THE NEW COMPANION TO URBAN DESIGN

The New Companion to Urban Design continues the assemblage of rich and critical ideas about urban form and design that began with the *Companion to Urban Design* (Routledge, 2011). With chapters from a new set of contributors, this sequel offers a more comparative perspective representing multiple voices and perspectives from the Global South.

The essays in this volume are organized in three parts: Part I: Comparative Urbanism; Part II: Challenges; and Part III: Aspirations. Each part contains distinct sections designed to address specific themes, and includes a list of annotated suggested further readings at the end of each chapter. Part I: Comparative Urbanism examines different variants of urbanism in the Global North and the Global South, produced by a new economic order characterized by the mobility of labor, capital, information, and technology. Part II: Challenges discusses some of the contemporary challenges that cities of the Global North and the Global South are facing and the possible role of urban design. This part discusses spatial claims and conflicts, challenges generated by urban informality, explosive growth or dramatic shrinkage of the urban settlement, gentrification and displacement, and mimesis, simulacra, and lack of authenticity. Part III: Aspirations discusses some normative goals that urban design interventions aspire to bring about in cities of the Global North and the Global South. These include resilience and sustainability, health, conservation/restoration, justice, intelligence, access and mobility, and arts and culture.

The New Companion to Urban Design is primarily intended for scholars and graduate students interested in cities and their built environment. It offers an invaluable and up-to-date guide to current thinking across a range of disciplines including urban design, planning, urban studies, and geography.

Tridib Banerjee holds the James Irvine Chair in Urban and Regional Planning at USC's Price School of Public Policy. His research and writings focus on the design and planning of the built environment and their human consequences. *Urban Design: Critical Concepts in Urban Studies* (a four-volume edited collection, 2014) is his most recent publication.

Anastasia Loukaitou-Sideris is Professor of Urban Planning and Associate Provost for Academic Planning at UCLA. She is the author of multiple articles on urban design, coeditor of *Jobs and Economic Development in Minority Communities* (2006), *Companion to Urban Design* (2011), and *Informal American City* (2014), and coauthor of *Urban Design Downtown* (1998), *Sidewalks* (2009), and *Transit-Oriented Displacement or Community Dividends?* (2019).

THE NEW COMPANION TO URBAN DESIGN

Edited by Tridib Banerjee and Anastasia Loukaitou-Sideris

Routledge
Taylor & Francis Group
LONDON AND NEW YORK

First published 2019
by Routledge
4 Park Square, Milton Park, Abingdon, Oxon OX14 4RN

and by Routledge
605 Third Avenue, New York, NY 10017

First issued in paperback 2022

Routledge is an imprint of the Taylor & Francis Group, an informa business

Publisher's Note
The publisher has gone to great lengths to ensure the quality of this reprint but points out that some imperfections in the original copies may be apparent.

British Library Cataloguing-in-Publication Data
A catalogue record for this book is available from the British Library

Library of Congress Cataloging-in-Publication Data
A catalog record for this book has been requested

ISBN 13: 978-1-03-247552-3 (pbk)
ISBN 13: 978-1-138-30212-9 (hbk)
ISBN 13: 978-0-203-73193-2 (ebk)

DOI: 10.4324/9780203731932

Typeset in Bembo
by Apex CoVantage, LLC

CONTENTS

CONTRIBUTORS

Daniel B. Abramson is Associate Professor of Urban Design and Planning and a China Studies faculty member at the University of Washington, Seattle. He received his doctorate from Tsinghua University, Beijing, was Founding Secretary of the International Association for China Planning, and has published broadly on planning and design in China.

Bruce Appleyard is Professor of City Planning and Urban Design at San Diego State University. He works at the intersection of transportation, land use, and urban design where he helps people and agencies make more informed decisions about how to achieve a range of sustainability, livability, health, and equity outcomes.

Tridib Banerjee holds the James Irvine Chair in Urban and Regional Planning at USC's Price School of Public Policy. His research and writings focus on the design and planning of the built environment and their human consequences. *Urban Design: Critical Concepts in Urban Studies* (a 4-volumed edited collection, 2014) is his most recent publication.

Rachel Berney is Assistant Professor of Urban Design and Planning in the College of Built Environments at the University of Washington, Seattle. Her work focuses on everyday expressions and contestations of the right to the city. She is the author of *Learning from Bogotá* and editor of *Bicycle Urbanism.*

Vinayak Bharne is joint Adjunct Associate Professor of Urbanism at the University of Southern California School of Architecture and Price School of Public Policy. He is also Principal and Director of Design of the architecture and planning firm Moule & Polyzoides. His books include *The Emerging Asian City: Concomitant Urbanities & Urbanisms*, and the *Routledge Companion on Global Heritage Conservation*

Alexander Bigazzi is Assistant Professor at the University of British Columbia in the Department of Civil Engineering and the School of Community and Regional Planning. He received his PhD from Portland State University in 2014. His primary research areas include transportation-related emissions and active travel behavior.

Stefano Bloch is Assistant Professor in the School of Geography and Development at the University of Arizona. His work looks at changing neighborhoods, transgressive subcultures, and anti-gang policing. His forthcoming book and autoethnography with University of Chicago Press is titled *Going All City*.

Bianca Bosker is an award-winning journalist and the bestselling author of *Cork Dork* and *Original Copies*. Her work has appeared in publications including the *Wall Street Journal, Fast Company, The Atlantic, Food & Wine, The New York Times, Far Eastern Economic Review*, and *The Oregonian*.

Anna Botta is an architect. She took part in the European Project GREAT Med, analyzing the Mediterranean area and proposing a model of "sustainable floating architecture." She is currently completing her PhD at Sapienza University (in Rome) focused on spontaneous Muslim worship places in Italy.

Pietro Calogero holds a PhD in city and regional planning from the University of California, Berkeley. He helped Allan Jacobs complete *Great Streets*, and then worked for the office of Peter Calthorpe. He has been working and researching intermittently in Kabul since 2003. In 2017–2018 he worked as Senior Policy Advisor to the urban planning sector of the Afghan government.

Surajit Chakravarty is Associate Professor of Urban Studies at the Indian Institute of Management, Calcutta. He holds a PhD in policy, planning, and development from the University of Southern California. His research interests include affordable housing, mobility, informality, and civic engagement in multiethnic societies. He has worked internationally in a variety of political-economic contexts.

Manish Chalana is Associate Professor in the Department of Urban Design and Planning at the University of Washington. His research focuses on historic preservation planning, planning history, and international planning and development. He is the author of multiple articles and the co-editor of *Messy Urbanism: Understanding the "Other" Cities of Asia*.

Felicity H.H. Chan teaches geography and urban planning at the Nanyang Technological University, Singapore. Her research interest lies at the intersections of the formation of social life in cities, sociocultural diversity and the planning/design of the urban built environment, particularly employing mapping as a method of inquiry.

Suzanne Lanyi Charles is Assistant Professor in the Department of City and Regional Planning and the Baker Program in Real Estate at Cornell University in Ithaca, New York. Her research addresses redevelopment, neighborhood change, and gentrification in older inner-ring suburban neighborhoods.

Francesco Chiodelli is Associate Professor of Urban Planning at the Gran Sasso Science Institute, Italy, where he is the Vice-Coordinator of the PhD Program in Urban Studies and Regional Science. His research focuses on social externalities of spatial regulation, in particular with reference to questions of religious diversity, pluralism, and informality/illegality.

Im Sik Cho is Associate Professor in the Department of Architecture at the National University of Singapore, where she serves as the Leader of Urbanism Research Cluster. Her research addresses the challenges and opportunities that Asian cities face with accelerating social change, focusing on the social dimension of sustainable development.

Diane E. Davis is the Charles Dyer Norton Professor of Regional Development and Urbanism and Chair of the Department of Urban Planning and Design at Harvard's Graduate School of Design. Her research interests include comparative urban governance, socio-spatial practice in rapidly urbanizing locales, and the coping and adaptation strategies adopted by citizens and authorities to push back against violence.

David Dewar is Emeritus Professor and Senior Research Scholar at the University of Cape Town. He is the author or coauthor of nine books and over 280 monographs or articles on issues relating to city and regional planning. He consults widely in Southern and Eastern Africa and has received numerous awards for both his academic and professional work.

Kim Dovey is Professor of Architecture and Urban Design at the Melbourne School of Design, University of Melbourne. He is the Director of *InfUr* (Informal Urbanism Research Hub) and has published widely on social issues in architecture, urban design, and planning, including the books *Framing Places, Fluid City, Becoming Places, Urban Design Thinking*, and *Mapping Urbanities.*

Fábio Duarte is Research Scientist at the Senseable City Lab at the Massachusetts Institute of Technology, and Professor of Urban Management at the Pontifícia Universidade Católica do Paraná, Brazil. He is also a consultant for the World Bank in urban planning and mobility, and the author of *Unplugging the City* (with Rodrigo Firmino, 2018).

Renia Ehrenfeucht is Professor of Community and Regional Planning at the University of New Mexico. She has two research focuses: everyday life on sidewalks and other public spaces, and shrinking cities or how people and places adapt to population loss. She has written two books and numerous journal articles.

Ramzi Farhat is a senior lecturer in Urban and Regional Planning in the American University of Beirut and Cal Poly Pomona. His research has appeared in journals such as *Journal of Urban Design, Urban Studies, and Planning Theory*. He teaches classes on comparative urbanism, urban design, and land use planning.

Rebecca Ferdman is a graduate student at UCLA pursuing dual master's degrees in urban planning and public health. She holds undergraduate degrees in environmental studies and neuroscience from USC. She has worked as a green building consultant in Los Angeles and as Scheduler to the Chair at the White House Council on Environmental Quality.

Scott Fisher is Associate Dean of Research and Professor of Media Arts + Practice at the USC School of Cinematic Arts. He is an interaction designer focusing on mobile media, interactive environments, and technologies of presence, and the Director of USC's Mobile & Environmental Media Lab and cofounder of the Emergent Cities Research Group.

René Fleschurz is a research assistant and PhD candidate at the Department of International Planning Systems at the Technische Universität Kaiserslautern's faculty of Spatial and Environmental Planning. His research focuses on urban shrinkage, demographic decline in rural regions, participatory planning and planning cultures, and comparative research.

Sabrina Förch holds a bachelor's degree in urban planning from Technische Universität Kaiserslautern. She currently pursues a career in urban and environmental planning.

Lawrence Frank is the Bombardier Transportation and Public Health Professor at the University of British Columbia. He specializes in the interaction between land use, travel behavior, air quality, and health. He was among the first to study links between planning, environment, and health collectively, and he coined the term "walkability."

Sandra Frem is an urbanist, architect, educator, and the founder of platau | platform for architecture and urbanism. She holds a SMArchS in architecture and urbanism from the Massachusetts Institute of Technology, where she was the recipient of the SMArchS award, and a DES in architecture from the Lebanese University in 2004 summa cum laude.

Alexander Garvin has combined a career in urban planning and real estate with teaching, architecture, and public service. He has been an Adjunct Professor of Urban Planning and Management at Yale University, his alma mater, where for the past 51 years, he has taught courses in architecture, city planning, and real estate development.

D. Asher Ghertner is Associate Professor in the Department of Geography and Director of the South Asian Studies Program at Rutgers University. He has published widely on aesthetic governmentality, gentrification, and urban informality in India and is the author of *Rule by Aesthetics: World-Class City Making in Delhi* (2015).

Kian Goh is Assistant Professor of Urban Planning at UCLA's Luskin School of Public Affairs. She researches the relationships between urban ecological design, spatial politics, and social mobilization in the context of climate change and global urbanization.

Kelly Gregg completed her PhD in planning at the University of Toronto in 2019. Her research is positioned at the intersection of urban design, landscape architecture, and urban planning, focusing specifically on postwar "modernist" and contemporary street designs including pedestrian spaces, shared streets, and Complete Streets.

Susan Handy is a Professor in the Department of Environmental Science and Policy and the Director of the National Center for Sustainable Transportation at the University of California, Davis. She is known for her research on the relationships between transportation and land use and on the concept of accessibility.

Lawrence A. Herzog is Professor Emeritus at the Graduate Program in City Planning, School of Public Affairs, San Diego State University. He has written or edited 11 books on urban planning, design, and global/cross-border development, including *Global Suburbs: Urban Sprawl from the Rio Grande to Rio de Janeiro* (2015).

Paul Hess is an Associate Professor in the Department of Geography and Planning at the University of Toronto, where he teaches urban design and planning for active transportation. His research interests include pedestrian environments and design, streets as public space, transport equity, and suburban form and transformation.

Max Hirsh is a Professor at the University of Hong Kong, where he teaches courses on infrastructure, transportation, and urban design. He is the author of *Airport Urbanism: Infrastructure and Mobility in Asia* (2016).

Justin B. Hollander is Associate Professor of Urban and Environmental Policy and Planning at Tufts University. His research and teaching are in the areas of shrinking cities, Big Data, brownfields, and the intersection between cognitive science and the design of cities. He hosts the iTunes podcast *Cognitive Urbanism.*

Yosef Jabareen is on the Faculty of Architecture and City Planning at the Technion University in Israel. He focuses on the nexus between planning theory and practices in different realms: justice, urban rights and minorities, and climate change and sustainability. He is the author of *The Risk City* (2015).

Richard J. Jackson is Professor Emeritus at UCLA's Fielding School of Public Health. A pediatrician, he has served as California's State Health Officer, and as the Director of the CDC's National Center for Environmental Health in Atlanta. He is a member of the Institute of Medicine of the National Academy of Sciences, and a recipient of a Presidential Distinguished Service award.

Annette Kim is Associate Professor of Public Policy at the University of Southern California. She founded and directs SLAB, the Spatial Analysis Laboratory, that advances the visualization of the social sciences for public service, in order to find inclusive and humane ways to design and govern the 21st-century city.

Zachary Lamb is a Mellon Fellow in Urbanism and the Environment at Princeton University. His research focuses on the intersection of design, planning, and the politics of flood risk in cities in the Global North and Global South. He has graduate degrees in planning and architecture from MIT.

Anastasia Loukaitou-Sideris is Professor of Urban Planning and Associate Provost for Academic Affairs at UCLA. She is the author of multiple articles on urban design, coeditor of *Jobs and Economic Development in Minority Communities* (2006), *Companion to Urban Design* (2011), and *Informal American City* (2014), and coauthor of *Urban Design Downtown* (1998), *Sidewalks* (2009), and *Transit-Oriented Displacement or Community Dividends?* (2019).

Bryce C. Lowery is Assistant Professor of Regional and City Planning at the University of Oklahoma's Christopher C. Gibbs College of Architecture. He researches the spatial inequality of physical form with a focus on public health and community engagement. More can be found at brycelowery.com.

Scott A. Lukas has taught anthropology, sociology, and American studies at institutions in the United States and Germany. He is the author/editor of numerous monographs and collections, including *The Immersive Worlds Handbook*. He was recognized with the McGraw-Hill Award for Excellence in Undergraduate Teaching of Anthropology by the American Anthropological Association.

Darshini Mahadevia is Professor at Ahmedabad University. Previously, she had served as the Director of the Centre for Urban Equity established by her in 2009 at CEPT University. She researches equity issues in housing, transport, water and sanitation, municipal finance, and climate change resilience and mitigation.

Stefano Moroni is Professor of Planning at Milan Polytechnic University. He works on planning theory and on ethical and legal issues. Recent publications include *Ethics, Design and Planning of the Built Environment* (with C. Basta, 2014), *Cities and Private Planning* (with D. Andersson, 2014), and *Space and Pluralism* (with D. Weberman, 2016).

Jeremy Németh is an Associate Professor of Urban and Regional Planning at the University of Colorado, Denver. His research looks at how planners, designers, and citizens can help create more socially and environmentally just places. All of his work can be found at http://jeremy-nemeth.com.

Mary Ann O'Donnell is an artist-ethnographer. She lives and works in Shenzhen, where she has co-founded the Handshake 302 Art Space. Other projects include coediting *Learning from Shenzhen: China's Post Mao Experiment from Special Zone to Model City* (2017), blogging at Shenzhen Noted, and her ongoing participation in the Shenzhen-Hong Kong Bi-City Biennale of Urbanism/Architecture.

Dietmar Offenhuber is Associate Professor at Northeastern University in Art + Design and Public Policy. His research focuses on the relationship between design, technology, and governance. He is the author of the award-winning monograph *Waste is Information* and has published books on the subjects of urban data, accountability technologies, and urban informatics.

Karina Pallagst is Professor for International Planning Systems at Technische Universität Kaiserslautern's faculty of Spatial and Environmental Planning. Her research focuses on international comparative urban development, shrinking cities, urban growth, planning cultures, and planning theory. She is also a cofounder of the Shrinking Cities International Research Network (SCiRN).

Balaji Parthasarathy is Professor at the International Institute of Information Technology Bangalore, and cofounder of the institute's Center for Information Technology and Public Policy. His teaching and research examine the implications of globalization and technology design for social change. One aspect focuses on urbanism, urbanization, and the political economy of urban form.

Richard B. Peiser is the Michael D. Spear Professor of Real Estate Development in the Department of Urban Planning and Design at the Harvard Graduate School of Design. He is the author of *Professional Real Estate Development: The ULI Guide to the Business*. His research focuses on new towns, affordable housing, suburban regeneration, and the urban development industry.

Rebecca S. Popowsky is a landscape architect and Research Associate at OLIN in Philadelphia. She leads the firm's external research initiatives, including work with academic and scientific institutions and allied professionals. In addition to practice, she teaches design studios, representation, and field ecology at the University of Pennsylvania School of Design.

Elena Porqueddu is an architect and an independent researcher. She received her PhD from the Polytechnic University of Milan, with a thesis developed in part at the University of Melbourne. Her research focuses on the role of design in emergent urban systems.

Carlo Ratti is Professor of Practice and Director of the Senseable City Lab at the Massachusetts Institute of Technology. He is also a founding partner of the international design and innovation office CRA-Carlo Ratti Associati. He is the author, among other publications, of *The City of Tomorrow* (with Matthew Claudel, 2016).

Jenny Roe directs the Center for Design + Health and the Healthy Cities Lab in the School of Architecture, University of Virginia. She is an environmental psychologist who specializes in how urban design can support mental health including access to urban green space.

Yodan Y. Rofè is a Senior Lecturer of Urban Planning and Design at Ben-Gurion University, Israel, and founder and former board member of the Movement for Israeli Urbanism (MIU). He holds a master's in architecture and a PhD in city and regional planning from the University of California, Berkeley. His interests include urban morphology, sustainable street design and cognition, and feeling in the built environment.

Ananya Roy is Professor of Urban Planning, Social Welfare, and Geography and Director of the Institute on Inequality and Democracy at the UCLA Luskin School of Public Affairs. Trained as an urban planner, she is critical of ideas and practices that perpetuate urbanisms of inequality and separation. Her most recent book is *Encountering Poverty: Thinking and Acting in an Unequal World* (2017).

Brent D. Ryan is Associate Professor of Urban Design and Public Policy at MIT. His research examines emerging urban design paradigms, particularly in postindustrial, global cities. His most recent book is *The Largest Art* (2017), and his first book was *Design after Decline* (2012). He has lectured, practiced, and written widely on urban design.

Hashim Sarkis is the Dean of the School of Architecture and Planning at MIT and the principal architect of the firm Hashim Sarkis Studios. Prior to MIT he was the Aga Khan Professor of Landscape Architecture and Urbanism at Harvard's Graduate School of Design. His architectural and planning projects include affordable housing, institutional buildings, and town planning throughout the globe.

Brinda Sastry is Adjunct Faculty member in the M.Arch (Urban Design) program, RV College of Architecture, Bangalore. Her practice and research interests lie at the intersection of spatial planning, urban governance and urban design policy. Specifically, she seeks to develop methodologies to embed place-based design strategies within the planning framework.

Lisa A. Schweitzer is a Professor in the Department of Urban Planning and Spatial Analysis at the University of Southern California Sol Price School of Public Policy. She writes about urbanism at lisaschweitzer.com.

Brettany Shannon has a PhD in urban planning and development from USC. She studies how digital communications intersect with urban and social placemaking. She is the USC Bedrosian Center for Governance Scholar-in-Residence and part-time faculty member at California State University, Northridge. Her coedited volume, *Planning for AuthentiCITIES*, was published in 2018.

Jennifer Stein is a design researcher examining the implications of emerging technologies on everyday life by designing for the near future. Her research and speculative design work explores interactive architecture, ambient storytelling, and virtual and augmented reality.

Frederick Steiner is Dean and Paley Professor at the University of Pennsylvania School of Design and codirector of the Ian L. McHarg Center for Urbanism and Ecology. His publications include, most recently, *Making Plans: How to Engage with Landscape, Design, and the Urban Environment*.

Fran Tonkiss is Professor of Sociology at the London School of Economics, where she teaches in the Cities Programme, an interdisciplinary graduate center for the study of city design. Her works include *Space, the City and Social Theory* (2005) and *Cities by Design: The Social Life of Urban Form* (2013).

Jared Ulmer, AICP, MPH, is the Climate and Health Program Coordinator at the Vermont Department of Health, where he promotes better health through climate change mitigation. Previously, he had worked for Urban Design 4 Health, Inc., on research and applications that demonstrate the health impacts of transportation system and community design.

Lawrence J. Vale is Ford Professor of Urban Design and Planning at MIT and director of MIT's Resilient Cities Housing Initiative (RCHI). He is the author or editor of ten books examining urban design, housing, and planning, including four prize-winning volumes on American public housing.

Ryan Whitney is an urban sustainability planner and PhD candidate in planning in the Department of Geography and Planning at the University of Toronto. His current research focuses on the uptake of urban sustainability best practices in the context of equitable urban development in Latin America.

Li Xiangning is Professor in Architectural Theory and Criticism at Tongji University. He has published widely on urbanism and architectural culture of China, Asia, and the world. He is the curator of several major biennales in China and 2018 Venice Biennale Chinese Pavilion. He is the founder and editor-in-chief of the international magazine *Architecture China.*

Su-Jan Yeo is Postdoctoral Fellow in the School of Community and Regional Planning at the University of British Columbia. Her research on cities and urban transformation brings into focus the interplay between public space use, design/planning, and public policy vis-à-vis global processes.

INTRODUCTION

Challenges and aspirations of urban design

Although titled *The New Urban Design Companion*, this volume is a sequel to our earlier project, *Urban Design Companion*, which was published in 2011. In the introduction of the earlier publication, we discussed how the field of urban design, while having roots in architecture, landscape architecture, and urban planning, has become increasingly interdisciplinary, as appropriate for a field that focuses on the urban form and the socio-spatial organization of cities. Indeed, in addition to technical and aesthetic matters of form, urban design's content and praxis are influenced by public policy, relevant social sciences, and also the arts and humanities. As an intellectual endeavor intended to inform practice, urban design scholarship is interested not only with aesthetic issues but also with the social, political, cultural, and economic forces that affect the built environment and its residents, as well as the human and environmental consequences of design interventions. Accordingly, this *New Companion* continues the assemblage of rich and critical ideas about urban form and design from a set of new contributors. Purposefully, not all authors invited have design backgrounds; nevertheless, they all have interest in the contemporary urban experience, are engaged in critical analysis of the urban outcome, and are willing to discuss normative implications of their critical arguments for urban design.

The *New Companion to Urban Design* attempts to correct some of the lacunae of the first volume. One important omission in the first volume was its general lack of a comparative perspective, as it primarily concentrated on issues and concerns in the United States, Canada, and Western Europe. A related second important omission was the absence of voices and perspectives from the Global South. However, at a time when major transformative urban development is taking place in the Global South, it is important to focus on consequential urbanism in a comparative perspective. Therefore, a number of topics are discussed in this volume as they appear in different geographic contexts, in cities of both the Global North and the Global South. Thus, in contrast to the dominance of Anglo/US/Canada viewpoints that characterized the earlier volume, this new compilation of works presents additional perspectives and assessments from the Middle East, Persian Gulf, South and East Asia, Southern Africa, and Latin America, as well as Australia, Europe, and North America. Indeed, each section has multiple set of voices from both the Global North and the Global South.

While the topic categories in the original *Companion* were more foundational, substantive, and archetypal, in this sequel we have conceptualized the topics as aspirational ("opportunities") or as those that remain stubborn "challenges." Several substantive areas of practice and application of urban design that were missing from the first volume now occupy specific sections devoted to them. These include such topics as historic preservation/restoration, obtaining justice through urban design, and the role of urban design in achieving sustainability and resilience in an era of climate change, among others.

Since the publication of the first *Companion*, several new trends that concern urban design are now apparent. First, the landscape of some cities has changed, as urban development has intensified in many parts of the world in response to rapid urbanization, ascendant economies, and the rise of information and communication technologies. Second, while many cities have experienced explosive growth and/or densification, other cities continue to witness disinvestment and shrinkage. Third, the transnational flow of populations between countries and regions has intensified, frequently leading to diverse spatial claims and conflicts, and enhancing divisions between natives and immigrants or refugees, between formal and informal settlements, and with heightened levels of inequity and injustice. Fourth, new technologies and the brave new world of digital social media are transforming urban life, influencing mobility and travel patterns, social interactions and communications, and also the spatial organization of cities. Many of these issues and topics are addressed in specific sections or chapters of this volume.

More specifically, the essays in this volume are organized in three parts: Part I: Comparative Urbanism; Part II: Challenges; and Part III: Aspirations. Each part contains distinct sections designed to address specific themes, and each section consists of two to five chapters. At the end of each chapter is a list of suggested further readings, annotated by the respective author.

Part I: Comparative Urbanism examines different variants of urbanism in the Global North and the Global South produced by a new economic order characterized by the mobility of labor, capital, information, and technology. We asked our contributors to consider the challenges for urban design in the context of these evolving and mutant urbanisms shaped by the tension between globalism and localism as well as the interactions, influences, and flows between places in the Global North and the Global South. The first section of Part I, titled "Arguments and Observations," begins with an examination by Fran Tonkiss of the matter of "design translations." Rejecting standardization and the export model of design ideas from the Global North to the Global South, Tonkiss focuses on design practices instead of issues of form. She argues that some ordinary practices of urban design – informality, incrementalism, improvisation, impermanence, and insurgency – common in low-income cities can be transferred and translated to high-income cities. In the same section, Ananya Roy calls for urban design to resist its colorblindness and Eurocentrism, shift away from its pre-occupations with aesthetic improvements, act in solidarity with subaltern and displaced groups, and exercise a "decolonial imagination of placemaking" that learns from the South, as found not only in the cities of the Global South but also in some cities of the Global North, such as Los Angeles, which she interprets as a "post-colonial city."

In the second section of Part I, titled "Regional Experiences," five authors discuss the emerging urbanisms and spatial relations in different regions of the world as influenced by the forces of globalization. Lawrence A. Herzog focuses his attention on Latin American cities, their evolution, and their contemporary urbanism characterized by their "global suburbs" and "transfrontier" metropolitan regions – both examples of a transnational urbanism. David Dewar examines the urbanism of the cities of Southern Africa. Referring to the UN-Habitat report that finds these cities among the most inequitable in the world, he proposes a set of criteria – ecological justice, generative capacity, access, livability, and choice – for evaluating spatial justice. He calls for radical changes in the design and spatial planning practices of Southern African cities, as well as state intervention in their urban land markets. Ramzi Farhat examines how neoliberal policies of development in Eastern Mediterranean cities are changing the fine-grain urban fabric and appearance of historic cores in Amman, Beirut,

Cairo, and Damascus, threatening displacement of existing residents and exacerbating gentrification trends. In the last chapter of this section, Surajit Chakravarty discusses how the rapid growth of cities of the Arabian/Persian Gulf, framed by poor urban design, has led to a human-unfriendly urban form, which he calls the "un-city." He argues for urban design interventions that instill back humanity in these cities, creating spaces that invite a diversity of users and uses, and counteracting tendencies for segregation and exclusion.

Part II: Challenges discusses some of the contemporary challenges that cities of the Global North and the Global South have to face. These include spatial claims and conflicts; challenges generated by urban informality; explosive growth or dramatic shrinkage of the urban settlement; gentrification and displacement; and mimesis, simulacra, and lack of authenticity.

New waves of migration – both domestic and foreign – have led to new claims on space and the "right to the city" (to invoke Henri Lefebvre (2008)). The global economic order and the international political discourse have triggered both the mobility of labor and also the exodus of displaced refugees trying to escape oppression, torture, hunger, and death. Cities are the primary recipients of the new migration, which has challenged the conventional norms and uses of urban settings. How can urban designers respond to the emerging ethnoscapes (cf. Appadurai 1996) of diaspora that bring different religions and cultures in close spatial proximity to one another? This is the overarching question of the first section of Part II, titled "Claims and Conflicts." This question is addressed by Stefano Moroni, Francesco Chiodelli, Elena Porqueddu, and Anna Botta in their chapter. They examine the spatial claims of Muslim immigrants in Rome for their places of worship, and the attendant opposition from local communities and authorities. The authors are critical of design arguments that privilege homogeneity of the urban form, linkage to past typologies, and building traditions of the past, arguing instead for urban design and spatial planning that can better accommodate the new diversity in cities. In the next chapter, Felicity H.H. Chan takes us to the city-state of Singapore where different spatial claims emerge, particularly from high-skilled migrant workers from the Global North, resulting in the development of a "cosmopolitan grid" with amenities for the transnational elites that are distinct from the spatial settings for the migrants of the Global South. In the last chapter of this section, Jeremy Németh uses Barcelona's well-known multicultural neighborhood of *el Raval* to demonstrate how its urban form and physical features have successfully accommodated the integration of a diverse public of immigrants and locals and suggests this as a best practice.

Informal urbanism represents a reality and sometimes a challenge for cities of the Global South but also of the Global North. Yet the formal response to informality has been either authoritarian, exclusionary, or, at best, a benign tolerance (Mukhija and Loukaitou-Sideris 2014). Nevertheless, the phenomenon of informality involves urban space and form, in the cracks, fissures, nooks, and niches of the built city. How should the scope of urban design be defined to better address the challenges and opportunities of informality in cities? Kim Dovey, who opens the second section titled "Informality," argues that traditional design and planning approaches can eliminate informal urbanism. Instead, urban designers dealing with informal urbanism should be guided from an "assemblage approach" that examines the nexus between formal and informal sectors including conflicts and synergies. This can better help them understand the logic of their location, form, image, and identity. In the following chapter, Annette Kim argues that urban design practice should address the needs of all city residents, including those she finds hidden in plain sight, in informal housing accommodations embedded in the formal sector housing in Beijing and Shanghai. In the next chapter, Su-Jan Yeo turns our attention to the informal spatial practices as accommodated at certain "nightscapes" of East Asian cities. She asks urban designers to guard against the elimination of these nightscapes by softening boundaries between formal-informal spaces, by finding the spaces in-between, and by enabling "conditions of sociability and conviviality." In the next chapter, Manish Chalana uses two informal settlements from Mumbai and Delhi in India to highlight the importance of incremental design changes that are meaningful for residents, arguing that design "has a greater role to play in urban informality." In the last chapter of this section, Rachel Berney notes that "unsettledness" because of war, natural or economic disaster, or displacement has become a frequent occurrence as a condition of exception in cities globally. She articulates the design strategies that can help respond to this "urbanism of exception."

Rapid urbanization, accompanied by explosive growth fueled by a global economy, has significantly transformed many cities in the Global South, often causing irreversible changes in their historic structure, environmental degradation, and at times exacerbating disparities between landscapes of wealth and landscapes of poverty. But the opposite phenomenon – shrinkage – is also affecting some cities of both the Global North and the Global South, as a result of human-made and natural disasters, wars, terrorist acts, economic downturns, product cycle changes, and the like that wreak devastations, obsolescence, and abandonment of housing or infrastructure. How can urban design anticipate such damages and imagine cities that can bounce back effectively? How might urban design address issues of recovery and adaptation or manage decline? In the first chapter of the section "Explosive Growth Versus Shrinkage," Dan Abramson ponders the challenges of dramatic urban growth in the context of Chinese cities. He argues that the role of urban design is to integrate instead of separating the new from the old urban enclaves through incremental betterment and strategies of integration, rather than following a *tabula rasa* approach. In the next chapter, Im Sik Cho details how creative urban design strategies can generate new typologies and activate hybrid public spaces in the high-density urban environments of East and Southeast Asian cities. Examining city shrinkage and responsive urban design strategies in different parts of the world, Brent D. Ryan finds that these vary based on the larger context, economy, and governance system of cities and nations. Similarly, Karina Pallagst, René Fleschurz, and Sabrina Förch detail two different "paths of shrinkage" and responsive revitalization strategies, as applied by the US city of Cleveland and the German city of Herne. Justin Hollander's chapter that follows is also comparative, detailing different design and policy approaches to manage decline as followed in different US cities. He argues that within the US context, urban design for shrinking cities should first "focus on small impactful efforts that can then be rolled into a larger, more systemic change."

A type of intensive growth, albeit one that may be more concentrated in specific development sites, comes in the form of large-scale development projects. In some instances, these projects become mini-cities, as is the case of Battery Park City in New York – the 92-acre residential community at the southern tip of Manhattan. Because of their scale and complexity, mega-development projects entail not only economic but also physical challenges, such as how to best integrate the new development to the rest of the urban fabric or activate its public and commercial spaces. Two chapters in the section "Large-Scale Development" discuss some of these complexities. Alexander Garvin discusses the development story of three landmark mega-developments in New York City – Grand Central Station, Rockefeller Center, and the World Trade Center – showing the importance of good urban design for the eventual commercial success of large-scale development projects. Sandra Frem and Hashim Sarkis discuss the complications in the development of another megaproject, this one in a city of the Global South: Byblos in Lebanon. Rather than designing an isolated landmark building, the opportunity arose to integrate it in a wider and more coherent urban design framework that reestablished and strengthened the site's connection to the city's waterfront and historic core.

Flows of private capital and public investments in existing neighborhoods and redevelopment of the urban stock may lead to residential and commercial gentrification and displacement of existing residents, in cities of both the Global North and the Global South. What is the role of urban designers in such a conundrum, if the physical improvements, revitalization, and upgrading sought by them also bring along these negative byproducts? What strategies can help ameliorate these growing concerns related to land and property development? Gentrification has typically plagued central city neighborhoods, but newly built suburban retrofit developments may also raise gentrification concerns. In the first chapter of the "Gentrification and Displacement" section, Suzanne Charles and Richard B. Peiser examine if this is the case in a suburban retrofit development in Lakewood, Colorado. While they find that this project has not resulted in displacement impacts in the surrounding neighborhoods, they caution that retrofitting suburbia into denser, walkable, and aesthetically attractive locales may bring along exclusionary gentrification, as the market price far exceeds the ability of current households to purchase property and live there. Displacement is another outcome discussed in the next chapter by Darshini Mahadevia, who examines the Indian state's housing efforts as part of its Smart Cities initiative. This has involved displacement of previously centrally located informal housing and the relocation of residents to formally built housing in the urban periphery. Mahadevia shows that the peripheral location and modernist urban design of the new housing projects fail to serve the residents' needs and result in their diminished well-being.

Many emerging forms of new urban development seem to lack authenticity and identity. "Mimesis and Simulacra" – the topic of the next section – abound in both new residential and commercial environments in the Global South. Some call it "Disneyfication"; however, this phenomenon is not only a characteristic of countries of the Global South. Why is it so widespread? Is this an inexorable outcome of globalization that leads to a lack of authenticity? Scott A. Lukas puts this phenomenon in its larger context, drawing from examples from the Global North. He explains how mimetic "urbanscapes" may represent a city's efforts for "urban remaking." Nevertheless, the design of such spaces would do well if it privileges the desires of their users. Picking up the topic of Disneyfication, Li Xiangning shows in the next chapter how the design and programming of contemporary themed spaces in China and other countries, following a profit motive, try to incorporate local cultural elements. The themed new towns of China are also the focus of the next chapter by Bianca Bosker. Revisiting them one decade after their creation, she finds that, far from being empty shells, some of these towns have filled out and become "more Chinese in spirit" and with "a Chinese soul." If there is a common message in all these three chapters, it is that even environments that started as simulacra, in time, become authentic by the lived experiences of their inhabitants. Lastly, the fourth chapter of this section by D. Asher Gertner discusses another type of mimicking or "improvised urbanism," created by planners and developers of unauthorized settlements in India. This mimicry projects an ideal condition (conformance with planning regulations, housing occupancy) as a way to formalize informality or to avoid demolition. According to Ghertner, these practices "mimic the formal city but in doing so, simultaneously make a mockery of it."

Part III: Aspirations discusses some normative goals that urban design interventions aspire to bring about in cities of the Global North and the Global South. These include resilience and sustainability, health, conservation/restoration, justice, intelligence, access and mobility, and arts and culture.

In the context of global warming and climate change, sustainability has come to the forefront of public policy debate. In the first section of Part III, titled "Resilience and Sustainability," authors examine how sustainability goals are addressed in the practice of urban design. Are there some normative positions that urban designers should offer? Are there inherent conflicts and contradictions in meeting sustainability goals? What are the best practices in these areas? In the first chapter of this section, Zachary Lamb and Lawrence J. Vale call for resilience practices that are guided by concepts of environmental justice. Using examples from cities of the Global North and the Global South, they outline principles that seek to refocus urban design towards this end. In the next chapter, Diane E. Davis draws from specific design interventions in Latin American cities to explore the potential for design to reduce vulnerabilities in marginalized communities, especially those experiencing chronic violence. Yosef Jabareen, in the next chapter, brings us to New York City, discussing how city planners motivated by perceptions of risk are creating and implementing plans and designs to respond to climate change. In the last chapter of this section, Frederick Steiner and Rebecca S. Popowsky present the new rating system SITES that guides and encourages designers and developers to create more sustainable landscapes.

The public health community, concerned about growing obesity and other pathologies, has long urged planners and urban designers to design cities that promote active living, walkability, opportunities for exercise, access to good food, and the like. In parts of the Global South, access to clean water is crucial for health. Issues of public health also include the uneven exposure to pollution and its damaging effects on older adults, children, and residents of low-income neighborhoods. The authors of the next section, titled "Health," broadly interrogate how urban design can help create healthy cities. To this end, Richard J. Jackson and Rebecca Ferdman articulate public health priorities for different components of the built environment – housing, health-care facilities, schools, transportation systems, and green infrastructure – and propose an action agenda to achieve them. Along similar lines, Jenny Roe in the next chapter offers a framework for achieving what she calls "new health urbanism," by improving through design the characteristics of living streets, urban green and blue spaces, and creating cities that are more compact and enabling of healthier and better lives. In the last chapter of this section, Lawrence Frank, Jared Ulmer, Bruce Appleyard, and Alexander Bigazzi review the literature to show how the design of one component of the built environment – streets – can have health-related impacts.

The imperatives of rapid economic development have taken a toll on the historic cities and districts around the world. Preservation of important buildings and places, restoration of valued landscapes, and urban conservation, more broadly, remain major challenges in managing change and continuity. In the next section, titled "Conservation/Restoration," authors ponder the role of urban design practice in achieving such goals and detail some good practices from cities in different parts of the world. In the first chapter of this section, Vinayak Bharne discusses three heritage sites in three different countries – Japan, Iran, and India – to illustrate how an urban design approach that views these sites in their larger contexts can act synergistically with goals of conservation, ecology, and social policy. The next chapter takes us to Shenzhen, China. Mary Ann O'Donnell details how local architects and planners, who were part of the preservation movement Hubei 120, have managed to preserve Shenzhen's historic city market, despite earlier city attempts to redevelop this site. In the last chapter of this section, Pietro Calogero details how urban design has been used historically to "instill modernity" in Kabul but has also been used to express various disparate ideals and ideologies in this conflict-ridden city.

Addressing inequality of opportunity and freedom and obtaining a just city remain two important normative goals in achieving a good city. Many of the injustices encountered in urban locales of the Global North and the Global South have spatial implications. How urban design can respond to such challenges and promote spatial justice was one question that we posed to the authors of the next section, titled "Justice." For Bryce C. Lowery and Lisa A. Schweitzer, who authored the next chapter, the answer lies in participatory design approaches and the co-creation of the built environment with residents. Similarly, Kian Goh, who interrogates the concept of "just design" within the context of Jakarta, finds that designers should align with community advocates to initiate counterproposals to dominant spatial initiatives such as "showcase" projects that do not adequately protect vulnerable groups from the whims of climate change.

In the rapidly changing landscape of digital technology, Internet of Things (or IOT), social media, and "big data," talk about "smart cities" is commonplace. Yet, what are the urban design implications? How can urban designers respond to such challenges? What would be the scale and scope of their involvement? These are some questions that the authors in the next section, titled "Intelligence," address. In the first chapter, Dietmar Offenhuber helps us make sense of different urban technologies and their implications for urban form, grouping them in four distinct categories based on their values, propositions, and shortfalls. In the following chapter, Jennifer Stein and Scott Fisher review four citizen-led projects in the Global North and the Global South to detail what urban designers can learn from the examples of their participatory platforms and citizens' use of smart city infrastructure. In the third chapter of this section, Fábio Duarte and Carlo Ratti use four projects from their own lab to show how digital technologies can provide critical data to urban designers, who can then use them to create more fitting and adaptable urban forms. While the previous two chapters pose an optimistic view of the potential of new technologies to contribute to good design, the last chapter in this section takes a more cautionary stance. Drawing from the experiences of India's Smart Cities initiative, Balaji Parthasarathy and Brinda Sastry argue that new technology would do little to improve the quality of urban living and placemaking if it seeks to prescribe and determine human behavior.

Access and mobility are intrinsically linked to the urban form and spatial structure of cities. Yet, in certain socio-spatial contexts, some groups (women, older adults, lower-income and minority groups) enjoy less access and mobility than others. How can urban design play a central role in enhancing access and mobility for all in cities? In the first chapter of the section "Mobility and Access," Susan Handy contemplates "what it means to design for accessibility." She offers a best practice: the design and layout of the city of Davis, California, where design and policy have encouraged alternative modes of transportation to the car, increasing life quality for residents, and particularly those who are carless. Similarly, Yodan Y. Rofè discusses the various design interventions for urban streets that, in the late 20th and early 21st centuries, seek to counteract the automobile-oriented city design and increase livability without sacrificing mobility in cities. Paul Hess, Kelly Gregg, and Ryan Whitney, in the next chapter, also discuss and critique street design initiatives in the North American context (complete streets, parklets, pavement-to-plaza initiatives) that seek to increase the pedestrian friendliness of cities. In the 21st century, mobility is also largely defined by air travel, even in countries like China where, a few decades ago, only a miniscule segment of the public was traveling by plane. In the last chapter of this section, Max Hirsh examines the urban form impacts of airport urbanism, namely, the building of mega airports in East and Southeast Asia.

Increasingly, arts and culture are playing an important role in influencing urbanism and public life in cities. The influence of the "creative class" in economic development and local placemaking continues to capture the imagination of many planners and policymakers. What role may urban design play in integrating local expressions of arts and culture, while avoiding displacement and exclusion? Responding to this question, Renia Ehrenfeucht, who offers the first chapter in the section "Arts and Culture," describes four urban design strategies that can help integrate arts and culture in urban street life, without excluding certain social groups. In the next chapter, Stefano Bloch shows us the value of insurgent artscapes in cities, reminding us that urban designers in the city may not be only those who are professionally trained as such. In the last chapter of this section, and of this volume, Brettany Shannon shows how urban designers can mine digital archives such as those of Instagram to understand what the public thinks about design or arts interventions in cities.

References

Appadurai, A. (1996). *Modernity at Large*. Minneapolis: University of Minnesota Press.
Lefebvre, H. (2008). "The Right to the City," in *Writings on Cities*. Oxford: Blackwell Publishing.
Mukhija, V. and Loukaitou-Sideris, A. (2014). *The Informal American City: Beyond Tao Trucks and Day Labor*. Cambridge, MA: MIT Press.

PART I
Comparative urbanism

PART I.I

Arguments and observations

Introduction

The term urbanism has become quite commonplace these days, and in many ways a leitmotif for thinking about cities and urbanization across disciplines. Indeed, the term has offered a medium of conversation across multiple and adjacent disciplines interested in understanding and explaining the transformative changes in human habitats across the globe. Within the fields of planning and urban design, the term urbanism is conjoined with multiple qualifiers – more than 60, according to one inventory (see Barnett, 2011). But our hunch is that the number of various "urbanisms" is still growing. This is quite remarkable given that a few decades ago, the term was not as commonly used in the Anglo-American planning and design literature. The common usage today in the United States in particular, we believe, may have been triggered by the concept of "New Urbanism" and its subsequent popularity.

Remarkably, the scholarly definition of urbanism was first introduced by sociologist Louis Wirth (1938), who defined the concept as "a way of life," different from the rural lifestyle common in the United States in the first half of the 20th century. Several decades later, with the spread of the postwar suburban growth, sociologist Herbert Gans (1973) claimed that suburban living represents a distinct departure from the original notion of urbanism and should also be considered a way of life. In the early 1990s, the protagonists of "New Urbanism" advanced some normative notions about the design of the built environment and an attendant lifestyle harkening back to the urban form and lifestyle of late-18th- and early-19th-century small US towns. Today, with the continuing expansion of urbanization worldwide, we might consider even "peri-urbanism" as a way of life.

The contemporary relevant literature has linked the concept of urbanism to multiple realities of our urban experience, encompassing the sensorium, form, social ecology, habitus, and identity of the city (Banerjee, forthcoming). Writers on urbanism have often seen urban design as a catalyst for change from the ground up, or as an agent of power in shaping the built environment to serve the interests of global capitalism. Consequently, urban design is viewed as squarely engaged in matters of social and environmental justice. This places significant pressure on the practitioners of urban design, and its scholarly inquiry. In the two chapters included in this section, Fran Tonkiss and Ananya Roy, accordingly, propose to redefine the charge of urban design in a comparative armature.

In her chapter, Tonkiss argues against a standardized – "one size fits all" – model of urban design remiss of the cultural, economic, historical, geographical, and social differences. Rejecting the "export model" of urban design in a globalizing world, where concepts, paradigms, and techniques are hatched in the Global North to be exported to the Global South, she emphasizes the role of endogenous design practice that translates such ideas across economies and cultures to fit local circumstances. The innovative practice and process of urban design, rather than the resulting urban form, is more significant here. She proposes five modes of innovative practices and processes of design in engaging alliteration – informality, incrementalism, improvisation, impermanence, and insurgency – and elucidates these concepts with appropriate and relevant examples. Her discussion of these concepts clearly suggests an alternative to the top-down planning and urban design that has shaped much of contemporary urbanism and the urban experience more generally.

The chapter by Roy expands the errands and challenges of urban design even farther by referring to the new debate on the global understanding of urbanism in a comparative perspective involving both theory and methodology. Resisting the current trend of globalizing notions of urban design and development based on Western, Eurocentric theories and methods, she proposes an alternative mode of interrogation based on postcolonial theories and black geographies. Instead of focusing on the cities of the Global South, Roy chooses Los Angeles as an example of displacement and dispossession of the communities of color, not unlike the "alterity" and subaltern status common in the colonial history of the Global South. In interpreting Los Angeles as a postcolonial city, she argues that the long history of occupation and expulsion in this city has led to resistance and social movements that may now offer a new vision of placemaking.

This is not the first time Los Angeles has been examined in the context of the Global South, or as a postmodern city. Journalist David Reiff (1991) has argued that the multicultural urbanism of Los Angeles makes it a candidate for being "The Capital of the Third World." Earlier, Scott Ridley's (1982) portrayal of a dystopic Los Angeles in his movie *Blade Runner* also suggested an emergent urbanism of the Global South. In the 1980s, geographers Edward Soja (1986) and Michael Dear (1998) – scholars of the then-emergent Los Angeles School – argued that the city's polycentric, polyglot, multicultural urbanism is the new reality of a global city rejecting the earlier hegemony of the Chicago School model of urban spatial structure. In a similar vein, querying the positivist models of urban reality, Banerjee and Verma (2001) have argued that the sense of the existential reality of the "soft metropolis" can be best captured through appropriate metaphors, some of them ostensibly drawn from the Global South.

In her chapter, Roy proposes still another perspective, albeit a critical one, to capture an aspect of Los Angeles urbanism, previously neglected, in the context of a globalizing world. Here she suggests that this view of Los Angeles urbanism is relevant to urban design in three ways: to understand the racialized nature of displacement and dispossession in urban development; as an alternative perspective to the Eurocentric universalizing tendencies in contemporary urban development; and to appreciate the creative imaginations and practices of the solidarity and social movements of subaltern social groups. She implores urban designers to learn from what she sees as the "decolonial" imaginations of placemaking.

References

Banerjee, T. (forthcoming). *In the Images of Development: City Design in the Global South*. Cambridge, MA: MIT Press.

Banerjee, T. and Verma, N. (2001). "Probing the Soft Metropolis: Third World Metaphors in the Los Angeles Context." *Planning Theory and Practice*. 2(2): 133–148.

Barnett, J. (2011). "A Short Guide to 60 of the Newest Urbanisms." *Planning*. 77(4): 19–21.

Dear, M. and Flusty, S. (1998). "The Iron Lotus: Los Angeles and Postmodern Urbanism." *The American Academy of Political and Social Science.* 551: 151–163.

Gans, H. (1973). "Urbanism and Suburbanism as a Way of Life: A Re-Evaluation of Definition." In *American Urban History: An Interpretive Reader with Commentaries.* New York: Oxford University Press. pp. 507–521.

Reiff, D. (1991). *Los Angeles: Capital of the Third World.* New York: Simon and Schuster.

Soja, E. (1986). "Taking Los Angeles Apart: Some Fragments of a Critical Human Geography." *Environmental and Planning D: Society and Space.* 4(3): 255–272.

Scott, R. (1982). *Blade Runner.* Producers: The Ladd Company.

Wirth, L. (1938). "Urbanism as a Way of Life." *The American Journal of Sociology.* 44(1): 1–24.

1
COMPARATIVE URBANISM
Design in translation

Fran Tonkiss

Introduction

Thinking about cross-border movements in urban development and design frequently focuses on the "high end" of global urban processes – the typical forms created by the reproduction of capital across transnational space. A new international style of corporate headquarters, cultural centers, shopping malls, gated housing, and chain hotels marks the production of dispersed spaces in easily recognizable forms; whether in the premium brands of a coterie of "starchitects" or the banal gestures of the behemoths of Big Architecture. Less prominent spatial types play out around the morphology of back offices, call centers, and assembly and processing plants, where the routine service and production work of a transnational economy gets done. And at another level, the precarious infrastructures of sweatshops, workers' camps, and waste landfills sketch out the geographies of exploitation and despoliation that are produced on the jagged edges of an international economy of "flows." The iconic, the identikit, and the insecure are all spatial expressions of globalized urbanism. This discussion departs from such versions of homogenized, offshored, and extruded spaces to approach urban design as an art of incomplete translation. In their prevailing mode, orthodox urban schemes offer a kind of trade language for international design: limited in vocabulary, simplified in grammar, and geared to commercial exchange. These are powerful and pervasive models of spatial design and development, and their impact in shaping urban environments can hardly be understated. But it is also possible to think in different ways – and at different scales – about a transnational traffic in urban design that works through more idiomatic forms, and more minor practices.

In considering how spatial design travels between urban contexts, it is obvious enough to think about the duplication of dominant formats across disparate settings: from the common-language commercial developments seen in the serial reproduction of shopping malls, retail parks, corporate plazas, or gated residential compounds; to the signature moves of high-end international architects in museums and concert halls, stadium constructions, or airport buildings. There is nothing especially new in the replication of design styles in diverse locations, as histories of colonial architecture or of international modernism will suggest. But there *is* something particular about the scale, spread, and speed of contemporary design translations, enabled in various ways by digital technologies in design, the electronic outsourcing of routine design processes, the extended circulation of design imagery, and the increasingly transnational organization of the design, construction, and development industries (see McNeill, 2009; Sklair, 2017). Against such a backdrop, it can be professionally challenging to think about smaller-scale and lower-key exchanges of design ideas and idioms that are less capitalized and less high-profile, but nonetheless make urban space in inventive and persistent ways.

A critical impetus for focusing on these more muted design translations comes from a growing body of work in the field of "comparative urbanism." The latter has emerged from postcolonial interventions calling for urban studies to work across an expanded and more diverse range of contexts; to redress a long-standing skew towards work on (and generalizations from) cities in the Global North; and to engage with the particular features, as well as the passing resemblances, of dispersed urban processes and forms. It follows that the tasks of urban comparison may have less to do with narrowly defined and tightly controlled contrasts between discrete urban *cases*, than with an engagement with urban connections, overlaps, and divergences open to a range of translations (see, among many others, Jacobs, 2012; McFarlane, 2010; Myers, 2014; Peck, 2015; Robinson, 2011, 2016a, 2016b; Roy, 2009b). Taking up such an approach to urban design means being sensitive to local and minor practices in specific spatial settings; to the circulation of techniques and typologies through less orthodox channels of urban exchange; and to the array of actors involved in the purposeful making of urban environments. It means being alert, moreover, to movements of design discourse and practices which interrupt the direction of urban travel from north to south, the rolling-out or trickle-down of corporate and cut-and-paste forms across transnational space.

My aim, in what follows, is to suggest several ways of thinking about urban design in this more open comparative frame that draw not on the design rhetorics of global urban development, but on the design idioms of less privileged spatial strategies. Rather than examining how iconic or airport or expat architectures come to be copied across very different urban sites in a kind of "design dumping," I am interested in design practices which are less obvious in their physical outcomes and often involve translations from low-income to higher-income cities. Such migrations disrupt the standard logic that repeats, across a leveled-out "global" geography, object forms conceived in the design centers of the Global North. These alternative design principles have less to do with how built forms and spaces *look*, much less with the recognizable mark of any architectural style or corporate designer, than with the conditions of their production and use. This is to foreground design as a mode of practice rather than as primarily a question of form. In this chapter, I set out five ways for thinking about such design in translation; in terms of spatial practices that are informal, incremental, improvised, impermanent, and insurgent.

Informality

It is easy enough, within contemporary design discourse, to focus on major players, spectacular forms, and grand designs. But it is important to temper such an approach to spatial design and development with a sense of the more mundane and certainly more typical ways in which urban spaces are made and remade. The global architectural styles with which I began this discussion may take a disproportionate share of capital investment in built environments, as well as dominating mainstream cultural representations of urban design, but they account for only a limited – if powerful – field of spatial production. A very substantial share of urban form emerges from the work of nameless "designers," unplanned development, and the everyday investments that shape informal urbanism. From the provision of urban housing to spaces of urban economic exchange to configurations of urban infrastructure: to a significant degree, basic morphologies of contemporary urban life emerge from non-expert practices of design, unregulated processes of development, and unofficial acts of planning (Roy, 2005; Roy and AlSayyad, 2004). This is well established in relation to poor-world urbanism; the necessary work of self-help spatial production that creates physical forms and meets urban needs in sites forgotten, ignored, or forsaken by the state and overlooked by mainstream markets. But urban informality is also characteristic of cities where state capacities are more developed and vigilant, capital more abundant, and planning systems more extensive.

Informality is endemic of contemporary urbanization: whether in the movement of populations, the building of physical environments, or the reproduction of urban economies, a major share of urban growth in the early 21st century is taking place outside the range of formal planning systems and the reach of government regulation. While it conventionally has been understood as an urbanism of the poor, informality "has become a primary avenue for home ownership for the lower-middle and middle classes" in cities of the Global South (AlSayyad, 2004: 20), as well as a key resource for a growing rentier class in any number of cities globally. The provision of housing, infrastructure, commercial, and common spaces through more or less informal means is a less visible but still pervasive element of cities in the Global North; long traditions of squatting in Northern and Western European cities run alongside patterns of irregular settlement in the transition economies of Eastern Europe, while informality is threaded through urban environments in the United States, notwithstanding "one of the world's strongest legal and governance regimes" (Durst and Wegmann, 2017: 294; see also McFarlane and Waibel, 2012; Mukhija and Loukaitou-Sideris, 2014; Vasudevan, 2015a, 2017; Wegmann, 2015). Echoing accounts of informal settlement in poor-world cities, Durst and Wegmann (2017: 283) trace the ways in which "informal housing in the US is characteristically geographically uneven, or highly variable from location to location; interwoven with formal housing and land ownership; and hidden, both in a figurative, and sometimes literal sense." Informal shelter solutions in these highly developed planning systems reveal numerous cracks in the regulatory field: in vehicle living, tent cities, squatting, occupations and repossessions, *colonias* and *chabolismo*, informal and "wildcat" subdivisions, subletting, multiple occupation, beds in sheds, unpermitted uses, and an emerging economy of petty rentier capitalism enabled by room- and home-sharing digital platforms.

Incrementalism

A second principle for thinking about a common grammar of design is in the forms of incremental development that underpin emergent housing strategies across a range of urban cases. The design of building cores and provision of basic collective infrastructure, which allow residents to rework or build out their homes over time as their needs and their means change, represents a reinterpretation – often with a stronger accent on design – of the site and services model deployed since the 1970s in many low-income housing contexts, and particularly in South Asia (see Wakely and Riley, 2010). Such interventions are motivated by variable factors: in large part, they respond to financial conditions in which public subsidies, collective equity, or individual finance are more or less severely limited; but they are also driven by a commitment to user engagement and co-production in the making of urban living space. Incrementalism, after all, is at the core of what a great many people do who stay in the same home for any length of time (from getting rid of the landlord's paint-job or enclosing the porch, to adding a lean-to or building another story); this approach to housing development builds in such a principle at the design stage. Again, the gradual building-out and regular remaking of basic housing structures is a taken-for-granted element of the urbanism of poor-world cities. But this mode of spatial practice can also be seen in diverse housing histories, and in current strategies in higher-income regions, where more and more people are squeezed out of overheated housing markets – from the precedent provided by the Mutual Self-Help Housing program in California, whose modest housing designs allowed for incremental expansion over time by householders (Mukhija, 2014); to the architect-led schemes of Berlin's *Baugruppen*, which combines design innovation with adaptable interiors and co-housing politics (see Ring, 2013); or the bare-bones designs of Naked House in London, which seek to further reduce costs by infill building on underutilized or void public land, including garages and parking spaces on council housing estates (see Wainwright, 2017).

The Chilean practice *Elemental*'s approach to incremental design – to draw on a well-known example in the Latin American context – has been much discussed in the contemporary design literature. Their low-income schemes provide "half a house" so as to allow residents to fill-in and enlarge the basic structure over time, redistributing the work of design – and the social relations of design production – between architect and inhabitants (see Aravena and Iacobelli, 2012; Iacobelli and Aravena, 2008). Just as notable is this model's reliance on the incremental co-financing of housing through combining government subsidies, private investments, and sweat equity. In projects in Iqique and Santiago in Chile and in Mexico City, for example, low-cost housing developments have pooled government funding to households to acquire land and finance collective infrastructure, and then provided the structural shell and basic facilities of a "middle-income house half-built" for lower-income residents to build out over time (Iacobelli and Aravena, 2008: 352). *Elemental*'s practice – like that of the *Baugruppen* in Germany's rather different development regime – is exemplary of a mode of design in translation, which draws variously on a low-cost site-and-services model, self-help housing strategies, the politics of architectural co-production, and an on-trend, camera-ready style. The mix of poor-world urbanism and design hipsterism is not to all tastes – *Elemental*, in particular, has been subject to a range of criticisms; whether for its incremental projects being too limited in scale and impact, or for taking too much corporate money and winning too many prizes. But architectural strategies such as these mark out a space for design between basic housing and infrastructure provision by development agencies in poor-world sites, and sub-standard or over-priced developer-led housing in rich-world cities – underlining how incrementalism might be understood as a common grammar of urban design in cities everywhere.

Improvisation

Both informal and incremental approaches to the design of urban space have an improvisational quality – in the sense of acts and interventions that have not been planned, and in getting by with resources and materials that come to hand. Orthodox modes of design tend to provide little room for improvisation, but reinvention is a common theme of everyday spatial practice. It includes the appropriation and adaptation of existing spaces for new or alternative uses – producing places of commerce, congregation, and civility in sites laid out for other purposes and to different designs. The work of design in these cases is most often undertaken by users themselves, repurposing shopfronts, garages, lock-ups, homes, and offices as places of worship, markets, dormitories, and social centers. Design, in this mode, concerns the way that spaces are made through practices of use, rather than how they are planned and produced around projections of future uses.

Acts of everyday urbanism (Chase et al., 2008), mixing opportunism, infiltration, and occupation in various parts, work within the formal morphology of the city to make spaces of commerce, care, shelter, information exchange, and social support. We might think, to take an example that travels widely across transnational urban contexts, of the improvisations of minority, migrant, and marginal religious groups in clandestine or commonplace spaces: unofficial mosques and prayer rooms in stores or restaurants; pop-up churches in strip malls or shacks; informal synagogues in private homes or – indeed – public bomb shelters (for the latter, see Aharon-Gutman and Ram, 2017; see also Finlayson, 2017; Hatziprokopiou and Evergeti, 2014; Kuppinger, 2011, 2014; Ribbens and De Beer, 2017). Similar strategies appear in relation to trade and commerce, as enterprising practices of design carve out retail and work spaces from holes-in-the-wall, underneath railway arches and overhead tracks, in garages, and on pavements. Money might be made where spaces can be created, and urban entrepreneurs in very disparate settings are adept at the annexations, adaptations, and accretions that support the generation of income through the production of space. Asef Bayat (2010: 15) points to the processes by which rural-to-urban migrants in large Middle Eastern cities "take over street sidewalks and other desirable public spaces to spread their vending businesses . . . turning the public streets into parking spaces for private gains, or use sidewalks as sites for outdoor workshops and other businesses." In her study of a "super-diverse" local street in south London, meanwhile, Suzanne Hall traces the many internal subdivisions of standard shopfronts to form "hybrid interiors" accommodating a range of trading and service activities: makeshift emporia combining mobile phone shops, money transfer, grocery stores, nail bars, and barbers (Figure 1.1). Attempts to map such ad-hoc economies, she notes (Hall, 2015: 31), were confounded by the manner in which they simply "defeated the standard land-use and retail categories."

Everyday improvisations such as these provide a low-key counterpoint to the design-savvy tactics at the boutique end of DIY urbanism, with its repertoire of urban parklets, book booths, and guerrilla gardens. If the informal mosque or the impromptu emporium represent a kind of tactical urbanism "from below," these more "vogueish" urban interventions are typically the tactics of a certain fraction of the middle. Douglas (2018: 14) notes that the "privileged

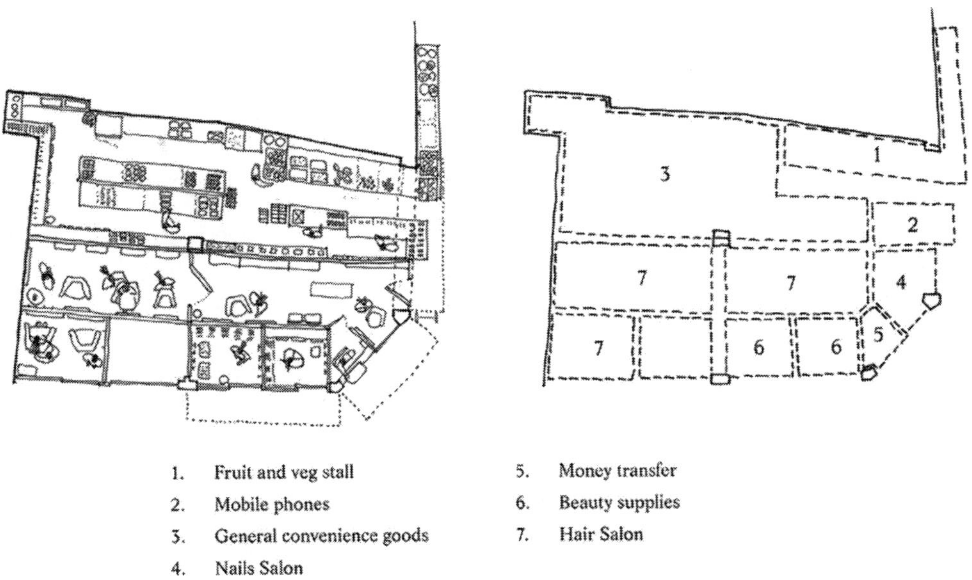

Figure 1.1 The improvised emporium: hybrid store interior in South London.

Source: Hall (2015); image by Nicolas Palominos (2015)

do-it-yourselfers" he studied, mainly in US cities, often "employ sophisticated knowledge of professional planning and scholarly urbanism" (as well as decent design software). Moreover, they may be more ready to risk the sketchy legality of closing a street or occupying a vacant property than others who feel more vulnerable in making claims to public space, and in their interactions with urban authorities; who are less middle class, more female, or less white (Douglas, 2018: 13). This is not to dismiss the spontaneous civic urbanism of AstroTurf-ed parking spaces, pallet pavilions, or stealth greening, but a preoccupation with these photogenic typologies – which are often intended as spatial *statements* – can obscure the mundane and more muted urban tactics that routinely remake space but show up less readily in the design literature. Different modes of improvised urbanism, however – while they vary markedly in terms of their aesthetic, their uses, their demographic profile, and their social media reach – share certain basic elements. Most simply, they put into question a distinction between the designers and the users of space, as use itself becomes the instrument (rather than the outcome) of urban design. This is to think about design as a form of direct action in the city; one in which agency is distributed and expertise may be democratized (see, for example, Deslandes, 2013; Douglas, 2014, 2018; Finn, 2014; Iveson, 2013; Talen, 2015).

Impermanence

Urban improvisations work to their own timescales. The kinds of tactical urbanism that transform city streetscapes with rogue signage, seed bombing, or pavement art appear suddenly and last for as a long as they can get away with it. The many makeshift forms in which everyday users rework space as places of congregation, worship, livelihood, or residence have a more durable – if still precarious – presence in the urban environment. The temporalities of tactical design are uneven and often unstable, but they share impermanence as a basic condition. This has a number of senses. Urban designs may be time limited as a matter of intent: figurations in space that are also interventions in time. In unofficial road closures, flash retail, mobile gardens, or urban beaches, temporal formats are key to producing spatial forms. For design tactics such as these, making space is itself a *matter* of time. In a slightly different mode, impermanent design is about using up empty time as well as occupying vacant space. The spatial projects associated with interim or "meanwhile" uses offer a temporal infill on underutilized land, in derelict buildings, or in becalmed development sites. Such spatial appropriations, despite their debt to histories of urban squatting (see Vasudevan, 2015a), have become preferred tactics of both city governments and private developers, who are happy to see free-issue or low-cost design provide services, secure property, clean up lots, and generate buzz on underused public land and dormant developments (see Lydon and Garcia, 2015; Madanipour, 2017a, 2017b; Tonkiss, 2013). In these design strategies, impermanence is a matter of the brief. For the greater part of everyday urbanism, however, the provisional nature of design and use is a common feature of spatial life. Where uses are unlicensed, adaptations are unpermitted, and tenures are irregular, impermanence may be a long-term condition. This is less about "meanwhile" uses than *mean-to-stay* uses, which are embedded in the urban environment but subject to legal precarity and temporal contingency as standard states.

If these designs on space – some more visible and some more illegal than others – have various timeframes, they all run counter to conventional models of urban design, planning, and development that are geared to the long term. Viewed from another standpoint, however, impermanence appears as the *normal* urban condition. Rahul Mehrotra has written suggestively of Indian urbanism, and that of Mumbai in particular, in terms of a "kinetic city"

Figure 1.2 The "city in motion" – Kolkata: the kinetic city is characterized by movement, in contrast to the static city of monumental and architectural form.

Source: Image by Claire Mookerjee

(Figure 1.2); temporary in character and inventive in its use of recycled materials and repurposed spaces (Mehrotra, 2008, 2011). Such a concept refers to the dynamic transformations of urban space that take place during festivals and ritual processions, but also to the more regular modes of city-making created by bodies on the move – the street vendors, delivery runners, cart pullers, rickshaw drivers, and waste-pickers, who compose mobile systems of everyday infrastructure to serve a complex and expansive city. This "city in motion" is produced around the *static city* of fixed material form: the latter is the city that shows up on official plans and is mappable in two dimensions. The kinetic city is not so easily captured in this way; it is a "three-dimensional construct of incremental development" (Mehrotra, 2008: 206), which takes spatial chances and builds where it can, with what it has, for as long as it holds. Its primary physical domain, in this sense, is the field of urban housing, given the share of populations in developing cities (estimated to be from one third to almost half in urban India) living in informal settlements or unauthorized structures. The fact that so much housing in contemporary cities might be classed as a kind of temporary shelter underlines the fact of impermanence as an urban condition, the provisional nature of everyday urbanism, and the extent to which it is persistent use and sustained occupation – as much as built form – that establishes degrees of permanence in the city.

Insurgency

The point, however, is not simply to celebrate the contingency of spatial form in the city, given the precarious hold so many city dwellers may have on their urban environments and livelihoods. Mehrotra's account of the kinetic city, while alert to its adept and inventive qualities, is clear about the ways in which colonial and modernist planning legacies intersect with the deepening inequalities of a neoliberal urban economy to produce the conditions of kinetic existence. The spatial sorting and functional divisions carved into the urban landscape by these earlier planning histories are overlaid by economic segmentations, splintering infrastructures, and environmental inequities as more people seek to make their lives and livelihoods in increasingly polarized and poorly served cities. Improvised and interim designs on space are not simply the signal moves of experimental designers in more privileged urban contexts, but part of a much broader, more "ordinary" and less expert set of spatial practices that cut across distinctions between low- and high-income urban settings. In situations of widening disparity, state withdrawal, or punitive austerity, urban actors in many rich-world cities also draw on the repertoires of poor-world urbanism to advance claims to space in insistent and insurgent ways. (Holston, 1998, 2009; see also Bayat, 2010; Dikeç, 2018; Hou, 2010; Miraftab, 2009; Roy, 2009a; Vasudevan, 2015b).

In recent years, some of the most striking reframings of urban space have occurred in sites of protest, as movements in response to political repression, financial crisis, and worsening inequality have taken shape in street-based demonstrations and spatial occupations in numerous cities. Much of this spatial activism has reworked the architectural formats of urban order – claiming parks and plazas as arenas of occupation, using roads and roundabouts as "infrastructures of insurgency" (Davis and Raman, 2013). These movements on the streets and the squares have also translated into more sustained incursions in urban space. Anti-austerity protests in European cities – in Spain, Italy, and Greece, most notably – are the visible expression of more dispersed movements, which have occupied vacant property and "activated" public space to provide infrastructures of information, informal welfare provision, social clinics and pharmacies, housing, migrant services, education, and social care. Against a backdrop of ongoing crisis and government incapacity, the protests in the squares of cities such as Athens and Thessaloniki have connected up with networks of social centers, providing local bases of solidary provision and self-organization (see, for example, Arampatzi, 2017; Hadjimichalis, 2013). In the Italian case, such initiatives build on a radical urban tradition of *Centri Sociali*, which frequently make use of the abandoned architecture of the state, and employ spatial tactics of trespass and occupation with the aim of "producing rights by means of illegality" (Mudu, 2018: 450; see also Montagna, 2006; Mudu, 2004). The politics of urban occupation are seeded through these urban environments, composing a social architecture of welfare, services, and support; making spatial claims through appropriation and persistent presence, as well as through larger statements of protest in public space (see Vasudevan, 2015b; see also Benjamin, 2008). Insurgent urbanisms in this sense involve different designs on space. They are most visible when they bring the politics of dispossession, exclusion, and inequity into public sites of occupation and demonstration. But they rework the spatial fabric of the city in more stubborn and often more surreptitious ways, disrupting urban logics of over-regulation, privatization, or dereliction as they make spaces of autonomous action, common access, and resourceful reuse (Figure 1.3).

Figure 1.3 Je So' Pazzo (EX-OPG), Naples: In this *centro sociale*, which occupies a former secure psychiatric hospital, a visiting room has been remade as a library/reading room.

Source: Image by Fran Tonkiss

Conclusion: the arts of incomplete translation

This discussion has been concerned with how precepts and practices of spatial design travel across urban settings. It has tried to "aim off" the most dominant and highly capitalized movements of international urban design and development, placing emphasis less on the ways that mobile capital remakes space at a distance, than on how mobile bodies reconfigure it *in situ*. Its focus has been on design migrations that subvert a conventional direction of travel from Global North to Global South, and with seeing resemblances across cities based on suggestive connection rather than strict comparison. This is to approach design in terms of an iterative rather than an "imitative urbanism" (Clarke, 2012; Robinson, 2006), parsed not in the simple "idiom of emulatable cities" (Ong, 2011: 17) but in borrowed phrasings and partial translations. Two critical points emerge from such an approach: a focus on urban design centering on practices rather than forms; and an extended sense of the actors involved in the work of designing the city.

How might more formal design practice and pedagogy respond to these ways of making urban space? Let me pose four propositions which extend design thinking towards these everyday interventions without simply incorporating them into the field of professional practice. The first is to treat design as provisional. There is no contradiction in seeking to create forms that are durable, sustainable, and robust while at the same time assuming these will be open to translation, adaptation, and addition at different scales and over various timeframes. The sustainability of built forms, indeed, depends in large part on the degree to which they lend themselves to adaptations over time. The formal work of the designer represents a phase in the production of urban space that necessarily remains unfinished, as the work is taken up by other actors who augment, transform, or undo it through processes of use and reuse. It follows, second, that collaboration is open-ended. Collaborative principles are fundamental to urban design as a field that mediates architecture, planning, engineering, and the social sciences; they are also crucial to a design politics that seeks to engage potential users in processes of spatial design. Such collaborations go beyond the design and build stages, and the formal repertoires of user consultation or co-production: collaborative work continues long after the expert designers have left the scene, as everyday practitioners inhabit, build out, and re-fit their spaces of work, residence, commerce, culture, and encounter. The third proposition is to see planning in terms of affordance rather than prescription. This is to focus on the kinds of spatial interventions and occupations that urban planning systems promote or permit – whether as part of a positive planning agenda to encourage temporary uses, interim structures, *pro tem* leases, *de facto* tenures, or informal exchange; or on the basis of a permissive planning attitude that tolerates or overlooks such practices, due to a liberal orientation or to a lack of regulatory capacity (Tonkiss, 2013: 313–314). The making of spatial forms, fourth, is about creating conditions of possibility rather than composing fixed spatial scripts. Urban planning and design, understood in these terms, seek to provide opportunities for use rather than to prescribe them. This is particularly relevant to the design of open and shared spaces, which seek to accommodate a diversity of uses and users at different times and for various purposes: the idea of design as affordance means that physical forms support and suggest a range of possibilities for use, but do not over-determine them. The political thinker Michael Walzer (1986: 470) once wrote in simple terms of the qualities of "open-minded" urban spaces that are "designed for a variety of uses, including unforeseen and unforeseeable uses, and used by citizens who do different things and are prepared to tolerate, even take an interest in, things they don't do." More recently, the sociologist Richard Sennett (2018) has argued for urban design geared to the principles of an "open city" in which spatial forms (like the city itself) remain permeable, porous, and incomplete.

The design "principles" that have been outlined above – informality, incrementalism, improvisation, impermanence, and insurgency – are not objectives to be laid out in advance of making a plan or conceiving a space. They arise, rather, from spatial practice; as means of achieving certain spatial ends – accessing shelter, enacting faith, setting up shop, finding refuge, claiming identity (for diaspora in particular), exercising freedom of expression, opposing hegemony of power – rather than ends in themselves. These modes of design are not typically articulated around proper names, but via practical idioms. Although more "expert" designers may play a critical role in translating spatial practices between urban contexts – and while city governments, planning agencies, and private developers may seek to appropriate the uses of "guerrilla" tactics within their own development strategies – the place of everyday social actors in the making and remaking of urban space puts into question any narrow sense of design as a specialist field or an authorized activity. Concepts of informality or insurgency have been strongly associated with the urbanism of cities of the Global South and low-income populations, but these design idioms run across disparate economic and political situations, and they underscore the ways in which poor-world urban strategies speak to the design dilemmas found in many rich-world cities. Consider, for example, low-tech or low-cost design, incremental housing solutions, de-motorized streets, and unconventional trading spaces: rather than being urbanisms of last resort, these means of making space are premised on alternative circuits of capital and a design language that departs from the dominant rhetoric of "global" urbanism. Such practices are *minor*, not necessarily because they are small in scale (informal housing and irregular economies, after all, take a significant share of urban settlement and livelihoods), but because they are marginal within mainstream design discourses. Yet, in the colloquial common language of informality, the piecemeal grammars of incremental development, the urban slang of improvised solutions, and the battle cries or quiet witness of insurgent urbanism, more people may get to have a spatial say in the making of their cities.

References

Aharon-Gutman, M. and Ram, M. (2017) "Objective Possibility as Urban Possibility: Reading Max Weber in the City," *Journal of Urban Design*. DOI:10.1080/13574809.2017.1377065

AlSayyad, N. (2004) 'Urban Informality as a "New" Way of Life', in Roy, A. and AlSayyad, N. (eds.) *Urban Informality: Transnational Perspectives from the Middle East, Latin America, and South Asia*. Lanham, MD: Lexington Books. 7–30.

Arampatzi, A. (2017) "Territorializing Social Movements," *City* 21(6): 724–736.

Aravena, A. and Iacobelli, A. (2012) *Elemental: Incremental Housing and Participatory Design Manual*. Berlin: Hatje Cantz Verlag.

Bayat, A. (2010) *Life as Politics: How Ordinary People Change the Middle East*. Amsterdam: Amsterdam University Press.

Benjamin, S. (2008) "Occupancy Urbanism: Radicalising Politics and Economy Beyond Policy and Programs," *International Journal of Urban and Regional Research* 32(2): 719–729.

Chase, J., Crawford, M. and Kaliski, J. (eds.) (2008) *Everyday Urbanism* (Expanded edition). New York, NY: The Monacelli Press.

Clarke, N. (2012) "Actually Existing Comparative Urbanism: Imitation and Cosmopolitanism in North-South Interurban Partnerships," *Urban Geography* 33 (6): 796–781.

Davis, D. and Raman, P. (2013) "The Physicality of Citizenship: The Built Environment and Insurgent Urbanism," *Thresholds* 41: 60–71.

Deslandes, A. (2013) "Exemplary Amateurism: Thoughts on DIY Urbanism," *Cultural Studies Review* 19 (1): 216–227.

Dikeç, M. (2018) *Urban Rage: The Revolt of the Excluded*. New Haven, CT: Yale University Press.

Douglas, G. C. C. (2014) "Do-It-Yourself Urban Design: The Social Practice of Informal 'Improvement' Through Unauthorized Alteration," *City & Community* 13 (1): 5–25.

Douglas, G. C. C. (2018) *The Help-Yourself City: Legitimacy and Inequality in DIY Urbanism*. New York: Oxford University Press.

Durst, N. J. and Wegmann, J. (2017) "Informal Housing in the United States," *International Journal of Urban and Regional Research* 41 (2): 282–297.

Finlayson, C. (2017) "Church-in-a-Box: Making Space Sacred in a Non-traditional Setting," *Journal of Cultural Geography* 34 (3): 303–323.

Finn, D. (2014) "DIY Urbanism: Implications for Cities," *Journal of Urbanism* 7 (4): 381–398.

Hadjimichalis, C. (2013) "From Streets and Squares to Radical Political Emancipation?" Resistance Lessons from Athens During the Crisis," *Human Geography* 6 (2): 116–136.

Hall, S. M. (2015) "Super-Diverse Street: A "Trans-ethnography" Across Migrant Localities," *Ethnic and Racial Studies* 38 (1): 22–37.

Hatziprokopiou, P. and Evergeti, V. (2014) "Negotiating Muslim Identity and Diversity in Greek Urban Spaces," *Social & Cultural Geography* 15 (6): 603–626.

Holston, J. (1998) "Spaces of Insurgent Citizenship," in Holston, J. (ed.) *Cities and Citizenship*. Durham, NC: Duke University Press. 155–174.

Holston, J. (2009) "Insurgent Citizenship in an Era of Global Urban Peripheries," *City and Society* 21 (2): 245–267.

Hou, J. (ed.) (2010) *Insurgent Public Space: Guerilla Urbanism and the Remaking of Contemporary Cities*. London: Routledge.

Iacobelli, A. and Aravena, A. (2008) "Elemental: Housing as an Investment not a Social Expense," in Ruby, I. and Ruby, A. (eds.) *Urban Trans Formation*. Berlin: Ruby Press. 344–357.

Iveson, K. (2013) "Cities Within the City: Do-It-Yourself Urbanism and the Right to the City," *International Journal of Urban and Regional Research* 37 (3): 941–956.

Jacobs, J. (2012) "Commentary: Comparing Comparative Urbanisms," *Urban Geography* 33 (6): 904–914.

Kuppinger, P. (2011) "Vibrant Mosques: Space, Planning and Informality in Germany," *Built Environment* 37 (1): 78–91.

Kuppinger, P. (2014) "Flexible Topographies: Muslim Spaces in a German Cityscape," *Social & Cultural Geography* 15 (6): 627–644.

Lydon, M. and Garcia, A. (2015) *Tactical Urbanism: Short-Term Action for Long-Term Change*. Washington, DC: Island Press.

Madanipour, A. (2017a) *Cities in Time: Temporary Urbanism and the Future of the City*. London: Bloomsbury.

Madanipour, A. (2017b) "Temporary Use of Space: Urban Processes Between Flexibility, Opportunity and Precarity," *Urban Studies* 55 (5): 1093–1110.

McFarlane, C. (2010) "The Comparative City: Knowledge, Learning, Urbanism," *International Journal of Urban and Regional Research* 34 (4): 725–742.

McFarlane, C. and Waibel, M. (eds.) (2012) *Urban Informalities: Reflections on the Formal and Informal*. Abingdon: Ashgate.

McNeill, D. (2009) *The Global Architect: Firms, Fame and Urban Form*. New York: Routledge.

Mehrotra, R. (2008) "Negotiating the Static and Kinetic Cities: The Emergent Urbanism of Mumbai," in Huyssen. A. (ed.) *Other Cities, Other Worlds: Urban Imaginaries in a Globalizing Age*. Durham, NC: Duke University Press. 205–218.

Mehrotra, R. (2011) "The Static and the Kinetic," in Burdett, R. and Sudjic, D. (eds.) *Living in the Endless City*. London: Phaidon Press. 108–115.

Miraftab, F. (2009) "Insurgent Planning: Situating Radical Planning in the Global South," *Planning Theory* 8 (1): 32–50.

Montagna, N. (2006) "The De-commodification of Urban Space and the Occupied Social Centres in Italy," *City* 10 (3): 295–304.

Mudu, P. (2004). "Resisting and Challenging Neoliberalism: The Development of Italian Social Centres," *Antipode* 36 (5): 917–941.

Mudu, P. (2018) "Introduction: Italians Do It Better? The Occupation of Spaces for Radical Struggles in Italy," *Antipode* 50 (2): 447–455.

Mukhija, V. (2014) "The Value of Incremental Development and Design in Affordable Housing," *Cityscape: A Journal of Policy Development and Research* 16 (2): 11–20.

Mukhija, V. and Loukaitou-Sideris, A. (eds.) (2014) *The Informal American City: Beyond Taco Trucks and Day Labor*. Cambridge, MA: The MIT Press.

Myers, G. (2014) "From Expected to Unexpected Comparisons: Changing the Flows of Ideas About Cities in a Postcolonial Urban World," *Singapore Journal of Tropical Geography* 35 (1): 104–118.

Ong, A. (2011) "Introduction: Worlding Cities, Or the Art of Being Global," in Roy, A. and Ong, A. (eds.) *Worlding Cities*. Oxford: Blackwell. 1–26.

Peck, J. (2015) "Cities Beyond Compare?" *Regional Studies* 49 (1): 160–182.

Ribbens, M. and De Beer, S. F. (2017) "Churches Claiming a Right to the City? Lived Urbanisms in the City of Tshwane," *HTS Teologiese Studies/Theological Studies* 73 (3).

Ring, K. (2013) *Self-Made City: Self-Initiated Urban Interventions and Architectural Interventions*. Berlin: Jovis.

Robinson, J. (2006) *Ordinary Cities: Between Modernity and Development*. London: Routledge.

Robinson, J. (2011) "Cities in a World of Cities: The Comparative Gesture," *International Journal of Urban and Regional Research* 35 (1): 1–23.

Robinson, J. (2016a) "Thinking Cities Through Elsewhere: Comparative Tactics for a More Global Urban Studies," *Progress in Human Geography* 40 (1): 3–29, 60–182.

Robinson, J. (2016b) "Comparative Urbanism: New Geographies and Cultures of Theorizing the Urban," *International Journal of Urban and Regional Research* 40 (1): 187–199.

Roy, A. (2005) "Urban Informality: Toward an Epistemology of Planning," *Journal of the American Planning Association* 71 (2): 147–158.

Roy, A. (2009a) 'Strangely Familiar: Planning and the Worlds of Insurgence and Informality," *Planning Theory* 8 (1): 7–11.

Roy, A. (2009b) 'The 21st Century Metropolis: New Geographies of Theory', *Regional Studies* 43 (6): 819–830.

Roy, A. and AlSayyad, N. (eds.) (2004) *Urban Informality: Transnational Perspectives from the Middle East, Latin America, and South Asia.* Lanham, MD: Lexington Books.

Sennett, R. (2018) *Building and Dwelling: Ethics for the City.* London: Allen Lane.

Sklair, L. (2017) *The Icon Project: Architecture, Cities and Capitalist Globalization.* Oxford: Oxford University Press.

Talen, E. (2015) "Do-It-Yourself Urbanism: A History," *Journal of Planning History* 14 (2): 135–148.

Tonkiss, F. (2013) "Austerity Urbanism and the Makeshift City," *City* 17 (3): 312–324.

Vasudevan, A. (2015a) 'The Makeshift City: Towards a Global Geography of Squatting," *Progress in Human Geography* 39 (3): 338–359.

Vasudevan, A. (2015b) "The Autonomous City: Towards a Critical Geography of Occupation," *Progress in Human Geography* 39 (3): 316–337.

Vasudevan, A. (2017) *The Autonomous City: A History of Urban Squatting.* London: Verso.

Wainwright, O. (2017) "Strip Club," *RIBA Journal*, 25 May. www.ribaj.com/buildings/ommx-naked-house-london

Wakely, P. and E. Riley. (2010) "The Case for Incremental Housing." *CIVIS* Special Issue, World Urban Forum 5, March.

Walzer, M. (1986) "The Pleasures and Costs of Urbanity," *Dissent* 33 (4): 470–475.

Wegmann, J. (2015) "Research Notes: The Hidden Cityscapes of Informal Housing in Suburban Los Angeles and the Paradox of Horizontal Density," *Buildings & Landscapes* 22 (2): 89–110.

Further reading

Davis, D. and Raman, P. (2013) "The Physicality of Citizenship: The Built Environment and Insurgent Urbanism," *Thresholds* 41: 60–71. This piece includes examples from Cairo, Manama, and New York to consider how physical spaces in the city provide sites for "insurgent collective action," both in the architecture of public squares and in the infrastructure of roads and roundabouts.

Mehrotra, R. (2008) "Negotiating the Static and Kinetic Cities: The Emergent Urbanism of Mumbai," in Huyssen, A. (ed.) *Other Cities, Other Worlds: Urban Imaginaries in a Globalizing Age.* Durham, NC: Duke University Press. 205–218. An examination of the idea of the "kinetic city" in relation to Mumbai, exploring impermanence as a basic feature of urban forms and everyday urban life.

Miraftab, F. (2009) "Insurgent Planning: Situating Radical Planning in the Global South," *Planning Theory* 8 (1): 32–50. This piece draws on anti-eviction struggles in South Africa's Western Cape to consider how critical planning approaches might respond to insurgent citizens' movements and spatial claims.

Roy, A. (2005) "Urban Informality: Toward an Epistemology of Planning," *Journal of the American Planning Association* 71 (2): 147–158. This piece addresses informality as a general mode of contemporary urbanization and examines how urban planning and policy might work with conditions of informality in the interests of distributive justice in the city.

2
FISHBOWL CITY
Postcolonial Los Angeles and the philosophy of the urban

Ananya Roy

Introduction

In 2016, the American hip-hop collective A Tribe Called Quest released the song "We the People," which was immediately hailed as a fierce critique of the Trump presidential campaign and presidency. *Rolling Stone* named it one of the "13 great anti-Trump protest songs" (Dolan et al. 2017), and the group's 2017 Grammy performance of the song, with Anderson Paak, Consequence, and Busta Rhymes, portrayed Trump as "President Agent Orange" and ended with the call to resist. The song's refrain speaks to displacement and expulsion: "And all you poor folks, you must go/Muslims and gays/Boy, we hate your ways/So all you bad folks, you must go." But the song is not simply about the current historical conjuncture of Trumpism and its distinctive constellation of class power, xenophobia, racism, and homophobia; it is also a commentary on the persistent practices of urban exclusion and displacement for which the term gentrification has become shorthand. In addition, I read the song as a critique of practices of representation, those that turn marginalized neighborhoods into a fishbowl and yet "overlook street art." The appropriation of space thus often goes hand in hand with narratives that silence and suppress forms of knowledge produced in these communities. With this in mind, the title of this essay is drawn from "We the People": "Niggas in the hood living in a fishbowl/Gentrify here, now it's not a shit hole."

In this essay, I take the provocations presented by A Tribe Called Quest as the starting point for an exploration of urban theory and research. In particular, I am interested in drawing attention to how theorizations of displacement and dispossession must contend with the long and continuing history of multiple forms of colonialism, including settler-colonialism. Needless to say, this project of theorization cannot be separated from key questions of voice, representation, and narrative. With this in mind, I examine the entangled organization of space, power, and knowledge and seek to pinpoint epistemologies and methodologies that challenge coloniality. I view these two interrelated endeavors, one of reinterpreting urban exclusion and expulsion and the other of refusing extractive and exploitative research, as together constituting a philosophy of the urban. Such a philosophy relies heavily on postcolonial theory, from Said's (1978) seminal exposition of the relationship between knowledge and power, to Spivak's (1999) sharp critique of epistemic violence, to Mignolo's (2010) insistence on decolonial thinking. I also hold in simultaneous view a growing body of work titled "black geographies," which is not only an analysis of racialized spaces and histories but also a call to remake the practices of theory and research. In doing so, I contend that the black radical tradition, from Du Bois (1935) to Fanon (1961) to Gilroy (1993) to Robinson (1983), has always been concerned with world-systems of racial capitalism, with the far reach of Euro-American imperialism, and with the necessity for a decolonial imagination of placemaking. It is thus potentially productive to explicitly link the worldliness of black geographies with postcolonial theory and its epistemologies of worlding (Roy and Ong 2011).

I argue that such a philosophy of the urban serves as an epistemological reorientation for urban design. First, while urban design, in keeping with the rest of urban studies, expresses dismay over gentrification, there is often less attention to the long histories of colonial occupation and racialized dispossession. This type of colorblindness can serve to maintain the racial status quo and can elide the role of spatially oriented professions, including urban design, in producing landscapes of racial segregation and separation. Second, to challenge colorblindness, urban design has to confront its Eurocentrism. The problem with such Eurocentrism is not only that it obscures from view the complex urban transformations underway in many different parts of the world but also that it whitewashes the history of cities in the West. The universals that thus circulate, whether in urban theory or in urban design practice, often rehearse and consolidate urban experiences of whiteness, reducing racial-ethnic difference to exotic diversity or identity politics. Third, there is no design solution to such forms of coloniality. As postcolonial theory reminds us, the recuperation of subaltern voices and histories often engenders epistemic violence. As the black radical tradition reminds us, themes of suffering and death often proliferate, diminishing black personhood and obscuring a black sense of place. But, as I argue in the conclusion of this essay, it may be possible for urban design to learn from, and act in solidarity with, decolonial imaginations and practices of placemaking. Emerging from sites of social struggle, these frameworks exemplify a philosophy of the urban, which highlights racialized expropriation and resists extractive and exploitative forms of research. It is to these epistemologies and methodologies that I now turn.

The philosophy of the urban

In urban studies today, there is much debate about the possibilities and limits of "theory from the South." That mandate has often been misunderstood as populating urban theory with case studies of cities in the Global South. Similarly, postcolonial urbanism has often been reduced to the documentation of cities in formerly colonized countries. Theory from the South, like postcolonial theory more broadly, is a theoretical approach, which devises concepts derived from the historical specificity of the Global South. As I have emphasized in prior work, not only are these concepts generalizable but they also have particular relevance to the understanding of urban processes in the Global North (Roy 2009a). Especially pertinent here is the conceptualization of the Global South by anthropologists Jean and John Comaroff. In their influential text, *Theory from the South: Or How EuroAmerica Is Evolving toward Africa*, they argue that it is in the Global South where "radically new assemblages of capital and labor are taking shape" and that these "prefigure the future of the global north" (Comaroff and Comaroff 2012: 12). Theory from the South, the Comaroffs (2012: 7) note, is not about narrating modernity from its "undersides," but rather revealing the "history of the present" from the "distinctive vantage point" that these frontiers of accumulation offer. In this sense, Los Angeles must be understood as a postcolonial city, a space of "colonial returns," to borrow a phrase from Jacobs's (1996: 64) brilliant interpretation of London as the "edge of empire." Such returns, as Jacobs (1996: 59) argues, can be nostalgic deployments of a colonial past, catalyzing racialized struggles over territory and memory, for example in conservation practices. Or they can mark an unsettling of imperial boundaries, a return by colonized groups, those either kept afar or expelled.

A theoretical approach, which positions Los Angeles as a postcolonial city, challenges Eurocentrism. In his incisive analysis of Eurocentrism, postcolonial theorist Mignolo (2010: 159) demonstrates how epistemic dominance is maintained through "a detached and neutral point of observation" or the "hubris of the zero point." As he notes, it is from here that "the knowing subject maps the world and its problems, classifies people and projects into what is good for them." It is from here that the "hood" appears as a fishbowl, its people merely objects of study or derision as delineated by A Tribe Called Quest. This neutral point of observation, this knowing subject, these systems of classification, constitute the epistemological and methodological foundations of universalism. Indeed, the power of Eurocentrism lies in the capacity to produce and disseminate universal models – of urban form, urban development, urban meaning, urban identity. Such universals emerge from quite particular experiences, be it the Italian piazza or small-town America, and yet circulate as neutral and detached. Inevitably, such universalism suppresses historical difference, relegating the place histories of the subaltern other to either obscurity or exotic diversity.

Postcolonial theory has long grappled with how to deal with the problem of Eurocentrism and subaltern histories. The study of subalternity, Spivak (1999) reminds us, cannot be undertaken via conscientious ethnography. Instead, it requires an examination of the principles and procedures of our own disciplines; it requires a remaking of the forms of representation that repeatedly suppress and subordinate subaltern histories; it requires facing up to our complicity, as scholars, in the creation of archives and narratives that "overlook street art," or even worse, replace it with street art authorized by the knowing subject.

If we are to think of this as an epistemological reorientation of our disciplines and professions, then two imperatives immediately come to mind. First, the spatially oriented disciplines often demonstrate an amnesia about their role in producing and legitimizing racialized landscapes of exclusion and expulsion. As Çelik (1997) and many other architectural historians have shown, the norms and forms of European modernism were crafted in the crucible of colonialism; for example, under French occupation in the cities of Algeria and Morocco. In the United States, urban planning has played a crucial role in underwriting what Lipsitz (1995) has famously called "the possessive investment in whiteness," ensuring white unity through programs and policies such as residential segregation. These disciplines and professions are also steeped in what Ranganathan (2016) has called "racial liberalism," an unshakeable faith in integration despite persistent histories of exclusion and abandonment. Ranganathan (2016: 6) thus calls for attention to the "paradigmatic practices of racial liberalism," which she defines as "conjoined processes of racialized property making and property taking." In other words, an epistemological reorientation requires a reckoning with the role of urban planning, urban design, and related fields of expertise in producing and maintaining racialized expropriation.

Second, it is not enough to acknowledge expulsion and dispossession. Scholars of black geographies insist that theory and research must narrate black life, not just black death. Thus, Woods (2002: 62) boldly states that "predictions of the death of impoverished and actively marginalized racial and ethnic communities are premature." Even more boldly, he asks: "Have the tools of theory, method, instruction, and social responsibility become so rusted that they can only be used for autopsies?" (Woods 2002: 63). In homage to Woods, I argue that a philosophy of the urban must learn from both prominent and less-known social movements, those that are committed to decolonial imaginations and practices of placemaking. Often steeped in anti-colonial and anti-racist struggles, these movements exhibit a worldliness that urban theory and research can only hope to embody. Often cast in language that is fierce and raw such as that in "We the People," they tell an undeniable truth about core urban practices such as property-making and property-taking. The intention here is not to romanticize subaltern practices. Rather, in keeping with the legacies of Fanon and other philosophers of freedom, it is to consider how the human can be reclaimed in the face of dehumanization and how in turn the urban might be reclaimed as spaces of freedom and collectivist history.

"Shit's about to pop"

In cities across the United States, there are intense struggles being waged by communities against gentrification. In Los Angeles, a key site of contestation is Boyle Heights, a Latino working-class neighborhood east of downtown Los Angeles. A *Newsweek* article provided the following socioeconomic profile: "Of the neighborhood's 92,000 residents, 94 percent are Latino, 33 percent live in poverty, 89 percent rent, 95 percent do not have a four-year college degree, 17 percent are undocumented immigrants" (Nazaryan 2017). As Loukaitou-Sideris et al. (2017) note, while the most visible and established measures of gentrification are not immediately apparent in Boyle Heights, representatives from community-based organizations express widespread anxiety about "cultural tourism" as well as about possible displacement caused by a new transit station. One example of the new aura surrounding Boyle Heights is its description, provided by Airbnb, as an "on-the-border neighborhood" complete with "mariachi, mariscos, and murals" and with "convenient access to livelier locales like Downtown and the Arts District" (www.airbnb.com/locations/los-angeles/boyle-heights).

But such representations are being vigorously resisted by organizations and movements based in Boyle Heights. From protests against hipster coffee shops such as Weird Wave Coffee (Vives 2017) to rallies against and boycotts of art galleries such as PSSST Gallery, activists have used blunt language and direct action to reveal and challenge the coalitions between developers, city officials, nonprofit organizations, and the art industry that produce what the Boyle Heights Alliance Against Artwashing and Displacement (BHAAD) calls "art-washing" or the "beautification" of "violent gentrification" (http://defendboyleheights.blogspot.com/). In particular, they have diagnosed such processes as the forced removal of people of color, thereby placing contemporary urban transformations in a long history of displacement and dispossession enacted through state violence. This includes the Pico Aliso public housing demolitions and evictions of 1996 that launched the formation of Union de Vecinos. But it also includes what Sánchez (2017: 165–167) describes as the "tangled history of forced removals in Boyle Heights": the repatriation of Mexican residents in the 1930s, organized by the Los Angeles County Board of Supervisors; the internment of Japanese Americans during World War II; and large-scale urban renewal to build public housing and freeways in the 1950s. Indeed, as several scholars, including Avila (2014), have demonstrated, a prolonged period of highway construction tore up neighborhoods in east Los Angeles and, as Sánchez (2017: 167) notes, ultimately devoured "almost 15 percent of the land in Boyle Heights by the time construction was completed." What is at stake in Boyle Heights, then, is what McKittrick (2011: 951) has called "urbicide," "the deliberate death of the city and willful place annihilation." As she notes, the term urbicide, often used by urban theorists to indicate geographies of imperialism, must also be extended to geographies of/at home, from the plantation to displacement.

The current struggles against gentrification in Boyle Heights have to be understood in this context. They also provide important analytical lessons towards building a philosophy of the urban. Take, for example, the frames used by Defend Boyle Heights to describe the "state of war." Eschewing the usual language of neoliberalization, the group analyzes gentrification as "white supremacy" and calls out the ruse of diversity and integration staged by what it pinpoints as "white art." Vividly expressed in blog posts and protest graffiti, Defend Boyle Heights produces narratives of anger on behalf of "low-income working class renters of color." Here is one glimpse of this politics of representation, from the Defend Boyle Heights blog post of April 29, 2017.

I quote at some length from it because such statements of boycott, divestment, and refusal are rare in urban studies. But this precisely is the epistemological reorientation for which I am arguing. This is the rejection of death and expulsion that Woods hopes for.

> Yes, PSSST left because they were scared, terrified. They were confronted by members of the community – where they imposed their neighborship to a community just a few hundred feet away who have the fight against displacement flowing through their veins, still a fresh memory from the mass displacement and demolition of the Pico Aliso/Aliso Village housing in the late 1990s. . . . *To be frank, PSSST Gallery left because of the dangerous atmosphere they created for themselves, attracting rage to their building, their staff, their board of directors and their supporters.* Finally, it was clear that Boyle Heights was not a safe space for PSSST Gallery. . . . Because the most marginalized, the most oppressed have everything to lose if Boyle Heights loses the war to gentrification, and that right there, that conclusion, that understanding, has the power, when put into action, to do remarkable things. . . . Boyle Heights is not a safe space for gentrifiers, for white hipster-artist scum who arrogantly parade around our streets, our hood, our barrio, like contemporary colonizers, walking their dogs at Mariachi Plaza or Hollenbeck Park, taking pictures of all the authentic, supposedly "exotic" Chicanx/Mexican culture like poverty tourists, like zoologists. . . . In short, shit's about to pop.
>
> *(http://defendboyleheights.blogspot.com/; emphasis in original)*

Postscript: The struggles against gentrification in Boyle Heights have generated controversy in various spheres of knowledge and power, including media, urban design and planning, and academia. Explicit in their use of frames such as urban colonialism and targeted in their protest tactics, such as the graffiti "Fuck White Art," Boyle Heights activists have insisted that shit will pop. In 2017, a group of graduate students from the Department of Urban Planning at UCLA formed a year-long working group, "Our Hoods, Our Stories," to learn from and collaborate with anti-gentrification movements and organizations. Realizing soon that it was not their role to narrate the "state of war," the students organized a panel held at UCLA with representatives from Defend Boyle Heights, Union de Vecinos, and the Boyle Heights Neighborhood Council. Toward the end of the evening, an adjunct member of the UCLA faculty commented that each time he or one of his colleagues takes a group of their students to Boyle Heights for a field trip, they face hostility. Why is that? Why can't it be different, he asked? Implicit in his questions is the fishbowl city, unmediated and unchecked access to the lives of the subaltern and the spaces of the hood. Needless to say, Boyle Heights residents are not making field trips to the manicured whiteness of Westwood and Bel-Air amidst which the UCLA campus sits. That evening at UCLA, Boyle Heights activists patiently explained why their community is not a laboratory or an exhibition. They made the case for an end to practices of research that exploit communities and extract their narratives for academic commodification. They sought to tell their own stories about their own hood.

Ways of seeing

Recently there has been considerable discussion in urban studies about displacement. Stretching well beyond understandings of gentrification as class struggle or evictions as symptoms of neoliberalization, scholars have called for attention to "old and new urban colonialities" that produce a permanent state of displaceability for certain social groups (Yiftachel 2017). In doing so, they have insisted, as have social movements in neighborhoods such as Boyle Heights, that contemporary displacement be understood in relation to histories of settler-colonialism and in particular to what Ranganathan has described as property making and property taking. As Hern (2016: 19) puts it, "under all is the land," and this in turn requires an account of "the colonial accumulation our cities are constructed on and the kinds of labor that built them." Thus, Porter and Yiftachel (2017: 2) delineate the following task for urban studies inquiry: "holding colonization as a contemporary phenomenon that not only endures through time in settler-colonial cities, but also reconstitutes new dimensions of oppression and new possibilities for decolonization." In my own work, I have argued that it is necessary to shift from a theorization of displacement to that of dispossession (Roy 2017). My interest in doing so is not only to highlight the role of state violence in gentrification, evictions, and urban exclusion, but also to assert the constitutive role of racial hierarchy in structures of dispossession.

Let me highlight two theoretical implications of such a philosophy of the urban. The first is the denaturalization and deconstruction of liberalism's foundation, property. As Blomley (2004: 15) argues, instead of asking "what is property?" we must ask "what is to *count* as property?" (emphasis in original). I have extended Blomley's work by asking an additional question: who can count as the subject who can claim home and land? If certain subjects are always necessarily dispossessed, or constituted as property owned by others, how do they claim property? (Roy 2017). To rethink the "natural" character of property is also to necessarily rethink the law. As Goldstein (2008: 836) has shown, it is legal reason that has built "proprietary regimes," underpinning settlement as "an entitled and possessive relation to place" and casting indigenous populations as "supposedly unsettled." Thus Blomley (2004: 107) argues the law was the West's "instrument of empire," turning territories into "exploitable property." Possession, then, Nichols (in Goldstein 2017) argues, must be understood not as preceding dispossession but rather as its effect. "Differential racialization," Goldstein (2017) emphasizes, is necessary for this colonial capacity to possess.

We are surrounded by instantiations of dispossession and possession, of settlement and movement. Yiftachel (2009: 88–89) describes these as "gray spaces," "positioned between the 'whiteness' of legality/approval/safety, and the 'blackness' of eviction/destruction/death." He notes that these spaces are tolerated and managed but "while being encaged within discourses of 'contamination,' 'criminality' and 'public danger' to the desired 'order of things.'" In contrast, gray spaces created by "powerful or favorable interests" are "laundered." Similarly, I have shown how in India, the planning functions of the state valorize, and even produce, urban forms of elite illegality and criminalize subaltern informality (Roy 2009b). A striking example of such state violence comes from the West Bank, a context that Yiftachel (2009) describes as "urban colonialism." This is Area C, a jurisdiction encompassing 61% of the West Bank and created by the 1995 Oslo II Accord, where Israel has full control of civil and security affairs, including planning, construction, infrastructure, and development. In the foreground of this photograph (Figure 2.1) is an "unrecognized" Bedouin village. Displaced from the Negev desert, these Bedouins face the constant threat of forced removal (Figure 2.2). On the horizon is the sprawling Jewish settlement of Maale Adumim, illegal by international law but "laundered" by Israeli occupation and settler-colonialism. While the illegal Jewish settlement enjoys urban amenities and is connected to Jerusalem through premium infrastructure, the unrecognized village of Abu Nuwar faces eviction, with the Israeli government threatening to relocate them to a nearby toxic garbage dump to make way for the planned expansion of Maale Adumim. At the kindergarten self-built and self-run by the Bedouins, children draw their urban hopes and aspirations. They draw homes that resemble those that they see every day on the horizon, in a settlement that they can never enter, the suburban homes of Maale Adumim (Figure 2.3).

To think from Boyle Heights or from Abu Nuwar is to theorize from the South. The South is not a location, it is a relation. It is a structural relation – of space, power, and knowledge – produced

Figure 2.1 An unrecognized Bedouin village in Area C, West Bank.

Source: Photo by the Author

Figure 2.2 Threat of forced removal in Abu Nuwar.

Source: Photo by the Author

Figure 2.3 Kindergarten drawings of Maale Adumim, a Jewish settlement.

Source: Photo by the Author

and maintained in the crucible of racial capitalism on a global scale. To see from the South is to see EuroAmerica differently. It is to see expulsion and dispossession.

"If I Was President"

In the video accompanying their song, "If I Was President," the East Los Angeles–based band Las Cafeteras imagines a society constituted through free education, clean water, the end of incarceration, mutual aid, community history, and demilitarization. Performed live on *Democracy Now!* on August 1, 2017, the song starts with a commentary on the illegalization of black and brown workers: "Señor President le vengo avisar / No tengo papeles para trabajar / Señor Presidente le pido porque / Matan al Moreno con piel de café." But it is also a bold plan for redistribution and collectivism, a remaking of undocumented status into what band members have described as "a wild imagination about self-determination" (www.democracynow.org/2017/8/1/las_cafeteras_in_conversation_performance_on). Inspired by the Eastside Café, a Zapatista-inspired space of autonomy and community in East Los Angeles, which itself has been threatened by gentrification, the song is meant to express how those not recognized by a government can "govern themselves in a way that they find holds dignity."

I read "If I Was President" as decolonial imagination and practice, one rooted in the contested terrain of East Los Angeles. The dreams expressed in the song are very much in keeping with what Gibson-Graham (2014) have celebrated in their account of "post-capitalism": diverse economies with a wide range of social relations ranging from bondage, guilt, and self-exploitation to trust, cooperation, solidarity, and stewardship. Gibson-Graham (2014: S147) call for a revised understanding of the economy, one that is able to take account of these "diverse determinations." Such an approach is crucially important as well for a philosophy of the urban. But they also call for the active practice of post-capitalism, for "making revolution by pulling threads of hope through ongoing crises" (Gibson-Graham 2014: S151). Such indeed is the case of the undocumented Morena who dares to imagine herself as president.

But the question neither posed nor answered by theories and dreams of post-capitalism is why some communities are able to make revolution from crisis. I argue that, for this, we must turn to the long histories of colonial occupation and racialized expropriation. In a compelling analysis of multiracial communities in Hawai'i, Cape Town, and Los Angeles, Sánchez (2017: 162) develops the concept of "interracial radicalism." He tells the history of neighborhoods such as Boyle Heights as one shaped through segregation and exclusion, places where those kept out of "lily-white suburbs" were forced to live and where racially mixed communities thrived (Sánchez 2017: 157). Sánchez interprets these communities, such as District Six in Cape Town, as stories of "unification," where one can find "a progressive culture of political coalitions and democratic living."

Sánchez's analysis leads us to what McKittrick (2011: 947) calls a "black sense of place." As a key interlocutor of the theories and methodologies increasingly known as black geographies, McKittrick, like Woods, argues that "black suffering and naming racism" should not be the "sole conceptual schemas through which to 'understand' or 'know' blackness or race." I find striking resonance between McKittrick's theoretical and methodological interventions and the project of postcolonial theory. Taken together, they return us to questions of epistemic violence and provide important direction for a philosophy of the urban.

In her path-breaking critique of postcolonial reason, Spivak (1999) calls into question the efforts of postcolonial historiography to recover suppressed histories of the subaltern. For her, subalternity marks the limits of representation. The subaltern is the absence in the archives. She is the fishbowl subject who is represented through ethnography. She does not speak. She is spoken for. The subaltern is thus constituted through epistemic violence. Similarly, McKittrick (2011: 948) argues that the focus on the "suffering black body" can serve to maintain the equivalence of whiteness with freedom and blackness with dispossession and can "reify racial-colonial categories." As Mohanty (1991) reminds us of how "Third World women" are constructed "under Western eyes," so McKittrick warns that "inhabitants of spaces of absolute otherness" can be "discursively colonized by our intellectual investigations." In its place, she calls for "a collective history of encounter" that is based on "a difficult interrelatedness." The concepts of encounter, interaction, and interrelatedness are also central to postcolonial theory and its challenge to Eurocentrism. Thus, Mitchell (2000: 2) emphasizes a methodology that foregrounds "sites of interaction": "If modernity had its origins in reticulations of exchange and production encircling the world, then it was a creation not of the West but of an interaction between West and non-West." Thus, the slave ship and sugar plantation were not elsewheres or exteriors of Western modernity but rather constitutive of it. Postcolonial theorists remind us that such sites of interaction and forms of constitution are obscured from view by dominant epistemologies. Mignolo (2010: 166) describes this as "the epistemic privilege of the First World." McKittrick, too, is concerned with forms of knowledge that obscure a black sense of place. She writes:

> Instead of encounter, in fact, our present system knowledge, inherited from enlightened colonialism and Eurocentric modernity, repetitively constitutes blackness as a discreet (and hostile) racial category that routinely "troubles" an already settled whiteness (Morrison 1992). This paradigmatic perspective on race and blackness, in its denial of an entangled racial history produced through geographies of encounter, normalizes practices of colonization as it naturalizes overdevelopment, accumulation, and land ownership as identifiable – seeable locales of emancipation.
>
> *(McKittrick 2011: 950)*

Central, then, to postcolonial theory and black geographies is a revolution in epistemology. Jazeel (2017: 335) makes a distinction between postcolonial and decolonial methodologies, noting that "the decolonial shift is a project of de-linking, whereas postcolonial theory is a project of scholarly transformation within the academy." Indeed, Mignolo (2010: 161) delineates "decolonial freedom" as "de-linking from the magic of the Western idea of modernity." But if we take the collective history of encounter seriously, if we center the difficult interrelatedness of racial and spatial histories, then such de-linking is always out of reach. I see decolonial freedom as a normative horizon, the ever-present task of transforming the global university and its distinctive practices of racial capitalism. As Banfill et al. (2017) note in their reflections on the project of creating an urban humanities pedagogy at UCLA, including work by students with community organizations in Boyle Heights, a decolonized curriculum is often a "liminal zone," inflected with the very "spatial contestations and tensions" that are present in the neighborhoods and cities we hope to study, design, and transform.

In my work I have argued that a key part of decolonizing the university has to be the sustained challenge to the Eurocentrism of our disciplines; that Eurocentrism is the "sanctioned ignorance" (Spivak 1999) of colonial histories and racial oppression. Against such "unbearable whiteness" (Derickson 2017), I am interested in staging what Mignolo (2010) calls "epistemic disobedience." In urban studies, scholars have recently described this as an "unlearning," which includes decentering universal theories, locating and situating these theories in specific geographies, and limiting the reach of their explanations (Lawhon et al. 2016). What might such unlearning and disobedience entail for urban design? Let me suggest two possibilities.

The first is the epistemological task of thinking from the South. As I have argued earlier in this essay, to think from the South is to rethink the universal history of urbanism. It is also to recognize the revolutionary potential of subaltern histories and spaces. The South, this South, is also in the United States. It is postcolonial history. It is black geography. Thus, in their recent book, *Chocolate Cities* (2018: 3) Hunter and Robinson call for a new map of black life, one that sees the "geography of the Black American experience . . . as existing within and across varying versions of 'The South.'" For them, the South is "one large territory, governed by a historically rooted and politically inscribed set of practices of racial domination"; but they also see the South as a "frame," as "cultural forms" that travel across the United States. The South, then, is epistemology and experience; it is territory and history; it is plantation, prison, colony, but it is also migration, movement, and freedom. The South is a philosophy of the urban.

The second is the political task of abandoning racial liberalism and making a commitment to racial and spatial justice. This in turn requires that urban design shifts from ideologies of aesthetic improvement to dreams of freedom, from civic engagement to the fight for civil rights. Many of the concepts discussed here, such as interracial radicalism or a black sense of place or a wild imagination about self-determination, have an urban materiality to them. Social movements know this already. They seek to build urban futures amidst colonial returns. An urban design that is in solidarity with such movements would seek to give form to these decolonial imaginations. In doing so, it would both confront and redesign its own collective history of encounter and the difficult interrelatedness that lies at the heart of the disciplines and professions of late modernity.

References

Avila, E. (2014). *The Folklore of the Freeway: Race and Revolt in the Modernist City*, Minneapolis: University of Minnesota Press.

Banfill, J., Presner, T., Zubiaurre, M. (2017). "Urban Humanities Pedagogy: The Classroom, Education, and New Humanities." https://boomcalifornia.com/2017/01/23/urban-humanities-pedagogy-the-classroom-education-and-new-humanities/

Blomley, N. (2004). *Unsettling the City: Urban Land and the Politics of Property*, New York: Routledge.

Çelik, Z. (1997). *Urban Forms and Colonial Confrontations: Algiers Under French Rule*, Berkeley: University of California Press.

Comaroff, J., Comaroff, J. (2012). *Theory from the South: Or, How Euro-America Is Evolving Toward Africa*, Boulder: Paradigm Publishers.

Derickson, K. (2017). "Urban Geography II: Urban Geography in the Age of Ferguson." *Progress in Human Geography* 41(2): 230–244.

Dolan, J., Shteamer, H., Esposito, S. (2017). "13 Great Anti-Trump Songs." *Rolling Stone*. www.rollingstone.com/music/lists/13-great-antidonald-trump-protest-songs-w467532, February 21.

Du Bois, W.E.B. 1935 (2007). Gates, L.H. (ed.) *Black Reconstruction in America: An Essay Toward a History of the Part Which Black Folk Played in the Attempt to Reconstruct Democracy in America, 1860–1880*, Oxford: Oxford University Press.

Fanon, F. (1961). *The Wretched of the Earth*, New York: Grove Press.

Gibson-Graham, J.K. (2014). "Rethinking the Economy with Thick Description and Weak Theory." *Current Anthropology*, 55(S9): S147–S153.

Gilroy, P. (1993). *The Black Atlantic: Modernity and Double Consciousness*, Cambridge: Harvard University Press.

Goldstein, A. (2008). "Where the Nation Takes Place: Proprietary Regimes, Antistatism, and U.S. Settler Colonialism." *South Atlantic Quarterly*, 107(4): 833–861.

Goldstein, A. (2017). "The Ground Not Given: Colonial Dispositions of Land, Race, and Hunger." Unpublished paper.

Hern, M. (2016). *What a City Is For: Remaking the Politics of Displacement*, Cambridge, MA: The MIT Press.

Hunter, M., Robinson, Z. (2018). *Chocolate Cities: The Black Map of American Life*, Berkeley: University of California Press.

Jacobs, J. (1996). *Edge of Empire: Postcolonialism and the City*, New York: Routledge.

Jazeel, T. (2017). "Mainstreaming Geography's Decolonial Imperative." *Transactions of the Institute of British Geographers*, 42: 334–337.

Lawhon, M., Silver, J., Ernston, H., Pierce, J. (2016). "Unlearning (Un)Located Ideas in the Provincialization of Urban Theory." *Regional Studies*, 50(9): 1611–1622.

Lipsitz, G. (1995). "The Possessive Investment in Whiteness: Racialized Social Democracy and the 'White' Problem in American Studies." *American Quarterly*, 47(3): 369–387.

Loukaitou-Sideris, A., Gonzalez, S., Ong, P. (2017). "Triangulating Neighborhood Knowledge to Understand Neighborhood Change: Methods to Study Gentrification." *Journal of Planning Education and Research*, Online, 1–16.

McKittrick, K. (2011). "On Plantations, Prisons, and a Black Sense of Place." *Social & Cultural Geography*, 12(8): 947–963.

Mignolo, W. (2010). "Epistemic Disobedience, Independent Thought, and Decolonial Freedom." *Theory, Culture & Society*, 26(7–8): 159–181.

Mitchell, T. (2000). "The Stage of Modernity," in Mitchell, T., (ed.) *Questions of Modernity*, Minneapolis: University of Minnesota Press.

Mohanty, C.T. (1991). "Under Western Eyes: Feminist Scholarship and Colonial Discourses," in Mohanty, C.T., Russo, A. and Torres, L., (eds.) *Third World Women and the Politics of Feminism*, Bloomington: Indiana University Press.

Morrison, T. (1992). *Playing in the Dark: Whiteness and the Literary Imagination*. New York: Vintage Books.

Nazaryan, A. (2017). "The 'Artwashing' of America: The Battle for the Soul of Los Angeles Against Gentrification." *Newsweek*. www.newsweek.com/2017/06/02/los-angeles-gentrification-california-developers-art-galleries-la-art-scene-608558.html, May 21.

Porter, L. and Yiftachel, O. (2017). "Urbanizing Settler-Colonial Studies." *Settler Colonial Studies*, Published Online, 1–9.

Ranganathan, M. (2016). "Thinking with Flint: Racial Liberalism and the Roots of an American Water Tragedy." *Capitalism Nature Socialism*, 3: 17–33.

Robinson, C. ([1983] 2000). *Black Marxism: The Making of the Black Radical Tradition*. University of North Carolina Press.

Roy, A. (2009a). "The 21st Century Metropolis: New Geographies of Theory." *Regional Studies*, 43(6): 819–830.

Roy, A. (2009b). "Why India Cannot Plan Its Cities: Informality, Insurgence, and the Idiom of Urbanization." *Planning Theory*, 8(1): 76–87.

Roy, A. (2017). "Dis/Possessive Collectivism: On Property and Personhood at City's End." *Geoforum*, 80: A1–A11.
Roy, A. and Ong. A., (eds.) (2011). *Worlding Cities: Asian Experiments and the Art of Being Global*, Chichester: Wiley-Blackwell.
Said, E. (1978). *Orientalism*, New York: Pantheon Books.
Sánchez, G. (2017). "Why Are Multiracial Communities So Dangerous? A Comparative Look at Hawai'i, Cape Town, South Africa, and Boyle Heights, California." *Pacific Historical Review*, 86(1): 153–170.
Spivak, G.C. (1999). *A Critique of Postcolonial Reason: Toward a History of the Vanishing Present*, Cambridge, MA: Harvard University Press.
Vives, R. (2017). "A Community in Flux: Will Boyle Heights Be Ruined by One Coffee Shop?" *Los Angeles Times*. http://beta.latimes.com/local/lanow/la-me-boyle-heights-gentrification-coffee-20170621-story.html, July 18.
Woods, C. (2002). "Life after Death." *The Professional Geographer*, 54(1): 62–66.
Yiftachel, O. (2009). "Theoretical Notes on 'Gray Cities': The Coming of Urban Apartheid." *Planning Theory*, 8(1): 88–100.
Yiftachel, O. (2017). "Displaceability: A Southeastern Perspective." *New Directions in Displacement Research: A Digital Blog Symposium*. http://mitdisplacement.org/symposium-oren-yiftachel/

Further reading

The three texts that follow examine how postcolonial theory and black geographies make possible readings of the urban which are attentive to racialized dispossession.

Hunter, M., Robinson, Z. (2018). *Chocolate Cities: The Black Map of American Life*, Berkeley: University of California Press.
Jacobs, J. (1996). *Edge of Empire: Postcolonialism and the City*, New York: Routledge.
Roy, A. (2009). "The 21st Century Metropolis: New Geographies of Theory." *Regional Studies*, 43 (6): 819–830.

PART I.II

Regional experiences

Introduction

The arguments and observations presented in the previous chapters by Tonkiss and Roy about urban design and comparative urbanism reverberate in the chapters presented in this section, although each chapter presents perspectives on emerging urbanism from different corners of the world. In each instance, we find regional experiences of what can be broadly described as the urban outcomes of globalization (Banerjee 2008). The specific development trajectories are of course varied, shaped by the different history, geography, and culture of places. Yet, in all of these regional perspectives, there are some consistent leitmotifs that concern the production of space and the resulting urbanism shaped by the imperatives of global economy and local power, exacerbating inequality, displacement, fragmentation, and alterity.

In the chapter entitled "Globalization, Resiliency, and Change," Lawrence A. Herzog addresses the emergent urbanism in Latin American cities. The scope of this chapter is quite comprehensive, as Herzog uses a rather long historical armature to place the current trends in a broader context. He begins by reviewing the three phases of Latin American city design, beginning with the colonial era and the Laws of the Indies, leaving behind the endogenous and pre-colonial traditions of the Mayans, Incas, and Aztecs. He argues that the early colonial city from 1500 AD onward was quite compact and efficient; its design largely shaped by the Laws of the Indies that dictated settlement policies. Beginning around 1800, with increased European immigration, Latin American cities expanded along corridors of upper-class settlements, with the poor living in emergent slum neighborhoods. The contemporary form of Latin American cities has largely resulted from the massive rural-urban migration that began in the 1940s. Finally, the era of globalization has brought about new developments reminiscent of, and influenced by, the model of the North American suburbia. In the rest of his chapter, Herzog focuses prospectively on three phenomena common in Latin America today: the rise of "slow urbanism" as a counterpoint to the fast lifestyle of the global economy; the challenges of the US style "global suburbs" and the attendant problems of sprawl and anomie; and finally, the new urban design prototype of the "transfrontier metropolis," exemplified in the US-Mexico border, which offers a "living laboratory of globalization and urban design," according to the author.

In the next chapter in this section, David Dewar focuses on the settlements of Southern Africa, a region of long colonial history and alterity famously enforced in the apartheid policies of the recent past. Drawing from a recent UN-Habitat report that characterizes the urban form and spatial organization of the Southern African cities as "the most inequitable and unjust in the world," Dewar suggests multiple criteria for evaluating spatial justice – ecological, generative capacity, access, livability, and choice – and finds these cities lacking in all accounts. He argues that the principal obstacle to redress some of these inequities and injustices is the current planning system, which tends to be programmatic and without any normative vision for how such issues of justice and inequity can be addressed. He makes a case for what he calls "structural-spatial approaches" to planning and urban design but concedes that such changes may not be enough to redress inequity and injustice; it will require aggressive intervention in the land market from the state.

In "The Fading Pulse of Place," Ramzi Farhat presents a critical panorama of the recent urban development efforts and outcomes in four important cities of the Middle East: Amman in Jordan, Beirut in Lebanon, Cairo in Egypt, and Damascus in Syria. These cities are well known in the West and globally because of their critical role in regional economy and politics, and in recent times, because of their political strife and upheavals of national and international scope. These cities have been deeply embedded in their regional and international political history, and they have suffered from major damages to their urban form, infrastructure, and urbanism. What many observers may not realize is that despite the provisional nature of peace and reconciliation, considerable urban design efforts are underway to rebuild and restore the historic urban form and urbanism rooted in the cities' culture. In reviewing the recent urban development efforts in these four cities, Farhat sees a "neoliberal agenda" shaped by overlapping imperatives of communalism, and by motives that are mainly political, and profit-oriented. This neoliberal agenda, he laments, threatens the traditional urbanism derived from a trade economy, a close-knit community character, and fine-grain social integration. He notes how the ongoing "downtown" redevelopment efforts to achieve such traditional urbanism rooted in cultural practices can get mired in an "illiberal development policy" leading to displacement and gentrification. However, Farhat sees redemption in decentralized and bottom-up approaches that obtain the kind of organic urbanism rooted in the cultural practices of these cities by regenerating "hip-town" districts and the unique sense of various localities. He sees the success of these initiatives depending on whether the traditionally authoritarian state can rid itself of the hegemony of real estate and other political interests.

The last chapter in this section, with the title "Un-Cities" by Surajit Chakravarty, focuses mainly on the emergent urbanism of Abu Dhabi, associated with "the rapidly growing cities in the Gulf region" that includes the entire United Arab Emirates. In a way, the theme of this chapter is parallel to what sociologist Harvey Molotch (2017) refers to as "cities without urbanism" in reference to Dubai, the other major city in the UAE. Chakravarty begins his chapter reviewing the spectacular transformation of the Gulf countries and their desert landscape – from small desert settlements to some of the most modern and striking cities of the world, fueled by oil revenue and massive infrastructure investments on ports, airports, highways, transit system, and the like. These cities aspire to be globally prominent in scope, creating consumption opportunities for visitors from the West and also for the newly emergent bourgeois class of the Global South. Consequently, luxury hotels, shopping malls, vacation homes, and other tourist attractions abound. The urbanism – or lack thereof – of these transformed landscapes brings multiple liabilities, and these remain the principal focus of Chakravarty's critical assessment. He refers to four "uniquely dysfunctional" characteristics of the cities of the Gulf region. These cities and the supporting infrastructure are essentially car-oriented, with all of the associated problems of urban sprawl: low-density development, traffic congestion, and pollution. A "fragmented urban fabric" is the expression the author uses to characterize the disjointed character of project-based development involving artificial islands and peninsulas without any comprehensive vision of the urban form. Residents are separated and segregated by income and identity, especially the large number of expatriates, many of them from the Indian subcontinent, earning low wages and living in what is tantamount to labor camps. Chakravarty also points to what he refers to as "people-last" architecture – signature architectural pieces, some from prominent Western designers – that symbolizes power and profit but not an urbanism of place and community.

It should be evident that the arguments by Tonkiss and Roy in the previous section continue to remain the leitmotif in these four chapters of regional perspectives of contemporary urbanism in different sociocultural contexts.

References

Banerjee, T. (2008). "Urban Outcomes of Globalization: Theory, Research, and Practice." *Journal of Urban History*. 34(6): 1044–1054.

Molotch, H. (2017). "Cities Without Urbanism: Lessons of Money and Power in the Gulf Region," Lecture Given at the Department of Urban Planning and Spatial Analysis Seminar, University of Southern California. March 2.

3

GLOBALIZATION, RESILIENCY, AND CHANGE

Latin American urban design in the 21st century

Lawrence A. Herzog

Introduction: changing Latin American urban space

A driving force in the way urban space is valued and used in Latin America is the historic emphasis on the traditional center, or downtown. This cultural prioritization of "centrality" can be linked to colonial town planning. The Spanish royal family coordinated plans for the design of its colonial cities via an imperial system of governance and control. The planning rules were laid out and then rigorously enforced through the Laws of the Indies in the early 1500s. More specifically, a subset of these laws, called the "Royal Ordinances for the Laying Out of New Towns," dictated the first urban planning and engineering rules for cities in the Americas. (Stanislawski 1947). Using ideas borrowed from the Greeks and Romans, Spanish town engineers were instructed to build cities in a "grid-iron" configuration, anchored by a central square (*plaza mayor*) set within a rectangular street grid. The ordinances stated that the key institutions – the church, the government palace, the market – would be sited on or near the main square. Larger cities would have a number of secondary nodes in the form of plazas geometrically dispersed and equidistant from the main square.

The main plazas both literally and symbolically encapsulated the powerful role of church and state. Centrality was enhanced by locating not only the main institutions and social infrastructure (markets, church, etc.) on or near the core, but also the elite members of society, whose homes were adjacent to the center, alongside key facilities of the colonial era – the customs house, royal granary, royal tobacco warehouse, and royal treasury. Further, the royal ordinances made it very clear that indigenous and subaltern members of society were to be kept at a distance or pushed completely away, as revealed in these words from the original language of the ordinances: "All settlers, with greatest possible haste, are to erect jointly some kind of palisade, or dig a ditch around the main plaza so that the Indians cannot do harm" (Nuttall 1922: 252).

From an urban design perspective, Latin American cities evolved through three phases prior to the 21st century. In the first phase, from around 1500 to 1800, cities were compact, and the status of their residents was defined by the proximity of their homes to the main plaza. In the second phase, in the 19th century, European immigration increased, and linear corridors extending from the downtown core defined new zones of high status and upper-income residence. Meanwhile, the poor lived in emerging "slum quarters," close to the city center or nearby industrial districts. During the third phase, from the 1940s onward, giant waves of rural-to-urban migration brought millions of landless peasants to the cities. This eventually produced a large-scale socio-spatial polarization of rich and poor, with the poor forced to find shelter on the least desirable land in squatter settlements and shantytowns, mainly isolated in districts on the urban periphery. Meanwhile, the rich concentrated in a cone-shaped area roughly defined by downtown and an elite corridor that connected the city center to a second high-status zone on the periphery. These two zones for the wealthy were served by streetcars, and later by automobiles (Griffin and Ford 1980).

For most of the second half of the 20th century, as land closer to the center became scarce, urban growth was pushed towards the urban periphery and was dominated by *asentamiento irregulares* (irregular settlements) – squatter settlements and shantytowns that evolved as poor rural migrants began arriving in cities in the 1950s. Indeed, to speak of urbanization in Latin America, is to recognize that most of the outward urban expansion was due to unplanned invasions of land, and subsequent informal settlement construction by the most disadvantaged citizens.

The plight of the urban poor still remains a critical issue in Latin American urbanization (see Mangin 1967; Cornelius 1975). A dominant narrative of this continuing plight has been that of "uncertainty." Migrants to the city were caught in a vortex of perilous circumstances. Unable to survive from subsistence agriculture in their rural lands, they arrived in large cities and faced high unemployment, lack of capital, and discrimination. Their only choice was to create their own "spontaneous" settlements, often on precarious and unwanted land: near toxic pollution sources or urban nuisances like airports or prisons, on unstable flood-prone hillsides, or at the bottom of steep canyons and floodplains (Davis 2006). Further, these city dwellers often found shelters only through land invasions or fraudulent property deals with less scrupulous investors.

During the fourth phase of urban evolution, which began in the first decade of the 21st century, globalization began to reshape the spatial form of Latin American cities. The forces of globalization and the neoliberal model of economic development brought the elements of "fast urbanism" to the urban periphery: mass-produced, master-planned neighborhoods for the upper and middle classes, the expansion of shopping spaces and business parks, and heightened consumerism, advertising, and high technology. These forces combined to create metropolitan regions characterized by fragmentation and the formation of what has been termed an "archipelago" of gated residential spaces and fortified centers, all served by the automobile and all disconnected from each other (Sabatini, Caceres, and Cerda 2001). These closed-off "islands" of activity included gated suburbs, malls, private schools, and social and recreational clubs. As in the United States, these increasingly spread-out developments became overwhelmingly dependent on automobiles and, even in some cases, on private toll roads that allow the wealthy to get to work more quickly. These patterns can be seen across the Americas, especially in Argentina, Chile, Mexico, and Brazil (Coy and Pohler 2002).

An urban design crisis on the periphery: global suburbs in the Americas

Since the 1990s, portions of the North American suburban model have been exported to other parts of the world. These "global suburbs" often employ the cultural branding of the US prototypes through their place names: One of the most affluent suburbs outside Cairo, Egypt, is called "Beverly Hills." An upper-class enclave in the outskirts of Beijing is named "Orange County." A wealthy peripheral residential district in Hong Kong calls itself "Palm Springs." Similar global suburbs have risen outside Jakarta, Manila, Lagos, and Johannesburg.

In Latin America, the elite or middle-class developments built on the periphery of cities are partially influenced by the North American suburb of the 1950s, though they do not necessarily look exactly like their low-density, single-family US counterpart. They manifest in different forms adapted to their cultural settings. For example, in Brazil, elegant high-rise luxury suburban towers, or *condominios fechados* (closed condominiums) define the new suburbs. In Mexico, there are both mid-rise and high-rise suburban apartment towers for the wealthy, but also spread out, low-density subdivisions for the working and middle classes.

While the built landscapes may vary, what remains consistent across the planet is the "socio-cultural design narrative" that has come to define global suburbs. This suburban narrative is embodied by social exclusivity, an increasing desire for private over public space, the fear of crime, a preference for predictable, homogenous built environments, and a greater emphasis on consumerism and shopping within artificially constructed malls, as opposed to traditional corner stores or street-oriented shops in pedestrian-scale neighborhoods (Herzog 2015). Scholars have also suggested that Hollywood and American cinema may have influenced the construction of American-style suburbs south of the border; they argue that many Hollywood films unwittingly set up a tension between the dark, dangerous landscapes of inner cities and the idealized, peaceful, healthy image of the suburbs (Pizarro 2002).

The evolution of global suburbs in Latin America illustrates an early-21st-century fundamental reorganization of space in the urbanized regions of the Americas. Wealthy and poor are ever more separated geographically, via technology, walls, fences, or politics. The term "global suburb" thus captures both the socio-spatial reorganization that has spread across the globe, and the growing cultural/psychological importance to urban dwellers of retreating or escaping into protected enclaves within the metropolis. These new social and cultural hybrids are part of an evolving social model of cities that may well be reinforcing the conditions of inequality, allowing wealthy classes to further distance themselves from the realities of urban life. As US-inspired global suburbs proliferate in the periphery of cities in Latin America, and more generally, the Global South, they end up competing for space with the *favelas*, *colonias*, and other squatter communities that house the growing population of urban poor across the planet (Figure 3.1).

Figure 3.1 Suburban development in the Global South. A case from Tijuana.

Source: Photo by L. Herzog

Two design and spatial trends are notable in the shifting structure of the Latin American periphery in the 21st century. First, gated communities and fenced suburbs are now spreading into zones that were previously occupied by only squatter communities. The new gated suburbs in Latin America are called "*barrios privados*" (private communities) in Argentina, *condominios* in Chile, *conjuntos* or *urbanizaciones cerradas* in Ecuador, *condominios fechados* (closed condominiums) in Brazil, and *fraccionamientos cerrados* in Mexico. In Buenos Aires, middle- and upper-class suburban enclaves are now surrounded by illegal settlements and garbage dumps. In general, on the Latin American periphery, we are seeing luxury residential quarters served by highways juxtaposed against squatter communities – literally enclaves of wealth sitting alongside vast swaths of poverty (Borsdorf, Hidalgo, and Sanchez 2007).

This trend alludes to a dramatic shift in the scale of segregation. In the 20th century, large-scale segregation was the hallmark of urban form, with rich and poor profoundly separated in space. Today we are seeing what one scholar terms "small-scale segregation," in which wealthy and upper-middle-class enclaves are spread across the metropolitan region (Sabatini, Caceres, and Cerda 2001). These enclaves, often adjacent to squatter neighborhoods, are fortified with fences and walls and are built on plots that become available depending on topography, land markets, and other factors.

The second recent trend in the Latin American periphery lies in the tendency to build mega-projects that are not merely gated communities, but rather gated cities – giant developments of 30,000 and more inhabitants – in Chile, Brazil, Mexico, and elsewhere (Jones and Moreno-Carrasco 2007). They are often products of direct foreign investment and influenced by foreign design and marketing strategies. The original mega-suburb was probably Alphaville, on the edge of São Paulo, Brazil; its construction began in 1974. Scholars point out that Alphaville and other projects in São Paulo were influenced by North American ideas about suburbs and marketing (Caldeira 2000). Another example is the elite suburb of Santa Fe, on the western outskirts of Mexico City. Its development involved US and other international investors as well as the building of foreign-owned commercial, industrial, and office developments (Valenzuela 2007).

Both these early-21st-century trends in the Latin American periphery underscore the overwhelming role of exclusion as the emerging structuring force in the social and geographical construction and design of urban space. This exclusion is not only observed in the formation of gated elite and upper-middle-class residential zones, but also in the rise of highly protected, often fortified, privatized business centers, schools, leisure and recreation spaces, and shopping malls. Exclusion, isolation, and fragmentation also hint at a corresponding loss of community and public life resulting from the steady decline of walkable, community-oriented public spaces. This kind of new urban sprawl has also been environmentally destructive, unleashing more pollution from automobiles, trucks, and buses using the new peripheral roads, and eradicating forests and other natural ecosystems as these suburbs push further into exurban spaces.

Sense of place and "slow urbanism" in Latin America

If the suburban model is an example of design influences flowing from north to south in the Americas, there is also evidence of an opposite force (flowing south to north) that might alter American cities – what might be termed "Latino slow urbanism." Western cultures have become habituated to "fast living" on a variety of levels. In the cities of North America (United States and Canada), residents are accustomed to traveling fast, typically in their cars and often on mammoth systems of freeways. The average American travels 12,000 miles per year, almost double the distance traveled by citizens in most other industrialized nations. Americans also suffer from a time crunch, by working longer hours than other global citizens. North America's urban dwellers live at an accelerated pace within a built environment (spread-out suburbs, freeways) that exacerbates this pattern. This ethos of fast living has both environmental and ethical costs that cut across every major layer of people's lives – food, clothing, shelter, and ecology. This fast-urbanist development pattern adds more traffic, air pollution, and noise.

But alternatives to the "fast urbanism" paradigm are emerging. The "Slow Food" movement, born in Italy in 1986, pushed back against the globalization of fast food, fast life, non-sustainable food production, and the eroding of local economies. Slow Food later gave way to the "Slow Cities" movement, or *Cittaslow*, also born in Italy. The Cittaslow charter argued that globalization wipes out uniqueness and generates mediocrity (Cittaslow 2009).

Both the Slow Food and Slow City movements are largely about protecting sense of place and local culture (Mayer and Knox 2006). The idea is to preserve both local ecology (agriculture, vegetation, grown in local soil, water, etc.) and human tradition (cultural practices), both of which allow diverse parts of the planet to generate their own unique foods, styles of consumption, and ways of life. This is then extended to urban buildings and spaces in the form of conservation, green spaces, walkability, appealing public spaces, improved public transit, and eco-friendly design (Knox 2005).

There are qualities of urbanism and urban design in Latin America that inherently embrace this "slow urbanist" perspective. Latin American slow urbanism qualities include memory, ritual, art/design, walkability, and the cultural power of streets and public spaces. Of course, one must be careful not to reduce this idea of "Latino slow urbanism" to a cultural stereotype. The term "slowness" here refers only to the pace of change in urban design.

To speak of memory in Latin American urban design is to acknowledge the collective force of three powerful historic layers that define the heritage of the urban built environment: indigenous, colonial, and modern. These three periods have left an indelible and lasting imprint on the physical form of cities; for example, some national governments have even gone so far as to legislatively preserve that "cultural patrimony" as a valuable part of their heritage (Herzog 1999; Scarpaci 2005). In Mexico, for example, for more than five decades, the national government has created legal protections for its cultural patrimony, as well as at least two cabinet-level agencies to manage historic properties (Monnet 1995). More recently, national governments have been employing historic preservation policies to jump-start economic development in old city centers in Cartagena, Montevideo, Buenos Aires, Havana, Mexico City, Quito, Valparaiso, Santiago, and Rio de Janeiro, to name a few salient examples (Scarpaci 2005).

In today's global economy, historic buildings and public spaces have immense value in urban design for redevelopment, and in stimulating economic sectors such as tourism, entertainment, design arts, and cuisine. The experience of living in and using historic spaces involves a set of rituals that are inherently connected to a "slow" urban lifestyle, one that scientists argue contributes to higher levels of public health, since it involves walking, socializing, and being part of a community (Pucher and Buehler 2015).

A number of scholars have discussed the rituals surrounding behaviors in or near public squares and plazas in Latin America (Figure 3.2) (see Arreola and Curtis 1992; Scarpaci 2005; Herzog 2006). These rituals involve everything from casual meeting, discussions of politics, eating, dating practices among young teens, children playing, and mothers interacting, to having one's shoes shined, reading, or gambling (Low 2000). More formal rituals include the weekly flag ceremony on the *zocalo* in Mexico City, or the annual "*grito*" (shout) of "Viva Mexico" on the same *zocalo*, the evening before Mexico's independence day (Herzog 2006).

Art itself is important to Latin American urbanism. The legacy from the building of cities remains embedded in both the built environment (in the form of buildings, streets, plazas, gardens, patios, and other spaces from the past), as well as in the artwork (paintings, murals, sculpture,

Figure 3.2 Plaza in Rio de Janeiro during a spring carnival "*bloco*" (block party). Plazas define Latin American urban design and are still utilized today.

Source: Photo by L. Herzog

photography, film, and so forth) that celebrates culture and place in Latin American cities. This is seen, for example in the proliferation of street murals, public art, outdoor craft markets (Figure 3.3), and high-profile architectural commissions as part of the overall renaissance and redevelopment of historic urban centers in cities like Buenos Aires, Valparaiso, Montevideo, Rio de Janeiro, and Santiago. Valparaiso, Chile, stands out as one of the critical examples of a city where progressive politics, murals in public spaces with populist themes like social justice or housing rights, and street art have become embedded in the aesthetic landscape of the historic center, part of the reason why it was declared a UNESCO World Heritage Site in 2003 (Palmer 2008).

In historic centers and downtown areas of Latin America, there has been a revitalization of slower forms of transportation – trolleys, streetcars, bicycling, and, of course, walking. The city of Curitiba, Brazil, has been celebrated as a cutting-edge example of the implementation of sustainable urban practices. One of Curitiba's great accomplishments has been the successful construction of a Bus Rapid Transit (BRT) network throughout the metropolitan region, a system that is based on creating dynamic, high-density centers near bus stations, or along linear transit lines, with walkable neighborhoods along the corridors and around the stations. This model is being copied in other cities in Latin America, including Bogotá, Columbia; Rio de Janeiro, Brazil; Santiago, Chile; Guatemala City; and São Paulo, Brazil. An estimated 32 cities in Latin America have adopted some form of BRT lines, amounting to 20 million weekly users (Schlaiker 2011).

Globalization has also found its way to the historic core of many downtown districts, like that of Mexico City. The attention to redevelopment through the lens of protecting historic spaces

Figure 3.3 Street Market, San Telmo, Buenos Aires. Street markets are part of the "slow urbanism" that defines Latin American urban space.

Source: Photo by L. Herzog

is a lesson that other regions of the world can point to. For example, UNESCO's designation of Mexico City's historic center as a World Heritage Site provided crucial international recognition of the historic infrastructure of downtown, and it became a focus of opportunity for global investors soon after the North American Free Trade Agreement (NAFTA) was signed in 1993. This had the additional impact of attracting international journalists to cover Mexico, and thus redressing the negative stereotypes of Mexico City spread by some foreign media sources. For much of the late 20th century, international news coverage in Mexico City focused on negative events: political corruption, the devastation of the 1985 earthquake, the problems of air pollution in the early 1990s, and reports of crime against foreign visitors, particularly in the form of taxi cab kidnappings. UNESCO's designation reminded the world design community that the central core of Mexico City is a rich assembly of historic buildings, streets, patios, and plazas, and that there are important stories to be told about their histories (see Malkin 2005).

A new urban design prototype: the transfrontier metropolis

Globalization in the 21st century has produced new forms of spatial relations in the Americas, including the emerging prototype of a unique kind of global urban space – what might be termed a "transfrontier metropolis" (Herzog 1990, 1999). The gradual acceleration of transnational banking, offshore manufacturing, multination trade blocs, global communications, digital technology, and the international division of labor have all shaped a profound "internationalization" of urban space. No longer are cities merely artifacts shaped by local stakeholders; increasingly, urban centers have become containers developed and transformed by international actors, from corporate investors and transnational financial interests to transborder marketing entities and cross-national governmental organizations. To accommodate the forces of globalization, the city has been forced to reinvent itself; sprawling "edge cities" and dense high-tech corporate business districts are just two recent responses to the globalization of the metropolis. But the transfrontier metropolis is a unique form of globalized urbanism in the Americas, especially on the Mexico-US border.

Transfrontier metropolitan regions typically consist of two or more settlement core areas located around an international boundary. Over time, these areas fuse together to form a single ecological and functional city/region, whose socio-economic, cultural, and ecological integration is a hallmark of the ways in which 21st-century global processes transcend national boundaries. More than ten million people today live in transfrontier metropolitan regions that blanket the 2,000-mile boundary from Matamoros-Brownsville to Tijuana-San Diego (Figure 3.4). Citizens on either side of the boundary are increasingly drawn together into a web of north-south relations, where the Global North vs. Global South and developing vs. developed dichotomies are cast aside, as urban neighbors become part of a common transnational living and working space.

Since the rise of the nation state in the 19th century, cities usually evolved in locations entirely inside (and often distant from) the boundaries of sovereign nations. Yet, during the late 20th century, this pattern began to loosen in some regions across the planet, and notably in North America, as population, economic resources, and infrastructure migrated towards the edges of nations, leading to the formation of city regions that sprawl across international boundaries. By the second decade of the 21st century, this pattern has become well entrenched.

Cross-border metropolises are urban configurations that can potentially benefit from the interplay between the flows of goods and services in a globalizing economy and the proximity of territorial borders. Open borders offer opportunities for cross-border metropolises to reinforce their position at the heart of global economic networks and affirm their autonomy as cross-border regional entities (Sohn 2014). And the process of what has been called *debordering* allows cross-border metropolises to use the border as a resource, which allows the bordering nations to benefit from the comparative advantages on either side (for example, less expensive

Figure 3.4 The Tijuana-San Diego metropolitan region.

Source: Photo by permission of Keith Pezzoli, San Diego

labor on one side of the border, technological expertise on the other; specialized micro-level skill sets and products on one side, and demand for those outputs on the other).

But globalization has also produced an opposite trend that runs counter to de-territorialization and debordering; we can call this trend *rebordering*. Rebordering deals with the securitization of state borders. Although the potential security threat that borders represent was already a concern before September 11, 2001, it has become a defining paradigm for many governments after the terrorist attacks. As a matter of fact, this ideology has had a major impact on the policy shift towards greater border security, particularly along the US frontier (Andreas and Biersteker 2003). In the name of the "global war on terror," a sharp increase of controls (at the borders and potentially everywhere) has unfolded, along with the implementation of networked systems of surveillance and, in some cases, the erection of walls and security fences. National borders are seen as a line of defense against various threats. Beyond the issue of terrorism, the securitization discourse also serves to control immigration and reduce the number of migrants from the South entering the North, mainly for economic reasons.

One of the important qualities of the emerging transfrontier metropolis is that it serves as a prototype of an urban place that softens rather than hardens the meaning of the international boundary line. One can argue that the debordering character of a cross-border metropolis should be the goal along the entire Mexico-US border. Indeed, this argument was made long ago by well-known urban designers Kevin Lynch and Donald Appleyard in their "Temporary Paradise" study of the San Diego-Tijuana region (Lynch and Appleyard 1974)

The Tijuana metropolitan region at the California-Mexico border is, indeed, a case in point. It is projected to grow from 1.8 million in 2015 to 2.7 million in 2025 and will need some 300,000 new housing units to accommodate its expanding population. Urban designers are employing the virtues of Latino slow urbanism and placemaking to redesign downtown Tijuana. They are drawing upon Mexican design traditions of public space, lively streets, music, and local food, all of which have become critical tools for rebranding and rebuilding the future of downtown Tijuana.

One of the tenets of the new redevelopment vision for downtown Tijuana will be shifting the construction of housing away from sprawling suburbs that many residents have abandoned (see Herzog 2015), towards the downtown core, possibly building up to 10,000 housing units over the next decade and adding about 30,000 new residents to downtown (Urban Land Institute 2013). The strategy builds around salvaging or upgrading the unique urban design elements of the center of Tijuana: the street-friendly small galleries, business ventures and shops in the abandoned spaces along or near Revolution Avenue; the nearby *pasajes* (covered spaces) that had been created as markets more than five decades ago.

Tijuana's advantages are arguably built around the power of centrality and the vibrancy of spaces near the plazas and streets in the older part of town. A place-based design and economic development strategy are arguably being connected to art, design, regional cuisine, locally manufactured beer and wine, and boutique stores selling Baja California-produced food, beverages, clothing, crafts, or other folk products. Innovative experimental projects have already sprung up; they include a new office building in a previously run-down section near downtown, with spaces for galleries and design studios, a new design college with a rooftop public space, micro-breweries, a food truck space called Telefonica Gastro Park, remodeled restaurants, banks, and a redeveloped former government complex called Estacion Federal, close to the boundary. Local cuisine in Tijuana is now attracting international attention, thereby adding gourmet food spaces as part of the overall redevelopment and design strategy.

Tijuana's center already had historic buildings, from the original California Mission Revival-style Jai Alai building to old cantinas, dance halls, movie theaters, and stores that could be refurbished to create a sense of place and local identity and attract the "creative class" back into the center.

Tijuana's downtown also has potentially valuable public spaces – including the previously mentioned "*pasajes*" (covered market spaces) as well as several historic plazas (Plaza Santa Cecilia, Parque Teniente Guerrero) and the widened pedestrian tourism promenade, Revolution Avenue. These spaces have historically served as valuable gathering places. The thinking is now to re-insert them into the "mental maps" of residents, investors, and visitors alike. Further, the design strategy also includes the creation of mixed-use developments that will bring more residents and density to the downtown core. These new projects are being marketed to millennial families and designed around a future, more sustainable urban model of walkable space that partially eliminates the wasteful use of cars. This echoes the global "new urbanism" and "smart growth" movements, but with an uniquely Mexican "slow urbanist" flair, anchored by the plazas, *pasajes*, and vibrant street life of downtown, and by nearby neighborhoods along the border.

Conclusion

This chapter has been aimed at rethinking scholarly views on Latin American urban design. It began with a review of the legacy of colonial urbanism, the European period of the 19th century and the first half of the 20th century, and the pattern of socio-spatial polarization that dominated the second half of the 20th century. All of these historic influences have shaped Latin American urban space and set the stage for the immense change that followed.

Latin American cities find themselves in the midst of a new and transformative era of change – mainly driven by globalization, though filtered by patterns from previous eras. First, we have witnessed the burgeoning impact of the formation of "global suburbs" – elite enclaves of either gated residential space or "mega projects" like giant malls, office complexes, and large universities, all clustered in master-planned, automobile-centric, North American-style suburbs on the urban peripheries, from Mexico to the southern cone nations of Argentina, Chile, and Brazil. These elite suburbs are gradually displacing some squatter communities on the periphery, leaving what scholars are describing as an "archipelago" of fragmented rich and poor enclaves, a fundamental reorganization of urban space in this part of the world.

However, the chapter also suggests a different way of thinking about Latin American "design urbanism" and the legacy of visceral physical spaces created over centuries. It views these spaces as tied to what scholars term "slow urbanism." From indigenous ceremonial architecture to Spanish colonial city planning, the southern cultures of the Americas embraced the notion of centrality – discrete, civically engaged gathering spaces, including plazas, market squares, gardens, patios, alleys, and promenades, as well as pedestrian-scale streets. Rituals became common civic practices that celebrated both the public spaces and the buildings around them. More recently, street murals, festivals, and other cultural activities have also enriched public spaces. These spaces, and the larger idea of civic nodes of engagement, are part of the historic legacy that defines Latin American slow urbanism. This legacy sets a standard that should be of interest to urban designers across the planet. While we often think of Latin America as negatively impacted by the more dominant cultural power of the United States or by the problems of social polarization, "slow Latino urbanism" is an example of a cultural force that might very well flow from south to north.

The very act of using public spaces suggests a way of life that is healthy for communities and for human engagement. Plazas and pedestrian-scale streets inherently create spaces for human activity, for slower movement (walking, cycling), and for meaningful social encounters both among known participants and among strangers experiencing democracy in a "third place" (Oldenburg 1999). Despite the political and economic challenges facing nations today in southern regions of the Americas, the "slow urbanist" legacy holds great promise in allowing governments to design more resilient cities, built around local/community rights, preservation of historic spaces, traffic calming in the streets, less automobile use and more walking/cycling, and counteracting consumerism by creating authentic spaces for meaningful social interaction.

The chapter concluded by reviewing another form of global urban space critical to Latin American urban design in the global era: the US-Mexican "transfrontier metropolis" that now defines parts of the 2,000-mile international border. This metropolitan spatial form encapsulates both the positive and negative elements of globalization. It represents a daily, cross-border lived space whose patterns – commuter flows, journeys to school, exchange of commodities, cross-border industrial assembly, and tourism flows – can be viewed as part of an evolving trans-border functional metropolitan space that could become a critical 21st-century urban prototype with global significance. At the same time, citizen fears of terrorism, smuggling, immigration, and crime have energized a movement of "rebordering," whereby some interests want to see frontiers walled-up and closed off. The United States, the giant force that lies at the gateway to Latin America, finds itself in a time of resurgent nationalism since 2016. As cities like Tijuana undergo innovative urban growth and redevelopment, whether they can build ongoing connections with their "twin city" partners in the United States remains a question, amidst uncertainty and battles fought in the political arena. The outcome speaks to the future of Latin American cities, which, on many levels, are more and more interconnected with North American economies and societies.

References

Andreas, P. and Biersteker, T.J. (2003). *The Rebordering of North America: Integration and Exclusion in a New Security Context.* New York and London: Routledge.

Arreola, D. and Curtis, J. (1992). *The Mexican Border Cities.* Tucson: University of Arizona Press.

Borsdorf, A., Hidalgo, R., and Sanchez, R. (2007). "A New Model of Urban Development in Latin America: The Gated Communities and Fenced Cities in the Metropolitan Areas of Santiago and Valparaiso, Chile." *Cities.* 24 (5): 365–378.

Caldeira, T. (2000). *City of Walls.* Berkeley: University of California Press.

Cittaslow. (2009). www.cittaslow.net/download/DocumentiUfficiali/2009/newcharter%5B1%5D.pdf.

Cornelius, W. (1975). *Politics and the Migrant Poor in Mexico City.* Stanford: Stanford University Press.

Coy, M. and Pohler, M. (2002). "Gated Communities in the Latin American Megacities." *Environment and Planning B: Planning and Design.* 29: 355–370.

Davis, M. (2006). *Planet of Slums.* New York: Verso.

Griffin, E. and Ford, L. (1980). "A Model of Latin American City Structure." *Geographical Review.* 70: 397–422.

Herzog, L. (2015). *Global Suburbs: Urban Sprawl from the Rio Grande to Rio de Janeiro.* New York: Routledge.

Herzog, L. (2006). *Return to the Center: Culture, Space and City Building in a Global Era.* Austin: University of Texas Press.

Herzog, L. (1999). *From Aztec to High Tech.* Baltimore: Johns Hopkins University Press.

Herzog, L. (1990). *Where North Meets South.* Austin: University of Texas Press and CMAS.

Jones, G. and Moreno-Carrasco, M. (2007). "Mega-Projects: Beneath the Pavement, Excess." *City.* 11 (2): 144–164.

Knox, P. L. (2005). "Creating Ordinary Places: Slow Cities in a Fast World." *Journal of Urban Design.* 10 (1): 1–11.

Low, S. (2000). *On the Plaza.* Austin: University of Texas Press.

Lynch, K. and Appleyard, D. (1974). *Temporary Paradise: A Look at the Special Landscape of the San Diego Region.* San Diego: City of San Diego and Marston Foundation.

Malkin, E. (2005). "A Modern Space in Mexico City's Historic Center." *New York Times*, December 21. www.nytimes.com/2005/12/21/arts/design/a-modern-space-in-mexico-citys-historic-center.html.

Mangin, W. (1967). "Latin American Squatter Settlements: A Problem and a Solution." *Latin American Research Review.* 2: 65–98.

Mayer, H. and Knox, P. (2006). "Slow Cities: Sustainable Places in a Fast World." *Journal of Urban Affairs.* 28 (4): 321–334.

Monnet, J. (1995). *Usos e Imagenes del centro historico de La Ciudad de Mexico*. Mexico: DDF.
Nuttall, Z. (1922). "Royal Ordinances Concerning the Laying Out of New Towns." *Hispanic American Historical Review*. 5: 249–254.
Oldenburg, R. (1999). *The Great Good Place*. New York: Marlowe and Co.
Palmer, R. (2008). *Street Art Chile*. London: Eight Books Limited.
Pizarro, R. (2002). "Exporting the Dream: Hollywood Cinema and Latin American Suburbia," in Heikala, E.J. and Pizarro, R. *Southern California and the World*. New York: Praeger.
Pucher, J. and Buehler, R. (2015). "Walking and Cycling for Healthier Cities." *Built Environment*. 36 (4): 391–414.
Sabatini, F., Caceres, G., and Cerda, J. (2001). "Residential Segregation in Major Cities in Chile." *EURE*. 27 (82): 21–42.
Scarpaci, J. (2005). *Plaza and Barrios*. Tucson: University of Arizona Press.
Schlaiker, E. (2011). "Latin American Bus Rapid Transit Boom Offers Lesson for the U.S." *World Resources Institute-City Fix*. March 28.
Sohn, C. (2014). "The Border as a Resource in the Global Urban Space: A Contribution to the Cross-Border Metropolis Hypothesis." *International Journal of Urban and Regional Research*. 38 (5): 1697–1711.
Stanislawski, D. (1947). "Early Spanish Town Planning in the New World." *Geographical Review*. 37: 94–105.
Urban Land Institute (ULI). (2013). *Downtown Tijuana, B.C., Mexico Revitalization Concept Plan*. Downtown Tijuana Technical Assistance Panel. San Diego: Urban Land Institute.
Valenzuela, A. (2007). "Santa Fe: Megaproyectos Para Una Ciudad Dividida." *Cuadernos Geograficos*. 40: 53–66.

Further reading

Caldeira, T. (2000). *City of Walls*. Berkeley: University of California Press. An ethnography of the changing social relations and polarization of social space in Brazilian cities, with an emphasis on the case of São Paulo, Brazil's largest metropolis.
Herzog, L. (2015). *Global Suburbs: Urban Sprawl from the Rio Grande to Rio de Janeiro*. New York: Routledge. A comprehensive study of "global suburbs" in Latin America, including an analysis of the "socio-cultural design narrative" influence of US suburbs, and the ecological challenges created by urban sprawl for the case studies of Mexico, Brazil, and the US Southwest border region.
Herzog, L. (2006). *Return to the Center: Culture, Space and City Building in a Global Era*. Austin: University of Texas Press. An historic, cultural geographic and urban design analysis of public spaces in Spain and Latin America, including Mexico and the US-Mexico border.
Lynch, K. and Appleyard, D. (1974). *Temporary Paradise: A Special Look at the Landscape of the San Diego Region*. An urban design report on the future of the San Diego region's growth and planning, including an analysis of the advantages posed by the rich cross-border connections between landscape, topography, and urban design.
Scarpaci, J. (2005). *Plaza and Barrios*. Tucson: University of Arizona Press. A study of land use, buildings, urban design, and downtown revitalization in Latin American cities, both past and present.

4

SPATIAL JUSTICE AND URBAN DESIGN

The case of Southern African settlements

David Dewar

Introduction

Political analyst Naomi Klein (2014) reminds us that our world is changing, whether we want it to or not. A question for us, as urban designers and spatial planners, is how might we participate in enabling a world that is more just for most rather than for an elite few? Such a question becomes particularly relevant when one considers that urban design interventions tend to focus on "select prime spaces in the city, [thereby] perpetuating urban inequalities" (Loukaitou-Sideris and Mukhija 2016: 577), while spatial planning interventions tend to be beset by investor-friendly policies, entrepreneurial strategies, and desires for economic competitiveness as catalysts for urban change (Fainstein 2010). Furthermore, if we consider that we do, in fact, "collectively make our cities, and that we do so on a daily basis, then why is it that we somehow end up creating public domains that few of us particularly like and most find unjust?" (Graeber 2013: 222). In this chapter, I will address this pertinent question by revisiting the normative idea of spatial justice. I also hope to stimulate further debate on what spatial justice might mean for those of us who are engaged in reimagining our cities.

Thus, I begin by suggesting that spatial justice includes concerns for the performative aspects of human habitats as a whole. With this suggestion, I am not arguing for a return to "rational comprehensive planning," or "the master plan" as a finalized product with ready-made solutions. Rather, I am suggesting that spatial justice needs to be considered in conjunction with other desirable performance qualities, such as environmental and socioeconomic sustainability, spatial integration, a sense of place, choice, and resilience.

The contextual focus of this chapter is the urban settlements in Southern Africa, which are often referred to as cities of the Global South. There are, of course, real dangers in generalizing about these Global South cities, for as we know, different geopolitical, socio-economic, cultural, and other forces shape cities across diverse world regions. Nevertheless, Southern African cities resemble spatial manifestations derived from shared histories of colonialism and imperialism (Watson 2016), and their underlying structural characteristics are, in fact, very similar. Southern African settlements (in particular settlements found in South Africa, Botswana, Namibia, and Zimbabwe), also share characteristics commonly associated with other cities of the Global South, including: high rates of population growth and increasing rates of urbanization; high levels of poverty, unemployment, and inequality; low levels of literacy and skills; relatively low life expectancy; high levels of informality, both in the economy and in terms of shelter; and limited public resources relative to the demands made on public coffers. However, in contrast to many countries in Latin America and Asia, urbanization in Africa is a relatively recent process, and while rates are increasing, levels of urbanization remain low. (South Africa is 60% urbanized, the highest level in Africa, but lower than that in more developed countries). It is against this background that I turn to the concept of spatial justice.

The concept of spatial justice

We can conceptualize spatial justice as having a number of components: some of these overlap but are not the same as the desired performance qualities identified by Kevin Lynch (1981) in his *Theory of Good City Form*. These are:

Ecological justice Responsible urban design is based on respect for ecological principles and "putting nature first." The first design decision is determining where development should not go.

Generative power Spatial justice implies access to opportunities. Settlements can generate a rich range of opportunities (economic, social, cultural, and recreational) for inhabitants. However, they do not automatically do so. Two interrelated characteristics are central to the concept of generative power. The first is relatively high aggregate densities in local areas, since a precondition for small business to flourish is a vibrant local market. The second characteristic is deep, hierarchical orders of access, defined in terms of scale and pace of movement. More continuous routes have the power to break down economic dependence of local areas. They can be used to link many local areas, and local businesses can benefit from the aggregate purchasing power of multiple local areas.

Equity of access Equity of access refers to the ability of all people to access the opportunities that are generated in a settlement. Access relates to mobility. The most equitable form of access is universal access, which includes movement of people with disabilities and movement on foot: people should be able to carry out most of their daily activities on foot. When foot movement is not possible, because distances become too great to walk comfortably, non-motorized transport and public transport are the next most equitable forms of transportation, as affordable to most people. Equity of access is better enabled by a considerable amount of mixed uses and, in larger settlements, a movement network that promotes polycentricism. Equity also relates to access to resources. While the distribution of city resources is never equal, settlement structure should contribute to narrowing the gap between the rich and the poor. At the very least, it should not worsen poverty and inequality.

Livability All people should be able to live with dignity. Central to this is the quality of the public environment, particularly in contexts of high poverty. When people are poor, they often cannot carry out many of their daily activities in private spaces. Most of their daily living takes place in public spaces (such as on sidewalks, streets, squares, and other forms of public spaces). These are the places where people socialize and play, where lovers flirt, where students go to study when the house is overcrowded. When the quality of the public space is good, in the sense that the space is well-defined, humanly scaled, multifunctional, comfortable (with places of shade and informal seating) and naturally surveilled (by other space occupants), then it enhances the enjoyment of the activities which take place within it. These spaces give dignity to a settlement, regardless of how much or how little is invested in built structures. Conversely, when public space quality is poor, the entire environment remains hostile.

Choice In just environments, all people have choices that are enabling, not imposing. While many types of choices are desirable, the most fundamental one is the choice of lifestyle along a continuum ranging from very public or exposed in very accessible locations, to very private or embedded in quiet local areas.

Southern African settlements and spatial justice

When Southern African settlements are assessed in terms of these criteria, it is clear that they are performing very poorly. At the heart of the problem is the spatial planning system itself, which in Southern Africa has been strongly informed by the urban tenets of modernism, perhaps most clearly articulated in the 1934 Athens Charter of the CIAM (Gold 1998). These included:

Separation of uses This particularly includes the separation of live, work, move, and play, as well as vehicular from pedestrian movement in order to reduce conflict. Further, in Southern Africa, separation frequently (and historically) has had racial (and thus class) overtones.

The belief that technology would free society Modernist cities were not designed for pedestrians but for automobiles, which allowed for relatively unrestricted lateral spread (it was assumed erroneously that all households would own a vehicle), while elevators allowed for a vertical expansion of the urban form, particularly in city centers.

The object building The basic building block of settlement was seen as the individual free-standing building (the "object building"), whereas historically, it was public open space – the space between buildings – that composed the fundamental element of the city's structure (Gehl 2011). From the 1930s onward, design was primarily expressed in architectural, and not urban, forms. Indeed, modernism consciously de-emphasized the structural and spatial role of the street: the great Swiss architect, Le Corbusier, one of the fathers of modernism, is on record as saying, "we must kill the corridor street."

Belief in industrialization There was a belief in the use of mass production processes and materials for the creation of standardized forms, believed to be better suited to the future. According to functionalism and the international style, the world could, and should, look increasingly the same. One manifestation of this is mass-produced low-income housing schemes, which litter the landscape in Southern Africa.

A new planning model Modernism and functionalism also introduced a new planning model, which was enthusiastically embraced in Southern Africa: the rational comprehensive model. This model first took root in Europe and the United States but was rapidly transported around the world. The exported model contained two additional ideas, which were appropriated from other paradigms. The first was the "neighborhood unit" concept, first introduced in the United States by Clarence Perry (1927). In this concept, larger urban settlements were seen to be composed of a collection of smaller settlements, evocative of the rural village, in the belief that this form could capture the best of both rural and urban living. Each neighborhood was supposed to be relatively self-sufficient in terms of commercial and social facilities. Internally, the circulation network of each neighborhood cell was orientated inward, focused on a grade school at the center. Other facilities were located towards the periphery, with a radius of 0.25–0.5 meters from the center, a distance defined by the optimal walking distance of children (Banerjee and Baer 1984). The neighborhood model was based on the belief that this form would promote a sense of "community" and neighborliness.

The second concept was the promotion of suburbia, as the good life for all. Originally, the image underpinning suburbia was the Palladian image of the spacious, well-landscaped pavilion.

However, this same model has prevailed in many parts of the world, including in Southern African cities, even though, in terms of poorer households, lot sizes and housing units have been reduced to the degree that there are few benefits of urbanity or nature.

Superimposed on these planning ideas was the political ideology of separate development (termed apartheid in South Africa). Apartheid embraced the concepts outlined earlier but grotesquely distorted them. Particularly, the concept of separation was expanded to include separation on the grounds of race, and, since there is a strong correlation between race and class in Southern Africa, it has been the poorest people who have been forced to live the furthest from urban opportunities.

The dominant spatial patterns, which have resulted from these planning and political ideologies acting in tandem, are sprawl, separation, and fragmentation. These patterns in turn have had devastating impacts on the five different components of spatial justice, as will be explained next.

Ecological justice

The low-density and sprawling form of development has had severe negative ecological consequences. In Cape Town, for example, where aggregate net densities are about 30 dwelling units per hectare and have been declining over the last 30 years, sprawl has driven roughshod over agricultural and wilderness landscapes, destroying potentially productive land and wilderness areas at an alarming rate. Although environmental impact assessments are required for all major new developments, they have proved to be ineffective tools in promoting ecological justice. Further, the low-density, sprawling pattern generates large amounts of traffic but is inimical to the provision of cheap, viable public transportation. Thus, most movement is by private automobiles, which contribute to air pollution and greenhouse gas emissions (Figure 4.1).

Figure 4.1 Cape Town, showing the dominant spatial patterns of low-density sprawl, fragmentation, and separation, underpinning many of the city's structural problems.

Source: P Louw

Generative capacity

The generative capacity of settlements in the region is very low. Generally, densities are too low to support adequate levels of social facilities and other amenities. In terms of the widely employed neighborhood unit concept, which promotes embedding (as opposed to exposing) public facilities, each neighborhood is dependent on its own internally generated thresholds to support social and commercial activities and facilities. When densities are very low, and the population of the cell is almost ubiquitously poor, the purchasing potential is too low to support and sustain local facilities, and it is very difficult for people to access facilities embedded in other neighborhoods. Additionally, as the demographic structure of the neighborhood changes over time, lifecycle needs and the demand for facilities change. Unsurprisingly, a feature of the urban landscape in Southern African settlements is the phenomenon of some facilities (such as schools) being overcrowded, while others are underutilized.

The low-density patterns also have severely negative economic effects. In the face of increasing globalization and structural unemployment in the formal economy, many people have no option but to attempt to engage in (largely survivalist) activities in the informal sector. However, a precondition for a flourishing micro-economy is increasing economic specialization and diversification (Dewar and Todeschini 2013), and these, in turn, require high densities to create vibrant local markets: in most cases, these simply do not exist. A striking feature of the informal economy in the region is its relatively small scale and lack of diversification.

Additionally, because densities are too low to support efficient, viable public transportation, there is a strong emphasis on transportation planning on automobile-based movement systems and limited access routes. Indeed, transportation planning in Southern Africa has been, and still is, powerfully influenced by the North American model (the American Highway Code remains the standard reference for transportation planners), as many South African transportation planners have been trained in the United States. Their central concern remains mobility (enabling cars to move as easily and rapidly as possible), as opposed to accessibility (reducing aggregate amounts of movement by ensuring a close relationship between urban opportunities and places of residence). The perception that problems of access can be resolved by increasing mobility remains widespread amongst transportation planners in Southern Africa and has resulted in an overemphasis on limited access routes, which act as building walls across the city: what happens on one side of the route has nothing to do with the other side, since activities are not able to respond directly to the flows along them. Effectively, when these barriers, together with above-ground train lines, surround an area on all sides, they create a system of isolated "boxes," or "cells," destroying permeability and increasing fragmentation. When the "boxes" are too small, integration becomes impossible (Dewar 2017). As an indication of the scale of the problem, the average size of the "box" in Cape Town is 2 kilometers by 2 kilometers (Figure 4.2). By contrast, in Shanghai, China, it is 7 kilometers by 12 kilometers.

From an economic perspective, a further implication of the overreliance on limited access routes is that the ensuing land use pattern is coarse-grained. Access to and egress from them occurs relatively infrequently (usually at vertically separated intersections with other limited access routes). Further, points of high access occur not on but rather off the route, at the foot of connecting ramps. This coarse-grained pattern of access creates conditions of spatial monopoly: the points of higher access are controlled by higher rent-paying activities (for example, shopping centers, industrial parks, office parks, and so on), thus excluding the poor. It is very difficult to access various life-supporting amenities in Southern Africa without owning an automobile.

Lastly, when large-scale facilities (such as regional shopping centers) are located at the edge of lower-income areas (as they frequently are), they have devastating impacts on the local economies

Figure 4.2 The barrier-like effect of limited access routes fragments the city.

Source: P Louw

of the poorer areas as they suck out of the local area the thresholds necessary to support local business.

Equity of access

Southern African settlements also perform poorly in terms of equity of access. The patterns of sprawl, separation, and fragmentation generate enormous amounts of movement at great cost in terms of household budgets, productive time, infrastructural investment, inefficient freight movement, and air pollution. As mentioned previously, densities are too low to enable the provision of cheap, efficient, and viable public transportation. Most of the movement generated is automobile-centric, which results in high levels of greenhouse gas emissions (ASSAf 2011), traffic congestion, and traffic-related fatalities. Ironically, it is the poor who have to move the furthest. Indeed, the choices facing poor people are stark. On the one hand, they are virtually forced to purchase a car, but transportation costs assume a disproportionately high proportion of their household budgets, negating alternative investment opportunities such as housing. The vehicles that the poor can afford are the oldest (frequently breaking down) and the least efficient in terms of fuel consumption and greenhouse gas emissions. On the other hand, many households are unable to afford the purchase price of even the oldest vehicles and are trapped in space. They are unable to travel even to seek employment or pursue the opportunities of the city. They become the target of abject exclusion.

In a very real sense, therefore, settlement structure and form significantly aggravate the key social problems of poverty, unemployment, and inequality, which underpin most developmental challenges in the region.

Livability

Central to the concept of livability is the quality of the public realm. The quality of public spaces in Southern African settlements is almost ubiquitously sterile, particularly in the poorer areas. Because of the modernist belief in the importance of land use separation, entrenched by zoning, there are high levels of mono-functionality, which is inherently sterile. Large tracts of subsidized public housing have been produced by capital intensive, mass-housing programs. Efficient delivery demands a high level of uniformity in lot size and configuration as well as in the form of the housing units. The consequence is large blankets of identical, small, single-storey units located in the center of the plot, and not defining the street (Figure 4.3).

Figure 4.3 Delft in Cape Town, illustrating the sterility of mass-housing schemes, lack of urban design, and a circulation system of "roads" as opposed to "streets."

Source: Author

The rational comprehensive planning model introduced by modernism has almost no concern with spatial or aesthetic quality; its focus is on land use only. The concept of using buildings to define and make public space does not form part of the planning regulations. Indeed, the reverse is true: the regulations entrench buildings as free-standing objects.

Most movement infrastructure (which accounts for some 25% of open space in South African cities), takes the form of "road" (a highly engineered conduit or "pipe" for automobiles), as opposed to the "street" (a linear public space which can accommodate a range of human activities, including movement). There is an enormous amount of leftover space, often in the form of excessive road reserves, excessive space standards for community facilities, and land reserved for particular activities which have never materialized. Rather than amenities, these spaces generally become dangerous "rape spaces" and dumping grounds for garbage. Not uncommonly, backyards front onto streets and other forms of public space, removing informal surveillance and creating dead-edges.

Choice

Southern African settlements offer a very limited range of lifestyle choices. The overwhelmingly dominant model is that of low-density suburbia: the free-standing, usually single-story, unit on its own lot. Higher density and associated urbanism occur in small pockets only, usually in the larger cities. The problem with the suburban model is that it forces dependence on the automobile. A high proportion of vehicle trips undertaken daily (particularly by higher-income households) are from parents taking their children to and from school and extra-mural activities. Households are faced with "either-or" choices: either access to green space with very little convenience, or easy access to urban opportunities and convenience but with almost no access to greenery.

In terms of all the indicators of spatial justice, therefore, Southern African settlements are highly inequitable and unjust. Southern Africa is not alone in facing problems associated with sprawl, separation, and fragmentation. However, the scale of problem is quite unique here. A United Nations-Habitat review of world cities found that, measured in terms of the Gini Co-efficient (a measure of inequality ranging from 0 to 1, with 0 describing perfect equality and 1 describing absolute inequality), Southern African cities are among the most inequitable in the world (UN-Habitat 2009). In 2014, the co-efficient in South Africa was 0.66 in terms of expenditure and 0.69 in terms of income (Stats SA 2015).

This situation is clearly unsustainable, a fact that is widely recognized rhetorically. For example, the National Development Plan for South Africa (RSA, NDP 2015) emphasizes the imperative of restructuring settlements. However, the patterns have worsened over the last 20 years since the advent of democracy. The primary reason for this is that little planning is occurring at the level of settlement as a whole. Rather, the growth of towns and cities is shaped by four main forces. One is the deeply entrenched nature of the suburban model, which remains dominant. A second is the developer-driven speculative sprawl, whereby developers seek to privatize amenities for the wealthy. The ensuing form is "leap-frog" sprawl, with development occurring in discrete pockets, further and further beyond the urban edge. A third force is the policy for low-income housing in the region, which largely takes the form of capital-intensive mass-housing projects. Their "success" depends on two factors: large parcels of cheap land, which can be found only beyond the urban edge; and standardization and uniformity, usually in the form of the single unit on its own lot, creating mind-numbing monotony and sterility (Figure 4.3). A fourth force is the rapid growth of informal settlements. Frequently, the primary factors determining their location are availability of land (usually public) and a desire to avoid harassment by authorities. These four forces all contribute to lateral spread and fragmentation of the settlements.

Thus, a great challenge facing policymakers is how to restructure Southern African towns and cities, at a variety of scales, to make them more efficient, just, sustainable, and resilient. This, in turn, has two interrelated dimensions: controlling sprawl and making settlements more compact; and intensifying, densifying, and making settlements more mixed-use, in order to optimize the benefits of urbanity (such as promoting expansion and diversification of small and micro-enterprises, establishing viable and efficient public transportation, and so on). It is neither possible nor desirable, however, to intensify everywhere: it is necessary to think strategically and structurally – to intensify around strategic corridors and nodes (for example, social housing around existing agglomerations of urban opportunities; reinforcing corridors carrying public transportation (Dewar and Todeschini 2004); transit-orientated development; and so on). But for significant improvements in spatial justice to be achieved in the region, radical adjustments to mainstream spatial planning practices must take place, since the planning system itself is a major obstacle to sustained improvements.

The spatial planning system and the nature of the plan

To explore this issue further, it is necessary to distinguish between two different approaches to plan formulation: programmatic and structural-spatial approaches. The rational comprehensive planning model, which was introduced into the region by the modernists and which still dominates today, is a programmatic approach. Its focus is on land use (despite the fact that, ironically, in capitalist societies planning cannot control land use): it can prevent uses from happening in a particular place but cannot create them – only market demand can do that. In the rational planning system, an idealized land-use pattern is conceived and neatly separated. The land parcel is quantified (planning is seen as a quasi-science), future space demands are projected (on the basis of range and threshold), and a land-use schedule or budget is generated (x number of households can support y number of schools, z square meters of commercial space, and so on). Planning and urban design then become a more-or-less rational distribution of land uses, buildings, and spaces. In this concept, settlement-making is seen as a rational, comprehensive, highly controlled process leading to balance and stability (the suburban image of the good urban life for all). The central problem with this form of planning is that it inevitably results in sterility. There are a number of reasons for this.

One reason is that too few minds are involved in this process. A hallmark of good environments is that they are complex. Complexity, however, cannot be designed: it results from process (Alexander 1972). A second reason is that, in this approach, there is a great deal of planning around the parts, but nothing holds the whole together. As a consequence, there is a high degree of orderliness, but little order. The third reason is that the system is based on the practice of forecasting space requirements and reserving land. However, the science of prediction is notoriously unreliable. If the prediction is wrong, large amounts of vacant residual space make the settlement look permanently incomplete, and by reducing thresholds, the practice effectively ensures that the planned facility is never built.

Structural-spatial approaches to plan formulation are different from programmatic approaches in a number of important respects. First, they are driven by a concern about the performance of the whole, not the maximization of any part. They are based on the recognition of the truism that, for the whole to work well, no part can be maximized, because compromises are required. Second, their focus is not on land use, but on public structure and space. Third, they are based not on a conception of "the good life for all," but on the creation of rich choices. In this approach, plans are enabling, not prescriptive. Fourth, the starting point of design is not assumptions about access to technology or personal resources, but rather about the lowest common denominator: the needs of nature and of people on foot.

The concepts of structure and space are central to these approaches. The main elements of public structure (green space, circulation space for all modes, public institutions, public space, utility and emergency services) are harmoniously brought into association with each other to create a geometry of points, lines, cross-overs, and grids. This geometry in turn, creates a logic to which all activities, large and small, formal or informal, can respond in their own self-interest. The key to understanding the spatial logic of structure lies in the concept of access. The geometry generated in this way creates an "accessibility surface" across the landscape – a reference grid of line and points of greater or lesser accessibility. The system is a hierarchical or differentiated one.

Each activity and use in urban systems has its own logical requirements in terms of accessibility, ranging along a continuum from very public or exposed to very private and embedded. Users seek to satisfy these to the greatest degree possible through the competitive processes of the urban land market. It follows from this that the deeper the hierarchy of access, the richer the range of locational choices. Conversely, the flatter the hierarchical surface, the fewer the viable locations and the greater the tendency for the more accessible locations to be appropriated by the strongest players requiring exposure, to the exclusion of all others: this characteristic promotes spatial monopoly.

In this way of thinking, the "glue" that holds the settlement together, as it historically always has been, is public space. In the same way that it is possible to create a hierarchy of access, it is possible to create a hierarchy of space. In structural-spatial approaches, all public space (including the street) is social space – there is no residual or leftover space. Positive public spaces have a number of characteristics: they are well defined, partially sheltered from the elements, humanly scaled, multifunctional, under natural surveillance (by other people in the space), and usually landscaped. Surrounding buildings should help define the public space, in the configuration of their footprints. In other words, design is the creative integration and coordination of these different hierarchies (hierarchies of movement, public institutions, public space, and utility services) in ways that respond sensitively to landscape, hydrology, and climate.

It is important to stress that the outcomes of the approach are not comprehensive plans but guiding frameworks. These frameworks are minimalist or partial. The challenge is to identify the minimum strong actions necessary to give direction, while allowing for the ingenuity and creativity of many others to play a role. There are three reasons why this is important. The first is adaptability; that is, in conditions of volatility and uncertainty, such frameworks help to deal with the unexpected. The second reason is qualitative; that is, preventing sterility and monotony that are the inevitable consequences of a design process dominated by very few people, regardless of how talented they may be. The third reason is greater welfare; that is, the benefits from the economic processes associated with settlement-making can be spread over a wider area.

Positive plans are plans of process. Figure 4.4 diagrammatically illustrates a plan of this kind (Dewar and Todeschini 2004). The movement network takes the flexible form of a grid, which is highly permeable. The point of highest access is where the two most continuous routes intersect. The hierarchy of public space and public institutions largely corresponds to the hierarchies of access. The variations in access send clear signals to the land market that all activities can find a place in the system.

An important dimension of this way of thinking is to think consistently across scales. Commonly, therefore, a "package of plans" approach is utilized. Decisions taken at higher-order scales provide the first level of interventions for the increasingly detailed and refined scales below. This way of thinking is not new. It resonates with David Crane's (1960, 1964) long-standing argument

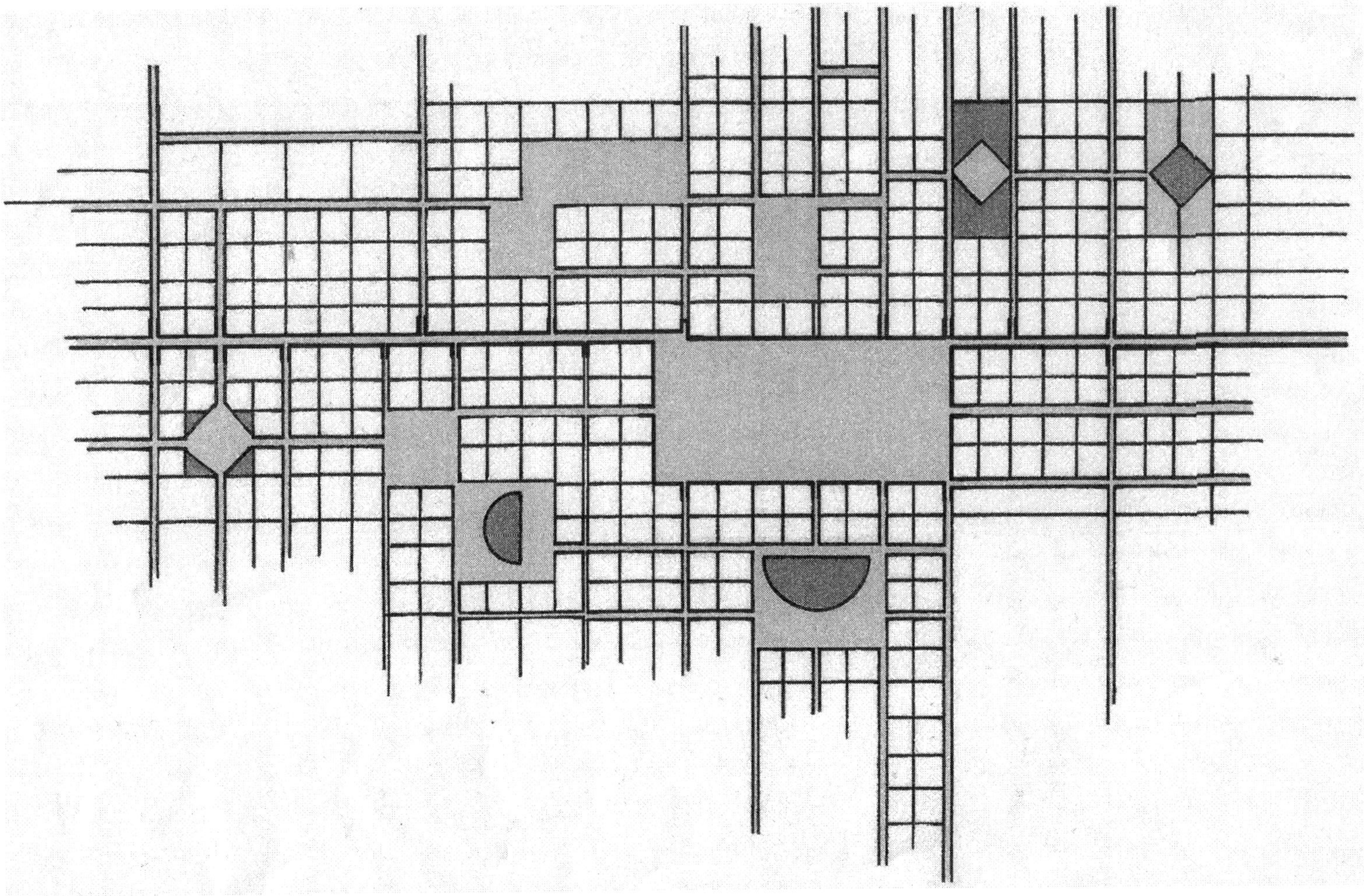

Figure 4.4 A plan without a program.

Source: Dewar and Todeschini (2004)

for the creation of a "capital web." For Crane, this entails identifying and consolidating design decisions around the main structuring elements of an urban context, in order to establish a "skeletal continuity" of public buildings and public places throughout the city that are spatially, economically, and symbolically linked to each other. The capital web then becomes "a unified system of civic activities" that allow for the integration of urban parts into "the whole" without foreclosing every aspect of the urban context by re-inscribing only privileged positions (Crane 1960: 282). Instead, Crane (1960: 283, 285) is calling for "a city of a thousand designers" by promoting spatial frameworks that are robust enough to guide future investment and design interventions, and also flexible enough to accommodate changing needs, choices, and, in the case of Southern African cities, growth. The framework serves not only to "convey a memorable sequence of public-events and activities, a systemic rhythm of symbolic intelligibility," it also allows individuals to "build freely in response to a unified whole" (Crane, 1960: 288). In so doing, it evokes the idea of "adaptability with some sense of permanence," which is arguably of particular relevance to Southern African cities (Ibid.).

From a more contemporary standpoint, this approach resonates with Dewar and Todeschini's (2004) argument for planning partially, as well as with aspects of Gunder and Hillier's (2007) "utopia" strategic planning. All of these allow for design practices that are "incomplete," since urban contexts are ever changing, fluid, and contingent in nature. They "accept struggle as an inevitable and necessary aspect of urban environments" that negates possibilities at arriving at fixed and inflexible design solutions (Gunder and Hillier 2007: 478). In so doing, they desist from establishing a "finalized and fully colored-up map of a 'good' spatial form." Rather, they anticipate multiple futures and realms of freedom via design practices which are critical, inclusive, and dynamic (Gunder and Hillier 2007: 480, 467)

Finally, it is important to recognize that, even if spatial planning and urban design practices are changed radically, greater spatial justice will not be achieved until the state, at all levels, is prepared to intervene aggressively in the urban land market. Once inequitable spatial patterns have taken root, they become entrenched through the land market. Writing down the cost of public land, so that it becomes accessible to poor households, is a necessary response to inequity. The legal instrumentation for this to occur is in place: what is lacking currently is the political will to do so.

Conclusion

This chapter showed that spatial injustice – defined in terms of ecological injustice, a weak generative capacity, inequity of access, lack of livability, and limited choices – is deeply entrenched in Southern African towns and cities. The chapter also showed how the structure and form of settlements in the region are actively and severely worsening the structural and social problems of poverty, inequality, and unemployment. Finally, the chapter discussed an avenue of change, which is necessary for significant improvement: a different approach to planning and urban design, which has, at its heart, structure and space.

References

Academy of Science of South Africa (2011). *Towards a Low Carbon City: Focus on Durban.* Academy of Science Press, Pretoria, South Africa.

Alexander, C. (1972). "The City Is Not a Tree." In Bell, G., J. Tyrwhitt, (eds.) *Human Identity and the Urban Environment.* London: Penguin Books.

Banerjee, T. and Baer, W.C. (1984). *Beyond the Neighborhood Unit: Residential Environments and Public Policy.* New York: Plenum Press.

Crane, D. (1960). "The City Symbolic." *Journal of the American Institute of Planners.* 26(4): 281–292.

Crane, D. (1964). "The Public Art of City Building." *Annals of the American Academy of Political and Social Science*. 352: 84–94.

Dewar, D. (2017). "Transportation Planning in South Africa: A Failure to Adjust." In *Transactions on the Built Environment*. Vol. 36. WIT Press, UK.

Dewar, D., and Todeschini, F. (2004). *Rethinking Urban Transportation after Modernism: Lessons from South Africa*. London and New York: Ashgate Publishers.

Dewar, D., and Todeschini, F. (2013). "Local Area Planning and Design: Some Lessons from the Old City of Ahmedabad, India." *Architecture SA*. 63, (September–October): 36–45; Cape Town.

Fainstein, S. (2010). *The Just City*. Ithaca, NY: Cornell University Press.

Gehl, J. (2011). *Life Between Buildings*. Washington, DC: Island Press.

Gold, J. R. (1998). "Creating the Charter of Athens: CIAM and the Functioning City, 1933–43." *Town Planning Review*. 69(3): 225–247.

Graeber, D. (2013). "It Is Value That Brings Universes into Being." *HAU: Journal of Ethnographic Theory*. 3(2): 219–243.

Gunder, M. and Hillier, J. (2007). "Planning as Urban Therapeutic." *Environment and Planning A*. 39: 467–486.

Klein, N. (2014). *This Changes Everything: Capitalism vs. the Climate*. Penguin, London.

Loukaitou-Sideris, A., and Mukhija, V. (2016). "Responding to Informality Through Urban Design Studio Pedagogy." *Journal of Urban Design*. 21(5): 577–595.

Lynch, K. (1981). *A Theory of Good City Form*. MIT Press, Cambridge, MA, US.

Perry, C. (1927). *Regional Plan for New York State*. New York: *Regional Planning Association of New York*.

The Republic of South Africa (2015). *The National Development Plan*. Government Printers, Pretoria, South Africa.

Stats SA (2015). *Statistics South Africa*. www.statssa.gov.za/

United Nations-Habitat (2009). *State of the World's Cities Review*. Harmonious Cities, Earthscan Publications, London, UK.

Watson, V. (2016). "Shifting Approaches to Planning Theory: Global North and South." *Urban Planning*. 1(4): 32–41.

Further reading

Dewar, D. (2018). "Rethinking the Meaning of Livable Cities: The Case of Cape Town, South Africa." In Wagner, F. and Caves, R.W. (eds.) *Livable Cities from a Global Perspective*. Ashgate Publishing, Farnham, UK. This paper identifies criteria for defining urban livability and applies the matrix to South African cities, finding that, in large part, they are highly unlivable.

Dewar, D. and Kiepiel, J. (2012). "Spatial Development Frameworks on a Broader Scale: An Integrative Approach." *Town and Regional Planning*. 61: 30–35. This paper discusses inter alia, the issues of spatial justice at regional and metropolitan scales.

Dewar, D. and Todeschini, F. (2017). "South Africa." In Pojani, D. and Stead, D. (eds.) *The Urban Transport Crisis in Emerging Economies*. Springer Publishers, USA. This paper provides a comprehensive overview of the form and performance of the transportation system in South Africa and argues that poor transportation planning practices have contributed significantly to the current dysfunctionality of South African cities.

5

THE FADING PULSE OF PLACE

Eastern Mediterranean cities in the neoliberal era

Ramzi Farhat

Introduction: broadening and localizing gentrification discourse

Across cities of the Eastern Mediterranean, the events of the past few years have been seismic. As cities from Cairo to Baghdad convulsed, popular uprisings often referred to as the "Arab Spring" laid bare the long-simmering tensions between rich and poor, locals and immigrants, and liberals and conservatives. While political agitation has subsided, these rifts remain largely unaddressed. At this pivotal moment, this chapter examines patterns of contemporary placemaking and emerging challenges of gentrification using Amman (Jordan), Beirut (Lebanon), Cairo (Egypt), and Damascus (Syria) as examples. The broader discourse on the neoliberal restructuring of the state frames this discussion (Harvey 2005). The examples aim to illustrate how the state, special interests, and the citizenry contest over the articulation of the cultural and economic rights to the city and influence urban design schemes. Based on an examination of projects in the region, the chapter also explores how gentrification theory can be nuanced.

The traditional Western debate on gentrification is notable for its recognition of the distinction between governments and market actors. While Smith (2006) and others have highlighted the influential roles of financiers and developers, subsequent discussions on neoliberal and state-led gentrification tug the debate in different directions. The cooptation of the state by monied interests in a process that erodes the protection of marginalized populations is one interpretation of the neoliberal thesis (Harris 2008). On the other hand, the examples of state-led gentrification help us to better understand how public policy can be channeled to catalyze the regeneration of disinvested places (Cameron and Coaffee 2005). To build on this approach, any invocation of the role of the state in the Eastern Mediterranean must be attuned to the fact that the state's regulatory role is often conditioned by its authoritarian qualities, with varying degrees of respect for the rule of law, protection of property rights, and adherence to liberal tenets of freedom of expression or contract. Under such circumstances, the distinction between neoliberal and state-led gentrification is often blurred.

Studies of gentrification have also acknowledged the role of cultural motives in determining which areas are targeted for investment. For example, Zukin (1987) contends that culturally validated neighborhoods underpin the sense of collective identity that some upwardly mobile groups desire. Smith (2006) has noted that the consumption of urban experiences contributes to the construction of class identity. That identity may also be defined around a particular urban outlook. Caulfield (1989), for example, frames the renewed interest in urban living as a rejection of the cultural conformity of the suburb. In the Eastern Mediterranean, however, the cultural motives of gentrification are also often reflective of ethnic, religious, and tribal tensions. While this might manifest itself in symbolic place-claiming, it might in more extreme cases also result in violent internal displacement of one ethnic or religious group by another. While the discussion on gentrification in a Western context can be qualified by attention to the degree to which public policy is biased, the liberal deficit in statecraft in the Eastern Mediterranean makes alliances between state and social actors particularly synergetic.

Finally, the traditional theory of gentrification presupposes that economic motives are formulated in a context where the state fulfills certain responsibilities, ranging from the provision of basic services to the protection of rights to private property. This is not always the case in the region, where the state has often failed to reliably provide basic services such as power (e.g. Lebanon), waste management (e.g. Egypt), and open public spaces (most other countries). Additionally, the state's commitment to protecting property rights and expanding those rights to the citizenry has also often been unsatisfactory (Ghertner 2015). Expropriation schemes on questionable legal basis are not uncommon (e.g. Jordan and Syria), as is institutionalized discrimination in housing. These considerations, in addition to social contestation, should inform any assessment of economic motives contributing to gentrification. The next section will illustrate these points through examples of specific state, cultural, and economic influences on urban design schemes. I will refer to these as "projects," and they include the state seeking legitimation through urban preservation and modernization, the efforts of certain social groups to assert rank and privilege by appropriating culturally significant urban environments, and the determination of elites to build enclaves of privilege in response to the failure of the state to provide basic services and public goods.

Gentrification: projects and the regeneration toolkit

In the neoliberal era, the state in the Eastern Mediterranean region continues to negotiate a precarious balance between promoting economic competitiveness by broadening corporate freedom and managing political legitimation. In the arena of placemaking, this is often manifested in both historic preservation and urban modernization. For example, the old towns of Cairo and Damascus exemplify how historic preservation not only promotes tourism but legitimates the state as it signals its fulfillment of obligations in the stewardship of sites and artifacts significant to world heritage (Sudermann 2013). As illustrated by the financial district planned for downtown Beirut and the al-Abdali district in Amman, legitimation may alternatively take the form of a project of economic modernization. Legitimation here derives directly from a commitment to economic competitiveness as the state promises to usher once provincial urban centers into the service-oriented, 21st-century global economy (Hourani 2014; Parker 2009). Across the region, the urgency of this project is necessitated by the failing prospects of monopoly capitalism, where key enterprises in the agricultural, industrial, and service sectors are controlled by politically connected elites.[1]

Cultural projects, on the other hand, often express a loss of rootedness in the transition from nationalist to populist models of citizenship, and more recently, from the latter to the reactionary conservative fundamentalist model that has flourished in the neoliberal era. Historically, the traditional mercantile bourgeoisie was one of the first social groups to be aggrieved by the rise of authoritarian governments. As the reclamation of the Haussmann-inspired Khedive Cairo and the development of a new downtown in Amman reflect, neoliberal reforms have allowed this group to cement associations with regional real estate investors and leverage these ties to regain stature. Not to understate the multifaceted nature of the process, these efforts partially interface with the desire of natives to reclaim the urban cores from rural migrants, whose move to the city was made possible as original residents abandoned a deteriorating building stock, and as populist housing policies encouraged relocation. Increasing attention to questions of identity politics and the newfound appreciation for historic neighborhoods (both as status symbols and commercial concerns) contribute significantly to this trend, as the renovation of old homes, bed-and-breakfasts, and boutique hotels in old Damascus illustrates (Totah 2014). More recently, secular elites and upwardly mobile liberals have been emboldened to assert their rights in a social context that has become increasingly more conservative. The Rainbow Street district in Amman and the Mar Mikhael district in Beirut are two examples of efforts to carve out spaces for a multicultural, forward-looking lifestyle.

Furthermore, economic projects are often articulated in serving the interests of the emerging elite. While the neoliberal tide has washed over the region with alarming ferocity, the gated communities on the outskirts of Cairo and Amman have offered traditional elites and an emergent class of upwardly mobile businessmen the opportunity to segregate themselves from their poorer compatriots. In traditionally territorial Beirut, gated towers have offered similar conveniences. These types of environments are also popular amongst middle-class residents due to the state's inability to adequately supply basic services, such as clean water, power, green spaces, and pollution control. While they deliver where the state often fails, the often-exclusive developments are gradually disarticulating the traditionally walkable urban fabric (Alaily-Mattar 2008). As walls rise, storefronts that typically lined the ground floors of buildings are also sacrificed, exacerbating physical barriers and the segregation of uses in a trend that recalls sprawling development in the United States and elsewhere.

In realizing these projects, various agents employ a spectrum of tools, including district upscaling, infrastructure upgrading, and, in extreme scenarios, wholesale upzoning and population uprooting. For agents acting through private markets, the restoration of older buildings often contributes to the upscaling of a district. Upgrading, alternatively, is an important tool given that urban design remains a relatively peripheral state concern. While sidewalks are generally provided in urban areas outside of the informal settlements, an investment in high-quality materials, urban furniture, or open spaces is reserved for tourist or trendy areas. That said, the upgrading of the public realm is often a potent catalyst in the marking of certain areas for gentrification (Figure 5.1). The stark difference between the quality of the urban infrastructure in downtown Beirut and the rest of the city is illustrative. The commitment to district-wide historic preservation is a more recent concern that is synergistic to these efforts. When Damascene authorities detailed renovation standards for old city homes, for example, a large segment of the housing stock became unaffordable to lower-income residents (Copertino 2014).

Urban infrastructure upgrading is often coupled with potentially more disruptive tools that throw into sharp relief the power and reach of the state. In many contexts, politically connected real estate interests have influenced the passage of laws that raise development densities to levels that create powerful incentives to tear down low-rise historic buildings (Sakr-Tierney 2017). For example, most of the structures officially listed on the historic building registrar in Beirut have been lost over the years, despite public outcry over the state's unwillingness to stand up to politically connected developers. Alternatively, public-private real estate redevelopment corporations remain popular when the wholesale reconfiguration of the urban fabric is desired. Unfortunately, many urban redevelopment projects in the region feature the sort of weak commitment to property and tenure rights that has plagued 20th-century urban renewal efforts elsewhere. The uprooting of local residents to realize the al-Abdali district in Amman is one example (Daher 2008). To further illustrate these dynamics, the next section will discuss how the aforementioned projects and tools have impacted the redevelopment of select old town, downtown, and "hip-town" districts.

Figure 5.1 Muizz Street in old Cairo, the pedestrianized spine through the historic core.

Source: Ahmed Al Badawy

Regenerated cityscapes

Old towns

The regeneration of old towns presents a classic example of the legitimation dividends that the state will earn from a political project centered around historic preservation. Successive efforts to revitalize old Cairo are illustrative. In the late 19th century, the area around the Bab Zuweila, one of the few remaining gates in the walls of the old city, was renovated in an effort to restore the old core to an Islamic identity (Hampikian 2005). More recent attempts to expand this effort to the entirety of the old city have faced challenges, however, since building and use patterns in the 20th century further compromised the district's authenticity. Jarring transitions between older structures of modest scale and more recently built multi-story apartment buildings abound. All the while, light industrial activities continue to be interspersed amongst historic buildings with little attention to land use planning. Given this situation, recent plans have largely abandoned the ambition to achieve a district-wide consistency. Current efforts, dating to the UNDP plan of 1997, prioritize heritage corridors and monument clusters, especially around the al-Azhar, Ibn Tulun, and Qalawun religious and funerary complexes (Sutton and Fahmi 2002). The upgrading efforts, which included closing off the historic Muizz Street and al-Azhar Square to traffic, as well as restoring the Ayyubid medieval city wall, attempt to achieve the state's agenda in a more pragmatic fashion.

The old city of Damascus, a UNESCO World Heritage Site, is instructive of how political, cultural, and economic projects overlap. During the last two decades, the old city has gradually been upscaled as hundreds of traditional historic courtyard homes have been repurposed as restaurants, coffee shops, and later boutique hotels, with names such as Beit Ramza and Qasr Shahbandar (Copertino 2013) (Figure 5.2). This project of historic preservation was accelerated after a pivotal political succession in the year 2000, with the new leadership committing to a policy of economic and cultural liberalization. If commercializing the built heritage creates investment opportunities that contribute to political legitimation, politically connected monied interests also stand to benefit (Copertino 2014; Sudermann 2013). Alternatively, a local cultural project of the urban elites features as prominently. For the original residents of Damascus (the shameen), wrestling back control of traditional courtyard homes in the old city from the rural migrants (the muhajireen) remains a prime goal (Jacobs 2010; Totah 2014). In this case, upscaling and restoration contribute to a project of redeeming cultural standing lost during socialist and populist eras.

Figure 5.2 Bait Al-Moalem, a recently renovated boutique hotel in the Jewish Quarter of Damascus.

Source: Kareem Mayan

Downtowns

Across the region, overlapping projects have determined the dynamics through which a civic and corporate downtown was imposed upon the bustling city of tradesmen and merchants. In Beirut, a state project of modernization informed the rebuilding of the war-torn downtown starting in the mid-1990s. While the area around the boulevards radiating from the Place de L'Etoile was upscaled and upgraded to French-mandate-era elegance, a zone by the waterfront was designated as Beirut's new office district. It was envisioned that high-rises would house corporate headquarters that would reclaim the city's reputation as a regional business hub (Nagel 2000). The redevelopment of the downtown area, however, also channeled a sectarian cultural project. During the conflict, the political influence of the Sunni Muslim community had waned considerably. In spearheading development efforts, leading figures of a Sunni oligarchy, with the financial backing of regional sponsors, reclaimed political and cultural standing as agents of reconstruction. The economic project of designing an upscaled and upgraded high-end district in the center of the city cannot be discounted, either. While the district master plan sensibly addressed pedestrian experience issues, it imposed very exacting architectural standards, especially concerning the quality of materials. The pedestrianized area around Foche and Allenby Streets is illustrative (Gavin 2015). This fully serviced enclave of privilege at the center of an impoverished urban core was oriented to an elite monied class and regional investors alike.

Amman's al-Abdali district, similarly, sought no less than to endow the sprawling, hill-hugging city with a downtown structured around the Parliament Building, the High Court, and the King Abdullah Mosque. The wholesale upzoning and upgrading of the area resulted in the uprooting of a local community. The mid-rise, block-form downtown is organized around a central spine named "The Boulevard," designed as a pedestrian promenade and anchored by a high-end enclosed mall. The Lebanese-Saudi Hariri family were instrumental in the realization of both the al-Abdali and the downtown Beirut projects. In the al-Abdali district, the neoliberal influences can be further traced to the privileges developers were afforded to realize the vision of the master plan through negotiated agreements on large-scale projects. As a state project of modernization, the al-Abdali district provides Amman with the center it sorely lacked, a significant achievement during the reign of King Abdullah. The development also serves as an economic project of the traditional capitalist elite, an oligarchy whose control over the banking sector was expanded after the relaxation of currency controls in the 1980s (Parker 2009). When a later, mid-1990s law made real estate investment tax-free, and trade agreements from the early 2000s facilitated foreign investment by mediation of this elite, their relationship with the monarchy was further cemented (Hourani 2014).

Hip-towns

Challenging this establishment politics are the intellectuals, progressives, and creative professionals shaping the hip-towns of the city. In Cairo, we see short-lived glimpses of this effort in the regeneration of Khedive Cairo, the 19th-century "Paris on the Nile." Once associated with a cosmopolitan bourgeoisie, the district later reverted to an aspiring class of officer corps after the military coup of 1952, before being abandoned for better-serviced master-planned developments in the desert fringes of the city (el-Sorady, in press). More recently, state agencies teamed up with banks in an attempt to infuse life into the district. Tepid measures such as the painting of the district's tired façades failed, however, to catalyze a wider regeneration. During the upheavals of 2011, area galleries and cafés attracted intellectuals, the liberal youth, and the upwardly mobile creative class to the district. This trend was reversed when the state regained its footing, as it was feared that these cultural claims would prove to be subversive. In response, the state mobilized a project of modernization that was synergistic to a bourgeois project of cultural consumption. Notable accomplishments to date include the upgrading and pedestrianization of downtown streets and the upscaling of modernist apartment buildings through investment activities by regional investors.[2] As in the example of Damascus, regeneration was geared towards sophisticated urbanites (el-Shahed 2015).

In Amman, the Rainbow Street district on the hills west of the al-Abdali district continues to grow into a compact-growth, inner-city neighborhood. An upgrading plan for the area developed with community input; it featured traffic calming, public furniture, open spaces, and symbolic signage (Zada 2014). The upscaling and renovation of the Rainbow Cinema, the city's oldest movie theater, was an important catalyst, and it was followed by the opening of an array of artsy restaurants and boutiques (Parker 2009). Young professionals, unsatisfied with a suburban lifestyle in the sleepy city, steered the endeavor towards a project of community building. The physical regeneration is complemented by the staging of periodic events, such as flea and farmers' markets, that animate the side streets. As in Cairo, competing interests put the area at a crossroads. The rising stature of the street has precipitated a more conventional process of real estate investment, as mainstream developers continue to move in with plans to upscale the area (Parker and Debruyne 2012).[3]

The Mar Mikhael district east of downtown Beirut offers the most vivid illustration of how transformative these efforts can be once they take root. The charm of this formerly working-class enclave derives from its narrow streets, walk-up apartment buildings, and traditional businesses. The district is a classic example of pioneer-driven gentrification and continues to attract the creative, liberal, and forward-thinking denizens of Beirut. The in-migration of design firms led the way more than a decade ago, and area buildings continue to be upscaled as it attracts more bistros, bars, and restaurants.[4] Progressively, the district grew into a nascent creative class community, a hospitality destination, and later a community-building project that attracts young, liberal professionals away from the various sectarian enclaves of the city (Krijnen and De Beukelaer 2015). Higher-end developments, such as the AYA residential-commercial complex, constitute a larger share of more recent construction, and signs of gentrification are easy to identity as the imprint of the working-class past continues to fade.

The city reclaimed and its fading pulse

These trends pose several placemaking challenges. Until recently, the city of bustling souks and busy commercial streets has survived the type of planning that adapted Western cities to the demands of the modern industrial economy. While the region was late to develop, the current economic agenda carries a decidedly neoliberal flavor, with emphasis on hospitality, professional services, and knowledge-based occupations. Invariably, the livelihood of many is jeopardized. For example, while tourism has increased employment in the hospitality industry, no effective plans were put in place to account for the displacement of other occupations, such as light industry, that were deemed incompatible to the new agenda. Those employed in the traditional service sectors, now shut out of the higher-end new-build districts, suffered as well. Furthermore, investments in more specialized and volatile sectors also carry significant risks. For example, after years of political instability, old Cairo has yet to fully recover the tourism it once attracted. Across the Sinai Desert, the early hopes that relatively small, poor cities would attract global corporate interest have largely faded. Today, few expect the corporate skyline that the plan for downtown Beirut envisioned would materialize.

These trends also pose cultural challenges. Across the region, a lifestyle traditionally focused on the social life of the street is slowly being compromised. Ill-considered urban design schemes that aggressively define and police property rights are compromising city life defined around casual street conversations, sidewalk cafés, and street vending and hawking. The pedestrianized areas around Allenby Street in downtown Beirut, sanitized of all competing claims on space, evoke little of the lively urbanism of the old city, while the upscale Beirut Souks mall is an almost cynical simulacrum on its namesake, the traditional souks (marketplace) of the pre-war city. Similarly, the ritual of the promenade that the Boulevard in Amman hopes to popularize stands in contrast with the everyday strolling down the city's streets. Regrettably, the new mall attracts high-end commerce away from the city's main streets.

One may also criticize whether the newly created or rehabilitated built environments undergird cultural identities. For example, most of the restored courtyard homes in Damascus do not have a public function. The freedom that developers have in interpreting the urban design plan for the al-Abdali district in Amman is also questionable. The design language of the mid-rise housing can be best described as an exercise in "restrained modernism." Where the design is context-sensitive, quality materials are used, craftsmanship is high, and the dignified buildings inspire a certain civic gravitas (Abed 2004). But when glass surfaces dominate, especially if commercial-grade cladding systems are used, the language is often too generic, as exemplified in downtown Beirut. While these are missed opportunities to further develop a local design language, one should commend the Amman downtown plan for its coherent and restrained high-rise skyline, in contrast to many of the skylines in the Arabian Gulf, which feature attention-seeking, leaning, and twisting towers of questionable design quality.

Notwithstanding these challenges, the emergent landscape of old towns, downtowns, and hip-towns is agreeable by many standards of urban design, including walkability, mixed-use development, and imageability. The plan for Beirut, for example, was developed by a team led by Angus Gavin, formerly involved in the redevelopment of the London Docklands and notable for his focus on the street experience, volumetric controls, and form-differentiated districts. For the al-Abdali district, a Beirut-based regional consultancy (LACECO) produced a master plan that focuses the visitor experience along prime routes, invests heavily in urban amenities, and distributes the corporate high-rises to prevent nighttime dead-zones that result from their clustering (Figure 5.3). While formal planning frameworks are in effect citywide, the level of control they

Figure 5.3 Work continues on the al-Abdali district, Amman's new "downtown."

Source: Eurovaran

impose over built form, uses, and public space does not compare to the more elaborate and stricter regulations that govern construction in the master-planned communities. It is the verve of that special place – a city that flourishes on the unplanned, unexpected, and unsanctioned – that is at stake. If such informalities survive the worst of the current design agenda, then something of the traditional economic and cultural ways of life may, too.

Gentrification, a looming threat

Questions remain as to whether these trends will result in further gentrification. Notable in this respect are laws encouraging foreign investment that have been introduced starting in the 1990s. For example, a Jordanian law dating to 1995 made profits from real estate tax-free (Hourani 2014). Syria followed suit by rewriting its laws on foreign investment and property ownership in the early 2000s (Clerk and Hurault 2010). Foreign investors were also in mind when a new wave of plans ushered in an era of an "informalized" regime of development regulation. Examples include the 2008 Amman master plan (Daher 2013), the 2010 "Cairo 2050 Development Plan" (el-Sorady in press), and relaxed development intensity ceilings in the Damascus and Beirut master plans (Clerk and Hurault 2010; Krijnen and Fawaz 2010). These master plans often imply the relocation of lower-income populations to edge regions, for which the plans for Amman and Damascus have been criticized (Clerk and Hurault 2010; Parker 2009). In the desert on the outskirts of Cairo, along the Amman airport road, and on Beirut's last remaining large lots, a new landscape of megaprojects is taking shape, one where entitlements are often negotiated with developers (Daher 2013; el-Shahed 2015).

In special districts, specifically downtowns, the transfer of planning authority to development corporations has compromised the public's ability to hold developers of large-scale projects accountable (Schmid 2006). Examples include the Solidere corporation developing downtown Beirut and the al-Abdali corporation developing downtown Amman. The most worrisome recent development, however, is the regionwide reform of rent control laws. Jordan led the way in 1994, and rent liberalization proved instrumental in accelerating redevelopment in the downtown area (Zada 2014). In Syria, the reform of rental laws in 2004 rendered the renovation of historic buildings financially feasible by making evictions easier to obtain (Totah 2014). In Lebanon, laws expected to become fully enforced in the next five years are feared to result in a mass exodus to the poorer suburbs (Krijnen and De Beukelaer 2015). While the old rent laws in Egypt continue to be enforced, plans for a decade-long process of liberalization are included in legislation currently under consideration (el-Sorady in press).

In the old towns, the effects of gentrification have generally been troubling. The upgrading of Old Cairo is a case in point. The tunnel under the newly pedestrianized al-Azhar Square disrupted the livelihoods of many in the local workshop-based economy. Redevelopment efforts in the Gamalia district, including the emptying out of the cemetery and the renewal of Bab el-Shaaria Square, resulted in evictions (el-Daidamony and Shetawy 2016; Fahmi 2012). In Damascus, the renovation of old courtyard homes accelerated after the capping of eviction compensation (Totah 2014). The laws allowed property owners to raise rents and deny automatic tenure transfer to descendants, previously a bulwark against dislocation. The ability to proceed with evictions attracted investment in the repurposing of the older structures.

In the downtown districts, the situation has been even more dire. The social toll exacted in the rebuilding of downtown Beirut was cataclysmic. After the passing of legislation allowing the state to exercise district-wide eminent domain, all outstanding property interests were liquidated into shares in the newly formed public-private real estate corporation. Consequently, almost none of the original residents were able to reestablish domicile in the hyper-gentrified district

(Nagel 2000). In the summer of 2015, a months-long disruption in trash collection evolved into a broader crisis of political legitimacy. The events that ensued exposed this unfortunate reality as activists used sit-ins, camp-ins, and bazaar-like markets to breathe life into a district of struggling hotels, unoccupied apartment buildings, and seldom-frequented luxury boutiques. Lower-income populations did not fare any better in the al-Abdali district. The old multimodal transit hub, utilized by a lower-income population, was relocated to the edge of town, as was the stall market frequented by the residents of the now-displaced Za'amta neighborhood (Daher 2013).

Hip-towns, finally, remain at a crossroads. In Khedive Cairo, the buzz created by the involvement of institutional investors is threatening to enable the type of corporate-initiated gentrification seen in Beirut and Damascus. The social impact of urban change in the hip-town of Mar Mikhael has so far been mixed. Many landlords and tenants with rent-controlled apartments or commercial spaces who could afford to vacate their units have benefited handsomely from the rising stature of the area (Figure 5.4). Local tradesmen and lower-income families, however, have

Figure 5.4 "Aya," One of many developments in Beirut's Mar Mikhael district that integrates with older buildings (visible at the base of the tower).

Source: Ramzi Farhat

faced an affordability crisis. With a larger share of the population being elderly, long-term renters, fear of social displacement remains unaddressed.

Conclusion: spaces of hope and the liberal state

The neoliberal tide, set in motion decades ago, has washed over the Eastern Mediterranean (Harvey 2005; Swyngedouw et al. 2002). In studies of recent development trends, it might be intuitive to frame the recent regeneration of old towns in terms of the state's promotion of tourism, the revitalization of downtown areas as an initiative of corporate interests, and the organic development of trendy "hip-towns" as indicative of the emergence of a local creative class. However, regional particularities must be taken into consideration to qualify the arguments on placemaking and gentrification in the broader discourse. In any political consideration, the notable resilience of the authoritarian and clientelist forms of government must be reckoned with (Carothers 2002; Nathan 2003). Alternatively, and while cultural themes have featured in global studies of gentrification, the region is also notable for a resurgent identity politics as traditional urban elites position to reclaim lost privileges (AlSayyad 2001). Economically, it is necessary to acknowledge local forms of wealth accumulation and the state's failure to provide basic urban services as an alternative economic motive. The various state, cultural, and economic projects alluded to in this chapter illustrate this notably more complicated reality of synergy and contestation.

In old Damascus and downtown Amman, for example, the state's project of legitimation and modernization was synergetic to elite-urban and elite-bourgeois projects of communal self-assertion and economic enclave-building, respectively. In Beirut's Mar Mikhael district, progressive liberal and elite bourgeois projects continue to be in contention. These experiments, which currently include Khedive Cairo and Amman's Rainbow district, are at a crossroads of progressive liberal and classist establishment projects. It is critical that these fissures are exploited if the bustling city of the Eastern Mediterranean, that uniquely vibrant place, should survive. To that end, promising experiments such as in Mar Mikhael offer lessons on how local entrepreneurship and social mobilization can be harnessed. While the footprint of the workshop economy has been reduced significantly, new service businesses have flourished as the local economy continues to be invigorated. Through grassroot initiatives, long-neglected public stairways and alleys have a new lease on life as social spaces. In interesting responses to the challenge of marrying heritage preservation to denser development, older building façades and courtyards have been salvaged and integrated into taller new construction.

Seriously addressing the challenge of increasing social polarization, however, requires that the state asserts its regulatory roles that have been compromised as policies of neoliberal flavor continue to be advanced. At the most basic level, the state must balance the priorities of corporate and real estate interests with those of the citizenry at large. It must also temper the competing cultural claims to urban neighborhoods by committing to the equal treatment of all citizens under the law. Lastly, the adequate provision of basic urban services can be an important first step in addressing the tendency of those of means to seek them in exclusive and gated environments. As urbanization patterns across the region portend, the state's failure to meet the organic adaptation of culturally meaningful places with adequate regulatory commitments can have unfortunate consequences. In Cairo, Damascus, Amman, and Beirut, the regeneration of the cores as exclusive developments and the proliferation of gated communities in the surrounding desert, farmland, and once-sleepy villages threaten to radically transform the traditionally mixed-use, walkable, pedestrian-scaled city. If the project of statecraft in the region remains unfinished, spaces of hope await in a more liberal future.

Notes

1 Various factors have contributed to the necessity of reform, including weak economic opportunities for a younger, better educated, and growing population. Although the chapter presents examples of flagship efforts, the bet on real estate is widespread and geographically diffuse. The Waterfront City in Dbayeh (Beirut) and the Maspero Triangle mixed-use project in Cairo are two other examples of how large-scale development initiatives are changing the face of the city in the region.

2 Khedive Cairo is named after Khedive Ismael, who commissioned French architects to design the "belle epoque" radial plan in the spirit of Haussmann's Paris and was the city's 19th-century downtown. This chapter references the more recent period – the last two decades – during which its fate was contested between competing projects: the first exemplified by the progressive "Townhouse Gallery" and the "Contemporary Image Collective," which hoped to channel that history into a sustained cultural renaissance; and the second exemplified by a more traditional project financed by investors such as the Saudi-financed Isma'iliya corporation, which aimed to capitalize on its architectural marketability.

3 The Rainbow Street district illustrates of how deep-pocketed investors are rapidly gentrifying neighborhoods in which upwardly mobile professionals show initial interest. For the past decade, developers led by the "Ahya Amman" group have amassed a considerable portfolio of strategic properties in the area. Developers have also proposed planning regulations that would make the realization of their plans possible. In this case, a developer-initiated neighborhood plan submitted to the municipality includes high-end mixed-use development as well as a shopping mall complex.

4 Artists, designers, and creative professionals in Beirut have established, abandoned, and reestablished foothold in various neighborhoods over the years, as they pioneer districts and later move when institutional investors follow suit. When the "Papercup" design bookshop and the "MGD" design firm set foot in Mar Mikhael, they were at the vanguard of a migration from the nearby Gemmayze district, which flourished as the center of Beirut counter-culture for a decade or so before rising market-rate rents triggered an eastward relocation.

References

Abed, G. (2004). "The Art of Selling Cities." In Abed, G. (ed.) *Architecture Reintroduced: New Projects in Societies in Change.* The Agha Khan Award for Architecture.

Alaily-Mattar, N. (2008). "Beyond Gated Communities? Detachment and Concentration in Networked Nodes of Affluence in the City of Beirut." *Urban Design International*, 13(4): 263–271.

AlSayyad, N. (2001). *Consuming Tradition, Manufacturing Heritage: Global Norms and Urban Forms in the Age of Tourism.* New York, NY: Routledge.

Cameron, S. and Coaffee, J. (2005). "Art, Gentrification and Regeneration: From Artist as Pioneer to Public Arts." *European Journal of Housing Policy*, 5(1): 39–58.

Carothers, T. (2002). "The End of a Paradigm (Political Transition)." *Journal of Democracy*, 13(1): 5–21.

Caulfield, J. (1989). "Gentrification and Desire." *Canadian Review of Sociology and Anthropology-Revue Canadienne De Sociologie Et D'Anthropologie*, 26(4): 617–632.

Clerk, V. and Hurault, A. (2010). "Property Investments and Prestige Projects in Damascus: Urban and Town Planning Metamorphosis." *Built Environment*, 36(2): 34–47.

Copertino, D. (2013). "Reinterpreting and Reforming the City: Patrimonialization, Cosmopolitanisms, and the Ethnography of the Heritage-scape in Damascus." *Paysages Patrimoniaux*, 35(2): 101–128.

Copertino, D. (2014). "The Tools of the Trade: The Materiality of Architecture in the Patrimonialization of 'Arab Houses' in Damascus." *Journal of Material Culture*, 19(3): 327–351.

Daher, R. (2008). "Amman: Disguised Genealogy and Recent Urban Restructuring and Neoliberal Threats." In el-Sheshtawy, Y. (ed.) *The Evolving Arab City: Tradition, Modernity, and Urban Development.* London: Routledge, 37-68.

Daher, R. (2013). "Neoliberal Urban Transformations in the Arab City: Meta-narratives, Urban Disparities and the Emergence of Consumerist Utopias and Geographies of Inequalities in Amman." *Environnement Urbain/Urban Environment*, 7: 99–115.

el-Daidamony, M. and Shetawy, A. (2016). "Gentrification Indicators in the Old City of Cairo." *Procedia: Social and Behavioral Sciences*, 225: 107–118.

el-Shahed, M. (2015). "The Prospects of Gentrification in Downtown Cairo: Artists, Private Investments, and the Neglectful State." In Lees, L., Shin, H. and Lopez-Morales, E. (eds.) *Global Gentrifications: Uneven Development and Displacement.* Bristol: Policy Press., 121-143.

el-Sorady, D. (in press). "Revitalisation of Downtown Khedive Cairo: A Contested Sovereignty." *Cities*.
Fahmi, W. (2012). "Sustainable Tourism and the Rehabilitation of Cairo's Historical Districts: The Case of the Bazaar Area and the Cities of Dead" Paper presented at the 2nd World Sustainability Forum, November 1–30.
Gavin, A. (2015). "The Corporate Discourse: Learning from Beirut's Central Area Renewal." In Saliba, R. (ed.) *Urban Design in the Arab World: Reconceptualizing Boundaries*. London: Ashgate.
Ghertner, D. (2015). "Why Gentrification Theory Fails in 'Much of the World'." *City*, 19(4): 552–563.
Hampikian, N. (2005). "Medieavalization of the Old City as an Ingredient of Cairo's Modernization: Case Study of Bab Zuwayla." In AlSayyad, N., Bierman, I. and Rabbat, N. (eds.) *Making Cairo Medieval*. Lanham, MD: Lexington Books, 201–234.
Harris, A. (2008). "From London to Mumbai and Back Again: Gentrification and Public Policy in Comparative Perspective." *Urban Studies*, 45(12): 2407–2428.
Harvey, D. (2005). *A Brief History of Neoliberalism*. Oxford: Oxford University Press.
Hourani, N. (2014). "Urbanism and Neoliberal Order: The Development and Redevelopment of Amman." *Journal of Urban Affairs*, 36(S2): 634–649.
Jacobs, J. (2010). "Re-branding the Levant: Contested Heritage and Colonial Modernities in Amman and Damascus." *Journal of Tourism and Cultural Change*, 8(4): 316–326.
Krijnen, M. and De Beukelaer, C. (2015). "Capital, State, Conflict: The Various Drivers of Gentrification in Beirut." In Lees, L., Shin, H. and Lopez-Morales, E. (eds.) *Global Gentrifications: Uneven Development and Displacement*. Bristol: Policy Press., 285-310.
Krijnen, M. and Fawaz, M. (2010). "Exception as the Rule: High-end Developments in Neoliberal Beirut." *Urban Environment*, 36(2): 117–131.
Nagel, C. (2000). "Ethnic Conflict and Urban Redevelopment in Downtown Beirut." *Growth and Change*, 31(2): 211–234.
Nathan, A. (2003). "Authoritarian Resilience." *Journal of Democracy*, 14(1): 6–17.
Parker, C. (2009). "Tunnel-bypasses and Minarets of Capitalism: Amman as Neoliberal Assemblage." *Political Geography*, 28: 110–120.
Parker, C. and Debruyne, P. (2012). "Reassembling the Political Life of Community: Naturalizing Neoliberalism in Amman." In Künkel, J. and Mayer, M. (eds.) *Neoliberal Urbanism and its Contestations: Crossing Theoretical Boundaries*. Basingstoke and New York: Palgrave Macmillan.
Sakr-Tierney, J. (2017). "Real Estate, Banking and War: The Construction and Reconstructions of Beirut." *Cities*, 69: 73–78.
Schmid, H. (2006). "Privatized Urbanity or a Politicized Society? Reconstruction in Beirut After the Civil War." *European Planning Studies*, 14(3): 365–381.
Smith, N. (2006). "Gentrification Generalized: From Local Anomaly to Urban 'Regeneration' as Global Urban Strategy." In Fisher, M. and Downey, G. (eds.) *Frontiers of Capital: Ethnographic Reflections on the New Economy*. Durham: Duke University Press, 191–208.
Sudermann, Y. (2013). *Gentrification and Urban Heritage Under Authoritarian Rule: The Case of Pre-war Damascus, Syria*. Dissertation, The University of Edinburgh.
Sutton, K. and Fahmi, W. (2002). "The Rehabilitation of Old Cairo." *Habitat International*, 26: 73–93.
Swyngedouw, E., Moulaert, F., and Rodriguez, A. (2002). "Neoliberal Urbanization in Europe: Large-scale Urban Development Projects and the New Urban Policy." *Antipode*, 34(3): 542–577.
Totah, F. (2014). "'Nothing has Changed': Social Continuity and the Gentrification of the Old City of Damascus." *Anthropological Quarterly*, 87(4): 1201–1227.
Zada, J. (2014). "How Did Neoliberalism Steer the Rainbow Street Regeneration Project in Amman?" International Urbanism Conference, Universität Stuttgart.
Zukin, S. (1987). "Gentrification: Culture and Capital in the Urban Core." *Annual Review of Sociology*, 13(1): 129–147.

Further reading

el-Sheshtawy, Y. (ed.) (2011) *The Evolving Arab City: Tradition, Modernity, and Urban Development*. London: Routledge. Historical and critical essays on Arab capitals from the MENA region, with a focus on global questions, 37–68.
Lees, L., Shin, H., and Lopez-Morales, E. (eds.) (2015) *Global Gentrifications: Uneven Development and Displacement*. Bristol: Policy Press. Reader on global gentrification, contains several contributions from the Arab world region, 121–143.
Saliba, R. (ed.) (2016). *Urban Design in the Arab World: Reconceptualizing Boundaries*. London: Ashgate. Contains essays on political and cultural determinants of urban design practice in MENA region.

6

UN-CITIES

The urbanism of rapidly growing cities in the Gulf region

Surajit Chakravarty

Introduction: key features of the urban context of the Gulf region

Abu Dhabi has witnessed phenomenal growth and development over the last decades. Various aspects of Abu Dhabi's urban experiment have been reported in the existing literature. Published work includes studies on sustainability (Chakravarty, 2017), public transit and social exclusion (Chakravarty and Qamhaieh, 2015), gentrification (Qamhaieh and Chakravarty, 2017), the Masdar eco-city project (Caprotti, 2014; Cugurullo, 2015; 2013), the lives of taxi drivers (Mohammad and Sidaway, 2012), a critique of megaprojects (Ponzini, 2011), economic trends (Davidson, 2009), and Abu Dhabi's global ambitions (Elsheshtawy, 2010; 2008). In addition to discussing the urban context, these publications also feature accounts of the social, political, and economic context of the city and region. While it is not possible to discuss here all aspects of the region's development and institutions, a few points are worth repeating. The following four key features of the region's political economy establish a context for the arguments made in this chapter.

First, countries in the region are non-democratic, but governments are typically sensitive to the needs and complaints of the locals (also known as nationals). Thus, municipal governance is efficient, and technological advancements are usually adopted promptly. Abu Dhabi, for example, makes no claims to guaranteeing basic democratic freedoms even to its citizens, far less to expatriates. There is no right to protest, no right to go on strike (even if salaries are withheld for the entire staff of a large firm), no right to free speech, no right to collective legal action (class-action), and no right to free association. Moreover, the city is under the constant surveillance of a network of cameras, and plain-clothes policemen looking for anything out of the ordinary – even something as banal as taking pictures of non-touristic places.

Second, the United Arab Emirates (UAE) and other countries in the region rely heavily on foreign workers, including a large number of low-wage workers, mainly from South Asia (for construction, maintenance, driving, etc.), and the Philippines (domestic helpers). Foreign workers are commonly known as "expatriates" ("expats," for short) because they have no route to citizen-ship, permanent residence, or tenure of any kind. In most urban centers in the region, 75%–80% of the residents are expats.

Third, the state maintains a façade of competition, free markets, and, significantly, the separation of public and private sectors. But most major sectors of the economy are characterized by oligopolistic setups, tight control and overall ownership by a small elite group of nationals. For urban development, this means artificially set rents, a real estate sector controlled by a few corporations (which are ultimately in the hands of a few people), and urban planning oriented towards setting the stage for megaprojects.

Fourth, the economies of the region depend on oil, although not equally so. Similarly, the local "taste" or "feel" of each city in the region is distinct. Although this chapter employs a broad brush (with arguments based primarily on the experience of Abu Dhabi), it is important to remember that even Abu Dhabi and Dubai, both in UAE and only 130 km apart, vary in terms of their growth trajectories and economic goals, and also in terms of their culture and urbanism. The objective in this chapter is not to flatten the diversity of the region, but rather to provide an overview of tendencies and predispositions that are common to its various cities.

Abu Dhabi's urban form and urbanism

Early attempts to plan Abu Dhabi (in the 1970s and 1980s) were motivated by a vision of modernity that glorified the personal automobile. This led to flaws in the design of city blocks, in terms of connectivity, walkability, and ultimately, sustainability. Further, peculiarities of the political culture and economic system have fostered an urbanism where people and the environment receive the lowest priority. Specifically, four aspects of the city reveal a planning culture that is at the heart of its urbanism: car-oriented outlook to urban design, fragmented urban fabric, "people-last" architecture, and layers of segregation.

Effects of a car-oriented outlook on urban design

The city building, from the time oil wealth began flowing into the nation, was an act of compensation for the hardships of the past. The city held the promise of comforts and conveniences that had been missing. Prime amongst these was the speed, reach, and comfort of the personal automobile. The passenger car was to be an integral element of the future of the rapidly modernizing state. The original street grid laid out in the late 1960s assumed widespread car ownership. With oil priced well below international market rates, the dependence on personal automobiles was advanced further.

As engineers prioritized cars and vehicular movement, pedestrians and cyclists were eventually forgotten. Abu Dhabi's superblocks (as large as 1 km x 1 km, although not all blocks are of the same size) typically feature 12- to 15-story buildings along the periphery, and 2- to 4-story blocks (or villas) within. Some downtown blocks are comprised entirely of taller structures. When this pattern was designed in the 1970s (Kyriazis et al., 2002), the city was a blank slate, marked by only a few gathering points of social significance. Those who did not have cars did not have far to go. As the city filled in the grid, it quickly became clear that the form created distances in everyday life that could not be covered without automobiles. Thus, Abu Dhabi embarked on a series of stopgap measures to bring a minimum mobility to low-income residents, mainly to allow them to reach their places of employment. To begin with, shared (fixed-route) taxis were introduced as the main means of transport for those who could not afford to buy a car. By the late 1990s, taxis were upgraded, and shared taxis were slowly eliminated. Low-wage residents were again priced out of the principal accessibility option. At the same time, it became common practice for larger employers to provide shuttle services for their employees. It was not until 2008 that a bus system was introduced in the city, which to this day continues to be stigmatized, as well as flawed from the perspectives of equity and sustainability (Qamhaieh and Chakravarty, 2017).

The absence of a high-quality rapid transit system makes life extremely difficult for hundreds of thousands of individuals, in a prohibitively expensive city, where even taxis are not affordable. In addition to diminishing the quality of life, the absence of reliable and affordable mass transit also effectively excludes people from the city and its amenities. Not least, it exposes pedestrians to the elements. Pedestrians must brave high temperatures (that do not fall below 40 degrees Celsius for several months, and approach 50 degrees at the peak of summer) as they go about their day, buy groceries, and catch buses. Late into the night, the heat continues to radiate back out of the concrete.

One poignant issue is that of crossing streets. While cars breeze through, pedestrians find their already stressful and unpleasant walks unduly lengthened by the need to go to the nearest intersection and then walk back, when they only needed to cross the street. Although a few midblock crossings (some signalized) and overhead walkways have been constructed in recent years, the problem of crossing 6- to 10-lane carriageways can still be seen all over the city.

Probably the most profound impact of auto-dependence is the fragmented urban fabric of Abu Dhabi. This fundamental outlook towards city building has not changed even in recent times, even as the city makes tall claims of promoting sustainability (Chakravarty, 2017). Planners and designers of the city have continued to create pockets of growth (by way of megaprojects in attractive locations), rather than weaving a continuous fabric. This aspect is discussed further in the following section.

Fragmented urban fabric

The city was built in pieces. Distant suburbs were soon added to the original grids; large lots (typically 60 m x 60 m or larger) accommodated the needs of large families. Suburbs have kept growing over the decades as local population has grown, and families have become relatively smaller. Medium-density options were not considered for locals, who had been promised a plot of land and a home by the generous first president and father of the nation, the late Sheikh Zayed. One of the founding promises of state welfare, the claim to free land, continues to be made to young Emiratis starting families. And it is a commitment the state continues to honor.

Meanwhile, the rest of the city seems to have been assembled as a collection of discreet megaprojects, sporadically built in various parts of the city. Masdar City, the new airport (and its additions), Reem Island, Yas Island, Saadiyat Island, Souwah Island, the Grand Mosque, the World Trade Center and mall, Bain al Jissrain, Al Reef, the Etihad complex, Two-Four 54 (the Ministries Complex) etc., are all scattered far and wide. Abu Dhabi's original city blocks now exist in the shadows of these megaprojects. These blocks, too, are being gentrified posthaste.

At the same time, small settlements between Abu Dhabi and Dubai (for example, Al Shahama, Al Bahia, Al Rahba, Bani Yas, etc.) have grown into peri-urban entities, featuring suburban villas for locals, and a limited number of medium-density projects targeted towards expats (at significantly lower rents than in the city). There is a chronic dearth of civic facilities in some of these areas, especially in comparison with central Abu Dhabi. Similarly, Musaffah – mainly known for light industry and automotive repair, is also home to several large "labor villages" and affordable medium-density housing. These are exceptional spaces on the outskirts of the city, functioning (at least for their lower-income residents) as spaces disconnected from downtown Abu Dhabi.

Overall, the urbanization of Abu Dhabi seems to be led by various ambitious, albeit disjointed visions, all jostling to realize successive increments of their grand desires. There is no published plan that explains the logic with which these spaces have been planned, or how the various projects constitute a coordinated growth strategy. It is not clear why these megaprojects are so far away from each other, or why so much undeveloped land has been left vacant in between. It could take decades to fill in the empty gaps that exist between the dots that are the prestigious projects of Abu Dhabi.

When it comes to urban form, Abu Dhabi seems to use the same model of decision-making as Dubai – finance first. Dubai is notorious for "off-plan" constructions, allowing almost any

Figure 6.1 "Earth mover parked in downtown Abu Dhabi."

Source: Author

development that can show a profit-making business model (Husnein and Chakravarty, 2015). Abu Dhabi, though not quite as laissez-faire in its approach, does not seem overly concerned with controlling its overall urban form, either. For the foreseeable future, long drives and travel times are going to be the norm in Abu Dhabi. And the problem will continue to be interpreted as a question of speed and highway capacity, rather than density and land use.

"People-last" urbanism

A "People-last" urbanism prioritizes the imperatives of power, profit, and image and cares very little for the diversity of resident perspectives and needs. Planning, in Abu Dhabi, has not been considerate to all sections of society. The city operates for profit – and, at a deeper level, as a carefully curated mirage that lends an image of normality to a perverse and exploitative system. People are incidental. The original grids – as discussed earlier – were laid out for cars, with the assumption that this would lead to greatness of the nation. The adverse consequences for residents have already been discussed.

One hears of "affordable housing," but it means something quite different in Abu Dhabi from the usual understanding of the term. Affordable housing policies are targeted towards middle- and higher-income individuals, who are single and need studios or one-bedroom apartments. These apartments are not affordable for those who really need affordable housing and who are being priced out of the city center with few alternatives (Chakravarty and Qamhaieh, 2015). The inadequacy of public transportation, as mentioned earlier, only makes matters worse.

All concerns for social issues and questions of welfare and equity have been ironed out of planning documents. Planning has given way to an incremental real estate development process, with minimal intervention from the planners. The construction industry is controlled by a few state-owned enterprises (Davidson, 2009; Ponzini, 2011; Chakravarty and Qamhaieh, 2015). The first priority of development belongs to megaprojects (museums or other state-funded prestige pieces), profitable mixed-use developments (typically featuring multiple skyscrapers), and entertainment-retail centers. Not only do these projects derive rent from land, but they also help in marketing the city in magazines and catalogs.

Social facilities are funded to the extent that they are useful to locals and the higher-income groups. Sports facilities, for example, are not affordable for most residents. The basketball and tennis facilities that *are* available for free are too few, and, as a result, overcrowded. People with cars have several options for accessing parks, waterfronts, and other open spaces. Few public spaces, save the Corniche, are shared by diverse sections of society. Those without cars, especially in low-income blocks of "downtown" or in "labor housing," have to make do with either paved spaces between buildings or sparse patches of grass on some sidewalks. There is only one public library, and it is located inside a cramped refurbished office building.

Planning has also become isolated from the lived experiences of residents. The idea of public participation, recently introduced into the planning process, remains an exclusive domain of state agencies, locals, and big businesses. Formal planning has become increasingly oriented towards codes and standards. These are mostly copied from abroad, enshrined in glossy manuals, and incorporated into local urban planning curricula. Graduating planners increasingly are expected to speak the same language of numbers and regulations. Consequent urban development creates spaces that are devoid of contestation and vitality. Rather than seeing the city as an imperfect but compassionate work in progress, the system approaches it as an exercise in mimesis – to be achieved by imported numbers and purchased artifacts (Chakravarty, 2017).

Layers of segregation in space

Socio-spatial segregation in Abu Dhabi is multi-layered. The most stark and conspicuous is the segregation of the locals from expats. Seemingly, this cleft has deepened in the past decade or so. With rents sky-high, low-income expats are already living in dense shared accommodations (or "labor villages" well outside the city), which keep them away from the suburban "villa" communities inhabited by expats. With the development of Reem Island and a few other high-rise projects, middle- and higher-income expats have also been segregated from the local population. Furthermore, a crackdown on the leasing of villas (and "divided villas"), to expats of all income groups, has ensured a clean separation of locals from "Others." At the same time, the "Emirati Housing" projects have created enclaves of locals within luxurious gated communities (Chakravarty and Quamhaieh, 2015).

Furthermore, expats are segregated by income. While low-income expats live in labor housing outside the city, the lower-middle-income group lives in the city, and the upper-middle-income professionals live on Reem Island (and a handful of other such projects). Of course, there may be a few exceptions, but the broader trends are consistent. Those in lower-middle-income groups – often sharing accommodations (with 12 or more occupants to a room) – may be further segregated – by religion, nationality, ethnicity, and employment status. Figure 6.2, for example,

Figure 6.2 "Rooms and 'bed spaces' advertised by ethnicity."

Source: Author

shows advertisements for rental accommodations posted in the public sphere in Abu Dhabi. To be fair, this segregation is not enforced by compulsion, and functions only for apartment rentals. There are no "ethnic enclaves" amongst the low-income groups. But the prevalence of matching identities (national, ethnic, or religious) of potential roommates reveals insecurities – of both landlords and tenants – in a section of society where precarity, transience, and marginality are the norms.

As mentioned earlier, the bus system contributes to the marginalization of lower-income residents of the city. In this context, one might also mention that the presence of a women's beach (in addition to separate beaches for "families" and "bachelors"), and women-only spaces (not only seats, but also standing space) in the front of public buses, point to an uncomfortable tension between traditional values and new realities. In fact, the perverse fascination with segregation has infiltrated institutions that supposedly uphold liberal values. Several university campuses (including those of New York University Abu Dhabi and the Sorbonne University Abu Dhabi) are gated and allow entry only to guests with appointments.

It would not be accurate, however, to say that Abu Dhabi does not have an urbanism. Meaningful spaces still exist – such as the Madinat Zayed Shopping Center, Hamdan Street, Al Markaziya, and the Tourist Club area – that are still hubs of activity. In fact, to an extent, these are the last few areas where the everyday life of low-income residents intersects with lower-middle- and middle-income groups. The sidewalks are abuzz and vital. People live in high densities all around these centers. The world's languages can be heard here: Arabic dialects mix with Tagalog, Malayalam, Hindy, Urdu, Bengali, Sinhalese, English, and French. Engagement across lines of identity, however, is predominantly transactional – not only between locals and non-locals, but also within the expat population.

There are no malls in these spaces. Storefronts are directly on the sidewalk. The smells of doner, falafel, and freshly baked bread fill the air. Even in the summer, patches of grass outside Madinat Zayed are occupied late into the night (albeit by men only). These are the vestiges of the Abu Dhabi of the 1980s, now in the crosshairs of urban planners. This was an organic urbanism that had evolved in the region until for-profit urbanization completely took over. Rather than build on this strong template, planners and decision makers got busy trying to mimic Vancouver (Khirfan and Jaffer, 2014) and import symbols from Paris (the Louvre and the Sorbonne) and New York (Guggenheim Museum and NYU). Meanwhile, aging parts of the city have been left to those people who cannot buy into the swankier new developments. These remnant spaces are shrinking fast as buildings fall to waves of urban renewal. Other cities in the region have similar stories to tell.

The un-city: civic spaces in uncivil times

So how can we characterize Abu Dhabi's urbanism? I will attempt to describe (rather than define) the "un-city." If this effort takes on allegorical shades, I hope that it will be accepted as a necessary instrument, for empathetically comprehending the experiences and lived spaces that have led to this neologism. I expect this concept, although based on the state of affairs in the Gulf region, to be broadly applicable to other parts of the Global South, where contemporary cultural and political vectors are having a similar impact on cities.

Abu Dhabi is a city that does not have any geist, spirit, continuity, rendition, or contribution. It serves industrial purposes, but not a human purpose. The city reads like pieces of a puzzle, which, if somehow put together, would still constitute but a question mark. As a modern metropolis, Abu Dhabi is an afterthought – an awkward appendage to an invisible industry. The oil rigs are either offshore or far inland – unseen and unheard. But the wealth and power they generate needs an almighty spatial fix – a whole city that can store and reproduce value – but also one that can reflect that value to the world. Abu Dhabi is that city – at once a world unto itself and a mirage to the world.

Henri Lefebvre (1991) has argued that the values, socioeconomic relations, and hierarchies of a society are inscribed in the spaces it produces. Urban planning and design, and the places and forms they create, constitute a record, an archive of the politics and contestations of the times – both ideological and at the level of everyday life. Frequent or large-scale flaws in the form and process of urban development provoke critical analysis of the prevailing planning culture, and related socioeconomic issues or conflicts. The question "whose city is it?" (Sassen, 1999; Zukin, 1995) has animated critical thinking with regards to cities in the 21st century, particularly in the context of contracting rights to the city and practices of neoliberal urban governance. In the "instant" (Bagaeen, 2007; Salama, 2008) cities of the Gulf region, however, where such an inquiry ("whose city?") can be answered objectively with a short list of names and statutes, a more fundamental question – *what* is a city? – appears unresolved and daunting.

Lefebvre (2003: 7) saw urbanization as a transformation of "pure nature" into a social space with agrarian, political, economic, and industrial functions. For Lefebvre, the first transition from an "agricultural" society to an "industrialized" one is followed, eventually, by a "critical phase" (2003: 15). In the critical phase, Lefebvre postulates that:

> a second transition occurs, a second reversal of direction and situation. Industrialization, the dominant power and limiting factor, becomes a dominated reality during periods of profound crisis. This results in tremendous confusion, during which the past and the possible, the best and the worst, become intertwined.
>
> *(Lefebvre, 2003: 16)*

Understanding the nature of the "critical phase" is an ongoing challenge for scholars of space and society. Abu Dhabi exhibits some of the contradictions of the critical phase. The oil-based economy that sustains Abu Dhabi has upended the assumption of industrialization. Abu Dhabi made the transitions postulated by Lefebvre (2003) in a very short period. The transformation, from pearling and fishing to oil extraction, happened so rapidly that the trajectories of social and economic development had to part ways. Moreover, neither agriculture nor industry "dominated" the landscape. Abu Dhabi did not have to address the spatial challenges of industrial social relations. Extreme variation, in terms of class and rights, did create similar problems to those faced by early industrial societies (e.g. lack of affordable housing, environmental excess, segregation, and exploitation of labor). But these were managed through radical interventions ill at ease with the norms of democratic politics.

Although a wealthy society and a prosperous economy, Abu Dhabi does not guarantee freedoms of expression, dissent, and association. Ordinary urban practices – exploration, flanerie, photography, social research, participation in local planning, etc. – are monitored and controlled. People rarely engage each other in open conversation. Art faces pressures of conformance. Dignity, inclusion, and equal access to opportunities (including housing and mobility) are distant dreams for many. This is one likely future in the critical phase of cities.

Most people live in a city voluntarily. But if a city does not nurture the spaces, institutions, and attitudes that provide verve and vitality to urban life, and instead either replaces them with faux alternatives or eliminates them altogether, then is it still a city? Is such a society "urban" in a Lefebvrian sense? One possible extension of Lefebvre's continuum, past the critical phase, could lead to the "un-city" – a hollow, apolitical, faux city – a dystopia of consumerism, climate carnage, and impersonal gesellschaft (the spatial expression of an overbearing polity) – an image to which the real is subjugated. From a Lefebvrian perspective, the un-city exposes the emptiness of our times and our urbanism – the ascendant phase of the global consumer society.

The un-city does not contribute to our collective human progress. It exists to generate profit, mainly from rent – derived from natural resources, from buildings and assets, and finally from exploitation of human labor. The un-city has buildings and roads, universities and shopping malls, but it has no objective apart from extraction. The un-city is full of exuberant displays of wealth and the prosperity of a few but remains fundamentally silent on the condition of human work and habitation. The urbanism of the "un-city" is not confined to just the Gulf region; it is incipient in other parts of the world, especially the Global South.

Planning in the un-city considers only its wealthy and powerful constituents. Others are neither consulted nor considered substantively in plans and projects. Their needs are considered only as far as they are subservient to the needs of the ruling class. Pedestrians, cyclists, public transit riders, children, the elderly, those with special needs, and those belonging to low-income groups all find themselves subjected to the apathy of the un-city. Further, diversity – of gender, faith, or ethnicity – is accepted within the limits of profit, efficiency, and political correctness. When it comes to the welfare (economic, cultural, or psychological) of the least well off, the maximum conceivable is never an iota more than the minimum required.

Un-city urbanism is characterized, most palpably, by a tension between violence and survival. Violence is manifested in many forms – denial of rights, segregation, marginalization (particularly in terms of mobility, housing, and access to other resources). But perhaps most significantly, violence is embedded in policymaking that dehumanizes individuals and communities – viewing them only as hours of productivity and statistics of profits and liabilities. The spaces of the wealthy and those of the deprived drift farther apart in an effort to eliminate every intersection of their everyday lives. Surviving the violence of un-cities is a process of compromise. Out of concern for their families and livelihoods, individuals internalize the trauma of everyday life in the un-city. They become accustomed to violence and accept it as a normal. Aware of their provisional and ephemeral status, they walk in malls, eschew political thought, speak guardedly, self-censor, and rarely engage with "others" beyond interactions at work or through transactions. Concern for community and conviviality is restricted to micro-communities of people of the same class, or property owners in the same neighborhood, or residents of the same gated community, or parents of a certain school district.

The un-city does away with politics – it obviates contestation, controls spatial practice, and produces spaces that are deeply uncomfortable in an ethical sense. The un-city prefers the language of legality and compliance over discourse and engagement. When participation is attempted, usually initiated under a clause of compliance, it remains an elitist exercise. The un-city can punish and reward behavior, but seldom does it solicit grounded and innovative solutions to local problems. Un-cities operate on discipline – not only in terms of compliance with codes, but also in terms of control of human agency. Flanerie and street photography may be frowned upon, while organizing groups or asking questions may be outright illegal. Freedom of expression is tightly controlled. Fear, self-censorship, acute self-interest, paranoia, mistrust, and distancing the "Other" contaminate space, infiltrate everyday life, and infect the very essence of human relations.

Conclusion: finding a way back to the city

Ash Amin (2006) has argued that the "good city" is not part of the lived experiences of most residents of today's megacities. Although there always were many who were less well off, the extent of absolute poverty, deprivation, and dispossession has reached extreme proportions today. The situation is made worse by a decimation of the rights that were once taken for granted in liberal democratic polities and were widely expected to spread globally over the 21st century. However, with the rise of the Gulf region and several other influential states outside the democratic framework, and the rightward tilt in politics in many democratic countries as well, the challenge we face is pernicious. It is not limited to the subjugation of urban space and community to capital through "revanchist" (Smith, 1996) policies and ideological "designscapes" (Julier, 2005). What we face today is the transformation of the very meaning of the city. The un-city is a collection of dispersed megaprojects, entertainment-retail centers for the elite, and gated communities, juxtaposed against pieces of an old town. The "urban design manifesto" forwarded by Jacobs and Appleyard (1987) recognizes the problems of "giantism," "large-scale privatization," "placelessness," and "injustice," amongst others. Three decades and mountains of literature later, the problems are more entrenched, and have spread to all corners of the world.

To a large extent, poor urban design is a result of decisions taken by real estate corporations, financial institutions, and state agencies. These decisions have little to do with urban planning or form. Urban planners and urban designers may or may not be involved in the process. But they certainly play an enabling role. The un-city is not *produced by* flawed urban design (deliberate or otherwise). Rather, the flawed visions and values of the un-city are reflected (or inscribed) in the design of space. The un-city cannot but suppress (actively or passively) urban design that is truly inclusive and democratic, because such spaces do not fit with the objectives of those who own and operate the un-city.

As discussed earlier, this reading of Abu Dhabi contributes to our understanding of the Lefebvrian idea of the "critical phase" (Lefebvre, 2003: 16) – an urbanity in search of purpose and orientation. Abu Dhabi, with fewer agencies of mitigation, reveals, in stark contrast, the values that underpin the current "planetary society" (Lefebvre, 2003: 17). Many countries, corporations, and communities are complicit in the widespread practice and acceptance of these values.

The colonization of space, and through it, of liberal values, requires that urban design confines itself to aesthetic-decorative endeavors. And this trend can spread quickly, particularly to cities in the developing world. In response, urban design – as a field of study and as a profession – must not bow out of the realm of the political. Urban design heuristics must integrate political thought. Education in the field must acknowledge and study the material influence of power on design practice. Students have to know that design choices are political decisions. Power relations and social hierarchies may be inscribed in space through various means, including site selection, scale, participation, communication, symbolism, uses, restrictions, memorializing/forgetting, etc.

Abu Dhabi also illuminates the arena for action – the initial milestones on the path towards "de-alienated" space (Purcell, 2014; Knierbein, 2014). In this context, what can urban design accomplish? Urban design may not be able to resolve the causes of the un-city in all contexts, but it must strive to be therapeutic to the extent possible. Further, in the face of power, urban designers must be committed to small local "insurgencies" – i.e. interventions that aim to return spaces to users. Urban design must resist developments that lead to exclusion, displacement, and segregation. When possible, urban designers must attempt to bring down barriers, open up spaces, and make places that invite a diversity of uses and users. This will help urban design find a way back to the city.

References

Amin, A. (2006). "The Good City." *Urban Studies, 43*(5–6): 1009–1023.

Bagaeen, S. (2007). "Brand Dubai: The Instant City; Or the Instantly Recognizable City." *International Planning Studies, 12*(2): 173–197.

Chakravarty, S. (2017). "Buying (into) Sustainability: Technocratic Environmentalism in Abu Dhabi." In Caprotti, F. and Yu, L. (eds.) *Sustainable Cities in Asia*. London: Routledge.

Chakravarty, S. and Qamhaieh, A. (2015). "City Upgraded: Redesigning and Disciplining Downtown Abu Dhabi." In Lees, L., Shin, H. B., and Lopez-Morales, E. (eds.) *Global Gentrifications: Uneven Development and Displacement*. Policy Press, 59-80.

Caprotti, F. (2014). "Eco-urbanism and the Eco-city, Or, Denying the Right to the City?" *Antipode*, *46*(5): 1285–1303.

Cugurullo, F. (2015). "Urban Eco-modernisation and the Policy Context of New Eco-city Projects: Where Masdar City Fails and Why." *Urban Studies*, 0042098015588727. Online before print.

Cugurullo, F. (2013). "How to Build a Sandcastle: An Analysis of the Genesis and Development of Masdar City. "*Journal of Urban Technology*, *20*(1): 23–37.

Davidson, C. (2009). "Abu Dhabi's New Economy: Oil, Investment and Domestic Development." *Middle East Policy*, *16*(2): 59–79.

Elsheshtawy, Y. (2010). *Dubai: Behind an Urban Spectacle*. London: Routledge.

Elsheshtawy, Y. (2008). "Cities of Sand and Fog." In Elsheshtawy, Y. (ed.) *The Evolving Arab City*. London: Routledge., 258–304.

Husnein, A. and Chakravarty, S. (2015). "Planning the Recovery: Dubai's Search for a New Model of Development." In Wagner, F., Mahayni, R. G. and Piller, A. G. (eds.) *Transforming Distressed Global Communities*. Burlington: Ashgate., 265-276.

Jacobs, A. and Appleyard, D. (1987). "Toward an Urban Design Manifesto." *Journal of the American Planning Association*, *53*(1): 112–120.

Julier, G. (2005). "Urban Designscapes and the Production of Aesthetic Consent." *Urban Studies*, *42*(5–6): 869–887.

Khirfan, L. and Jaffer, Z. (2014). "Sustainable Urbanism in Abu Dhabi: Transferring the Vancouver Model." *Journal of Urban Affairs*, *36*(3): 482–502.

Knierbein, S. (2014). "Public Space as Relational Counter Space: Scholarly Minefield or Epistemological Opportunity?" In Tornaghi, C. and Knierbein, S. (eds.) *Public Space and Relational Perspectives: New Challenges for Architecture and Planning*. Chicago: Routledge, 62–84.

Kyriazis, A., Almehairi, A. and Al Seri, I. (2002). "The Morphology of Abu Dhabi's Superblocks: A Modernist Vision." *Urban Morphology*, *1*: 21–32.

Lefebvre, H. (2003). *The Urban Revolution*. Minneapolis: University of Minnesota Press.

Lefebvre, H. (1991). *The Production of Space*. Oxford: Blackwell.

Mohammad, R. and Sidaway, J. D. (2012). "Spectacular Urbanization Amidst Variegated Geographies of Globalization: Learning from Abu Dhabi's Trajectory Through the Lives of South Asian Men." *International Journal of Urban and Regional Research*, *36*(3): 606–627.

Ponzini, D. (2011). "Large Scale Development Projects and Star Architecture in the Absence of Democratic Politics: The Case of Abu Dhabi, UAE." *Cities*, *28*: 251–259.

Purcell, M. (2014). "Possible Worlds: Henri Lefebvre and the Right to the City." *Journal of Urban Affairs*, *36*(1): 141–154.

Qamhaieh, A. and Chakravarty, S. (2017). "Global Cities, Public Transportation and Social Exclusion: A Study of the Bus System in Abu Dhabi." *Mobilities*, *12*(3): 462–478.

Salama, A. M. (2008). "Doha: Between Making an Instant City and Skirmishing Globalization." *MEI-Viewpoints of the Middle East Institute*, 42–45.

Sassen, S. (1999). "Whose City Is It? Globalization and the Formation of New. In Foo, A. F. and Yuen, B. K. P. (eds.) *Sustainable Cities in the 21st Century*. Singapore: Singapore University Press, 145–162.

Smith, N. (1996). *The New Urban Frontier: Gentrification and the Revanchist City*. Psychology Press.

Zukin, S. (1995). "Whose Culture? Whose City?" In Lin, J. and Mele, C. (eds.) *The Urban Sociology Reader*. London: Routledge, 281–289.

Further reading

Chakravarty, S. (2017). "Buying (into) Sustainability: Technocratic Environmentalism in Abu Dhabi." In Caprotti, F. and Yu, L. (eds.) *Sustainable Cities in Asia*. Abingdon, UK: Routledge. Argues that sustainability in Abu Dhabi has been reduced to commoditized/purchasable artifacts while substantive aspects are overlooked.

Chakravarty, S. and Qamhaieh, A. (2015). "City Upgraded: Redesigning and Disciplining Downtown Abu Dhabi." In Lees, L., Shin, H. B., and Lopez-Morales, E. (eds.) *Global Gentrifications: Uneven Development and Displacement*. Policy Press, 59–80. An account of the gentrification of downtown Abu Dhabi, a large-scale but undeclared project leaving thousands with no affordable housing options in the city.

Qamhaieh, A. and Chakravarty, S. (2017). "Global Cities, Public Transportation and Social Exclusion: A Study of the Bus System in Abu Dhabi." *Mobilities*, *12*(3): 462–478. Argues that the bus system in Abu Dhabi reflects the elitist visions of the city and marginalizes the less privileged sections of society, while also undermining the sustainability agenda.

Mohammad, R. and Sidaway, J. D. (2012). "Spectacular Urbanization amidst Variegated Geographies of Globalization: Learning from Abu Dhabi's Trajectory through the Lives of South Asian Men." *International Journal of Urban and Regional Research*, *36*(3): 606–627. An in-depth study of the lives and struggles of South Asian men in Abu Dhabi, most of whom are stuck in low-wage jobs.

Ponzini, D. (2011). "Large Scale Development Projects and Star Architecture in the Absence of Democratic Politics: The case of Abu Dhabi, UAE." *Cities*, *28*: 251–259. Questions the validity of megaprojects and star architecture in a non-democratic context, particularly in light of tie-ups with the Louvre and Guggenheim Museum.

PART II
Challenges

PART II.I

Claims and conflicts

Introduction

Contemporary growth and development of cities in a globalizing world are in many ways transformative. In an earlier era, urbanization and growth of cities resulted largely from rural-urban migration within national boundaries. Today, the new urban growth is resulting from unprecedented flows of international labor migration that lead to significant changes in the composition of population in cities, involving new socioeconomic and demographic patterns. In short, most cities connected to the global economy are becoming multiethnic, multiracial, and multicultural. This transnational migration is both an agent and an outcome of globalization. Anthropologist Arjun Appadurai (1991) has referred to this new dynamic change as a case of emergent ethnoscapes. Scholars like Leonie Sandercock (1998, 2003) have long followed the emergent multicultural ethnoscapes of cities and the claims and conflicts associated with this dynamic. Sandercock has discussed at length the planning problems in what she calls "cosmopolis" or "mongrel cities." Mohammad Qadeer (2016), a Canadian planner, has analyzed these problems and challenges in the arena of practice and suggested approaches practitioners might take. While some initial progress has been made in defining the scope of multicultural planning, according to these authors, the appropriate perspective for urban design has yet to be defined. This is why we have chosen to include this theme in the larger topic of "Challenges."

There is a general agreement that one of the challenges of planning and design in multicultural cities is to negotiate multiple claims and conflicts over space and urban form, as different cultural groups seek to negotiate their identities while trying to assimilate in the larger expectations and imperatives of assimilation (see Banerjee, Chakravarty, and Chan, 2017). What is the role of urban design in facilitating the dual processes of assimilation and identity negotiation? What are the urban form implications of managing such processes while maintaining the continuity of the host population and the associated spatial form? The three chapters included in this section attempt to address these broader questions with case studies from different corners of the world – two from the Global North and one from the Global South.

While most countries of the Global North are facing the challenges of assimilation from the diaspora of the Global South, the opposite can be also true, as Felicity H.H. Chan points out, drawing from the case of Singapore. In this instance, a national policy of encouraging multinational immigration to boost Singapore's talent pool and economy has drawn immigrants not only from the Global South, but from the Global North as well, particularly migrants form the United States, France, and Germany. Apparently, and not unlike the Global South diaspora in the cities of the Global North, these skilled and prosperous migrants have chosen not to assimilate spatially but instead form their own enclaves isolated from the multicultural host population of Singapore. Although Chan does not overtly mention this, clearly there is a whiff of the colonial tendency in the postcolonial context. But unlike the segregated spaces of the colonial era separated by *cordon sanitaire*, she discusses how the fabric of the Singaporean social space has now been overlaid with a "cosmopolitan grid" from deliberate planning and design intervention. So, here is a clear case of urban design intervention in response to different claims to everyday urban space and the inevitable tension that arises from it.

In their chapter, Stefano Moroni, Francesco Chiodelli, Elena Porqueddu, and Anna Botta focus on a particular aspect of the claims of diaspora – the space for the practice of religion. In particular, they consider the tension that arises from the location and design of mosques in Italy, where Muslim diaspora pray. Their case city is Rome, an ancient historic city with a palpable firmament of the Catholic religion. But this long Catholic religious tradition in the whole country, not just Rome, does not necessarily, and not surprisingly, lead to a catholicity of religious practices as the diaspora of the Global South demands. So, is there a role for urban design in accommodating this claim to spaces of different religious practices and the resulting conflicts? Moroni and his colleagues suggest in their chapter that creative responses in the design of such spaces might minimize, if not eliminate, the tension that comes from the multicultural reality.

Jeremy Németh begins by asking the question that we have posed in a different form for this section: "Can urban design affect immigrant integration?" In this chapter, he focuses on the possible role of what he calls "arrival neighborhoods" paralleling the earlier concept of "entrepôt," popular among economic geographers. But unlike a trading center, Németh's concept focuses on people, especially immigrants, and the places and neighborhoods that host them for their transition to full assimilation. His case example is the very successful *el Raval* neighborhood of Barcelona. His arguments are based on his field study of this neighborhood as participant observer. His conclusions about what takes to create such neighborhoods are the three familiar concepts for urban designers – concentration of activities, availability of public space, and accessibility of the place. His larger aim in this chapter is to develop a conceptual framework to understand the immigrant experience and uses of urban space.

References

Appadurai, A. (1991). "Global Ethnoscapes: Notes and Queries for a Transnational Anthropology." In Fox, R. (ed.) *Recapturing Anthropology: Working in the Present.* Santa Fe: School of American Research, 191–210.

Banerjee, T., Chakravarty, S., and Chan, F. (2017). "Negotiating the Identity of Diaspora: Experience of the Southeast Asian Communities in Los Angeles." In Moroni, S. and Webberman, D. (eds.) *Space and Pluralism.* Prague: Central European Press, 173–199.

Qadeer, M. (2016). *Multicultural Cities: Toronto, New York, and Los Angeles.* Toronto: University of Toronto Press.

Sandercock, L. (1998). *Towards Cosmopolis: Planning for Multicultural Cities.* Chichester: John Wiley & Sons.

Sandercock, L. (2003). *Cosmopolis II: Mongrel Cities of the 21st Century.* London and New York: Continuum.

7

IMMIGRANTS, MOSQUES, AND RELIGIOUS PLURALISM

Challenges for urban design and planning

Stefano Moroni, Francesco Chiodelli, Elena Porqueddu, and Anna Botta

Introduction

In recent years the steady flow of immigrants from other lands and cultures has greatly increased social pluralism (points of view, values, and needs) in cities across Europe, creating new challenges for urban design and urban planning. This chapter examines the immigrant influx from a thematic viewpoint – namely, the placement and construction of buildings of a religious nature – and from the specific geographical viewpoint of Italy. In this country, the provision of proper places of worship for some immigrant groups has become an increasingly pressing problem, as it has triggered heated conflicts in numerous Italian cities (Triandafyllidou 2002). Accordingly, the chapter first presents the Italian case and then offers suggestions on how to rethink urban design and urban planning as a means for responding to this emerging issue.

Migration and religious diversity in European and Italian cities

The history of migration in European countries in the years after the Second World War varies from country to country but nonetheless is characterized by several general trends across the western European countries (Fassmann & Münz 1994; Kocsis et al. 2016). During the last two decades, migration in Europe has become more dynamic and complex, with further complexities during the last few years because of two main factors: the political crises in the neighboring areas of Asia and Africa – which have led to an increase of illegal immigration, refugees, and asylum seekers – and the economic crisis, which has increased the intra-European migration (Trenz & Triandafyllidou 2017).

Against this backdrop, Italy has its own challenges. First, international immigration flows to Italy started to become significant only in the 1990s – at least two decades later than in other western European countries, such as France and Germany, for example. In 1991, the number of foreigners residing in Italy was around 350,000 (about 0.6% of the population). By 2017, this figure had risen to five million (8.3% of the overall population), with the majority of immigrants coming from countries outside the European community (mainly Morocco, Albania, and China) (ISTAT 2017); if we add the number of immigrants present in Italy without legal documents, estimated around 400,000 (Fondazione ISMU 2016), the total number amounts to 9% of the overall population.

Second, for centuries Italy was a country dominated almost exclusively by the Catholic religion, which therefore played a central and pervasive role in the social, cultural, and political spheres of the country. Only in the last few decades – owing to the varied religious beliefs of the many different immigrants – has Italy started to experience a shift from an overwhelming Catholic majority towards a tapestry of diverse religions (Pace 2013). For instance, about one third of the migrants residing in Italy come from Muslim countries, and another third come from mainly Christian Orthodox regions, while the size of Hindu groups is also growing (Centro Studi e Ricerche IDOS 2017). This has come as a shock to some Italians, who were not used to this religious diversity causing changes in the composition of Italian society, but also affecting the country's diverse urban environments. In fact, Italian cities are experiencing development of places characterized not only by a different ethnicity (Grandi & Tanzi 2007), but also by a different religious character. Some of these places directly serve religious purposes, such as places of worship and burial grounds; others are just indirectly related to particular practices influenced by religion, such as *halal* butchers. These spaces still represent a minority in the overall Italian urban landscape. Nevertheless, they are now emergent in major cities such as Milan, Turin, Rome, Bologna, and Florence, characterized by a significant presence of immigrants. Limited incidence notwithstanding, the perceptions of "alien" spaces and their impact on the public opinion are significant, as evidenced by the increasing political attention and related policies addressing this issue.

The case of Islam is particularly significant in this regard, because of both the magnitude of the Muslim presence and the perceived and actual "cultural divide" between Muslims and the native population. In Italy, as in many other European countries, Islam is the most common religion practiced by the immigrant population and *de facto* the country's second religion after Catholicism, with around 1.5 million worshipers (Introvigne & Zoccatelli 2013). For several reasons, many Muslim migrants consider not only their ethnicity, but also their religion as a central pillar on which their identity as immigrants is formed (Allievi 2010; Saint-Blancat 2002). In what follows, we will focus on Muslim places of worship, considering their relevance and how they are addressed by urban policies in general, and planning regulations in particular.

Muslim places of worship in Italy

In Italy, as well as in many migrant-receiving countries, there are two main types of places where Muslims gather to worship: formal "purpose-built mosques" (i.e., mosques in total compliance with the law and with all the requisite building permits, which are intended to be permanent) and informal prayer halls (i.e., spaces only in partial compliance with the law, or totally illegal, but accepted by worshipers only as provisional solutions).

Formal purpose-built mosques are iconic buildings, usually characterized by a dome, minarets, and Arabic scripts and symbols (Allievi 2010); they are built purposefully, following all the planning rules and building procedures. Most of them are built thanks to the financial support of organizations (e.g., foundations) and foreign governments. Currently (2018), fewer than ten purpose-built mosques exist in Italy, usually located outside the city center. They are quite isolated and often difficult to reach.

Informal prayer halls are small and simple prayer rooms. They are frequently located in former apartments, shops, or warehouses and have no recognizable external signs identifying their function. They are mostly temporary and informal places. In the majority of cases, they respond to the Muslims' daily prayer needs – even if they are frequently too small for necessities related to special occasions, such as Ramadan. Due to their informal status, no accurate data are available on their number; however, it is estimated that about 1,000 such prayer halls exist in Italy (Chiodelli

2015). As Allievi (2009, 25–26) argues convincingly, these figures "may seem surprising, given the widespread assumption that Muslim places of worship are few in number. Therefore, there is no problem of a lack of places of worship. . . . The problems that arise . . . are of a qualitative, not a quantitative, nature." This point is crucial: the overall availability of Muslim places of worship in Italy is quantitatively quite adequate for the number of worshipers. Hence, in Italy (as well as in other European countries), we cannot say that there is scarcity *per se* of Muslim places of worship. However, there is scarcity of *formal* and *appropriate* Islamic places of worship. This is not the case in France, where about 200 purpose-built mosques have been built (Allievi 2009). But in Italy, almost all mosques are informal prayer rooms, which do not satisfy the Muslims' need for public recognition, visibility, and dignity, which the building of a solemn formal mosque would provide. As has been argued: "For many Muslims in the diaspora the importance of the mosque has . . . come to rest in this symbolic role as a mark of their presence. Sacred or not, mosques increasingly represent Islam in the West to Muslims and non-Muslims alike" (Metcalf 1996, 17–18). Furthermore, there are also practical problems linked to the informal character of prayer halls. Such places have a precarious status; hence they are always at risk of being closed down. In many cases they are too small; often they have been established in inappropriate locations, from both an architectural viewpoint (e.g., in basements) and from the point of view of building and zoning codes (e.g., in residential areas without parking lots, which may cause conflicts with the local population).

The case of Rome is particularly interesting in this regard. In fact, Rome hosts the first purpose-built mosque constructed in Italy, inaugurated in 1995. This mosque is, at present (2018), the largest in the country, and it can hold around 12,000 worshipers (Figure 7.1). However, this mosque is not sufficient to satisfy the varied religious needs of the Islamic population residing in Rome, currently around 120,000 people, who belong to different ethnic groups (mainly

Figure 7.1 The formal mosque in Rome.

Source: Photo by Anna Botta

Bangladeshis and Egyptians). For this reason, several informal prayer rooms have mushroomed in the city. Our research discovered more than 30 informal prayer rooms in the city (but the actual number could be around 40–50).

Most of these prayer halls are located in the southeastern part of the city – although a few are also located in the central areas (close to St. Peter's Basilica and around the central railway station, two areas characterized by a high concentration of immigrants). Some of these prayer halls are quite old. Others were founded several years ago but changed their location over the years in order to find better premises, as is the case of *Dar al Salaam*, in Laurentino. However, for the large majority of cases, informal prayer halls in Rome are recent establishments. Most of the informal prayer halls are located in "residual" spaces, such as former garages or basements in residential buildings. They are usually small, with insufficient ventilation and poor lighting.

In regard to the organization and design of their internal spaces, these are also quite similar: They are composed of a main room for prayer, an area for ablution,[1] and a wardrobe. The main difference is that longer-established prayer halls are usually larger and more structured, as they may include an office, a room for language and Koran courses, etc. Such spaces, although not directly connected to prayer, are nonetheless important. In fact, Muslim prayer halls play the typical role of many gathering places for minorities in foreign countries; hence, they provide help and social assistance for newcomers and people in need and offer cultural and social activities for children and young people (Germain & Gagnon 2003). In terms of their insertion in the urban fabric, in most of the cases prayer halls are located in peripheral neighborhoods, characterized by a low availability of public transportation and parking lots. They have no recognizable external signs indicating their function, apart from small plaques with Arabic scripts or symbols. As a consequence, their presence is usually not recognized by non-Muslims (Figures 7.2 and 7.3).

Figure 7.2 Main entrance of Rome Muslim Center.

Source: Photo by Anna Botta

Figure 7.3 Main entrance of Dar al Salam mosque.

Source: Photo by Anna Botta

Regulation of mosques in Italy

One of the main reasons for the lack of formal and permanent purpose-built mosques in Italy is often the opposition of local council authorities (Chiodelli & Moroni 2017; Saint-Blancat & Schmidt di Friedberg 2005). This opposition is due to different reasons, such as security concerns, cultural hostility, or even Islamophobia. In almost all cases, it materializes mainly through planning and building controls. In this regard, it is worth stressing that Italy shares many of these controls with several other European countries. Despite some differences in specific legal frameworks, local authorities in many other European countries have the *de facto* power, through planning regulations and controls, to block or hamper the construction of certain places of worship (Eade 1996; Gale 2005; Gale & Naylor 2002).

In Italy, the legislative framework governing places of worship is quite complex, and comprises four normative levels, as follows.

The constitutional level

The Italian Constitution defends the freedom of individuals and groups to profess their own religion. The availability of a place of worship is considered intrinsic to this constitutional right. In a nutshell, the principle of religious freedom cannot be reduced to inward prayer, but must also be expressed through collective prayer in a devoted place of worship (Bettetini 2010).

The national level

National laws impose no specific restrictions on the construction of new places of worship, whatever the religion. New places of worship are subject to ordinary planning regulations, as is any other building.

The regional level

In Italy, town planning legislation is mainly a regional prerogative. Hence, each region has its own planning legislation – which, obviously, on paper should be congruous with the national laws and the Constitution. However, certain regions have recently passed planning laws that discriminate against several minority religions (Islam in particular), with reference to building places of worship. Lombardy is a blatant case in this respect. Lombardy planning law no. 2/2015 introduced several clauses that hamper religious minorities, and Islam in particular, in building their places of worship. These obstacles, for instance, concern the requirement for new places of worship to respect the "particular characteristics" of the Lombardy landscape, the requirement to draw a specific plan – the "Plan of Religious Facilities" – for (new) places of worship only, and the possibility to conduct a public municipal referendum on a proposal to build new places of worship. Basically, the law introduces "fuzzy" planning and building criteria that give leeway for discretionary decision-making in each council area. At the same time, it introduces special requirements that apply exclusively to the religious facilities of religious minorities and open the door to bureaucratic delays and complications (Chiodelli & Moroni 2017). It is worth stressing that Lombardy planning law no. 2/2015 is an extreme case of over-discriminatory planning laws in Italy. There are also several regions that, on the contrary, do not set any specific constraints to the construction of minority places of worship. This is the case, for instance, in the Region of Lazio, where Rome is located. However, this does not change the fact that purpose-built mosques are usually not present in these territories. The reason is related to the municipal level of government that is discussed next.

The municipal level

Planning activities at the municipal level must take place within the framework of the regional planning law. In some cases, as in the case of Lombardy planning law no. 2/2015, the regional framework sets explicit clauses that allow local authorities to directly impede the construction of (or conversion of an existing building into) a place of worship for minority religions; hence, if for any reason a local municipality does not welcome the construction of a mosque, it can block it through planning measures.[2] Even in regions where the regional planning law does not allow local authorities to implement such discrimination, municipalities have several technical devices for reaching the same goal. This occurs, in particular, through the biased application of planning regulations, sluggish administration, bureaucratic holdups, and extra red tape reserved for "special interest" groups. Several cases of this kind have been documented in many Italian regions (Bettetini 2010; Roccella 2008). If we combine the possibility of discretional treatment by local authorities with an often overt popular hostility towards Muslims (Wike et al. 2016), it is easy to understand why purpose-built mosques do not spread in Italy, despite the desire and many attempts of Muslim organizations. At the same time, municipal authorities also have the possibility to close informal prayer rooms for "technical reasons." As mentioned, in fact, most of the informal prayer rooms are located in former garages or cellars, often without proper authorization and permits. According to planning and building regulations, these places are normally not adequate to host collective activities – such as collective prayer, for instance – because of their insufficient ventilation and poor lighting, or their lack of emergency exits. Hence, local authorities can easily close these spaces down, in particular if they are pressured by local residents or political parties, as has occurred several times in Rome (for instance, a prayer room in Tor Pignattara neighborhood was shut down in October 2017 for violation of planning and building regulations).

Rethinking urban design and urban planning

The difficulty faced by Islamic places of worship in Italy is primarily the outcome of a widespread social and political opposition. Therefore, in order to change the situation, it is necessary to embark on a social and political battle. However, there are several crucial challenges also for urban design and urban planning. In particular, we want to discuss and criticize three beliefs that are still influential in urban design and urban planning, which often end up, sadly, endorsing certain socio-political trends.

First, there is the tendency to believe that when typological, formal, dimensional, and decorative elements are homogeneous, they necessarily and unequivocally strengthen the identity of a place. This belief emerges in the aforementioned regional law of Lombardy, according to which the design of religious buildings must provide "architectural and dimensional congruity of the places of worship with the general and peculiar characteristics of the landscape of Lombardy, as provided for in the Regional Territorial Plan."[3] The text of the law reveals the fear that alien architecture different from that "typical of Lombardy" may undermine the visual characteristics that identify the local landscape.

Actually, Lombardy's landscape is already in itself heterogeneous and diverse, similar to the Italian landscape in general. In fact, its uniqueness is the fruit of countless historic stratifications among various architectural, construction, and decorative traditions ascribable to different places and cultures. Italy in its whole, for example, is studded with architectural masterpieces originating exactly by its encounter with Arabian and Byzantine architecture and art. In Venice, for example, the blending of Arabian and Byzantine elements with Lombardy-Romanesque and International Gothic allowed the creation of a peculiar and original synthesis: the church of San Marco is indeed an extraordinary union between the East and the West. Also, the stylistic elements of the Doge's Palace recall Arabian and Byzantine architecture (Fontana 2012). Innumerable other cases of such blending are found in Sicily, and even in Lombardy (Fontana 2012).

The variegated architectural and urban landscape in Italy highlights how diversity is not necessarily a threat for the identity of a place. Identity originates, in fact, not by tracing boundaries and by excluding what is considered the "other" and is beyond those boundaries. On the contrary, the uniqueness of a place often originates from the fact that it becomes a point of intersection where people and cultures coming from different places actually meet (Dovey et al. 2009; Massey 1994).

Second, there is the tendency to think that the sense of place is necessarily and principally linked to the past. For example, the current Landscape plan of the Region of Lombardy envisages regulations that aim "to protect the structural elements of historical memory" (Regione Lombardia 2010: 23); in this perspective, new buildings must "respect the traditional formal characters and building techniques (typology, materials and building details equivalent to those of the original nucleus)" (24). Indeed, on the one hand, there is an architectural and artistic patrimony of great value in Italy, which must be protected. On the other hand, though, vital places have developed over time and the desire to codify and crystallize their character *rigidly* means to interrupt their inevitable evolution and adaptation. The sense of place is not reducible to something rigid and immutable (Massey 1994) but transforms together with the life of its inhabitants.

Moreover, it is possible for the character of a building or of an urban place to remain unchanged, even if its spaces are modified. Actually, transformations can even highlight the character of a space: its formal logic, its DNA. Architect Rafael Moneo (1985) notes how the construction of a building is only the beginning of its life; the desire to maintain it rigidly in all its aspects (as it was realized originally) means to crystallize it in a precise moment – and therefore interrupt its "lifecycle." On the contrary, if the initial construction and geometric principles are clear and readable, a building can maintain its character despite undergoing innumerable changes throughout its history.

Third, certain laws and cultural norms tend to suggest that the identity and character of a place are reducible to its architecture or its physical form, and that it can be codified and protected by a series of rules involving morphological elements. Actually, the character of a place is difficult to codify because it emerges from the continuous interaction between the people that inhabit it – or pass through it every day – and its physical structure (Dovey 2010). As highlighted by Dovey et al. (2009, 2612–2613),

> while planning codes and consultants' studies generally try to reduce character to a set of formal elements, the ways it is experienced in everyday life tend to resist attempts to separate the social from the physical. Struggles to prevent the wrong kinds of building can easily slip into the exclusion of the wrong kinds of people. . . . A serious threat to character may lie in the desire to reduce it to a series of fixed features which turn character into caricature.

If we truly start to consider places as complex living systems – whose identity emerges from the mutable interaction between various spaces and people, and from a continuous exchange with other places – we will be able to assert that the insertion of a mosque in the architectural landscape of an Italian city does not necessarily constitute a threat for the characteristics that identify the local landscape. Indeed, it can be a transformation even capable of emphasizing the character of the place, contributing towards its evolution and uniqueness.

Therefore, the problem is not to rigidly *confirm* the historical memory, standardizing colors and materials or codifying all the stylistic and formal elements in great detail. The challenge for architecture and for urban design consists in knowing how to insert new elements in the lifecycle of a given place, how to cooperate with the (social-spatial) emerging transformations, and how to satisfy the many needs continuously evolving (Ouf 2001). The pertinent questions are thus totally different: for example, what is the spatial principle or structure that allows a place, a complex of buildings, or a single building to keep its integrity, while accommodating countless transformations made by history? This question concerns the design of new buildings and their ability to insert themselves in a specific urban context and to adapt to the spatial and social changes in that context over time. In this regard, open incremental (urban) structures are of particular interest, as they are designed with the aim to welcome subsequent unforeseeable transformations without losing their founding characteristics.

These considerations invite us to deeply rethink not only the culture of architecture and urban design, but also the planning regulations (and bureaucratic procedures) that affect the built environment. With reference to the latter aspect, according to our opinion, it is a matter of adopting a more impartial perspective towards transformations and possible uses (Moroni 2007, 2010, 2015) in at least three scopes of action. First, the traditional and persistent idea that local planning must define the purpose of *use* of areas and spaces – their *function* – should be abandoned in favor of more general and flexible regulations that rather indicate which *negative externalities* any land use must not produce anywhere. In other words, the regulatory attention should abstain from determining the possible *uses* (residential, commercial, religious, etc.) in predefined areas, and concentrate instead on the *undesired effects* of any use whatever (Chiodelli & Moroni 2017). For instance, it is more advisable to establish a maximum limit to noise emitted outside, a threshold that must not be surpassed (by discos, bars, religious buildings, etc.), instead of trying to coordinate the urban fabric as a whole by defining and localizing all uses, top down. By this logic, religious buildings should be treated like any other building. Second, excessively detailed and meticulous regulations and standards should be avoided as they interpret the goal of respecting context as an unimaginative reproduction of the existing urban fabric. Third, obtaining a building permit should become automatic if the proposed building transformations comply with all the prevailing laws, thereby eliminating any form of administrative discretionary power.

This strategy could be brought into play by shifting the focus of attention onto what has been termed the *right to the city*; the version of this that we prefer here, however, involves primarily the right whereby no obstacles are posed towards certain *basic liberties* (closely linked to the civic experiences, such as freedom of worship, to join associations of one's choice, etc.). These basic liberties would be assured through a radical idea of *equal treatment by the law* (Chiodelli & Moroni 2014).

It is worth noticing that relaxing the planning requirements for the building of various places of worship could better respond also to the frequent concerns for security of the kind highlighted in Lombardy's regional law, for example. In fact, in all likelihood, this would drastically reduce the quantity of "invisible" places and would rather foster their diffusion as visible and recognizable to all citizens and municipal governments. Moreover, to move certain faiths from the informal and distant places to which they are currently consigned, making them instead a visible and integral part of the urban reality, would foster a desirable daily and ordinary perception of their presence.

Conclusion

The aforementioned urban design and policy actions have become particularly urgent in Italy as a means to effectively and justly address the challenges that are currently emerging and respond to the claims and conflicts brought about by immigration. Significant changes in the culture of urban design and regulatory approaches seem indispensable also in other countries. The challenge of pluralism and multiculturalism has become, unavoidably, the main challenge to which design and planning should respond (Chiodelli & Moroni 2017).

Notes

1 Ablution (*wudu*) is the ceremonial act of washing parts of the body (hands, forearms, face, etc.), to mark a break from normal life and enter a state of devotion suitable for worship.

2 The Italian Constitutional Court examined the constitutionality of Lombardy planning law no. 2/2015. The ruling of the Court (Italian Constitutional Court, ruling 63/2016) declared unconstitutional certain requirements; by contrast, some other provisions of the law were not condemned (e.g., the respect for the local regional landscape and the necessity of a specific Plan for religious facilities). After the ruling of the Constitutional Court, a part of the Lombard law is still valid and operative. Moreover, the regional authority may enact a new law on the issue.

3 In the regional law (no. 2/2015), *this* architectural congruity requirement is mentioned explicitly only with reference to minority places of worship. (Clearly, requirements for congruity with the context exist for other types of building, but this specific formulation was introduced in this law exclusively regarding places of worship.)

References

Allievi, S. (2009). *Conflicts Over Mosques in Europe: Policy Issues and Trends*. London: Alliance Publishing Trust.

Allievi, S. (2010). "Immigration and Cultural Pluralism in Italy: Multiculturalism as a Missing Model." *Italian Culture*, 28(2): 85–103.

Bettetini, A. (2010). "La Condizione Giuridica dei Luoghi di Culto Tra Autoreferenzialità e Principio di Effettività." *Quaderni di diritto e politica ecclesiastica*, 1: 3–26.

Centro Studi e Ricerche IDOS (2017). *Dossier Statistico Immigrazione*. IDOS: Roma.
Chiodelli, F. (2015). "Religion and the City: A Review on Muslim Spatiality in Italian Cities." *Cities*, 44, 19–28.
Chiodelli, F. and Moroni, S. (2014). "Public Spaces, Private Spaces, and the Right to the City." *International Journal of E-Planning Research*, 3(1): 55–65.
Chiodelli, F. and Moroni, S. (2017). "Planning, Pluralism and Religious Diversity: Critically Reconsidering the Spatial Regulation of Mosques in Italy Starting from a Much Debated Law in the Lombardy Region. *Cities*, 62: 62–70.
Dovey, K. (2010). *Becoming Places: Urbanism, Architecture, Identity, Power*. New York: Routledge.
Dovey, K., Woodcock, I., & Wood, S. (2009). "A Test of Character: Regulating Place-identity in Inner City Melbourne." *Urban Studies*, 46(12): 2595–2615.
Eade, L. (1996). "Nationalism, Community, and the Islamization of Space in London." In Metcalf, B.D. (ed.), *Making Muslim Space in North America and Europe*. Berkeley: University of California Press, 217–233.
Fassmann, H. and Münz, R. (1994). *European Migration in the Late Twentieth Century*. Aldershot: Edward Elgar.
Fondazione ISMU (2016). *Ventunesimo Rapporto Sulle Migrazioni 2015*. Milano: Angeli.
Fontana, V. (2012). Orientalismi all'origine dell'architettura veneziana. *MDCCC 1800*, 1: 43–54.
Gale, R. (2005). "Representing the City: Mosques and the Planning Process in Birmingham." *Journal of Ethnic and Migration Studies*, 31(6): 1161–1179.
Gale, R. and Naylor, S. (2002). "Religion, Planning and the City." *Ethnicities*, 2(3): 387–409.
Germain, A. and Gagnon, J.E. (2003). "Minority Places of Worship and Zoning Dilemmas in Montréal." *Planning Theory and Practice*, 4(3): 295–318.
Grandi, F. and Tanzi, E. (eds.) (2007). *La Città Meticcia*. Milano: Angeli.
Introvigne, M. and Zoccatelli, P. (2013). *Enciclopedia delle Religioni in Italia*. Torino: Elledici.
ISTAT (2017). *Cittadini non Comunitari: Presenza, Nuovi Ingressi e Acquisizioni di Cittadinanza*. Available at: www.istat.it/it/archivio/immigrati. Accessed December 2017.
Kocsis, K., Molnár Sansum, J., Kreinin, L., Michalkó, G., Bottlik, Z., Szabó, B., Balizs, D., and Varga, G. (2016). "Geographical Characteristics of Contemporary International Migration in and into Europe." *Hungarian Geographical Bulletin*, 65(4): 369–390.
Massey, D. (1994). "A Global Sense of Place." In Massey, D. (ed.), *Space, Place and Gender*. Minneapolis: University of Minnesota Press, 146–156.
Metcalf, B. D. (1996). "Introduction: Sacred Words, Sanctioned Practice, New Communities. In Metcalf, B.D. (ed.), *Making Muslim Space in North America and Europe*. Berkeley: University of California Press, 1–27.
Moneo, R. (1985). "La Vida de los Edificios." *Arquitectura COAM*, 256: 26–36.
Moroni, S. (2007). "Planning, Liberty and the Rule of Law." *Planning Theory*, 6(2): 146–163.
Moroni, S. (2010). "Rethinking the Theory and Practice of Land-use Regulation: Towards Nomocracy." *Planning Theory*, 9(2): 137–155.
Moroni, S. (2015). "Complexity and the Inherent Limits of Explanation and Prediction: Urban Codes for Self-organizing Cities." *Planning Theory*, 14(3): 248–267.
Ouf, A. M. S. (2001). "Authenticity and the Sense of Place in Urban Design." *Journal of Urban Design*, 6(1): 73–86.
Pace, E. (2013). *Le Religioni nell'Italia che Cambia*. Roma: Carocci.
Regione Lombardia (2010). *Piano Pesaggistico Regionale* (Landscape plan). Available at www.regione.lombardia.it (accessed December 2018).
Roccella, A. (2008). "Edifici di Culto Nella Legislazione Regionale." *Jus: Rivista di scienze giuridiche*, 55(2–3): 505–560.
Saint-Blancat, C. (2002). "Islam in Diaspora: Between Reterritorialization and Extraterritoriality." *International Journal of Urban and Regional Research*, 26(1): 138–151.
Saint-Blancat, C. and Schmidt di Friedberg, O. (2005). "Why Are Mosques a Problem? Local Politics and Fear of Islam in Northern Italy." *Journal of Ethnic and Migration Studies*, 31(6): 1083–1104.
Trenz, H. J. and Triandafyllidou, A. (2017). "Complex and Dynamic Integration Processes in Europe." *Journal of Ethnic and Migration Studies*, 43(4): 546–559.
Triandafyllidou, A. (2002). "Religious Diversity and Multiculturalism in Southern Europe: The Italian Mosque Debate." *Sociological Research Online*, 7(1). Available at: www.socresonline.org.uk/7/1/triandafyllidou.html. Accessed December 2017.
Wike, R., Stokes, B., & Simmons, K. (2016). "Europeans Fear Wave of Refugees Will Mean More Terrorism, Fewer Jobs." *Pew Research Center*. Available at: www.pewglobal.org/2016/07/11/europeans-fear-wave-of-refugees-will-mean-more-terrorism-fewer-jobs/ Accessed December 2017.

Further reading

Burayidi, M. (ed.) (2000). *Urban Planning in a Multicultural Society*. Westport: Praeger. This book collects contributions from leading scholars concerning planning in multicultural societies; the chapters focus on both theoretical and empirical issues.

Burayidi, M. (ed.). (2015). *Cities and the Politics of Difference: Multiculturalism and Diversity in Urban Planning*. Toronto: University of Toronto Press. Several leading scholars discuss, from both a theoretical and a practical viewpoint, ways in which urban planning can enhance and support diversity and inclusiveness in cities.

Moroni, S. and Weberman, D. (eds.) (2016). *Space and Pluralism: Can Our Cities Today Be Places of Toleration?* Budapest: CEU Press. This book discusses how to interpret and foster tolerance in contemporary cities.

Qadeer, M. A. (2016). *Multicultural Cities*. Toronto: University of Toronto Press. Discusses planning for multiculturalism, offering a tour of three of North America's premier multicultural metropolises (Toronto, New York, and Los Angeles).

8

CLAIMING ORDINARY SPACE IN THE "COSMOPOLITAN GRID"

The case of Singapore

Felicity H.H. Chan

Introduction

As a postcolonial city-state embedded in the global networks of capital and human mobility, the contemporary cultural landscape of Singapore reflects its long affair with immigration and its relentless pursuit for international relevance. At the height of the Asian financial crisis in 1997, Singapore embarked on an aggressive pursuit of becoming a global city by courting transnational elites, who were highly skilled and well connected to the global flows of information and knowledge. The former Prime Minister of Singapore, Mr. Goh Chok Tong, announced during his 1997 National Day Rally Speech the vision for Singapore "to become a cosmopolitan, global city, an open society where people from many lands can feel at home" because "in the information age, human talent, not physical resources or financial capital, is the key factor for economic competitiveness and success" (cited in Yeoh and Chang 2001, 1029). As Yeoh and Chang (2001, 1030) noted further, this vision was "quickly translated into strategic policy thrusts and implicit 'best practices' for Singapore, even though the debates it had sparked in public discourse had yet to reach a clear resolution."

Between 1998 and June 2017, the foreign workforce[1] had more than doubled, from about 600,000 to 1.4 million (Yap 2014; Ministry of Manpower Statistics), and the overall proportion of foreign-born in Singapore in 2017 was close to 50% of its 5.6 million population, according to the United Nations population estimate (United Nations Population Division 2017). The structure of the foreign workforce in Singapore follows the contours of Sassen's (1991) findings of labor polarization in global cities. Global cities concentrate transnational professionals and managers, as well as low-wage labor-intensive workers who provide the services demanded by these elites. In the case of Singapore, the foreign workforce is similarly "bifurcated" into groups of foreign talents and foreign workers (Yeoh and Chang 2001; Yeoh 2006). This rapid rise of foreign-born in Singapore and the growing labor bifurcation has stoked social tensions between social groups. The citizens in this small 280-square-mile island city-state felt crowded out by the valorizing of foreign-born, who also competed for employment opportunities and affordable living space. The depth of the public anger was witnessed in 2013, when the release of the controversial *White Paper on Population* that recommended further population growth through immigration to support economic growth, led to a public furor and political backlash.

The necessity of foreign labor to augment the aging population and the economic restructuring of Singapore is a common rationalizing mantra for this survivalist and pragmatic nation. In his speech at the Ministry of Manpower Committee of Supply debate 2017,[2] the former Minister of Manpower, Mr. Lim Swee Say, justified the need for foreign labor and re-framed the rhetoric of a bifurcated labor market as a pragmatic transactional relationship of the locals needing the foreign-born. According to Mr. Lim, Singapore "can never be self-sufficient in having enough local workers, both in number and diversity of expertise" because foreign-born work in physically demanding jobs like construction "that few locals want to do" (Ministry of Manpower). In addition, they provide the additional capacity to meet the demand for professional skills in areas of nursing and engineering, for example; and lastly, foreign global companies that hire foreign-born create new job opportunities for locals.

With one of the world's highest GNP per capita, public discussion of the impacts of global immigration has focused largely on the one million low-wage foreign workers on work permits in Singapore,[3] whose numbers and proximity have challenged the comfort zone of the locals. In addition, its highly circumscribed consciousness of multiculturalism according to the ethnic categorizations of Chinese-Malay-Indian-Others has also been destabilized by the emergence of Chinese and Indian foreign talents alongside the arrival of foreign workers from China and various parts of South Asia such as India and Bangladesh, whose presence challenges the accepted notions of race and social status in the city-state.

In contrast, the Western foreign-born and foreign talents from other developed nations, such as Japan and Korea, have received less attention, likely due to their smaller numbers, self-sufficiency, and lack of societal expectation for their social integration, the latter particularly since they do not fit into any of the official ethnic groups. These foreign-born, especially the Westerners, are readily accepted as part of the "Others" category in a societal consciousness that remains entangled with the colonial past. Similarly, the Japanese and the Koreans, whose culture of food and film are particularly popular among locals, find themselves welcomed into the broad and tolerant "Others" category. These foreign-born from the developed nations are frequently associated as expatriates who are highly educated, employed in multinational corporations usually on short- to medium-term assignments, and fitting well into the economic pragmatism and meritocratic mores of the city-state (Chang 1995). Their lack of social integration is also unquestioned and socially acknowledged as the "expat bubble." Central districts like Orchard Road, Tanglin, Holland Village, River Valley, and Bukit Timah, with their fashionable malls, large private houses, expensive condominiums, and international schools, form the nucleus of this bubble, where well-to-do locals also live (Chang 1995; Yeoh and Khoo 1998). Thus, in more persistent ways than intended, the social space of early-21st-century Singapore bears an echo of its past as a colony of the British Empire, where Western elites lived and socialized separately in enclaves of privilege.

Social space reconfigured

Two dichotomous socio-spatial concepts have been introduced and reified in Singapore that mirror the contours of Castells's (1996 [2000]) conceptualization of the *spaces of flows* and the *spaces of places*, corresponding to the spatial circuits of those who are transnational elites and those who are not, respectively. Through his National Day Rally Speech in 1999, the former Prime Minister Goh Chok Tong reconfigured urban space in Singapore into cosmopolitan and heartland landscapes. According to Mr. Goh:

> As Singapore becomes more international, two broad categories of people will emerge. One group I call the "cosmopolitans," because their outlook is international. They speak English but are bilingual. They have skills that command good incomes, banking, IT, engineering, science and technology. They produce goods and services for the

> global market. Many cosmopolitans use Singapore as a base to operate in the region. They can work and be comfortable anywhere in the world. The other group, the "heartlanders," make their living within the country. Their orientation and interests are local rather than international. Their skills are not marketable beyond Singapore. They speak Singlish. They include taxi-drivers, stallholders, provision shop owners, production workers and contractors.
>
> *(National Day Rally Speech 1999, National Archives of Singapore)*

With these aspirations to host "cosmopolitans" and the capital (financial and cultural) they bring with them, Singapore is set towards a determined path to further adapt its social and built environment to transnational design sensibilities with amenities and infrastructure typical of a major financial center in the developed West. Works of signature architects like Zaha Hadid, Moshe Safdie, and Daniel Liebeskind dot the skyline of the city. As Yeoh (2004, 2435) noted, the efforts to make Singapore cosmopolitan was deliberate, "highly coordinated and planned." Thus, land use, economic, and social policies plan and extend in concert what I call "the cosmopolitan grid" – a transnational space comfortable for elites and expatriates in Singapore to inhabit.

The "cosmopolitan grid" is an infrastructural and amenity space that enables the transnational elites to have the quality of life to feel at home while on the move. Like the immigrant enclaves that have helped to incorporate new immigrants over the centuries, the "cosmopolitan grid" provides comfort zones for new transplants. Unlike the enclosed spaces of enclaves that take on the culture of the group that inhabits it, the grid is planned to enable expansion and locational freedom akin to the electrical grid that allows for mobility and ability to plug and play. For this purpose, the designers of the "cosmopolitan grid" in a multicultural setting like Singapore avoid any references to particular cultural groups (save for the culture of transnationalism) but instead cast the grid in universally accepted design parameters of safety and comfort for all ages. In this way, the "cosmopolitan grid" becomes a universal space for multiple cultures to connect, albeit designed mainly according to the perceived needs of transnational elites.

The "cosmopolitan grid" is composed of a network of quotidian "home-work-play" spaces.[4] For example, in Beaverstock's (2011) research of the British expatriates in Singapore, he illustrated the significant role of social "expatriate" clubs in helping British talents settled in Singapore. Quoting Andi, a recent German transplant whom I interviewed, Singapore's greatest asset is that it is the "most Western city in Asia!" He described the availability of goods and services to support the Western lifestyle as abundant and likened Singapore to a comfortable transnational "aircraft carrier" with the safety and familiarity of a retreat, where internationals can meet and exchange culturally. This is the amenity, according to Andi, that attracts many Germans and other Western expatriates to settle their families in Singapore, while they travel around the region to work during the week.

What is, thus, ironically attractive about the "cosmopolitan grid" is not exactly its hype of *global* footlooseness, but its promise of securing universal goods such as safety, predictability, and familiarity in a *local* place so that ordinary life routines, such as food shopping, education, work, and recreation, can be fulfilled.[5] Creating the value of ordinariness is a craft that many cities striving to become global hubs of talents *par excellence* are trying to achieve. However, ordinariness is a highly contested value insofar as whose ordinariness is being valued, and how ordinariness may vary among different cultural groups. In their study of Amsterdam and Eindhoven, Musterd et al. (2016) found that employment is a primary factor to attract international highly skilled migrants, but amenities such as schools, neighborhood, public transportation, shopping spaces, and supply of serviced and rental apartments are a critical complementary infrastructure that cities also need in order to attract and retain these talents.

In contrast to the cosmopolitan spatial circuit, Singapore's heartland is presented as occupied by those who are less educated and place-bound. In popular parlance, the heartland refers to places outside the city center, i.e., districts where the everyday landscape is characterized by largely public housing, neighborhood parks, and local coffee shops, rather than landscapes of private condominiums, Fitness First, acai bowls, and Starbucks.

While the opposing characteristics of these two socio-spatial circuits suggest parallel universes in the discourse, the reality is more dynamic and overlapping. Over the last two decades, some socio-spatial characteristics of the "heartland" have gradually become similar to the "cosmopolitan" landscape. "Heartland" settings are being remodeled with the international style architecture of glass, steel, and concrete to exude a "cosmopolitan" flair. Urban design guidelines by the Urban Redevelopment Authority of Singapore (URA) – such as controlling for uses at the basement, first-story, and second-story levels; outdoor refreshment areas; building form and height; pedestrian network; and public spaces – are increasingly being extended from the city center to the heartland. This change is particularly prominent in the planned regional centers that are undergoing a new spurt of growth, as decentralization is being aggressively pursued to accommodate the rising population. For example, planning and urban design guidelines made up two thirds of the conditions for the land sale at Jurong East Regional Center in 2010. These guidelines emphasized activity-generating uses, pedestrian connectivity, and building edges to "create pockets of lush permanent planting at street level" (6) (Urban Redevelopment Authority Singapore 2010). A 2016 request for proposal (RFP) for the creation of a master plan for the Jurong Lake District required that "urban design guidelines, landscaping and public space strategies, and plans to create, improve, and integrate green and blue spaces" were included prominently in the proposal (Urban Redevelopment Authority 2016).

As physical space in the city center becomes increasingly limited and expensive, deliberate and strategic urban planning policies have steered the "heartland" to become a new destination for the burgeoning number of international schools in Singapore that are being built to meet the demands of the children of foreign talents. Indeed, the "heartland" has become the workplaces and residences for the majority of the 1.4 million foreign-born in Singapore. Concurrently, "heartland" settings have turned into spaces of "ordinary transnationalism" (Beaverstock 2011), where foreign-born form comfort zones and create new contact zones, where daily intercultural encounters with the locals are even more probable (Pratt 1991; Amin 2002; Wood and Landry 2008). How do these ordinary spaces of transnationalism look like? Do they have unique spatial formations? What can the locations of these spaces tell us about how the foreign-born use and claim space in a planned city, where every inch of land is accounted for?

Spatial claims in a planned global city

With rising numbers of foreign-born low-income and high-income migrants working as domestic helpers in homes, as workers in construction sites, and also in Singapore's banks, universities, and technology-media firms, the everyday cultural landscape of the city-state has been palpably reshaped. The global flows of migrants are far from being deterritorialized and frictionless (Appadurai 1996; Gupta and Ferguson 1997; Tsing 2005). Over the past two decades, new spatial clusters of foreign-born have emerged within the ever-expanding planned "cosmopolitan grid": "Little America," "Little Paris," "Little Manila," "Little Thailand," and "Little Indonesia," to name just a few. While the trajectory of spatial cluster formation and change in Singapore unfolds within the planned parameters of the state's urban planning and housing policies, none of these new landscapes are given official status like the ethnic enclaves in North American cities (Banerjee et al. 2016). The only officially recognized ethnic enclaves in Singapore are the historic Chinatown, Little India, and Malay Village (see Figure 8.1 for a map of places mentioned in the chapter).

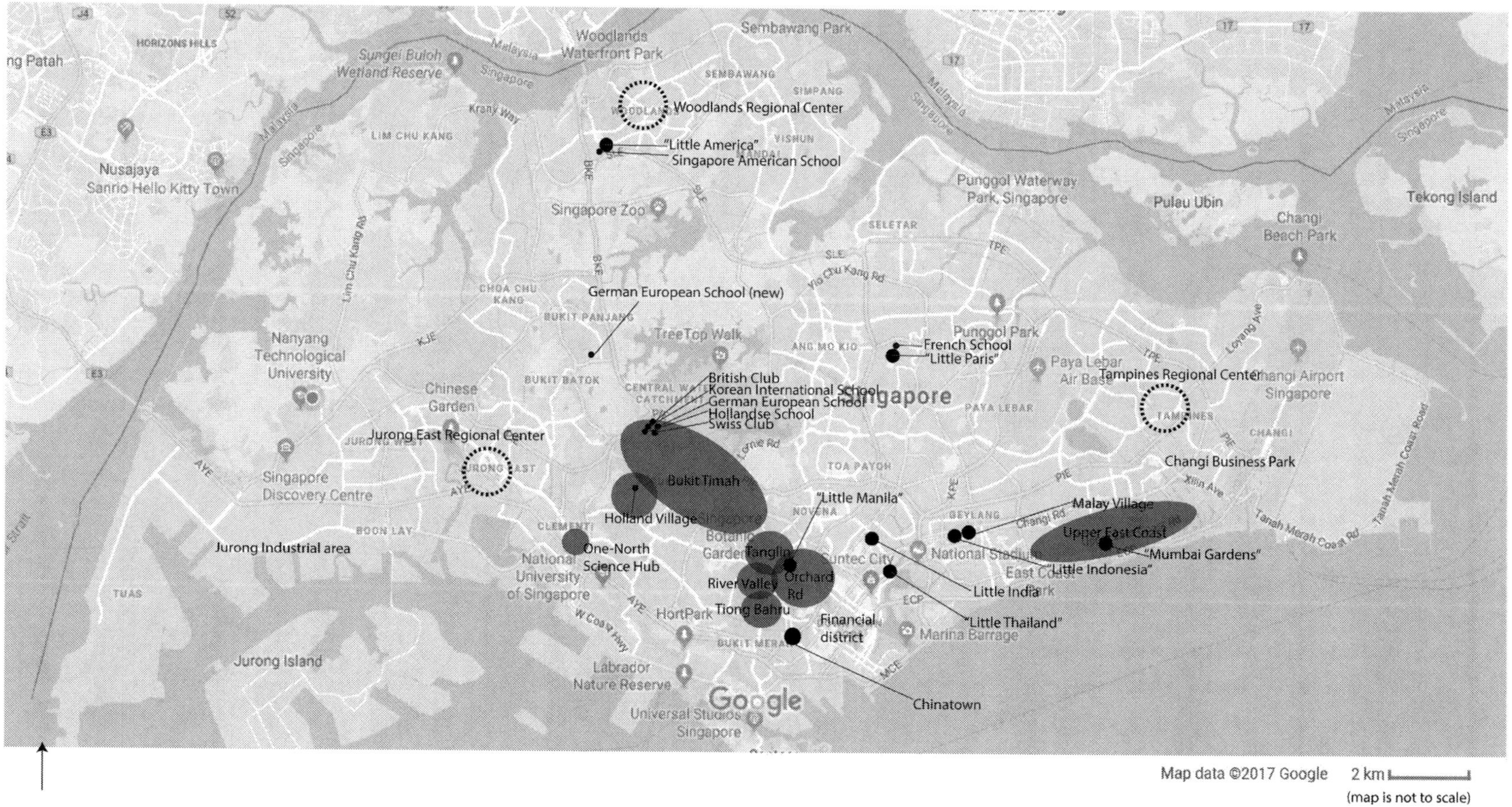

Figure 8.1 Map of contemporary cultural landscape in Singapore.

Source: Author using Google Maps

As Peter Marcuse (2005, 15) observed, "Spatial clustering seems to be an inevitable accompaniment of urban life." It can be voluntary or involuntary, as well as hierarchical or nonhierarchical with respect to wealth, status, or power. The foreign-born spatial clusters in Singapore straddle between the voluntary and involuntary, containing hierarchies of wealth and status. Unlike the spatial claims of ethnic enclaves, described by the urban ecology model of the Chicago School, where poor immigrants cluster in the low-rent areas of the city center and, subsequently, suburbanize, upgrade, and assimilate into the White American way of life, spatial clusters in Singapore emerge in different urban contexts, serve different purposes, and take different forms.

These spatial clusters reflect the importance of territories in globalization and express claims to occupy, craft, and use space in ways that disrupt the status quo. Where people live, socialize, and recreate are key markers and mediums of transnational mobility and claims of identity. These spaces of ordinary life are critical to our understanding of how global immigration translates, replaces, and/or reconstructs the urban form and fabric of cities. In what follows, I will discuss some of the socio-physical characteristics of migrant spatial clusters in Singapore.

Global North in the suburban heartland[6]

"Everything was the school!" recalled Renee Blackburne about her experience living in a suburban neighborhood known as "Little America" next to the Singapore American School (SAS) (Ng 2000). In 1996, the SAS relocated from the city center to a 36-acre land in the outskirts to accommodate its growing ambition and enrollment. Like many other American families, Renee exchanged her centrally located condominium at Orchard Road for a large house in Woodlands, so that her children could participate fully in sports after school and during weekends, without having to spend two hours on the school bus traveling to and from school. During those years, Renee said that 85% to 90% of her neighbors were Americans, and the residential streets were lined with rows of basketball hoops. Every day between 4 and 7 p.m., children would play in the streets, and evening street parties became a norm, as American families organized programs to overcome the lack of recreation options in Woodlands. Over the decade, even as Woodlands acquired new shopping amenities, most American families continued to drive to the city center for their shopping and recreation needs, where the American Club, the American Association, and the rest of their social networks were based. According to Renee, the vibrancy of "Little America" has waned over the last decade, as only one in two households are now American.

However, a few unique imprints remained in the built environment around "Little America" that are uncommon in other suburban residential neighborhoods in Singapore. For example, along one residential street, there are multiple basketball hoops and bicycles in almost every home. Quite a number of houses have their fences opened, with parked SUVs, patio furniture, and swimming pools, even though the yards are not large. As my visit was near Christmas time, many homes were decorated for Christmas. In deep contrast to the openness and porosity of the residential streets and nearby public housing, the SAS is gated by a high steel fence with a row of barbed wire at its top. The school entrance is also under tight security. The guards conveyed that without prior appointments with the school administration, visitors are not allowed on its premises. Apart from these imprints, there is little to no observable presence of Americans using the surrounding public community and retail spaces.

Like "Little America," "Little Paris"[7] in the northeastern suburb of Ang Mo Kio-Serangoon Gardens came on the heels of the opening of the new facility of Lycée Français de Singapour in 1999 when the French School outgrew its former location at Bukit Timah. I was informed by Maria and Lily, two French mothers who I met at a community fundraising event, that over a thousand French are living in rented condominiums and single-family homes in the surrounding

Figure 8.2 Scenes of "Little America" and "Little Paris."

Source: Photos taken by Author

neighborhoods around the new French school. They said that the French like to live close to the school, so that their children could travel to school independently by walking or cycling. Indeed, at the school's bicycle garage, there were as many as a hundred bicycles parked inside. The imprints of the French in the built environment were somewhat different from "Little America." Instead of expressing cultural difference through the private space of homes and the highly secured design of the American school, "Little Paris" is a space formed by visible activities of the French population in the public space. For example, adults cycle on the streets to do their daily routines of picking up food and kids; children cycle between home and school; families eat weekend brunch at the local cafés; and a French cheese shop (Le Petit Boutique) and a Parisian bakery (Maison Kayser) opened in 2015 and 2017, respectively. These expressions have a cumulative effect in reshaping the use and formation of public space in the suburb (see Figure 8.2).

Expat-talent abode and condo huddle

The Bukit Timah district is the emerald gemstone of Singapore. Located adjacent to the central district of Orchard Road and the largest nature reserve in the land-scarce city, its single-family homes and high-end condominiums sit amid lush tropical greenery made possible by strict density zoning guidelines. It is home to multiple top local public schools, international schools, and the residences of the powerful and wealthy. A local real estate website described Bukit Timah as the traditional and top "expat enclave" for the "wealthier expatriate crowd" (Ong 2016). Shops, restaurants, and services in the vicinity cater to different Western sensibilities that sell an unusually large selection of meat, produce, and products from Europe, North America, Australia, and New Zealand. In 2010, Bukit Timah also became home to the Singapore Korean International School. Within the last decade, a new cluster of Korean eateries has replaced a stretch of local restaurants in the vicinity (Kok 2011). The combination of global and local infrastructure available in Bukit

Timah through public and private investments makes this district an exceptionally sought-after comfort zone to conduct ordinary life in the space of flows.

While there is no known "Little Germany" in Singapore despite a long and growing presence of German population, my interview with Karl Göbel, a German who has lived in Singapore for several decades, suggested that the Bukit Timah area has been an abode for expatriate German families for at least the past 30 years. Since the early 1980s, this district housed the German Association and the German European School Singapore (GESS) (formerly known as the Deutsche Schule Singapur). Karl explained that the residential decision of Germans is usually shaped by the location of their workplace and their children's school. As many German expatriates work in German companies located in the nearby Jurong industrial area, this neighborhood has remained the choice residential location for most Germans. For this reason, the GESS chose among the options provided by the government to build their new multimillion-dollar campus in the fringe area of Bukit Timah. The land is on lease from the Singapore government for 30 years.

The propensity for foreign-born to spatially and socially cluster is also an outcome shaped by overt housing and land policies by the Housing Development Board and the Singapore Land Authority. These policies place a non-citizen quota on renting out public housing to foreign-born, as well as restrictions on ownership of public housing and private land. As a result, the majority of foreign talents in Singapore, who are not naturalized citizens or permanent residents, would usually rent private homes or rent or buy private condominium units. Thus, popular expatriate residential enclaves can be found in Tiong Bahru, Tanglin, Upper East Coast, and River Valley (Ong 2016), where there is a large stock of private condominiums, and in areas close to major universities and research facilities, such as the One-North science hub. In Upper East Coast, for example, the popularity of the district's condominiums among Indian foreign talents, who work in the nearby financial district and technology firms, has resulted in a spatial clustering of foreign Indians. This has ruffled the feathers of other condominium owners. One such example is a condominium complex named Mandarin Gardens that became known as "Mumbai Gardens," as many foreign Indians buy and rent there. The formation of the "condo huddle" of foreign talents enlarges the expatriate bubble and increases the social distance between foreign talents and the majority of locals, who live in public housing.

"Global South" malls

It is common knowledge that "Little Manila" refers to Lucky Plaza, a shopping mall located along the central spine of Orchard Road; "Little Indonesia" refers to City Plaza, and "Little Thailand" to the Golden Mile Complex, located at the eastern fringe of the city center. These three buildings are centrally located malls built in the 1970s that have fallen out of fashion. Built for mixed commercial and residential uses, the first six stories of Lucky Plaza are occupied by a bazaar-style shopping mall with mostly small and non-chain businesses interspersed with eateries, while the higher floors are reserved for business offices and condominium units.

Anecdotal evidence suggests that even as early as the 1990s, Filipino domestic workers started patronizing the stores and services at Lucky Plaza mall during the weekends as an additional gathering space to the open space across the road, where they would picnic on Sundays. The open space was replaced by a new luxury mall in 2009, and Lucky Plaza became increasingly acknowledged as the commercial hub of the Filipino population in Singapore. The seven floors (including the basement) of the mall are occupied by businesses and services targeting Filipino clientele, who are a mix of foreign workers and professionals. During my visit, I noticed remittance businesses catering to Filipino and Indonesian clients, Filipino banks, shipping services, Cebu Air travel and tours, small shops selling picnic snacks and canned food, the Filipino fast-food chain Jollibee,

Figure 8.3 Images of "Little Manila" and "Little Indonesia."

Source: Author

employment agencies, hair salons, and small eateries serving Filipino cuisine and operated by Filipinos. In comparison, City Plaza, or "Little Indonesia" is less vibrant. Businesses patronized by the Indonesian clientele appeared to be limited to the first two stories, and these included multiple remittance businesses, women's fashion stores, employment agencies, and a couple of halal eateries operated mainly by locals. There are also many other retail stores, travel agents, non-halal eateries, and wholesale garment stores that service a local crowd.

On Sundays, when many foreign domestic workers get their mandatory day off, the walkways, staircases, and open space around the two malls are appropriated as "public" spaces for gathering of friends over food and drinks. At City Plaza, I saw many groups picnicking outdoors, or chatting and snacking along the mall walkways, especially along quiet or empty storefronts. There were also significant numbers of people sitting on stairways with their hands-free sets chatting quietly on phones and having video chats, likely with loved ones. According to Yanni, an Indonesian domestic helper who lives and works in one of the large single-family homes in Bukit Timah, City Plaza has always been a meeting point to connect with her friends and to remit money home. Every Sunday morning, she brings along a potluck dish on her bus ride from Bukit Timah to City Plaza. Usually Yanni hangs out with her friends picnicking at the outdoor area outside City Plaza from noon to 6 p.m. and rarely patronizes the shops or eateries in the mall. Instead, she travels to "Little Thailand," about 20 minutes away by public transport, to shop for cheaper products (see Figure 8.3).

Conclusion

Portes and Manning (1986, 156) noted that occupationally skilled immigrants *do not* normally "give rise to spatially concentrated communities." Instead, their settlement pattern is characterized by dispersal, as they follow career opportunities. The case of Singapore has demonstrated different patterns. Foreign talents have formed enclaves in part guided by a slew of land, housing, immigration policy, and land use planning that place constraints on the residential options of the foreign-born. These policies have also enlarged the "cosmopolitan grid" so that the comfort zone for transnational elites has expanded to areas outside the city center, creating a more spatially dispersed settlement pattern than two decades ago. More than before, transnational elites can now easily "plug and play" in Singapore, without needing to create their ordinary spaces to live, work, and play. According to the annual HSBC Expat Explorer Survey (2017, 10), "For the third year running, expats have crowned Singapore the best overall destination, ranking consistently high across a range of economic, experience and family criteria."

In contrast, foreign workers from the Global South, who are normally dispersed during the weekday in private homes, on construction sites, and in dormitories around the island, have to recreate their comfort zones within the interstices of the "cosmopolitan grid." They assert a temporary claim to centrality by occupying and gathering in centrally located public and semipublic spaces of sidewalks and old malls. The foreign workers' visible claim for public space contrasts with the relatively invisible claim for private space by the foreign talents through the expansion of the public-initiated "cosmopolitan grid." Likewise, while the spatial claims by the low-income foreign workers are ephemeral, temporary, and disputed (with the exception of "Little Manila" at Lucky Plaza that has evolved into a viable ethnic business enclave), the spatial claims by foreign talents are durable and accepted.

The growth of the "cosmopolitan grid" has enabled and enhanced the spaces of ordinary life for some, making assimilation and integration easier for certain social and cultural groups and harder for others, thereby risking a deepening of social tensions arising from unequal access to resources. The 21st-century Singapore stands at the cusp of designing a city that enables human connections to grow across differences. The challenge for Singapore is inherently an intercultural one. Amin's (2002, 967) succinct argument for an intercultural approach to address "living with diversity" bears repeating here, as the city-state struggles with its multicultural consciousness and its desires of being cosmopolitan:

> The term "intercultured" is used to stress dialogue, to contrast with versions of multiculturalism that either stress cultural difference without resolving the problem of communication between cultures, or versions of cosmopolitanism that speculate on the gradual erosion of cultural difference through inter-ethnic mixture and hybridization.

The extension of the "cosmopolitan grid" into the heartland over the last two decades, through the addition of new metro lines, new condominiums, new international school campuses, multiple retail options, carefully designed public spaces, and new urban design guidelines, has expanded the "ordinary" grounds for encounters between the foreign talents and locals outside the central districts and the workplace. However, the verdict is still out on whether the smarter, better-networked, and increasingly transnational urban grid in Singapore can substantively converge the spaces of ordinary life of the local and foreign-born beyond physical bounds.

Notes

1 "Foreign" is the preferred term of use in Singapore rather than "migrant."
2 Committee of Supply refers to an annual session organized by each ministry to debate about its expenditure plans for the new Financial Year.
3 According to MOM statistics, in December 2017, the total foreign workforce is 1.368 million, with 187,700 on employment pass (professional class), 184,400 with S Pass (mid-skilled skilled professional who earns at least SGD $2,400 a month) and the remaining 965,200 with a work permit.

4 I am creatively borrowing the title of a book published by the Urban Redevelopment Authority of Singapore in 1999 (Tan 1999) that outlined a vision to make Singapore "become a top Asian Business and financial hub, as well as the most liveable city" that "will hopefully appeal to Singaporeans" (foreword by former Minister of National Development).

5 I am borrowing a concept of "ordinary transnationalism" from Beaverstock (2011).

6 I have removed the contextual details of the interviews and the situational particularities of my interviewees to protect their privacy. All the names used in the text are pseudonyms.

7 The "Little Paris" name was mentioned by one of the interviewees at the French School. However, when asked if she thought that Serangoon Gardens felt like "Little Paris," she shrugged her shoulders.

References

Amin, A. (2002). "Ethnicity and the Multicultural City: Living with Diversity." *Environment and Planning A* 34: 959–980.

Appadurai, A. (1996). *Modernity at Large: Cultural Dimensions of Globalization.* Minneapolis: University of Minnesota Press.

Banerjee, T., Chakravarty, S. and Chan, F.H.H. (2016). "Negotiating the Identity of Diaspora: Ethnoscapes of the Southeast Asian Communities in Los Angeles." In Moroni, S. and Weberman, D. (eds.) *Space and Pluralism: Can Contemporary Cities Be Places of Tolerance?* Budapest: Central European University.

Beaverstock, J.V. (2011). "Servicing British Expatriate 'Talent' in Singapore: Exploring Ordinary Transnationalism and the Role of the 'Expatriate' Club." *Journal of Ethnic and Migration Studies* 37 (5): 709–728.

Castells, M. ([1996] 2000). *The Rise of the Network Society: The Information Age: Economy, Society and Culture Vol. I.* Cambridge, MA and Oxford, UK: Blackwell Publishers.

Chang, T.C. (1995). "The 'Expatriatisation' of Holland Village." In Yeoh, B.S.A. and Kong, L. (eds.) *Portraits of Places: History, Community and Identity in Singapore*, Singapore: Times Edition.

Gupta, A. and Ferguson, J. (1997). "Beyond 'Culture': Space, Identity, and the Politics of Difference." In Gupta, A. and Ferguson, J. (eds.) *Culture, Power, Place: Explorations in Critical Anthropology.* Durham and London: Duke University Press.

Housing Development Board, Singapore. Renting out your Flat: Eligibility. www.hdb.gov.sg/cs/infoweb/residential/renting-out-a-flat-bedroom/renting-out-your-flat/eligibility (accessed 29 December 2017).

HSBC. (2017). *Expat Explorer: Broadening Perspectives Global Report.* London: HSBC Holdings.

Kok, M. (2011). Korean Boom in Bukit Timah. *The Straits Times* 28 August.

Marcuse, P. (2005). "Enclaves Yes, Ghettos No: Segregation and the State. In Varady, D.P. (ed.) *Desegregating the City: Ghettos, Enclaves, and Inequality.* New York: State University of New York Press, 15–30.

Ministry of Manpower Singapore. Foreign Workforce Numbers. www.mom.gov.sg/documents-and-publications/foreign-workforce-numbers (accessed 29 December 2017).

Ministry of Manpower Singapore. Main Speech by Mr Lim Swee Say, Minister of Manpower at Committee of Supply 2017. www.mom.gov.sg/newsroom/speeches/2017/0306-main-speech-by-min-lim-at-cos-2017 (accessed 29 December 2017).

Ministry of Manpower Statistics. Singapore. www.mom.gov.sg/documents-and-publications/foreign-workforce-numbers (accessed 15 March 2019).

Musterd, S., Bontje, M., and Rouwendal, J. (2016). *Skills and Cities: Implications of Location Preferences of Highly Educated Workers for Spatial Development of Metropolitan Areas.* London and New York: Routledge.

National Archives of Singapore. National Day Rally Address by Prime Minister Goh Chok Tong 22 August 1999 First-World Economy, World-Class Home. www.nas.gov.sg/archivesonline/speeches/view-html?filename=1999082202.htm (accessed 29 December 2017).

Ng, S.L. (2000). "Little America." *The Straits Times* 17 September.

Ong, R. (2016). 5 "Unofficial Expat Enclaves in Singapore," 24 June 2016. www.99.co/blog/singapore/unofficial-expat-enclaves-singapore/ (accessed 29 December 2017).

Portes, A. and Manning, R.D. (1986). "The Immigrant Enclave: Theory and Empirical Examples." In Lin, J. and Mele, C. (eds.) *The Urban Sociology Reader (2005).* London and New York: Routledge.

Pratt, M.L. (1991). "Arts of the Contact Zone." *Profession (Modern Language Association)*, 33–40.

Sassen, S. (1991). "The Global City." In Fainsten, S. and Campbell, S. (eds.) (1996) *Readings in Urban Theory.* Malden and Oxford: Blackwell Publishers.

Singapore Land Authority. Foreign Ownership of Properties. www.sla.gov.sg/Services/Restriction-on-Foreign-Ownership-of-Landed-Property (accessed 29 December 2017).

Tan, S. (1999). *Home Work Play.* Singapore: Urban Redevelopment Authority.

Tsing, A.L. (2005). *Friction: An Ethnography of Global Connection*. Princeton, NJ and Oxford: Princeton University Press.

United Nations Population Division. (2017). International Migrant Stock 2017, Department of Economic and Social Affairs. www.un.org/en/development/desa/population/migration/data/estimates2/estimates17.shtml

Urban Redevelopment Authority of Singapore Website. Urban Design Guidelines. www.ura.gov.sg/Corporate/Guidelines/Urban-Design (accessed 15 March 2019)

Urban Redevelopment Authority of Singapore. (2010). "Sale of Site for White Site Development Land Parcel at Jurong Gateway Road: Technical Conditions of Tender dated 26 April 2010. www.ura.gov.sg/sales/JEast-St13/tenderdocs/JE13-TCOT_CA.pdf (accessed 15 March 2019)

Urban Redevelopment Authority of Singapore. (2016). Jurong Lake District Media Release. Published on 11 July 2016. www.ura.gov.sg/corporate/media-room/media-releases/pr16-45 (accessed 15 March 2019)

Wood, P. and Landry, C. (2008). *The Intercultural City: Planning for Diversity Advantage*. London and Sterling, VA: Earthscan.

Yap, M.T. (2014). "Singapore's System for Managing Foreign Manpower." In Adams, R. Jr. and Ahsan, A. (eds.) *Managing International Migration for Development in East Asia*. Washington, DC: World Bank Group.

Yeoh, B.S.A. (2004). "Cosmopolitanism and Its Exclusions in Singapore." *Urban Studies* 41 (12): 2431–2445.

Yeoh, B.S.A. (2006). "Bifurcated Labour: The Unequal Incorporation of Transmigrants in Singapore." *Tijdschrift voor Economische en Sociale Geographie* 97 (1): 26–37.

Yeoh, B.S.A. and Chang, T.C. (2001). "Globalising Singapore: Debating Transnational Flows in the City." *Urban Studies* 38 (7): 1025–1044.

Yeoh, B.S.A. and Khoo, L.M. (1998). "Home, Work and Community: Skilled International Migration and Expatriate Women in Singapore." *International Migration* 36 (2): 159–186.

Further reading

Vertovec, S., (ed). (2015). *Diversities Old and New: Migration and Socio-spatial Patterns in New York, Singapore and Johannesburg, Global Diversities*. London: Palgrave-Macmillan. A comparative study of how global immigration has shaped the use of urban spaces with a case study on Jurong West in Singapore.

Yap, M.T., Koh, G., and Soon, D., (eds.) (2015). *Migration and Integration in Singapore: Policies and Practice*. Oxon and New York: Routledge. An overview of the debates and challenges of global immigration in Singapore.

Yeoh, B.S.A. and Huang, S. (1998). Negotiating Public Space: Strategies and Styles of Migrant Female Domestic Workers in Singapore. *Urban Studies* 35 (3): 583–602. A critical discussion of the dynamics of foreign workers' claim on space with a case study of Filipinos' negotiation of space at Lucky Plaza.

9

DESIGNING FOR DIFFERENCE IN BARCELONA'S *EL RAVAL*

Jeremy Németh

Introduction

Plaça de Pedro in the heart of Barcelona's *el Raval* neighborhood contains a *pastisseria* (pastry shop), *xarcuteria* (deli), *Halal carnisseria* (butcher), tailor, print shop, phone store, and a café, nearly all owned or run by Middle Eastern or South Asian immigrants who live in nearby flats. Heading north up a single block of *Carrer d'Erasme de Janer* brings one past *Fundácio Comparte*, a non-profit working on Latin American issues; the Centre de'Estudis Africans, a resource for African immigrants; a Filipino barber; an Indian *perruqueria* (hair salon); the Shani Hindu temple; a money transfer store popular with Ecuadorian residents sending remittances; and the Minhaj Mosque and Islamic Center. A December 2017 Google search of the neighborhood reveals four Christian churches, four mosques, three Hindu temples, and a Sikh *gurdwara*.

El Raval is a quintessential "arrival neighborhood" in Barcelona, a dense jumble of streets and buildings to which immigrants have flocked for centuries (Degen 2017). In the late 1800s, this marginal working-class neighborhood in the *Ciutat Vella* (Old City) was pejoratively coined *Barri Xino* (Chinatown) and has long been known for prostitution, drugs, and crime. Since the 1980s, however, *el Raval* has gentrified considerably, partly as a result of efforts by the *Ajuntament* (Municipality) to attract newcomers and tourists to the neighborhood, partly through major investments in cultural facilities and the demolition of well-known hotspots for drug dealing and prostitution (Degen 2003). Tourists now know the neighborhood as a bohemian destination replete with organic groceries and craft beer bars (Arbaci & Tapada-Berteli 2012).

In spite of these gentrifying socioeconomic changes in *el Raval*, the percentage of its foreign-born residents has grown explosively since the early 1990s and has remained constant since the early 2000s (see Table 9.1). Over 125 nationalities live within the neighborhood's 0.42 square miles, an area just slightly larger than New York City's Greenwich Village (Ajuntament de Barcelona Department d'Estadística 2017).

El Raval's persistence as an arrival neighborhood makes one wonder why it remains such a welcoming place for immigrants in the face of such upscale change. Portes and Rumbaut (2014) use the term "contexts of reception" for the combination of opportunities and challenges that shape an immigrant's acceptance and integration into a host community. For decades, scholars have debated the factors that make contexts of reception more or less likely to result in successful immigrant integration, but nearly all of these studies have ignored the role of the built environment in shaping the immigrant experience.

Table 9.1 Foreign-born population (%) by year in *el Raval* and Barcelona

	El Raval	*Barcelona*
1991	4.8%	1.4%
1996	9.1%	1.9%
2000	23.1%	4.9%
2004	42.8%	13.0%
2008	48.0%	17.3%
2012	49.2%	17.5%
2016	48.5%	16.6%

Source: Ajuntament de Barcelona Department d'Estadística (2017)

In this chapter, I explore the role that urban design and planning can play in the economic, sociocultural, and civic integration of immigrants. This question is particularly pressing given our increasingly multicultural future and the contemporary refugee crisis reshaping cities around the globe. First, I outline several different forms of immigrant integration and describe how contexts of reception help us conceptualize the immigrant experience. Next, I present three common urban design concepts – spatial concentration, public space, and accessibility – and describe how these might shape such contexts. I then turn to *el Raval* to examine how these concepts play out in a thriving arrival neighborhood and conclude that place matters for immigrant integration.

Throughout this chapter, I rely on a combination of observational research, including personal field notes taken from September 2015 to June 2016 and several interviews with long-time residents of *el Raval*; data from the *Department d'Estadistica* of the *Ajuntament* (Municipality) *de Barcelona*; and a review of newspapers, professional reports, and academic sources, many of which contain original quotes and firsthand accounts of neighborhood dynamics in *el Raval*. Importantly, I see this single "revelatory" case study not as a source of explanation but as a site of exploration (Yin 2009). Taken together with contemporary scholarship on immigrant integration, my observations help me develop a conceptual framework for understanding how the built environment can shape contexts of reception.

Living with/in difference

Given the intense socio-economic, cultural, racial, ethnic, and attitudinal diversification of the population, researchers and policymakers are increasingly seeking the most appropriate ways of living together (Tasan-Kok et al. 2013). Long-standing debates often center on whether assimilationist or multiculturalist approaches are the most appropriate methods of cohabitation.

Assimilationists support a "melting pot" approach that narrows the differences between immigrants and receiving populations and asks immigrants to adopt the social, political, and cultural customs of the host country (Bloemraad et al. 2008). They believe that absorption into host communities can reduce exclusion from resources, capabilities, and other rights that can lead to health and quality-of-life inequities (Popay et al. 2008). Assimilation critics, however, claim that newcomers should not have to abandon their deeply held cultural foundations, identities, practices, and heritage in order to conform to dominant, mainstream societal norms (Young 1990).

Multiculturalists support a "mosaic" approach, wherein immigrants retain their cultural customs and ethnic heritage but contribute to the societal whole, like individual tiles that together form a mosaic (Bloemraad 2006). Proponents believe that multiculturalism can build deeper social ties; strengthen racial, social, and political identity; increase community satisfaction; and reduce distrust of diverse others (Bader 2016; Schmid et al. 2014). But critics assert that multiculturalism facilitates segregation, discourages loyalty to the host country, and hinders the development of shared cultural norms such as language and family structure (Bissoondath 1994).

Yet as Bloemraad (2006) argues, these coarse characterizations can obscure more than illuminate realities of the immigrant experience. In her work comparing immigration policy in the "melting pot" United States and "mosaic" Canada, she finds that immigrants and native-born residents in both countries agree on immigration policy more than they differ. In the United States in particular, beliefs on assimilation have come a long way since their early-20th-century roots and now explicitly encourage and celebrate the maintenance of diverse cultural and social beliefs of immigrants and non-immigrants (Zolberg & Long 1999). Such reformulations tend to "undermine an oft-posited dichotomy" regarding these positions (Bloemraad et al. 2008, 163).

For these reasons, I use the more general term "integration" to establish a middle ground between assimilation and multiculturalism. My use of the term implies that both group differentiation *and* racial and cultural integration are deeply valued. Successful immigrant integration, I infer, can enrich daily life; infuse communities with creativity, innovation, and prosperity; support diverse sources of knowledge and experience; and prepare immigrants and receiving communities for an increasingly globalized future (National Academies of Sciences, Engineering, and Medicine 2015).

Modes of immigrant integration

A recent report by a panel of immigration policy experts suggests several interrelated forms of immigrant integration, which I collapse into just three: economic, sociocultural, and civic (National Academies of Sciences, Engineering, and Medicine 2015).

Economic integration receives the bulk of attention in immigration scholarship. Studies focusing on personal and structural mechanisms for integration find that individual characteristics such as education level, work experience, language proficiency, and age can influence integration just as much as the character of the host labor market or citizenship policies of the receiving government (Portes & Rumbaut 2014). Mechanisms for measuring immigrant economic integration include: proximity to reliable sources of capital (Portes & Zhou 1992), access to diverse labor markets (Wilson & Portes 1980), and clustering in enclaves that provide opportunities mobilized through ethnic social networks (Valenzuela-Garcia et al. 2017).

Sociocultural integration is shaped by a number of mechanisms: norms and customs including language acquisition, religious inclusion, similarities between home and host cultures, and familial traditions. A common measure of sociocultural integration is the extent to which immigrant attitudes and opinions on social issues or political ideologies approximate those held by native-born populations (National Academies of Sciences, Engineering, and Medicine 2015). Language is another factor influencing sociocultural integration, although cultural values and identifications can be just as critical in shaping contexts of reception (Castillo & Caver 2009; Schwartz et al. 2014). Primary mechanisms for sociocultural integration include: access to social centers such as schools, churches, and recreation centers; participation in cultural institutions such as markets, cultural fairs, or parades; and regular interaction with other immigrants and long-term residents.

Civic integration is a highly localized social process that involves regular engagement of immigrants in debates, groups, practices, and decision-making of the receiving society (Bloemraad 2006). It foregrounds informal modes of engagement and differs from political integration, which relies on formal mechanisms such as voting behavior, acquisition of formal citizenship, or running for elected office. Primary mechanisms for measuring civic integration include:

engagement with local community groups and organizations; participation in political mobilization efforts; and immersion in interpersonal networks of family, friends, and neighbors (Bloemraad 2006).

In sum, mechanisms of immigrant integration center on social interaction, access to economic opportunity, and civic and political engagement. In this chapter, I argue that all of these have a clear spatial component and are laudable goals for any urban project, not just for arrival neighborhoods. Although a welcoming institutional environment, easy acquisition of legal citizenship, and access to a strong low-skilled labor market clearly contribute to positive contexts of reception, I suggest they fail to tell the complete story for arrival neighborhoods like *el Raval.*

The built environment and contexts of reception

This chapter starts with the premise that buildings, streets, public spaces, and neighborhood location can contribute to, or detract from, immigrant integration or isolation, as "social relations are inevitably correlated with spatial relations" (Park 1952). Thus, I ask not whether the built environment can *create* positive contexts of reception for immigrants, but whether immigrant integration "thrives better, or can be sustained longer, under certain physical conditions that planners [and designers] may have some control over" (Talen 2006, 242).

Attempts to sequester newcomers in segregated immigrant enclaves can breed isolation, increase inequality, and even radicalize extremists who feel cut off from mainstream society, as some believe to be the case in the Parisian *banlieues* (see Saunders 2010). If poorly conceived built environments can reduce the chances of positive integration, can well-designed environments do the opposite? Put another way, how might urban designers cultivate social interaction, access to opportunity, and civic and political engagement in arrival neighborhoods? In what follows, I return to *el Raval* to show how we might achieve these goals through placemaking endeavors that emphasize three concepts central to urban design theory and practice: spatial concentration, public space, and accessibility.

Spatial concentration

El Raval is one of the densest neighborhoods in Europe. At 174 residents/acre, the neighborhood is nearly three times denser than greater Barcelona (64 residents/acre), 17 times denser than Los Angeles, and almost as dense as Dhaka, Bangladesh, the world's densest city (180 residents/acre) (Ajuntament de Barcelona Department d'Estadística 2017; United Nations Human Settlement Program [UN-Habitat] 2017). One of the neighborhood's *seccions censales*, akin to a US Census tract, has a density of 541 residents/acre, and several more are in the 500+ range (Figure 9.1). Part of this is due to the size of residences: the share of small flats (below 656 sq. ft.) in the neighborhood is 56.4%, compared to a 30.7% citywide average, and several tracts contain more than 70% of these small flats (Ajuntament de Barcelona Department d'Estadística 2017).

El Raval residents hail from 125 different countries; 21.4% of foreign-born residents come from Pakistan, 17.4% from the Philippines, and 11.3% from Bangladesh (Ajuntament de Barcelona Department d'Estadística 2017). Subirats and Rius (2006) maintain that few, if any, ethnically clustered residential zones exist in Barcelona. But a walk through the neighborhood reveals several distinct nodes of immigrant *commercial* activity. As of 2015, for example, there were 244 Pakistani-run businesses in the neighborhood, with many immigrant-run shops clustered tightly along corridors, such as the 15 Pakistani-operated mobile phone shops along *Carrer de Sant Pau* alone (Güell 2016). Pakistani entrepreneurs also run Internet cafés, supermarkets, groceries, restaurants, hairdressing salons, butcher shops, travel agencies, tailor shops, and jewelry stores (Molina et al. 2015).

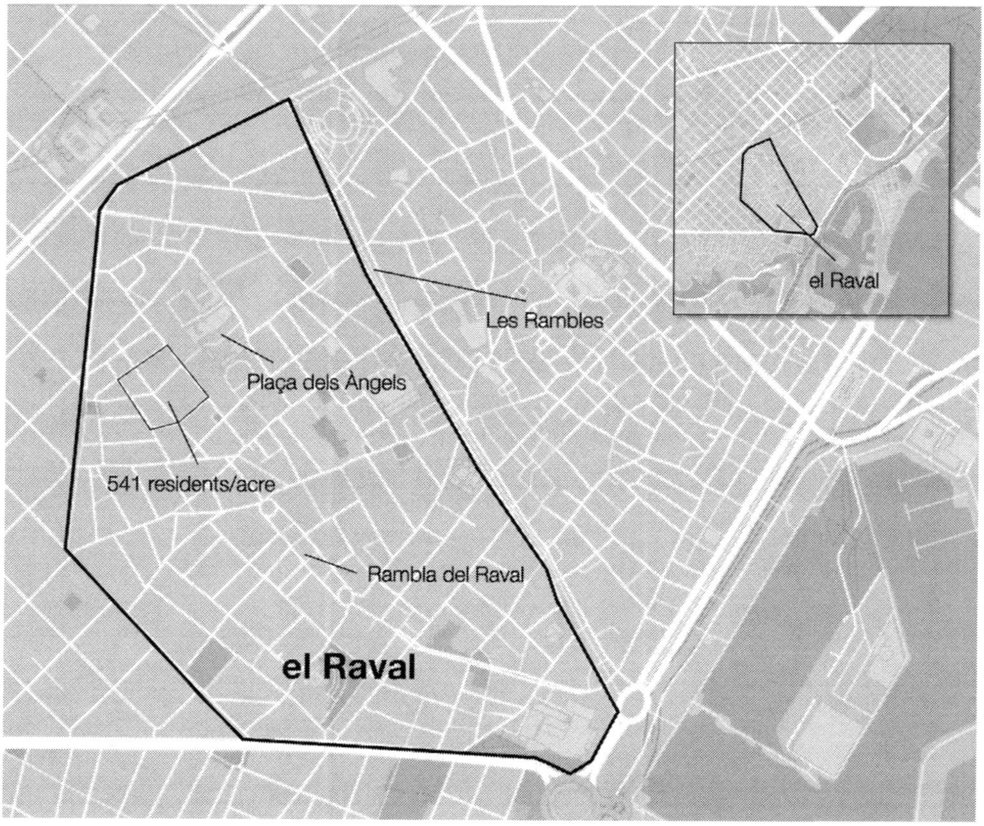

Figure 9.1 Map of *el Raval.*

Source: Author and Google

Co-ethnic business concentration is a distinctive feature of ethnic enclaves, giving marginal communities advantages they might not otherwise enjoy when integrating directly into traditional, mainstream economies (Molina et al. 2015; Portes & Manning 1986). Indeed, a recent study of Pakistani employees in *el Raval* revealed that clustering of these businesses has created high levels of social capital among shop workers and business owners as manifested in the sharing of a co-ethnic workforce, reliance on trusted production and supply chain contacts, and use of shared recruitment and training resources. Another study showed that life satisfaction among Pakistani women living in *el Raval* partly stems from the dense concentration of immigrant-serving shops and the ability to live with neighbors that share their language and cultural background (Ortiz et al. 2004; Valenzuela-Garcia et al. 2017).

Commercial enclaves thus represent important "pull factors" that attract immigrants to arrival neighborhoods due to efficiencies in staffing by local community members; vertical and horizontal integration in supply and production chains; information circulation within ethnic networks; access to credit, loans, and financial resources; and deep networks of social capital via engagement with community associations, businesses, places of worship, and other centers of socialization (Güell 2016; Molina et al. 2015; Portes and Rumbaut 2014). Put another way, commercial concentration can help cultivate positive contexts of reception for immigrant economic integration.

El Raval also contains an intense concentration of cultural and community organizations. Subirats and Rius (2006) show that from 1992 to 2002, the number of cultural and arts-related businesses increased from 229 to 571, while the neighborhood houses three nursery schools, five primary schools, two secondary schools, four libraries, nearly a dozen health centers serving primarily underprivileged children and elderly residents, and dozens of other active social services and neighborhood-serving organizations. The nonprofit *Fundació Tot Raval* is a prominent voice on social issues in the neighborhood and claims to bring together more than 60 community groups, many of them geared towards immigrants such as the Islamic Cultural Council of Catalunya (Fundació Tot Raval 2017).

These observations in *el Raval* tend to mirror those in other arrival neighborhoods. Price et al. (2011) show that a mixed-use concentration of commercial and residential spaces in such neighborhoods increases socialization opportunities, and high population densities tend to encourage exchanges among neighbors due to physical propinquity. Immigrant communities' reliance on dense and diverse institutional networks can also "constitute an important resource for mobilizing and sustaining place and civic engagement," a critical mechanism for immigrant civic integration (Price et al. 2011, 199; see also Mohan & Mohan 2002). In his study of a Dutch arrival neighborhood, Saunders (2010, 240) maintains that "neighborhoods that work best as urban neighborhoods and arrival cities – two- to five-story structures with direct access to the road and small businesses below – tend to be very high density."

Thus, urban designers and planners of multicultural neighborhoods can support the potential for socialization opportunities across racial and ethnic lines by requiring higher minimum residential densities, lower maximum street widths, and mixed-use ground-floor retail, all well-known examples of design standards that encourage social mixing.

Public space

Much of life in *el Raval* is lived outdoors, due to the mild Mediterranean climate, the high number of small apartments with no outdoor space, and, as one long-time resident claimed, a deeply community-oriented culture. Public spaces are busy at all hours, and buskers, touts, and late-night beer sellers are regular appearances in the neighborhood's crowded streets, alleys, and plazas.

The neighborhood's largest and most significant public spaces include the *Boqueria* market, as much a tourist attraction now as a fresh food market; the *Rambla del Raval*, a large linear pedestrian space that resulted from the demolition of five blocks of residences in the late 1990s; and various large public spaces around the *Casa de Caritat* cultural complex that includes the Museu d'Art Contemporani de Barcelona (MACBA) and its *Plaça dels Àngels* forecourt, which attracts skateboarders from all over the world. These last two spaces were proposed by the *Ajuntament* as strategies to "increase light and air" in the dense neighborhood, but many believe the real intent was to increase tourism and reduce the perception of disorder in what had been a "no go" neighborhood for so many years (Degen 2017). Former Head of Planning Oriol Bohigas developed a philosophy of targeted interventions in the neighborhood called *esponjamiento* (literally "hollowing out"), which involved demolishing dense, decaying parts of the urban fabric and rebuilding them in order to "bestow dignity on the area and its residents" (Arbaci & Tapada-Berteli 2012, 293). Invoking a metaphor of colonialism, Degen (2003, 873) argues that these public space interventions were meant to populate *el Raval* with outsiders and "civilize the existing population with their new practices and values." Indeed, the immigrant population grew from 23.1% to 42.8% between 2000 and 2004, and these popular public spaces are now successful tourist destinations partly responsible for attracting 18.8 million visitors to the neighborhood each year (Rius Ulldemolins 2014).

Iconic spaces such as these ideally serve a *bridging* function, bringing together socially heterogeneous groups to exchange information and ideas and build consensus among them (Putnam 2000). For example, the *Rambla del Raval* was built to function as a meeting point for neighborhood residents and to host the many cultural festivals put on by community organizations and associations (Horta 2010). The plaza is ringed by several *Halal* butchers and a half dozen kebab shops, many serving late-night tourists heading home after a night at one of dozens of neighborhood bars and clubs.

In contrast, the *Plaça dels Àngels* at MACBA (Figure 9.2) plays a bridging role. According to Degen (2017):

> In the morning dog-walkers let their pets defecate on the square. Groups of homeless [people] set up their belongings for the day and start to play cards, chess or even on their Playstation. I see how the private security guard of the museum greets them. The homeless [people] respond and they have a brief chat. Later, at around midday and early evening the square is buzzing. Young Filipino children are cycling around, some training with cross bikes; groups of skateboarders take over the ramps of the museum – their rattling rhythms mixing with the chattering voices from young tourists and locals that gather with sandwiches and beers.

Figure 9.2 Plaça dels Àngels.

Source: Author

Immigrant civic integration stems partly from interactions between families, friends, and strangers, and the stage for such activities is public space. As Sennett (2009, n.p.) notes:

> Gathering together strangers enables certain kinds of activities which cannot happen, or do not happen as well, in the intimate private realm. . . . In public, people can discuss and debate with people who may not share the same assumptions or the same interests. This promise of turning a fresh personal page among strangers has lured many migrants to cities.

Many of the neighborhood's smaller, local-serving spaces tend to function as *bonding* spaces, where people from socially homogeneous groups band together to share common concerns, claims, and interests (Putnam 2000). These simple spaces often lack ornamentation and are adapted by different groups for different purposes. *Plaça del Pedro* and its surrounding streets, where this chapter began, is the site of several bustling immigrant-run produce markets. Tourists, expats, and their children are the primary users of nearby *Plaça de Vicens Martorell*, which contains a popular children's playground and is located on the far eastern edge of the neighborhood abutting *Les Rambles*. In the northern reaches of the neighborhood, the *Plaça de Castella* is full day and night with large groups of students from the *Universitat de Barcelona*. And at the foot of the notorious red light district sits *Plaça de Salvador Segui*, a space filled with parents watching their children on the playground, filmgoers catching a show at the new *Filmoteca*, and prostitutes and drug dealers still plying their trade in full view of the police who park their cars in the *plaça*.

Tertiary spaces include some of the everyday spaces of interaction, such as streets and street corners, sidewalks (Loukaitou-Sideris & Ehrenfeucht 2011), and other "sites of banal transgression" such as supermarkets, community gardens, recreation centers, libraries, and childcare centers (Amin 2002). One of the most vibrant of such spaces is *Carrer de Joaquín Costa*, a multi-ethnic haberdashery of butcheries, hipster bars, thrift stores, spice shops, produce markets, and hair salons serving the broader neighborhood and city (Figure 9.3).

Figure 9.3 Carrer de Joaquín Costa.

Source: oh-barcelona.com via flickr.com/CC by 2.0

In addition to interaction, a critical component of immigrant civic integration is civic engagement, or "the involvement in communal activities that have some purpose or benefit beyond a single individual or family's self-interest" (Ramakrishnan & Bloemraad 2008, 16). In *el Raval*, civic events nearly always take place in outdoor public space, especially in the *Plaça dels Àngels*, the *Plaça de Joan Coramines*, or the *Rambla del Raval*, each of which hosts popular multicultural festivals, performances, parades, and concerts throughout the year (Ortiz et al. 2004; Fundació Tot Raval 2017).

It is worth noting that much of *el Raval*'s political mobilization not only takes place *in* public space but is a result of conflict *over* public space. Redevelopment projects in the 1990s and 2000s together resulted in the displacement of around 10,000 long-time residents, and dozens of scholarly articles, books, reports, and popular editorials chronicle the plight of the neighborhood and the deployment of cultural regeneration strategies to "clean it up" by opening it to outsiders (Fernández 2014). These projects have resulted in deep conflicts over the meaning, design, function, and identity of the neighborhood (Subirats & Rius 2006). In this way, public space has fostered the very political mobilization that it now hosts.

Designers seeking to support multicultural neighborhoods can take several lessons from the public spaces of *el Raval*. First, they can advocate for the development of primary, secondary, and tertiary spaces serving different kinds of populations, rather than focus only on the production of large, iconic public spaces. Infrequent contact across diverse groups is unlikely to increase respect and empathy across racial or class differences. Indeed, Amin (2002, 967) argues that there are dangers in reifying co-presence and that the power of visibility and encounter between strangers has been overstated: "the depressing reality is that in contemporary life, urban public spaces are often territorialized by particular groups . . . or they are spaces of transit with very little contact between strangers." Sandercock (2003), Fincher and Iveson (2008), and Valentine (2008) agree that designers and planners should focus their energies not only on large public spaces but also on schools, libraries, supermarkets, sports clubs, community gardens, and other everyday spaces that can serve for social contact and encounter, and sociocultural exchanges.

However, designers must recognize that there is a natural inclination, especially among immigrant groups, towards social group differentiation, something that Young (1990, 47) sees as "both an inevitable and a desirable aspect of modern social processes." Anguelovski (2014) shows that small, homogeneous bonding spaces can help strengthen the formation of racial, socioeconomic, and political identities and help protect against discrimination, persecution, and retribution in the broader public realm. These important spaces can become home to what Nancy Fraser (1990, 67) calls *subaltern counterpublics*, or places where "members of subordinated social groups [like immigrants] invent and circulate counter discourses to formulate oppositional interpretations of their identities, interests, and needs." In *el Raval*, designers can recognize the importance of providing stages for both bridging and bonding to occur, from the iconic spaces of spectacle and protest to the local card clubs and produce stands.

Urban design professionals can achieve these goals not by over-regulating design standards and requirements, but by providing greater flexibility in how spaces are designed and managed. Creating simpler, more adaptable, more flexible, and "looser" spaces that can be appropriated by different groups for alternative uses can make them more sustainable, inclusive, and "open-minded" (Franck & Stevens 2006; Walzer 1986).

Accessibility

Residents of arrival neighborhoods must be connected in real and perceived ways to mainstream society in order to access economic opportunities and foster civic and political engagement. Accessibility can be thought of in both geographic and economic terms. The location of *el Raval* near the city's port made it Barcelona's original arrival neighborhood and the first port of entry for immigrants (Degen 2017, 145). The neighborhood is also adjacent to *Les Rambles* and the *Barri Gotíc*, so tourists stay in and traverse the district regularly, although mostly along its main corridors. This strategic location, Valenzuela-Garcia et al. (2017) argue, allows immigrant entrepreneurs to market their products to a broader clientele rather than only to their own ethnic community.

Proximity to a strong tourist- and service-based economy is critical for an arrival neighborhood like *el Raval*. Compare this to the situation in many immigrant neighborhoods located on the outskirts of European or American cities, many along poorly serviced transit routes. Saunders (2010, 290) quotes a resident of one such place that "only after a few weeks in Slotervaart [Amsterdam], I knew there was something seriously wrong. It had become a dumping ground for migrants, cut off from everything." Many recent immigrants are increasingly settling in disconnected and outlying suburban contexts in Europe and the United States, some by choice, but others due to "back to the city" movements that bring middle- and upper-class gentrifiers back to the center, while pricing out lower-income, immigrant residents. In this context, urban designers and planners must work hard to address the long-standing challenge of spatial mismatch and better connect displaced residents to socioeconomic opportunity.

Economically, *el Raval* also provides key accessibility advantages for recent immigrants and refugees from the Global South – who have historically represented around three of every four immigrants to the neighborhood – due to lower rents there than in comparable areas of the city, although these too have risen in recent years (Ajuntament de Barcelona Department d'Estadística 2017; Subirats 2006). Median rent in 2016 was €664 ($800), or around 24% lower than the €874 ($1050) average just across *Les Rambles* in the *Barri Gotíc* (Ajuntament de Barcelona Department d'Estadística 2017). This brings up a key challenge for arrival neighborhoods: they must not become too successful lest they risk gentrification and ultimately displacement of long-time residents.

As such, designers can promote what Curran and Hamilton (2012, 2018) have coined "just green enough" interventions, or initiatives that are small scale, neighborhood-driven, and accompanied by anti-gentrification protections, as opposed to grand civic projects that geographically concentrate resources, use "conventional urban design formulae," and can spur gentrification (Wolch et al. 2014, 241). Indeed, some of the most thriving arrival neighborhoods have high poverty rates, because if residents "are able to leave within a generation for more prosperous middle-class homeowner districts, the neighborhood will be constantly refilled with new migrants from poor rural regions" (Saunders 2010, 315). In the process, urban designers should promote urban infill projects that foster the connectivity and density that *el Raval* offers and encourage the clustering of residential and commercial uses in existing immigrant-rich pockets while seeking to blur neighborhood edges and boundaries to encourage integration across ethnic lines (Ellin 2006; Young 2000).

Conclusion

This chapter explored the role that urban designers and planners can play in shaping immigrant integration in arrival neighborhoods. I used the case of Barcelona's *el Raval* and found that spatial concentration, public space, and accessibility – three concepts familiar to urban designers – can promote positive contexts of reception for immigrants, especially as they foster social interaction, access to opportunity, and civic and political engagement.

I provide a number of recommendations for urban designers seeking to foster immigrant integration in arrival neighborhoods. These include supporting higher residential and commercial densities, a mix of uses, and a pedestrian-oriented public realm. I argue that immigrant integration in *el Raval* relies heavily on the availability of a mix of public spaces that serve different populations, with some appropriated by recent immigrants and others by long-time residents or tourists. This confluence of bridging and bonding spaces affirms the value of frequent, respectful, and unmediated engagement with "the Other" but also celebrates natural affinities towards group-based differentiation and clustering and the maintenance of cultural heritage and practices (Young 2000). Urban designers can also learn from the many simple spaces around *el Raval* that are flexible and adaptable enough to be adopted by different groups at different times for different purposes. Urban design professionals should focus their energy not only on large iconic projects but also on the micro-spaces of everyday encounter and the smaller-scale, scattered, "just green enough" interventions that are less likely to induce gentrification and displacement of immigrant populations. Urban designers can also support immigrant integration by concentrating retail and residential uses in ethnic clusters but blurring boundaries and promoting more porous and permeable edges. Resource and population concentration is perhaps even more vital in the less-dense contexts where so many newcomers to Global North cities are now settling due to the suburbanization of immigration (Lung-Amam 2017).

This chapter is not about how well *el Raval* works as an arrival neighborhood. Some features I discuss might work well for immigrants but create resentment by long-time residents who feel they have lost control of their old neighborhood (Ortiz et al. 2004). In addition, the neighborhood is still struggling with issues of displacement, gentrification, discrimination, and persistent poverty. So, while in recent years, *el Raval* "certainly got a facelift and new city and community infrastructures, the problems that the population were suffering: health, education, and employment were not attended to . . . [and] life expectancy is five years less than in the rest of the city" (Degen 2017, 150). Instead, my intent in this chapter is to develop a framework for thinking through the relationship between the built environment and immigrant integration.

Exploratory work like this inevitably presents many challenges. With the exception of several semi-structured interviews, I had to rely primarily on nonsystematic observations. Such observations do not allow for direct transfer of prescriptions to other contexts. Nevertheless, the historical trajectory, political-economic forces and imperatives, and property development regimes shaping *el Raval* are not dissimilar to those facing arrival neighborhoods in global cities such as Los Angeles, Paris, or Berlin. In addition, my focus on the neighborhood scale is based on the notion that although it is "the most meaningful scale at which individuals participate, belong, and believe (or not) in cities, nations, and globally . . . there is little empirical information with respect to contemporary immigrant dynamics" (Price et al. 2011). Nonetheless, one might argue that we live in "spaces of flows," and just because people live or work in a certain neighborhood does not mean their lives are bound up within that space or that they derive satisfaction, opportunity, or meaning from neighborhood experiences alone (Castells 1989). Finally, and perhaps most importantly, immigrants are much more than this singular label ascribes. A more intersectional account of immigrant life is outside the scope of this chapter but is imperative to move beyond one-dimensional accounts of the human experience. If we believe that place matters in shaping arrival neighborhoods, then urban designers and planners must recognize the critical role they play in the lives of immigrants.

References

Ajuntament de Barcelona Department d'Estadística. (2017). Visualitzador de dades estadístiques. Retrieved from http://estadistica.bcn.cat

Amin, A. (2002). Ethnicity and the Multicultural City: Living with Diversity. *Environment and Planning A*, *34*: 959–980.

Anguelovski, I. (2014). *Neighborhood Reconstruction, Place Remaking, and Environmental Justice in the City.* Cambridge, MA: MIT Press.

Arbaci, S., and Tapada-Berteli, T. (2012). "Social Inequality and Urban Regeneration in Barcelona City Centre: Reconsidering Success." *European Urban and Regional Studies, 19*: 287–311.

Bader, M. D. M. (2016). *Diversity in the D.C. Area: Findings from the 2016 D.C. Area Survey.* Washington, DC: Metropolitan Policy Center and the Center for Latin American and Latino Studies, American University.

Bissoondath, N. (1994). *Selling Illusions: The Cult of Multiculturalism in Canada.* Toronto: Garamond Press.

Bloemraad, I. (2006). *Becoming a Citizen: Incorporating Immigrants and Refugees in the United States and Canada.* Berkeley, CA: University of California Press.

Bloemraad, I., Korteweg, A., and Yurdakul, G. (2008). "Citizenship and Immigration: Multiculturalism, Assimilation, and Challenges to the Nation-state." *Annual Review of Sociology, 34*: 153–179.

Castells, M. (1989). *The Informational City: Information Technology, Economic Restructuring, and the Urban-regional Process.* Oxford: Basil Blackwell.

Castillo, L., & Caver, K. (2009). "Expanding the Concept of Acculturation in Mexican American Rehabilitation Psychology Research and Practice." *Rehabilitation Psychology, 54*: 351–362.

Curran, W., and Hamilton, T. (2012). "Just Green Enough: Contesting Environmental Gentrification in Greenpoint, Brooklyn." *Local Environment, 17*: 1027–1042.

Curran, W., and Hamilton, T. (eds.). (2018). *Just Green Enough: Urban Development and Environmental Gentrification.* New York: Routledge.

Degen, M. (2003). "Fighting for the Global Catwalk: Formalizing Public Life in Castlefield (Manchester) and Diluting Public Life in el Raval (Barcelona)." *International Journal of Urban and Regional Research, 2*: 867–880.

Degen, M. (2017). "Urban Regeneration and "Resistance of Place": Foregrounding Time and Experience." *Space and Culture, 20*: 141–155.

Ellin, N. (2006). *Integral Urbanism.* New York: Routledge.

Fernández, M. (2014). *Matar al Chino: Entre la Revolución Urbanística y el Asedio Urbano en el Barrio del Raval de Barcelona.* Barcelona: Virus Editorial.

Fincher, R., and Iveson, K. (2008). *Planning and Diversity in the City: Redistribution, Recognition, and Encounter.* London: Macmillan.

Franck, K., and Stevens, Q. (2006). *Loose Space: Possibility and Diversity in Urban Life.* New York: Routledge.

Fraser, N. (1990). "Rethinking the Public Sphere: A Contribution to the Critique of Actually Existing Democracy." *Social Text, 25–26*: 56–80.

Fundació Tot Raval. (2017). *Fundació Tot Raval.* Retrieved from http://totraval.org/

Güell, B. (2016). "The Backstage of Pakistani Businesses in Barcelona: Unravelling Strategies from the Ground." *South Asian Diaspora, 8*: 15–30.

Horta, G. (2010). *Rambla del Raval de Barcelona.* Barcelona: El Viejo Topo.

Loukaitou-Sideris, A., and Ehrenfeucht, R. (2011). *Sidewalks: Conflict and Negotiation Over Public Space.* Cambridge, MA: MIT Press.

Lung-Amam, W. (2017). *Trespassers? Asian Americans and the Battle for Suburbia.* Berkeley, CA: University of California Press.

Mohan, G., and Mohan, J. (2002). "Placing Social Capital." *Progress in Human Geography, 26*: 191–210.

Molina, J., Valenzuela-Garcia, H., García-Macías, A., Lubbers, M., and Pamaplona, J. (2015). "Social Capital in Ethnic Enclaves: Indians in Lloret de Mar and Pakistanis in Barcelona." In Y. Li (Ed.), *The Handbook of Research Methods and Applications on Social Capital.* Cheltenham, UK: Edward Elgar Publishing, 377–398.

National Academies of Sciences, Engineering, and Medicine. (2015). "The Integration of Immigrants into American Society: Panel on the Integration of Immigrants into American Society." In Waters, M.C. and Pineau, M.G. (eds.), *Committee on Population, Division of Behavioral and Social Sciences and Education.* Washington, DC: The National Academies Press.

Ortiz, A., Garcia-Ramon, M., and Prats, M. (2004). "Women's Use of Public Space and Sense of Place in the Raval (Barcelona)." *GeoJournal, 61*, 219–227.

Park, R. (1952). *Human Communities: The City and Human Ecology.* Glencoe, IL: The Free Press.

Popay, J., Escorel, S., and Hernandez, M. (2008). *Understanding and Tackling Social Exclusion: Final Report to the WHO Commission on Social Determinants of Health.* Social Exclusion Knowledge Network.

Portes, A. and Manning, R. (1986). "The Immigrant Enclave: Theory and Empirical Examples." In J. Nagel and S. Olzak (eds.), *Ethnicity: Structure and Process.* New York: Academic Press, 47-68.

Portes, A. and Rumbaut, R. (2014). *Immigrant America: A Portrait.* Berkeley, CA: University of California Press.
Portes, A., and Zhou, M. (1992). "Gaining the Upper Hand: Economic Mobility Among Immigrant and Domestic Minorities." *Ethnic and Racial Studies, 15*: 491–522.
Price, P., Lukinbeal, C., Gioioso, R., Arreola, D., Fernández, D., Ready, T., and de los Angeles Torres, M. (2011). "Placing Latino Civic Engagement." *Urban Geography, 32*: 179–207.
Putnam, R. (2000). *Bowling Alone: The Collapse and Revival of American Community.* New York: Simon & Schuster.
Ramakrishnan, S. and Bloemraad, I. (2008). *Civic Hopes and Political Realities: Immigrants, Community Organizations, and Political Engagement.* New York: Russell Sage Foundation.
Rius Ulldemolins, J. (2014). "Culture and Authenticity in Urban Regeneration Processes: Place Branding in Central Barcelona." *Urban Studies, 51*: 3026–3045.
Sandercock, L. (2003). *Cosmopolis II: Mongrel Cities of the 21st Century.* London: Continuum.
Saunders, D. (2010). *Arrival City: How the Largest Migration in History Is Reshaping Our World.* New York: Pantheon Books.
Schmid, K., Al Ramiah, A., and Hewstone, M. (2014). "Neighborhood Ethnic Diversity and Trust: The Role of Intergroup Contact and Perceived Threat." *Psychological Science, 25*: 665–674.
Schwartz, S., Unger, J., Lorenzo-Blanco, E., Des Rosiers, S., Villamar, J., Soto, D., Pattarroyo, D., Baezconde-Garbanti, M., and Lourdes Szapocznik, J. (2014). 'Perceived Context of Reception Among Recent Hispanic Immigrants: Conceptualization, Instrument Development, and Preliminary Validation.' *Cultural Diversity and Ethnic Minority Psychology, 20*: 1–15.
Sennett, R. (2009). *Quant: The Public Realm.* Retrieved from richardsennett.com.
Subirats, J., and Rius, J. (2006). *From the Xino to the Raval: Culture and Social Transformation in Central Barcelona.* Barcelona: Centre de Cultura Contemporània de Barcelona.
Talen, E. (2006). *Design for Diversity: Exploring Socially Mixed Neighborhoods.* Oxford, UK: Architectural Press.
Tasan-Kok, T., van Kempen, R., Raco, M., and Bolt, G. (2013). *Towards Hyper-diversified European Cities: A Critical Literature Review.* Utrecht: Utrecht University, Faculty of Geosciences. Retrieved from www.urbandivercities.eu/wp-content/uploads/2013/05/20140121_Towards_HyperDiversified_European_Cities.pdf
UN-Habitat. (2017). *Data Explorer.* Retrieved from http://urbandata.unhabitat.org/explore-data/
Valentine, G. (2008). "Living with Difference: Reflections on Geographies of Encounter." *Progress in Human Geography, 32*: 323–337.
Valenzuela-Garcia, H., Parella, S., and Güell, B. (2017). "Revisiting the 'Ethnic Enclave Economy': Resilient Adaptation of Small Businesses in Times of Crisis in Spain." *International Journal of Anthropology and Ethnology, 1*: 5.
Walzer, M. (1986). "Public Space: Pleasures and Costs of Urbanity." *Dissent, 33*: 470–475.
Wilson, K., and Portes, A. (1980). "Immigrant Enclaves: An Analysis of the Labour Market Experiences of Cubans in Miami." *American Journal of Sociology, 86*: 295–319.
Wolch, J., Byrne, J., and Newell, J. (2014). "Urban Green Space, Public Health, and Environmental Justice: The Challenge of Making Cities 'Just Green Enough'." *Landscape and Urban Planning, 125*: 234–244.
Yin, R. (2009). *Case Study Research: Design and Methods.* Thousand Oaks, CA: Sage Publications, Inc.
Young, I. (1990). *Justice and the Politics of Difference.* Princeton, NJ: Princeton University Press.
Young, I. (2000). *Inclusion and Democracy.* Princeton, NJ: Princeton University Press.
Zolberg, A. and Long, L. (1999). "Why Islam Is Like Spanish: Cultural Incorporation in Europe and the United States." *Politics and Society, 27*: 5–38.

Further reading

Degen, M. (2008). *Sensing Cities: Regenerating Public Life in Barcelona and Manchester.* New York: Routledge. Details ethnographic research conducted in *el Raval* for many years.

Saunders, D. (2010). *Arrival City: How the Largest Migration in History Is Reshaping Our World.* New York: Pantheon Books. A fascinating tour of dozens of arrival cities and neighborhoods told through the eyes of the people who live, work, and play there.

Subirats, J. and Rius, J. (2006). *From the Xino to the Raval: Culture and Social Transformation in Central Barcelona.* Barcelona: Centre de Cultura Contemporània de Barcelona. One of the most extensive English-language reports to provide empirical insights into the recent changes occurring in *el Raval.*

Valentine, G. (2008). "Living with Difference: Reflections on Geographies of Encounter." *Progress in Human Geography, 32*: 323–337. A literature review that provides valuable empirical evidence on the potentialities and pitfalls of encounter in public space.

PART II.II

Informality

Introduction

Urban design has always been an instrument for the transformation, revitalization, and marketing of real estate in prime locations of the formal city. In contrast, the spaces of informal economy remain beyond the ken of urban design practice. Many urban designers may assume that informal activities require policy rather than design intervention; while others may regard informality as a phenomenon marginal at best and illegal at worst, and not relevant to the scope of their work or expertise (Loukaitou-Sideris and Mukhija 2015). However, in the last decade, many scholars have argued that far from being marginal, informal urbanism is quite omnipresent and visible in many cities, and it is not only characteristic of the Global South, but also present in the Global North. Informal activities and settings may differ from city to city, but far from being distinct and rigidly separated, informal spaces are often extensions of or even linked to formal ones (Roy 2005; Mukhija and Loukaitou-Sideris 2014). It then behooves urban designers to include informal urbanism within their scope of urban interventions. But how? This is the central question that unites the chapters in this section.

For Kim Dovey, who defines urban informality as "a mode of production of public space and therefore a form of urban design," urban designers must start their inquiry by first getting a deeper understanding of informal settlements. This exploration should involve three dimensions: (1) *location*, or why informal settlements and activities are located where they are; (2) *morphogenesis*, or how the architecture and urban design of these settlements co-evolves; and (3) *image and identity*, or how these settlements are perceived by insiders and outsiders, and how these perceptions and images motivate positive (upgrading) or negative (eradication) intervention. Dovey calls for a new direction of urban design thinking, one that follows an "assemblage approach" that examines the interconnections, conflicts, and synergies between formal and informal practices and spaces, and which views not only the deficits of informality but also its assets, or, in his words "capacities that enable residents to live a life, acquire an income, raise their children and build prospects for a better future." Thus, urban designers seeking to upgrade or formalize informal settings should not take away but rather build on their benefits of location (that allows them better connections to jobs) and opportunities for adaptation (that enables the accommodation of extended families) – and thus realize their occupants' right to the city.

Illustrating that informality can be literally an extension of formal settlement, Annette Kim finds thousands of informal housing units hidden in plain sight: In Beijing, they occupy the basements of formal housing, some of which served as the city's former air defense shelters. In Shanghai, formal apartment housing is being converted informally into group housing. While the condition of such housing may be characterized as subpar, because of lack of light and air and overcrowding, the central location and affordability of these units makes them desirable. To respond and better accommodate informal urbanism, Kim views the role of urban design as "reforming the city building process" in ways that help achieve spatial inclusion. She sees this as happening in two interrelated ways: (1) *reframing urban narratives* to include informality as a legitimate focus of urban design; and (2) *rewriting property rights* by redrafting building codes and regulations, urban design guidelines, and taxation for value capture finance with the goal of making informal settings better connected, with better accommodation and services, and safer, more livable, and less stigmatized.

In yet another evidence of the coexistence and interrelationship between formality and informality, Su-Jan Yeo examines informal urban practices during nighttime in one of the most highly regulated urban spaces in the world – the city-state of Singapore. She finds that even in such a context, nighttime often allows the possibility of relaxation of formal social norms and accommodation of informal spatial practices, which are often forbidden during the formal working hours. Night markets in a number of East Asian cities are prime examples of how the formal streetscape in some neighborhoods is transformed in the late evening and night hours by street vendors, tourists, and city residents, who interact over street food and commerce. Yeo examines Holland Village, a mixed-use residential and commercial neighborhood in Singapore that has become a night favorite for locals and tourists of different incomes, who come to eat, stroll, and spend some time there. The popularity of the neighborhood is behind the impetus for its redevelopment and possible gentrification, and Yeo wonders how the enduring qualities and inclusivity of the neighborhood can be safeguarded in its redevelopment. She urges multiple economic strategies, such as: supporting small businesses that offer low-cost goods and services; spatial strategies that soften the boundaries between formal and informal spaces; and temporal strategies that take advantage of the permissiveness of nighttime to foster a more inclusive and vibrant public realm.

Manish Chalana uses India, a country where informal urbanism is extremely prominent, to continue the investigation about the role of urban design in responding to informality. He is critical of the planners' various efforts to formalize and modernize informal settlements. Such efforts have often drawn from a vision of "high modernism" to impose an alien and culturally insensitive spatial form on local residents, and have typically sought to displace and scatter them to cheaper sites on the urban periphery (see also Mahadevia's account in Chapter 23). Using as case studies the informal settlements of Dharavi in Mumbai and Kathputli in Delhi, Chalana also observes their assets: These two informal neighborhoods manage, despite their infrastructural shortcomings, to "serve residents' needs in myriad ways that the proposed redevelopment plans are unlikely to re-create." Chalana calls for "ordinary design" interventions that recognize the specific needs of residents and work with them to instigate spatial improvements that can make their lives easier. But, as the aftermath of this chapter poignantly demonstrates, power and profit, rather than noble design intentions, have unfortunately determined the fate of one of the settlements, razing Kathputli Colony to the ground and pushing its residents to a camp.

Such a form of "unsettledness," observes Rachel Berney in the next chapter, is unfortunately a prominent characteristic of contemporary urbanism. This "urbanism of exception" has become not exceptional at all in cities globally, as homeless denizens and refugees try to survive in not just homeless and refugee camps, but also on city streets, under freeways, and on public lands. What should be the urban design response to such unsettledness and "temporary" settlement? Berney urges urban designers to acknowledge the reality of "temporary spaces that are no longer temporary" and to seek to plan and design them in ways that do not perpetuate stigmatization and "othering," but instead understand and respond to the daily needs of the people who have no other option but to live in them. She suggests the use of public lands to host temporary housing and the creation of collective infrastructures in cities that can help mitigate the plight of unsettledness.

References

Loukaitou-Sideris, A. and Mukhija, V. (2015). "Responding to Informality Through Urban Design Studio Pedagogy." *Journal of Urban Design*. 21(5): 577–595.

Mukhija, V. and Loukaitou-Sideris, A. (Eds.) (2014). *The Informal American City: Beyond Taco Trucks and Day Labor*. Cambridge, MA: MIT Press.

Roy, A. (2005). "Urban Informality: Toward an Epistemology of Planning." *Journal of American Planning Association*. 71(2): 147–158.

10

INFORMAL SETTLEMENT AS A MODE OF PRODUCTION

Kim Dovey

Introduction

Much has been made of the fact that most of the global population is now urban; cities are primary centers of economic opportunity, and rural-to-urban migration continues unabated. It is not so often noted, however, that most of this new urban population has been accommodated through the expansion of informal settlements in cities of the Global South. Under circumstances where the urban poor cannot secure access to formal housing, they settle wherever they can in the interstices and on the margins of the formal city. Such settlements are never informal in every way and are always enmeshed in complex relations with the formal structures of state planning. Depending on how such settlements are defined, they now house over a billion people.

The disciplines of architecture, urban design, and planning have traditionally focused on the formal order of the city – on formal protocols, procedures, plans, and design outcomes through which the city is transformed and governed from the top down. The challenge of engagement with urban informality requires fundamentally different modes of thought and practice. While the construction, services, and open space are often of low quality (by formal design criteria), and occupation may be unauthorized, we need to move beyond the labeling of informal settlements as "slums" or "squatter" settlements. Urban informality is a mode of production of public space, and therefore a form of urban design. In general terms, informal settlement involves self-organized production of affordable housing for a labor force that is otherwise unable to gain access to urban housing markets. While informal settlements are marginalized in many ways, they are not marginal to the cities that have attracted them (Perlman 1976). The most realistic future is one of *in situ* upgrading, utilizing the very self-organizing forces that produce informal settlement in the first place.

Some important distinctions need to be made between "informality" as a mode of production, the "slum" as a particular urban condition, and "squatting" as a particular kind of land tenure. A quick analysis of Google Ngrams will show that the use of the terms "slum" and "squatter" have declined over recent decades, while the use of "informal settlement" has increased, often as a replacement. A "slum" is defined by the UN in terms of a range of housing conditions, such as overcrowding and low-quality construction, generally accompanied by a lack of services, sanitation, and open space (UN-Habitat 2006). Yet "slum" is a "slur" word (like "slob," "slut," "sly" and "slump") with a history that hinders any accurate understanding of living conditions (Mayne

2017). As Mayne argues, it evokes a stereotype that obscures a diversity of living conditions along with the energy and resilience of residents. It suggests a neighborhood in a parasitic and illicit relationship with a host city; and the language of the slum is used to justify coercive action such as eviction and demolition. While there are good reasons to avoid the use of the term "slum," "informal settlement" is not a synonym – informality is not a condition, but a process.

The term "squatter" also needs to be used carefully, less because it is derogatory than because it is rarely used accurately. "Squatting" invokes a binary division between ownership and unauthorized occupation, yet most informal settlements fall into the ambiguous zone between these categories. Illegal subdivisions that take place on private land with the collusion of the owners are often called "pirate" developments, where residents "own" the land without a formal title. Squatters on public land often acquire some form of de facto tenure after a period of occupation – the right known as "squatter's rights" or "adverse possession."

Informal settlement, then, is a set of practices that need to be defined and understood separately from "slum" conditions and "squatting" practices. Having drawn these distinctions, however, there are overlaps and ambiguities. Many informal settlements are unhealthy, overcrowded, and poorly serviced with insecure tenure. Many are vulnerable to forced displacement at the hands of a state determined to "cleanse," modernize, and formalize the city, often in league with the interests of capital to raise land values and claim key urban sites for redevelopment.

In contrast to the stereotypes of "slums" and "squatting," I want to suggest that we understand informal settlement as a mode of production. Here, I return to the original definition of the informal sector as that part of economic production that falls outside the purview of the state and is not captured in measures such as GDP (Hart 1973). Informal settlement includes all forms of urban development that emerge outside the formal controls and regulations of the state. Thus, informality is largely defined by its relations to formality – its urban design is neither organized nor controlled by the state. My focus here is limited to the self-organized production of urban space and excludes those forms of industrial production that can take place within informal settlements (Brugmann 2009).

The phrase "mode of production" has a particular history in Marxist thought, where it refers to the ways economic production is organized, incorporating both labor and the means of production (capital). The capitalist mode of production is one where a ruling class owns the means of production; the current mode of neoliberal capitalism involves an alliance of the state with the interests of that class (Harvey 2007). For Davis (2006: 17), the policies produced by neoliberal capitalism are "an inevitable recipe for the mass production of slums." While informal settlement is insinuated into this formal framework and can work in synergy, it involves a very different mode of organization – it is self-organized, and the means of production are primarily controlled by residents. Informality is not simply a pre-capitalist mode of production, but part of a complex co-dependency between formal and informal modes of production: organized capital clearly benefits from the low cost of labor that is enabled by informal settlements (Moser 1979: 1062). While politicians concerned with law and order may wish to erase such settlements, capital markets depend upon this cheap labor. Thus, informality is a mode of production that services the neoliberal state, even as it maintains autonomy and is targeted for eradication. In the case of informal settlements, what is primarily being produced is urban space.

While it is the poorer cities of the Global South that have high levels of informal settlement, informality and formality have reciprocal relations in all economies and cities. We need to understand this relation not as a binary opposition, but as an assemblage where informal and formal intersect to form complex alliances and possibilities. There is no simple transition from formal city to informal city; rather, the formal city becomes informalized as informal settlements become formalized. The primary contrast with the formal city is that informal settlements are self-organized rather than hierarchically organized, community-based, and bottom-up rather than top-down.

Assemblage thinking

I have argued elsewhere that an effective theoretical framework for understanding informal settlements, and urban design more generally, might be called a "complex adaptive assemblage approach" (Dovey 2012; 2016: 263). This approach seeks to join an assemblage approach from philosophy and social theory (Deleuze and Guattari 1988) with complex adaptive systems or resilience thinking (Walker and Salt 2006). A brief introduction to assemblage thinking will stress that an assemblage is a whole, whose properties emerge from interconnections, interactions, flows, and synergies between its different parts (DeLanda 2006). The places, buildings, neighborhoods, and identities that are the most visible parts of cities are the emergent effects of these processes. Assemblage is a way of understanding the city as produced by multiple desires and at multiple scales. It seeks to understand morphogenic processes through which forms come into being – a philosophy of "becoming" rather than a stable sense of "being" or fixed identity. Any urban morphology embodies a set of immanent capacities that are linked to the possibilities of what it might become. Assemblage is thus a way of analyzing and mapping the city that connects research to design practice and urban imagination (Dovey et al. 2018).

The concept of "assemblage" is a translation of the French word "agencement," which also means "arrangement," "alignment," and "agency." It has roots in the Foucauldian critique of power as embodied in micro-practices, as distributed rather than simply held by human agents. Assemblage thinking exploits this revolution in thinking about power for its emancipatory potential, as well as its critique of discipline and hierarchy, with a focus on power *to* (power as capacity) in addition to power *over* (power as control) (Lukes 1974). An axiom of assemblage thinking is that power is distributed and embodied in material spatial arrangements. This is not a case for environmental determinism, nor is it displacement of human agency; it is a recognition that power as capacity is produced and practiced through the materiality and spatiality of the city.

The roots of assemblage thinking in urban design can be seen in the work of Jacobs (1961) and Alexander (1996) from the mid-20th century – the value of urban density, mix, and walkable access networks, the seminal insight that "a city is not a tree." Assemblage thinking brings a focus on the key distinction between the horizontal rhizomic connectivity between parts and the vertical tree-like hierarchies that are used to organize the city. Jacobs also suggested that the then-emerging work on complexity theory held potential for a better understanding of the synergistic effects of cities that cannot be reduced to cause-effect relations. This is the work that has evolved into complex adaptive systems, emergence theory, and resilience thinking (Johnson 2001; Walker and Salt 2006) with considerable influence in urban analysis. While I have reservations about any reduction of urban design to a science (Dovey and Pafka 2016), such work has many parallels with assemblage thinking. Assemblage thinking is anti-reductionist, resisting reduction to essence, text, or materiality. An urban neighborhood is neither a thing nor a collection of things. While people, buildings, practices, streets, and so on all comprise the neighborhood, it is the assembled connections between them and to other neighborhoods that are crucial. An assemblage is dynamic – it is the flows of life, traffic, goods, and money that give a neighborhood its intensity and emergent sense of place identity. From an assemblage perspective, the city is a dynamic field of differences – the identities of places, neighborhoods, and districts emerge from this field. Finally, it can be helpful to understand assemblage thinking as a conceptual toolkit. Indeed, theory is the means rather than the end of assemblage thinking; the goal is to transform the ways we think about informal settlements.

Informal production

As a mode of production, informal settlement can be understood through three primary dimensions focused on location, morphology, and identity. Where and why do such settlements emerge, how are they formed, and what kinds of place identity are constructed? First, informal settlements produce a new kind of urban spatial structure to generate new synergies of functional mix within the wider metropolis, particularly walkable connections between housing and employment. Second, they produce the material infrastructure of buildings and street networks through informal development practices, morphogenic processes, and new forms of urban governance. Finally, they produce particular forms of place identity within the larger urban imaginary. These are complex and interlaced issues. What follows is much more an agenda for research than a definitive account.

Locating informal settlements

First is the question of where and why informal settlements emerge. The simple answer is that they are where they need to be in order to gain access to the jobs and opportunities that the city offers. Any particular answer will be mediated by the availability of land together with topography, urban morphology, politics and history, among other factors. I have previously suggested a typology that includes escarpments, waterfronts, infrastructure easements, and accretions attached to the formal architecture of the city (Dovey and King 2011). Informal settlements emerge both in the interstices of the formal city and on its fringe. Interstitial encroachments are often juxtaposed or mixed with industry, retail, recreation and transport functions.

Figure 10.1 shows a mapping of urban informality in four cities drawn from an earlier study that included 12 cities (Dovey and Kamalipour 2017). These are based on data from Google Earth and are limited to those morphologies that are visible and identifiable in aerial photographs, excluding those forms of informality that occur within the shell of existing buildings. This is, of course, a problematic task because there are different morphologies that can be identified as informal, and there is no simple continuum of greater or lesser degrees of informality. While a settlement may begin as entirely formal or informal, it often becomes mixed over time, as formal buildings are added to an informal morphology, as informal buildings are formalized, and as informal buildings encroach within a formal street plan. There are also degrees of informality of the street and laneway networks, as semi-organized encroachments produce semi-formal subdivisions. Fully informal networks also become formalized over time with laneways widened, straightened, and so on. These maps are simplified and show urban areas divided into formal (dark gray), mixed (mid gray) and informal (light gray). This distinction is exemplified in Figure 10.2 and will be discussed in the following section.

The most obvious finding here is that the areas of mixed formal and informal settlements (mid gray) are far more pervasive in most cities than the stereotypic informal morphologies characterized by labyrinthine laneways (light gray). Another is that informal settlement patterns differ markedly from city to city and are strongly mediated by topography and politics. Cities on the flat and wet coastal plains of Southeast Asia, such as Manila, Bangkok, and Jakarta, show highly distributed patterns of small settlements and extensive areas of mixed informality. Jakarta is a most difficult city in which to draw any line between formal and informal (Simone 2014). Latin American cities such as Lima, Caracas, Rio de Janeiro, and Medellín are strongly characterized by encroachment onto steep escarpments. The legacy of apartheid in South Africa ensured that informal settlements were pushed to the margins of the formal (dark gray) city, as is evident in the Cape Town settlements. Cairo is characterized by vast areas of semi-formal development

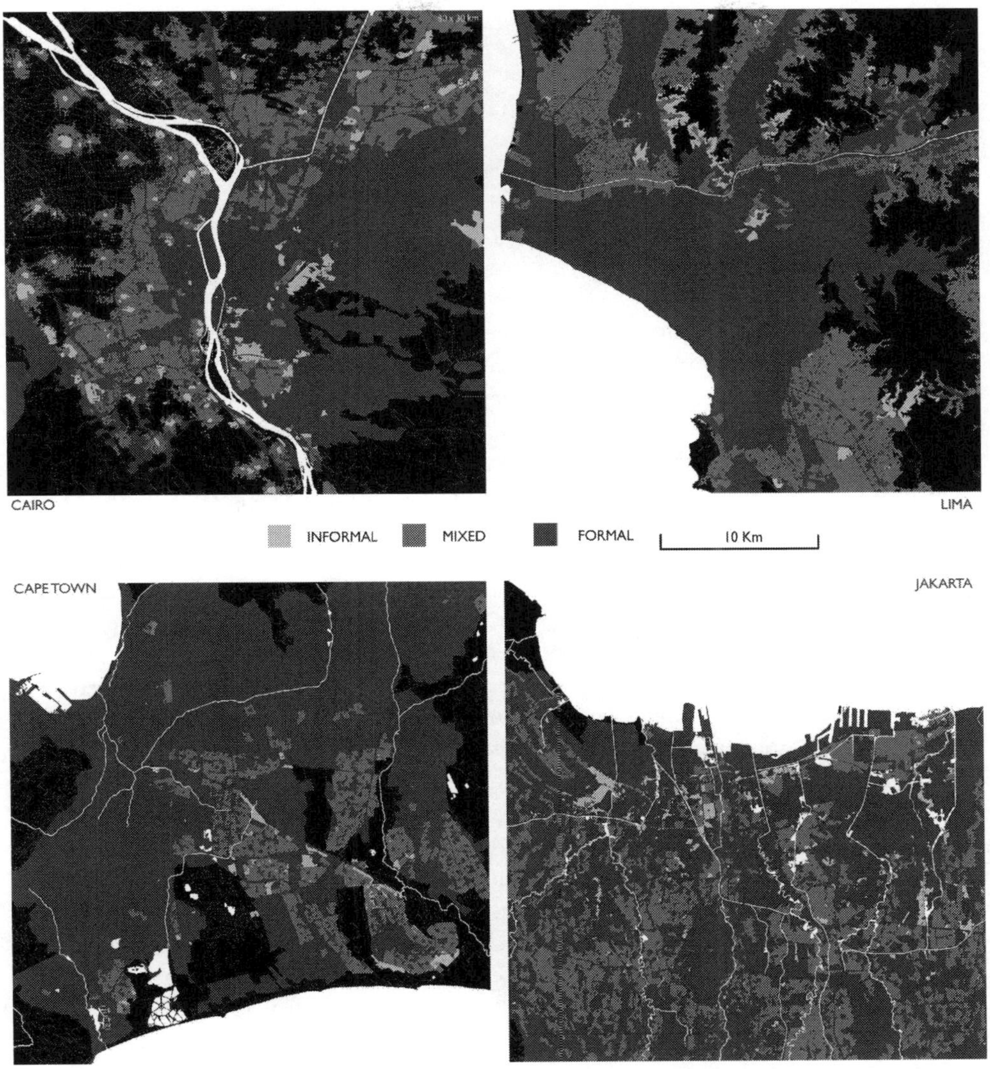

Figure 10.1 Metropolitan mapping of informal and mixed settlements in four cities.

Source: Dovey and Kamalipour (2017)

on agricultural land, ironically produced in part by the state's plan to protect such land from development (Sims 2010).

Such a mapping of informality enables a diagnosis of how such settlements integrate with the formal city, or not, and how they might be better integrated. It enables a more transparent judgment of the conditions under which they might need to be relocated for resident safety or public benefit. And it facilitates a form of anticipatory design and planning – preventing encroachment where it will be dangerous for residents or harmful to the city or environment; constructing key infrastructure in advance of settlement to render sites safe and well serviced (Gouveneur 2014).

Morphogenesis

The second key dimension of informal settlement as a mode of production lies in the processes of morphogenesis, of how built form and urban design emerge. I noted in the earlier section a key distinction in this regard between the more stereotypic informal settlements and the larger range of mixed morphologies as mapped in Figure 10.1 and illustrated in Figure 10.2. In a formal morphogenic process, the street layouts with building plots always precede the construction of buildings – urban design precedes architecture. The informal morphogenic process by contrast is relatively incremental and self-organized – the architecture and urban design co-evolve.

While there are variations of informal morphologies, the most informal of settlements have a relatively similar morphogenic process that produces a relatively consistent pattern of access networks that is evident in many very different cities, climates, and topographies (Figure 10.2, left column). The single room of a span no more than about four meters (13 feet) is the most common spatial increment that may be extended horizontally or vertically to a limit of about four rooms. Incremental construction is geared to income flows, to capacities to store materials, and to the need for building materials to be transported and assembled by hand. Many houses are in a permanent state of becoming – a yard becomes a room as walls are extended and then roofed. A roof becomes a terrace when a stairway access is added, and a roof terrace becomes another room as walls are added and roofed. Smaller increments such as walls, roofs, balconies, verandas, and stairways are deployed in a myriad of ways to add to, divide, or adapt different spaces over time.

At the same time, a public space network emerges and co-evolves with the architecture. Buildings may be constructed along existing pathways, but pathway networks also emerge incrementally between constructed buildings – a co-production of architecture and urban design. Some access ways will become more integrated with the network, and parts of houses fronting them may be converted to shops; others may be encroached and blocked, but walkable access will be maintained to all buildings. Thus, informal settlements embody an informal order or code that can be described as vernacular design (Rapoport 1988; Mayne 2017). While the morphologies we see in Figure 10.2 (left column) emerge from complex local negotiations between neighbors over issues of density and access, the consistencies between them are global. While it can be tempting to describe this as an "organic" process, these informal codes are geared to socially mediated desires for space, access, privacy, and security that produce any neighborhood. Under conditions of poverty, however, such informal codes are often insufficient, and we see the result of a nasty version of the "tragedy of the commons," where incremental overdevelopment starves the public realm of space, light, and air.

The right-hand column of Figure 10.2 shows a range of mixed morphologies that are much more varied. Here plots and streets have often been established in advance of informal construction. The informalization of the formal city generally involves the previously described incremental informal processes emerging within, around, and on top of the architecture of the formal city (Figure 10.2, Cape Town). The informal design of the street layout may be undertaken in anticipation of formal tenure and incorporation into the formal city – an attempt to shadow the forms of the formal city in terms of street widths and lot sizes. In other words, the codes that shape the formal city can also shape the informal city in an informal manner. Figure 10.3 shows the results of recent informal morphogenesis as visible from Google Earth data in four cities. The morphogenic patterns shift markedly, from the highly informal in Accra and Manila to the semi-organized (or mixed) in Lima and Delhi.

Our understanding of informal morphogenesis remains very weak, and this is a crucial field for future research. Perhaps the best insight comes from the work of Hakim (2014) on the informal codes that traditionally produced Mediterranean urbanism and became formalized over time.

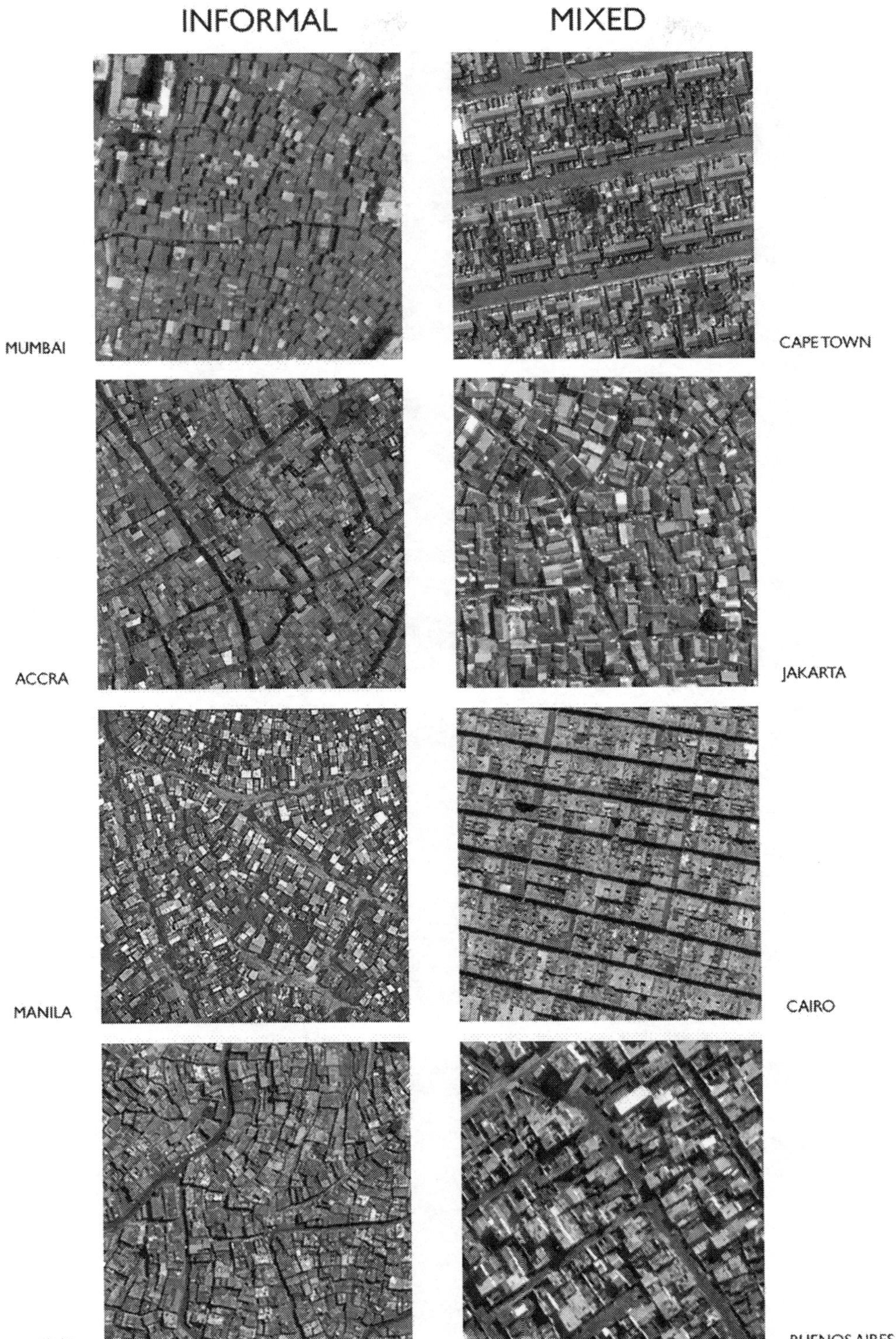

Figure 10.2 Informal and mixed morphologies.

Source: Google Earth/Digital Globe

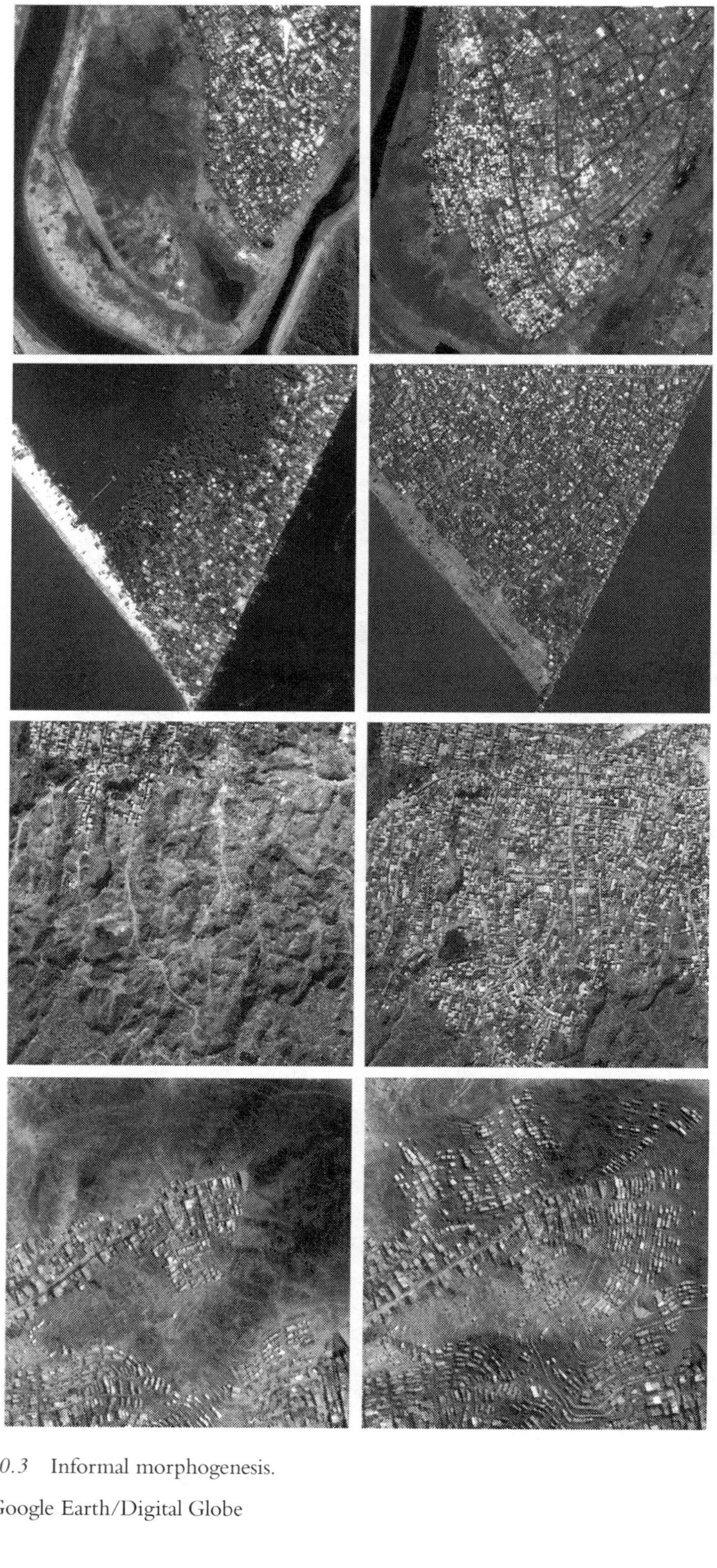

Figure 10.3 Informal morphogenesis.

Source: Google Earth/Digital Globe

He suggests that such codes embody proscriptions about encroaching on rights-of-way, limits to the height and to the extent of cantilevered balconies and rooms; they generally respect existing usages and include the preservation of views and privacy. The urban design codes that prevail in informal settlements often fail to protect much more than minimal rights of access; yet in other cases, encroachment is contained, and open space is protected. To understand such morphogenic processes provides insight into the emergence of informal urban design codes and the escalations of density into slum conditions. The challenge is to develop such informal codes into more formal codes where the escalation of encroachment is contained or reversed. Any newly formalized codes that emerge need to sustain the productivity, amenity and sociality that is already embodied in the place. This all depends on a better understanding of informal morphogenesis.

Image and identity

In her longitudinal study of *favelas*, Perlman (2010: 30) comments:

> Perhaps the single persistent distinction between *favelas* and the rest of the city is the deeply rooted stigma that adheres to them and to those who reside in them. . . . Even after the extensive . . . upgrading programs . . . there is little doubt as to where the *asfalto* (pavement) ends. . . . The visual markers of each are unmistakable, whether viewed from above or on street level.

Variations of this are true in all cities of the Global South – images of informality generally have negative symbolic capital for local populations. This is the third key dimension of how images of informality produce certain kinds of place identity that impact on the image of the city and political struggles over the right to the city. How are urban informal settlements perceived within the discursive field of the city and then geared to political practices and policy development? Since the image and identity of informal settlements are generally seen as negative, their removal can be seen as a form of urban cleansing – rendering the city modern and formal. Images of law and order are strongly geared to political legitimacy. While it can seem odd to focus on design imagery when dealing with issues of poverty, this is a crucial and under-researched dimension of urban informality that plays a key role in political struggles over the right to the city.

It is, however, a most complex problem. Informal settlements are often hidden from the gaze of the formal city, do not appear on municipal maps, and Google Streetview rarely penetrates them. At times the state will act to hide informal settlements, generating a façade of law and order; residents may also pursue a deliberate blurring of boundaries between formal and informal, a form of urban camouflage. On the other hand, there are many examples where arts projects are designed to produce a new identity and build local pride. Communities may be torn between the need to remain invisible in order to avoid eviction and the prospect of enhancing their visibility as a means to the same end. The rise of "slum tourism" in some cities meets a global market which is also a complex mix of voyeurism, aesthetic attraction, and a desire to learn more about such settlements (Dovey and King 2012).

Informal agency

These three dimensions of informality as a mode of production are inextricably intermeshed in the production of urban neighborhoods with particular metropolitan locations, morphologies, and meanings. While this involves a focus away from the discourse of "slums," there is no turning away from the pervasive conditions of poverty – informality is a means of both managing and potentially escaping poverty. As a mode of production, informal settlement is also a form of agency. Lister (2004) argues that the agency of people in poverty can be usefully understood in terms of four overlapping practices: "getting by"; "getting back"; "getting out"; and "getting organized." "Getting by" is subsistence activity – basic shelter, food, and sanitation. "Getting back" involves transgression of the formal codes of state control and informality as a form of everyday resistance. "Getting out" of poverty is more strategic – improving or extending the house, renting out a room, getting a job, educating children. "Getting organized" encompasses strategic and often political activities by community-based organizations. These categories are neither mutually exclusive nor all encompassing, but this is a useful framework for understanding urban informality as a means of managing and escaping poverty.

These forms of agency are the "weapons of the weak" as Scott (1985) puts it. This power is not simply held by residents but is embodied in the socio-spatial assemblage of the informal settlement and its many intersections with the formal city. Informal settlements can be understood as an assemblage of assets or capacities that enable residents to live a life, acquire an income, raise their children, and build prospects for a better future. The more we understand what these assets are and the ways in which they are geared to the urban morphology and spatiality, the more likely it is that policy and practice will be effective.

A first cluster of assets involves those embodied in the house and land. Informal settlements are geared to job markets and to the larger urban context – walkable access to local employment and public transport networks are crucial assets. Most informal settlements hold some form of ambiguous or *de facto* land tenure, often recognized by the state based on length of occupation or payment of land tax. Whether formally recognized as land tenure or not, the occupation of land is an asset that increases over time through the right of adverse possession. Regardless of land tenure, the buildings constructed within an informal settlement are owned by those who constructed, inherited, or bought them. The house often represents a long-term investment over more than one generation and the most substantial asset held by that household. The assets of house, land, and location are geared to each other because informal tenure is held by the house, and the value of the land lies in its location.

A second cluster of assets is embodied in the particular morphologies of informal settlements through their potentials for adaptation. The incremental construction process enables rooms to be added at a rate commensurate with income flows. Residents hold "air rights" to the space above the last room they have built, and this represents a prospect for additional space and possible rental income. Direct interface with a laneway network enables public space to be used as an extension of the private realm – relieving internal crowding, creating opportunity for socialization, childcare, and domestic production. It can be adapted to various kinds of production and exchange, or external access to an extra room and rental income. These adaptive capacities are embodied within the informal morphology and are generally extinguished in replacement housing.

These spatial and material assets need to be understood as crucial to any decisions residents might make about "upgrading" programs, which may require various tradeoffs: Will they have better housing, access to jobs, or open space? Will they be asked to convert from ownership to rental? Will they lose rental income? Can they build an extra room or open a shop? If any of these assets are taken away without equivalent benefit, then it will not be in the residents' interest to support such a program. Of course, they may negotiate a deal where they accept compensation and relocation but then sublet their allocated apartment and move back closer to employment – to another informal settlement where the cycle repeats, a common phenomenon in the history of managing informal settlements.

Urban design thinking

In this chapter, I have sketched an urban design knowledge base necessary for any attempt to understand or transform informal settlements. First, we need to know why they are located where they are, and how this location enables access to jobs via public or walkable transport. This is the larger social, political, and economic assemblage. For many settlements on urban fringes and in inaccessible locations, better access is the key to upgrading. Some settlements are unsustainable in current locations due to unstable conditions that cannot be mitigated. Under freeways, lining railway lines, and on sidewalks is no place to live; however, if the displacement is not to stimulate more informal encroachment, then replacement housing must be located with similar access to employment. Second, any attempt to upgrade settlements should be based on an understanding of the highly adaptive morphogenic processes through which they will continue to develop. In terms of morphology, we need micro-scale studies of building density, functional mix, access networks, building types, plot sizes, and public/private interfaces – the spatial assemblage. In terms of process, we need to understand local organization, construction increments, local materials, and supply chains. Any new design must consider the value of adaptability and ongoing change, even when designed at higher densities. Residents will rightly resist any form of relocation and reconstruction that does not acknowledge and at least match the current range of assets embodied in the informal morphology.

Third, any attempt at upgrading must engage with issues of image and place identity: areas of visibility from the surrounding neighborhood, access routes through it, the meanings attached, and the prospects for change. A key challenge is to incrementally blur the symbolic codes of social status that announce informal settlements as "slums" within the conceptual field of the metropolis. Will we continue to produce large housing estates with populations socially and symbolically segregated into ghettos that reproduce poverty? Or will we be prepared to engage with the incremental redevelopment of existing informal settlements as formally, socially, and functionally mixed districts of a more spatially just city? An effectively upgraded informal settlement can become an attractive place to live and work, by celebrating and developing the diversity and dynamism that is already present in the informal morphology. These are the same attractions that characterize the best of mixed-use, socially and formally diverse, inner-city neighborhoods of many formal cities.

An assemblage approach to such an understanding is useful because it brings a multiplicity of scales and approaches together; it entails prioritizing the relations between things, particularly the multiple interconnections between formal and informal practices, the ways informality works as both morphology and as discourse. The socio-spatial assemblage is at once a dynamic process and a formal configuration that reveals how informal settlements are always already in a process of upgrading and formalization. However, the take-home message of this chapter is not to make a theoretical point, but to understand how urban informality works to produce affordable housing for billions of people throughout cities of the Global South.

Current practice in the upgrading and redevelopment of informal settlements is far too diverse to cover here, and there are many examples of good practice in this regard (Huchzermeyer and Karam 2006; Hernandez et al. 2010; Smithsonian 2012; Gouveneur 2015). Most current thinking and practice, however, remain locked into a formalization framework, wherein knowledge developed within the formal city is applied, and the settlement becomes formalized top down. The challenge is not simply to replace top down with bottom up, but to understand how formal and informal intersect. This often entails dealing with a paradox in that the informality that produced the existing settlement is necessary for its effective formalization. Formal codes inevitably coexist and intersect with informal practices – a form of urbanism that works with and between different sets of rules and is irreducible to either, a double logic. Upgrading requires an engagement with both formal and informal logics, practices, and codes at the same time. While slum upgrading is a kind of formalization, it cannot be achieved without harnessing the informality.

Residents of informal settlements are not just getting by, getting back, and getting out, but they are also getting organized; so, will this ever be enough? Many good thinkers suggest or imply that the retention and transformation of informal settlements is pointless, that the key task is one of overturning global conditions of economic inequality and neoliberal capitalism that produce poverty. In his seminal book, Davis (2006: ch. 4) argues against what he terms "self-help" schemes that often fail or exacerbate the problem. He argues that funds leak through corruption; that owner-built housing is shoddy; that incremental construction is inefficient; and that NGOs can co-opt the interests of residents to their own. While there is some truth in all of this, there is also a blindness to the ways urban informality works as a mode of production. Informal construction produces affordable housing at a fraction of the cost per square meter compared to the formal production process (Sims 2010). Incremental corruption is of less consequence than in formal construction processes. While construction standards are often initially shoddy, in most settlements the majority of buildings can be incrementally and effectively upgraded *in situ* rather than replaced.

Davis (2006) also suggests that self-help programs divert the urban poor from political struggle. Yet informal settlement is already a form of political insurgency that produces new forms of urban governance and deep democracy (Appadurai 2001). The political struggle of informal settlement residents to remain in place and to upgrade their neighborhoods is being won in many parts of the world. While coercive eviction remains common, forced eviction and demolition of informal settlements has become the exception rather than the rule. This struggle is over not only eviction, but control of the settlements themselves; in many "informal" settlements, development is organized, and profit is extracted, by gangs or slumlords. Some house clusters of industrial production in sweatshop conditions that again demonstrate deep affinities between the deregulatory regimes of neoliberal capitalism and informality. These local and global struggles are interconnected – there is no need to choose between support for upgrading practices and for global political change. Indeed, informal settlement upgrading can be the most political form of urban design practice, because effective upgrading is at once a struggle for and a legitimation of the right to the city. Informal settlement is a mode of urban production involving the urban design of new neighborhoods and new forms of agency exercised by the urban poor. As I hope to have showed, it also requires some new directions for urban design thinking.

Acknowledgment

I want to thank David Week and Tanzil Shafique for their critiques of this chapter.

References

Alexander, C. (1996 [1965]). "A city is not a tree." In R. LeGates and F. Stout (eds.) *The City Reader*. London: Routledge, pp. 118–131.

Appadurai, A. (2001). "Deep democracy." *Environment and Urbanization, 13*(2), 23–43.

Brugmann, J. (2009). *Welcome to the Urban Revolution*. London: Bloomsbury.

Davis, M. (2006). *Planet of Slums*. London: Verso.

DeLanda, M. (2006). *A New Philosophy of Society*. New York: Continuum.

Deleuze, G. and Guattari, F. (1988). *A Thousand Plateaus*. London: Athlone.

Dovey, K. (2016). *Urban Design Thinking: A Conceptual Toolkit*. London: Bloomsbury.

——— (2012). "Informal settlement and complex adaptive assemblage." *International Development Planning Review, 34*(3), 371–390.

Dovey, K. and Kamalipour, H. (2017). "Informal/formal morphologies." In K. Dovey, E. Pafka and M. Ristic (eds.) *Mapping Urbanities*. New York: Routledge, pp. 223–248.
Dovey, K. and King, R. (2012). "Informal urbanism and the taste for slums." *Tourism Geographies*, *14*(2), 275–293.
——— (2011). "Forms of informality." *Built Environment*, *37*(1), 11–29.
Dovey, K. and Pafka, E. (2016). "The science of urban design?" *Urban Design International*, *21*(1), 1–10.
Dovey, K., Pafka, E. and Ristic, M. (eds.) (2018). *Mapping Urbanities*. New York: Routledge.
Gouverneur, D. (2015). *Planning and Design for Future Informal Cities*. New York: Routledge.
Hakim, B. (2014). *Mediterranean Urbanism*. Dordrecht: Springer.
Hart, K. (1973). "Informal income opportunities and urban employment in Ghana." *Journal of Modern African Studies*, *11*, 61–89.
Harvey, D. (2007). *A Brief History of Neoliberalism*. Oxford: Oxford University Press.
Hernandez, F., Kellett, P. and Allen, L. (eds.) (2010). *Rethinking the Informal City: Critical Perspectives from Latin America*. New York: Berghahn.
Huchzermeyer, M. and Karam, A. (eds.) (2006). *Informal Settlements*. Cape Town: UCT Press.
Jacobs, J. (1961). *The Death and Life of Great American Cities*. Harmondsworth: Penguin.
Johnson, S. (2001). *Emergence*. London: Penguin.
Lister, R. (2004). *Poverty*. Cambridge: Polity.
Lukes, S. (1974). *Power*. London: MacMillan.
Mayne, A. (2017). *Slums: The History of a Global Injustice*. London: Reaktion.
Moser, C. (1979). "Informal sector or petty commodity production." In R. Bromley (ed.) *The Urban Informal Sector*. Oxford: Pergamon, pp. 1041–1064.
Perlman, J. (2010). *Favela*. Oxford: Oxford University Press.
——— (1976). *The Myth of Marginality*. Berkeley: UC Press.
Rapoport, A. (1988). "Spontaneous settlements as vernacular design." In C. Patton (ed.) *Spontaneous Shelter*. Philadelphia: Temple University Press, pp. 51–77.
Scott, J. (1985). *Weapons of the Weak*. New Haven: Yale University Press.
Simone, A. (2014). *Jakarta: Drawing the City Near*. Minneapolis: University of Minnesota Press.
Sims, D. (2010). *Understanding Cairo: The Logic of a City Out of Control*. New York: American University in Cairo Press.
Smithsonian (2012). *Design with the Other 90%*. New York: Cooper-Hewitt National Design Museum.
UN-Habitat (2006). *The State of the World's Cities*. London: Earthscan.
Walker, B. and Salt, D. (2006). *Resilience Thinking*. Washington: Island Press.

Further reading

Dovey, K. (2016). *Urban Design Thinking: A Conceptual Toolkit*. London: Bloomsbury. An accessible assemblage-based approach to urban design, including a chapter on informal settlements.
Dovey, K. and Kamalipour, H. (2017). "Informal/Formal Morphologies." In K. Dovey, et al. (eds.) *Mapping Urbanities*. New York: Routledge. Proposes a typology of informal/formal morphologies that is mapped across 12 cities of the Global South.
Hakim, B. (2014). *Mediterranean Urbanism*. Dordrecht: Springer. An investigation of the informal and formal urban design codes that produce traditional Mediterranean cities.
Sims, D. (2010). *Understanding Cairo: The Logic of a City Out of Control*. New York: American University in Cairo Press. Perhaps the best account of both morphologies and processes of informal development within a single city.

11

THE NEW GEOGRAPHIES AND SPATIALITIES OF INFORMALITY

Annette M. Kim

Introduction

The nature of informal urbanism has changed over the last several decades physically and spatially. In the popular imagination, the idea of "informal" settlements still conjures up images of vast "slums" and "shantytowns," whose main defining characteristics are shoddy construction and being outside of laws and regulations. But more recently, informal housing has taken on astonishing permutations, in some places becoming more hidden and dispersed. Meanwhile, a global struggle for informal livelihoods vies for public space, openly contesting the spatial order and formal urban visions of planners and designers.

To elaborate the changing nature of contemporary informal urbanism, in this chapter I first outline the evolution of our understanding of "informality." This quick review lays the context to discuss some contemporary cases in East Asia, the site of the largest and most rapid urbanization processes (World Bank Group 2015). Because governments there have exercised strong planning and urban design powers over the built environment, these cases help to elucidate the spatial logic of the current state of informal urbanism and outline the challenging role that urban designers must play.

A hidden and dispersed constellation: contemporary informal spaces in the city

Earlier literature on informal urbanism focused on the vast settlements on the periphery of the city, often referred to as "slums," which were the target of the United Nations and its Millennium Development Goals (United Nations Human Settlements Program 2003). This earlier literature helped establish the fact that informality was not only a significant phenomenon, but in some cities of the Global South, it constituted the majority of new urbanization. Informal settlements were often large and distinct areas, even adjacent to formal urban development, thus providing a jarring juxtaposition of class difference among residents. The visibility of massive informal settlements and their disturbing squalor confronted the sensibilities of concerned citizens and policymakers and demanded attention. Urbanization was thus seen as a problem. Initially, slum clearance, through the razing of these settlements, was the norm. However, during the 1980s, protests and activism eventually made this practice politically untenable

(Asian Coalition for Housing Rights 1993), and a global discourse began around slum upgrading projects to improve the living standards of tenants *in situ* (Mukhija 2001).

While the physical conditions of un-serviced settlements presented public health hazards, such settlements also provided more affordable housing. As rural populations flocked to be near the economic opportunities found in central cities, residing in these peripheral zones was often close enough to participate in the urban economy. Figure 11.1 shows one typical example of rapid urbanization land use patterns in Ho Chi Minh City, Vietnam, in which the extent of the city expands as informal construction proliferates on the periphery. However, while these informal settlements multiplied, public budgets for housing projects decreased in response to the global economic recession of the 1980s, the debt crisis in the Global South, and the accompanying policy push to devolve and decentralize government.

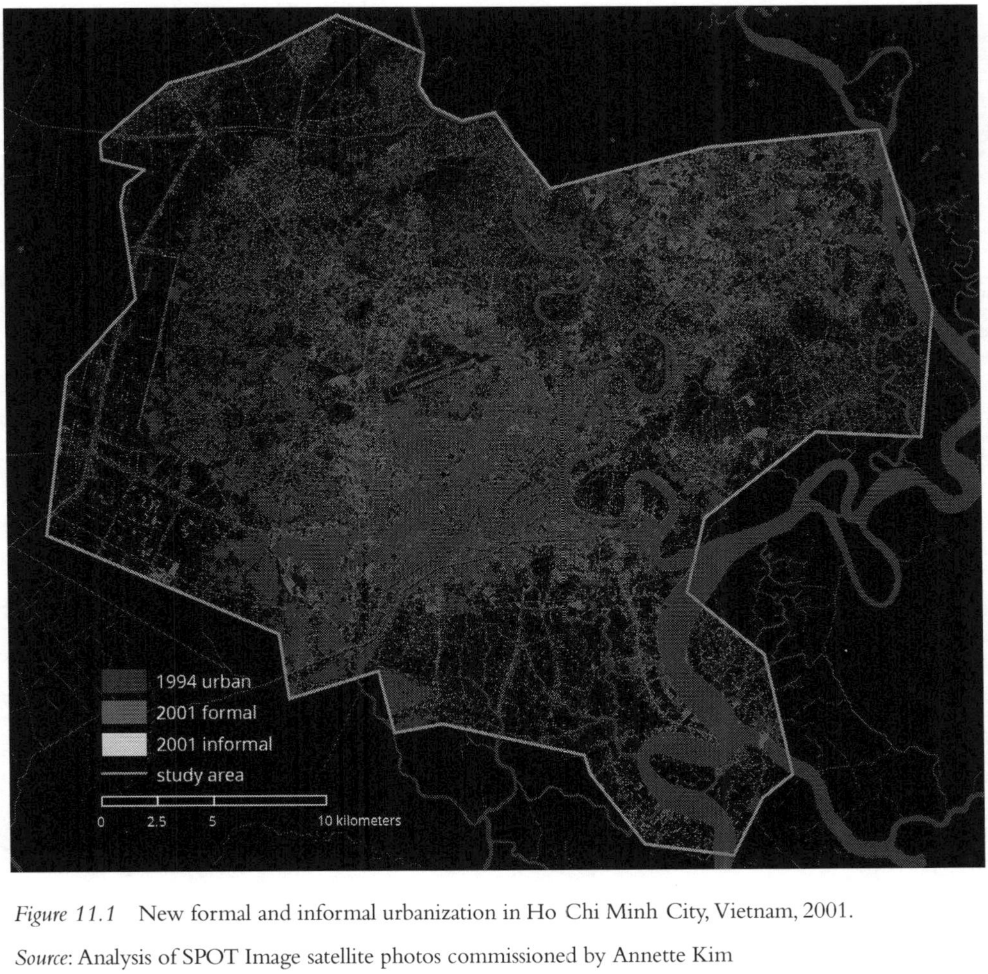

Figure 11.1 New formal and informal urbanization in Ho Chi Minh City, Vietnam, 2001.

Source: Analysis of SPOT Image satellite photos commissioned by Annette Kim

[Land use change from rural to urban land use was detected comparing images from 1994 and 2001. Formal construction was categorized by the brightness levels of the construction materials used. Informal construction had lower brightness levels but still had the texture patterns of urban construction that distinguished it from rural land use. This map does not detect the legal status of buildings and so informality is undercounted, if buildings were informally constructed with durable materials.]

The turn to a market-centered approach increasingly focused on investing in the capacity of institutions that support the market provision of housing (Farvacque & McAuslan 1992). In addition to administrative and financial institutions, there was particular focus on reforming legal and property rights institutions. The theory was that formal legal titles to land would remedy the insecurity of tenure that was holding back investment into the real estate market and therefore producing low-quality and insufficient amounts of housing. However, implementing titling projects was complicated, as these peripheral urban zones often had ambiguous property rights because of a variety of specific historical reasons, such as postcolonial, indigenous, customary, or collective agricultural land right systems that were being overlaid with or supplanted by modern, market-based, and bureaucratized land tenure (Doebele 1983; Payne 2001).

One of the major points that scholars have made, after decades of research, is that there is not a clear binary between formal and informal conditions but rather many permutations of non-conformance, such that a significant number, and sometimes the majority, of settlements could be called informal. If informality is the norm, then it is formality that needs as much reconsideration. Indeed, more than by accident, rapid urbanization may be supported by the interdependency of formality and informality: rapid market economic development and global trade through low-cost labor that is unentitled to public services (Chen 2006; Kudva 2009; Roy 2005; Song et al. 2008; Wang et al. 2009; Wu 2004). Clarifying informality's role in the national economy echoed the larger debates in international development strategies. The rise of China and other rapidly growing socialist market economies seemed to defy the Washington Consensus of phased reforms about the rule of law and trade liberalization. Rather than formal market institutions being a prerequisite for real estate development, these institutions may develop after the pioneering market transactions have claimed territories in order to enforce their property rights (Kim 2007, 2012).

After the turn of the millennium, human migration and urbanization growth rates are now peaking. Many cities, which have expanded in girth, appear to be reaching their limit of how far people can reasonably commute from the periphery. What was once the urban periphery has become valuable real estate in the metropolitan core, too expensive for laborers. In addition, the increased land values have supported the policy progression to enforce formal property rights more thoroughly in order to protect investments and maximize profit.

Therefore, as more areas of the city are becoming formalized, and property is increasingly protected, the migrant and lower-resourced underclass population, who cannot afford to participate in this institutionalized private real estate market, desperately have to seek out increasingly crowded spaces within the city (Myers et al. 1996). Informal housing has appeared on top of, between, behind, and under formal buildings (Canham & Wu 2010; Huang & Yi 2014). We discuss two such cases in this chapter: the subterranean housing market in Beijing, and the extreme crowding of the "group rental" housing market in Shanghai. Research into these phenomena helps clarify some of the root reasons for the emergence of new patterns of informal housing. Understanding these urban dynamics is necessary for designers as they endeavor to shape the physical manifestations of the socioeconomic and political processes of informality (Loukaitou-Sideris & Mukhija 2016).

The subterranean housing market in Beijing

> The example of China is like no other. It is possibly the only large country that has managed, so far, to urbanize rapidly without the creation of large slum areas or informal settlements.
>
> *(United Nations Human Settlements Program 2003, 126)*

As I researched the approximately one million people who live in bomb shelters and basements in Beijing, a phenomenon that captured my attention was the changing spatiality of contemporary informality (Kim 2016). While it is well known that Chinese cities have constructed an unprecedented number of buildings in the last several decades (Zhang & Song 2003), as the Human Settlements Program quote reflects, few know that a large supply of subterranean urban space is being used for housing. While some Chinese cities like Beijing are unusual in terms of the vast amount of underground space that they have, such space is not unusual in displaying how informal migrant housing may exist in any interstitial space that the formal market does not claim.

Underground spaces in Chinese cities are the result of a series of policies over the last six decades in which all new buildings were and still are required to construct air defense shelters and "common basements." The building construction codes specify how these must be built, which includes the provision of infrastructure such as electricity, water, and sewer. During China's earlier economic hardships, with its housing shortages and overcrowding, the official policy for 24 years encouraged the "economic" utilization of this space during "times" of peace. Up until 2010, one could apply for a Certificate of Use for these subterranean spaces, if they met the regulatory codes and passed inspections. While these uses could include commerce and storage, this underground space became increasingly subdivided into residential units, primarily housing migrants to the city. These subdivided rental spaces were so ubiquitous that thousands of advertisements for these units could be easily found on Internet classified ad sites, as well as on signs posted around the city. This market developed in response to the huge demand for private housing, as China has transitioned from a centrally planned economy, where the state used to provide all housing through employment work units, to a private housing market.

The underground housing market, which had been encouraged by official policy and swelled with historic migration, was one alternative to the urban villages. Official policies reversed in 2010, and Beijing made a three-year plan to evacuate residents from basements, with evictions starting to occur in 2011 in the inner-city districts. Still, during the time of my field research in 2014, there still existed a large market whose advertisements were visible on the Internet and on the street. My team collected data about this market from Internet ads in Beijing during October 2012–September 2013 and mapped their location, to identify the patterns of their spatial distribution.

Figure 11.2 shows that this underground housing market, rather than being on the periphery of Beijing's ring roads, appears in the central city. Such housing is evenly distributed throughout the city, without a specific concentration in marginalized city areas. Like the usual pattern described at the beginning of this chapter, land on the city periphery, which used to be rural, urbanized rapidly, with collectives building large "urban villages," which have been regularly razed by the government.

The map also confirms what demographic scholars have found in examining government census figures: roughly one half of Beijing's migrant population – approximately 4.3 million people (Liu et al. 2013) – actually live in the urban center, not in the periphery. They live in the city despite the fact that the migrants driving China's urbanization do not possess a household registration permit to live in the city (*hukou*). A holdover from central planning, individuals without urban *hukou* are not entitled to city public services such as education, health, and public housing.

This map also helps us to visualize the contemporary spatial pattern of informality: a constellation of dispersed informal units hidden under the formal city. These residential spaces can range from single units to large basements with hundreds of units, but essentially they are isolated from each other. Our fieldwork further found that entrances to the underground were kept separate from the main building. In interviews, residents of the formal units above expressed some fear

Figure 11.2 Map of underground housing in Beijing, China, 2013.

Source: Author

and anxiety about the underground tenants, wanting to maintain their invisibility and to keep their spaces separate (Lu & Kim 2018). One of the implications of this geography is that collective organizing and action of underground residents would be difficult, even if it were allowed legally. The spatial arrangement deters the incidental communication about shared interests and relationships built through spatial proximity, the usual foundational process for building social capital in traditional informal settlements.

This geography also implies that the traditional policy focus for dealing with informal housing, namely to provide better infrastructure, would not pertain in the same way. This constellation of informal units exists in the midst of the formal city rather than in areas without urban investment. Rather, it is the micro-level access and unit living conditions that require attention. In the case of Beijing, our fieldwork confirmed that generally these underground spaces have infrastructure connections, since the buildings above were built to meet building codes. They usually have common toilet and bathing facilities, electricity, and Internet service. Such spaces, however, were designed not for longer-term human habitation, but rather for emergency shelter. So, the main living quality issues are about crowding, ventilation, and enforcing safe egress and fire safety measures. Additionally, the dispersed and hidden nature of these informal units makes their management a challenge.

Because of the unusual situation of having vast amounts of underground apartments distributed throughout the city rather than low-income populations being segregated into ghettos, we can better study the migrants' difficult preference tradeoffs between housing characteristics such as rental price, location, or living space quality. Of course, if one had more resources, one would prefer to live above ground. We found that units that are halfway underground with a small window at the ceiling are more expensive than those completely underground. Still, this was less important for residents than other conditions such as heating and Internet availability. Furthermore, all these reasons were overshadowed by the combination of location and access variables. Being closer to the city center, to jobs near subway stops, and to particular city districts was much more important, and the reason these migrants live underground.

Thus, the most valued characteristics of Beijing's basement housing is its location; these underground units provide low-cost housing located closer to the city center than urban villages. When one considers the widespread availability of urban village housing on the urban periphery, it is clear that the choice that these renters are making is about location in the city center. This finding was further confirmed by interviews with underground residents and property managers in 2014 (Lu & Kim 2018). These interviews revealed that underground residents particularly valued proximity to a subway station. However, they did not use the subway for transportation, as the modest 2 RMB fare is still above their willingness to pay. The majority interviewed walked or bicycled to work but chose to live near the subway stations because of the many more employment opportunities in Beijing's transit-oriented districts. While there are cheaper housing units on the periphery, people would rather choose to live underground and not have to spend time or money to commute.

Crowding and the market for a bed in Shanghai

After completing the Beijing study, we turned to examine China's other premier city, Shanghai. We wanted to determine if Shanghai also has an underground housing market. But while there were many advertisements for renting underground spaces in Shanghai, we soon found that such spaces were rented mostly for extra storage and not for housing. The city had already effectively evicted the former underground residents. However, we came across a different and even larger informal housing market: group rental housing (Harten et al. 2018). Essentially, this is the market for a bed in a room full of bunk beds with strangers, much like a crowded college dormitory. This group rental housing is located in the city center, within formal buildings. For example, during our fieldwork visiting group rentals, we entered a high-rise commercial office building in the central city, and behind the door of a two-room office, we found 37 beds.

The phenomenon of Shanghai's group rental housing market further illustrates what happens when formal regulations come in to halt one form of informality (underground housing). The strong demand for migrant housing creates another kind of market that makes use of residual urban space, increasingly hidden within the formal built environment. Shanghai's group rental market had proliferated so much that we were able to scrape a large housing dataset of over 33,000 Internet ads during February–June 2016. Additionally, we complemented and verified this web data with fieldwork.

Groundtruthing revealed how hidden this market is. Invariably, brokers that we contacted through the ads would ask to meet us in the street and take us to a place that was different from the unit advertised in the original listing. The hiddenness of this market was emphasized by the behavior of the brokers. They would meet us surreptitiously and nervously, ask us to be quiet as

Table 11.1 Mean rent by crowding level for Shanghai group rental housing (n = 747)

Number of People per Room	*Frequency*	*Mean Rent (in Yuan)*
1	4	1250
2	35	1007
4	153	1078
6	249	931
8	188	668
10	86	625
12	32	589

Source: Fieldwork, June 2016

we entered buildings, and communicated the need to be discreet. Meanwhile, we would encounter groups of young people also visiting group rentals in the same building.

Because of discrepancies with Internet ads, we decided to collect original, real estate market data. We recruited a research team of five Chinese graduate students, who, along with an American doctoral student, visited roughly 190 group rental apartments during two weeks in July 2016, documenting location, price, crowding levels, and unit amenities, among other factors.

Not surprisingly, we found that actual units were more crowded, farther away, and more expensive than what was advertised. Still, the units shown were not too different from what was advertised, to the point that it would deter a transaction. We found that brokers were managing several units, and that these beds rented in a matter of days. So, there may have been no incentive to update advertisements with details as long as they were similar enough to attract inquiries and transactions. But we also found that the actual rentals were consistently more expensive than what was advertised.

Again, similar to what we found in Beijing's underground housing, these units were widely distributed throughout the central city. Indeed, the group rentals form a constellation of hidden units within the formal city. These units are very crowded. Table 11.1 indicates that the majority of these housing units involved 20 people in a unit, with 4–10 people per room. We also found that the market rationally prices the disamenity of crowding at the room level: the more people in the room, the lower the rent. Unlike the surprising example of the beds in the office building described earlier, about 80% of group rentals were within residential buildings. However, most often the kitchen was used as a bedroom, and cooking was not allowed.

Crowding was worse than suggested by Internet ads. Only 40 ads (or 30.3% of 132) mentioned crowding levels at all. However, all the sites visited had multiple people per room. The average number of persons per room in the ads was 5.5, but in reality it was 6.6. Perusing advertisements would suggest that rooms with more than 8 persons are rare (only one ad mentioned 10 persons sharing a room). In contrast, we saw 118 rooms with over 10 persons each, with additional people sleeping in hallways and living rooms. Real sites also tended to have fewer bathrooms than advertised.

Invisibility and public space conflict

As the Beijing and Shanghai cases indicate, migrants come to urban centers and live in such cramped spaces in the central city, even if urban planning visions do not provide space for them, because of employment opportunities. Formal and informal businesses in the city are employing them, and urban centers also provide opportunities for micro-entrepreneurialism.

Therefore, another major way that contemporary informal urbanism shapes the city, beyond where migrants sleep, is through people's informal spatial practices in pursuing their livelihoods. Indeed, the literature on informal economies in public space is large, and the cases are plentiful because informal practices, such as street vending, are not new (Bhowmik 2005; Bostic et al. 2016; Bromley 2000; Chen 2006; J. Cross & Morales 2007; Loukaitou-Sideris & Ehrenfeucht 2009). Rather, because of the large and rapid migration, such informal activities are now more prevalent, visible, and politically consequential in cities (Hou 2010).

In our current historic process of urbanization, as millions of people migrate to the city for jobs and opportunities, lower-income people need space to recreate and socialize. So, many go to the open and free public spaces in the city. Furthermore, given the high and rising rents in cities, many low-income people are using public spaces such as streets and sidewalks to earn a living. But high-density and multiple claims to public space produce tensions and challenges (A. Brown 2006; J. C. Cross 2000). As the social controversies and policy debates in cities around the world attest, we are in an active period of rewriting our public space rules.

The sheer bodily presence of the migrant underclass is controversial. This issue of their visibility is core to the conflict and directly touches urban designers' practice. Critical urban theorists have shown how the current global aesthetic and approach to urban space, designed for modern-day consumer flâneurs, follows a historical trajectory of propagating classist, sexist, and racist visions of urban space (Chase et al. 2008; Crawford 2015; Hayden 1995). Formal plans and blueprints, drawings, and renderings help build society's cognitive frames of who is legitimately in the city and the ways of living that will be supported (Kim 2015).

Rebuilding the global city: urban design's responsibility to contemporary informality

One of the things we learn from the two cases of informal Chinese urbanization is that, as much as government planning might try to erase the migrant labor population from urban centers, migrants have shown remarkable persistence in coming to and staying in the city, and so are only driven underground, sometimes literally. So, today, in addition to the large districts of informality on the periphery, there is a wide hidden web of dispersed informal units within the formal city. While the particular spatial manifestations depend on the particular context (Brown et al. 2017; Mukhija & Loukaitou-Sideris 2014; Palta 2017), the geography of contemporary informality of a hidden and widespread constellation of dispersed spaces in the central city is true for the global cities of both the North and South, since the fundamental dynamics of immigration, crowding, and privatization are all essentially global.

These cases demonstrate how desperately people still want to be in the city center. This presents a spatial challenge. While some call for a fundamental, universal right to the city, not everyone can fit in. So, by what system will we determine spatial entitlements? Systems that rely primarily on the market lead to the current outcome of a formal city interlaced with hidden informality. Can we do better? What will it take?

Every 20 years, the United Nations holds a major global conference to take stock of the world's housing situation and to promote an international policy dialogue. I attended Habitat 2 in Istanbul, Turkey, in 1996, and Habitat 3, 20 years later in Quito, Ecuador, in 2016. In the latter conference, the policy framings had rightly shifted from a focus on building more houses to the processes of urbanization. The discourse was more interdisciplinary, including environmental sustainability, technology, design, etc. But, perhaps the most notable change was the shift towards greater inclusion. In Habitat 2, nongovernmental organizations were placed at a site separate from the main conference. In Quito, the many conference sessions included the expertise and perspectives of a mix of constituencies: multinational, national, and local government; private sector; NGOs; activists; and academics. Of particular symbolic significance was the fact that the newly formed civil society General Assembly of Partners (GAP) helped to edit Habitat 3's New Urban Agenda statement. The language of the Agenda now discussed building a more inclusive city. Still, amidst this global dialogue shift towards inclusive urbanization, there was hardly ever an explicit mention of migrants/immigrants as people to be included, planned, and designed for, even though one-seventh of the world's urban population are migrants.

Soon afterward, a global, populist anti-immigrant/migrant sentiment became apparent with the 2016 US elections, Brexit in Europe, and in many countries of the Global South increasing sentiments of animosity against fellow citizens born in rural areas. As our cities come under stress with overcrowding and increased diversity, there is a mounting fear and demonization of fellow urban residents that we need to address and counter. The literature on institutional change suggests that achieving spatial inclusion requires the change of two fundamental and related topics, and urban designers can play an important role in each.

Reframing urban narratives

The narrative of who is legitimate to be present in the city relates to how we imagine that the city should look and be experienced. Urban planning practice, particularly in the Global South, has a physical, architectural approach to the city, and these are constituted and communicated by drawings, renderings, and visions of the future city (Loukaitou-Sideris & Mukhija 2016; Roy 2004). However, most of these renderings show oddly under-populated scenes in megacities of singles and couples shopping at high-end stores (Chase et al. 2008; Kim 2015). Contemporary urban design practice that seeks to be inclusive and deal with the actual city would need to depict all of its residents and their spatial practices.

Rewriting property rights

The vision of an inclusive city must also be institutionalized by real-world mechanics. These include not only the property laws and adjudicating courts, but the many plans and regulations that govern land use, public space, and building codes (Garnett 2009), as well as the civil servants who enforce these on the ground (Blomley 2011; Lipsky 1980; Martin 2007). Reforming rights to urban land and space is fundamental to achieving an inclusive city.

Globally, we are in the midst of rewriting property rights. To make practical space out of the right for spatial inclusion means a broader host of strategies that may require changing:

a) legal codes to include the legitimacy of people who are outside the formal documentation system
b) land use plans in order to share urban space between different users and uses
c) regulations, building codes, and design guidelines to allow for smaller, more affordable spaces
d) taxation for land value capture to help finance urban transformations

We can be much more creative about creating new kinds of spaces and property rights. Civil society is already agitating for these changes. The only way to make adequate space is for government, along with civil society and business, to create a new paradigm of property rights.

Designers are part of this international division of labor. We work inside the government at the international, national, and local levels helping to update codes, redraw plans, and create new developments. Communities of activists, artists, and academics continue to work with and outside government institutions to create a world of more spatially inclusive cities.

References

Asian Coalition for Housing Rights. (1993). *Environment and Urbanization*, *5*(2): 153–165. doi:10.1177/095624789300500213

Bhowmik, S. K. (2005). "Street Vendors in Asia: A Review." *Economic and Political Weekly*, *40*(22): 2256–2264.

Blomley, N. (2011). *Rights of Passage: Sidewalks and the Regulation of Public Flow*. New York: Routledge.

Bostic, R. W., Kim, A., and Valenzuela, J. Abel. (2016). "Contesting the Streets: Vending and Public Space in Global Cities." *Cityscape*, *18*(1): 3–10.

Bromley, R. (2000). "Street Vending and Public Policy: A Global Review." *The International Journal of Sociology and Social Policy*, *20*(1/2): 1–28.

Brown, A. (2006). *Contested Space: Street Trading, Public Space, and Livelihoods in Developing Cities*. Rugby, UK: ITDG Publishers.

Brown, A., Mukhija, V., and Shoup, D. (2017). "Converting Garages into Housing." *Journal of Planning Education and Research*, 0739456X17741965. doi:10.1177/0739456X17741965

Canham, S. and Wu, R. (2010). *Portraits from Above*. Berlin: Peperoni Books.

Chase, J., Crawford, M., and Kaliski, J. (2008). *Everyday Urbanism: Updated and Expanded*. New York: Monacelli.

Chen, M. A. (2006). *Rethinking the Informal Economy: Linkages with the Formal Economy and the Formal Regulatory Environment*. New York: Oxford University Press.

Crawford, M. (2015). "From the "Feel Good" City to the Just City." Retrieved from www.youtube.com/watch?v=lEGliu4P-Mk

Cross, J. and Morales, A. (2007). *Street Entrepreneurs: People, Place, & Politics in Local and Global Perspective*. Taylor & Francis.

Cross, J. C. (2000). "Street Vendors, Modernity and Postmodernity: Conflict and Compromise in the Global Economy." *The International Journal of Sociology and Social Policy*, *20*(1/2): 29–51.

Doebele, W. A. (1983). "Concepts of Urban Land Tenure." In Dunkerley, H.B. (ed.), *Urban Land Policy: Issues and Opportunities*. Washington, DC: World Bank, 63-107.

Farvacque, C. and McAuslan, P. (1992). *Reforming Urban Land Policies and Institutions in Developing Countries*. Washington, DC: World Bank.

Garnett, N. (2009). *Ordering the City: Land Use, Policing, and the Restoration of Urban America*. New Haven: Yale University Press.

Harten, J., Kim, A. and Brazier, C. (2018). *Hidden Informality in Urban China: A Mixed Method Approach to Shanghai's Market for Overcrowded Housing*. Retrieved from Los Angeles.

Hayden, D. (1995). *The Power of Place: Urban Landscapes as Public History*. Cambridge: MIT Press.

Hou, J. (ed.). (2010). *Insurgent Public Space: Guerrila Urbanism and the Remaking of Contemporary Cities*. New York: Routledge.

Huang, Y. and Yi, C. (2014). "Invisible Migrant Enclaves in Chinese Cities: Underground Living in Beijing, China." *Urban Studies*, *52*(15): 2948–2973.

Kim, A. M. (2007). "North Versus South: The Impact of Social Norms in the Market Pricing of Private Property Rights in Vietnam." *World Development*, *35*(12): 2079–2095.

Kim, A. M. (2012). "Seeds of Reform: Lessons from Vietnam About Informality and Institutional Change." *International Economic Journal*, *26*(3): 375–390.

Kim, A. M. (2015). *Sidewalk City: Remapping Public Space in Ho Chi Minh City*. Chicago: University of Chicago Press.

Kim, A. M. (2016). "The Extreme Primacy of Location: Beijing's Underground Rental Housing Market." *Cities*, *52*(2016): 148–158.

Kudva, N. (2009). "The Everyday and the Episodic: The Spatial and Political Impacts of Urban Informality." *Environment and Planning A: Economy and Space*, *41*(7): 1614–1628. doi:10.1068/a41101

Lipsky, M. (1980). *Street-Level Bureaucracy: Dilemmas of the Individual in Public Services*. New York: Russell Sage Foundation.

Liu, X., Wang, M., Cai, J., and He, M. (2013). "An Analysis on the Spatial Structure of Non-Native Permanent Population of Beijing Metropolitan Area in 2000–2010." *Urban Development Studies*, *20*(10): 86–93.

Loukaitou-Sideris, A. and Ehrenfeucht, R. (2009). *Sidewalks: Conflict and Negotiation Over Public Space*. Cambridge: MIT Press.

Loukaitou-Sideris, A. and Mukhija, V. (2016). "Responding to Informality Through Urban Design Studio Pedagogy." *Journal of Urban Design*, *21*(5): 577–595. doi:10.1080/13574809.2015.1071650
Lu, B. and Kim, A. (2018). *Excavating the Subterranean City: The Living Conditions, Livelihood Strategies, and Governance of Beijing's Underground Housing Population*. Retrieved from Los Angeles.
Martin, J. (2007). "A Reasonable Balance of Law and Sentiment: Social Order in Democratic Taiwan from the Policeman's Point of View." *Law & Society Review*, *41*(3): 665–698.
Mukhija, V. (2001). "Upgrading Housing Settlements in Developing Countries: The Impact of Existing Physical Conditions." *Cities*, *18*(4): 213–222.
Mukhija, V. and Loukaitou-Sideris, A. (2014). *The Informal American City: Beyond Taco Trucks and Day Labor*. MIT Press.
Myers, D., Baer, W. C., and Choi, S.-Y. (1996). "The Changing Problem of Overcrowded Housing." *Journal of the American Planning Association*, *62*(1): 66–84.
Palta, R. (Producer). (2017, January 15, 2018). "Can't Afford an Apartment in Los Angeles? Rent an RV" [article]. Retrieved from www.scpr.org/news/2017/11/27/78011/cant-afford-an-apartment-in-los-angeles-rent-an-rv/
Payne, G. (2001). "Urban Land Tenure Policy Options: Titles or Rights?" *Habitat International*, *25*(3): 415–429.
Roy, A. (2004). "Transnational Trespassings: The Geopolitics of Urban Informality." In Roy, A. and Alzayad, N. (eds.), *Urban Informality: Transnational Perspectives from the Middle East, Latin America, and South Asia*. Lanham, MD: Lexington Books, 289–313.
Roy, A. (2005). "Urban Informality: Toward an Epistemology of Planning." *Journal of the American Planning Association*, *71*(2): 147–158.
Song, Y., Zenou, Y., and Ding, C. (2008). "Let's Not Throw the Baby Out with the Bath Water: The Role of Urban Villages in Housing Rural Migrants in China." *Urban Studies*, *45*(2): 313–330.
United Nations Human Settlements Program. (2003). *The Challenge of Slums: Global Report on Human Settlements*. UN Habitat Retrieved from Nairobi, Kenya.
Wang, Y. P., Wang, Y., and Wu, J. (2009). "Urbanization and Informal Development in China: Urban Villages in Shenzhen." *International Journal of Urban and Regional Research*, *33*(4): 957–973.
World Bank Group. (2015). *East Asia's Changing Urban Landscape: Measuring a Decade of Spatial Growth*. Retrieved from Washington, DC: https://openknowledge.worldbank.org/handle/10986/21159
Wu, W. (2004). "Sources of Migrant Housing Disadvantage in Urban China." *Environment and Planning A*, *36*: 1285–1304.
Zhang, K. H. and Song, S. (2003). "Rural: Urban Migration and Urbanization in China: Evidence from Time-Series and Cross-section Analyses." *China Economic Review*, *14*(4): 386–400.

Further reading

Kim, A. M. (2014). "Hidden City: Beijing's Subterranean Housing Market." *Land Lines* (October), 2–7. Provides a journalistic version of my underground Beijing research accompanied by the photo-journalism of award-winning photographer's Sim Chi Yin portraits of the underground spaces and residents.
Kim, A. M. (2015). *Sidewalk City: Remapping Public Space in Ho Chi Minh City*. Chicago: University of Chicago Press. Provides alternative critical cartographic approaches to the public space of the sidewalk and its many everyday informal uses.
Loukaitou-Sideris, A. and Mukhija, V. (2016). "Responding to Informality Through Urban Design Studio Pedagogy." *Journal of Urban Design*, *21*(5), 577–595. doi:10.1080/13574809.2015.1071650. Discusses how urban design education needs to expand its scope and teach students about informality through experiential knowledge, negotiation and dialogue with communities.

12
URBAN INFORMALITY AND THE CITY AT NIGHT

Su-Jan Yeo

Introduction

The duality between day and night has created a pervasive gap in our understanding of the city (Gallan and Gibson 2011), such that the city at noon vis-à-vis the city at midnight are viewed as two worlds apart – the former evokes notions of productivity, work, and safety, while the latter signals cautions of inactivity, incivility, and danger. It is, then, not surprising that daytime norms and practices have dominated the discourse on public space and urban life; consequentially, constraining the scope of urban design along the spatial and temporal trajectories of daytime city users.

However, in the 21st-century capitalist city – one that is incessantly plugged into 24-hour global and digital networks – urban rhythms carry on well into and beyond the night (Kreitzman 1999). Here, the city at night takes on an exceptional space-time dimension of obscurity, either real or imagined, which subverts controls and rules imposed on urban public spaces by formal systems of planning and policing as well as by societal virtues. During nighttime, "there exists an informal city located just beneath and in the interstices of the formal city" (Laguerre 1994: 24), as certain places in the city after dark provide fertile ground for informal urban practices. Moreover, this nocturnal sphere can illuminate the presence or absence of certain segments of the public, such as nightshift workers, older adults, women, and children, for whom the city at night could either lessen or heighten their "vulnerability" – a factor that is very much contingent on how urban nightscapes are effectively planned and designed to be inclusive of everyone.

As more and more contemporary cities in the Global North and Global South embrace neo-liberal urban strategies towards economic growth, it becomes increasingly necessary to rethink the challenges and opportunities for planning and designing the nightscape. Such rethinking should not merely serve as a means to an end in promoting economic vitality, but, more crucially, should safeguard inclusivity of participation in the nighttime life and economy. Formally planned and "branded" entertainment zones are, today, major sites in the construction of the nighttime economy (Hollands and Chatterton 2003), yet there is also much to be explored in nightlife culture of the mundane and ordinary kind. For it is here where informal but routine modes of urbanism are enacted amongst nighttime denizens through their everyday interactions with the built environment in the public realm. In turn, these nightly interactions help foster a sense of place and community in the current epoch of "worlding" projects and practices (Ong 2011).

Taking the city-state of Singapore as a case study, I examine the characteristics of urban informality as manifested temporally at night and spatially within a commercial enclave by the name of Holland Village. Situated approximately ten kilometers from Singapore's central business district, Holland Village embodies a certain quality of nightlife that sets it apart from the entertainment destinations in the city center. However, as Holland Village undergoes an impending plan for future extension and development, I pose this question: what might be the policy implications for urban nightlife, and how can the boundaries between established spaces of informality and new spaces of formality be "softened" (Loukaitou-Sideris 2012)? By exploring these connections through the temporal lens of the night, this chapter seeks not only to address the changing landscape of urbanism in a globalized economy, but also to explore how the urban design field can play an influential role towards engendering a vibrant yet inclusive public realm at night.

Night and nightscapes in cities

Time is simultaneously an abstract notion and a functional construct. Time unifies geographically distant societies and structures human activities according to a universally accepted temporal rhythm. In this way, time affects our experience of places and shapes our collective attitude towards work, rest, and leisure. This collective attitude is evolving vis-à-vis advancements in information and communication technologies (ICT) that enable us, from any place at any time, to plug into "one network of quasi-simultaneous interaction that brings together processes, people, buildings, and bits and pieces of local areas, in a global space of interaction" (Castells 2002: 372). In this "space of flows," our lifeworlds need not be restrained by the clockwork of time (Ibid.), thus influentially restructuring the temporal dimension of human activities within cities that are very much embedded in 24-hour global and digital networks. Long before the advent of ICT, however, human civilizations had created a regimented relationship with time.

During the pre-industrial age, for example, the toll of bells in many cities indicated the end of a day and signaled inhabitants to return to their homes for rest and slumber. Hindered by culturally derived connotations of darkness as simultaneously sacred and dangerous, the night was subjugated by disciplined routines of domestic nature, which reinforced familial and kinship connections as well as religious or mystic rituals of devout practice, which deepened attachments to either the church-state or spirit world (see Palmer (2000) for an expanded historical scholarship on night cultures of liberation and transgression). Public participation in activities outside the private conclaves of hearth and home was discouraged, if not prohibited, at night. Over time, as new markets and trade opportunities in a globalizing economy would expand beyond city boundaries, night curfews would grow increasingly less restrictive, making certain types of work acceptable or even required to be conducted at night. One such type of work is trading, which connects to the opening and closing hours of Wall Street or other stock markets around the world.

More often than not, however, work related to a local economy would be carried out by the poor working class, who toiled in the factories, workshops, and streets at the onset of darkness. Nocturnal labor emerged during the industrial age and escalated with the rise of early capitalism. It was also during this period in the mid-19th century that a night-based public realm, one that was associated with neither the home nor the workplace, contributed to new perceptions of the night as a time (and space) for conviviality and conspicuous consumption. From alehouses to opera houses, the night-based public realm introduced different class stratifications of leisure and promoted the nightscape as an inviting environment for urban recreation.

The invention of electric lighting in the late 19th century and the subsequent systematic installation of street lighting further advanced the notion of night as a new space-time "frontier" that expanded the social and economic potential of cities (Melbin 1987). Similarly, the ostentatious illumination of the city – showcased through world expositions and amusement parks – drew attention to how public space at night could be utilized for revelry, celebration, and merriment. The festive qualities of night lighting drew people into the city for purposes that extended beyond work. As artificial illumination improved over time, new kinds of public spaces and urban experiences emerged alongside new imaginaries of the city afforded by the night. For example, the "Great White Way" aesthetic, exemplified by New York City's theater corridor on Broadway, involved the strategic use of lighting to make commercial districts attractive after dark (Jakle 2001) and so "became a model for the modern American night," which spurred replications elsewhere in cities across the United States (Baldwin 2002: 594). Over time, the illuminated urban nightscape significantly implicated the economic dimension of the city through the valorization of land and private-sector investment in urban projects.

In this vein, time is viewed as a commodity and the urban nightscape a terrain for mass consumption. It is, then, not surprising that the urban nightscape has the capacity to captivate the interest of architects, planners, developers, and city authorities. Contemporary cities in today's globalized economy are actively being imagineered and shaped, through public policy and urban design, into symbolic expressions of modernity, wealth, and power (Ooi 2004; Yeoh 2005). The prevalent "24-hour ethos" (Kreitzman 1999) is illustrative of this paradigmatic shift in city building, which aims not only to transform the physical landscape and service infrastructure towards such global aspirations, but also to manipulate the temporal dimension in order to create a vibrancy that is often associated with urban living in a "city that never sleeps." Here, nighttime becomes a ubiquitous resource for social and economic exploitation, as formal mechanisms such as zoning ordinances permit extended hours of service for certain businesses and public events. Such mechanisms effectively allow public space after dark to be commodified and controlled in formal ways with the objective of attracting capital and conspicuous consumption – for example, pedestrianization of streets at night to promote vitality in the downtown core after typical office hours, and staging of "night festivals" to enhance art and cultural offerings at various city venues.

Thus, in the postindustrial urban context, the "night-time economy" is characterized by two types of formal processes. First, local governments are investing in "city branding" (see Lang 2011), employing strategies of civic boosterism to promote urban life after dark, as well as to join the ranks of a growing global network of international events (such as the "Nuit Blanche" festival and Earth Hour movement's "Lights Out"). Second, there is increasing participation of corporations in the production of nightlife spaces of consumption, and consequentially, the market segmentation of these spaces through advertising tactics that target specific consumer groups (Lovatt and O'Connor 1995; Hollands and Chatterton 2003).

It is crucial to acknowledge the potentially homogenizing effect of the nighttime economy, particularly in urban districts that are deliberately programmed and thematically branded with venues and events that promote a standardization of nightlife culture. On the other hand, to assume that all nighttime economies operate in this manner would be to disregard alternate nightscape narratives that do not fit neatly into such designated categories as prime streets, shopping districts, and entertainment zones. Alternate nightscapes may not be necessarily any less formally planned or designed, but they represent a nightlife culture that exudes a recognizably informal sensibility, expressed through the relaxation of social norms and the ordinariness of spatial practices. Yet, such mundane nightlife – or "everyday (night)life" (see Yeo *et al.* 2012) – can have important contributions towards placemaking. The habitual nature of quotidian nightlife, characterized by the commute home after work or the last-minute grocery run to purchase items for dinner, for example, give "shape to spaces" and "weave places together" (De Certeau 1984: 97).

Over time, these nocturnal rhythms and routines, enacted individually and amassed collectively, not only shape the affective character of a neighborhood, but in doing so create a sense of community for those to whom the neighborhood is a place of livelihood, leisure, and lifestyle.

Everyday nightlife and the public realm

Quotidian nightlife involves the public realm where habitual routines and repetitious activity patterns are performed on a recurring basis at night. For participants of everyday nightlife, these nocturnal patterns of "mundanity" can be all-too-familiar and insignificant. By deconstructing the taken-for-granted rituals of everyday nightlife, we can better understand the hidden possibilities of public space at night – that is, possibilities for meaningful social interaction, economic enterprise, and identity formation – for it is precisely the whys and hows by which people expend and allocate their time that bring to view the variegated, nuanced, and geographic specificities of the nightscape. Indeed, many cities at the forefront of today's global age are socially diversifying and physically densifying at a remarkable rate and scale. The challenges posed by this ensuing scenario require greater attention and consideration towards urban spatial outcomes that are not only efficient and functional, but also meaningful and useful for multiple publics.

But multiple publics result in plural spheres of interests and values, and intervening city authorities are more likely to establish formal controls that regulate and manage public space, in order to uphold acceptable forms of civility and order. Such formal controls, however, are effective when there is the collective will to observe and enforce them. In other words, the oscillation between informal and formal modes of public life is played out through the dynamics of the urban citizenry such that what might be deemed socially acceptable in one context is potentially unacceptable in another – for example, youth playing street hockey on a residential neighborhood street versus a commercial plaza. However, this dynamic may be influenced by the framework of time, so that youth playing street hockey in an open plaza of a commercial precinct might very well be accepted after office hours or on weekends. As Crawford (1999: 13) asserts, "[b]y recognizing and building on these understandings of time, we can explore new and barely acknowledged realms of urban experience." Arguably, quotidian nightlife is an urban experience characterized by the relaxation of social norms and spatial practices, otherwise eschewed during the discernably formal and "polite" daytime hours. In this regard, quotidian nightlife unsettles the dualism between formal and informal – instead, it illuminates the pertinent coexistence of both spheres, which may entail vibrant and inclusive public spaces after dark.

A perspective from the Global South

As the 24-hour city gains momentum on a global scale, there is a need to better understand not only the policy, planning, and urban design implications of nightlife, but also, and equally importantly, the international facets of this burgeoning phenomenon. For some time, scholarship on the nighttime economy tended to privilege Euro-American notions of nightlife with case examples from cities in the Global North (see, for example, the special issue of *Planning Practice and Research* (Montgomery 1995) that includes select articles stemming from a 1993 conference in Manchester on "The 24-Hour City"). Subsequent nighttime economy scholarship from the Global North demonstrated an increasingly dominant narrative of the city after dark that regularly features the bar-pub-club scene as a "monocultural" landscape of young party-revelers and drinkers and critiqued the exclusionary outcomes of these nightscapes for being "less intergenerational, less intercultural" (Landry 2006: 125). In the last two decades, new insights about the nighttime economy have emerged from research that unveils the plurality of nightlife as experienced and documented in varying contexts of the Global South (see, for example, Heng and Chan 2000; Chiu 2013; Tadie and Permandadeli 2015; Song *et al.* 2016). Within contemporary Asian cities, for example, the depth and breadth of nightlife geographies can range from the neon-illuminated commercial precinct of Tokyo's Shibuya district to the lantern-lit food thoroughfare of Jalan Alor in Kuala Lumpur. More interesting, perhaps, is the overt juxtaposition of formal and informal modes of nightlife as demonstrated in cities such as Bangkok, Hong Kong, Shanghai, and Taipei, to name a few, where high-brow shopping boulevards and open-air night markets coexist simultaneously as well as proximately.

Indeed, nightlife narratives from the Global South are rich and diverse but, in being so, also present a case for multi-city comparative investigations. Consequentially, as the collection of single-city nightlife case examples from the Global South continues to expand in numbers, (dis)similar thematic strands and theoretical connections are likely to surface that warrant comparative studies. As a starting point, contemporary Global South cities at the nexus of global and cultural influences commonly exhibit a particular experience of nightlife that is often described as "cosmopolitan" – that is, a "fluid multifaceted urbanism" (Yeoh 2004: 2443) – that allows for the old and new, local and foreign, informal and formal to persist in a state of coexistence. The following section, a case study of Holland Village in Singapore (Figure 12.1), illustrates the convergence of urban informality and nightlife in what is regarded to be one of the most highly regulated and formally planned cities in the Global South.

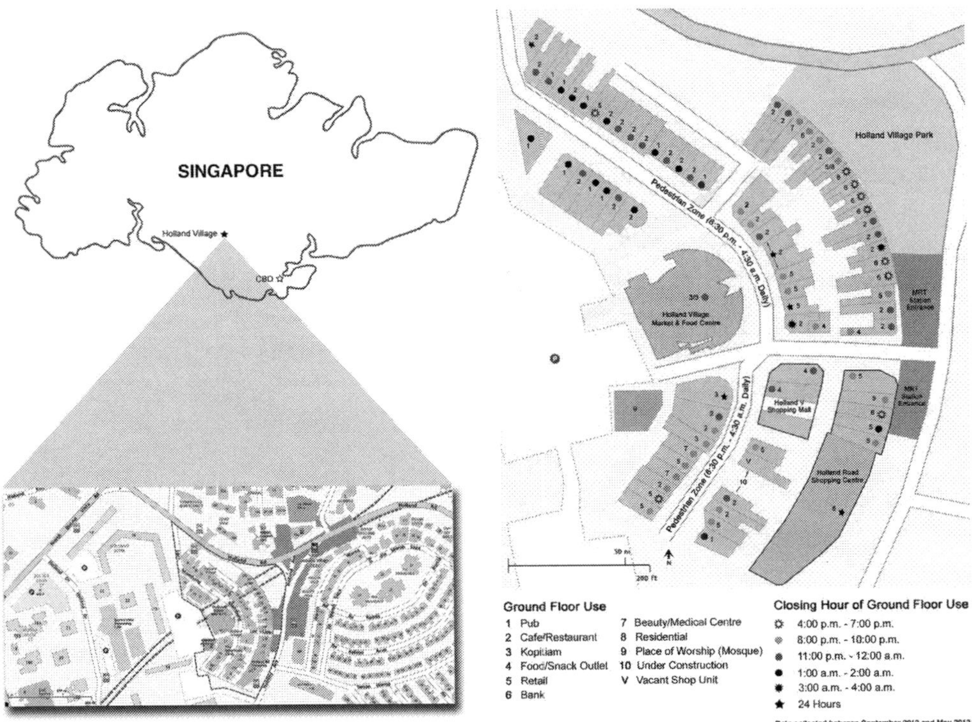

Figure 12.1 Geographic location and temporal mapping of Holland Village from data collected between September 2012 and May 2013.

Source: Image by Su-Jan Yeo

Case study: Holland Village, Singapore

Site context

In the mid-1940s, and under British colonial planning, Holland Village was developed as a modest commercial enclave of two-story shophouses (with shops on the ground floor and residences on the upper floor), accommodating trades and businesses that provided amenities to British soldiers and their families living in military bases nearby (Chang 1995). These shophouses continue to characterize the urban fabric of Holland Village today. Another significant feature of Holland Village during that time was a makeshift tin- and zinc-roofed market constructed in 1946 and populated by independent vendors that supplied fresh produce, poultry, seafood, and dried sundries such as rice, noodles, garlic, and onions. In the 1960s, a new extension was added to the market that expanded its space and offered proper facilities that included partitioned food stalls, lavatories, and an administration office.

From the 1970s to the present, Holland Village has undergone a succession of developments beginning with the insertion of two modern shopping malls that stand in contrast to the vernacular shophouse streetscape, yet are still in keeping with its low-rise urban form. In 1972, a new public housing estate – developed and managed by Singapore's public housing authority – emerged adjacent to Holland Village and boosted the residential population in the area. In the following decades, Holland Village evolved in a relatively organic manner to become an eclectic commercial hodgepodge of restaurants and coffee shops, retail shops, medical clinics, and banks. More interesting, however, is the broader geographic context surrounding this commercial enclave – Holland Village occupies a peculiar niche flanked by private housing on one side and public housing on the other. The juxtaposition between privatized and subsidized residential communities surrounding Holland Village is striking from a planning standpoint. Being situated at this nexus, Holland Village is an important case study of urban informality with everyday nightlife as a potential driver of sociability, conviviality, and inclusivity amidst socioeconomic diversity.

As the novelty of Holland Village continues to grow amongst locals and expatriates alike, several public projects were initiated to enhance its place identity and economic vitality. First, the market was demolished and rebuilt in 2006 to become the present-day Holland Village Market & Food Centre. Second, the construction of a mass rapid transit (MRT) station at Holland Village was completed in 2011, thus improving transportation accessibility, and a new park opened in 2013. At around this same time, the Singapore's Urban Redevelopment Authority announced plans to "extend" Holland Village on an adjacent 2.3-hectare site. This entailed the clearance of residential blocks comprising the original HDB public housing estate erected in 1972, and which had been targeted by the government for demolition since 2005. Additionally, the extension eliminated an existing open-air public parking serving Holland Village and a mosque that was constructed there during the 1970s, and which ceased operations in 2014, when it was demolished in order to make the land available for redevelopment.

The Holland Village extension plan capitalizes on the place character of a distinct commercial enclave; it aims to inject a new mix of residential and commercial developments through private-sector investment by way of the government's land sale program. In May 2018, Stirling Land Holdings was awarded the development of the Holland Village extension on a 99-year leasehold, as a result of its concept proposal and submission of the highest bid (see Channel News Asia 2018). This begs the question: How might this policy-guided and design-oriented neighborhood transformation implicate the existing place identity and sense of community in Holland Village? In other words, how can the boundaries between established spaces of informality and new spaces of formality be "softened" to safeguard enduring qualities of everyday nightlife in Holland Village? The following sections will unpack these questions through the temporal lens of the nightlife with findings from my fieldwork from September 2012 to May 2013.

Night rhythms and placemaking

The transition to nightlife in Holland Village begins at approximately 6:30 p.m. on a daily basis, when two of the three internal roads serving the commercial enclave are gated and closed to vehicular traffic. With the conversion of these roads to pedestrian use, social and economic activities not only expand temporally into the nighttime hours, but also spatially onto the roadway. At nightfall, eateries begin to maximize their seating capacity by setting up tables, chairs, and signboards that spill outdoors. This operational ritual is a nightly spectacle performed by the nightshift service staff as they deftly arrange various "props," transforming the physical space as well as the ambience of Holland Village. Here, bistro chairs and clothed tables rub alongside barstools and aluminum tables to create an atmosphere that is at once formal and informal (see "a" in Figure 12.2). The street activity reaches a lively rhythm by 7:30 p.m. as crowds comprising of families, couples, students, and office workers from the vicinity and elsewhere stream into Holland Village for dinner. For many, the Holland Village Market & Food Centre is an affordable and unpretentious setting for a meal. Some patrons are seen dressed in immaculate office attire, others in t-shirts, shorts, and open-toe sandals; locals eat side by side with foreigners; nonresidents blend inconspicuously with residents; and teenagers as well as older adults chat with their friends.

The former Masjid Holland Kampong mosque, which was present and operating at the time of my fieldwork, added to the eclecticism and distinctiveness of Holland Village. It was not uncommon, especially on Fridays, for the mosque to be filled to capacity with almost 500 Muslim devotees, as it was conveniently located *en route* after work or school to the congregants' homes. The location of the mosque over its 40-year tenure within the commercialized landscape of Holland Village may seem, at first glance, jarring against this context; yet a closer examination reveals the enduring bustle that once emanated from the mosque as it converged in sync with the everyday and nightlife rhythms of Holland Village. From this perspective, the nighttime vivacity of Holland Village was enacted in part by the ebb and flow of crowds that patronize place-based institutions ranging in ambience from the reticent quality of the mosque to the vociferous character of the Holland Village Market & Food Centre.

Such place-based institutions help to create a sense of community amongst regular patrons and visitors through their routine participation in the everyday nightlife of Holland Village. This collective sentiment is shaped more prominently on Fridays and weekend nights, as Holland Village becomes the "in" place to be, thereby replacing the traditional role and common perception of the city center as a nightlife destination. Indeed, the overt urban informality that is manifested in the public realm at night is part of the appeal, which continues to draw crowds to Holland Village. The nightly road closure, for example, extends the availability and accessibility of the public space to pedestrians, thereby creating a spatially negotiated terrain for urban life that straddles the informal-formal and public-pri-

Figure 12.2 The everyday nightlife character of Holland Village, illustrating multiple uses and informal appropriations of public space after dark.

Source: Photos by Su-Jan Yeo

vate binaries through street activities that range from curbside snacking to alfresco dining; from playful rollerblading and skateboarding to purposeful strolling; and from passive resting/loitering to active people-watching. Such "flexing" of social norms and "loosening" of spatial boundaries at night open up possibilities for informal urbanism, which, in turn, facilitate the convergence of multiple publics in Holland Village as people from diverse walks of life – students, office workers, families with children, retirees, tourists, food hawkers, retailers and vendors, and service industry employees – facilitate the production of everyday nightlife and, in this way, the making of place with a sense of community.

More specifically, the routines and rituals of everyday nightlife foster place identity through the habitual practices and repetitive patterns enacted by people performing ordinary activities in the informal space-time continuum afforded by the nighttime activities. From dining and shopping to people-watching and socializing, the informal modes and arrangements by which these banal acts of urban leisure unfold after dark reveal the capacity of everyday nightlife to create distinct places of rhythmic vibrancy and diversity. Such quotidian milieus can enrich the lived experiences of and attachments to place, as they expose the extraordinariness of ordinary settings in the display of sensorial aesthetics – "a kind of visual excitement, a quality of electric invigoration" (Lofland 1998: 87). In short, the informal urbanism that characterizes everyday nightlife helps to sustain the rhythmic intensity of the public realm and, hence, enhances the vitality and identity of place.

Informal spatial practices

Cobblers, locksmiths, newspaper peddlers, and confectionery vendors commonly occupied the street corners of Holland Village in the past. At the time of my fieldwork, there were three cobblers and a newspaper/magazine merchant operating regularly on the pavements of Holland Village. One of the cobblers, an elderly man who had been plying his trade in Holland Village for the past 30 years, "sets up shop" at his preferred street corner from late morning until around 9:00 p.m. His equipment consists of a sitting stool and the portable tools of his trade. Situated across the street corner from the cobbler is a newspaper/magazine merchant with a sustained presence in Holland Village. This long-standing family business occupies a miniscule "hole-in-the-wall" shopfront. Utilizing a narrow stretch of passageway, the merchant displays swathes of print merchandise, which spread laterally and vertically against the building façade in order to optimize space. His display creates congestion along this passageway, especially on weekend nights as pedestrians meander and pass those who are leisurely browsing the magazines, temporarily oblivious to the crowding (see "b" in Figure 12.2).

The street corners and pedestrianized internal roads in Holland Village at night, therefore, provide an outdoor stage for the making of urban life after dark; in this regard, public space is akin to the "front region" of a theatrical stage (see Goffman 1959). Here, everyday "actors" knowingly or unknowingly participate in the scripting of place through their individual and collective gestures, interactions, and trajectories of travel from one point to another – all of which produce a dynamic performance. The performance, in and of itself, is also a spectacle, often consumed by the very actors creating it. On Friday nights, for example, Holland Village serves as a gathering place for teenagers of both local and foreign nationalities. The front region affords impressionable youth an informal platform to express their social identities, while at the same time gazing as spectators (see "c" in Figure 12.2). The public realm of streets and sidewalks provides opportunities for identity formation amongst individuals; this is where "contesting or deviating from rules, stereotypes, and norms of appropriateness" (Loukaitou-Sideris and Ehrenfeucht 2009: 56) is amplified in the hours after dark with greater social acceptance and less social policing.

Conversely, the "back region" of alleys that serve the shophouse businesses from behind provides a setting for very different forms of informal spatial practices. Here, there is a significant reduction in the noise from the crowds and the ambient music in the "front region." The back region provides service industry staff a relatively quiet and intimate setting to take their breaks, individually or in small groups. Some restaurants arrange tables and chairs in the alleys for their employees to sit, smoke, and have their meals. In this way, the alleys are informally "colonized," with visual markers indicative of the type of ownership and degree of "publicness" that can be encountered there (Ho and Lim 1992). Hidden from the public eye, alleys further promote the easing of social norms and loosening of spatial practices, thereby fostering a sense of community amongst the users that share these common areas.

The back region of Holland Village is also a place for the staging of festivities intended for community stakeholders and not necessarily open to all members of the public. The annual Hungry Ghost Festival is one such event. In Singapore, the Hungry Ghost Festival is a month-long occurrence in August, during which large celebrations and events are commonly organized in neighborhoods outside the city center. The main setup typically involves the erection of a massive tent under which a makeshift stage provides the platform for live performances, entertainment, and auctions. Throughout the duration of the festival, rituals such as incense and paper burning, prayers, and food offerings are practiced. At night, tables fill the tent as attendees dine together in a feast that could vary from a moderate meal to an elaborate banquet for over 100 people. The Hungry Ghost Festival is popular amongst the business community, and the tradition continues in present-day Holland Village, where it still holds meaning for older generations of shopkeepers and hawkers. Although the festival may no longer enjoy the same degree of extravagance and scale as it once did, its ongoing practice is an indication of the importance of spaces of everyday nightlife in enabling informal spatial practices with cultural and community significance.

Designing the urban nightscape: challenges and opportunities

"Eclectic" and "cosmopolitan" are descriptions often applied to the commercial enclave of Holland Village to express not only the aesthetics of its physical environment, but also the affective quality and novelty of its nightlife scene, vis-à-vis the purposefully planned entertainment districts in the city center. The peculiarity of the urban nightscape in Holland Village is further compounded by its geography – a locale situated at the margins of the local and the global. This "in-betweenness" provides the niche for urban informality to flourish, temporally as well as spatially. For example, the instituted nightly closures of the internal roads have extended the availability of the public realm to pedestrians and businesses, so that public space is used simultaneously for a variety of formal and informal activities: dining in upmarket restaurants; snacking from the pavements; "pubbing" on the street side; people-watching from the coffee shops; rollerblading and freeform playing on the roads; and staging of prayers and traditional rites. More significantly, the confluence of locals and foreigners, young and old, is heightened at night as people come to Holland Village to participate in nighttime leisure. This quotidian commercial enclave, outside the aura of the city center, sustains a vivid public life, where strangers cross paths, providing a meaningful opportunity for sociability, conviviality, and inclusivity.

However, challenges in safeguarding these ideals are common when neighborhoods undergo transformative urban redevelopment, as is the current situation with the impending extension project in Holland Village. It is therefore necessary to critically unpack the interrelation between planning and design deliberations, on the one hand, and policy implications on the other. First, a large-scale physical transformation of the built environment could potentially compromise effective, though perhaps no longer "fashionable," expressions of scale and aesthetics that have enabled certain places to endure. How can comprehensively planned redevelopment schemes integrate new expressions of architectural styles and public realm treatments in a manner that supports, rather than abandons, the nuanced sensibilities of the existing environment? Second, a deliberate effort to further the economic transformation of a mature neighborhood, through the injection of new infrastructure to stimulate residential and commercial development, may create a gentrified retail landscape that privileges established brands and disenfranchises small businesses. What initiatives can be exercised to accommodate a range of goods and services at different price points that would address the gamut of individuals and groups living, working, and playing in the neighborhood and enhance, or at least maintain, its diversity and distinctiveness? Third, and related to the previous two points, an urban transformation of significant physical and economic impact may consequentially alter the social dynamics of place. Herein lies the potential of the urban nightscape to bring together people from different walks of life in a frame of time that celebrates relaxation and revelry – but only if the planning and design strategies are explicitly articulated to enable conditions of sociability, conviviality (see also Banerjee 2001), and inclusivity to flourish.

In the process of planning and designing a good-quality urban nightscape, it is important to consider how the "formal" facilitates the "informal," and how the existence of one helps to define the other. More often than not, planning and design strategies separate the spheres of formality and informality. In fact, what may be crucial for diverse and varied nightlife is the softening of formal-informal boundaries through spatial and temporal practices that enable the creation of in-between spaces. The informalization of the public realm after dark – as manifested in nocturnal routines and informal spatial practices – creates a rare place in time without necessarily removing formal norms and systems, but rather loosening their restraints in certain spaces and at specified hours. After all, urban life is spontaneous, flexible, and dynamic. And for urban life to flourish in a global city, it becomes ever more important to transcend the duality of formal-informal and, instead, reconcile the messiness and delight that transpire when communities find inventive means to reclaim and reconceptualize the public realm in the restless hours of the night.

References

Baldwin, P.C. (2002). "'Nocturnal Habits and Dark Wisdom': The American Response to Children in the Streets at Night, 1880–1930." *Journal of Social History*, 35(3): 593–611.

Banerjee, T. (2001). "The Future of Public Space: Beyond Invented Streets and Reinvented Places." *Journal of the American Planning Association*, 67(1): 9–24.

Castells, M. (2002). "The Culture of Cities in the Information Age." In I. Susser (ed.) *The Castells Reader on Cities and Social Theory*. Malden; Oxford: Blackwell Publishers. 367–389.

Chang, T.C. (1995). "The 'Expatriatisation' of Holland Village." In B.S.A. Yeoh and L. Kong (Eds.) *Portraits of Places: History, Community and Identity in Singapore*. Singapore: Times Editions. 141–157.

Channel News Asia (2018, May 16). "Stirling Land Gets Holland Road Tender with S$1.21b Bid." Retrieved from www.channelnewsasia.com/news/business/holland-village-stirling-land-tender-awarded-retail-dining-10239744

Chiu, C. (2013). "Informal Management, Interactive Performance: Street Vendors and Police in a Taipei Night Market." *International Development Planning Review*, 35(4): 335–352.

Crawford, M. (1999). "Introduction." In J. Chase, M. Crawford, and J. Kaliski (Eds.) *Everyday Urbanism*. New York: Monacelli Press. 8–15.

De Certeau, M. (1984). *The Practice of Everyday Life*, transl. by S. Rendall. Berkeley: University of California Press.

Gallan, B. and Gibson, C. (2011). "New Dawn or New Dusk? Beyond the Binary of Day and Night." *Environment and Planning A*, 43(1): 2509–2515.

Goffman, E. (1959). *The Presentation of Self in Everyday Life*. New York: Anchor Books.

Heng, C.K. and Chan, V. (2000). "The Night Zone Storyline: Boat Quay, Clarke Quay, and Robertson Quay." *Traditional Dwellings and Settlements Review*, 11(2): 41–49.

Ho, K.C. and Lim, V.N.E. (1992). "Backlanes and Contested Regions: Construction and Control of Physical Space." In B.H. Chua and N. Edwards (eds.) *Public Space: Design, Use and Management*. Singapore: Singapore University Press. 40–54.

Hollands, R. and Chatterton, P. (2003). "Producing Nightlife in the New Entertainment Economy: Corporatization, Branding and Market Segmentation." *International Journal of Urban and Regional Research*, 27(2): 361–385.

Jakle, J.A. (2001). *City Lights: Illuminating the American Night*. Baltimore: John Hopkins University Press.

Kreitzman, L. (1999). *The 24 Hour Society*. London: Profile Books.

Landry, C. (2006). *The Art of City-Making*. Sterling: Earthscan.

Laguerre, M.S. (1994). *The Informal City*. New York: St. Martin's Press.

Lang, J. (2011). "City Branding." In Banerjee, T. and Loukaitou-Sideris, A. (eds.) *Companion to Urban Design*. London and New York: Routledge. 541–551.

Lofland, L.H. (1998). *The Public Realm: Exploring the City's Quintessential Social Territory*. New York: Aldine de Gruyter.

Loukaitou-Sideris, A. (2012). "Addressing the Challenges of Urban Landscapes: Normative Goals for Urban Design." *Journal of Urban Design*, 17(4): 467–484.

Loukaitou-Sideris, A. and Ehrenfeucht, R. (2009). *Sidewalks: Conflict and Negotiation Over Public Space*. Cambridge, MA: MIT Press.

Lovatt, A. and O'Connor, J. (1995). "Cities and the Night-time Economy." *Planning Practice and Research*, 10(2): 127–133.

Melbin, M. (1987). *Night as Frontier: Colonizing the World After Dark.* New York: Free Press; London: Collier Macmillan.

Montgomery, J. (ed.) (1995). "Urban Vitality and the Culture of Cities." *Planning Practice and Research*, Special Issue, 10(2).

Ong, A. (2011). "Introduction: Worlding Cities, or the Art of Being Global." In Roy, A. and Ong, A. (Eds.) *Worlding Cities: Asian Experiments and the Art of Being Global.* Chichester and Malden, MA: Wiley-Blackwell. 1–26.

Ooi, G.L. (2004). *Future of Space: Planning, Space and the City.* Singapore: Eastern Universities Press.

Palmer, B.D. (2000). *Cultures of Darkness: Night Travels in the Histories of Transgression.* New York: Monthly Review Press.

Song, H., Pan, M. and Chen, Y. (2016). "Nightlife and Public Spaces in Urban Villages: A Case Study of the Pearl River Delta in China." *Habitat International*, 57: 187–204.

Tadie, J. and Permandadeli, R. (2015). "Night and the City: Clubs, Brothels and Politics in Jakarta." *Urban Studies*, 52(3): 471–485.

Yeo, S.J., Hee, L. and Heng, C.K. (2012). "Urban Informality and Everyday (Night)life: A Field Study in Singapore." *International Development Planning Review*, 34(4): 369–390.

Yeoh, B.S.A. (2004). "Cosmopolitanism and Its Exclusions in Singapore." *Urban Studies*, 41(12): 2431–2445.

Yeoh, B.S.A. (2005). "The Global Cultural City? Spatial Imagineering and Politics in the (Multi)cultural Marketplaces of South-east Asia." *Urban Studies*, 42(5/6): 945–958.

Further reading

Bunnell, T. and Harris, A. (2012). "Re-viewing Informality: Perspectives from Urban Asia." *International Development Planning Review*, Special Issue, 34(4). A collection of articles that extend the theorization of "informality" through empirical cases in Asia.

Kreitzman, L. (1999). *The 24 Hour Society.* London: Profile Books. A seminal book that unpacks the "24 hour city agenda" from a broad urban studies perspective.

Laguerre, M.S. (1994). *The Informal City.* New York: St. Martin's Press. Conceptualizes and contextualizes "informality" as everyday urban practices that emerge from within the fabric of the formal system.

Van Liempt, I., Van Aalst, I., and Schwanen, T. (eds.) (2015). "Geographies of the Urban Night." *Urban Studies*, Special Issue, 52(3). A collection of articles that explore how cities at night manifest in different international contexts.

13

URBAN DESIGN AND INFORMAL URBANISM

Case studies from Delhi and Mumbai

Manish Chalana

Introduction

Indian cities contain large pockets of informal settlements (often labeled as "slums") that are homes and places of work for tens of millions of poor and low-income individuals and families. These settlements are the main source of affordable housing in the country's urban centers, especially in megacities like Delhi and Mumbai. Informal settlements provide a launchpad for new migrants arriving daily in search of employment, offering the kinship networks and support systems needed for transitioning into urban life. However, most of these communities lack access to even the most basic urban infrastructure and services. For this and many related reasons, these settlements are stigmatized and often dismissed by design and planning practices. When they are considered, it is usually as part of formalization attempts that are grossly inadequate for residents.

In this chapter, I present two informal neighborhoods in India: Dharavi, Mumbai, and Kathputli Colony, Delhi. Both are "illegal" squatter settlements on city land, with residents lacking tenure rights. They exhibit high-density, low-rise, compact, and organic urban form, with eclectic architectural styles and building traditions, and ingenious use of materials and vernacular technology. They also occupy prime real estate that their respective cities are eying for redevelopment. I explore both settlements, focusing primarily on their urban form, and how it has been shaped to efficiently integrate the various demands of lives and livelihoods. I argue that, in their current forms and despite all infrastructural shortcomings, these settlements serve residents' needs in myriad ways that the proposed redevelopment plans are unlikely to recreate. I present alternative design solutions for informal settlements elsewhere that have tried to bridge the gap between technocratic urban planning and lived experiences of inhabitants. These practices have fallen out of favor in recent decades as neoliberal planning has taken root in India, but they might still hold vital lessons for today's informal settlements.

Urban informality in India

Large swaths of land in urban India are informally developed without involvement from professional designers and planners or other "experts," and largely without regulations. These areas are settlements of low-income migrants seeking work, which is relatively easy to find in India's massive informal economy. However, affordable housing is a rare commodity, especially in megacities like Delhi and Mumbai, where most people are priced out of the real estate and rental markets. With no other options, poor migrants settle in informal settlements (*jhuggi jhopri* or *basti* in the local parlance). Often they select a settlement where extended family or friends from their home community already reside. They may live with them until they are able to rent their own room (or *kholi*) or to buy or build one when land becomes available in the informal market. Most informal settlements are maximally built out, so the most recent wave of immigrants have to rely on cohabitating, renting, or purchasing. (Building up is sometimes an option, although this can be expensive as it requires sturdier materials, and the more established settlements are generally near their vertical limits as well). The result is a tightly packed built environment, often lacking basic services such as electricity, water, and indoor plumbing. Despite the benefits of low cost and family networks, residents must deal with the unhealthy living conditions that come from overcrowding and minimal urban infrastructure.

Urban informality has begun to receive greater attention from academics in recent decades, including those in design and planning disciplines. Some urban planning scholars have focused on the production and evolution of informal settlements and their spatial and economic relationships to the formal city. For instance, Roy and AlSayyad (2004) argue that planners see urban informality as irregular, despite being the predominant form of urbanization in the Global South; this results in misplaced efforts at formalizing it. Others have critiqued planning practice towards urban informality; the common theme is that professional planners have overly focused on creating a city that matches their narrow vision of a visually orderly and efficient settlement. Scott (1998) terms such failed attempts at ordering the built environment to improve the human condition "high modernism," which deems everything traditional or nonscientific as obsolete. In the case of India, attempts at formalizing informal settlements date back over a century, beginning with colonial British planning institutions like the Improvement Trusts to "modernize" Indian cities (Chalana & Rishi 2015). Despite calls from Patrick Geddes to consider alternate approaches to informality, which he termed "conservative surgery," practice grounded in early modernism prevailed. Modernism reached its height in India during the mid-20th century postcolonial period. Le Corbusier, the master modernist who planned Chandigarh, was "virtually offended by the disarray and confusion" of the pre-modern city (Scott 1998). This high modernism ideology paralleled approaches to slum clearance, urban renewal, and development of housing projects in the United States and elsewhere. Despite the failures of high modernist ideology in all these arenas, it continues to shape contemporary urban planning vis-à-vis slum redevelopment in India. Elsewhere, I have argued that what modernism undervalues and perceives as "messy" can have multiple layers of ordering within the apparent visual disorder, which together integrate nested community networks and economic systems in a way that benefits both community and city (Chalana & Hou 2016).

On the other end of the design and planning disciplinary spectrum, architects and urban designers have rarely focused on urban informality. Even as the domain of urban design has extended beyond the formal city to include vernacular neighborhoods and concerns of affordable housing and livability, the work of urban designers largely neglects the "Informal City." Even without direct involvement from design professionals, however, interventions grounded in urban design skills have influenced physical improvements and upgrading of informal settlements, but they remain at the fringes of the field and are seldom discussed (Mukhija 2011). These include small projects, such as creating open spaces and community facilities, as well as larger efforts at infrastructure upgrading, including: access to water and sewer connections; integration of the informal and formal city by building shared infrastructure; and concentrated efforts for job creation, economic development, and affordable housing (Ibid.).

A few design professionals have worked directly on improving the built environment of informal settlements and developing solutions to improve the lives of residents. Most are motivated by a sense of social responsibility and a desire to contribute towards an inclusive city, developing adaptive solutions that challenge the imagination of "glittery" "world-class" cities. These can take the form of large-scale redevelopment, discussed in more detail in this chapter. But designers have also explored an array of smaller creative strategies. For example, in the San Rafael Unido section of Caracas's La Vega Settlement, designers developed an innovative network of stairs and public landings that incorporated water and sewer systems, improving sanitation while still allowing families to remain in place. In Iquique, Chile, the design team for the Quinta Monroy project created half-built homes, with empty space between units on the first and third floors for residents to build on their own. This provided an outlet for residents' entrepreneurship and creative expression (Smith 2007). In Freedom Park, Cape Town, a design team created low-cost homes using sandbags held by timber frames, which do not require specialized labor and are energy efficient and climate-appropriate (Ibid.). In Lagos, architect Kunlé Adeyemi designed the Makoko Floating School as an amphibious teaching facility for a fishing village turned informal settlement. A resident team constructed the school using locally grown bamboo and wood scraps from a local sawmill; the school was powered by rooftop solar panels and floated on 250 plastic drums (Makoko Floating School, NLE 2013).

In addition, some designers have attempted to understand urban informality in more complex ways than it is usually credited for. This includes recognizing the ingenuity, creativity, and resourcefulness exhibited in informal settlements and extracting lessons for wider practice. In India, such ingenuity has been called "*Jugaad*" (Hindi for "improvising shrewdly with available resources"), with some designers recognizing how "the rise in India's enterprising spirit can be an inspiration and a catalyst for architects, designers, and urban planners around the world," particularly its focus on creative recycling and frugal innovations (Jeffrey & Harriss 2014: 63).

Admiration for *Jugaad Urbanism* and other forms of informality, however, can be critiqued as a colonial fantasy removed from the harsh reality of slum-dwelling. Rem Koolhaas, for example, famously declared the urban conditions in Lagos as ahead of their times and a harbinger of urban futures (Koolhaas 2007). In addition, designers who have worked on informal settlements have been criticized for not fully understanding the complexity of urban informality, and their efforts have been dismissed as unsuccessful in making a significant long-run difference for residents. This critique argues that only the settlement's built environment and physical amenities are improved, but not "people's capacities or livelihoods" (Roy 2005). This emphasis on physical environment over economic and political capacities is what Roy (2004) terms the "aestheticization of poverty"; it is also rooted in the people-versus-place debate in urban policy trying to address socioeconomic inequities such as poverty, housing affordability, and unemployment. Neoliberal economists often emphasize investment in people over place-based strategies, arguing that the former allows people greater mobility to better their lives. However, the people-versus-place debate is rather incomplete and misleading, as it downplays the inherently dual nature of community economic development (Crane & Manville 2008).

One counter-argument to those who critique designers' focus on urban informality is that the urban form of informal settlements exhibits many of the characteristics designers and planners value and try to achieve by spending enormous amounts of time and effort. Ignoring the lessons these places have to share is both top-down and arrogant. Informal settlements have high density and use value, encourage walkability, and reuse construction materials. They also house the robust social networks that design and planning professions try to promote elsewhere for sustainability and resilience. And while urban informality is indeed far too complex a condition for good design alone to be a panacea, I would argue that urban design can play a critical role in shaping the built environment of informal settlements, making incremental and positive differences in residents' lives (Mukhija 2011). However, to do so requires the practice of urban design to recalibrate, using both indigenous sensibilities and a postcolonial analysis that disrupts the formal/informal binary (Boano, Lamarca, & Wade 2014). Such a recalibration has the potential to generate development alternatives or forms of alternative development (McFarlane 2004); however, this process has thus far been indolent, as the next section demonstrates.

Dharavi and Kathputli Colony: urban form

Since the early-1990s economic liberalization and globalization, India has been dealing with rapid urbanization. The country's megacities are increasingly positioning themselves to benefit from global capital and commerce and aiming to reinvent themselves as modern, "world-class" cities. Subsequent decades have seen rapid expansion and restructuring of metropolitan areas, with new investments in infrastructure and urban form. However, these transformations are often at the expense of low-income populations, who reside and work in the city's informal settlements. The most massive of these transformations are in India's two largest metropolises: Mumbai and Delhi.

Long India's financial and entertainment capital, Mumbai is the country's second-most populous city, with a regional population of over 21 million. Since colonial times it has been a port city with an active textile industry, which has remained central to Mumbai's economy into the postcolonial decades. Given its peninsular geography, much of Mumbai's growth has been from south to north into suburban zones engulfing non-urban land uses, including rural settlements and fishing villages. Mumbai has numerous informal settlements within its fold, but no other nearly as expansive as Dharavi. Indeed, by some accounts, Dharavi is the largest "slum" in Asia, and one of the two largest in the world.

The land that would eventually become Dharavi was located on the fringes of Colonial Bombay in the swampy lowlands along Mahim Creek. Here informal settlements grew around a *koliwada*, or fishing village, where they were tolerated. These have since coalesced and grown into a 530-acre (2.1 sq. km; 0.83 sq. mi.) settlement with a largely low-income population of over 700,000 people, and by some estimates considerably more. This puts its population density around six times that of *daytime* Manhattan, but all in low-rise construction. Today Dharavi is located in the city's heart, along suburban railway lines and near Mumbai's international airport and a tech center. Meanwhile, Mumbai has become the most expensive real estate market in India, on par with New York and London. This places enormous pressure on the city to develop further, leading the city to demarcate a "golden" triangular area, which includes Dharavi, for redevelopment. Mumbai has considered several possibilities for redeveloping Dharavi over two decades, but none has yet materialized.

At over 22 million, the Delhi metropolitan area is one of the world's largest. Like Mumbai, it is a comparatively prosperous city by Indian standards, with a diversified economy based in government, manufacturing, tourism, and education. Situated on a plain with low-lying hills, Delhi's postcolonial growth has followed a radial (as opposed to Mumbai's axial) pattern. The growth has centered around two established cores: Old Delhi (Shahjahanabad), the 17th-century Mughal capital city on the west bank of the Yamuna River; and New Delhi, the early-20th-century British capital city some two miles south from it. Today, Delhi and its suburbs spread far across on the east bank as well. The city has its share of informal settlements; however, they are smaller, more scattered, and generally less visible than Mumbai's. Multiple urban villages that were engulfed by Delhi's expansion are now zones of urban informality, since development is exempt from building by-laws and zoning regulations within the villages' *lal dora* – literally "red line," created by British planners to mark the habitable zone boundaries. Informal settlements are essentially absent in New Delhi, the area most visited by global politicians, bureaucrats, and tourists.

Kathputli Colony is one such squatter settlement. It began in the 1960s as an encampment west of the city for nomadic artisans and street performers along their traditional migratory route from Rajasthan. Artists and other migrant workers from around India eventually joined them, and over time, the settlement grew to 13 acres. Today, with Delhi having expanded in all directions, Kathputli Colony lies well within the urban core, a mere 10-minute Metro ride from the colonial city center. Much smaller than Dharavi, it also occupies prime land that Delhi has long sought to redevelop. Local authorities have begun relocating residents from Kathputli Colony to a site of temporary accommodations (known as a transit camp) to start the redevelopment project.

Both Kathputli's and Dharavi's urban forms are defined by distinct districts organized by regional, religious, caste, linguistic, and occupational affiliations. Most are marked by an organic spatial organization comprising a complex network of narrow streets and winding lanes along which rows of structures (two to four stories, sometimes taller) are tightly packed (Figure 13.1). The districts are primarily mixed use, with most homes also functioning as work spaces. This tight urban form is interrupted only by places of worship and education, medical dispensaries, and occasional open spaces, which are rare outside of streets and intersection nodes that serve some public uses. A typical home is small (100–200 sq. ft.), without a water closet and often housing large, extended, or even multiple families. It is built incrementally using "found" or recycled building materials. The majority of homes comprise one multifunctional, often rented, room used as a living space during the day and at night for sleeping on foldout mattresses. Courtyards and roof terraces expand the indoor spaces. While the homes are spatially constrained, they have high use value, especially those shared among extended family and doubling as workshops (Figure 13.2).

Figure 13.1 Dharavi's tightly packed, mixed-use, and pedestrian urban form.

Source: Author

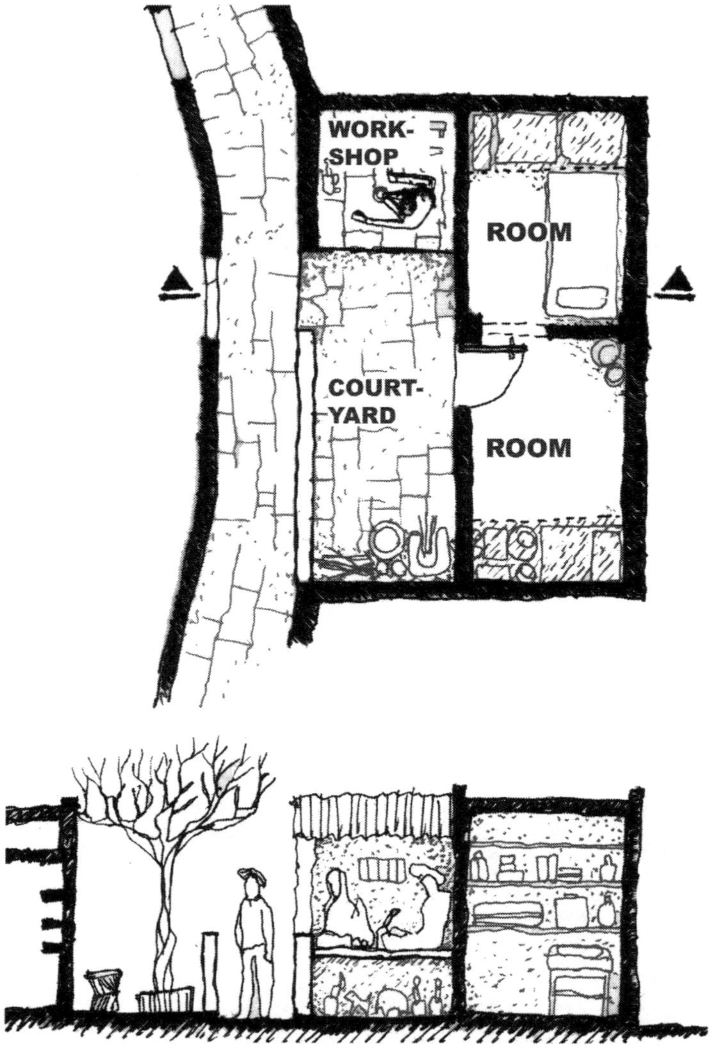

Figure 13.2 Self- and incrementally built dwelling unit with workshop and courtyard shared among joint family. Plan and section.

Source: Sketch by Amit Ittyerah

In both Dharavi and Kathputli, bazaars comprise a major form of public space and community interaction. The Chamra ("leather") Bazaar is a side street that forms the edge of the Dharavi *koliwada*. It hosts a row of small shops run by Muslim residents specializing in leather products. The living spaces above the shops double as workshops and storage. The bazaar and adjacent mosque were the site of major inter-communal violence during the 1992–1993 Mumbai riots. Several shops were damaged or destroyed by arson and vandalism but later rebuilt in much the same way. In Kathputli, the main bazaar divides the neighborhood's similarly sized artist and non-artist sections. The bazaar is lined by shophouses constructed on high plinths; the shops sell items of daily use and products made by artists (Chalana & Rishi 2016). In addition, both bazaars provide spaces for informal commerce.

Given the high population density, larger open spaces are rare in the informal settlements. In Dharavi, the largest open space is the Holi Maidan, used for celebrating weddings and festivals, political rallies, and cricket matches, and home to a large fish market. Another area – the former shores of Mahim Creek, defined by stones and steps – serves as a public space for local residents. In Kathputli Colony, the hardscaped open space around the local *darga* (Muslim shrine), with its associated mosque and madrassa, is the main hangout for neighborhood youth outside the bazaar. Additionally, the open space of the workshop of the *Bhule Bisre Kalakar* (a co-op of artisans) and a school run by an NGO (Kalakar Trust) provide places to socialize and host community events. However, both are just outside of the settlement, with controlled access.

Despite varying along linguistic or occupational lines, most districts in Dharavi and Kathputli share similar urban form characteristics; however, there are exceptions to this rule. For instance, in Dharavi, the *koliwada* – the original 10-acre fishing village that was later urbanized and incorporated into Dharavi – exhibits a highly distinct urban form. It is largely a residential and retail neighborhood of majority Hindu population with a sizeable Christian minority. Its basic urban form constitutes a fenced cross (believed to date to 1850) and the Holi Maidan mentioned earlier. The *koli* women continue to host a long-standing fish market in the Holi Maidan, even as reliance on fishing has dwindled due to urban encroachment into Mahim Creek's riparian zone and resulting water pollution. The district overall retains its village-like spatial organization, including remnants of timber-framed houses with porches, with a sense of openness compared to Dharavi's other districts.

Another district, Khumbwara (roughly 12.5 acres), was settled by potters originally from Gujarat, relocated here by the British in the early 20th century from a site developed as a military fort. Since then, potters have shaped their live-work spaces by creating a workshop in the back for the production of pottery and selling their finished products from the shop-room or road in front of their homes (Figure 13.3). Common areas behind homes contain shared pottery kilns.

Figure 13.3 Khumbwara: Potters' village in Dharavi.

Source: Author

Both the *koliwada* and Khumbwara retain aspects of rural building typologies and spatial configurations in their urban fabric. Both also have larger homes with more open spaces and better infrastructure, despite having densities far above those of most major world cities. Other zones of Dharavi with unique morphologies include the recycling district ("13th Compound") and goldsmith district (*Sakina bai Chawl*).

In Kathputli, the artists' section has clusters of homes arranged around a series of interlocking courtyards, shared among members of extended families. Originally, the homes were built out of temporary materials using discarded tents on earthen plinths; eventually they were made permanent, first using earth and thatch, then brick and concrete to allow for more vertical expansion. The courtyards have distinct climatic, functional, and social benefits in facilitating traditional ways of living in the Indian context. They function as outdoor rooms, extending the living and work spaces by accommodating a variety of uses, including socializing, craft making, sleeping, and for women cooking over wood-fire stoves and bathing with some degree of privacy. While the size and shape of courtyards vary, most homes in the artists' section apply a similar spatial typology.

Attempts at formalization

Both Mumbai and Delhi have in recent decades been trying to formalize their many informal settlements using an eviction-demolition-peripheral relocation model. However, this has largely been unsuccessful, especially in the absence of efficient and affordable transportation networks. More recently, Mumbai has developed an alternative approach using a public-private partnership model of slum redevelopment that controls for displacement, as residents are rehabilitated *in situ*. After attaining consent from a majority of residents, the private developers can develop the property using cross-subsidies to build apartments for qualifying residents in exchange for excess FAR for market-rate developments on a portion of the site. This approach has been widely used in formalizing sites of informal settlements across the city.

Kathputli Colony recently became the first case in Delhi to adapt Mumbai's public-private partnership model of slum redevelopment. The land has been sold to a developer, who proposes to use over 60% of the site for high-end uses, including a 54-story residential tower – the city's tallest – alongside high-end offices and luxury retail. In the interim, residents who qualified for rehabilitation verified by residency on site prior to the cutoff date (originally 2009, later extended to 2015) were asked to relocate to one-room prefabricated dwelling units in a transit camp three miles away, where they must live until the project is complete.

Mumbai has tried formalizing Dharavi since the 1980s. This includes the creation of social housing by the Slum Rehabilitation Authority (SRA) in the form of high-rise apartment blocks on the periphery that required relocating displaced residents to a temporary relocation site within Dharavi, which has since become permanent. More recently, an attempt at redeveloping Dharavi, led by a US-based developer and touted as the "Opportunity of the Millennium," fell through. The project imagined the neighborhood as a high-rise city organized in five distinct zones containing skyscrapers, open spaces, and shopping malls, along with apartment blocks for residents of Dharavi. Units were to be 250 square feet each, with eligibility based on a residency cutoff date. The apartment blocks would have taken less than half of the area, increasing the density on that portion of the site even more than current levels and adding to other problems of traffic and congestion. The project was discontinued for this and other reasons, including a recognition that it lacked even the most basic considerations for residents' ways of living and of generating livelihoods from the informal economy (Chalana 2010).

Both the Kathputli Colony and Dharavi rehabilitation proposals are characterized by fundamental misunderstandings about the social and economic networks in these settlements. They continue to further the ideals of high modernism and reliance on formalization as a panacea to all problems facing informal settlements. This is despite the fact that elsewhere in India and beyond, designers as early as the 1980s experimented with multiple alternatives grounded in traditional urban forms. These have lost popularity in India in recent decades since urban planning in the post-economic liberalization decades has become increasingly entrepreneurial, using market-based finance, land acquisitions, and slum evictions to restructure urban space prioritizing business interests. Each case differs in details and context, for sure; nonetheless, these efforts can provide guidance in what alternatives might be possible for sites like Dharavi and Kathputli Colony.

Urban design and urban informality: efforts and limitations

In Belapur in Navi Mumbai (a 1970s planned expansion of Mumbai), architect Charles Correa developed a low-income settlement using a traditional urban form of housing clusters organized around courtyards to create micro-neighborhoods that are able to maintain their social networks (Brown 2012). Correa derived this paradigm from courtyard dwellings in South India, which share many similarities with those the artists created in Kathputli Colony. The interlocking courtyard dwelling is better suited to accommodate practices of everyday life in a low-rise, high-density built form compared to high-rise projects. The site interprets Turner's "sites-and-services" approach (Turner 1968) (Figure 13.4) to achieve a spatial network of high-density, low-rise clusters around courtyards and streets, integrating different socioeconomic and religious groups (Davidson & Serageldin 1995).

Likewise, the Aranya housing project in Indore, by architect B.V. Doshi, shares many common elements. Aranya and Belapur have achieved densities of 650 and 500 persons/hectare, respectively, compared to the current density of Dharavi, which is about 1,000 persons/hectare. Following in this tradition, architect and planner Revathi Kamath created a plan for an artist village at Kathputli Colony called *Anandgram*, which never materialized due to budget constraints and reluctance by the Delhi Development Authority (DDA) to allocate the entire site for artists to the exclusion of non-artist residents (Dubey 2016). Kamath studied the spaces the artists had generated in Kathputli Colony and their understanding of space from their lives in the village, and sought to reproduce them in the new design using a mix of modern and traditional materials and techniques (Figure 13.5). She engaged the residents to document social networks and kinship ties to ensure they be maintained in the new design. The residents would have built their own homes using the design Kamath generated with their input. The proposal would have retained the spatial patterning of interlocking courtyard clusters (including scale and massing) as well as the social fabric that residents desired (Interview with Kamath 2013).

Yet efforts like Belapur and Aranya that explore alternative models for low-income housing and slum rehabilitation have received their own share of criticism; they have been labeled utopian, based in nostalgia, and pastoral, even as they have received national and international accolades (Datta 2008; Verma 2002). Much like with the aestheticization-of-poverty critique advanced by Roy (2004), these efforts are seen as piecemeal and superficial. Most importantly, they are seen as mere stopgap measures, doing little to prevent the production of slums and inefficient in overturning the massive structural barriers to meaningful social advancement for India's poor. While these critiques possess merits, I argue that doing nothing to alleviate the deeply unhealthy conditions of the urban poor, while waiting for all structural inequalities to be eliminated, consigns multiple generations to lower opportunity for health and well-being. Moreover, with the development pressures that Delhi and Mumbai both face, doing nothing in the short term is no solution at all – it simply leaves too much room for the forces of global capital to push their own

Figure 13.4 Belapur incremental housing, Navi Mumbai. Image from 2014; sketch of the micro-neighborhood module showing a grouping of seven units around a courtyard. Redrawn by Amit Ittyerah from Charles M. Correa, The New Landscape.

Source: Author and Correa (1989: 56)

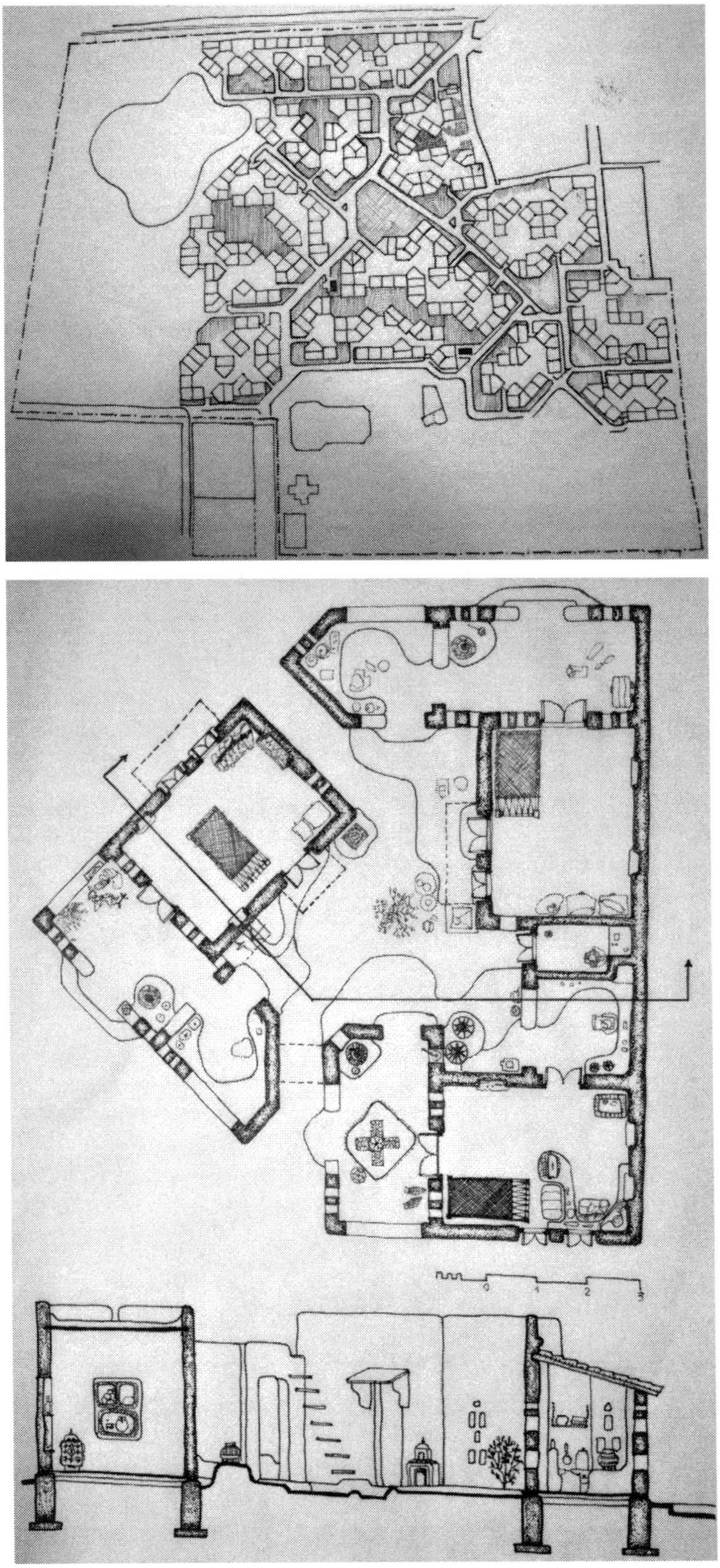

Figure 13.5 Anandgram: proposed site plan and cluster plan and section.

Source: Redrawn by Sreya Sreenivasan from original drawings from "Kamath Design Studio – Architecture, Planning and Environment." Used by permission solutions that are worse for the residents of informal settlements. While approaches like Belapur and Aranya – which rejected the modernist paradigm and deliberately used traditional forms to explore slum housing – may not have served as a panacea for urban disparities, they have been more successful than the modernist housing projects produced by SRA in Mumbai and DDA in Delhi.

It is easy to imagine that the abandonment of these types of approaches is due solely to economic forces; that in the current neoliberal climate of India, cities like Delhi and Mumbai want only to maximize the financial return they can get from redeveloping informal settlements. This, of course, occurs through the sort of approaches that are currently in vogue – relocating residents to far-out high-rise projects or resettlement colonies and freeing up all in-city real estate for high-end uses; or, as a compromise, poorly constructed *in situ* apartment blocks for residents on part of the property, while freeing the bulk for high-end development. This focus on return-on-investment that emerges from public-private partnerships undoubtedly explains much of the current trend. But it would seem that, less obviously, design professionals themselves might also share some responsibility. Informal settlements exemplify "everyday urbanism," reflecting a "fabric of space and time defined by a complex realm of social practices – a conjuncture of accident, desire, and habit" (Chase, Crawford, & Kaliski 2008: 6). But attempts to understand the internal logic of these informal settlements are often dismissed by other planning professionals and academics as romanticizing or fetishizing poverty. While some designers – most notably Correa and Doshi – have found reward in their work in this realm, it remains far too uncommon and not much respected within the field. If the design fields fetishize anything, it is the idea that grand design is more powerful than "ordinary" design, even in terms of providing opportunity for the economic, health, and social well-being of the vast numbers of urban poor. But that vision has always been more about the wishes of the designer than the reality on the ground. At the other end of the spectrum is the belief that only efforts at structural change have any meaningful long-term impact – but this consigns multiple generations to urban conditions that are more marginal than they need to be. In between lies a host of "ordinary design" approaches that work with residents through incremental but meaningful spatial improvements that have the potential to improve the living conditions of the many millions living in informal settlements in India and beyond. In the words of Mehrotra (2007), "we cannot jump from our present situation to some ideal condition. The city here is not about grand design but about grand adjustment."

Aftermath

During the publication process for this volume, the remaining residents of Kathputli Colony, Delhi – one of the two case studies in this chapter – were forcefully evacuated from their homes under the order of the Delhi Development Authority, who then razed the settlement to the ground. Along with the physical fabric, all of the complex community networks and livelihoods were forever disrupted. The population was left isolated and even more vulnerable, trying to make sense of their lives, now in a transit camp. This occurred too late to update the chapter in its entirety. The underlying lessons of the work remain the same, however, and are indeed only accentuated. The fate of Kathputli is emblematic of countless other informal settlements, where residents have to negotiate their lives as permanent transients dealing with the heavy-handedness of the state. Even then, they cautiously invest in their built environments with little municipal support and with no prospect of building secure long-term real estate equity. The negative impacts of this repeated dislocation on their lives, livelihoods, families, communities, social networks, and economic security, in the present and down through the generations, are immeasurable.

References

Boano, C., Lamarca, M., and Wade, A. (2014). "Deconstructing and Recalibrating Urban Design in the Global South." In Carmona, M. (ed.), *Explorations in Urban Design: An Urban Design Research Primer.* London: Routledge, 25–34.

Brown, R. M. (2012). "Revivalism, Modernism and Internationalism: Finding the Old in the New India. In Costa, A. P. D. (ed.), *A New India? Critical Reflections in the Long Twentieth Century.* London: Anthem Press.

Chalana, M. (2010). "Slumdog vs. Millionaires: Balancing Urban Informality and Global Modernity in Mumbai, India." *Journal of Architectural Education, 63*(2): 25–37.

Chalana, M., & Hou, J. (2016). *Messy Urbanism: Understanding the "Other" Cities of Asia.* Hong Kong: Hong Kong University Press.

Chalana, M., and Rishi, S. (2015). "Twilight in Delhi: The Failure of Slum Rehabilitation Programs in Creating a Healthier City." In Wagner, F.W., Mahayni, R.G. and Piller, A.G. (eds.), *Transforming Distressed Global Communities: Making Inclusive, Safe, Resilient, and Sustainable Cities.* London: Routledge, 245–264.

Chalana, M. and Rishi, S. (2016). "Making Sense of the Order in the Disorder in Delhi's Kathputli Colony." In Chalana, M. and Hou, J. (eds.), *Messy Urbanism: Understanding the "Other" Cities of Asia.* Hong Kong: Hong Kong University Press, 155–175.

Chase, J., Crawford, M., and Kaliski, J. (2008). *Everyday Urbanism* (Expanded ed.). New York: Monacelli Press.

Correa, M. C. (1989). *The New Landscape.* London: Butterworth Architecture.

Crane, R., and Manville, M. (2008). "People or Place? Revisiting the Who versus the Where of Urban Development." *Land Lines, 20*(3): 2–7.

Datta, A. (2008). "Architecture of Low-income Widow Housing: 'Spatial Opportunities' in Madipur, West Delhi." *Cultural Geographies, 15*(2): 231–253.

Davidson, C., and Serageldin, I. (1995). *Architecture Beyond Architecture.* London: Academy Editions.

Dubey, S. (2016). "Understanding Participation in a Heterogeneous Community: The Resettlement of Kathputli Colony." In Chakravarty, S. and Negi, R. (eds.), *Space, Planning and Everyday Contestations in Delhi.* New York: Springer.

Jeffrey, C. and Harriss, J. (2014). *Keywords for Modern India.* Oxford: Oxford University Press (UK).

Koolhaas, R. (2007). *Lagos: How It Works.* Zurich: Lars Müller.

Makoko Floating School. NLE. (2013). *Architectural Review, 234*(1402): 68–69.

McFarlane, C. (2004). "Geographical Imaginations and Spaces of Political Engagement: Examples from the Indian Alliance." *Antipode, 36*(5): 890–916.

Mehrotra, R. (2007). "Urban Age: A Worldwide Investigation into the Future of Cities." www.urban age.net/10_cities/07_mumbai/_quotes/mumbai_quotes.html Retrieved August 3, 2009.

Mukhija, V. (2011). "Urban Design for a Planet of Informal Cities. In Banerjee, T. and Loukaitou-Sideris, A. (eds.), *Companion to Urban Design.* London: Routledge, 574–584.

Roy, A. (2004). "Transnational Trespassings: The Geopolitics of Urban Informality." In Roy, A. and AlSayyad, N. (eds.), *Urban Informality: Transnational Perspectives from the Middle East, Latin America, and South Asia.* Lanham, MD: Lexington Books, 289–317.

Roy, A. (2005). "Urban Informality: Toward an Epistemology of Planning." *Journal of the American Planning Association, 71*(2): 147–158.

Roy, A. and AlSayyad, N. (2004). *Urban Informality: Transnational Perspectives from the Middle East, Latin America, and South Asia.* Lanham, MD: Lexington Books.

Scott, J. C. (1998). *Seeing Like a State: How Certain Schemes to Improve the Human Condition Have Failed.* New Haven: Yale University Press.

Smith, C. E. (2007). *Design for the Other 90%*: Cities. Cooper Hewitt, Smithsonian Design Museum.

Turner, J. C. (1968). "Housing Priorities, Settlement Patterns, and Urban Development in Modernizing Countries. *Journal of the American Institute of Planners, 34*(6): 354–363.

Verma, G. D. (2002). *Slumming India: A Chronicle of Slums and Their Saviours.* New Delhi and New York: Penguin Books.

Further reading

Chalana, M., and Hou, J. (2016). *Messy Urbanism: Understanding the "Other" Cities of Asia.* Hong Kong: Hong Kong University Press. This work examines a range of cases in Asia to explore the social and institutional politics of urban informality and the contexts in which this "messiness" emerges or is constructed while raising questions about the production of cities, cityscapes, and citizenship.

Imparato, I. and Ruster, J. (2003). *Slum Upgrading and Participation: Lessons from Latin America.* Washington, DC: World Bank Publications. Formal housing markets around the world are unable to meet the needs of poor urban residents; this work explores successful examples of participatory upgrading from Latin America that improved the living conditions of residents.

Smith, C. E. (2011). *Design with the Other 90%: Cities.* New York: Cooper Hewitt. Professional designers have traditionally focused on the 10% of the world's population that can afford their goods and services; this work instead explores ways that design could improve the lives of poor and marginalized communities around the world.

Weinstein, L. (2014). *The Durable Slum: Dharavi and the Right to Stay Put in Globalizing* Mumbai. Minneapolis: University of Minnesota Press. Despite efforts by Mumbai to erase and redevelop, Dharavi has persisted. This work documents Dharavi's durability despite increasing pressures of urban development and globalization.

14

THE EVERYDAY EXCEPTIONALISM OF TEMPORARY SETTLEMENTS

The role of urban design response

Rachel Berney

Introduction

> Everyone has the right to a standard of living adequate for the health and well-being of [themselves]and [their] family, including food, clothing, housing and medical care and necessary social services, and the right to security in the event of unemployment, sickness, disability, widowhood, old age or other lack of livelihood in circumstances beyond [their] control.
>
> *(United Nations General Assembly, Universal Declaration of Human Rights (1948), Article 25)*

Currently, 68.5 million people throughout the world are displaced from their homes (UNHCR 2017a). An additional 1.6 billion people are considered inadequately housed (Chamie 2017), and 200 million people have been affected by natural disasters in the first decade of this century – 98% of them in the Global South (Aquilino 2011: 7). The numbers of those who are unsettled globally make "unsettledness" a prominent characteristic of contemporary urbanism. Al-Abweh (2017: 3) defines unsettledness as frequently changing the place in which one dwells or lives; this is one of the characteristics of those who are houseless as well as refugees and others who move by force or by choice. Those who are unsettled are often also described as occupying spaces not considered rightfully theirs and perceived as "the other."

Unsettledness takes different forms, including refugee camps, emergency shelters, and growing conditions of homelessness. It is typically born out of critical events or circumstances – natural disasters, conflicts, climate change, and/or economic peril. Aquilino (2011: 287) broadens this understanding to include all who are "vulnerable and impoverished" throughout the world, maintaining that "entire countries, entire populations live precariously, day-to-day, in a perpetual state of emergency." The conditions that lead to this precariousness are not limited to the Global South. The overwhelming nature of displacement and homelessness in US cities and counties has caused no fewer than ten of them since 2015 to declare a "state of emergency" (Weber & Mulvhill 2017).

This chapter explores urban design responses to unsettledness. It focuses on those who become unsettled and the consequences of displacement in the built environment, including rapid growth, shifts in population location, changing needs for services, and conflicts over public space. The chapter provides an overview of unsettledness, its relationship to everyday life, and the rise of an "urbanism of exception" experienced in human settlements globally. It then presents the challenge for urban design and a path forward.

Experiencing unsettledness

People experience different ways of being unsettled, from refugee camps and emergency shelters, to those displaced into houseless camps, to squatters and pavement dwellers, among others. In this chapter, I focus on the overarching condition of unsettledness and the role of urban design to positively address the growing issue of urban unsettledness. While some settlements and camps are located in rural or border regions, the reality is that more than 60% of the world's refugee population and 80% of internally displaced persons under UNHCR mandate live in urban environments (UNHCR n.d.), and a majority of the US houseless population lives in urban areas as well (Misra 2017).[1]

The task of aiding those who are unsettled around the globe is monumental because of their sheer numbers. Nation states are losing the ability to close "temporary" settlements and are growing less capable of delivering long-term, "durable solutions" that can help people rebuild their lives, such as allowing refugees to settle in countries in which they are currently living, returning to their countries of origin, or resettling in other countries (UNHCR 2017b). On the other hand, informal encampments are always under threat of removal. Houseless populations in the United States struggle to keep their camps in place, even if they are a "de facto waiting room for affordable and accessible housing" (Donovan 2010: 6).

Why should urban design be concerned with unsettledness? The immense number of people who are unsettled begs the questions of where they can go, and how cities can better accommodate them. How do we support people's access to a dignified life and a standard of living that reaches towards that described in Article 25, quoted at the start of the chapter? Urban design can help redress the precariousness and vulnerability felt by many who are unsettled. Acknowledging that temporary settlements are no longer temporary creates an urgency to plan and design cities to accommodate them. Caring about people's everyday experiences is key: opposing the "othering" of people and places should be a fundamental aspiration of contemporary urban design.

Everyday life and an "urbanism of exception"

Being unsettled can mean living with significant uncertainty and vulnerability. Mendel (2011: 157) characterizes this as a "'dual reality,' in which the houseless are part of the common world, but because they are located on its margin, they are also beyond it, and they look at it from a distance." Those who are unsettled frequently occupy spaces of "exception." For them, exception can be understood as a series of paradoxical circumstances with which they contend daily, including belonging and not belonging, temporary and permanent, legal and extralegal, and so forth.[2] I argue that we should broaden our thinking on the experience of exception and the exceptional – to include not only the extreme situations and spaces that emerge from human and natural disasters, but also the exceptional circumstances in everyday life. Because, after all, spaces created by exceptional circumstances are also everyday places, shaped by the daily needs, movements, behavior, and hopes of the people residing within them.

In the work surrounding the concept of "the right to the city," I find a compelling counter response to the multiple exceptions that those who are unsettled experience. First expressed by Lefebvre (1968), this right is strongly tied to how people survive and make use of the spaces they come to inhabit, and how they engage in the "social production of space" (Lefebvre 1991) – producing places based on personal and community values, actions, and meanings. Harvey (2008: 1) further characterizes a right to the city as a fundamental right, which, I have argued, is connected to first- and second-generation rights (Berney 2013: 154–155). Harvey (2012: xv) also calls out the collective nature of pursuing a right to the city, calling the process a "means" rather than an "end" that is always evolving through negotiation and contestation. Velásquez Carrillo (2004: 3) further emphasizes the right to the ability to construct one's own life via "equal access to the city and its resources, to exercise full citizenship, and to be provided the capacity to construct his or her life and to participate in the equitable development of the city."

It is through the potential of and struggles for a right of each person to "the city" – literal or symbolic – that everyday life is practiced. Highmore (2002: 2) points out that in modern life "the everyday has been continually invaded." An interest in everyday life in relation to unsettledness can take many turns, but Highmore (2002: 5) highlights that the most central concern as one of a duality of resistance and/or power. He asks: "Is the everyday a realm of submission to relations of power or the space in which those relations are contested (or at least negotiated in relatively interesting ways)?" That negotiation suggests a "blurring" between camps and cities and also within camps themselves, as they become more like cities (Murray 2017: xii, 234–235). I argue that this blurring allows for a diversity of spaces and experiences and is a form of engaging a right to the city.

This diversity is echoed in Murray's (2017) description of contemporary urbanism as an "urbanism of exception," one driven by enclaves and heterogeneous settlement patterns, different logics, and experiences. This view of urbanism allows temporary settlements to be seen as another manifestation of this process, in which rapidly accreted territories with differing levels of service and investment take on regional urban form, especially in areas with rapid population growth and/or a rapid rise in the cost of living (Figure 14.1). In Murray's (2017: x) estimation,

Figure 14.1 This tent camp in Seattle is an enclave within a neighborhood. It is tolerated but not legal, and it lacks services.

Source: Rachel Berney

the driving logic of the form of cities is a continuation of the creation of neoliberal, socially acceptable, privately managed spaces that arose in the late 20th century and that is shaping cities around the globe. Whether formal or not, places for those unsettled are part of a "disaggregated urbanism" composed of "assemblages, work in progress, and enclaves."

Over the last decades, one of the most pervasive urbanism trends is what Murray (2017: 9) calls the "breakdown of centrally-managed public resources," including the breakdown of benevolent planning on behalf of the public good. With little oversight of the process by which communities rise or fall in terms of infrastructural inputs and levels of service, the result is what Marvin and Graham (2001) have called "splintering urbanism" – the differential provision of infrastructure based on a community's ability to pay or on its strategic importance. Splintering urbanism reflects and perpetuates the "othering" of places and creates and/or exacerbates disparities between the people who reside in those places and others.

Exceptional and every day at the same time, these places and people are frequently viewed as difficult, different, and/or dangerous. This perception is partially driven by the fact that those who are unsettled display private behavior in public places. Indeed, for people living in camps or who are houseless living in public spaces, public and private activities often overlap. Watson (2006: 161) describes private activities as those "framed with reference to bodies and corporeality," such as sleeping, eating, personal care, and working.

Al-Abweh (2017: 19) describes the life of the US houseless population through an interpretation of Ray Oldenburg's (1989) "third place" by contrasting Oldenburg's tidy classification of "first" (home) and "third" (public) places with the reality that "the houseless community conducts both its private and public activities in the public realm and spends most of their time there as well."[3] This represents a conflation of Oldenburg's "first" and "third" places. Oldenburg (1989: 16) writes that third places "host the regular, voluntary, informal, and happily anticipated gatherings of individuals beyond the realms of home and work." These spaces, thought to help create the stage for participatory politics, social mixing, and social equity, now must contain a wider range of activities, especially private behaviors that might be perceived negatively by others. Normalizing this shared experience is a new challenge for Lisa Peattie's (1998) notion of "conviviality," which emphasizes common experiences, informal social connections, and a sense of community in public spaces. A lesson can be learned from the Latin American norm that has traditionally allowed for a broad spectrum of private and public use in public space (Low 2000), though challenges must be negotiated. My work in Bogotá (Berney 2011, 2017) revealed that when various city actors believed the scale has tipped too far towards private use of public space, corrections occurred in the form of vendor and houseless harassment and relocation.

The challenge for urban design and the path forward

How can urban design help address the situations of those who are unsettled in a manner that does not diminish people or place? This is a challenging question. Placemaking is often limited to issues of quality of place rather than quality of life; two conditions that are often conflated to the detriment of community well-being (Berney 2017: 129). Instead, we need to acknowledge that placemaking is a political activity that can either contribute to or lessen just outcomes and that "making place" is fundamentally about shaping the spaces that provide for all people. Many of these actions fall under the rubric of "public-interest" design.

A subset of environmental design professionals engages in public-interest work. However, designers do not regularly figure into long-term solutions for crises, despite the great impact that urban conditions and the design of places (or lack thereof) may have on unsettled populations. Aquilino (2011: 7) notes that architects promulgate "a range of creative strategies to improve social, environmental, and economic equity . . . but in extreme circumstances, in crises, architects offer no coherent response." Urban designers, likewise, lack an articulate, action-based, transformative agenda. Urban design is no panacea, but since the field first emerged as one that is sensitive to social and political concerns in human settlements, it should engage its own credo now.

For urban design to address unsettledness at the scale at which it is now present in the world, designers, who are often more comfortable designing small and intimate spaces, need to embrace the metropolitan scale in their work. At the same time, government and its agencies need to promote a vision and support for design. Urban design should become relevant at the metropolitan scale and through publicly backed visioning (Berney 2017). These goals could be met through the efforts of public actors and public-private partnerships robust enough to provide visioning, financing, and management.

However, urban design is currently confronted with several challenges to its relevance, including the need to expand its "scope and perspective" (Loukaitou-Sideris 2012: 476). Loukaitou-Sideris and Banerjee (1998) have noted a diminishing role for urban design in the public sector as well. Murray's (2017) description of the loss of benevolent public management echoes this. Another great challenge for urban design is how to handle rapid change in and near cities. This includes continuing to rethink and address how urban design deals with displacement, unplanned growth, and extending a right to the city to all residents – new and old. The spatial logic and function of places should be reinterpreted towards more just outcomes.

What I propose is a stretching of the public realm in a deliberate, designed manner to better support everyone living in cities, prioritizing those who are unsettled and need it most. By making use of publicly owned land in a more adaptive manner, urban design can include those unsettled, improve neighborhoods, and provide housing. This also includes creating emergency and disaster-management plans for temporary use of formal and programmed open space.

In "Environmental Adaptability," Kevin Lynch (1958: 379) critically examined the idea of flexible environments. He outlined two possibilities for flexible cities – a wide variety of choices and the ability to shape one's own environments – before landing on adaptability – "general adjustment of an environment or artifact, with minimal effort, to future changes of use," as an appropriate response to the problem of "*unspecified* future change." Lynch's recommendations focus on concentrating essential supportive structures into cores, with surrounding functional areas at a rather coarse grain or scale, especially in areas that can be identified as likely and/or easy to change. He also notes the usefulness of extra capacity in any design to allow for growth and change over time, as well as the importance of interrelationship and communication among entities to be able to respond quickly to change. Lynch's analysis suggests performance characteristics for improving adaptability and resilience in cities. This includes providing functional cores for housing, parks, and other places that bring essential services to people, while allowing for expansion or remodeling at a later time. It also highlights the applicability of a metropolitan-scale urban design response that combines communication, organization, and vision.

A challenge and a moral obligation of urban design is to think of camp and settlement building as part of an adaptable approach to urban development, in which select publicly owned land in cities might be used to provide temporary housing for those who are unsettled. At the city scale, then, government-owned vacant land that is not immediately needed and is appropriate for housing should be made available to lease in the short or medium term. Other concerns at the metropolitan scale that would support these efforts include access to services and amenities, such as equitable access to safe transportation modes (Berney 2018: 1) and reasonable standards of affordable civic infrastructure.

Specifically, one way to move away from temporary settlements as privately managed spaces of everyday exception is to draw them into the public process of building and managing "collective infrastructures" within cities. Collective infrastructures are assemblages and systems of pubic goods that are created and managed as a system. These systems rely on public input and resident engagement, providing support for the civic and social life of the city.

Creating collective infrastructures is a new urban design strategy to link those who largely lack access to public resources. It is a process of increasing people's right to public space, mobility, and well-being in cities. For example, in Bogotá, Colombia, the rise of collective infrastructure occurred in the 1990s and early 2000s as bus and bicycle routes, libraries, and public spaces were developed and made available to all residents. Networks of public goods were intentionally put in places where lower-income people could use them and to support the mixing of different socioeconomic groups (Berney 2017). Visitors and vendors working in the city's public spaces indicated high levels of satisfaction with these changes (Berney 2017).

As this successful example from Bogotá indicates, urban design needs to look beyond the Global North for solutions. Murray (2017: 4) cautions that "western models are too often used to understand the trajectories of global urbanism." With an urbanism of exception and exclusion in the Global North and the Global South, there is opportunity to draw ideas for change from both contexts and for urban design to embrace models, examples, and hybrids via cross-pollination of practices and ideas. In what follows, I discuss some past and present examples from the Global North and the Global South of creative design responses to the challenge of unsettledness.

Adaptable cities

Public lands and public space have been used to house city populations after disaster. The 1906 earthquake and fire that leveled much of San Francisco, California, displaced more than half of the city's population. Displaced residents migrated towards the lesser-developed areas of the city, and 20,000 people were housed in US Army–run refugee camps located within the city's public parks, including the Presidio and Golden Gate Park (Bevk 2015). Longer-term shelter for those displaced came from an arrangement between the army, the San Francisco Parks Commission, and the San Francisco Relief Corporation, which constructed "tiny houses" on parkland (Figure 14.2).

Figure 14.2 A San Francisco tiny house and tenant.

Source: San Francisco History Center, San Francisco Public Library

Figure 14.3 Tiny house being moved to a new location.

Source: San Francisco History Center, San Francisco Public Library

Used to shelter 16,000 people, these cottages ranged from 10 feet by 14 feet (140 square feet/13 square meters) to 14 feet by 18 feet (252 square feet/23.4 square meters) (Bevk 2015).[4] A house costs $50 to buy through a $2 per month rent-to-own program.[5] The new owners of these tiny homes were responsible for relocating them to private property (Figure 14.3). Besides helping to replenish the housing supply, the tiny houses helped to create first-time homeowners from those of lower incomes (Bevk 2015).

Colombia has a history of "pirate subdivisions," or neighborhoods of homes built outside of planning regulations. While such informal and semi-formal growth has expanded in Colombian cities, especially in the second half of the 20th century, recent urban design efforts have connected these types of neighborhoods to the services and fabric of the formal city (Berney 2017). Bogotá's Patio Bonito neighborhood, a low-income, informally settled community, was connected in the 2000s to city systems via a water main inserted down a main street with a *paseo* (pedestrian and bicycle walkway) installed on top (Figure 14.4). The *paseo* allows people to move out of the muddy streets through which the cars must navigate and connects the community to a nearby TransMilenio bus rapid transit station. In Medellín, Colombia, transit investments in the form of gondolas and hillside escalators have linked lower-income settlements perched high above the downtown with jobs in the center, allowing people to easily and quickly reach their work.

Other contemporary examples are emerging from the use and design of public land in cities to accommodate those who are unsettled. Melbourne, Australia, is currently moving ahead on a project that uses a public-private partnership to supply 57 units of affordable temporary rental housing in its tight rental market on currently unused public land. The lease on the land, set at a peppercorn rent, is held by the Department of Health and Human Services. Launch Housing, a community-based organization providing services to the houseless, subleases the land from them.[6] The lease is set at an initial period of five years with a 12-month notice-to-vacate clause

Figure 14.4 Patio Bonito *paseo* under construction in Bogotá.

Source: Rachel Berney

that can be activated if the State of Victoria's Department of Transportation requires the land to be returned. Launch Housing guarantees the re-housing of tenants, if they need to be relocated in the future (Launch Housing 2017). According to them, the project "has the potential to grow and extend to other government and nongovernment bodies who hold blocks of vacant land that can be used to create high quality, relocatable housing to dramatically reduce the number of people experiencing homelessness in Melbourne" (Launch Housing 2017).

Other short- to medium-term solutions for providing shelter in public space include safe lot programs and tiny home villages. Seattle's 2016 Safe Lot program hosted people sleeping in vehicles, including RVs, on public land and provided sanitation, waste pickup, and social services (Beekman 2016). While the Seattle program struggled with costs (Jaywork 2017), the "Dreams for Change" privately run safe-lots program in San Diego shows a path forward towards safe, cost-effective solutions for vehicle homelessness and moving people into housing (Davila 2018). Tiny house villages are springing up in cities across the United States. Seattle hosts eight villages; the most recently opened is one for women only (Clark 2018).

In Los Angeles, the shortage of affordable shelter has led to a rental system of box trucks and RVs that sit on public streets. Owners move the vehicles they manage every 72 hours to avoid citations and text their tenants to tell them where they have moved their home (Palta 2017). While this is problematic for several reasons, a more formalized program could be created using tiny houses (with or without wheels) that could be located on streets, parking lots, and other publicly owned land.

In Chile, Alejandro Aravena, through his firm, Elemental, has designed the now well-known "Incremental Housing" project. In this model, the architect provides a small house core with the ability for owners to expand over time. According to Aravena: "Instead of designing a small house, we provided a middle-income house, out of which we were giving just a small part now" (in Franco 2017). This type of affordable housing can be used by anyone seeking to start small with what they can afford.

Brice Maryman, a Seattle-based landscape architect, has pursued the project HomeLandLab as an Innovation and Leadership Fellow of the Landscape Architecture Foundation, which combines his long-term interest in the ability and responsibility of designers to make change at the intersection of homelessness and public space. The project "explores the spatial manifestations of homelessness on the landscape, documents current management approaches, and intends to offer comprehensive, community-based spatial strategies to create better, more successful public spaces for all" (Maryman n.d.).

Cities can and must develop their own durable solutions. City leaders play a unique and growing role in people's quality-of-life options (Berney 2017: 140). Leaders should sanction the use of publicly owned land to house those who are unsettled and provide short- to medium-term security of tenure. City housing programs should focus on supporting alternative types of housing such as tiny houses, accessory dwelling units, and small units that have room for expansion and remodeling in the future. Progressive cities are those that offer collective infrastructure to all residents.

Examples of creative design thinking are happening around the world. Urban design can do much more by learning from them as well as by taking on a leadership role in helping cities prepare for future crises and disasters. How does urban design move beyond its current norms to include such resilient responses?

The need for an *urgentiste* agenda

We need critical questioning that broadens the scope of urban design and the role of urban designers. In the preface to *Expanding Architecture: Design as Activism*, Bell (2008: 14) notes that

> to make design more relevant is to reconsider what "design" issues are. Rejecting the limits that we have defined for ourselves, we should instead assume that design can play a positive role in seeking answers to many different kinds of challenges. We have limited our potential by seeing most major human concerns as unrelated to our work.

Aquilino (2011:7) calls for a corps of designers she calls *urgentistes* – those who respond to crisis – to be placed into aid agencies, governmental departments, NGOs, and so forth to provide design, visioning, and project-management skills for resilient responses to crisis and disaster. These design professionals would intervene at crucial moments in recovery processes to produce enduring solutions. Fisher (2008: 12) notes that "we need a career path, and possibly even a profession, of public-interest architecture, parallel to public health and public defense, that has its own educational requirements, practice models, financial support, and client base."

Urban designers need to take an interest in shelter as well as in the spaces and systems that accompany housing and are needed for the good and fair functioning of people's lives. This can be called "just placemaking". The need for shelter is universal, and shelter is "the greatest humanitarian challenge facing us all" (Architecture for Humanity 2006: 1). According to Fisher (2008: 9), "millions of American citizens, as well as billions of people around the world, battle the storms of inadequate services, unaffordable housing, and unsafe neighborhoods on a daily basis." Responding to disasters and crises necessitates creative means for providing shelter and, in the longer term, for supporting services and amenities. The need for community engagement is also important. When houseless residents are involved in the process of design, construction, and management, they feel more empowered and comfortable in the spaces designed for them (Al-Abweh 2017). By engaging the creativity of community members while working with needs and assets, shelter solutions – temporary or permanent – always benefit. As Aquilino (2011: 125) puts it, "It is cheaper and safer to build a neighborhood than a house."

Conclusion

This chapter lays out a framework for understanding unsettledness as a major 21st-century urban issue with which urban design must contend. Through the examination of aspects of unsettledness, it is possible to recognize current challenges to city design as well as important design futures. Contemporary examples of the use and design of public space in cities in the Global North and Global South show a path for urban design to better incorporate those who live in precarious circumstances into the city. These aspects and examples help demonstrate that those who are unsettled are part of an "urbanism of exception" and, at the same time, suggest less exceptional and more ordinary circumstances in which cities incorporate those who are unsettled into their planning and design considerations in more profound and meaningful ways.

In conclusion, I argue that urban design should act and expand its purview. More specifically, urban design has a role to play in responding to displacement and unsettledness, providing dignity to people, whether they are ending up in refugee camps, houseless encampments, RVs, or public spaces in urban neighborhoods. Resilient communities emerge from supportive cities. Adaptable and supportive neighborhood design helps accommodate different lifestyles and life cycles and better prepares residents to handle a disaster as well as welcome new residents. There is an urgent need for urban design to go down this path and meet this responsibility.

Notes

1 Drawing on Al-Abweh's (2017) work, I use her term "houseless" rather than "homeless" to acknowledge that being without a house is not synonymous with lacking a sense of home.

2 While I am discussing "exception" more broadly, I want to acknowledge the term "spaces of exception" used to describe the increase in power structures and management that governments employ in supposed times of crisis to justify increased government action in particular spaces (Agamben 2004).

3 In Oldenburg's framing, "third" places are communal spaces for public activities in which people with housing spend the least amount of time.

4 These are similar in size to today's micro-unit residential apartments and tiny houses, which typically measure 100–400 square feet.

5 In 2019 dollars, this would be $1,404 to purchase with a monthly payment of $56 (Official Data Foundation).

6 In the English tradition, an actual peppercorn was offered as the rent so that there would be no need to collect it.

References

Agamben, G. (2004). *State of Exception.* Chicago: University of Chicago Press.

Al-Abweh, R. H. (2017). *What Makes People Feel at Home: Reframing Home and Homelessness Through Stories from Seattle's Tent City 3* (Master's thesis). Retrieved from ResearchWorks Archive, University of Washington, Seattle.

Aquilino, M. J. (Ed.). (2011). *Beyond Shelter: Architecture and Human Dignity.* New York: Metropolis.

Architecture for Humanity (ed.). (2006). *Design Like you Give a Damn: Architecture Responses to Humanitarian Crises.* New York: Metropolis.

Beekman, D. (2016, Jan 9). "Seattle 'Safe Lots' to Hold Homeless Living in Vehicles." *The Seattle Times.*

Bell, B. (2008). "Preface." In Bell, B. and Wakeford, K. (eds.) *Expanding Architecture. Design as Activism.* New York: Metropolis, 14–17.

Berney, R. (2011). "Pedagogical Urbanism: Creating Citizen Space in Bogotá, Colombia." *Planning Theory, 10*(1): 16–34.

Berney, R. (2013). "Public Space versus Tableau: The Neoliberal Paradox in Bogotá, Colombia." In Samara, T.R. et al. (eds.) *Locating the Right to the City in the Global South.* London & New York: Routledge, 152–170.

Berney, R. (2017). *Learning from Bogotá: Pedagogical Urbanism and the Reshaping of Public Space.* Austin: University of Texas Press.

Berney, R. (2018). "Advancing Bicycle Urbanism." In Berney, R. (ed.) *Bicycle Urbanism: Reimaging Bicycle Friendly Cities.* London & New York: Routledge, 1–13.

Bevk, A. (2015, Feb 24). "Remembering Earthquake Shacks, San Francisco's Original Tiny Houses." *Curbed San Francisco.*

Chamie, J. (2017, July 13). "As Cities Grow Worldwide, So Do the Numbers of Homeless." *YaleGlobal Online*, MacMillan Center, Yale University. Retrieved from https://yaleglobal.yale.edu/content/cities-grow-worldwide-so-do-numbers-homeless

Clark, K.M. (2018, May 11). "Seattle's New Tiny House Village for the Homeless – Women Only." *Crosscut.* Retrieved from https://crosscut.com/2018/05/seattles-new-tiny-house-village-homeless-women-only.

Davila, V. (2018, May 3). "San Diego Did what Seattle Didn't: Give People a Safe Place to Sleep in Cars." *The Seattle Times.*

Donovan, N. (2010). *Tent Cities in America: A Pacific Coast Report.* Washington, DC: National Coalition for the Homeless.

Fisher, T. (2008). "Public-interest Architecture: A Needed and Inevitable Change." In Bell, B. and Wakeford, K. (eds.) *Expanding Architecture: Design as Activism.* New York: Metropolis, 9–13.

Franco, J.T. (2017, Jun 18). "The Construction Details of Elemental's Incremental Housing." *ArchDaily.* Retrieved from www.archdaily.com/873477/the-construction-details-of-elementals-incremental-housing.

Harvey, D. (2008). "The Right to the City." *New Left Review, 2*(53): 23–40.

Harvey, D. (2012). *Rebel Cities: From the Right to the City to the Urban Revolution.* London: Verso.

Highmore, B. (Ed.). (2002). *The Everyday Life Reader.* London: Routledge.

Jaywork, C. (2017, May 15). "With Seattle's 'Safe Lots' for RVs All but Gone, Campers Are Creating Their Own." *Seattle Weekly.* Retrieved from www.seattleweekly.com/news/last-year-seattles-safe-lots-for-rvs-all-but-disappeared-now-rv-campers-have-created-their-own/

Launch Housing. (2017, Jan 11). "A Road Map for Affordable Housing." Retrieved from www.launchhousing.org.au/road-map-affordable-housing/.

Lefebvre, H. (1968). *Le Droit à la Ville.* Paris: Anthropos.

Lefebvre, H. ([1974], 1991). *The Production of Space.* Translated by D. Nicholson-Smith, Oxford: Basil Blackwell.

Loukaitou-Sideris, A. (2012). "Addressing the Challenges of Urban Landscapes: Normative Goals for Urban Design." *Journal of Urban Design, 17*(4): 467–484.

Loukaitou-Sideris, A., and Banerjee, T. (1998). *Urban Design Downtown.* Berkeley: University of California Press.

Low, S. (2000). *On the Plaza.* Austin: University of Texas Press.

Lynch, K. (1958). "Environmental Adaptability." In Banerjee, T. and Southworth, M. (eds.) *City Cense and City Design: Writings and Projects of Kevin Lynch (1990)*, Cambridge: MIT Press, 379–395.

Marvin, S. and Graham, S. (2001). *Splintering Urbanism: Networked Infrastructure, Technological Mobilities and the Urban Condition.* Abingdon, UK: Routledge.

Maryman, B. n.d. "HomeLand Lab: About." Retrieved from www.homelandlab.com/about-1/.

Mendel, M. (2011). "Heterotopias of Homelessness: Citizenship on the Margins." *Studies in Philosophy and Education, 30*: 155–168.

Misra, T. (2017, Apr 7). "Mapping America's Intractable Homeless Problem." *CityLab.* Retrieved from www.citylab.com/equity/2017/04/map-america-homeless-crisis/522339/.

Murray, M.J. (2017). *The Urbanism of Exception: The Dynamics of Global City Building in the Twenty-first Century.* Cambridge, UK: Cambridge University Press.

Official Data Foundation. *In2013dollars.com.* Retrieved from http://www.in2013dollars.com/us/inflation/1906?amount=50.

Oldenburg, R. (1989). *The Great Good Place: Cafes, Coffee Shops, Community Centers, Beauty Parlors, General Stores, Bars, Hangouts and How they Get you through the Day.* New York: Paragon House.

Palta, R. (2017, Dec 19). "RVs Are Becoming an Alternative to High Rents in LA." *Marketplace.* Retrieved from www.marketplace.org/2017/12/19/economy/officials-see-emerging-economy-los-angeles-rvs-rented-out-homeless-people

Peattie, L. (1998). "Convivial Cities." In Douglass, M. and Friedmann, J. (eds.) *Cities for Citizens: Planning and the Rise of Civil Society in a Global Age.* London: John Wiley, 247–253.

United Nations General Assembly, Universal Declaration of Human Rights. (1948). Retrieved from http://www.un.org/en/universal-declaration-human-rights/

UNHCR (United Nations High Commissioner for Refugees). (2017a). "Figures at a Glance." *UNHCR: The UN Refugee Agency*. Retrieved from www.unhcr.org/en-us/figures-at-a-glance.html

UNHCR (United Nations High Commissioner for Refugees). (2017b). "Solutions." *UNHCR: The UN Refugee Agency*. Retrieved from www.unhcr.org/en-us/solutions.html

UNHCR (United Nations High Commissioner for Refugees). (N.d.). "Urban Refugees." *UNHCR: The UN Refugee Agency*. Retrieved from www.unhcr.org/urban-refugees.html

Velásquez Carrillo, F. (Ed.) (2004). *Ciudad e Inclusión: Por el Derecho a la Ciudad*. Bogotá: Fundación Foro Nacional por Colombia.

Watson, S. (2006). *City Publics: The (Dis)enchantments of Urban Encounters*. Abingdon, UK: Routledge.

Weber, C. and Mulvhill, G. (2017, Dec. 7). "America's Homeless Population Rises for First Time in Years." *Associated Press News*. Retrieved from www.apnews.com/47662ad74baf4bb09f40619e4fd25a94

Further reading

Agier, M. (2011). *Managing the Undesirables: Refugee Camps and Humanitarian Government*. Translated by Fernbach, D. Cambridge, UK: Polity Press. An in-depth study of refugee camps from an anthropological perspective.

Aquilino, M. J. (Ed.). (2011). *Beyond Shelter: Architecture and Human Dignity*. New York: Metropolis. An overview of architecture's response to humanitarian crises as well as specific case studies.

Loukaitou-Sideris, A. (2012). "Addressing the Challenges of Urban Landscapes: Normative Goals for Urban Design." *Journal of Urban Design*, *17*(4), 467–484. A contemporary manifesto for urban design that follows in the tradition of the CIAM *Athens Charter* and Jacobs and Appleyard's *Toward an Urban Design Manifesto*.

Marvin, S., and Graham, S. (2001). *Splintering Urbanism: Networked Infrastructure, Technological Mobilities and the Urban Condition*. Abingdon, UK: Routledge. An examination of the relationship between infrastructure investment and urban space.

Murray, M. J. (2017). *The Urbanism of Exception: The Dynamics of Global City Building in the Twenty-first Century*. Cambridge, UK: Cambridge University Press. An in-depth study of the constituent parts and overall whole of contemporary global urbanism.

PART II.III

Explosive growth versus shrinkage

Introduction

Explosive growth and shrinkage of urban settlements are polar opposite phenomena; yet both present significant challenges for urban design in cities of the Global North and the Global South. Explosive growth and development, experienced by some cities in South and Southeast Asia and the Persian Gulf, strain physical and economic resources and exacerbate such urban problems as housing affordability, traffic congestion, air and water pollution, and shortfall in city services. Growth is accommodated by densification of the urban core or spreading it horizontally, leading to leapfrog development and sprawl. Scholars find that the emergence of megacities and megaregions is associated with fragmented landscapes, which are relatively insulated and isolated from one another (Wheeler & Beebe 2011).

While the development of decentralizing megacities and megaregions appeared as a characteristic pattern of late-20th-century urbanization, the opposite phenomenon – shrinking populations and declining rates of urbanization – is also evident in some cities, and even in some nations. The reasons behind urban shrinkage vary. Deindustrialization and declining economic fortunes have led to the shrinkage of Rust Belt cities in the United States, the steel towns of the Ruhrgebeit in Germany, and the industrial cities of the former Soviet Union. Hurricane Katrina, a natural disaster, prompted the shrinking of New Orleans, while the rise of sea levels attributed to climate change and the subsequent loss of land are displacing residents of some US coastal towns. According to Brent D. Ryan (in this section), income growth and general prosperity may lead to changed lifestyles and reduced birth rates, and the eventual shrinkage of the population and settlement nationally, as is now happening in Japan. Regardless of the cause of shrinkage, it comes with an array of associated problems: abandoned houses, empty land, and vacant factories and commercial facilities.

How can urban design respond to explosive growth or shrinkage? The focus of most contemporary urban design interventions in many cities of the Global North has been at the level of the street, the block, or the urban neighborhood, such as the development of pedestrian-friendly spaces and outdoor rooms. Nevertheless, the challenges posed by explosive growth or shrinkage require urban design intervention at the metropolitan scale (Loukaitou-Sideris 2012). However, as the chapters in this section clearly point out, these challenges can also be viewed as opportunities for innovative design responses. The goal of this section, then, is not only to discuss the challenges encountered by the different locales experiencing explosive growth or shrinkage, but also to present some successful urban design responses. And while design may not be sufficient to reverse the impacts of demographic or climatic phenomena decades in the making, it can offer some innovative ideas about the built environment that complement responsive social and/or economic strategies

In his chapter about the "explosive urbanization" of China, Daniel B. Abramson identifies some of the negative social and environmental impacts or rapid growth and excessive overbuilding: displacement of existing communities, socio-spatial segregation, destruction of agricultural land, and unwarranted building construction (some of which remains unoccupied), which is frequently designed in a monumental and pedestrian-unfriendly scale. But he also finds "creative, locally responsive, and socially inclusive" spaces that appear unpredictably at the interstices between formally designed spaces. Examples are arts districts, riverfront embankments, self-built spaces in urbanizing villages, and parks. He also notes three recent positive governmental responses: (1) a shift towards a more "fine-grained approach" and incremental building in the place of clean-sweep planning and urban design; (2) a more open and flexible spatial organization of the urban form that discourages superblocks and gated communities and encourages small blocks and well-connected, multimodal streets; (3) a better balance between urban and rural development, and a renewed appreciation for local materials, local heritage, and community life.

Drawing from examples from Singapore, Tokyo, Hong Kong, and Seoul in her chapter, Im Sik Cho details how creative urban design can help overcome the negative externalities of very high urban density. In cities where only miniscule amounts of land remain unbuilt, she finds "elevated, underground and multi-level public spaces, intensified housing developments, mixed-use and transit-related infrastructural spaces." These settings – from the Cheonggye stream reclamation in Seoul, to the elevated parks (sky-gardens and sky-bridges) in Singapore, to the Central Mid-levels Escalator in Hong Kong – are being developed thanks to design ingenuity and regulatory flexibility. They provide much-needed recreational and social spaces in intense and high-density urban areas, and they point to the need to "synergize the various aspects of urban design, from spatial and functional to organizational dimensions."

Studying the impacts of decreasing rather than increasing urban densities, Brent D. Ryan also finds opportunities for successful urban design interventions to address shrinkage at the city and even national levels from the presence of "ample space, reduced competition for amenities, and even an enhanced quality of life." Detailing the experiences of the United States, Germany, Soviet Union, and Japan, he shows, however, that it is not only urban design but also policymaking and legacy of urban form that may determine the success or failure of interventions. Thus, Germany has been able to successfully repurpose industrial facilities into parkland and recreational space in the Ruhr, while Japan is in the process of a spatial reorganization that favors compactness as well as concentration and enhancement of urban amenities at town centers. On the other hand, urban design interventions in the United States and the cities of the former Soviet Union have often been hampered by a lack of funding and political will. Given the differentiation of the socioeconomic and political contexts in shrinking cities or nations, Ryan observes that different spatial approaches may be necessary to address conditions of shrinkage.

Following a similar line of thought, namely that "the socio-physical, economic, and political contexts of each shrinking city or region should influence the responding strategies," Karina Pallagst, René Fleschurz, and Sabrina Förch note that different shrinking cities have initiated different strategies for "shrinking smart." Strategies are defined by the larger political economy and planning framework of cities, and therefore employ different design responses. To illustrate these arguments, the authors contrast the strategies followed by the city of Herne in Germany to those of the city of Cleveland in the United States. In Cleveland, market-driven strategies and a weak and decentralized planning system promoted the development of flagship buildings by famous architects; urban design was seen as a means to attract private sector-initiated economic development and a substitute industry. In Germany, which has a strong multilevel planning system, the public sector assumed a much stronger role in funding and implementing urban greening strategies and green infrastructure.

In the next chapter, Justin Hollander elaborates the concept of "smart shrinkage"; that is, "local government policies that contribute to the conversion of land and buildings from residential to non-residential use." While urban designers are used to accommodate growth, he argues that urban design has an important role to play to manage decline in shrinking cities as well. For example, urban design can convert vacant structures into green public spaces; redesign residential neighborhoods for lower densities, as has happened in Baltimore (Maryland), Flint (Michigan), and New Bedford (Massachussetts); and even reconsider shrinking downtowns in ways that may accommodate vertical farming at the roofs of commercial buildings. He notes that "urban design in shrinking cities ought to focus on small, impactful efforts that can then be rolled into larger, more systemic change."

In conclusion, the common message of all five chapters in this section is that urban design can positively complement other social and economic strategies to accommodate growth or manage decline, but it should be tailored to the particular socioeconomic situations and political economy of each context. Design interventions for the benefit of the larger public in high-density or shrinking urban environments will help counteract some of the challenges that rapid growth or shrinkage entail.

References

Loukaitou-Sideris, A. (2012). "Addressing the Challenges of Urban Landscapes: Normative Goals for Urban Design," *Journal of Urban Design*, 17 (4): 467–484.

Wheeler, S. & Beebe, C.W. (2011). "The Rise of the Postmodern Metropolis: Spatial Evolution of the Sacramento Metropolitan Region," *Journal of Urban Design*, 16 (3): 307–332.

15

THE INSURMOUNTABLE OPPORTUNITIES OF EXPLOSIVE GROWTH

Urban design in China

Daniel B. Abramson

Introduction: the design of explosive urbanization

China presents an extreme case of "explosive" urbanization. Compared to other nations in the Global South that are also experiencing massive urban-rural migration, China has attracted an exceptional level of investment for large infrastructure projects, housing, and other developments (Angel 2012). It has done so throughout the country, in cities old and new, large and small, richer and poorer, and following a relatively high degree of central coordination, planning, and standardization. Somewhat paradoxically blended with the country's top-down bureaucratic governance is an entrepreneurial political culture that encourages local initiative (up to a point) and treats each development project as a potential national model, garnering fame and promotion for its backers. On the one hand, local officials and developers have considerable resources and incentives to create models of good practice and raise their political profiles and economic credit through innovative projects, encouraging designers to envision major improvements. On the other hand, the power of higher levels of government to promote and scale up local models holds out the promise that good design will have a widespread impact.

However, the very system of large-scale bureaucracy that enables the replication of successful projects also tends to strip them of their responsiveness to local conditions. For every ambitious local leader and developer who accepts the risk of investing in innovative design, there are countless others who take the safer path of hastily and uncritically adopting existing models for their own jurisdictions. In between, there are also countless shades of localization in which standard practices are adapted to site-specific geographic, historical, and socio-political conditions. Regardless of how many projects may ultimately prove sustainable, the speed, extent, and official intentionality of China's urbanization requires a specificity of design at a wide range of scales and grants urban designers an exceptional scope to influence the future of the built environment.

This chapter reflects on how China's outsized experience of explosive growth has expressed itself in urban design, and how some of the opportunities this growth presents have turned out to be illusory, even as others are now emerging and presenting the discipline with new challenges. These reflections are therefore partly a cautionary tale, and partly an attempt to direct professional designers to promising directions. First, however, some official statistics give a general sense of what "explosive" means, in relative and absolute terms. In 1975, China's urban population nearly equaled that of the United States, about 160 million. By 2015, it had more than quadrupled to nearly 780 million; India's had more than tripled, putting it in second place with nearly 420 million city dwellers (United Nations 2015). During the 1990s, China's total investment in completed urban infrastructure projects increased by 1,300% (National Bureau of Statistics of China 1999a); urban built-up land expanded by 69% (National Bureau of Statistics of China 1999b; National Bureau of Statistics Urban Social and Economic Survey Team 1990); the number of households increased three times faster than the population at large (Liu and Diamond 2005); and, from 1990 to 2003, the floor area of total urban housing increased by more than 250% (Ministry of Construction General Statistics Section 2000; National Bureau of Statistics Urban Social and Economic Survey Team 1990, p. 18). More recent estimates of urban expansion from 1992 to 2015 show a nearly fivefold increase in urban land cover, almost 2.5 times as rapid as the global average (Xu et al. 2016).

Intra- and interurban transportation infrastructure has also undergone a revolution. Total urban road area doubled between 1990 and 2000 (National Bureau of Statistics of China 1999c), but the number of cars owned by Chinese more than quadrupled in that time (Myers and Kent 2003). In Beijing, just from 2001 to 2003, the total length of roads increased by over 25% and the length of rail transit by about 50% (Beijing Municipal Statistical Bureau 2005). Over the past 20 years, Beijing's subway system expanded from only two lines to 21, with a total length that is second only to Shanghai's in the world. Car ownership, however, has increased even more quickly. It took the first 48 years of the People's Republic (1949–1997) for the number of automobiles registered in Beijing to reach 1 million; over the next six-and-a-half years, the number reached 2 million, of which 1.65 million are private. In 2005, the number increased by 1,000 vehicles each day (Liu 2005). China's expressway system is now the most extensive in the world, having surpassed that of the United States in 2011. The national high-speed rail (HSR) system emerged only in 2007 but, ten years later, accounted for two thirds of the world's commercially operating HSR service.

Although China's growth may be quantitatively unprecedented, it is not historically unique in qualitative terms. The growth of American cities in the late 19th and early 20th centuries similarly impressed the world with its explosiveness, and it inspired fundamental changes in the norms, forms, and techniques of urban design (Campanella 2010; Giedion 1941). To a smaller extent, but with similar speed and forms, the emerging urbanity of other East and Southeast Asian countries has also challenged some of the most basic ideas of what constitutes "good" urban design, as defined by the experience of earlier-urbanized societies (Marshall 2003; 2004; Rowe 2005, 2011). As Tom Campanella (2009) has argued, China's growth should inspire urban designers around the world to articulate new disciplinary principles and practices as bold and ambitious as those that accompanied the first wave of global industrial urbanization in the West, a hundred years ago. Instead, American- and European-influenced urban design remains "steeped in nostalgia," and paralyzed by a timid and small-minded focus on process and deliberation rather than on forward-looking visions.

To discover the inspirational value of China's growth for global urban design, it is important first to appreciate the actual "messiness" of that growth (Abramson 2016a), and to try to position it historically, with respect to China's longer-term internal development as well as the urbanization of other societies. Most features of China's current urban planning and design system are the peculiar products of the transition from Mao-era command-control economics, which emphasized national self-reliance, to Deng Xiaoping's global market-oriented Reform and Opening policies (Wu 2015). The transition featured a massive release of three decades of pent-up export-oriented industrial and trade capacity, punctuated by moments of politically high-profile "mega-events" attracting investment on a national scale, such as the hosting of the 2008 Olympic Games in Beijing, the 2010 World Expo in Shanghai, and the reconstruction after the 2008 Wenchuan earthquake in Sichuan (Abramson and Qi 2011; Broudehoux 2017). A megaproject and superblock approach tended to dominate design practice, which was in part a legacy of Mao-era, Soviet-influenced planning, but also simply the most efficient way to accommodate growth given a scarcity of designers and an embryonic regulatory system in the 1990s (Abramson 1997, 2010; Lu 2006).

Explosive growth and its discontents

The sudden increase in size and speed of projects in the 1990s and 2000s created an intoxicating atmosphere for urban designers. For some, growth held out the promise of restoring or adapting long-neglected historic quarters on a large scale, and reviving principles of urbanism that have been tested and evolved for centuries (Chen and Thwaites 2013). For others, it promised a blank slate on which to test the scaling up of cutting-edge ecological design technologies (Fraker 2013, n4,5; Lin 2014). Incentives to "think big," however, were matched by pressures to complete projects in a minimal amount of time and with an emphasis on visual impact, often at the expense of functional quality and innovation, and with minimal input from their ultimate users. The speed of construction and horizontal expansion of cities displaced existing communities, exacerbated socio-spatial segregation, and destroyed long-established, highly productive agricultural landscapes. These effects largely negated the energy- and population-accommodating efficiencies of new high-rise building construction and attendant urban forms, so that the net land use intensity and carbon-efficiency of development actually began to decrease (Chen et al. 2005; Güneralp and Seto 2010).

As a result, projects tended to express local identity and ecological function in only superficial ways, and with little relevance to the daily practices, needs, and values of their inhabitants (Gaubatz 2008; May 2011a; May 2011b). Monumental, supra-human-scaled designs tended to neglect pedestrian-friendly connectivity with the surrounding city, and they encouraged Corbusian architectural typologies, in which buildings sat relatively isolated in a neutral matrix of open space and mode-specific pathways for movement (Figure 15.1). New development tended to take the form of large blocks separated by wide avenues (Abramson 2008; Miao 2011), frustrating even conscious efforts by designers to improve pedestrian accessibility. In the Modern MOMA Linked Hybrid in Beijing, for example, a well-documented high-rise housing project that includes cutting-edge energy- and water-saving features such as geothermal wells and gray water recycling systems, the architect Steven Holl intended "to counter the current privatized urban developments in China by creating a new twenty-first century porous urban space, inviting and open to the public from every side" (Holl 2009; Rowe and Kan 2014; Steven Holl Architects 2009). The design included innovative elevated walkways that connected all the towers at the upper floors, enhancing the complex's internal connectivity. On the ground, however, the project like most other superblocks ended up being walled and gated, which combined with its location, hemmed in by elevated highways and other gated compounds on all sides, ultimately inhibited pedestrian activity (Figures 15.1 and 15.2).

Projects that better accommodated pedestrians and integrated better with the pre-modern, pedestrian-scaled urban fabric were largely limited to historic preservation districts or historicized "invented streets" that replaced whole communities (Banerjee 2001) and tended to become the domain of an extreme form of consumerism (Shin 2010). Shanghai's Xintiandi development was a pioneer of this model (He and Wu 2005; Wai 2006), but perhaps the most extreme example is Qianmen Street in Beijing (Figure 15.3). Located prominently on the imperial central axis

Figure 15.1 Partial view of model of Beijing at the city's Urban Planning Exhibition Hall. In the lower left is Steven Holl's Linked Hybrid project.

Source: Photo by author, 2010

Figure 15.2 The Modern Linked Hybrid project viewed from under the Airport Expressway, Beijing.

Source: Photo by author, 2010

Figure 15.3 Qianmen Street in Beijing, after its reconstruction and pedestrianization in 2008.

Source: Photo by author, 2010

outside the Ming dynasty central gate to the Inner City, just south of Tiananmen Square, this ancient market street was entirely rebuilt within a couple of years and pedestrianized for the 2008 Olympic Games, with early-20th-century-style shopfronts accommodating a tourist-oriented mix of famous old Beijing brands and global chain stores.

Meeting the challenges of large-scale design

In conditions of explosive growth, the most creative, locally responsive, and socially inclusive urban spaces often emerge unpredictably from processes that have only an indirect, and sometimes conflicting, relation with formal professional design and development practice and policy – e.g. arts districts that reoccupy and adapt former industrial zones, self-built spaces and structures in urbanizing villages, and other grassroots-initiated projects or compromises between official and unofficial interests (Abramson 2011; Al 2014). Some formally designed noncommercial projects have successfully become integrated with the daily life of their cities – especially when they involve large public spaces that all citizens enjoy, where people can gather in large numbers, and which have convenient connections to populous residential quarters and cultural centers. Many of these are urban riverfront embankments, such as Shanghai's Bund, Chengdu's extensive system of parks and promenades along the Jinjiang River, and Harbin's Songhua River embankment, which enhances and connects to Zhongyang Street, a highly successful pedestrianized street that balances conservation of authentic historic architecture and long-standing local institutions with newly designed urban amenities. Another category of successful new public spaces are those that combine park space with ecologically functional artificial wetlands and/or urban forests, such as

Beijing's Olympic Forest Park and Houtan Park in Shanghai, designed as part of the 2010 World Expo on the former Baoshan Steelworks site (Saunders 2013; Yu et al. 2008). Such projects require coordination on a vast scale, often involving the relocation of major institutions, and are a defining feature of explosive growth.

In other cases, large projects that originally failed to provide average citizens with richly amenable public space have since been improved by retrofits. The enormous circular elevated pedestrian overpass above the Lujiazui traffic roundabout in Pudong, Shanghai, has turned what used to be a forbidding automobile-dominated wasteland (Xue et al. 2011) into a legible and dramatic way for people to move between blocks. Transcending the mundane function of such overpasses, which are usually frustratingly crude ways to resolve pedestrian and vehicular traffic conflicts, Lujiazui's has become a destination in itself – a gathering place and platform from which large crowds may view the spectacle of high-rises that sprang up there over the previous two decades (Figure 15.4). Similarly, the new central subway station in Chengdu's Tianfu Square facilitates the meeting of people, as crowds ascend through a wide circular opening to the large plaza above, much as the main concourse of Grand Central Terminal in Manhattan, New York has facilitated the hurried movement and interaction of crowds of travelers since 1903.

While these examples of successful projects represent design at a large public scale, urban design during China's period of explosive growth often failed to support the interactions of communities of a few hundred up to 5,000 households. Government policy tacitly acknowledged this fact in pursuing its "community building" (*shequ jianshe*) programs beginning in the late 1990s (Shieh 2011). Traditionally, both under Mao and throughout earlier Chinese history, such

Figure 15.4 The central roundabout with elevated pedestrian walkway in Lujiazui, Pudong district of Shanghai.

Source: Photo by Stefan Wagener – Shanghai, licensed under Creative Commons by 2.0, https://commons.wikimedia.org/w/index.php?curid=48289652

communities were somewhat rigidly accommodated in well-defined quarters, often walled and gated – in other words, assuming a tight correspondence between social and spatial community, deeply at odds with the social and spatial mobility that characterizes modern capitalist cities (Bray 2005; Rowe et al. 2016). As society urbanized and became increasingly mobile and segmented by income levels, this earlier system of community identification and governance has given rise to a geographic social segregation of the city that exacerbates the tendencies of disconnection, exclusivity, extreme speculation, and waste in urban design (Huang and Li 2014; Miao 2003).

Circumstances are already changing in ways that suggest new urban design responses to these challenges. Although the global financial crisis of 2008 prompted a brief new surge in China's construction-oriented investment, overbuilding has since become a greater concern than at any point in the country's past 25 years, giving rise to serious questions about whether some of the new urban spaces will ever be occupied, and whether damage to ecological processes can ever be repaired. Although real estate speculation continues to balloon and deflate in cycles that never quite approach a national crisis, the country's largest design firms have undergone a persistent trend of retrenchment uncommon in the past three decades.

Urban design in China's "New Era"

Since 2013, President Xi Jinping's administration has heralded an economic "new normal" and a "new-type urbanization" for a political-cultural "New Era." While largely just slogans, the government's pronouncements coincide with three broad transitions increasingly evident in urban design over the past five years: (1) a shift from a *tabula rasa* approach to one that focuses on urban "betterment" and incremental change; (2) a shift from a view of urban structure as cellular, with a tight correspondence between social and spatial community, to a form of socio-spatial organization that is more open and fluid; and (3) a shift from an ideology that equates development with transition of rural to urban environments and livelihoods to a pursuit of "ecological civilization" based on the integration and balance of urban and rural development.

From tabula rasa *to urban betterment*

Starting about 2015, for the first time, cities across the country are experiencing a significant wave of investment in renovation and rehabilitation instead of wholesale clearance. These projects go beyond mere redecoration, beautification, and historic or postindustrial adaptive reuse; they include extensive efforts to infill or otherwise reconfigure interior and exterior space within and around existing buildings, most of which are mundane housing or office blocks built within the past 40 years. They also include substantial energy-conserving improvements to roofs, exterior materials, and openings. Similar to the investments in East German cities after the reunification in the 1990s, China's shift towards "city betterment" or "repair" (*chengshi xiubu*) represents a finer-grained approach to making socialist-era urban space more locally functional and adaptable, recognizing the embedded energy in existing buildings (China Academy of Urban Planning and Design 2016; Chow 2017).

In the case of villages enfolded into expanding cities, such projects have gone beyond the earlier efforts to regularize the informal construction of affordable housing for migrant workers, as was seen in Shenzhen over the previous two decades. These projects now, for the first time, envision long-term improvements of community space on the basis of the settlements' original layout and density. Thus, increasing overall built floor space is no longer always the prime motive in development; consequently, the relocation of villagers and urban residents has become increasingly difficult and less common.

Local governments increasingly acknowledge the need to maintain communities in place, involve residents in the gradual improvement of their own neighborhood space, and support them in improving and maintaining their own housing (Wang 2018). After some embryonic experiments in participatory design in the late 1990s and 2000s (Abramson 2007, 2016a), a small but growing number of cities now support resident-determined upgrading priorities with municipal budgets, modeled on Taiwan's community-building programs.[1] Beijing's historic *hutong* alleys and *siheyuan* courtyard housing are now for the first time undergoing preservation based on a gradual residential upgrading process, with existing households remaining in place, instead of commercialization or gentrification. In the East City District's Shijia Hutong, for example, the local subdistrict (*jiedao*) government has partnered with the Beijing Municipal Institute of Urban Planning and Design to establish a neighborhood history museum in an old courtyard. The museum also serves to exhibit neighborhood upgrading designs and technologies, including small, low-cost electric, chemical, and composting toilets suitable for infill, which do not require wholesale rebuilding of sewage lines. These technologies are funded by a national "toilet revolution" that has been part of the new environmental betterment approach since 2015 (Cheng et al. 2018). To participate in these projects, urban designers are learning new skills of communication, community engagement, integration of design with long-term management, and fine-scaled design compatible with already-built-up environments.

From the cellular city to a more open urban matrix

The second transition, from a cellular urban structure to a more open and fluid one, is the spatial corollary of the temporal shift in the pace of development described previously. As long as municipal governments were dependent on relatively few development firms to expand the city at a rapid pace, the most manageable design approach was one that enabled firms to undertake large superblock projects, and to do so with minimal coordination with surrounding projects. In February 2016, however, the Communist Party's Central Committee and the State Council jointly issued a set of Opinions on urban planning that called for a number of radical departures from this approach (Chinese Community Party Central Committee and State Council of the People's Republic of China 2016). Three aspects in particular have major implications for urban design: (1) discouragement of building more gated communities and even encouragement to break up existing ones into more permeable neighborhoods; (2) densification of the street network to form smaller blocks and provide greater choice of routes for through traffic; and (3) increasing public transportation capacity and improving multimodal use of roads and intermodal connections.

The Opinions are possibly the highest-level policy document addressing detailed urban design since the passing of the national Urban Planning Law of 2007. They are far more specific from a design perspective than the Law, and include such recommendations as "by 2020, the average road network density in urban built-up areas should increase to 8 [linear] kilometers per square kilometer of land area, and the area of road surface should increase to 15%." The Opinions do not specify the technical basis for these recommendations, nor the ultimate objective, other than to ease traffic congestion. Given the population densities of China's cities, no amount of road network density or surface area increase seems likely to reduce congestion without a concurrent reduction in private motor vehicle use.

To encourage more residents to use bicycles and transit, the Opinions also stipulate that all residences be within 500 meters of a transit stop. Meeting this fairly universal standard would probably require many large superblock housing districts to become more permeable for pedestrians, opening more gates in surrounding fences and laying out more direct pathways within the blocks. Diversifying the transportation modes accommodated within the existing super-wide rights-of-way that bound and separate large blocks would then be enough also to encourage transit ridership and to reduce travel times and possibly even congestion, especially if private automobile use is discouraged. Increasing the density of vehicular networks is thus actually not necessary in order to achieve the Opinions' stated goals. A more important focus for design would be to make existing streets more accessible from housing and more amenable, with wider sidewalks, comfortable furnishing, abundant shade trees, and convenient crossing points.

As Kan et al. (2017) point out, however, there are other reasons to consider breaking up the superblocks. In an era of slower growth, the grain of development is likely to decrease and lead to greater demand for smaller developable lots. Smaller blocks tend to accommodate a greater diversity of land uses and housing prices within a smaller area. They also provide more direct access for delivery of goods and services ordered online. Finally, a shift towards smaller blocks may also reflect trends in property management, community governance, and public security. New digital "grid management" and electronic surveillance technologies using personal biometrics provide the state with more direct policing options and are less reliant on walls, gates, and other "place-based" forms of social control such as *hukou*, the household registration system, which was established in the 1950s (Gong 2016). Somewhat counterintuitively, a more open city in terms of movement does not necessarily correlate with a more open society in terms of either state control or community self-governance. Meanwhile, the public remains attached to their residential compounds: they ensure a degree of stability in home values, and they represent the most important investment people make, as well as personal connection to the management of collective space, which is important given the low public trust in institutions.

From development as city building to "ecological civilization"

Another major transition in the developmental context of urban design is the blurring of urban and rural boundaries, and professional designers' engagement with the countryside (Abramson 2016b; Bray 2013). At one level, this shift is a straightforward expression of urbanization – cities' spatial expansion and rural residents' pursuit of urban livelihoods, lifestyles, and environments. After 2003, national programs to Coordinate Urban and Rural Development and promote Socialist New Countryside Construction brought a wave of completely new housing and infrastructure development to rural communities across the country. Most of these projects simply transferred urban models into the rural landscape, or addressed rural environmental character in only superficial ways. However, in tandem with the shift from *tabula rasa* to betterment approaches in the city, rural design is increasingly attending to the countryside's deeply coupled natural systems and human settlements that traditionally begot productivity and culture.

The new rural design is the product of evolving dialogues between local officials and farmers, who seek more stable livelihoods and an improved standard of living but increasingly experience migration to low-wage, dead-end urban jobs. Conversely, similar discontent abounds among diverse city-based actors – designers pursuing new clients and inspiration for aesthetic expression; developers seeking new opportunities for investment in a time of slowing growth in urban construction; consumers and activists concerned about environmental and food quality and the loss of rural beauty; and scientists researching ecological functions. At the architectural scale, many of China's highest-profile designers such as Wang Shu and Lu Wenyu, Ma Qingyun, Li Xiaodong, and Yung Ho Chang have undertaken projects that pursue new expressions of modern rurality based on local materials, climate, heritage, and community life (Stokols 2015). While still embryonic, new rural design is emerging as part of a cultural redefinition of "quality of life" that encompasses both rural and urban perspectives (Hu and Abramson 2015).

If, in its initial stages, rural design was modeled on a conventional urban pattern of development, more recently, designers' encounter with the rural is not only attempting to revive and redefine rural values, but also feedback into the design of cities, through projects that mimic or preserve the functioning of rural environments even in urban settings. Most such projects fall under the category of "sponge city" programs designed to manage stormwater and reduce demand on overwhelmed drainage systems – a problem directly associated with explosive growth. As with urban betterment projects, design that incorporates rural values sustainably must engage more closely with communities of users and managers, but it must also incorporate both traditional and scientific ecological knowledge. At their best, these projects transcend single engineering objectives and achieve complex multifunctional outcomes, accommodating species diversity and food production, and raising social awareness of environmental science.

In March 2018, one of the most significant reforms of the entire planning system in 30 years is likely to give added impetus to this trend: planning power is now consolidated in the Ministry of Natural Resources (formerly Ministry of Land and Resources), whereas before it was split between this ministry, the Ministry of Housing and Urban Development, and the National Development Reform Commission. It may therefore be no exaggeration to say that the national policy objectives of urbanization and economic development are now subsumed under the mandates of environmental protection and natural resource stewardship.

Conclusion

The greatest challenges that China's explosive growth presents for urban design have to do with achieving a balance of scales in three dimensions: the spatial, temporal, and social/organizational. How to coordinate development across very large jurisdictions of both urban and rural land, including tying them together with high-speed and high-capacity transport and communications infrastructure, while "stepping down" this connectivity to ensure that local places maintain their accessibility, identity, and socio-ecological viability? How to undertake projects in the short term that do not compromise the sustainability of long-evolved coupled human and natural systems, especially given densely settled and productive rural landscapes? How to accommodate and govern vast movements and settlements of people seeking out new livelihoods, while supporting existing small communities and enabling new ones? While these questions reach well beyond the scope of urban design, they all have implications for urban design and vice versa. "Urban" design may take its place among a newly enlarged array of disciplines, including not just community design but also rural design and regional design.

This chapter has outlined a number of observed trends. Their projection into the future is necessarily speculative. It is not at all certain that the era of explosive growth is slowing down for the long term, but even if it is, technological change continues to accelerate and transform social relations. At present, many of the obstacles that explosive growth and rapid change presented to long-term thinking and decision-making at the local scale are still in place. Local leaders still face pressure to favor quick fixes over long-term solutions. Data-sharing necessary for projects to incorporate environmental science into design is difficult, as is the organizational work needed for public participation in design. Nevertheless, designers have a clear mandate to reduce environmental impacts, while also helping communities and decision makers move forward under conditions of uncertainty. It seems reasonable to hope that the current "New Normal" provides more space and time to reimagine what a good, livable city is, at every scale.

Note

1 In July 2017, the author spent time in Quanzhou, Fujian Province, observing community engagement by designers and planners from Urbaneer, a Shanghai-based firm founded by graduates of National Taiwan University trained in community building, whom the Quanzhou Municipal Government commissioned to implement a new urban neighborhood upgrading program.

References

Abramson, D. B. (1997). "'Marketization' and Institutions in Chinese Inner-City Neighborhood Redevelopment: A Commentary on 'Beijing's Old and Dilapidated Housing Renewal' by Lü Junhua." *Cities*, 14(2): 71–75.

Abramson, D. B. (2007). "The Dialectics of Urban Planning in China." In Wu, F. (ed.), *China's Emerging Cities: The Making of New Urbanism*. London; New York: Routledge, 66–86.

Abramson, D. B. (2008). "Haussmann and Le Corbusier in China: Land Control and the Design of Streets in Urban Redevelopment." *Journal of Urban Design*, 13(2): 231–256.

Abramson, D. B. (2010). "A Trans-Pacific Planning Education in Reverse: Reflections of an American with a Chinese Doctorate in Urban Planning and Design." In Healey, P. and Upton, R. (eds.), *Crossing Borders: International Exchange and Planning Practices*. London; New York: Routledge, 289–312.

Abramson, D. B. (2011). "Places for the Gods: Urban Planning as Orthopraxy and Heteropraxy in China." *Environment and Planning D: Society and Space*, 29(1): 67–88.

Abramson, D. B. (2016a). "Messy Urbanism and Space for Community Engagement in China." In Chalana, M. and Hou, J. (eds.), *Messy Urbanism: Understanding the "Other" Cities of Asia*. Hong Kong: Hong Kong University Press, HKU, 215–237.

Abramson, D. B. (2016b). "Periurbanization and the Politics of Development-as-City-Building in China." *Cities*, 53: 156–162.

Abramson, D. B., and Qi, Y. (2011). "'Urban-Rural Integration' in the Earthquake Zone: Sichuan's Post-Disaster Reconstruction and the Expansion of the Chengdu Metropole." *Pacific Affairs*, 84(3): 495–523.

Al, S. (2014). *Villages in the City: A Guide to South China's Informal Settlements*. Hong Kong; Honolulu: Hong Kong University Press; University of Hawaii Press.

Angel, S. (2012). *Planet of Cities*. Cambridge, MA: Lincoln Institute of Land Policy.

Banerjee, T. (2001). "The Future of Public Space: Beyond Invented Streets and Reinvented Places." *Journal of the American Planning Association*, 67(1): 9–24.

Beijing Municipal Statistical Bureau. (2005). *Statistical Communiques for the Years 2000 through 2003 [English]*. City: Beijing Tongji Xinxi Wang (Beijing Statistical Information Net).

Bray, D. (2005). *Social Space and Governance in Urban China: The Danwei System from Origins to Reform*. Stanford, CA: Stanford University Press.

Bray, D. (2013). "Urban Planning Goes Rural: Conceptualising the 'New Village'." *China Perspectives*, (3): 53–62.

Broudehoux, A.-M. (2017). *Mega-Events and Urban Image Construction: Beijing and Rio De Janeiro*. Abingdon, Oxon; New York, NY: Routledge, an imprint of the Taylor & Francis Group.

Campanella, T. J. (2009). "The China Challenge." *Journal of Urbanism: International Research on Placemaking and Urban Sustainability*, 2(1): 3–4.

Campanella, T. J. (2010). "China: As We Once Were." *Journal of Architectural Education*, 63(2): 63–64.

Chen, F., and Thwaites, K. (2013). *Chinese Urban Design: The Typomorphological Approach*. Farnham, Surrey, England; Burlington, VT: Ashgate.

Chen, H., Ganesan, S., and Jia, B. (2005). "Environmental Challenges of Post-Reform Housing Development in Beijing." *Habitat International*, 29(3): 571–589.

Cheng, S., Li, Z., Uddin, S. M. N., Mang, H.-P., Zhou, X., Zhang, J., Zheng, L., and Zhang, L. (2018). "Toilet Revolution in China." *Journal of Environmental Management*, 216: 347–356.

China Academy of Urban Planning and Design. (2016). *City Betterment and Ecological Restoration: Catalysts in Transitional Development of China*. Beijing: China Architecture and Building Press [in Chinese].

Chinese Community Party Central Committee, and State Council of the People's Republic of China. (2016). "Zhonggong Zhongyang Guowuyuan Guanyu Jinyibu Jiaqiang Chengshi Guihua Jianshe Guanli Gongzuo De Ruogan Yijian [Certain Opinions of the CPC Central Committee and State Council Regarding Further Strengthening of Urban Planning and Construction Management Work] (February 6, 2016)." *Xinhua Wang*. New China News Agency. www.xinhuanet.com/politics/2016-02/21/c_1118109546.htm (Archived by WebCite® at www.webcitation.org/6yC7Ota0V), Accessed 25 March 2018.

Chow, R. Y. (2017). "Continuity and Change: Challenging the Disposable Chinese City." *Architectural Design (Special Issue: Loose-Fit Architecture: Designing Buildings for Change)*, 87(5): 114–121.

Fraker, H. (2013). *The Hidden Potential of Sustainable Neighborhoods: Lessons from Low-Carbon Communities.* Washington, DC: Island Press.

Gaubatz, P. (2008). "Commercial Redevelopment and Regional Inequality in Urban China: Xining's Wangfujing?" *Eurasian Geography and Economics*, 49(2): 180–199.

Giedion, S. (1941). *Space, Time and Architecture: The Growth of a New Tradition.* Cambridge; London: Harvard University Press; H. Milford; Oxford University Press.

Gong, Y. R. (2016). "Rental Housing Management as Surveillance of Chinese Rural Migrants: The Case of Hillside Compound in Dongguan." *Housing Studies*, 31(8): 998–1018.

Güneralp, B., and Seto, K. C. (2010). "Can Gains in Efficiency Offset the Resource Demands and Co2 Emissions from Constructing and Operating the Built Environment?" *Applied Geography*, 32(1): 40–50.

He, S., and Wu, F. (2005). "Property-Led Redevelopment in Post-Reform China: A Case Study of Xintiandi Redevelopment Project in Shanghai." *Journal of Urban Affairs*, 27(1): 1–23.

Holl, S. (2009). "Linked Hybrid." *Domus*, (928): 14–26.

Hu, J., and Abramson, D. B. (2015). "Visions of New Urban-Rural Relations and Alternative Definitions of Well-Being in Rapidly Urbanizing China: The Case of Chengdu, Sichuan." In Wagner, F., Mahayni, R., and Piller, A. (eds.), *Transforming Distressed Global Communities: Making Inclusive, Safe, Resilient, and Sustainable Cities.* Farnham, Surrey; Burlington, VT: Ashgate, 317–337.

Huang, Y., and Li, S. M. (2014). *Housing Inequality in Chinese Cities.* Abingdon, Oxon: Routledge.

Kan, H. Y., Forsyth, A., and Rowe, P. (2017). "Redesigning China's Superblock Neighbourhoods: Policies, Opportunities and Challenges." *Journal of Urban Design*, 22(6): 757–777.

Lin, Z. (2014). "Constructing Utopias: China's Emerging Eco-Cities." *ARCC Conference Repository.* Architectural Resource Centers Consortium. www.arcc-journal.org/index.php/repository/article/view/310/246, Accessed 15 March 2018.

Liu, J. (2005). "Beijing De Tudi Neng Chengzai Duoshao Qiche; 'Jing H' Sudu Shenzang Weiji [How Many Cars Can Beijing's Land Hold? The 'Jing H' License-Plate Number Capacity Is Reaching a Crisis Point]." Xinhua News Agency, 11 July 2005. http://auto.sina.com.cn/news/2005-07-11/0933127107.shtml (Archived by WebCite® at http://www.webcitation.org/77Xusyu5v) Accessed 10 April 2019.

Liu, J., and Diamond, J. (2005). "China's Environment in Globalizing World." *Nature*, 435: 1179–1186.

Lu, D. (2006). *Remaking Chinese Urban Form: Modernity, Scarcity and Space, 1949–2005.* London; New York: Routledge.

Marshall, R. (2003). *Emerging Urbanity: Global Urban Projects in the Asia Pacific Rim.* London; New York: Spon Press.

Marshall, R. (2004). "Asian Megacities." In Robbins, E. and El-Khoury, R. (eds.), *Shaping the City: Studies in History, Theory and Urban Design.* New York; London: Routledge, 194–211.

May, S. K. (2011a). "Ecological Urbanization: Calculating Value in an Age of Global Climate Change." In Roy, A. and Ong, A. (eds.), *Worlding Cities: Asian Experiments and the Art of Being Global.* Hoboken, NJ: John Wiley & Sons, 98–126.

May, S. K. (2011b). *Practices of Ecological Citizenship: Global Dreams for a Chinese Village.* PhD dissertation, University of California, Berkeley.

Miao, P. (2003). "Deserted Streets in a Jammed Town: The Gated Community in Chinese Cities and Its Solution." *Journal of Urban Design*, 8(1): 45–66.

Miao, P. (2011). "Brave New City: Three Problems in Chinese Urban Public Space Since the 1980s." *Journal of Urban Design*, 16(2): 179–207.

Ministry of Construction General Statistics Section. (2000). "2000 Urban Construction Statistics Public Report (2000 Nian Chengshi Jianshe Tongji Gongbao)." http://www.mohurd.gov.cn/download/tj/2000cjtj.doc. Accessed 10 April 2019.

Myers, N., and Kent, J. (2003). "New Consumers: The Influence of Affluence on the Environment." *PNAS (Proceedings of the National Academy of Sciences of the United States of America)*, 100(8): 4963–4968.

National Bureau of Statistics of China. (1999a). "Completed Fixed Asset Investment in Urban Construction (1999)." Beijing: National Bureau of Statistics of China.

National Bureau of Statistics of China. (1999b). "Urban Buildings and Housing for Each Area (1999)." Beijing: National Bureau of Statistics of China.

National Bureau of Statistics of China. (1999c). "Urban Public Works (1999)." Beijing: National Bureau of Statistics of China.

National Bureau of Statistics Urban Social and Economic Survey Team. (1990). *Zhongguo Chengshi Tongji Nianjian [China Urban Statistical Yearbook].* Beijing: China Statistical Publishing House.

Rowe, P. G. (2005). *East Asia Modern: Shaping the Contemporary City*. London: Reaktion.
Rowe, P. G. (2011). *Emergent Architectural Territories in East Asian Cities*. Basel; New York, NY: Birkhäuser; Distribution, ActarBirkhäuserD.
Rowe, P. G., Forsyth, A., and Kan, H. Y. (2016). *China's Urban Communities: Concepts, Contexts, and Well-Being*. Boston: Birkhäuser.
Rowe, P. G., and Kan, H. Y. (2014). *Urban Intensities: Contemporary Housing Types and Territories*. Basel; Boston: Birkhäuser.
Saunders, W. S. (2013). *Designed Ecologies: The Landscape Architecture of Kongjian Yu*. Basel/Berlin/Boston, DE: Birkhäuser.
Shieh, L. L. (2011). *Shequ Construction: Policy Implementation, Community Building, and Urban Governance in China*. PhD dissertation, University of British Columbia.
Shin, H. B. (2010). "Urban Conservation and Revalorisation of Dilapidated Historic Quarters: The Case of Nanluoguxiang in Beijing." *Cities*, 27(Supplement 1): S43–S54.
Steven Holl Architects. (2009). "Linked Hybrid." www.stevenholl.com/projects/beijing-linked-hybrid (archived by Internet Archive at https://web.archive.org/web/20180728220515/www.stevenholl.com/projects/beijing-linked-hybrid), Accessed 28 July 2018.
Stokols, A. (2015). "Village Acupuncture: Can Architects Restore Vitality to China's Countryside?" *China-File*, "Earthbound China" series of articles and multimedia reports. Asia Society Center on U.S.-China Relations. www.chinafile.com/reporting-opinion/earthbound-china/village-acupuncture (Archived by WebCite® at www.webcitation.org/6yXIIoJEy), Accessed April 8, 2018.
United Nations. (2015). *World Urbanization Prospects: The 2014 Revision*. New York, NY, USA: United Nations Department of Economics and Social Affairs, Population Division.
Wai, A. W. T. (2006). "Place Promotion and Iconography in Shanghai's Xintiandi." *Habitat International*, 30(2), 245–260.
Wang, J. (2018). "Tong Gui Zi Jian, Liu Shui Bu Fu: Beijing Laocheng Huifuxing Xiujan de Shiyi Lujing [Unified Planning and Self-Building; Flowing Water Does Not Putrefy: A Suitable Path for the Restorative Repair of Beijing's Old City]." *Guoji Chengshi Guihua Fazhan Baogao [International Urban Planning and Development Report]*, 23(4): 16–23.
Wu, F. (2015). *Planning for Growth: Urban and Regional Planning in China*. New York, NY: Routledge.
Xu, M., He, C., Liu, Z., and Dou, Y. (2016). "How Did Urban Land Expand in China Between 1992 and 2015? A Multi-Scale Landscape Analysis." *PLoS One*, 11(5): e0154839.
Xue, C. Q. L., Zhai, H., and Mitchenere, B. (2011). "Shaping Lujiazui: The Formation and Building of the Cbd in Pudong, Shanghai." *Journal of Urban Design*, 16(2): 209–232.
Yu, K., Zhang, L., and Li, D. (2008). "Living with Water: Flood Adaptive Landscapes in the Yellow River Basin of China." *Journal of Landscape Architecture*, 3(2), 6–17.

Further reading

Among English language publications:

Lin, Zhongjie, and Jose Gamez, eds. (2018). *Vertical Urbanism: Designing Compact Cities in China*. Abingdon, UK: Routledge. A collection of essays exploring in depth the relationships between density, intensity and form in Chinese urban design.
Rowe, P. G. (2011). *Emergent Architectural Territories in East Asian Cities*. Basel; New York, NY: Birkhäuser. A broad overview that encompasses other East Asian contexts as well as China, and commercial and mixed-use as well as residential projects.
Rowe, P. G., Forsyth, A., and Kan, H. Y. (2016a). *China's Urban Communities: Concepts, Contexts, and Well-Being*. Boston: Birkhäuser. Provides a more comprehensive background on the issues addressed in this chapter.
Saunders, W. S. (2013). *Designed Ecologies: The Landscape Architecture of Kongjian Yu*. Basel/Berlin/Boston: Birkhäuser. A collection of essays on the influence of Kongjian Yu on urban and large-scale landscape design, interspersed with descriptions of his projects.
Wu, F. (2015). *Planning for Growth: Urban and Regional Planning in China*. New York, NY: Routledge. A comprehensive presentation of China's overall planning system and developmental policies and trends as a context for urban design.

16

URBAN SPACE DESIGN FOR HYBRID AND HIGH-DENSITY ENVIRONMENTS

Im Sik Cho

Introduction

According to the United Nations' *World Urbanization Prospects* (2018), currently 55% of the world's population lives in urban areas, and this number is expected to increase to 68%, adding 2.5 billion to the world's urban population by 2050. While Africa and Asia are currently the least urbanized regions of the world, with their level of urbanization approaching 43% and 50%, respectively, it is expected that close to 90% of the world's urban growth will occur in these regions by 2050. Many of the fastest-growing urban agglomerations are located in Africa and Asia (United Nations 2018), while the majority of the current megacities with more than 10 million inhabitants are located in Asia. The world is projected to have 43 megacities by 2030, most of them in the Global South. Especially over the last three decades, Asian countries have experienced the most dramatic urban transformations, which is comparable to what the Western world experienced over the last two centuries.

The increased challenges in meeting the needs of growing urban populations, as the world continues to urbanize, inevitably lead to higher demands for environmentally, economically, and socially sustainable planning. As densities increase, the availability of open land decreases, and the provision of well-planned, well-designed, well-maintained, and well-regulated public spaces becomes a crucial factor in creating a livable urban environment (Amin 2006; Madanipour 2006). Under urban densification, public space often faces increasing typological and programmatic demands, and also intense negotiations among its diverse users, resulting in the emergence of new typologies of urban space (Cho *et al.* 2017). This chapter examines how the challenge of very high urban densities in East Asian cities has led to new trends and typologies in the production of urban space.

The term "urban space" that is adopted in this chapter, in contrast to the classical definitions of public space, acknowledges the emerging types of spaces that function as public space despite not being publicly owned or managed (Cho *et al.* 2016). This notion of urban space is similar to what Manuel de Solà-Morales (1992) coins as "collective space" that is neither public nor private, but both simultaneously.[1]

The quality of urban space in sustainable urban development becomes particularly critical in high-density contexts. Especially for cities characterized by scarcity of land (such as Tokyo, Hong Kong, or Singapore, among others), densification is an existential challenge. In these contexts, high-density and compact development is not a preferred choice, but an inescapable and pressing condition given the demands for rapid urban growth despite the topographical constraints. The specific political, spatial, economic, and sociocultural conditions of many East Asian cities have over time formed a unique circumstance in which various challenges have led these cities to embrace and create new hybrid modes of living and spatial or functional organizations that are different from conventional patterns of urban development (Cho *et al.* 2016).

Emerging urban space typologies[2]

In order to accommodate and respond to the diverse, contested, and unprecedented urban conditions shaped by rapid urbanization, it is imperative to better understand the "dynamics and complexities that characterize the emerging urban spaces in high-density contexts" (Cho *et al.* 2017: 614). Past research that "has predominantly focused on familiar models of public space," emphasizing the distinction between the public and the private, is insufficient to provide "consistent and clear consensus on what constitutes good urban space" in a high-density context, where the strict binary of public and private often disintegrates (Ibid.: 614). Consequently, there remains a lack of collective understanding regarding the emerging and increasingly complex types of urban spaces – such as elevated, underground, and multilevel public spaces, intensified housing developments, mixed-use and transit-related infrastructural spaces – which are typical of high-density and high-intensity development (Cho *et al.* 2017). While the "traditional" typologies still play a significant role in shaping the urban fabric and public realm, demands for higher density and intensified urban development are initiating a multiplicity of new urban space typologies and conditions, which call for a reexamination and reconceptualization of conventional approaches to urban design. These may require more flexible, context-sensitive, holistic, and comprehensive mechanisms, including "new, systematic-yet-flexible approaches, frameworks and instruments to understand, analyze, assess and guide the design and management of emerging urban spaces" (Cho *et al.* 2015: 148).

Addressing the issue of optimizing available spaces while retaining or enhancing their quality, some scholars have emphasized the role of integrating various functions and creating hybrid spatial conditions, giving rise to new vibrant urban environments with "thresholds, or places of intensity," where diverse people and activities may converge (Ellin 2006; see also Fenton 1985; Komossa 2011; Per *et al.* 2014). Specifically, in the Asian context, where high-rise development is now a predominant urban model (Hou 2012), the three-dimensional complexity and intensity of high-density developments are being recognized as key aspects of the emerging Asian urbanity, which is rarely encountered in other contexts. Exploring Hong Kong as a case, Shelton *et al.* (2011) have proposed a shift of focus from mere "vertical" to "volumetric" urbanism, in order to synergize the potentials of complex three-dimensional high-density environments based on proximity and connection; while Lau *et al.* (2005) have suggested multiple use of space and vertical intensification as suitable spatial practices to overcome constraints of land scarcity and provide high-quality, future city forms for the new millennium in Asia. In these hybrid and complex high-density environments, the conditions and qualities identified as desirable by theories for conventional cities – with less-dense urban forms and surface public spaces – have to be revisited and reinterpreted (Shelton *et al.* 2011).

Examples of the emerging urban space typologies include intensified residential developments, mixed-use developments, infrastructural transit-led spaces, recreational green hybrids, and hybrid urban voids, which are increasingly being adopted in high-rise and high-density developments across Asia. The previous list is based on the primary use of urban space and is not exhaustive;

rather, it seeks to capture the current trends of urban space development within today's hybrid and high-density urban contexts.

Intensified residential developments

The most apparent impacts of densely built and populated compact developments on the quality of living experiences are reflected in the intensified housing environments that often do not follow conventional housing schemes. To redress the lack of public space, emerging urban spaces in high-density, high-intensity residential developments are increasingly adopting various forms and functions (with an emphasis on sociability), catering to both the residents and the wider community, while creating new hybrid conditions for social exchange and spatial negotiation (Cho *et al.* 2016). Spatial features such as multilevel podiums, sky-courts, and sky-bridges, "which show a high level of complexity and interconnectivity, are some of the physical characteristics" of these new residential developments, while flexibility, "multifunctional programming and spatial negotiations are the main programmatic and operational conditions commonly found" in these cases (Cho *et al.* 2017: 630). These typologies offer alternative ways of space usage, especially in regard to incorporating publicly accessible communal spaces vertically, while at the same time ensuring the comfort of urban dwellers through alternative amenities and recreational spaces for social interaction (Yeh 2011, cited in Cho *et al.* 2016). Some of these cases include: Pinnacle@Duxton in Singapore, Shinonome Codan Court in Tokyo, and Dangdai Moma-Linked Hybrid and Jianwai Soho in Beijing.

Pinnacle@Duxton is a pioneering 50-story high-density public housing development in downtown Singapore, sited next to the business district. The development's location reflects Singapore's urban renewal strategy to inject vibrancy into its central area (Cho *et al.* 2017). The extensiveness of public housing throughout Singapore, where more than 80% of its residential population lives, implies the critical role of public housing in supporting national development but also in serving as an instrument to strengthen social cohesion and ethnic integration (Chiu 2008). As a landmark project, Pinnacle@Duxton was initiated by Singapore's public housing authority and designed by a local private firm whose winning competition scheme was selected from more than 200 international submissions. With a gross plot ratio (or floor area ratio) of 9.28, "three times higher than that of the typical public housing blocks in Singapore (3.00–3.65)," the Pinnacle@Duxton embraces the new vision of public housing amidst population growth in Singapore, which aims to challenge "the limits of density and scale without compromising the quality" of living (Cho *et al.* 2017: 621). The project experiments with unprecedented solutions for high-rise, high-density housing environments, such as integrating elevated public amenities with sky-gardens on the 26th and 50th levels, which not only connect seven residential towers but also provide "public spaces for residents and the general public, as a form of compensation for the detachment from the street level" (Ibid.: 622). The amenities on the 26th level sky-bridge include social and recreational facilities, which are typically located on the ground, such as residents' community center, community plaza, 800-meter jogging track, fitness corner for seniors, outdoor gym, and children's playground. The main nodes of activities at the 50th story rooftop, such as viewing decks, pavilions, and seating areas, provide the atmosphere of an "outdoor living room" with great panoramic view of the city center, which is often advertised as a major attraction of such high-rise development (Figure 16.1).

"Designs aimed at creating vibrant urban life with porous, open urban spaces and 24-hour access to the general public," (Ibid.: 630) such as the case of Pinnacle@Duxton, are often complicated by contested ownership between the residents and the general public. In this context, regulations and management intervene as inevitable negotiating mechanisms to ensure appropriate

Figure 16.1 Elevated public space – Pinnacle@Duxton, Singapore.

Source: Chen Xiuqi

uses and behaviors, as is evident in the various access and crowd control measures implemented. Nevertheless, stringent rules and control can restrict the use of space and limit the opportunity for social interaction (Hadi *et al.* 2018). Based on the negotiations among the stakeholders "concerning issues such as crime, safety and anti-social behaviour" (Ibid.: 109), different levels of publicness are imposed to secure the privacy of residents, building maintenance considerations, and control of vertical circulation, among others (Cho *et al.* 2016); these result in a "loss of freedom for adaptability and social activity" (Hadi *et al.* 2018: 109). Pinnacle@Duxton exemplifies the various potentials of an intensified residential development in the context of high-density cities, but, at the same time, poses its pertinent challenges in achieving a balance between accessibility and privacy; providing diversity of "uses without generating conflict among the users; encouraging openness of urban space without compromising safety; and ensuring security" in this high-density living environment (Cho *et al.* 2017: 630).

Mixed-use developments

Although heavily criticized for diminishing the public realm due to their primary orientation towards profitability and competitiveness (Fainstein 2009), mixed-use private developments or those based on public-private partnerships, "characterized by high diversity and intensity of users and activities" (Cho *et al.* 2016: 11), represent one of the dominant models of contemporary urban development. Arguably, compared to gated residential developments, emerging mixed-use megacomplexes in high-density contexts are often more concerned with providing quality urban spaces as a means to attract the public; as a consequence, they become focal points, creating new identities in urban districts (Cho *et al.* 2016). However, many of the urban spaces operating within the premises of commercial-oriented mixed-use developments are privately owned public spaces or pseudo-public spaces that emphasize individualism and consumption over public benefits and social equity; they can fall short of ensuring publicly accessible spaces and inclusiveness. In this respect, the role and level of commitment of the public sector, such as city or national governments, in protecting quality public spaces for civic interests becomes even more critical, "since it is at this level of urban planning where complex decisions are made pertaining to land use, by-laws and incentives on development issues" (Ibid.: xiii–xiv). On the other hand, the emerging trends in designing retail-focused megaprojects show considerable shift from the "conventional shopping mall" developments, by actively incorporating unconventional experimentations with hybrid urban space typologies and innovative uses (Ibid.: 11), with an emphasis on interactive relationships between the site and the surrounding context. Several examples such as Tokyo Midtown and Roppongi Hills in Tokyo, and Ion Orchard in Singapore are developed in city centers often near transport nodes, integrating large-scale buildings with various public facilities and urban spaces in high-density and high-rise built environments.

Tokyo Midtown is an urban infill project, a 569,000-square-meter mixed-use development comprising of offices, commercial facilities, residential buildings, a hotel, a world-class medical facility, and two museums, surrounded by green open spaces and interconnected by a network of pedestrian spaces. The project, which started in 2001 during the recession, was supported by the national government with extensive subsidies for public improvements (Urban Land Institute 2016), and was designed by a number of world-renowned architectural firms. Urban spaces within this dense vertical development show high diversity of activities surrounding its ample open or green areas, "encouragement for social interaction through the provision of formal and informal seating, art and culture" programs, guided tours, and events, as well as community engagement involving cleanup activities and neighborhood festivals and deepening local community interaction (Cho *et al.* 2015: 163). The preservation of the site's historical and environmental heritage, environmentally conscious design, and fostering of environmental awareness are valuable aspects of this megacomplex. The development is well integrated with a public park, facilitating a flow between the site and the indoor and outdoor environment. With various contemporary design elements blended with influences from traditional motifs, Tokyo Midtown explores the cultivation of a stronger urban space identity through cultural and artistic means, while "creating an environmentally sensitive and commercially active attraction that serves the entire neighborhood and adds economic value to nearby properties" (Urban Land Institute 2016). There is, however, room to improve "connectivity within the urban space and to make the space more legible," as well as to increase opportunities for "soft, non-intrusive security measures and informal activities" that enhance accessibility and inclusiveness (Cho *et al.* 2015: 163).

Infrastructural transit-led spaces

New transportation hubs (such as train stations) as well as reused older infrastructural systems or pedestrian facilities "increasingly adopt planning and design strategies that emphasize community, accessibility, pedestrian friendliness and culture as new important values" (Cho *et al.* 2016: 12). These transit-oriented spaces thus function not only as transit nodes that accommodate urban movement, but also as "urban regenerators" (Zacharias 2013) with rich, complex, and diverse functions at multiple levels, thus enhancing the economic and social viability of the surrounding built environment. Eminent examples of these spaces include Shinjuku subway stations in Tokyo, Kyoto Station in Osaka, and Central Mid-levels Escalator in Hong Kong.

The Central Mid-levels Escalator in Hong Kong is the longest outdoor covered escalator system in the world, crossing over 800 meters in distance and spanning more than 135 meters in elevation. The project was initiated by the local government as a means to alleviate the increasing pressure of massive traffic congestion in the Mid-levels area, "with its tortuous narrow road system and high residential densities, all arranged on a steep slope" (Zacharias 2013: 584). The new escalator system was an "out of the box" transportation solution designed by a group of civil engineers to provide better daily commute by linking various areas around all the elevated levels between the Central and Mid-level districts on Hong Kong Island. Despite the initial criticism, when the project was labeled a failure due to its overrun budget and inability to reduce traffic (Shelton *et al.* 2011), the elevator system has played an important role in regenerating its surrounding environment by spurring investment and encouraging high-intensity commercial development, as well as contributing to street vitality by attracting public-oriented activities (Zacharias 2013). It features a variety of spatial interventions at different levels that contribute to the diversity and intensity of uses in the vicinity. It has also become a memorable tourist attraction, increasing walkability and accessibility at the previously private and exclusive spaces on the upper levels, and opening up large tracts of intermediate levels above ("SoHo") and below ("NoHo") Hollywood Road to pedestrians and commerce (Cho *et al.* 2016). However, the escalator in itself is not fully integrated with its surrounding environment as it functions mainly as a pedestrian thoroughfare, thus not fully expanding its potential to form a synergistic network of vertical public spaces. Nonetheless, the Central Mid-levels Escalator exemplifies the possible impacts of such new types of infrastructural spaces for intensifying highly dense urban areas, while also enhancing and integrating the transitory, informal, and everyday experiences with facilities for pedestrian movement and circulation in highly built-up city centers (Cho *et al.* 2015) (Figure 16.2).

Figure 16.2 Elevated public infrastructure – the Central Mid-levels Escalator, Hong Kong.

Source: Ivan Kurniawan Nasution

Recreational green hybrids

Recreational green hybrids inject more vibrancy into city centers by integrating greenery such as parks, green promenades (linear parkways) or park-bridges (elevated parks) with the highly dense built environment. Having a positive impact on environmental benefits as well as human comfort, they often serve as urban respites or green oases for the surrounding urban developments. They also function as connectors, enriching the public space network of the city by reinforcing connectivity and accessibility at different levels. Some examples of recreational green hybrids include Namba Parks (or Nankai Namba) in Osaka and Henderson Waves in Singapore.

Namba Parks is a mixed-use recreational complex with rooftop parks, located adjacent to Namba Station, one of the most important transportation hubs in the Osaka region in Japan (Cho *et al.* 2016). The economic bubble burst in the early 1990s, and the persistence of the economic stagnation that followed led the government to restructure its urban policies and support new modes of public-private partnerships and instruments that use private finance and deregulation of strategic planning (Saito and Thornley 2003). As a result, integration of residential, commercial, and recreational urban spaces adjacent to transportation infrastructure, such as green transit-oriented development, emerged as a new approach to sustainable and comprehensive urban development, which aimed to rejuvenate the city center while maximizing the economic opportunities that the context provides. Namba Parks is an example of such joint development by private actors (the Nankai Electric Railway and the prominent department store chain Takashimaya), which was initiated in the midst of a prolonged economic recession, with governmental support (Urban Land Institute 2015).

The main conceptual premise of Namba Parks is that of a canyon coursing through an elevated urban park, while offering a combination of retail, entertainment, cultural, office, and residential spaces (Ibid.). Being half-open, half-enclosed, and thus well sheltered from the sun and rain, the canyon operates both as a commercialized pedestrian street and an urban connector. The roof park located at the uppermost roof level, surrounding the central opening, slopes upward, roughly eight levels from the main entrance (Cho *et al.* 2016). It consists of a number of outdoor common areas and a terraced amphitheater facing a flat stage area. Interspaced with green areas are circular plazas fronted by restaurants and equipped by a number of diverse seating amenities, an outdoor auditorium, a children's playground, a cinema, and entrances to the office tower and residential building (Reurbanist 2011). In such a way, Namba Parks functions as a natural intervention and urban retreat in Osaka's dense urban environment, while encouraging the interaction of people, culture, and recreation (Jerde Partnership 2017).

However, the elevated green park, a superimposition of a park on top of the shopping mall, is not easily accessible to pedestrians at the street level and is often criticized for its lack of permeability and legibility, displaying somewhat island-like and introverted characteristics. Despite its successful economic and environmental performance, based on quality green design, Namba Parks represents some of the impediments of a privately owned and managed urban development model, implying the importance of the public sector's role in integrating the public and private spheres to cater to the social values of the development.

Hybrid urban voids

Although they may be considered rather "conventional" urban space typologies compared to other cases introduced so far, urban voids acquire special attention in high-density contexts for their role in making highly dense contemporary cities more walkable and livable, and in improving the overall quality of life. Pedestrian-friendly city zones that emerge from initiatives, such as the reclamation of urban streets and other spaces often overtaken by cars or underutilized by pedestrians, are some of the examples of hybrid urban voids, which include Clarke Quay in Singapore and Cheonggyecheon stream in Seoul.

Cheonggyecheon stream, a 5.8-kilometer-long urban void in downtown Seoul, is a result of a massive urban renewal project initiated by then-mayor Myung-Bak Lee. In addition to improving the environmental and living conditions of the area, recovering natural and cultural heritage, creating vibrant public spaces, and addressing traffic problems (Cho and Križnik 2017), the project was also driven by the local government's political and economic interests to increase the city's global competitiveness and become an international tourist destination. The project involved the revival of the ancient Cheonggye stream, which used to have a significant historical and symbolic value for the city, by demolishing an elevated highway crossing downtown Seoul. It is "perhaps the most emblematic among large-scale public projects, which the local government has completed in Seoul over the past decade" (Cho and Križnik 2017: 51). The restoration of Cheonggyecheon greatly enhanced the local biodiversity by creating a sunken ecological corridor that functions as an urban oasis amidst the surrounding densely built environment, providing flood protection, better air quality, reduced urban heat, and restoration of flora and fauna (Figure 16.3). It has transformed the vehicle-oriented environment to a more pedestrian-friendly one that is conducive to walking and public transportation and offers various recreational opportunities along the way. The design of the stream incorporates both hardscapes and softer wetlands along three distinct zones that reflect the historical and cultural identity of the area while reconnecting the once-divided northern and southern parts of the city. The project contributed to not only

Figure 16.3 Sunken parkway – Cheonggyecheon stream, Seoul.

Source: Author

resolving the area's environmental degradation but also fostering its economic competitiveness, by creating new investment opportunities and reinforcing commercial development, especially in Seoul's northern part (Rowe 2010).

In such a way, the restored Cheonggyecheon, regarded as "the most attractive cultural civic space in downtown Seoul" (Cho 2010: 145), has positively affected the quality of everyday life for the citizens and boosted tourism in Seoul. However, the redevelopment also had social ramifications such as ongoing gentrification and decline of traditional industrial and service sectors (Kriňnik 2011), which resulted in the displacement of low-income residents, traditional merchants in the market, small independent retailers, and street vendors, who could no longer afford the exponential increase in real estate value as a result of the economic boost. A major problem during the planning and urban design process was the lack of civic participation in decision-making: The rapid process of redevelopment did not allow sufficient debate or consensus, benefitting the political agendas of some, rather than pursuing "ecological and historical authenticity," based on a careful restoration of the site's natural environment and historical heritage (Cho 2010: 163).

Conclusion

This chapter draws attention to the possibilities and challenges posed for urban design in the context of high-density East Asian cities, where the region's rapid pace of urbanization displays unique conditions for the development of new typologies of urban space. Although the different political, economic, and sociocultural factors of the region should be taken into account, the learning points gleaned are also relevant for other high-density contexts globally, such as Latin America and Africa, where similar trends of urban densification and intensification are emerging with the increasing levels of urbanization.

This chapter recognizes hybridization and intensification of urban processes as direct responses to the challenges of urban densification. New hybrid urban spaces are, therefore, seen as relevant units in the contemporary urban fabric; however, their potential to become vibrant and contribute more substantially to environmental and social sustainability has not been fully explored. Traditional ways of understanding, designing, and utilizing urban space are insufficient to fully understand the contemporary contexts of emerging new hybrid urbanities and modes of publicness. The emerging urban spaces are shaped by a dynamic process that aims to synergize the various aspects of urban design, from spatial and functional to organizational dimensions.

Once densities increase, space becomes a precious commodity, charged with intense competition, tension, and negotiation, as many of the case studies in this chapter have shown. The complex qualities of emerging urban spaces typically evoke contests around space ownership and privatization, emphasizing a continuum of "quasi-public" or "semi-public" spaces (Carmona 2010a, 2010b; Nissen 2008). There are also the pertinent difficulties of collaborative and timely decision-making, as well as adequately maintaining these spaces in practice (Mayer *et al.* 2005; Sandercock and Dovey 2002). Numerous challenges face these developments, which are triggered by the demands for higher building and population density, as well as for higher diversity and intensity of uses and users. The dynamic spatial and programmatic transformations that are involved in the production of new urban space typologies often cause confrontations and require negotiations among diverse groups of users and stakeholders. It is thus a challenging task to provide greater amounts and diversity of uses in high-density environments, as intensification generates not only vibrancy but also conflict, such as those among diverse user groups and subcultures over limited shared spaces, which can manifest as overcrowding, monopolization, or ghettoization of space (Cho *et al.* 2017).

The ever-evolving conditions that characterize the emerging model of urban space in high-density contexts necessitate more open, flexible, integrated, and holistic approaches to urban design and analyses that are capable of responding to change, hybridity, dynamism, and complexity. The role of the public sector (such as city managers or local governments) and private actors (such as urban designers and development firms) in creating inclusive and accessible urban spaces for collective interests becomes more crucial, where conditions that favor privatization and individualism prevail. Thus, openness to more collaborative and innovative design approaches is imperative for thinking beyond conventional regulatory instruments that are "based on rigid, static, ready-made and inflexible sets of design principles" (Cho *et al.* 2016: 237). Such new approaches are necessary to facilitate "decision-making for timely, flexible, intuitive and context-sensitive urban space interventions" (Ibid.: 237). This entails a shift from centralized, top-down approaches of planning and urban design to more decentralized, bottom-up processes; switching from master planning and excessive control to more strategic and spontaneous planning; and from singular and static design solutions to dynamic and pluralistic design processes (Ellin 2006; Meredith *et al.* 2008, cited in Cho *et al.* 2015), which can be instrumental for reconceptualizing urban space design for the hybrid and high-density environments of today and tomorrow.

Acknowledgments

The content discussed in this chapter is partially based on the research project "Urban Space Planning for Sustainable High-Density Environments" conducted at the Centre for Sustainable Asian Cities, School of Design and Environment, at the National University of Singapore, in collaboration with the Urban Redevelopment Authority, the National Parks Board, and the Housing and Development Board, all in Singapore. This work was supported by the Ministry of National Development – Singapore (grant number R-294-000-025-490).

Notes

1 For more elaboration on the definitions of public space, urban space and hybrid urban space, refer to Cho *et al.* (2016: 4–8).

2 For more discussion and illustrations on emerging hybrid urban space typologies, refer to Cho *et al.* (2016: 8–27).

References

Amin, A. (2006). "Collective Culture and Urban Public Space." *City: Analysis of Urban Trends, Culture, Theory, Policy, Action*, 12(1): 5–24.

Carmona, M. (2010a). "Contemporary Public Space: Critique and Classification, Part One: Critique." *Journal of Urban Design*, 15(1): 123–148.

Carmona, M. (2010b). "Contemporary Public Space, Part Two: Classification." *Journal of Urban Design*, 15(2): 157–173.

Chiu, R.L.H. (2008). "Government Intervention in Housing: Convergence and Divergence of the Asian Dragons." *Urban Policy and Research*, 26(3): 249–269.

Cho, I.S., Heng, C.K., and Trivic, Z. (2016). *Re-Framing Urban Space: Urban Design for Emerging Hybrid and High-Density Conditions.* New York, NY; Abingdon, Oxon: Routledge.

Cho, I.S. and Križnik, B. (2017). *Community-Based Urban Development: Evolving Urban Paradigms in Singapore and Seoul.* Singapore: Springer Nature Singapore.

Cho, I.S., Trivic, Z., and Nasution, I. (2015). "Towards an Integrated Urban Space Framework for Emerging Urban Conditions in a High-Density Context." *Journal of Urban Design*, 20(2): 147–168.

Cho, I.S., Trivic, Z., and Nasution, I. (2017). "New High-Density Intensified Housing Developments in Asia: Qualities, Potential and Challenges." *Journal of Urban Design*, 22(5): 613–636.

Cho, M.R. (2010). "The Politics of Urban Nature Restoration: The Case of Cheonggyecheon Restoration in Seoul, Korea." *International Development Planning Review*, 32(2): 145–165.

de Solà-Morales, M. (1992). "Public and Collective Space: The Urbanisation of the Private Domain as a New Challenge." In *A Matter of Things*. (2008). Reprint. Rotterdam: NAi Publishers.

Ellin, N. (2006). *Integral Urbanism*. New York: Routledge.

Fainstein, S. (2009). "Mega-Projects in New York, London and Amsterdam." *International Journal of Urban and Regional Research*, 32(4): 768–785.

Fenton, J. (1985). *Architecture Pamphlet 11: Hybrid Buildings*. New York: Princeton Architectural Press.

Hadi, Y., Heath, T., and Oldfield, P. (2018). "Gardens in the Sky: Emotional Experiences in the Communal Spaces at Height in the Pinnacle@Duxton, Singapore." *Emotion, Space and Society*, 28: 104–113.

Hou, J. (2012). "Vertical Urbanism, Horizontal Urbanity: Notes from East Asian Cities." In V. Bharne (Ed.) *The Emergent Asian City: Concomitant Urbanities and Urbanisms*. London; New York: Routledge.

Jerde Partnership. (2017). "Places: Namba Parks." *JERDE*. Accessed October 29, 2017. www.jerde.com/places/detail/namba-parks.

Komossa, S. (2011). "Researching and Designing GREAT; the Extremely Condensed Hybrid Urban Block." *Architecture & Education Journal*, 5: 27–36.

Kržnik, B. (2011). "Selling Global Seoul: Competitive Urban Policy and Symbolic Reconstruction of Cities." *Revija za sociologiju*, 41(3): 291–312.

Lau, S., Wang, J., Giridharan, R., and Ganesan, S. (2005). "High-Density, High-Rise and Multiple and Intensive Land Use in Hong Kong: A Future City Form for the New Millennium." In M. Jenks and N. Dempsey (Eds.) *Future Forms and Design for Sustainable Cities*. Oxford; MA: Architectural Press.

Madanipour, A. (2006). "Roles and Challenges of Urban Design." *Journal of Urban Design*, 11(2): 173–193.

Mayer, I.S., van Bueren, E.M., Bots, P.W.G., van der Voort, H., and Seijdel, R. (2005). "Collaborative Decisionmaking for Sustainable Urban Renewal Projects: A Simulation: Gaming Approach." *Environment and Planning B: Planning and Design*, 32(3): 403–423.

Nissen, S. (2008). "Urban Transformation from Public and Private Space to Spaces of Hybrid Character." *Sociologický Časopis/Czech Sociological Review*, 44(6): 1129–1149.

Per, A.F., Mozas, J., and Arpa, J. (2014). *This is Hybrid*. Vitoria-Gasteiz: A+t Architecture Publishers.

Reurbanist. (2011). "Namba Parks." *Reimagining Urban Environments: Using Retail to Remake Urban Mixed Use Environments*. Last modified March 9, 2011. http://reurbanist.com/2011/03/namba-parks/

Rowe, P.G. (2010). *A City and Its Stream: An Appraisal of the Cheonggyecheon Restoration Project and Its Environs in Seoul*. Seoul: Seoul Development Institute.

Saito, A. and Thornley, A. (2003). "Shifts in Tokyo's World City Status and the Urban Planning Response." *Urban Studies*, 40(4): 665–685.

Sandercock, L. and Dovey, K. (2002). "Pleasure, Politics, and the 'Public Interest': Melbourne's Riverscape Revitalization." *Journal of the American Planning Association*, 68(2): 151–164.

Shelton, B., Karakiewicz, J., and Kvan, T. (2011). *The Making of Hong Kong: From Vertical to Volumetric*. Abingdon, Oxon; New York, NY: Routledge.

United Nations. (2018). *World Urbanization Prospects: The 2018 Revision (Key Facts)*. Retrieved from: https://esa.un.org/unpd/wup/Publications/Files/WUP2018-KeyFacts.pdf.

Urban Land Institute. (2015). *ULI Development Case Studies: Namba Parks*. Retrieved from: https://casestudies.uli.org/wp-content/uploads/sites/98/2015/12/C035006.pdf.

Urban Land Institute. (2016). *ULI Development Case Studies: Tokyo-Midtown*. Retrieved from: https://casestudies.uli.org/wp-content/uploads/sites/98/2016/06/Tokyo-Midtown.compressed.pdf.

Zacharias, J. (2013). "The Central-Mid-Levels Escalator as Urban Regenerator in Hong Kong." *Journal of Urban Design*, 18(4): 583–593.

Further reading

Carmona, M., Tiesdell, S., Heath, T., and Oc, T. (2010). *Public Places Urban Spaces: The Dimensions of Urban Design (second edition)*. Oxford: Architectural Press. A thorough overview of urban design theories, practices, issues, and principles.

Shelton, B., Karakiewicz, J., and Kvan, T. (2011). *The Making of Hong Kong: From Vertical to Volumetric*. Abingdon, Oxon; New York, NY: Routledge. Exploration of new models of compact, high-rise, volumetric, dense, and intense urbanism based on a historical overview of Hong Kong's urban development.

Cho, I.S., Heng, C.K., and Trivic, Z. (2016). *Re-Framing Urban Space: Urban Design for Emerging Hybrid and High-Density Conditions*. New York, NY; Abingdon, Oxon: Routledge. Provides a comprehensive and accessible means to understand the critical properties that shape emerging hybrid typologies of urban space in high-density contexts.

17

SHRINKING CITIES, SHRINKING WORLD

Urban design for an emerging era of global population decline

Brent D. Ryan

Introduction: planet of slums, or planet of shells?

Humanity is one of the world's most successful animal species. *Homo sapiens*' ability to alter its environment through technology has improved comfort; provided abundant food, shelter, and safety; and set humanity on a seemingly ever-upward arc of population growth. One thousand years ago, the global human population numbered only around 300 million people (United Nations 1999). By 1800, humanity numbered 1 billion people; by 1960, 3 billion; by 1999, 6 billion; and by 2017, 7.6 billion (United Nations 2017). If the rate of growth that occurred between 1960 and 2000 were to continue, humanity would number an astonishing 12 billion by 2040, and an alarming 24 billion by 2080.

Humanity's rapid and seemingly inexorable increase, together with historic evidence of the deleterious effect of overpopulation on environmental quality, food availability, and overall well-being, has long spurred concern. Malthus's "iron law of population" (Gray 1946, 336), formulated just as the human population topped one billion, argued that inexorable increases in population would inevitably spur "distress," particularly among the poorest members of society (Malthus 1798, 19). This argument has retained its cautionary power in the two-plus centuries since Malthus formulated his law: Paul Ehrlich and Anne Ehrlich (1968) made a similar argument that the world's "population bomb," numbering over three billion at the time of their writing, would ultimately cause disaster. Meadows et al. (1972) made a similar argument in their well-known study on "the limits of growth." Today, sustainability thinking is motivated largely by the fears that continued population growth will exhaust our global ecosystem (Diamond 2011).

Cities mirror their nations, and nations with large, growing populations have typically had large, growing cities. Today, most of the world's largest cities are in the world's largest nations. The United Nations (2016) has calculated that all but 5 of the 25 largest world cities are in the world's 10 most populous countries. Megacities, with over 10 million inhabitants, are a particular challenge: these places face a host of problems including air pollution, traffic, low-quality housing, unemployment, violence, and poor public health (United Nations 2015). Mexico City is a typical example: as early as 1983, when the city had around 12 million inhabitants, a *New York*

Times article called it "a warning to the Third World," reasoning that "with a population that has doubled in the last decade," Mexico City had "woefully inadequate living conditions," with "appalling urban problems" caused by "uncontrolled population growth" (Riding 1983). Such alarmist language has continued to typify urbanists' discussion of megacities, from architect Rem Koolhaas (2000), who decried the problems of Lagos, Nigeria, to urbanist Mike Davis (2006), who warned of a "planet of slums."

Though not all megacity problems are real, many are, and many of these problems are in fact driven by growing populations. Few would deny the pressures that population growth has placed on slums in cities like Mumbai/Bombay in India or Medellín in Colombia. Although the word "slum" may be a pejorative and subjective word for some (Angotti 2006), one would be hard pressed not to find places like Dharavi, Mumbai's largest slum, to fit this term. Dense, humid, polluted, noisy, ringed by highways and sewage-ridden waterways, with housing constructed from concrete to sheet metal to plastic, Dharavi, whatever one chooses to call it, is the result not of historically high birth rates, but of high migration from rural areas. Dharavi is long established, but other slums are not; many in Medellín are quite recent, the result of internally displaced populations driven by civil war to the fringes of the metropolitan area.

It should come as a relief, therefore, to hear that growing populations in many parts of the world may soon be a thing of the past. At a local level, some cities in developed countries have been losing population for decades, and at a national level, many countries are beginning to taper off in population growth; while a few, most significantly Japan, Germany, and Italy, are beginning to lose overall population. The fact is that the world is in the early to middle stages of a profound demographic shift, or a demographic "tsunami" (*Wall Street Journal* 2018). Across the world, with some notable exceptions, populations are aging as women have fewer and fewer children, and as senior citizens have longer life expectancies. Why? Quite simply, economic growth, paired with technological advances, have improved human life tremendously, particularly since 1870, as Gordon (2016) has noted. As well-being improves, a variety of interrelated factors lead to both enhanced survival and reduced birth rates.

Women's decisions are critical: women with secure, prosperous lives and secure civil liberties get married later and have fewer children. If each woman in a population has fewer than approximately 2.1 children on average, that population will decrease. In order for a population to maintain itself, each woman must replace not only herself but a male, plus a fraction representing children who do not reach reproductive age. The map in Figure 17.1 shows how dramatically low world fertility rates have become by 2017. The number of countries with fertility rates less than 2.0 is startling; all of these countries are below replacement rate, some dramatically so (South Korea's fertility rate is only 1.2, Singapore's only 1.26, Japan's only 1.46). Without immigration, any country below replacement rate will eventually lose population. Even Brazil (1.75) and China (1.57) have fertility levels below replacement rate. Equally as startling, however, is the larger number of countries with fertility rates between 2.0 and 3.0. Equally as startling, however, is the larger number of countries with fertility rates between 2.0 and 3.0, meaning that they are just above replacement rate. All of Latin America and South Asia are in this category (except for those countries that are even lower). In fact, the only countries in the world with fertility rates above 3.0, with few exceptions, are in Africa.

The consequences for urban designers, and for probably every other profession, of this looming population decline will be tremendous. Although localized population decline has long been present in human society, these declines have typically occurred because of localized failures, such as those detailed by Jared Diamond (2005) in his book, *Collapse*. Widespread population decline, on the other hand, is due to success, not failure, of development because living standards have risen so high across much of the globe. A world with nations dropping

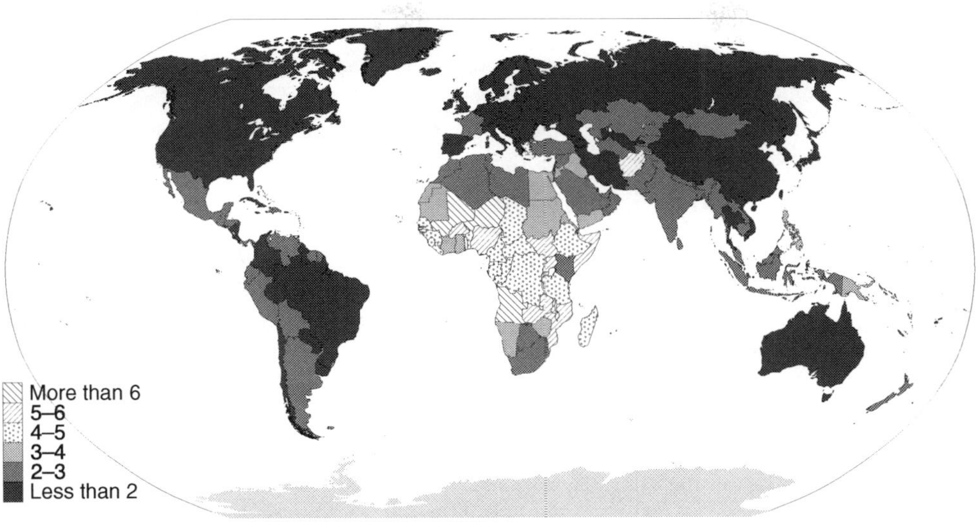

Figure 17.1 Map of Earth's fertility rates by nation, 2017.

Source: Wikipedia Commons, https://commons.wikimedia.org/w/index.php?curid=7248239

below the replacement rate is a world where more and more cities, regions, and nations will ultimately decline in population. Instead of the planet of slums feared by Mike Davis, Earth will increasingly become a planet of shells, with cities dotted with empty houses and empty lots, regions filled with empty villages, and entire nations with empty areas nearly devoid of habitation. How urban designers will confront this growing global population slowdown, occurring in many different nations and many different city regions, may ultimately become the discipline's greatest challenge.

This shift towards a planet of shells is already underway even in cities that were once growing rapidly. Mexico City, once "uncontrolled," is now a slowly growing city (the national fertility rate is 2.21, just above replacement), whose *favelas* have stopped expanding, and whose periphery is dotted with semi-abandoned settlements that are too far from downtown to be viable. There will be no further large-scale population influxes to Mexico City, as the country's rural-urban migration is close to completion. Mexico is 79% urban, meaning that there are not many rural inhabitants remaining available to migrate to Mexico City. In fact, much of the population increase still occurring in Mexico City comes not from rural Mexico, but from immigrants from poorer countries in Central America.

Shrinking cities and localized population decline

Human population dynamics reflect those of nature. Ecologists MacArthur and Wilson (1967), examining island species, identified a principle that they called R/K selection. R-selected species had many predators, short lifespans, and many children. Rabbits are an example of an R-selected species, with several litters a year, most of whose offspring do not survive. K-selected species had few predators, long lifespans, and few children. Blue whales are an example of a K-selected species; they breed every so often, and baby blue whales generally survive to adulthood. Though R/K selection has been criticized by later ecologists, the principle retains analogous value when looking at human populations; as human well-being, safety, and prosperity increase, we can understand humans as shifting from R-selected towards K-selected. In a growing number of countries, humans have few (or no) predators (other than themselves), long lifespans, and consequently, fewer children.

The shift from R towards K selection can be seen in the shift over time in the types and extent of shrinking cities. Prior to the 20th century, humans were more R-selected; family sizes were large and lifespans relatively short. In such populations with rapid replacement rates, local populations can be extinguished, but larger populations remain stable or growing. Thus, prior to 1900, cities occasionally shrank or vanished entirely due to localized catastrophes like volcanic eruptions, earthquakes, ecological exhaustion, epidemics, or invasions and warfare. These forces caused both the permanent extirpation of populations in places like Greenland (Diamond 2005), or the permanent abandonment of cities (Vale and Campanella 2002).

In more recent times, however, catastrophes have shifted in nature. Rather than natural or human-made catastrophes, both of which have diminished in frequency and in extent, factors driving population loss in the 20th century have been primarily technological, economic, and political. This shift has occurred in concert with humanity's shift towards a K-selected population model; in some cases, 20th-century factors driving decline have coincided with declines in fertility rates, and some of these factors may have driven fertility declines as well. Three examples illustrate shrinking cities' experience in an era of localized population decline: the United States, Germany, and the ex-Soviet Union.

The United States was early to industrialize and quick to modernize. By the early 20th century, American industry was at its apex, concentrated in the cities of the northeastern and midwestern United States. The primacy of these cities was guaranteed by the technological and social forces of the time, such as railroads, that obliged centralization (Rae 2003). Comparatively homogenous urban populations shared dense urban spaces by default, since transportation was limited to rail. After 1940, increased automobility and suburbanization devastated many American central cities; and later, globalization and a shift to a postfordist economy affected especially older industrial hubs. Rail declined, and automobiles permitted decentralization of residences, commerce, offices, and industry to the periphery. The economies of industrial cities declined as their enterprises relocated out of the city or went out of business. The migration of formerly rural populations of diverse races and ethnicities drove further decentralization as wealthier white populations suburbanized.

Gordon (2016) noted that the full effects of technologies are often felt many decades after they were first invented. The truth of this can be seen in American cities that continued to grow according to pre-automobile development patterns, even as the car made those patterns irrelevant. In Detroit, for example, automobile factories developed, from the 1910s until the 1950s, adjacent to rail lines in multi-story, large-windowed buildings that resembled 19th-century mills (Ryan and Campo 2013). The product they were constructing would make these factories obsolete. Contemporary automobile factories, of course, are quite different: oriented to the car and truck, they are windowless single-story buildings located adjacent to highways in low-cost suburban areas. Victimized by forces such as suburbanization and globalization that they could neither control nor fully understand, industrial American cities declined precipitously from 1960 onward (Rae 2003). Large shrinking cities included Philadelphia, Detroit, Cleveland, Baltimore, St. Louis, and, for some decades, Chicago, New York, and Boston. America's urban decline, however, was highly localized to the cities themselves; the suburban area of each declining city continued to thrive, and much of the rest of the United States grew rapidly, immune from the shocks affecting the older part of the country. This localized decline, with one part of the country (the Rust Belt) declining in both economic and demographic importance while other areas (mostly in the sunbelt) grew, made it difficult to achieve a national political consensus on how to address the problem.

Germany's industrial history was similar to that of the United States, but the country's 20th-century history was quite different. World War II damaged Germany badly, but the following 40 years of political division (1949–1989) had an arguably greater effect, splitting the country into a capitalist west, where economic trends were similar to those in the United States, and a communist east, where the state supported underperforming industries until 1989, when East Germany collapsed. Germany thus faced two types of shrinkage in its two regions: one in the west with technology-driven economic decline similar to that of the United States in places such as the Ruhr; and one in the post-1989 east, driven by the combined political, social, and economic shock of collapse and reunification. After 1989, many East German industrial cities emptied out, as populations moved west to obtain work, but other forces also drove the shrinking of German cities. Unlike the United States, suburbanization was less of a factor, as German populations were less socially divided than America's. But low fertility rates affected both areas of the country: with limited immigration, Germany's population grew only 13.5% between 1960 and 2016. Whereas America was characterized by steeply declining industrial cities in growing or stable surrounding regions,[1] western Germany was characterized by a patchwork of declining city regions in prosperous areas. In the declining east, the patchwork was reversed, with a few growing cities surrounded by a much larger shrinking territory. The built environment of German shrinking cities was characterized by partially vacant multifamily housing blocks, derelict industrial areas, and depressed, though still functional, business districts.

The decline of cities in the former Soviet Union mirrored that of Germany's to some extent, and to some extent that of the United States. Across the vast space of the Soviet Union, planners constructed industrial cities after 1928, according to national five-year economic plans designed to rapidly industrialize the formerly rural nation. They augmented existing cities like Kiev with abundant industrial facilities and created new cities from scratch in resource-rich, poorly populated areas, often with the help of convict labor (Hill and Gaddy 2003). As a result, by 1990, the Soviet Union was dotted with heavily industrial cities in widely dispersed locations, constructed according to an economic logic that became swiftly obsolete when the Soviet Union collapsed in 1991, just as East Germany had collapsed in 1989. Soviet cities, once swimming in state-directed economic activity, were left suddenly adrift.

One such city was Kherson, a small industrial city located after 1991 in independent Ukraine. Kherson had specialized in shipbuilding and cotton processing, the latter commodity coming from Soviet republics in Central Asia. Independence left both industries obsolete, making Kherson a city without an economic base. This situation was mirrored across the Soviet Union, where residents of newly obsolete economies often found themselves with little means of support, and with little means of moving elsewhere. Residents could relocate to the local capital, or to Moscow, but weak housing markets and low-value homes "trapped" many ex-Soviets in their declining cities. Unlike America, where cities shrank as former residents suburbanized, or Germany, where cities shrank as residents moved west or aged, ex-Soviet citizens mostly remained in their low-value homes. Fertility rates also dropped, as economic conditions became precarious and uncertain. The built environments of ex-Soviet shrinking cities were characterized by mostly occupied housing blocks in poor condition, large-scale abandoned or semi-functional industrial sites, and depressed center-city business areas.

Urban designers' response to shrinking cities

Prior to 1989 when East Germany collapsed, urban population decline was a comparatively rare event except in the United States, where decline was commonplace. Design responses to poverty, "blight," and abandonment began in American cities as early as the 1930s. Urban designers' response to population decline was heavily influenced by federal government policy and investment that initially funded the clearance and rebuilding of declining areas. Rebuilding peaked in the 1950s and 1960s with passage of federal acts that funded neighborhood clearance and infrastructure construction. Depressed residential neighborhoods and mixed industrial-commercial areas were acquired by city authorities and redesigned by city planners and architects, to encourage private investment. This funding collapsed in the 1970s with an economic slowdown and a corresponding shift in social attitudes towards government intervention.[2] Prior to 1970, urban design strategies were heavily influenced by Modernism, with standalone, abstractly designed buildings surrounded by open space and/or parking. Examples of such Modernist urban design included Lafayette Park in Detroit, Charles Center in Baltimore, and Lake Meadows in Chicago. After 1970, the Modernist ideology was tempered by an increasing appreciation of historic buildings and a preference for more representational (i.e. historicist) design (Ryan 2012). For the most part, post-1970s urban design strategies were smaller in scale, respectful of existing built environments, and more reliant on private market financing. This shift in strategies did not moderate the ongoing economic decline of cities like Detroit and Cleveland, which was driven by the exogenous factors detailed earlier. Abandonment in shrinking American cities thus continued, growing to epic proportions by the 1990s. As urban design involving city block reconfiguration, street pattern shifts, and large-scale construction of new structures requires capital, the absence of federal funding limited most declining American cities in the 1990s and early 2000s to constructing small-scale housing, rehabilitating older buildings in central locations, and repurposing vacant infrastructure and building sites as open spaces.

Urban design in American shrinking cities was hampered not only by a lack of funding and a lack of public-sector will, but by American cities' urban structure of rectilinear blocks subdivided into small building parcels, each occupied by a standalone or attached dwelling or commercial structure.[3] Since abandonment of such properties occurred on an individual basis, American building blocks assumed a piecemeal appearance, with a mix of empty parcels, vacated but still standing structures, and inhabited or occupied structures. Given that each parcel was typically individually owned, redevelopment of such city blocks, absent the large-scale government financing of the pre-1970 era, was effectively impossible.[4] Occasionally, cities were still willing to exert acquisition powers to redevelop large areas of declining neighborhoods, but such redevelopment was costly, as land and other issues like intervening properties and title problems made redevelopment even more challenging (Ryan 2012). For the most part, the urban design of American shrinking cities – even that which focused on landscape rather than the built environment (DeSimini 2014) – was dependent upon infill strategies accepting the existence of remaining inhabited structures.

Germany's experience with shrinking city urban design was very different. Unlike in the United States, federal investment in German cities did not evaporate, but remained high. As Jessen (2012) memorably said, Germans "have a rather high expectation concerning the role of the state and its interventions," thus supporting and accepting federal urban expenditures across the political spectrum. After 1989, as reunification and support of the depressed east was a national priority, the federal government funded the "Stadtumbau Ost" and "West" ("City Rebuilding East" and "West") programs. Whereas post-1975 American urban design was halting, small-scale, esthetically conservative, and driven by often myopic market forces, German urban design efforts in shrinking cities were nearly the opposite. Post-1990 German rebuilding was driven by well-funded federal programs, with a range of high-quality design approaches that incorporated varying amounts of demolition, rehabilitation, and conversion. The resulting projects were often splendid, particularly the Emscher Landschaftpark in Duisberg Nord (west) (Birkhäuser Architecture 2010) and the varied landscape, leisure, and reclamation-based projects of Furst-Puckler-Land in Lusatia (east) (Bauausstellung 2010). Both projects prioritized reuse of open spaces, with strategies that included preservation of industrial ruins, rehabilitation of nearby housing areas, construction of social and cultural facilities like community centers and museums, and leisure landscapes with ecological, ludic, and commercial programs such as concert facilities and office buildings.

Urban design in German shrinking cities, particularly in the east where many such cities were located, was "aided" by the region's socialist past. In the former East Germany, all land had been the property of the state, and most postwar apartments were in multifamily, prefabricated buildings whose inhabitants were renters, not owners. Unlike the ex-Soviet Union, where such flats were immediately privatized after 1991, Germany's socialist-era flats remained the property of housing cooperatives, and their inhabitants remained renters. With most housing occupying large, aggregated parcels of land, urban design proposals to consolidate and selectively demolish such housing was a practical proposition, particularly when backed by the Stadtumbau policy's funding. An urban design proposal such as that for Dessau, a shrinking city where designers proposed a greenbelt of recreational space on the site of selectively demolished socialist flats, could be implemented within just a few years of its proposal (Oswalt and Mittmann 2010) (Figure 17.2). Germany's shrinking city urban design thus offered much to learn from, but the particular qualities of Germany's urban landscape, and the country's high level of federal funding, made these urban design strategies difficult to emulate.

The former Soviet Union struggled perhaps the hardest to generate urban design strategies in its declining cities. Soviet cities before 1991 had been entirely shaped by state-funded, state-backed urban design, as housing, industry, and commerce were all state-controlled. But post-1991 ex-Soviet cities faced a host of challenges: bankrupt governments, corrupt oligarchies, and disempowered populations with low incomes. Civic groups were absent or underdeveloped, and nongovernmental organizations attempting to organize citizens sometimes found that citizens were unwilling or uninterested in improving even their own housing estates. Decades of state-driven city building, top-down decision-making, and suppression of citizen initiative had left city residents believing that improvement was a governmental responsibility requiring neither their consent nor their participation. This apathy, together with weak markets, depressed economies, and the massive abandoned industrial properties of ex-Soviet cities, resulted in a toxic combination.

Some ex-industrial areas were developed with retail by developers, often operating in combination with state authorities. But decaying Soviet housing was left mostly alone; it was often fully inhabited, and there was little latitude for action except for opportunistic infill with shops and kiosks. Urban design in the ex-Soviet Union thus remained an unfinished project. Large-scale state support like that of Germany was unlikely to materialize, and citizens had fewer options for relocation than did residents of shrinking cities in the United States. The challenges of urban design in shrinking ex-Soviet cities somewhat resembled those of American cities: in both countries, urban design in shrinking cities was limited in scale, rarely progressive in its design, and reliant upon the private sector for both initiative and funding. Ex-Soviet cities, with their disempowered citizens and corrupt governments, presented a dispiriting situation, even as they continued to slowly decline.

Figure 17.2 Greenbelt urban design proposal for the German shrinking city of Dessau. Housing is projected to be demolished in a belt around the historic downtown, reducing abandoned housing units and providing recreational space for citizens.

Source: Oswalt and Mittmann 2010, 591.

Shrinking nations: urban design and widespread population decline

Over the next 30 years, many Asian countries will reach their peak population, then begin a slow decline. Somewhere in the 2020s, China will reach peak population of around 1.4 billion people, barring a dramatic reversal in its fertility rate, or substantial immigration. Neither shift seems likely at present (Babones 2017). It is expected that India will outstrip China in population by the early 2020s (Katz 2015), but India, too, will begin declining in population around 2065, or even earlier: The *Times of India* (2018) estimates that some Indian states will begin hitting peak population as early as 2041. Asia as a whole is predicted to become a shrinking continent by the 2050s, though this decline will be heavily dependent upon China's and India's population dynamics. Should other trends remain the same, however, Asia's forthcoming population decline means that the human population in the 22nd century will be nearly half African. The implications of this coming "African century," as some have called it (Meynial 2018), will make the world in a hundred years a very different place from that of today.

While most urban literature on China has focused on the country's rapid urbanization, the pairing of this urbanization with the aging of China's population and the slowdown in its fertility rate means that many regions of China are beginning to shrink even as the nation's largest metropolitan areas continue to grow. Long and Wu (2016) found that 19,882 of China's 39,007 townships, or almost half, lost population between 2000 and 2010. Most of these townships were rural, reflecting village abandonment, but 1,147 of them were urban. Additionally, these authors identified 180 shrinking cities, which could be found in almost every province. While much of China's localized decline is due to rural-urban migration combined with dropping fertility, some decline is also due to resource exhaustion. Yichun, in Heilongjiang, is one of China's most rapidly shrinking cities. Yichun lost 14.8% of its population, or 111,000 people, between 2005 and 2015 (*Financial Times* 2017a) due to a ban on logging that ended the practice entirely in 2013 (Sixth Tone n.d.). Although its circumstances are to some extent unusual, Yichun provides an illustrative example of how urban design may unfold in China's shrinking cities.[5]

In the People's Republic of China, most land is nationalized, and urban policy is often dictated from the national level. Germany demonstrated that such national policy, though not privileging individual owners or residents, is a fertile one for urban design. The same is true in Yichun as well. Urban design in Yichun has focused on the rehabilitation or demolition of deteriorated single-family homes and the relocation of residents into higher-quality, multifamily housing. Many Yichun residents have taken the opportunity to relocate into flats, while others have chosen to rehabilitate their houses with aid from the state. Such rehabilitation is expensive, however, and poorer residents often cannot afford to rehabilitate; they must therefore move to the new flats, or remain in deteriorated older dwellings, often in semi-vacated areas where much housing has been demolished. The newer flats are themselves only semi-occupied, since many residents have decamped for cities with stronger economies, leaving their homes unoccupied. The result is a patchwork city, with districts of semi-occupied single-family homes (some rehabilitated, others demolished) and districts of semi-occupied multifamily flats. As Yichun continues to lose population due to demographic decline in addition to economic transition, this patchwork quality will likely grow in extent. The future urban form of Yichun will depend upon the central Chinese government's willingness to fund additional construction, or deconstruction, of declining districts, but such a fiscal commitment has not yet fully arrived.

Japan provides an indication of the coming urban design challenges in shrinking cities within shrinking nations. The nation is already past its peak population; it declined by 300,000 people in 2016 (*Financial Times* 2017b), and Japan is expected to lose 16.2% of its 2010 population by 2040 (Japan Foreign Policy Forum 2015). Two thirds of Japan's 1,727 municipalities lost more than 5% of their population between 2010 and 2015, and this number is itself misleading, because Japan reduced its municipalities from 3,232 prior to 2010 in order to accommodate population loss (Buhnik 2017). Japan is not unique in its nationwide population decline (Italy and Germany, together with many Eastern European nations, are also undergoing such nationwide declines), but Japan, uniquely, has instituted a national urban policy to reshape metropolitan areas as a reaction to nationwide population loss. In 2014, the national government stated that "measures for regional concentration and revitalization through compact city development" (Japan Foreign Policy Forum 2015) would be implemented to address population shrinkage. This strategy implies a kind of strategic retreat of urbanization, with populations receding to core areas and outlying areas abandoned. In a mountainous nation like Japan, where cities are surrounded by well-wooded mountains, this strategic shrinkage has a geographical logic.

The city of Toyama, a prefectural (provincial) capital on the north shore of Honshu, has taken early steps towards realizing such a compact city. In 2007, the city announced its first strategy towards "creating a compact city of the future." The key to Toyama's policy is infrastructure rather than housing; by constructing additional infrastructure around the city center, the municipality has declared success in attracting senior citizens to live closer to downtown, providing them with better access to other seniors, increased sociability, and increased access to urban amenities (Figure 17.3). Toyama's focus on city restructuring for the benefit of senior

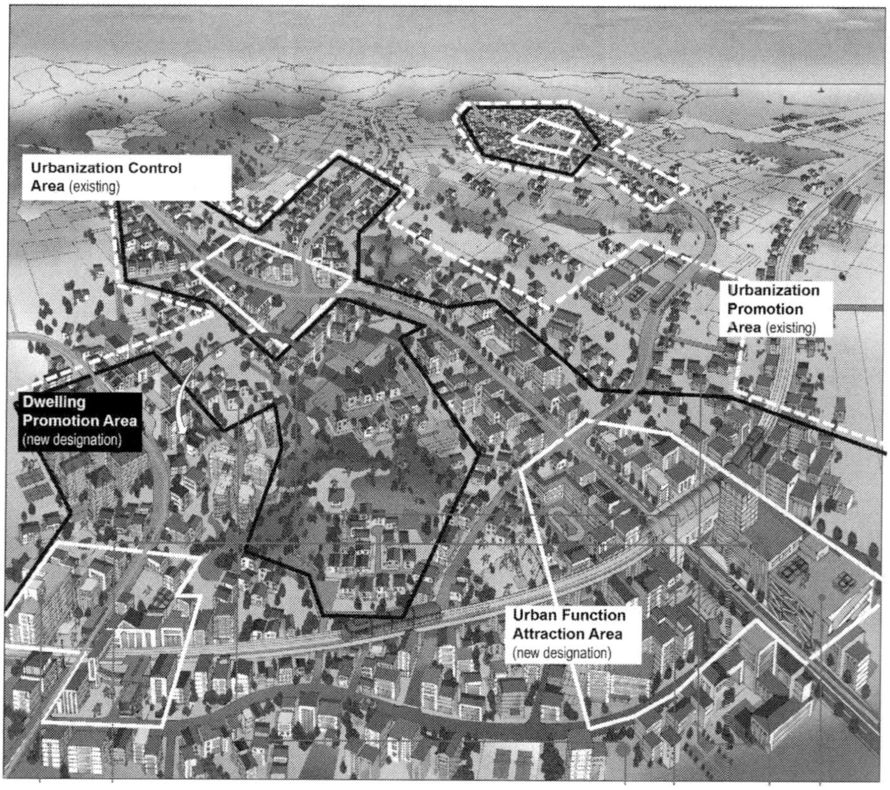

Figure 17.3 Japanese shrinking city concept. Shrinking cities in Japan are strengthening their centers in order to permit gradual diminution of population in outlying areas. The dashed white line denotes the current containment area of urban growth; the black and solid white denote newly designated "dwelling promotion areas" and "urban function attraction areas." Development in between the black and dashed white lines will now be officially discouraged.

Source: Japan Ministry of Land, Infrastructure, Transport and Tourism (2014).

citizens differs radically from previous shrinking city urban design strategies in Germany, which focused on poverty alleviation for low-income residents. As populations age further, however, Toyama's urban design for senior citizens may become an increasing, even the dominant, means by which cities in Japan are restructured. Of course, Japan's strategy also depends on abundant government funding for infrastructure and senior facilities, as well as on a national consensus to aid declining areas. Whether such funding will become more or less available as Japan senesces further is still unknown.

Conclusion: urban design in a shrinking world

This review of urban design efforts in shrinking cities and in shrinking nations indicates that the future for designers in these places presents both opportunities and challenges. The opportunities are numerous; shrinking cities can provide additional space, reduced competition for amenities, and even an enhanced quality of life. Such opportunities have motivated the successful urban design strategies of Germany, where industrial land has been repurposed for recreation and where social facilities have been rehabilitated and enhanced. Such opportunities also seem to be present in the emerging national scale urban design of Japan, where cities appear to be beginning a process of spatial reorganization to concentrate and enhance amenities.

The converse, however, may also be true: challenges for shrinking cities and shrinking nations may also be widespread. Experience demonstrates that shrinking cities often present problems that defy easy solutions. Both the United States and the former Soviet Union have shrinking cities that have found themselves politically isolated, left to their own economic devices, and riddled with social malaise and political dysfunction. While the United States is not likely to become a shrinking nation because of immigration, most post-Soviet countries are experiencing nationwide population declines but have little immigration and no national urban design policies. Will these countries marshal the political will for national strategies that can reshape their shrinking cities and regions? So far, the evidence is unclear.

Barring radical changes in human behavior, more and more nations will pass peak population and enter into gradual population decline in the decades to come. Accordingly, the potential and need for innovative urban design strategies at citywide and regional scales will only grow. Given that policymaking structures and urban forms differ widely from nation to nation and from region to region, there are likely to be a variety of urban design approaches proposed and implemented in shrinking nations. With its top-down governance, China demonstrates substantial potential for well-conceived, national-scale urban design strategies. The next decade or two will be critical in creating and testing such strategies. A few decades further on, India will provide additional opportunities for urban designers to participate in the physical restructuring of rural areas, then cities, in response to population decline. Even Indonesia and Brazil, unexamined in this chapter, will provide parallel opportunities for urban designers later in this century as their populations mature, age, and begin to decline. Ultimately, the urban design trajectories of shrinking cities in developed countries may provide only limited precedent for the coming shrinking of much of Asia and the Global South.

Notes

1 The St. Louis metropolitan region, for example, increased in population from 1,681,281 in 1950 to 2,698,687 in 2000 while its core city, St. Louis, declined from 856,796 to 348,189. Other industrial cities' metropolitan regions, such as Cleveland, Detroit, and Buffalo, declined slightly in population after 1970 as their core cities declined precipitously.

2 Elements of the relationship of American "urban renewal" to urban design are detailed in Ryan (2012), Klemek (2011), and Ammon (2016), among many others.

3 Though she does not treat urban decline or population loss, Vernez Moudon (1989), examining San Francisco, provides a detailed examination of a typical American city's rectilinear block and lot structure and its resistance to redevelopment.

4 I detail this dilemma in *Design after Decline* (Ryan 2012: ch. 3) and explore its manifestations in Flint, a typical American shrinking city, in Ryan (2015).

5 Additional detail on Yichun's planning and urban design measures is in Gao and Ryan (in press).

References

Ammon, F.R. (2016). *Bulldozer: Demolition and Clearance of the Postwar Landscape.* New Haven: Yale University Press.

Angotti, T. (2006). "Review of *Planet of Slums.*" *International Journal of Urban and Regional Research*, 30(4): 961–967.

Babones, S. (2017). "Has China Hits Its Peak? How Demography, Economics, and Geography Will Limit China's Growing Power in 2018." *Forbes* (December 31). www.forbes.com/sites/salvatorebabones/2017/12/31/china-max-how-demography-economics-and-geography-will-combine-to-limit-chinas-growing-power/2/#487d7cc81ad8

Bauausstellung, I. F. (2010): *Katalog: Neue Landschaft Lausitz.* New Landscape Lusatia. Berlin, Jovis.

Birkhäuser Architecture (2010). *Unter Freiem Himmel/Under the Open Sky.* Regionalverband Ruhr.

Buhnik, S. (2017). "Migrating Urban Decline through the Compact City? Reflections of 15 Years of Urban Recentralization Policies in Japan." *Metropolitics.eu* (December 7). www.metropolitiques.eu/Mitigating-urban-decline-through-the-compact-city-Reflections-on-15-years-of.html

Davis, M. (2006). *Planet of Slums.* New York: Verso.

Desimini, J. (2014). "From Planned Shrinkage to Formerly Urban: Staking Landscape Architecture's Claim in the Shrinking City Debate." *Landscape Journal*, 33(1): 17–35.

Diamond, J. (2011). *Collapse: How Societies Choose to Fail or Succeed.* Penguin.

Ehrlich, P.R. (1968). *The Population Bomb.* New York: Sierra Club/Ballantine Books.

Financial Times (2017a). "Shrinking Cities: Population Decline in the World's Rust-Belt Areas" (June 16). www.ft.com/content/d7b00030-4abe-11e7-919a-1e14ce4af89b

Financial Times (2017b). "Japan Suffers Record Decline in Population" (July 5).

Gordon, R.J. (2016). *The Rise and Fall of American Growth: The U.S. Standard of Living since the Civil War.* Princeton: Princeton University Press.

Gray, A. (1946, 1947). *The Socialist Tradition: Moses to Lenin.* Longmans, Green and Co., p. 336.

Hill, F. and Gaddy, C.G. (2003). *The Siberian Curse: How Siberian Planners Left Russia Out in the Cold.* Washington, DC: The Brookings Institute Press.

Japan Foreign Policy Forum (2015). "Japanese Regions in the Face of Depopulation and the Trend of Compact City Development" (May 7). www.japanpolicyforum.jp/archives/editor/pt20150507023249.html

Japan Ministry of Land, Infrastructure, Transport and Tourism (2014) "Pamphlet for Location Optimization Planning system under the Act on Special Measures Concerning Urban Reconstruction".

Jessen, J. (2012). "High Speed Suburban Decline: The Case of Cities in Eastern Germany." In Nan, S. and Gossop, C. (Eds.). *Fast Forward: City Planning in a Hyper Dynamic Age.* ISOCARP Review 8, pp. 164–177.

Katz, A. (2015). "India Will Have the Largest Population Earlier Than Expected." *Washington Post* (July 30). www.washingtonpost.com/news/worldviews/wp/2015/07/30/india-will-have-the-worlds-largest-population-earlier-than-expected-u-n-says/

Klemek, C. (2011). *The Transatlantic Collapse of Urban Renewal: Postwar Urbanism from New York to Berlin.* Chicago: University of Chicago Press.

Koolhaas, R. (2000). "Lagos." In *Mutations: Harvard Project on the City.* Actar.

Long, Y. and Wu, K. (2016). "Shrinking Cities in a Rapidly Urbanizing China." *Environment and Planning A*, 48(2): 220–222.

MacArthur, R. and Wilson, E.O. (1967, 2001 reprint edition). *The Theory of Island Biogeography.* Princeton: Princeton University Press.

Malthus, T.R. (1798). *An Essay on the Principle of Population.* Chapter 2, 19. Oxford World's Classics.

Meadows, D.H., et al. (1972). *The Limits to Growth: A Report for the Club of Rome's Project on the Predicament of Mankind.* New American Library.

Meynial, C. (2018). "Enquete: Le Siecle de l'Afrique." *Le Pointe Afrique* (February 22). http://afrique.lepoint.fr/economie/enquete-le-siecle-de-l-afrique-02-02-2018-2191521_2258.php

Oswalt, P. and Mittmann, E. (2010). *Less Is Future: 19 Cities-19 Themes: International Building Exhibition Urban Redevelopment Saxony-Anhalt.* Jovis Publisher.

Rae, D. (2003). *City: Urbanism and Its End.* New Haven, CT: Yale University Press.

Riding, A. (1983). "Problems of Mexico City: Warning the Third World." *New York Times* (May 15). www.nytimes.com/1983/05/15/world/problems-of-mexico-city-warning-to-third-world.html

Ryan, B. (2012). *Design After Decline: How America rebuilds shrinking cities.* Philadelphia: University of Pennsylvania Press.

Ryan, B. and Campo, D. (2013). "Autopia's End: The decline and fall of Detroit's automotive manufacturing landscape." *Journal of Urban History*, 12(2): 95–132.

Ryan, B. (2015) "Stadtschrumpfung als Determinante Städtischer Politik: Das beispiel Flint, Michigan, USA". *Forum Stadt* 3: 131–150.

Sixth Tone (n.d.). "Yichun: The Timber Town That Banned Logging." http://interaction.sixthtone.com/feature/2018/shrinking-cities/yichun.html

Times of India (2018). "India Will Start Shrinking from Kashmir and Kanyakumari" (February 2), p. 3.

United Nations (1999). *The World at Six Billion.* Population Division, Department of Economic and Social Affairs, United Nations Secretariat. www.un.org/esa/population/publications/sixbillion/sixbillion.htm

United Nations (2015). *2014 Revision of World Urbanization Prospects.* https://esa.un.org/Unpd/Wup/

United Nations (2016). *Demographic Yearbook 2016.* New York: UN: Statistics Division. https://unstats.un.org/unsd/demographic-social/products/dyb/dybsets/2016.pdf

United Nations (2017). *World Population Prospects.* https://esa.un.org/unpd/wpp/

Vale, L. and Campanella, T. (2002). *The Resilient City.* Oxford: Oxford University Press.

Vernez Moudon, A. (1989). *Built for Change: Neighborhood Architecture in San Francisco.* Cambridge, MA: MIT Press.

Wall Street Journal (2018). "Demographic Tsunami Menaces the World's Developing Economies: More Retirees, Fewer Young People Mean Little Money for Growth" (Monday, April 2).

Further reading

Oswalt, P. and Rieniets, T. (2006). *Atlas of Shrinking Cities.* Hatje Cantz. This early graphic investigation of shrinking cities as a global condition provides a variety of lively maps and graphics.

Richardson, H. and Bae, C.H. (2014). *Shrinking Cities: A Global Perspective.* London: Routledge. This edited volume provides concise overviews of the situations of shrinking cities in several global regions.

18

FROM ABANDONMENT TO URBAN QUALITIES?

Urban design strategies for shrinking cities in Germany and the United States

Karina Pallagst, René Fleschurz, and Sabrina Förch

Introduction: the phenomenon of shrinking cities

In recent years, planners have paid growing attention to the subject of shrinking cities as many countries, cities, and regions – including old industrialized "rust belts" and peripheral rural areas – have undergone structural transformations. Contemporary research on shrinking cities has referred to population loss as the main indicator of urban shrinkage. To fully understand the challenges of shrinkage, the term shrinking city must be adequately explained. According to Wiechmann (2006: 1), a "shrinking city" is defined as a densely populated urban area with a minimum population of 10,000 residents that has faced a population loss for more than two years and is undergoing economic transformation with some symptoms of a structural crisis. In addition to this academic definition of shrinking cities, in planning practice, shrinkage is also perceived in other ways. For example, in Northern Europe a large number of smaller cities are affected by shrinkage, and there are many causes of shrinkage such as climate change (that has led to the shrinkage of some Australian cities), or disasters (such as the Fukushima catastrophe in Japan). While in the last decades, a positive view of growth has dominated the discourses in both planning research and political debates, shrinkage has been viewed merely through its negative effects, such as decline in population and economic activities.

Shrinkage can show very different characteristics depending on national, regional, and local contexts (Cunningham-Sabot and Fol 2007). In the past, shrinkage was mostly the result of a few distinctive causes such as war, disaster, or famine (Benke 2004; Oswalt and Rieniets 2006). Today, many other factors influence the process, such as low fertility rates and consequent aging societies, increased mobility, migration, and suburbanization, in addition to political, economic, and demographic changes (Oswalt and Rieniets 2006). All these factors notwithstanding, increasingly, the shrinkage can be identified as a structural crisis caused by the loss in population and economic transformation, begot by globalization (Pallagst 2013). Meanwhile, shrinking cities have become a worldwide phenomenon today in many postindustrial societies, caused by postindustrial transformations due to the decline of manufacturing industries (Pallagst 2008).

Over the years, the persistence of shrinkage in urban development has been underestimated, although the problems are obvious: an economic basis in decline, out-migrating population, and ensuing housing vacancies. The overlay of sprawl and economic shrinkage has caused severe losses of population in some city centers, and even in entire regions. The problems of derelict sites, vacancies, and abandoned urban quarters are well known and include such social consequences as exclusion, poverty, and homelessness, which are happening to a much more dramatic extent in the United States than in European cities. Furthermore, it has been observed that shrinkages often have self-reinforcing attributes, leading some scholars to call this phenomenon "downward spiral" or "vicious circle" (Fritsche et al. 2007: 18).

Planning and design strategies for shrinking cities

In an earlier publication, Pallagst (2013) claimed that a shift in paradigm is taking place from growth-centered planning to shrinking smart (see also Hollander in this book, Chapter 19, for a definition of smart shrinkage). Yet it is still not clear if or in what ways planning paradigms, planning systems, planning strategies, and planning cultures are adapting when faced with urban shrinkage. Hollander et al. (2009) claim that "shaving the edges off Youngstown, or Detroit or Dresden, will no longer suffice," in particular when dealing with an issue as complex as shrinking cities. However, applying the right strategies when it comes to the future development of a shrinking city seems to be crucial.

But what are these "right" strategies? In search of planning strategies to deal with shrinkage, cities have initiated different policies, plans, and projects often in the frame of programmatic approaches such as urban redevelopment. These efforts try to tackle shrinkage-related causes to reverse or manage the downward-spiral process. These measures can serve as part of a planning strategy for expansion or as a strategy for managing decline. The common denominator is their focus on improving the quality of life in the respective area. Consequently, these cities are seeking to apply not only a new kind of urban planning, but also new forms of urban design.

In the introduction to the first volume of the *Companion to Urban Design*, Banerjee and Loukaitou Sideris (2011: 1) state that urban design showcases multiple scales of application. Cleary, one of them is shrinking cities. Mostafavi (2010: 220) observes in this respect that urban design can play a vital role in shrinking cities when it comes to "urban recycling of the remnants of the industrial city." These cities start from *tabula rasa*, and the new design can gain from the conditions of the territory or place when it comes to creating new developments. Madanipour (2014: 92) highlights the "regenerative role of urban design in economic change" and asks what role of urban design should play: one of innovation, or one of merely managing investments in land use? He discusses further if urban design could offer a "recipe for change" (Madanipour (2014: 119). However, the role of urban design in shrinking cities remains a contested ground, since it has frequently failed to revitalize them, as examples from Detroit and Philadelphia demonstrate (Ryan 2012), using rigid measures of urban renewal.

Still, planning and design approaches that help provide visions and guidelines for the future are essential (Ryan 2012). Indeed, shrinking cities offer the potential to trigger changes, reforms, and even innovations in planning and urban design. Recent research illustrates strong links between the perception or awareness of shrinkage, and the respective strategies, which are applied to deal with decline (Pallagst et al. 2017a). These may include: "decline as vicious circle" (this is a passive stage without specific goals); "expansive strategy" (in this stage communities expand residential areas in order to provide quality housing); "maintenance" (maintaining attractiveness and developing programs for target groups); or "planning for decline" (such as adapting a city to a smaller population and reducing its infrastructure).

Accordingly, these types of strategies might require different types of urban design responses. The following narrative presents an evidence-based investigation of how planning and urban design responses in two cities introduced strategies to deal with shrinkage. In this comparative research, we discuss two different cases: the city of Herne in Germany and the city of Cleveland (Ohio) in the United States. Both cities show comparable conditions in terms of shrinkage paths, population development, and economic structures; yet their planning contexts are quite different. Based on this comparative exploration, we highlight how shrinkage has affected these two cities and point at their struggle to address their shrinkage problems through planning and urban design.

Paths of shrinkage in Germany and the United States

In the new millennium, shrinking cities became a widely discussed topic in academia, politics, and society in Germany. This trend was fueled because after the German reunification, the cities in the eastern part of Germany (the "new Länder") started losing population due to out-migration to western Germany. A cause for this development was the transformation of former socialist economic structures, which triggered a massive deindustrialization in the eastern part of Germany (Glock and Häussermann 2004). However, even prior to the German reunification, and since the 1970s, economic transformations had already started to affect older industrial areas in Germany, most prominently in the Ruhr Area located in the state of North Rhine-Westphalia. In this metropolitan area, economic transformations were triggered by the declining coal and steel industries. The most obvious physical result of these shrinkage processes was an oversupply of buildings, vacant land, housing units, and commercial spaces.

Today, population development in German cities is strongly influenced both by economic factors and declining birth rates (Pallagst and Wiechmann 2005). The impacts of shrinkage in Germany are not much different from those in shrinking cities in other parts of the world: housing vacancies and derelict sites in urban areas. In addition, the provision of public infrastructure and services is becoming problematic for shrinking cities as the tax base keeps declining.

Much like in Germany, in the United States, shrinkage was caused by postindustrial changes going hand-in-hand with long-term industrial transformation processes (Pallagst 2012). However, unlike in old industrial regions of Europe, shrinkage in the United States is usually taking place in the urban core, while the surrounding suburban region continues to grow. In fact, early processes of shrinkage during the 1950s and 1960s were triggered by suburbanization. The sprawl pattern led to vast losses of population in the city centers. The problems of derelict sites, vacancies, and abandoned urban quarters are commonplace in some US cities. These are the problems and consequences of social exclusion, poverty, and homelessness, which are present to a larger extent in the United States than in European cities.

Despite the fact that US cities located in the Rust Belt have lost population since the 1960s, for a long time there was no active discussion of US shrinking cities (Pallagst and Wiechmann 2005). Only since the new millennium did the US media begin to pay attention to the issue of shrinkage, also marking the beginning of an academic discourse. The trigger for this attention was, arguably, the city of Youngstown's 2010 plan, which set off a public debate about shrinkage (Pallagst 2012). Youngstown was the first US city to publicly announce and actively seek to respond to shrinkage, and not follow the usual development path of growth. Indeed, the Youngstown 2010 plan laid the fundaments for a smaller city footprint, a strategy of "right-sizing," and a more sustainable development.

The following section discusses urban planning and urban design implications for shrinking cities by considering the strategies followed in Herne, Germany and in Cleveland, United States. Both cities were affected by changes or decline in one specific economic sector. Accordingly, they lost a significant amount of population following these economic shifts. In addition, both cities are located in industrial regions, also affected by shrinkage: Cleveland is located in the Rust Belt of the Midwest, while Herne is located in the Ruhr Area.

Revitalizing shrinking cities in Germany and in the United States: two different planning contexts

How do cities in different countries react when faced with shrinkage, and how might they utilize urban design to address this challenge? Before highlighting two examples, we briefly discuss the different planning contexts in which US and German cities are embedded.

The US planning culture has traditionally been market oriented, developer driven, and breeding ground for NIMBYism (Pallagst 2007). As it is widely known, sprawl is the dominant pattern. Several policies and tendencies have shaped today's sprawling land use in many US cities. In this respect, development decisions are driven by market forces and embedded in economic competitiveness. Property rights obtain an outstanding position in this interplay. In this regard, what planners do or consider relevant in terms of the quality of life of the community as a whole often stands in direct contrast to the property-driven interest of individuals (Hoch 1994).

In the United States, there is no national planning law, and land use planning at the state level differs greatly from state to state. At the regional level, regional planning agencies exist exclusively in metropolitan regions and often lack implementation powers (Porter 1997). For this reason, the local level formed by communities and counties retains a dominant role in the development of settlements. The single county or municipality, however, is often overextended by the challenges of accommodating growth or decline. Moreover, allocating profitable types of land uses is a tool of supporting a municipality's fiscal strategy. This involves promoting development that generates high amounts of taxes (Razin 1998).

The strong affinity towards growth, which characterizes US planning, makes it difficult for shrinking cities to break free from this context. Even though the Rust Belt has witnessed long-term decline, shrinkage has long been considered as a taboo (Pallagst et al. 2017a). Moreover, unlike in most European countries, shrinking processes in some US cities take place while the country is experiencing population growth at the national level. Against this background, the US discourse on shrinking cities started out with a time lag compared to Europe (Wiechmann and Pallagst 2012).

Germany is characterized by an elaborate multilevel planning system, where policies, plans, and strategies are released at every level of planning: nation, state, region, locality. In addition, guidelines set by the European Union influence the German planning practices. The länder (i.e. province) endorses spatial development plans and regional plans. At the local level, the municipality has the main responsibility for building and planning. There is no space within the country that is not covered by a national, regional, or local plan. Despite the fact that economic changes happened in many old industrial areas, such as in the Ruhr Area, in Germany since the 1970s, the debate regarding shrinking cities (and policymaking to address them) started with the eastern part of Germany suffering from the dramatic effects of post-socialist economic transitions and the vast decrease in fertility rates, combined with sustained processes of out migration. These led to a lively debate on the issues of shrinking cities after the year 2000 (Wiechmann and Pallagst 2012). For this reason, Germany can be seen as the birthplace of the international shrinking cities discourse.

As shown in the previous paragraphs, shrinkage is not a new phenomenon – neither for Germany nor for the United States. Consequently, shrinking cities have sought to address the problem for decades. But not every city can be the next tourism hub or high-tech center. It depends very much on the locale, the people (citizens, political leaders, city planners), the mode of governance, and many other factors (Pallagst 2017). The socio-physical, economic, and political contexts of each shrinking city or region should influence the responding strategies, but this complexity often makes it hard to envision what might work or not. In addition, a certain momentum might be necessary to leverage a specific approach in order to be successful. In an investigation of path dependencies of planning strategies, Sorensen (2015: 25f) calls these events "critical junctures."

In an earlier publication, Pallagst (2012: 31–32) characterized the process of coping with shrinkage in the United States during different decades, as follows:

- The 1950s: tear-downs and new constructions following modernity's rigid development principles (with new high-rise buildings and auto-oriented developments).
- The 1960s and 1970s: shift in trends towards preservation of historic buildings and enhanced citizen participation.
- The 1980s: diversification of the economy, revitalization at the level of neighborhoods by means of community building, and focusing on key projects and events such as sports stadiums (incrementalism).
- The 1990s: mixed-use developments in the city center (retail, housing, office, and hotel), and reconsideration of pedestrian-friendly spaces.
- The 2000s onward: while many of the aforementioned strategies are still applied, there is also stronger regional collaboration.

The United States and Germany seem to follow different strategies in dealing with shrinkage. In the United States, this often involves the development of flagship projects (such as Detroit's Renaissance Center of 1977, Baltimore's sports stadium of 1998, or Cleveland's sports complex of 1994), usually at the city center, in an effort to revive the local economy. It is, however, debatable whether the location and purpose of the new flagship projects have the potential to combat shrinkage in ways that can restore urban qualities. Nevertheless, such projects may offer new urban design qualities in shrinking cities.

While planning in the United States is very decentralized, with the federal government having little to no role in urban development in cities in the last decades, in Germany the central government has launched national programs to assist cities in their struggle of coping with urban decline. These programs – e.g. "Rebuilding the city East" and "Rebuilding the city West" have offered a structured and comparable methodology for the respective cities. For instance, in order to qualify for funding under these programs, the cities had to develop a comprehensive plan mapping out future goals for urban development for the entire city. Another key project, which has drawn international attention, was the IBA Emscher Park, carried out as a project of revitalization in the Ruhr Area initiated by the state of North Rhine-Westphalia in the 1990s (Höber and Ganser 1999). In the wake of the IBA, greening and green infrastructure were widely introduced as a means to improve the quality of life in shrinking cities. Greening has gained significance in recent years both in growing cities and in shrinking cities, assigning greening the status of a tool with strategic urban development potential (Reimer et al. 2015).

Greening has been used in Germany as a strategic approach that aims to develop green infrastructure along with supporting environmentally and socially acceptable practices at the national, state, regional, and local levels. Many cities have employed successful greening strategies. At the forefront is the aforementioned Emscher Park located in Germany's Ruhr Area, where a large-scale abandoned industrial space was turned into a landscape park for an International Building Exhibition. Greening in this respect is not a new planning concept. However, steering urban development in line with green spaces and facilitating linkages with the surrounding landscape is gaining momentum among local planners (Fox-Kämper et al. 2015). Greening is not only related to the built environment, but is of high relevance for the future development of cities as ecosystems and living spaces (Pallagst et al. 2017b). As a result, in Germany, urban design in shrinking cities often go hand-in-hand with greening. In addition, from an urban design perspective, Madanipour (2014: 64, 83) claims that as a form of "connective urbanism" urban design shows potentials for "linking the spatial fragments," which is a strong component of the IBA Emscher Park concept.

Revitalizing shrinking cities in Germany and in the United States: two different urban design strategies

In this section, we draw from research that we carried out in 2017 to highlight two examples of urban planning and design, which are representative of the different approaches in Germany and in the United States. For Germany, the redevelopment of the mining plant Mont Cenis located in the city of Herne is presented as a recent example of inducing design qualities in a shrinking city. For the United States, the development of University Circle in the city of Cleveland is presented as an example of a flagship project hoping to foster substitute industries in shrinking cities.

The German case can be seen as an example of the green dimensions of urban design. Moughtin and Shirley (2005: 192) particularly refer to ecological models of urban design when observing: "The quarter has the potential to support further the process of sustainable development, when its form is compatible with – and promotes – public transport, food production, waste and energy recycling and environmentally friendly systems of heating and power supply." On the other hand, the US case represents an approach to apply urban design as "means to attract economic investment" (Punter 2007: 169). In this context, local governments utilize urban design "to aesthetisize megaprojects" (Punter 2007: 169).

To understand the phenomenon of shrinkage in the respective cities, we examined a number of sources and conducted a review of the existing literature. We collected primary data through on-site investigations and stakeholder analysis, including qualitative interviews in each city with city government staff and NGOs. Additional sources of information came from archival records and planning documents in the two cities. The aim was to gather in-depth knowledge about shrinking city experiences and strategies that can steer the development of shrinking cities.

Greening a former mining city: Mont Cenis in the city of Herne

The city of Herne is located in the Ruhr Area of Germany. Since the 1950s, the city has lost population due to economic transitions in the coal and steel production of the area. The extent of depopulation was in part compensated by incorporating another town in 1975. Part of the city of Herne's industrial heritage is the former coal mining plant Mont Cenis. In the 1990s, the city was part of the project area of the IBA Emscher Park, a project that has influenced urban redevelopment in the cities of the Ruhr Area. Starting at the end of the 1990s, the area was revitalized in order to create a new location for the state of North Rhine-Westphalia's academy of continued education. The city of Herne was chosen as the new place for the academy because of its large unemployment and its strategic location in the center of the state of North Rhine-Westphalia.[1] Interestingly, this development is entirely funded by public funds, mainly from the state and the city of Herne, and from the European Union and the federal government.[2] The public actors installed the Mont Cenis development corporation to implement the project.

The project consists of the academy building embedded in a so-called "energy park." This development was shaped by two factors: (1) the need to offer new skills to the unemployed workers of the Ruhr Area to prepare them for a changed labor market; and (2) the desire to transition to environmentally friendly and energy-efficient uses.

The following three features of the project are of interest from an urban design perspective:

- Landscape design: The area of the park surrounding the building was designed in an elliptical shape with trees framing an alley. In addition, light effects were installed to highlight the alley, which was labeled the "oval light."
- Building design features: The building of the academy is covered by a glass shell of 20,000 square meters. This glass shell creates a specific Mediterranean microclimate in the building (Figure 18.1), which lowers the energy demand for heating.
- "Green" design: The building is part of the energy park. About 3,200 photovoltaic panels were installed on the building's roof, creating a large solar power plant. This solar power is used for heating not only for the building but also for the adjacent urban area that includes 250 housing units.

(Jourda and Hegger 2001)

Another part of the energy park is a block heating station that uses the mine gas, which is still leaking from the closed underground mine. Thus, the old industry is still supplying the new energy use of the area – but now in a more environmentally friendly way.

Figure 18.1 Interior of the academy building at Mont Cenis, Herne.

Source: Picture taken by Sabrina Förch in 2017

A flagship project in a former steel town: Cleveland's University Circle

Cleveland is a US shrinking city located in the country's Rust Belt. The city has lost 500,000 residents during the last decades, because of suburbanization and departure of the population from the city center, combined with the decline of the manufacturing sector. Consequently, there are many problems the city has to face. Today, many shrinking cities turn to the "eds and meds" sector seeking to develop educational and/or healthcare facilities as substitute industries (Harkavy and Zuckerman 1999). Following this path, the city of Cleveland started the Greater University Circle Initiative (GUCI). For many years, the University Circle neighborhood, located four miles east of downtown Cleveland, was struggling with problems caused by shrinkage and population losses. Nevertheless, the city and the institutions located in University Circle found a way to respond to the manifold challenges in their neighborhood through GUCI (Cleveland Foundation 2013), which brought a substitute industry –"eds and meds" – in the area. Since its founding in 2005, the initiative has already achieved a number of successes. Seventeen so-called "anchor institutions" are now located in this area, offering employment potential, with Cleveland Clinic and University Hospitals being the city's largest employers. In addition, a number of arts and cultural institutions have also located here.

GUCI came to life through a dialogue between the Cleveland Foundation and the institutions of University Circle. It is based on a multiple-anchor strategy engaging all the different institutions at University Circle. GUCI follows four different strategies to improve the situation in the area: institutional partnerships, physical development, economic development, and community engagement. In particular, in the area of physical development, urban design comes into play. The projects initiated under the physical development goal are highly acclaimed by private institutions, politicians, and residents, and are often designed by well-known architects. Our interviewees deemed this a positive effect, as it shows that "something is happening."

One example of a flagship project located in University Circle is the Frank Gehry-designed Peter B. Lewis Building, hosting the Weatherhead School of Management (Figure 18.2). With

Figure 18.2 Peter B. Lewis Building, Weatherhead School of Management, Cleveland.

Source: Picture taken by Pallagst (2017)

its distinctive design, this building creates an architectural landmark of high visibility for both University Circle and the school. The building, completed in 2002, also shows that private funds are a cornerstone of these developments: the construction was financed by entrepreneur Peter B. Lewis with a $36.9 million donation to the school. In this case, highly prominent design could be achieved in a shrinking city, also representing the new economic area in the "eds and meds" sector.

Conclusion: flagship projects versus greening: what can we learn for revitalizing shrinking cities?

As we have discussed in the previous paragraphs, the urban design approaches in German and US shrinking cities are different. These differences can be summarized as follows: German design efforts in shrinking cities have to a large extent focused around greening, whereas efforts in US cities are widely driven by the development of flagship projects, often sponsored by philanthropists and showcasing – as demonstrated in the city of Cleveland – star architects.[3] While one might argue that the Herne case is likewise a flagship project, its roots are embedded in the IBA Emscher Park's regional strategy of greening. This is in line, for example, with Spirn's (2011) arguments on ecological urbanism, as she pleads for a comprehensive frame for projects, which take regional requirements into consideration instead of carrying out incremental projects.

One might ask the question: why does Germany not induce comparable flagship urban design projects in shrinking cities to the same extent that US cities do? A possible answer might be that German design and building traditionally does not have a strong base in philanthropy. This phenomenon is almost absent in German urban development. This is evidenced by the fact that key projects in Germany such as the IBA Emscher Park are traditionally funded by public means. Not surprisingly, in both countries, different planning and design cultures come into play: On one side, a state driven planning system in Germany; on the other side, a market-oriented planning system in the United States (Pallagst 2010).

In the authors' opinion, the greening efforts generated in German shrinking cities by the IBA Emscher Park can also serve as best practices for US shrinking cities. Vice versa, leveraging private funds for key projects in shrinking cities would help German shrinking cities to overcome austerity and shape their image for the future in a more positive way. This will, however, necessitate the exchange of learning processes and best practices between the two planning and design contexts.

While we cannot argue that development of flagship projects automatically translates into more livability of shrinking cities, we see that – as demonstrated by the US and German examples – the demands for urban design are quite high in the context of shrinking cities. Moreover, different attributes of urban design are needed. It seems that considering a range of approaches would be advisable in order to offer flexible solutions for the future development of shrinking cities.

Acknowledgments

This work was supported by the German Research Association (Deutsche Forschungsgemeinschaft, DFG).

Notes

1 Interview with a representative of the management of the academy building, 2017.
2 Interview with a representative of the management of the academy building, 2017.
3 A number of highly prominent projects in US cities, especially those of medical or academic institutions, are sponsored by philanthropists.

References

Banerjee, T. and Loukaitou Sideris, A. (eds.) (2011). *Companion to Urban Design.* New York: Routledge.

Benke, C. (2004). "Historische Umbrüche. Schrumpfung und städtische Krisen in Mitteleuropa seit dem Mittelalter." *Städte im Umbruch – Das Online Magazin für Stadtentwicklung, Stadtschrumpfung, Stadtumbau & Regenerierung*, 1: 7-14.

Cleveland Foundation (2013). *Cleveland's Greater University Circle Initiative: Building a 21st Century City Through the Power of Anchor Institution Collaboration.* Cleveland: The Cleveland Foundation.

Cunningham-Sabot, E. and Fol, S. (2007). "Schrumpfende Staedte in Westeuropa: Fallstudien aus Frankreich und Grossbritannien." *Berliner Debatte Initial*, 18(1): 22-35.

Fox-Kämper, R., Lelong, B., and Sondermann, M. (2015). "Urbane Gärten als Teil grüner Infrastrukturen." *RaumPlanung*, 180(4): 23-29.

Fritsche, M., Langner, M., Köhler, H., Ruckes, A., Schüler, D., Zakirova, B., and Appel, K. (2007). "Shrinking Cities: A New Challenge for Research in Urban Ecology." In Langner, M. and Endlicher, W. (eds.) *Shrinking Cities Effects on Urban Ecology and Challenges for Urban Development.* Bern: Peter Lang. 17-33.

Glock, B. and Häussermann, H. (2004). "New Trends in Urban Development and Public Policy in Eastern Germany: Dealing with the Vacant Housing Problem at the Local Level." *International Journal of Urban and Regional Research*, 28(4): 919-929.

Harkavy, I. and Zuckerman, H. (1999). *Eds and Meds: Cities' Hidden Assets.* Washington: Brookings Institution.

Höber, A. and Ganser, K. (1999). *Industriekultur – Mythos und Moderne im Ruhrgebiet.* Essen: Klartext Verlag.

Hoch, C. (1994). *What Planners Do: Power Politics and Persuasion.* Chicago: Planners Press American Planning Association.

Hollander, J., Pallagst, K., Schwarz, T., and Popper, F.J. (2009). "Shrinking Cities as an Emerging Planning Paradigm." *Progress in Planning*, Special issue on emerging research agendas: Urban design and planning 2009. 223-232.

Jourda, F.H. and Hegger, M. (2001). *Mont Cenis. Lebendige Architektur.* Wuppertal: Müller und Bunsmann.

Madanipour, A. (2014). *Urban Design, Space and Society.* Basingstoke/UK: Palgrave.

Mostafavi, M. (2010). "Why Ecological Urbanism? Why Now? *Harvard Design Magazine* 32: Spring/summer: 124–135." Reprinted in Banerjee, T. (ed.) (2014). *Urban Design.* New York: Routledge. 215–228.

Moughtin, C. and Shirley, P. (2005). *Urban Design: Green Dimensions*, second edition. New York: Routledge.

Oswalt, P. and Rieniets, T. (Eds.) (2006). *Atlas of Shrinking Cities/Atlas der schrumpfenden Städte.* Ostfildern: Hatje Cantz.

Pallagst, K. (2007). *Growth Management in the USA Between Theory and Practice.* Aldershot: Ashgate Publishers.

Pallagst, K. (2008). "Shrinking Cities: Planning Challenges from an International Perspective." *Urban Infill*, 1: 6-16.

Pallagst, K. (2010). "Viewpoint: The Planning Research Agenda: Shrinking Cities: A Challenge for Planning Cultures." *Town Planning Review*, 81(5): I-IV.

Pallagst, K. (2012). "Shrinking Cities in the United States: Policies and Strategies." In Martinez-Fernandez, C. et al. (eds.) *Demographic Change and Local Development: Shrinkage, Regeneration and Social Dynamics.* Paris: OECD Local Economic and Employment Development (LEED) Working Paper Series. 41-46.

Pallagst, K. (2013). "The Interdependence of Shrinking and Growing: Processes of Urban Transformation in the USA in the Rust Belt and Beyond." In Pallagst, K., Martinez-Fernandez, C. and Wiechmann, T. (eds.) *Shrinking Cities: International Perspectives and Policy Implications.* New York: Routledge Publishers. 59-77.

Pallagst, K. (2017). "Substitute Industries: Panacea or False Hope for Shrinking Cities?" *DISP: The Planning Review*, 53(2): 82-83.

Pallagst, K., Fleschurz, R., and Said, S. (2017a). "What Drives Planning in a Shrinking City? Tales from two German and two American Cases." *Town Planning Review*, Special issue on shrinking cities. 88(1): 15-28.

Pallagst, K., Fleschurz, R., and Trapp, F. (2017b). "Greening the Shrinking City: New Sustainable Planning Approaches in the USA." *Landscape Research*, Special issue on Shrinking Cities: Expanding Landscapes. 82(6): 1-12.

Pallagst, K. and Wiechmann, T. (2005). *Shrinking Smart: Städtische Schrumpfungsprozesse in den USA*, Jahrbuch StadtRegion, Schwerpunkt Schrumpfende Städte. Wiesbaden: Verlag für Sozialwissenschaften.

Porter, D.R. (1997). *Managing Growth in America's Communities.* Washington: Island Press.

Punter, J. (2007). "Developing Urban Design as Public Policy: Best Practice Principles for Design Review and Development Management." *Journal of Urban Design*, 12(2): 167–202.

Razin, E. (1998). "Policies to Control Urban Sprawl: Planning Regulations or Changes in the 'Rules of the Game'?" *Urban Studies*, 35(2): 321–340.

Reimer, M., Othengrafen, F., and Rusche, K. (2015). Grüne Infrastruktur in Urbanen Rāumen. *RaumPlanung*, 4: 6–7.

Ryan, B.D. (2012). *Design After Decline: How America Rebuilds Shrinking Cities*. Philadelphia: University of Pennsylvania Press.

Sorensen, A. (2015). "Taking Path Dependence Seriously: An Historical Institutionalist Research Agenda in Planning History." *Planning Perspectives*, 30(1): 17-38.

Spirn, A. W. (2011). "Ecological Urbanism." In Banerjee, T. and Loukaitou-Sideris, A. (eds.) *Companion to Urban Design*. New York: Routledge. 600–610.

Wiechmann, T. (2006). *Types of Shrinking Cities: Introductive Notes on a Global Issue*. Paper presented at the International Symposium 'Coping with City Shrinkage and Demographic Change: Lessons from Around the Globe', Dresden, Germany. Retrieved from: www.schader-stiftung.de/wohn_wandel/966.php.

Wiechmann, T. and Pallagst, K.M. (2012). "Urban Shrinkage in Germany and the USA: A Comparison of Transformation Patterns and Local Strategies." *International Journal of Urban and Regional Research*, 36(2): 261–280.

Further reading

Hollander, J., Pallagst, K., Schwarz, T., and Popper, F.J. (2009). "Shrinking Cities as an Emerging Planning Paradigm." *Progress in Planning*, Special Issue on Emerging Research Agendas: Urban Design and Planning 2009. 223-232. This is one of the key publications in shrinking cities research, comparing US and European examples.

Pallagst, K. (2010). "Viewpoint: The Planning Research Agenda: Shrinking Cities: A Challenge for Planning Cultures." *Town Planning Review*, 81(5): I-IV. This article is one of the earlier publications on shrinking cities with a roadmap for further research on planning cultures.

Pallagst, K. (2012). "Shrinking Cities in the United States: Policies and Strategies." In Martinez-Fernandez, C. et al. (eds.) *Demographic Change and Local Development: Shrinkage, Regeneration and Social Dynamics*. Paris: OECD Local Economic and Employment Development (LEED) Working Paper Series. 41-46. This edited volume gives an overview on shrinking cities in different countries.

Pallagst, K., Fleschurz, R., and Trapp, F. (2017). "Greening the Shrinking City: New Sustainable Planning Approaches in the USA." *Landscape Research*, special issue on shrinking cities: Expanding landscapes, 82(6): 1-12. This publication discusses the link between green infrastructure and shrinking cities.

Ryan, B.D. (2012). *Design after Decline: How America Rebuilds Shrinking Cities*. Philadelphia: University of Pennsylvania Press. This is one of the few publications presenting shrinkage and urban design.

19

DESIGNING THE SHRINKING CITY

Justin B. Hollander

Introduction

Despite overall global population increases, not all places are growing. Whether due to natural or human-made circumstances, many cities and towns are in a state of population or employment decline (Oswalt and Rieniets 2006; Pallagst et al. 2014). With fewer people or jobs, demand for real estate decreases, leading to disinvestment and abandonment (Wilson and Kelling 1982; Wallace 1989; Glaeser and Gyourko 2005). This physical outcome of social, economic, demographic, or environmental conditions can be devastating to neighborhood stability and may further reinforce social inequalities (Hoyt 1933; Temkin and Rohe 1996; Dewar et al. 2014).

The conventional urban design and planning approach to decline has been to fight it; that is, to tackle the very forces contributing to depopulation or employment losses through firm recruitment strategies (Dewar 1998; Blakely and Leigh 2013). While this kind of economic development can be effective in certain circumstances and contexts, it is often unrealistic when demographic projections of future conditions suggest that future decline is likely (Hollander and Cahill 2011). Thus, a different approach of "smart shrinkage" acknowledges the possibility of decline and is increasingly being adopted by urban designers. Smart shrinkage strategies are not contingent on growth (Pallagst et al. 2014; Pallagst et al. 2017a; 2017b), and they can use urban design adjustments to the physical form of a shrinking place to better manage the typical problems of vacancy and abandonment associated with decline (Schilling and Logan 2008).

The section that follows offers a brief review of the mechanisms by which physical disorder results from decline and then turns to the question of how urban design can help impacted communities, examining urban design approaches to decline in residential neighborhoods and downtowns, and paying close attention to the specific policy dimensions of such approaches. More specifically, I discuss urban design interventions in shrinking residential neighborhoods in Baltimore (Maryland), Flint (Michigan), and New Bedford (Massachusetts). Some downtowns are also undergoing decline and face their own challenges, and I explore one possible design response. I conclude with reflections on the role and aims of urban design, in the face of shrinkage, and how designers can measure their successes in achieving those aims.

The process and impact of decline

Economists have discussed how municipalities offering different baskets of goods, in the form of government services, attract residents who make rational migratory decisions that satisfy their desired balance between services and costs (Tiebout 1956; Zodrow and Mieszkowski 1986). An extension of these assumptions suggests that, when employment opportunities are reduced in a locale, people are expected to act rationally, leaving that locale for other places with more abundant employment opportunities. Several problems arise when this happens, however. Crucial among these problems is the fact that, when residents leave a neighborhood, the physical form of the city does not naturally shrink. Glaeser and Gyourko (2005) studied the durability of housing in a time-series sample of 321 US cities and towns with at least 30,000 residents in 1970, finding that housing prices declined at a faster rate in depopulating cities than prices grew in growing cities. Their research suggests that the durability of housing poses a long-term threat to neighborhood stability, because houses do not disappear when people leave; instead, they remain and largely sit unoccupied for years.

Vacant houses impact neighborhood quality, not only because they progressively deteriorate, but also because increased abandonment and vacancy lead to higher levels of criminal activity. If housing does not disappear as quickly as people do, then those abandoned structures are likely to serve as havens for criminal activity (Wilson and Kelling 1982; Wallace 1989).

Furthermore, widespread past and present discrimination in hiring and in the housing market have systematically limited relocation options for African Americans and Latinos in the United States (Massey and Denton 1993; Sugrue 1995; Jargowsky 1997; McDonald 2015). During times of population decline, wealthier urban residents can easily relocate, leaving behind the poorest and most destitute residents. As a neighborhood loses jobs, its African American and Latino residents will have fewer relocation choices, causing further racial concentration and economic segregation in ghettos. It has been shown that a portion of racial disparities in employment outcomes can be attributed to a spatial mismatch between where a person lives and where potential jobs are located (Kain 2004; Stoll 2005). This is particularly concerning for minority and lower-income workers, who have fewer transportation options and less economic stability.

As housing demand declines in depopulating residential neighborhoods, housing units change from being high-rent (with affluent residents) to lower-rent (with less affluent residents). In a process described as "filtering," poorer economic conditions lead to shifts from homeowners to renters (Hoyt 1933; Temkin and Rohe 1996; Dewar et al. 2014). Specifically, "as most structures age, wear, and become obsolete, they filter down to lower-income occupants" (Baer and Williamson 1988; Bier 2001, 6). This is notably an uneven process whereby poor neighborhoods tend to depopulate more rapidly than wealthier neighborhoods (Guerrieri et al. 2012).

Access to housing, in terms of both tenure and neighborhood characteristics, has long been racialized, dating back to the earlier portions of the 20th century and the advent of an American society of homeowners (Metzger 2000; Crossney and Bartelt 2005, Lloyd et al. 2014). Ultimately, when demand sinks to a certain threshold level, owners tend to abandon their structures (Keenan et al. 1999; Henderson 2015). In time, many abandoned structures become derelict and may be subject to arson (Wilson and Kelling 1982). Thus, in a depopulating neighborhood, housing units become progressively unoccupied, resulting in derelict structures and vacant lots where fire often consumes the unit(s). This process suggests the pressing need for analyzing physical change through the lens of occupied housing density.

Rust's (1975) study of population and employment decline in 30 US metropolitan areas from the 1800s to the 1970s revealed much about why and how places decline. He found that these shrinking places experienced dramatic population loss and then "a long period of profound resistance to demographic or economic change which continues until the people, artifacts, and institutions which were assembled in the truncated growth era gradually erode away" (169). The very physical fabric of neighborhoods is expected to "erode away" in a period of decline.

This erosion process has been widely characterized as negative (Beauregard 2003); but population loss does not have to be a bad thing for a community. In my book, *Sunburnt Cities*, I explored how less people mean less congestion, which may translate to more open space and recreational opportunities (Hollander 2011). Less people in a city can mean lower student-to-teacher ratios in the classroom and faster response times from police and fire departments. A smaller population has a smaller environmental footprint, helping a city to reduce its impacts on global climate change. Fewer jobs mean fewer cars on the roads and improved air quality. Fewer factories mean less pollution and better water quality. While population and employment declines are never the aim of public policy and planning (and are usually associated with negative outcomes), such declines, if managed properly, can be potentially beneficial to residents and workers left behind.

However, cities know very little about what it takes to manage decline. More often, the loss in population is either caused by or contributes to a wide range of social, economic, and fiscal challenges. So, in a knee-jerk fashion, local governments simply work to reverse the process of depopulation – a skill that only a handful of cities have mastered.

But household size is shrinking as well (Ryan 2012). Some might argue that if a city drops from 50,000 to 40,000 people from one decade to another, but also witnesses a drop in household size (from five to four persons per household), then the demand for housing units might not change at all (there would be 10,000 households in the first period and 10,000 households in the second period). In theory, such a city with a smaller population, but smaller household size, might need the same number of housing units. This hypothesis was tested using a simulation model, and researchers were surprised with the result (Lauf et al. 2012). The study of the Leipzig region of Germany found that as the region contracted in population, modeling predictions showed that housing demand fell. The results showed that household size may not be shrinking enough to compensate for overall depopulation in some locales.

Rust's (1975) study also found that the effects of decline are "expected to be felt most strongly a generation after the cessation of growth and to persist for up to 50 years" (187). For this reason, it is important to look at both contemporary and historical planning approaches when studying a shrinking city.

The role of urban design in managing decline

How can urban design help manage decline and achieve "smart shrinkage"? Popper and Popper (2002, 23) define "smart shrinkage" as "planning for less – fewer people, fewer buildings, fewer land uses." Beauregard (2003: 242–243) explains the approach as:

> the privileging of quality-of-life concerns and neighborhood-based efforts over economic development. . . . The overriding goal is to make the city livable rather than to expand the economic base for the sake of growth. The model embraces the notion that the city might be better off smaller and that urban neighborhoods can become less dense and remain viable, thereby becoming more desirable.

Conventional urban design is growth-oriented; it is a tool to shape the built environment during new development processes. In already built-up urban areas, urban design aims focus on enhancing public spaces to make them more welcoming or inviting (Gehl 2010). But smart shrinkage provides a legitimate avenue for urban design to play a role in shaping the physical form of shrinking cities by turning vacant structures into public space. A number of urban design strategies in shrinking cities have proposed just that. For example, in Detroit the conversion of urbanized areas to greenery (Millington 2013) was described by Ryan (2012, 74) as "a beautiful, successional landscape of wildflowers." Clark (1989, 143) also called for an active promotion of greenery in vacated urban areas for "parkland and recreational spaces" – a suggestion also echoed by Schilling and Logan (2008). Others have looked into the potential for urban forestry in the abandoned spaces of shrinking cities (Kowarik and Körner 2005; Rink 2009; Haase 2013), while still others, including myself, have considered the potential use of urban agriculture in emptied-out land (Rugare and Schwarz 2008; Hollander 2011; Rosenman and Walker 2015).

A second urban design strategy for smart shrinkage is to redesign neighborhoods for lower densities. Armborst et al. (2006) have introduced the idea of widespread side yard acquisitions of vacant lots as a means of reducing housing density, a process they described as "blotting." Here, abutters to vacant lots either legally acquire or openly squat on the lots and begin to protect and maintain them through a variety of uses (grass, gardens, pools, extra parking). Armborst et al. (2006) found that the urban fabric of Detroit was changing on a daily basis, not by city plans or regulation, but by the actions of individual landowners expanding their lots to more closely mirror density patterns seen in suburbia.

Ryan's (2012) research into historic patterns of housing density in shrinking cities found quite a few examples of active redesigning of urban neighborhoods for lower densities. The 1980s-era Charlotte Gardens complex of detached single-family homes in Detroit (Michigan), for example, resulted in 6.8 housing units per acre, whereas the prior landscape had been dominated by five-story tenement buildings with a density of 290 units per acre. Likewise, the Jefferson Square 180-unit two-story apartment complex in Detroit was built in 1980 on a 3.5-block area that used to include 302 units of housing (Ryan 2012). This approach to managing depopulation was quite common in Detroit; between 1990 and 2001, a total of 15 single-family detached housing developments were built all over the city, effectively de-densifying Detroit (Ryan 2012).

With less land necessary for housing, the question becomes: what happens to the land that used to be devoted to residential use? In Detroit, Ryan (2012, 97) found that the additional land in the Jefferson Square project was largely devoted to "collective open spaces." But, certainly expanded parking areas were also a major land use. Fewer people and fewer cars, but the provision of additional parking helped address a previously pent-up demand where parking has historically been expensive.

What these smart shrinkage responses have in common is a commitment to nonresidential future uses for land and buildings that had previously been residential in nature. For the remainder of this chapter, I will use this as the core definition of smart shrinkage: local government policies that contribute to the conversion of land and buildings from residential to nonresidential use. That is a deceivingly simple definition, but it does capture a qualitatively unpopular and relatively uncommon practice in cities. Certainly, there are examples of homes being acquired for the expansion of industry (as was famously proposed by the city of New London in the Supreme Court case validating the use of eminent domain for economic development purposes; see *Kelo v. New London*). But in a city with stable population levels, rarely does the government kick people out of their homes to generate new, nonresidential uses. Such efforts may be different in exclusively residential versus downtown neighborhoods.

Designing shrinking residential neighborhoods

This section discusses examples of urban design and policy practices in three shrinking cities that have the explicit or implicit goal of managing decline in residential neighborhoods.

Baltimore's Growing Green Initiative (GGI)

Baltimore, Maryland, has seen consistent population decline since 1950 (Johnson et al. 2015). The city went from 949,708 residents in 1950 to 621,849 in 2015 (U.S. Census Bureau 2015a, 2015b). The result has been widespread housing vacancy and abandonment in a number of neighborhoods, predominantly in the city's East and West sections.

The city's mayor, Stephanie Rawlings-Blake, launched the Growing Green Initiative (GGI) in 2014 (Henneberry 2017), at a time when it was estimated that the city had about 16,000 abandoned buildings (Baltimore Department of Planning, Office of Sustainability 2015), and the program continues to this day. Through a series of policy interventions, the GGI reframes the conversation in Baltimore around the productive reuse of vacant and abandoned properties, focusing local funds on priorities that can be characterized as smart decline (Johnson et al. 2015). The goal of the GGI is to employ innovative, sustainable, and cost-effective methods for reusing vacant lots and mitigating the negative impacts of vacancy. Led by the city of Baltimore's Office of Sustainability, a small unit within the city's Planning Department, the program employs strategies such as growing food, greening of lots, and stormwater management and shapes much of the efforts of government and nongovernmental actors in the city along these themes. A *Green Pattern Book* written in 2015 offers guidance to residents and organizations on how to turn vacant and abandoned properties into green spaces, a quintessential smart decline activity (Baltimore Department of Planning, Office of Sustainability 2015).

This urban design effort has been undertaken largely through an expert-led, internal city process. Limited citizen engagement through a series of public meetings has generated some level of community buy-in, but the physical design solutions have been driven by city staff with some input by outside consulting groups (Nemeth et al. 2018). Motivations for various design and land use decisions have been grounded in the everyday functions of a city bureaucracy overwhelmed by the growing stock of vacant and abandoned property, with little room for concerns around social justice and participatory planning (Nemeth et al. 2018).

Flint, Michigan: Genesee County Land Bank

After the state of Michigan passed a law allowing land banks in 2003, the Genessee County Treasurer's Office quickly went to work to create their own land banking authority – in a county that encompasses Flint. Flint has witnessed massive hemorrhaging of its population over the last half-century and a concomitant decline in occupied housing (Hollander 2010). The Genesee County Land Bank took a new tack to dealing with the problem of abandoned properties. The conventional model (as typically dictated by state law) requires that any property, for which the owner fails to pay taxes or utility bills, is condemned by the local government and, after various waiting periods, is sold to the highest bidder at auction (with the hope that the local government will recover some delinquent taxes from the sale). Under that model, many absentee owners do not invest in these sites, instead just sitting on them or selling them to other absentee investors. The Land Bank legislation authors felt that a more strategic process would allow for a public entity to effectively plan for the reuse of these properties, consolidate them, package them, and market them to real estate developers. In Genesee County, planners have had much success in doing just that, repurposing hundreds of properties and partnering with real estate developers to rebuild in some cases. They are somewhat selective in screening for the most desirable properties and then working internally to reuse or land bank those properties (Genesee County Land Bank Authority 2014).

As in Baltimore, Genesee County urban designers have only modestly engaged community members in their spatial strategies (Schilling and Logan 2008; Pallagst et al. 2017b). Instead, they have responded to fiscal distress and pressing land management challenges associated with massive numbers of vacant and abandoned properties. Pallagst et al. (2017b: 724) found in their interviews with Flint area leaders that "community and neighbourhood groups [should] shape the city's new green spaces step by step." Beyond some tacit engagement, this was a call for full participation by these nongovernmental groups to design this abandoned landscape.

New Bedford, Massachusetts: smart shrinkage policies (un)masked

New Bedford is not a typical shrinking city. While historically the city has undergone depopulation, demographic conditions have stabilized, and between 2000 and 2010, the city even saw a jump in population. Not booming, but hardly shrinking, the city typifies the vast majority of all cities where there are forces pushing in both directions simultaneously, with declining and growing areas spatially distributed throughout the city.

New Bedford has instigated a number of policies to improve the pedestrian experience; develop parks and green spaces; foster tourism and the creative economy; promote transportation, sustainability, and historic preservation; and more (Hollander 2018). When viewed through the lens of smart shrinkage, a number of official city documents are also calling for a more right-sized infrastructure, improved land use policies, more recreational and agriculture uses, and a reimagining of cultural/heritage. While trying to play the conventional economic development game of chasing industry and betting on big "game-changing" projects, the city was also engaged in austerity urbanism, investing money and time in a genuine effort to decrease the city's housing density and find new non-housing uses for formerly residential structures and land. These approaches had some adverse implications for spatial justice and equity, as the most economically impoverished neighborhoods received minimum public investment, while middle-income neighborhoods and downtowns were well targeted with generous government support (Hollander 2018). As with the previous examples, there was not much community involvement in these design decisions, but instead the broader pattern was that vacant and abandoned buildings did largely get reused over a longtime horizon of 20–30 years. This urban absorption meant that most vacant parcels were reused without government involvement and instead through the hundreds of individual actions of property owners throughout the city, who took over the abutting lot and planted a garden, parked their car, or expanded their yard (Hollander 2018).

Designing shrinking downtowns

For many cities, towns, and regions, a central core downtown is a political, cultural, and economic center, and its decline can present unique challenges. This section reviews the problems and paradoxes in dealing with depopulation and decline in downtowns.

As discussed previously, shrinking cities have different ways to "right-size" their residential neighborhoods to accommodate smaller populations. By reprogramming formerly residential land and buildings for nonresidential purposes, these places can lower densities and assume a new identity, whether that means converting from urban to suburban, from suburban to rural, or something altogether different. But for most downtowns in postindustrial shrinking cities, this is not an option. A downtown is and always will be the cultural, governmental, and financial hub for the surrounding city and immediate region. There is a very real emotional and psychological need among communities to preserve that function of downtown. Thinning it out and making it largely forested is simply not an attractive option.

At the same time, the physical structures of a downtown are not easily demolished and "right-sized" in the way that residential neighborhoods can be. Downtowns certainly come in a range of densities, from 19th-century agrarian village centers as in Grafton, Massachusetts (see Figure 19.1), to suburban shopping-oriented downtowns such as Hanley (Stoke-on-Trent in the UK) (see Figure 19.2), to major downtowns of global cities like Tokyo (see Figure 19.3). For each density level, a unique arrangement of buildings, public plazas, government facilities, and parks is expected. But rearranging such disparate types of facilities to match a shrinking population is nearly impossible through the kind of right-sizing approach described earlier. For shrinking residential neighborhoods, the buildings are almost all residential, and reducing density (e.g. converting duplexes into single-family homes, or demolishing single-family homes to make larger yards for other homes) is quite straightforward and can be accomplished with little disruption to the overall neighborhood feel. For shrinking downtowns, however, the vast diversity of uses, including commercial, residential, institutional, government, and other uses, is one substantial barrier. Another challenge is the frequently planned arrangement of these uses along regular grids or around central plazas, whereby demolition or density reduction could have a destructive result. A different approach is needed.

Shrinking downtowns offer an opportunity to rethink the core of the central business district (CBD). Here, a new thinking around future uses can be valuable. Globally rich and economically vital agricultural regions often surround many shrinking cities. While urban agriculture has been widely seen as a prospective future use in residential areas of shrinking cities, CBDs would hardly seem to be compatible with agriculture. But with proper siting and considerations for preserving existing streetscapes, it might not be such a crazy idea.

Figure 19.1 Grafton, Massachusetts.

Source: Wikipedia; author: Daderot; public domain

Figure 19.2 Hanley Stoke-on-Trent, United Kingdom.

Source: Wikimedia; author: Leereyno; CC Attribution-Share Alike 3.0 Unported

Figure 19.3 Tokyo downtown district.

Source: Pexels; author: negativespace.co; CC0)

Figure 19.4 Rendering showing hydroponic container being inserted into a hollowed-out building.

Source: Justin Hollander/Daniel Camilletti

Downtown planners ought to consider the emerging urban practice of hydroponic, indoor (also known as vertical) farming. Using shipping containers retrofitted to support hydroponic farms, planners can work with property owners to preserve the historic façades and exteriors of CBD buildings and then "drop" in these pre-built, pre-configured, and module hydroponic containers. Efforts to do this in largely vacant industrial areas of Boston (near the airport) have been met with much success, and further exploration of this idea for shrinking downtowns is worth looking at (Abraham 2015; National Public Radio 2015). Often property owners want to find ways to preserve historic façades in downtowns in order to maintain the urban streetscape. Figure 19.4 presents a rendering of what the preserved façade might look like with a converted shipping container dropped into a demolished space (see Figure 19.4). These interim uses provide employment, food, and activity – all key to protecting and maintaining shrinking downtowns.

Conclusion

Designing growing cities is like baking a cake: add some sugar, flour, and eggs; mix together and put it in a pan; then bake at a high temperature. Designing shrinking cities is much like the opposite: taking a fully baked cake and then extracting the sugar, flour, and eggs. Clearly, it is a challenging task.

In this chapter, I have attempted to outline some key ideas and examples for how this might be accomplished – how and in what ways we might employ urban design in shrinking places. Uncracking the eggs is not simple. The political systems that operate in most capitalist, democratic cities are built off of a growth machine of real estate lawyers, appraisers, agents, and developers (Logan and Molotch 1987). This formidable group tends to wield great power and may resist urban design solutions that are not explicitly aimed to generate short-term property value appreciation. The ideas presented in this chapter may create great increases in property values, but a long-term horizon is helpful, as is patient money (Lucy and Phillips 2000), two qualities in short supply in most urban governance regimes.

Given this political reality, urban design in shrinking cities ought to focus on small, impactful efforts that can then be rolled into larger, more systemic change. Start with planting wildflowers on vacant lots. Invest in modest changes to local zoning codes to open up nonresidential uses in shrinking neighborhoods (like agriculture or light industry). While the examples presented provide little evidence of substantial community involvement in the urban design process in shrinking cities, the example of Genesee County suggests that community partnerships can be the answer, and the case of New Bedford shows how residents can play a key role in informally reusing and repurposing vacant and abandoned sites over time. As other contributors to this book have shown, urban design can shape people's lives, can change communities, and, when properly applied to a shrinking city, can help respond to deep and severe social, economic, and physical problems.

In closing, it is worth considering the question of success. My own research has explored what it means to do good urban design by operationalizing success as people's reported life satisfaction (Hollander 2010), their mental health (Sussman and Hollander 2015), or inferred well-being from social media posts (Hollander et al. 2016; Hollander and Renski 2017). For a shrinking city, conventional measures of success, like income or population growth, are less useful than an answer to the simple question: are the people of this city happy, and can urban design help?

References

Abraham, Y. (2015). "Couple Raises Crops in Recycled Freight Containers." *The Boston Globe*. April 26. www.bostonglobe.com/metro/2015/04/25/yvonne-abraham-couple-raises-crops-recycled-freight-containers/. Accessed October 10, 2015.

Armborst, T., D'Oca, D., and Theodore, G. (2006). "However Unspectacular." In Oswalt, P. (ed.) *Shrinking Cities. Volume 2: Interventions*, 324–329. Ostfildern, Germany: Hatje Cantz.

Baer, W.C., and Williamson, C.B. (1988). "The Filtering of Households and Housing Units." *Journal of Planning Literature* 3, no. 2: 127–152. doi: 10.1177/088541228800300201.

Baltimore Department of Planning, Office of Sustainability. (2015). *Green Pattern Book: Using Vacant Land to Create Greener Neighborhoods in Baltimore.* NRS-INF-32-15. Newtown Square, PA: U.S. Department of Agriculture, Forest Service, Northern Research Station: 109.

Beauregard, R.A. (2003). *Voices of Decline: The Postwar Fate of U.S. Cities*, 2nd ed. New York, NY: Routledge.

Bier, T. (2001). *Moving Up, Filtering Down: Metropolitan Housing Dynamics and Public Policy.* Washington, DC: Brookings Institution, Center on Urban and Metropolitan Policy.

Blakely, E.J., and Leigh, N.G. (2013). *Planning Local Economic Development.* Los Angeles: Sage.

Clark, D. (1989). *Urban Decline: The British Experience.* London: Routledge.

Crossney, K.B., and Bartelt, D.W. (2005). "Residential Security, Risk, and Race: The Home Owners' Loan Corporation and Mortgage Access in Two Cities." *Urban Geography* 26 (8): 707–736.

Dewar, M. (1998). "Why State and Local Economic Development Programs Causes So Little Economic Development." *Economic Development Quarterly* 12 (1): 68–87.

Dewar, M., Seymour, E., and Druță, O. (2014). "Disinvesting in the City." *Urban Affairs Review* 51 (5): 587–615. doi: 10.1177/1078087414551717.

Gehl, J. (2010). *Cities for People.* Washington, DC: Island Press.

Genesee County Land Bank Authority. (2014). *Celebrating the First Ten Years.* www.thelandbank.org/downloads/gclb_10th_anniversary_booklet.pdf

Glaeser, E.L., and Gyourko, J. (2005). "Urban Decline and Durable Housing." 113 (2): 345–375.

Guerrieri, V., Hartley, D., and Hurst, E. (2012). "Very Local House Price Dynamics." *American Economic Review: Papers & Proceedings* 103 (3): 120–126.

Haase, D. (2013). "Shrinking Cities, Biodiversity and Ecosystem Services." In *Urbanization, Biodiversity and Ecosystem Services: Challenges and Opportunities.* The Netherlands: Springer.

Henderson, S.R. (2015). "State Intervention in Vacant Residential Properties: An Evaluation of Empty Dwelling Management Orders in England." *Environment and Planning C: Government and Policy*, 33 (1): 61–82.

Henneberry, J. (2017). *Transience and Permanence in Urban Development*. Hoboken: John Wiley & Sons.

Hollander, J.B. (2010). "Moving Towards a Shrinking Cities Metric: Analyzing Land Use Changes Associated with Depopulation in Flint, Michigan." *Cityscape: A Journal of Policy Development and Research* 12 (1): 133–151.

Hollander, J.B. (2011). *Sunburnt Cities: The Great Recession, Depopulation and Urban Planning in the American Sunbelt*. London: Routledge.

Hollander, J.B. (2018). *An Ordinary City: Planning for Growth and Decline in New Bedford, Massachusetts*. New York: Palgrave Macmillan.

Hollander, J.B., and Cahill, B. (2011). "Confronting Population Decline in the Buffalo, New York, Region: A Close Reading of the Erie-Niagara Framework for Regional Growth." *Journal of Architecture and Planning Research* 28 (3): 252–267.

Hollander, J.B., Graves, E., Renski, H., Foster-Karim, C., Wiley, A., and Das, D. (2016). *Urban Social Listening: Potential and Pitfalls of Using Social Media Data in Studying Cities*. New York: Palgrave Macmillan.

Hollander, J.B., and Renski, H. (2017). "Measuring Urban Attitudes Embedded in Microblogging Data: Shrinking versus Growing Cities." *Town Planning Review* 88 (4): 465–490.

Hoyt, H. (1933). *One Hundred Years of Land Values in Chicago*. Chicago, IL: University of Chicago Press.

Jargowsky, P.A. (1997). *Poverty and Place: Ghettos, Barrios, and the American City*. New York: Russell Sage Foundation.

Johnson, M.P., Hollander, J.B., and Whiteman, E. (2015). "Data and Analytics for Neighborhood Development: Smart Shrinkage Decision Modeling in Baltimore, Maryland." In Geertman, S., Ferreira, J., Goodspeed, R. and Stillwell, J. (eds.) *Planning Support Systems and Smart Cities*. New York: Springer.

Kain, J.F. (2004). "A Pioneer's Perspective on the Spatial Mismatch Literature." *Urban Studies* 41 (1): 7–32.

Keenan, P., Lowe, S., and Spencer, S. (1999). "Housing Abandonment in Inner Cities: The Politics of Low Demand for Housing." *Housing Studies* 14 (5): 703–716.

Kowarik, I., and Körner, S. (2005). *Wild Urban Woodlands New Perspectives for Urban Forestry*. Berlin: Springer.

Lauf, S., Haase, D., Seppelt, R., and Schwarz, N. (2012). "Simulating Demography and Housing Demand in an Urban Region under Scenarios of Growth and Shrinkage." *Environment and Planning B: Planning and Design* 39 (2): 229–246. doi: 10.1068/b36046t.

Lloyd, C.D., Shuttleworth, I.G., and Wong, D.W. (2014). *Social-Spatial Segregation: Concepts, Processes and Outcomes*. Bristol, UK: Policy Press.

Logan, J.R., and Molotch, H.L. (1987). *Urban Fortunes: The Political Economy of Place*. Berkeley, CA: University of California Press.

Lucy, W.H., and Phillips, D.L. (2000). *Confronting Suburban Decline: Strategic Planning for Metropolitan Renewal*. Island Press.

Massey, D.S., and Denton, N.A. (1993). *American Apartheid: Segregation and the Making of the Underclass*. Cambridge, MA: Harvard University Press.

McDonald, J.F. (2015). *Urban America: Growth, Crisis, and Rebirth*. Abingdon, UK: Routledge.

Metzger, J.T. (2000). "Planned Abandonment: The Neighborhood Life-Cycle Theory and National Urban Policy." *Housing Policy Debate* 11 (1): 7–40.

Millington, N. (2013). "Post-Industrial Imaginaries: Nature, Representation and Ruin in Detroit, Michigan." *International Journal of Urban and Regional Research* 37 (1): 279–296.

National Public Radio, Here and Now Staff. (2015). "Freight Farms: How Boston Gets Local Greens, Even When Buried in Snow." *Here and Now*. www.npr.org/sections/thesalt/2015/02/23/388467327/-freight-farms-grow-local-flavor-year-round. Accessed May 12, 2016.

Nemeth, J., Hollander, J.B., Whiteman, E., and Johnson, M.P. (2018). "Planning with Justice in Mind in a Shrinking Baltimore." *Journal of Urban Affairs*. doi: 10.1080/07352166.2018.1454820.

Oswalt, P., and Rieniets, T. (2006). *Atlas of Shrinking Cities*. Ostfildern: Hatje Cantz Publishers.

Pallagst, K., Fleschurz, R., and Said, S. (2017a). "What Drives Planning in a Shrinking City? Tales from Two German and Two American Cases." *Town Planning Review* 88 (1): 15–28. doi: 10.3828/tpr.2017.3.

Pallagst, K., Fleschurz, R., and Trapp, F. (2017b). "Greening the Shrinking City: Policies and Planning Approaches in the USA with the Example of Flint, Michigan." *Landscape Research* 42 (7): 716–727.

Pallagst, K., Wiechmann, T., and Martinez-Fernandez, C. (2014). *Shrinking Cities: International Perspectives and Policy Implications*. New York: Routledge.

Popper, D.E., and Popper, F.J. (2002). "Small Can Be Beautiful: Coming to Terms with Decline." *Planning* 68 (7): 20–23.

Rink, D. (2009). "Wilderness: The Nature of Urban Shrinkage? The Debate on Urban Restructuring and Restoration in Eastern Germany." *Nature & Culture* 4 (3): 275–292.
Rosenman, E., and Walker, S. (2015). "Tearing Down the City to Save It? 'Back-Door Regionalism' and the Demolition Coalition in Cleveland, Ohio." *Environment and Planning A* 48 (2): 273–291.
Rugare, S., and Schwarz, T. (2008). *Cities Growing Smaller.* Cleveland, OH: The Cleveland Urban Design Collaborative, College of Architecture and Environmental Design, Kent State University.
Rust, E. (1975). *No Growth: Impacts on Metropolitan Areas.* Lexington, MA: Lexington Books.
Ryan, B.D. (2012). *Design after Decline: How America Rebuilds Shrinking Cities.* 1st ed. Philadelphia: University of Pennsylvania Press.
Schilling, J., and Logan, J. (2008). "Greening the Rust Belt." *Journal of the American Planning Association* 74 (4): 451–466.
Stoll, M.A. (2005). "Geographical Skills Mismatch, Job Search and Race." *Urban Studies* 42 (4): 695–717.
Sugrue, T.J. (1995). "Crabgrass-Roots Politics: Race, Rights, and the Reaction Against Liberalism in the Urban North, 1940–1964." *Journal of American History* 82 (2): 551–578.
Sussman, A., and Hollander, J.B. (2015). *Cognitive Architecture: Designing for How We Respond to the Built Environment.* London/New York: Routledge.
Temkin, K., and Rohe, W.M. (1996). "Neighborhood Change and Urban Policy." *Journal of Planning Education and Research* 15 (3): 159–170.
Tiebout, C.M. (1956). "A Pure Theory of Local Expenditures." *Journal of Political Economy* 64 (5): 416–424.
U.S. Census Bureau. (2015a). *Census of Population and Housing.* www.census.gov/prod/www/decennial.html. Accessed May 25, 2017.
U.S. Census Bureau. (2015b). "Population Estimates, July 1, 2015 (V2015)." *Baltimore City Maryland QuickFacts from the US Census Bureau.* www.census.gov/quickfacts/table/PST045215/2404000,00. Accessed May 25, 2017.
Wallace, R. (1989). "Homelessness, Contagious Destruction of Housing and Municipal Service Cuts in New York City: 1. Demographics of a Housing Deficit." *Environment and Planning A* 21: 1285–1603.
Wilson, J.Q., and Kelling, G.L. (1982). "Broken Windows: The Police and Neighborhood Safety." *The Atlantic.* March.
Zodrow, G.R., and Mieszkowski, P. (1986). "Pigou, Tiebout, Property Taxation, and the under Provision of Local Public Goods." *Journal of Urban Economics* 19 (3): 356–370.

Further reading

Hollander, J.B., Johnson, M.P., Drew, R., and Tu, J. (2018). "Changing Urban Form in a Shrinking City." *Environment and Planning B.* doi: 10.1177/2399808317743971. This article uses historic maps to explore how and why the urban form of Baltimore changed the way it did.
Ryberg-Webster, S. (2016). "Heritage Amid an Urban Crisis: Historic Preservation in Cleveland, Ohio's Slavic Village Neighborhood." *Cities* 58: 10–25. This journal article provides a rich case study of how urban design can be used to address the historic resources of a shrinking urban neighborhood.
Sussman, A., and Hollander, J.B. (2015). *Cognitive Architecture: Designing for How We Respond to the Built Environment.* London/New York: Routledge. The gist of the book is that humans evolved to experience their surroundings in a very particular and peculiar way that urban designers need to pay attention to.

PART II.IV

Large-scale development

Introduction

"One building is architecture, two buildings is townscape" – famously noted civic designer Gordon Cullen (1961: 9) once. While this has been the motto for many of his followers, the implicit reference to the collective production of urban space may not have been fully appreciated or understood. Presumably Cullen was trying to differentiate the scope of architecture from that of urban design and planning, emphasizing the collective, incremental, and organic nature of city design that is not limited to the caprices of a single actor or property owner. It is a result of multiple actors, agencies, and processes over time. Urban designer David Crane's (1960) notion that we live in a "city of a thousand designers" echoes this sentiment of pluralism and also the reality of the social production of urban space. In this *New Companion to Urban Design*, several essays urge for pluralistic design that empowers the subaltern and helps redress chronic inequalities and injustices. The opening chapters by Roy and Tonkiss are particularly noteworthy in this respect.

Such contemporary Aristotelian disposition towards city design notwithstanding, historically and presently the urban form has been subject to lumpy and large-scale redevelopment of older urban areas or greenfield development, such as new towns, most notably. Baltimore's Charles Center, Boston's West End, and Detroit's Renaissance Center are some noteworthy examples. Large-scale developments are particularly commonplace in the ascendant economies of the Global South: Dubai, Mumbai, Shanghai – the "ayes" seem to all have it. Whether they involve large-scale slum clearance, or building artificial islands with palm tree-like footprints, or new towns of ersatz design that mimic English, French, Dutch, German, or Italian cities of the past, large-scale developments financed by global capital and supported by local states are not only commonplace but a presumptive answer to a Faustian impatience. Furthermore, it can be argued that the scope of urban design was essentially defined as the large-scale project design undertaken by one, or more commonly, a consortium of architectural, engineering, and planning firms responsible to a public or private client with authority and autonomy. Public participation in the design process was neither required nor seen as necessary. So, it is appropriate to visit or revisit some of the large-scale developments already constructed or being planned.

Alexander Garvin's chapter, entitled "The Design of Large-Scale Development Projects," focuses on three major projects in New York, all located in Manhattan: Grand Central Station, Rockefeller

Center, and the World Trade Center. The leitmotif of Garvin's arguments that weaves through all three cases is that a successful public realm and seamless integration of the project with its immediate urban context are critical to its success. In these informative case studies, Garvin seeks to define the nexus between real estate and urban design that leads to market success. The history of Grand Central Station, the oldest of these three projects, reveals this nexus and the successful creation of one of the most spectacular indoor public spaces in the country. Next, he focuses on the design and development of Rockefeller Center and contrasts it with much later large-scale development of lower Manhattan, particularly the World Trade Center, a victim of the famous 9/11 attack. In this chapter, Garvin expounds the corporate and business perspectives of this development, the foresight (and the hindsight), and the challenges of land assembly in a prime location in the middle of Manhattan, and the basis for ultimate success. He argues that, to a large extent, planning for pedestrian and vehicular access and improvement of the larger public realm, as well as nudging the property owners to improve their properties, ultimately led to sustained market response and a place of attraction for the public and the tenants of Rockefeller Center. He notes that the World Trade Center, on the other hand, failed to create such a place mainly because its designers had ignored the pedestrian and vehicular flow and access. Successful public realm and public access and "ownership" of large-scale developments seem to be the key to success. Seemingly, the recently-opened multi-billion-dollar Hudson Yard project in New York has failed to follow Garvin's recommendations.

The chapter by Sandra Frem and Hashim Sarkis, entitled "Going from Town Hall to Town Plan: The Case of Byblos, Lebanon," offers another perspective about large-scale development in the context of a historic city in the Global South. While Garvin's chapter was retrospective of earlier developments, the chapter by Frem and Sarkis is prospective in tone, since the project has yet to be implemented. Their story is that of the process of designing a large-scale project that began from a competition for the design of the city's new town hall. The issue of the location of the town hall gradually expanded into an argument for developing a master plan that involved proposing a "coherent framework" for the residents and the tourists to negotiate the spaces of the historic city. Their story is not strictly one of a unitary project implementation with specific property boundaries, but really more about developing a script for development of the historic core, not unlike the way that Gordon Cullen had imagined designing the townscape.

References

Crane, D. (1960). "Chandigarh Reconsidered." *American Institute of Architects Journal*. May, 280–292.

Cullen, G. (1961). *Townscape*. New York: Reinhold.

20

THE DESIGN OF LARGE-SCALE REDEVELOPMENT PROJECTS

Alexander Garvin

Introduction

Since the 19th century, government officials and real estate developers have proposed major redevelopment projects intended to modernize and invigorate the downtown economy by attracting new businesses. In the 1860s in Milan, densely packed buildings around the Cathedral were displaced in order to create the Galleria Vittorio Emanuele II. Around the same time in Paris, a similarly dense section of the center of the city was cleared to make way for Les Halles, a vast new public produce market. During the 20th century, such redevelopment projects became even more common. Three of the most famous were in New York City: Grand Central Terminal (GCT), Rockefeller Center (RC), and the original World Trade Center (WTC). The first two were spectacularly successful, while the third failed to reverse the decline of lower Manhattan as a business center, even before the project itself became the victim of a terrorist attack.

In all three cases, the diagnosis of the pre-development condition (presence of an inadequate capital plant) and the prescription to address it (through infrastructure reconstruction, an improved public realm, and increased amounts of office and retail space) were sound. Their success or failure, however, was determined by how well the design of the public realm integrated the project with the surrounding district. That is why the redevelopment in and around the 48 acres occupied by GCT, and the 22 acres that became RC, were successful beyond anybody's dreams. It is also the reason why the now-destroyed original WTC failed to have much impact on the future of lower Manhattan.

Successful large-scale redevelopment projects provide a major attraction for the surrounding district. Their design, however, must also make pedestrian and vehicular circulation easier and faster, provide attractive and easy-to-use additions to the public realm, and induce property owners to make improvements to their property. Only then can they generate the widespread and sustained private market reaction that is the product of successful city planning and urban design.

Grand Central Terminal, Rockefeller Center, and the original World Trade Center shared similarities in their attractions, their circulation systems, and the additions they made to the public realm. A railroad terminal was the central attraction at GCT and the WTC. Tall skyscrapers with observation decks were additional attractions at RC and the WTC. These attractions bring millions of people to each site.

There is, however, a radical difference between the manners in which people use the two railroad terminals. On their way to and from the terminal at GCT, people pass through a system of streets and public spaces that includes other properties. This walk often prompts them to stop and purchase an item, making unintended expenditures, and boosting the economy of the surrounding area. In contrast, the destination for tens of thousands of people coming and going to the PATH terminal at the WTC was the project itself. Anything they wanted could be obtained on site. So, they never had to set foot outside the project area, thus obviating any contact with its immediate urban context.

There was a similar difference between the access to the observation decks at Rockefeller Center and the original World Trade Center. People going to RC have to pass through city streets. Those going to the original WTC could come and go directly from the PATH terminal or the subway stations within the project. Thus, as this chapter will explain, design of the circulation system and the arrangement of the components of the WTC, which did not integrate it well to the rest of the city fabric, also kept it from having much impact on lower Manhattan. In what follows, we detail the development and design characteristics of these three central city redevelopment projects.

Grand Central Terminal

Real estate professionals are fond of ascribing success to three factors: location, location, and location! When the current Grand Central Terminal opened in 1913, it may have been the single most valuable property on Manhattan Island (Roberts 2013). However, it was the design of the world's largest railroad terminal and adjunct rail yards, and not its location, that was the primary reason for its success.

In its first incarnation in 1871, the three-story Grand Central Terminal certainly was *not* the most valuable location in the city, as most of the city's businesses were concentrated south of Canal Street. But by 1898, when the original terminal was replaced by a different six-story terminal building, Midtown Manhattan had become a much denser neighborhood of four- to six-story residences. Nevertheless, Midtown property was still far less valuable than property in the city's prime business location in lower Manhattan, which contained 95.1% of the city's office workers.[1]

The New York Central Railroad had been planning the replacement of this recently built terminal when a serious accident forced it to accelerate the project. On the morning of January 8, 1902, a local train crashed into an express train in the tunnel under Park Avenue, filling it with smoke, dust, and debris. Fifteen people were killed and 36 were injured. Not long afterward, the New York State legislature passed a law requiring conversion from steam locomotives to electric power (Gray 2001).

The project envisioned by the railroad required more than just a grand new station, smoke-free rail yards, and tunnels made possible by switching to electric-powered trains. The site had to be expanded from 23 to nearly 48 acres, an area three times that of the World Trade Center. The expansion required massive redevelopment: demolishing more than 200 buildings, removing 1.6 million cubic yards of rock and 1.2 million cubic yards of earth, and reconfiguring below grade rail yards between 42nd and 50th Streets to accommodate two levels of tracks and platforms. Sections of Park Avenue, Vanderbilt Avenue, and nine east-west streets had to be erected over the tracks. All this had to take place while the trains continued to operate (Belle and Leighton 2000).

The St. Paul architecture firm of Reed & Stem won the design competition for what would be the world's largest rail terminal. In fact, Grand Central was a vast transportation hub for trains,

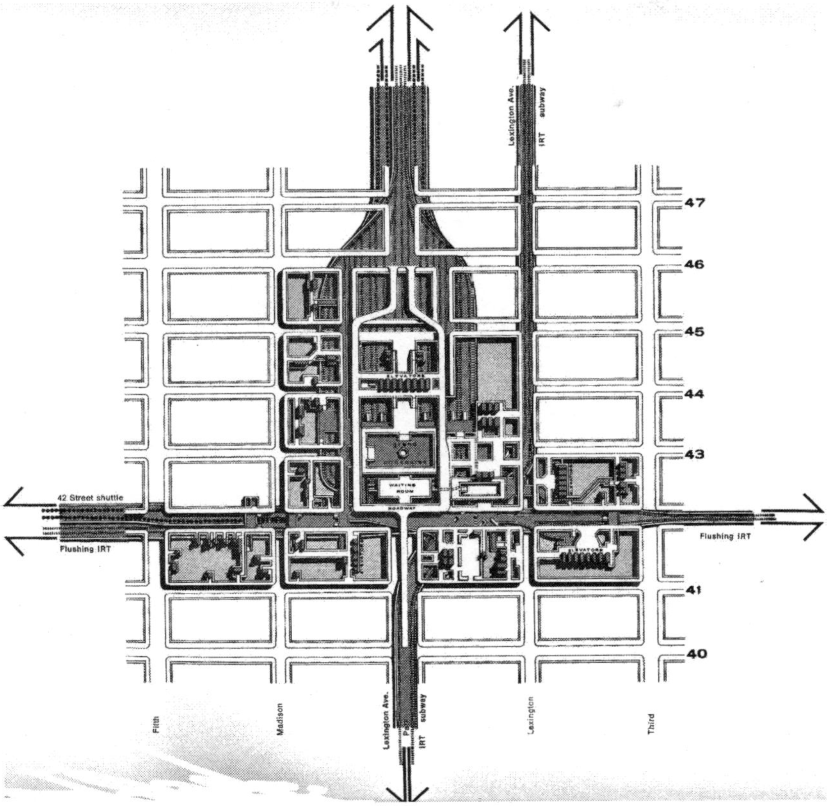

Figure 20.1 Grand Central Terminal site plan. Multilevel circulation in, out, and through Grand Central Terminal connects city sidewalks, streets, and subways directly to the train tracks.

Source: Alexander Garvin

subways, motor vehicles, and hundreds of thousands of pedestrians. It included 32 upper-level and 17 lower-level platforms, fed by 123 tracks, opening onto a nearly 38,000-square-foot main concourse with retail-lined passageways leading to two projected subway lines, to virtually all surrounding streets and to dozens of buildings and building elevators; an elevated roadway that allowed Park Avenue to continue around the terminal; and the structural foundation for what would eventually become a revenue-producing office building (Figure 20.1). Much of what was eventually built evolved from ideas initially put forward by William Wilgus, the railroad's chief engineer, and featured architectural details by Warren & Wetmore, the New York architectural firm that collaborated with Reed & Stem on the building's design (Roberts 2013).

Today, 100 years after its opening, some 700,000 people per day pass through Grand Central. This pedestrian circulation is higher than for any other retail complex in America. About 10,000 people come simply for lunch every day, even if they do not have to take a train or subway or purchase something in one of the retail stores (Carlson 2015). The spending generated by this flow of customers passing through the site by rail, subway, motor vehicle, and on foot is one of the reasons that Grand Central Terminal has continued to be the most valuable location in New York City (and the United States) for more than a century. A second reason is the rental (and eventually sales) revenue paid to the property owner (the Penn Central Railroad until it went bankrupt;

Figure 20.2 Grand Central Terminal interior.

Source: Alexander Garvin

developer Andrew S. Penson is the current owner) by the Yale Club, the Waldorf Astoria, the Commodore and Biltmore Hotels, and numerous other lessees of office buildings – a total of over 21 million square feet. Then, once the Grand Central became a designated landmark and was rezoned in 1968, the railroad gained a third source of revenue: the sale of unused development rights to nearby properties (Garvin 2014).

All three of these sources of revenue are the product of an inspired complex of public places and passages, which are located on the site (Figure 20.2), within privately owned property, and along the streets and sidewalks of New York City. This extraordinary public realm attracted developers who built office and apartment buildings, department stores, and hotels on Madison, Fifth, Park, Lexington, and Third Avenues, that created a vibrant urbanism and social and economic activity in the district, thus helping Midtown replace lower Manhattan as the prime business district of the New York metropolitan region.

Rockefeller Center

The New York Central Railroad understood that the projected opening of the subway lines running north-south on Lexington and Park Avenues and east-west on 42nd Street provided an opportunity to increase the value of its terminal by opening direct access to and from those subway lines from the new Grand Central Terminal. John D. Rockefeller Jr., who lived at 4 West 54th Street, perceived a similar business opportunity with the impending replacement of the Sixth Avenue elevated with a modern underground subway and the construction of an entirely new east-west subway across 53rd Street to Queens.

At that time "Sixth Avenue was a grubby collection of four- and five-story brick tenements, whose storefronts were occupied by cheap restaurants, furriers, barbershops, and drug companies" (Stern et al. 1987: 70, 107). Because these assets were underperforming, they could be purchased at relatively low prices. Additionally, lengthy, relatively secret negotiations were unnecessary because much of the territory between 49th and 51st Streets, and between Fifth and Sixth

Avenues, was in single ownership. The property had been given to Columbia University by the state of New York in 1814. When the university moved to its current location on Morningside Heights, Columbia divided the site into more than 200 properties and leased them to people who erected tenements, stores, theaters, and other small buildings.

A single purchase had the potential to transform this property into a large urban redevelopment project. Moreover, Rockefeller knew the leases expired between 1928 and 1931. Consequently, in January 1929, he leased the entire 12 acres from Columbia for $3.3 million per year, and thereafter bought out the outstanding leases for $900,000 (Okrent 2003). Soon thereafter, Rockefeller created a holding company that within two and one-half years had consummated 383 additional legal documents that enlarged the site sufficiently to make possible construction of most of those sections of Rockefeller Center that are between Fifth and Sixth Avenues (Okrent 2003).

To design what Rockefeller conceived of as a veritable city within a city, he assembled a talented team of designers that included the architectural firms of Hood & Fouilhoux and Corbett, Harrison & MacMurray. This first stage of Rockefeller Center took a decade to complete. It consisted of a private north-south street extending from 48th to 51st Street, a pedestrian plaza leading from Fifth Avenue to a sunken ice-skating rink that transformed into a restaurant when the weather warmed up, an underground shopping concourse, a six-level, 725-car garage, the 5,960-seat Radio City Music Hall, 14 office buildings totaling over 6.5 million square feet, and an off-street freight system that could handle between 700 and 1,000 trucks a day (Stern et al. 1987) (Figures 20.3 and 20.4).

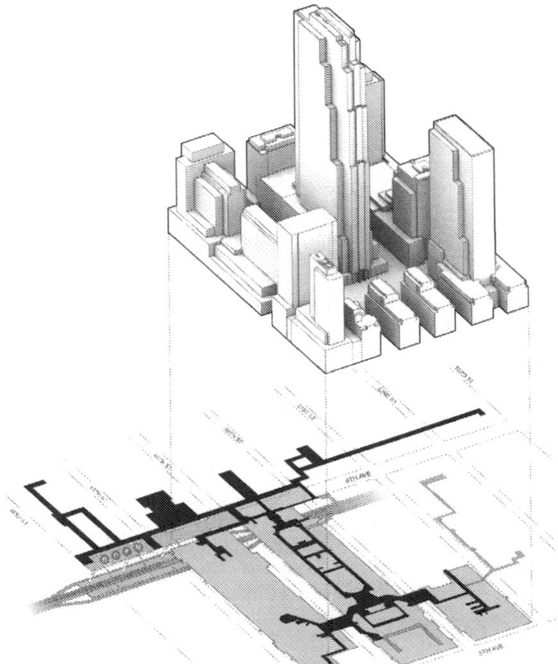

Figure 20.3 Rockefeller Center site plan. Access to Rockefeller Center was via city streets, one of which had been added to the city grid plan. A system of tunnels lined with a limited number of retailers provided access to the Sixth Avenue subway.

Source: Alexander Garvin

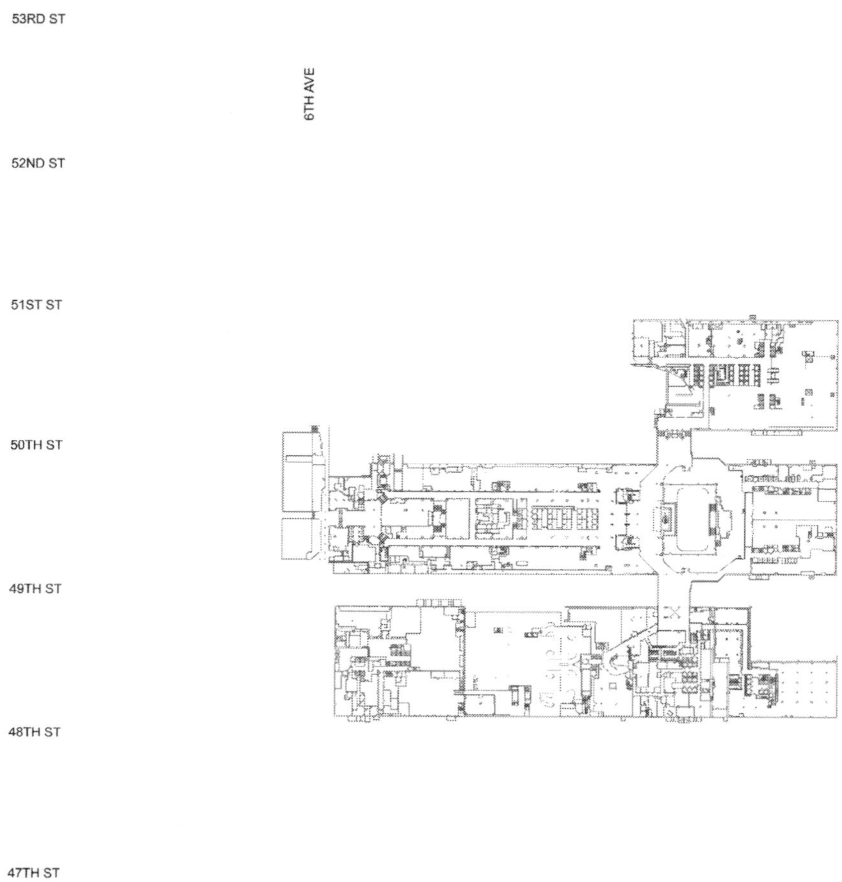

Figure 20.3 (Continued)

The centerpiece of Grand Central Terminal's pedestrian network was its vast concourse, through which hundreds of thousands of people passed every day. The centerpiece of Rockefeller Center, on the other hand, was an entire complex of public spaces: a network of public and privately owned streets and sidewalks, pedestrian promenades, sitting areas, and a plaza with its skating rink/restaurant, all crowned by an observatory with spectacular views of the city. That public space network seamlessly merged with the city's existing street grid.

Had all 6.5 million square feet of what became the initial section of Rockefeller Center opened on the same day, most of it would have stood vacant because, even in the best of times, it would have taken several years for the market to absorb that much expensive, new space. In fact, the 15 buildings in the original complex opened in stages over nearly a decade, thereby minimizing the cost to Rockefeller Center of carrying empty office space, and minimizing the impact of competing for tenants with other buildings in the vicinity that had available vacant space.

One reason for the lower-than-expected demand for space was that the section of the Sixth Avenue subway (providing service on the B, D, F, and M lines) was not open for service until 1940. Thus, what would eventually be understood to be a brilliant system of retail passageways from each of the buildings to the subway remained, for more than a decade, a tunnel network that "went from nowhere to nowhere" (Okrent 2003: 357).

Figure 20.4 Plaza at Rockefeller Center, 2005.

Source: Alexander Garvin

The project's first buildings had the added burden of opening at the start of the Great Depression. Consequently, in mid-1933, only 59% of the first four buildings had been rented. Two years later, however, occupancy had reached 75%. By 1940, Rockefeller Center was 90% occupied, and full rental was achieved in 1944 (Okrent 2003).

After World War II, Rockefeller Center was expanded to include four block fronts between 47th and 51st Streets on the west side of Sixth Avenue. When the project was completed, 65,000 people worked in its 15 million square feet of rentable floor area (The Rockefeller Group n.d.). Over the next half-century, new high-rise, high-density office buildings replaced the rest of the tenements on Sixth Avenue between 42nd and 57th Streets. Rockefeller Center had been the essential ingredient extending Midtown Manhattan west of Fifth Avenue, all the way to Times Square and the Theater District. Thus, like Grand Central Terminal, this was not merely a large-scale real estate investment, but one that triggered the widespread and sustained private market reaction that made it a successful redevelopment project.

The original World Trade Center

The original World Trade Center (WTC), unlike Grand Central and Rockefeller Center, was not primarily intended to be a business venture. It was conceived as a project that would spur the growth of lower Manhattan and enhance its economy. In fact, it failed to prevent the continuing slow but steady decline of what had once been the nation's largest business district.

In the decade between 1947 and 1956, 15.1 million square feet of new office space was erected in Midtown Manhattan, more than what existed in the Loop in Chicago, the nation's second-largest office district. During that time, only 1.1 million square feet had been erected in lower Manhattan in three modest office structures and one six-story addition (NYC Department of City Planning 1979). The situation was so serious that in 1952, the *Journal of Commerce* published articles that correctly predicted that businesses would soon be relocating from lower Manhattan to Midtown (Stern et al. 2009).

The first significant effort to retain businesses in lower Manhattan came in November 1955, when the Chase Manhattan Bank announced it would consolidate its 8,700-employee operations on a 2.5-acre site (one block east of Broadway in the middle of the Financial District) containing a new 60-story, 1.7-million-square-foot office slab, the existing 38-story building that had housed Chase headquarters since 1928, and a spacious new pedestrian plaza.

Between 1960, when the new Chase Manhattan Bank Building was ready for occupancy, and 1975, when the World Trade Center was completed, 47 buildings containing 39.1 million square feet of office space opened in lower Manhattan (NYC Department of City Planning 1979). These buildings, however, did little to enliven the district. A quarter of a century later, lower Manhattan still looked "almost the way it looked thirty years earlier," "down to the lapel cut of the suits worn by passing brokers." Those had been the very words used by the editors of *Fortune* magazine in describing lower Manhattan nearly *half a century earlier* (Fortune magazine 1956).

The World Trade Center was supposed to change all that. The idea for a complex of this sort began with the Downtown Lower Manhattan Association. This organization, which included the leadership of the corporations, financial institutions, and law firms located in lower Manhattan, was desperate to staunch the hemorrhaging of downtown business. Its members and most public officials mistakenly thought that the problem could be solved by redesigning the circulation system of a large portion of lower Manhattan and modernizing its antiquated physical plant. Consequently, they sought the demolition and replacement with up-to-date facilities of what they thought of as obsolete structures and, in 1961, requested the Port Authority of New York and New Jersey to consider the idea of a World Trade Center. The Port Authority believed that a project of this magnitude required integration with mass transit. Having just agreed to take over the nearly bankrupt Hudson & Manhattan Railroad (popularly known as the Hudson Tubes), the Port Authority was planning to modernize its underground tracks and tunnels and rebuild its 1908 terminal and office building. Consequently, the Authority devoted its attention to the blocks surrounding the Hudson Tubes.

At the time, however, the Port Authority was not authorized to engage in redevelopment. Therefore, in 1962, over the objections of Mayor Robert Wagner and a handful of public officials, the legislatures of the states of New York and New Jersey passed enabling legislation authorizing the Port Authority to acquire 16 acres of lower Manhattan for construction of a World Trade Center that (like Grand Central Terminal) would be a complex of mixed-use buildings built over a railroad terminal. It was planned by its designer, modernist architect Minoru Yamasaki, to be a virtually self-contained city three times the size of Grand Central. The complex included 95,000 square feet of retail businesses (a mere 22% of what the WTC would contain): 2.3 million square feet in the 35-story New York Central Building (now known as the Helmsley Building) that opened in 1929 and another 2.4 million square feet in the 59-story Met Life slab that opened in 1963 (Belle and Leighton 2000). Table 20.1 shows the distribution of space in the original WTC complex.

Unlike Grand Central, which already owned its site, or Rockefeller Center, which purchased most of its property on one day, the Port Authority had to assemble 14 blocks, demolish 164 buildings, and close five streets to create its 16-acre site (Jackson 1995). This removed businesses, jobs, retailers, and customers from lower Manhattan until the first building opened in 1970; the twin towers and a five-acre plaza enclosed by four seven-story structures opened in 1976. The last building of the complex was completed in 1988 (Jackson 1995) (Figure 20.5).

Table 20.1 Square footage of original World Trade Center

Land Use	*Building*	*Area*
Open space	Public plaza	5 acres
Office	Twin Towers (1 & 2 WTC)	9,500,000 sq. ft.
	4 & 5 WTC	1,000,000 sq. ft.
	Office Subtotal	**10,500,000 sq. ft.**
Civic/Govt.	US Customs Service	760,000 sq. ft.
Hotel	Host Marriott	600,000 sq. ft.
Retail	Shopping concourse	430,000 sq. ft.
Transit	PATH Terminal	420,000 sq. ft.
Operations	Service areas	810,000 sq. ft.
Parking	Parking structure	630,000 sq. ft.
	TOTAL	**14,150,000 sq. ft.**

Figure 20.5 World Trade Center site plan c. 2001: The original WTC consisted of seven buildings, built on a platform, that was separated from the rest of lower Manhattan. The platform was high enough on the western and northern edges that pedestrians were able to walk directly across a bridge to cross West and Vesey Streets. They had to go down stairs to get to the western edge of Liberty Street. The only direct, level connection with a surrounding sidewalk was from Church Street.

Source: Alexander Garvin

When Grand Central Terminal opened, more than ¾ of a century before the passage of the Americans with Disabilities Act, it was called the first great "stairless station," because gently sloping ramps allowed people to enter from 42nd Street, pass through the great concourse, purchase something in one of the stores, and stroll out to Lexington Avenue (Roberts 2013). At Rockefeller Center, pedestrians could do the same going from Fifth Avenue to the RCA Building, Radio City Music Hall, or any other structure at Rockefeller Center. The World Trade Center, however, was not nearly as pedestrian friendly (Figure 20.6).

To get from Church Street to the five-acre plaza at the heart of the World Trade Center required going up six steps, provided that was not during the five months when the plaza was closed to the public because of danger from falling icicles. Coming from Battery Park City required getting on a bridge 16 feet above West Street. A similar bridge crossed Vesey Street from the original #7 World Trade Center. To enter a store required descending one level to the dark, cavernous concourse that also provided access to the twin towers, the PATH Railroad, and the subway. To deal with the problem, the city even enacted the Greenwich Street Special Zoning District (expunged in the aftermath of 9/11) that required buildings south of the WTC to include bridge access to the Plaza.

Figure 20.6 World Trade Center, 1974.

Source: Alexander Garvin

The situation for motor vehicles was worse. Prior to the construction of the World Trade Center, Greenwich and Washington Streets were north-south arteries that carried trucks, buses, taxis, and private cars from the Battery to Canal Street. Fulton, Day, and Courtland Streets connected Battery Park City with Church Street and Broadway. Construction of the WTC disconnected Tribeca (the community north of the WTC) from everything south of the WTC.

These urban design flaws were minor compared with the World Trade Center's financial difficulties. Indeed, there was insufficient market demand for the complex's 14 million square feet when it opened in 1975. Thus, the Port Authority had no choice but to rent to government agencies and offer the vast amounts of modern office space at bargain prices. For more than two decades, the Port Authority operated the World Trade Center at what was essentially a loss – soaking up much of the demand for prime office space that might have further decreased the district's overall vacancy rate. Private developers, who were unable to withstand decades of losses, were forced to rent at similarly low prices, which they considered unfair. Finally, during the boom of the late 1990s, vacancies declined to a level that finally made World Trade Center a successful business venture.

Still worse, the World Trade Center was unable to prevent the outflow of jobs from lower Manhattan. In 1970, at the time the World Trade Center was under construction, lower Manhattan was home to nearly 15% of all workers in New York. By 2000, that had declined to just over 11% (Bloomberg 2002). At that time, more than 1 million people worked in 387 million square feet of office space in Manhattan, 76% in Midtown and 24% in lower Manhattan (Cushman & Wakefield 2002).

Why design matters

The design of large-scale redevelopment projects has a number of characteristics and challenges for developers and the city: It requires dealing with property assemblage, and also considering how to respond to market absorption on site and in the surrounding redevelopment area. It requires planning for, access to, and circulation within and outside the project, provision of public realm attractions, and consideration of the desirable level of integration or separation of the project from the surrounding area.

Most large-scale redevelopment projects (like the WTC, but unlike GCT) begin with assembling the property. Where this is true, it means eliminating market demand from the site until all the new buildings are fully leased, which can take a very long time (20 years in the case of the WTC). Such lengthy periods of absorption depress market demand in surrounding areas, which is exactly the opposite of successful redevelopment.

Urban and project design are critical in minimizing the problems of assemblage and absorption. Rockefeller Center, for example, consisted of separate buildings of different sizes, which opened at different times. The largest of them, the RCA Building, was half the size of one of the WTC twin towers. Thus, when the Port Authority put its 10 million square feet of office space on the market, it only exacerbated the declining fortunes of the rest of lower Manhattan.

Large-scale projects begin with a huge advantage over other redevelopment efforts: an enormous customer base. Grand Central and Rockefeller Center tenants (and the people coming there to do business) provided the spending for a widespread and sustained private market reaction in surrounding sections of Midtown Manhattan. The 50,000 people working in the World Trade Center every day should have had an even greater impact on lower Manhattan (Port Authority 2004). Instead, for at least two decades, the WTC soaked up customers, who could have been the tenants in other lower Manhattan locations, thereby depressing any private market reaction in the rest of lower Manhattan.

The customer base of large projects is not just people who work there. Tens of thousands of tourists went to the top of the World Trade Center to see the view. People from elsewhere in Midtown, as well as those using the railroad, purchase items in the food market and stores throughout Grand Central Terminal. Think of all the people who visit the stores in Rockefeller Center, go skating, see a show at Radio City Music Hall, or come to see the Christmas tree. Because of good urban design, at Rockefeller Center their activity is a seamless part of the Midtown street system. To enjoy the World Trade Center's attractions, on the other hand, required going inside and leaving lower Manhattan behind.

Moreover, the World Trade Center's design kept far too many building occupants inside when they could have been patronizing other businesses in lower Manhattan. Its vast underground retail complex provided all the goods and services that tenants might want, thereby making an off-site visit unnecessary, while the project's circulation system made the trip somewhat longer and more circuitous than necessary. Moreover, unlike at GCT or RC, its public areas were either above or below grade, and points of entry and exit were relatively restricted: thus, nobody ever walked through the WTC on their way to anywhere else.

In conclusion, urban design can and does affect how people use buildings and public spaces and is critical for the success of large-scale redevelopment efforts. As these three large-scale projects illustrated, however, one design characteristic that was particularly crucial for their public realm was or was not integrated with the surrounding city.

Note

1 Office space: Cushman and Wakefield (2002). Office workers: calculated from qualifying occupations in the 1900 and 2000 censuses.

References

Belle, J. and Leighton, M.R. (2000). *Grand Central: Gateway to a Million Lives*, New York: W.W. Norton & Company.

Bloomberg, M.R. (2002). "New York City's Vision for Lower Manhattan." *Speech Delivered*. December 12, 2002.

Carlson, J. (2015). "Everything You Never Knew You Wanted to Know about Grand Central Terminal." *Gothamist*. May 12, 2015. http://gothamist.com/2015/05/12/grand_central_facts.php

Cushman & Wakefield. (2002). *Office Space 2002*, New York: Cushman & Wakefield.

Fortune Magazine. (October, 1956). "'Downtown': A Last Look Backwards." *Fortune*, 54, New York, 157–162.

Garvin, A. (2014). *The American City: What Works, What Doesn't*, 3rd edition, New York: McGraw Hill Education.

Gray, C. (2001). "How a Deadly Train Accident Created an Elite Street." *The New York Times*, Section 11, p. 9, January 14.

Jackson, K.T. (1995). "World Trade Center." In *The Encyclopedia of New York City*, 2nd edition, New Haven: Yale University Press, p. 1415.

NYC Department of City Planning. (1979). *Manhattan Office Space Market 1960–78*, New York City: Unpublished Report.

Okrent, D. (2003). *Great Fortune: The Epic of Rockefeller Center*, New York: The Penguin Group (Viking).

Port Authority. (May 13, 2004). Briefing of the Author by Port Authority Staff.

Roberts, S. (2013). *Grand Central: How a Train Station Transformed America*, New York: Grand Central Publishing.

The Rockefeller Group. (n.d.). 1230 Avenue of the Americas, New York, N.Y., 10020–1579 Personal communication between the author and Rockefeller Group staff.

Stern, R.A.M., Gilmartin, G. and Mellins, T. (2009). *New York 1930*, New York: Rizzoli International Publications, Inc,

Further reading

Belle, J. and Leighton, M.R. (2000). *Grand Central: Gateway to a Million Lives*, New York: W.W. Norton & Company. A thorough historical review of the creation, construction, and impact of building Grand Central Terminal and the large-scale development complex on the rail yards behind the terminal.

Garvin, A. (2016). *What Makes a Great City*, Washington, DC: Island Press. A detailed explanation of the six characteristics of a great public realm that will attract and keep attracting the people who create great cities.

Gehl, J. (2001). *Life between Buildings: Using Public Space*, Copenhagen, Arkitekens Forlag: The Danish Architectural Press. An analysis of the way the spaces between buildings encourage walking, sitting, standing, seeing, and using the public realm.

Okrent, D. (2003). *Great Fortune: The Epic of Rockefeller Center*, New York: The Penguin Group (Viking). A comprehensive account of the activities and people who spent decades creating a large-scale development within the already-thriving heart of Midtown Manhattan.

21
FROM TOWN HALL TO TOWN PLAN
The case of Byblos, Lebanon

Sandra Frem and Hashim Sarkis

Introduction

The city of Byblos in Lebanon, Jbeil in Arabic, is one of the oldest continuously inhabited cities in the world with evidence of settlement dating back to about 8000 BC. Celebrated and protected by UNESCO as a World Heritage Site, the city has benefited from the visibility its status offers, and it has attracted funding and support from several European development and heritage agencies to help finance its public projects. Today, Byblos is the most visited city by tourists in Lebanon thanks to the variety and wealth of archaeological and heritage sites, its festivals and cultural activities, and its close proximity to Beirut. Yet the city's heritage status has not always been a blessing. During the French Mandate period (1920–1943), the historic core of the city was evacuated and destroyed in order to allow for archaeological excavations. This resulted in the initial migration of residents from the historic inner city out to its outer quarters. The constraints imposed by the Heritage status has also made it difficult for the property owners to develop their holdings. Consequently, they moved further out to find land to build.

Between 1975 and 1990, during the civil wars of Lebanon, Byblos was relatively calm and managed to attract many people seeking refuge. This led to chaotic growth in the immediate ring of surrounding villages producing a de facto suburban expansion of the city, now inhabited by many commuters who worked in the metropolitan Beirut area. Byblos's urban region now includes around 100,000 – only 40,000 of whom reside within the municipal boundaries. The recent addition of the new Lebanese American University campus to the east of the city has also boosted its economy and population. However, the agricultural sector, which has long supported the city and provided a green backdrop for its ruins, is being threatened by speculative real estate development of the surrounding farmland. Real estate development interests are now encroaching on the historic core and compromising the historic sites as well.

Thanks to a succession of enlightened mayors and to the continuity and energy provided by the municipal board, and also to a devolution of fiscal and administrative powers in the last decade from the Lebanese Ministry of Interiors to local authorities of cities and regions, the city has been able to take charge of many of its local problems. While roads, zoning, and other urban planning instruments continue to be centrally controlled, they are now subject to negotiation between local and national authorities. The increasingly vocal and empowered local population has also been able to exert pressure at both levels of governance in order to set priorities that are closer to the citizens' concerns.

The competition

In 2011, the firm of Hashim Sarkis Studios (or HSS) won a competition to design the new town hall of the ancient port city of Byblos. The competition was open, anonymous, and restricted to black-and-white presentations. The aim was to attract as many participants as possible; the anonymity was to ensure transparency in selecting the winning scheme; and the black-and-white restriction was meant to create a sense of equity and fairness for those who could not afford to produce and print colored images.

When HSS received the competition package, the team was taken aback by the site. The new was to be located in the middle of a highway interchange – a leftover island surrounded by multilevel roads that also serves as the main entrance to the city from Beirut (Figure 21.1a and Figure 21.1b). There have been several disparate attempts to fill the interchange with public activities, including a park on its eastern half, adjacent to the Municipality plot; and bus parking, an information center, and a small playground on its western side, aimed at tourists visiting the World Heritage city. Because of their isolation from the city by fast traffic and exit ramps, these public proposals have been challenged, if not considered unsuccessful. To place the main public building of the city on this highly inaccessible site felt like the city was only aggravating its already knotty traffic situation.

Figure 21.1a The municipality site, Byblos Central Interchange.

Source: Aerial photo courtesy of Wissam Chaaya

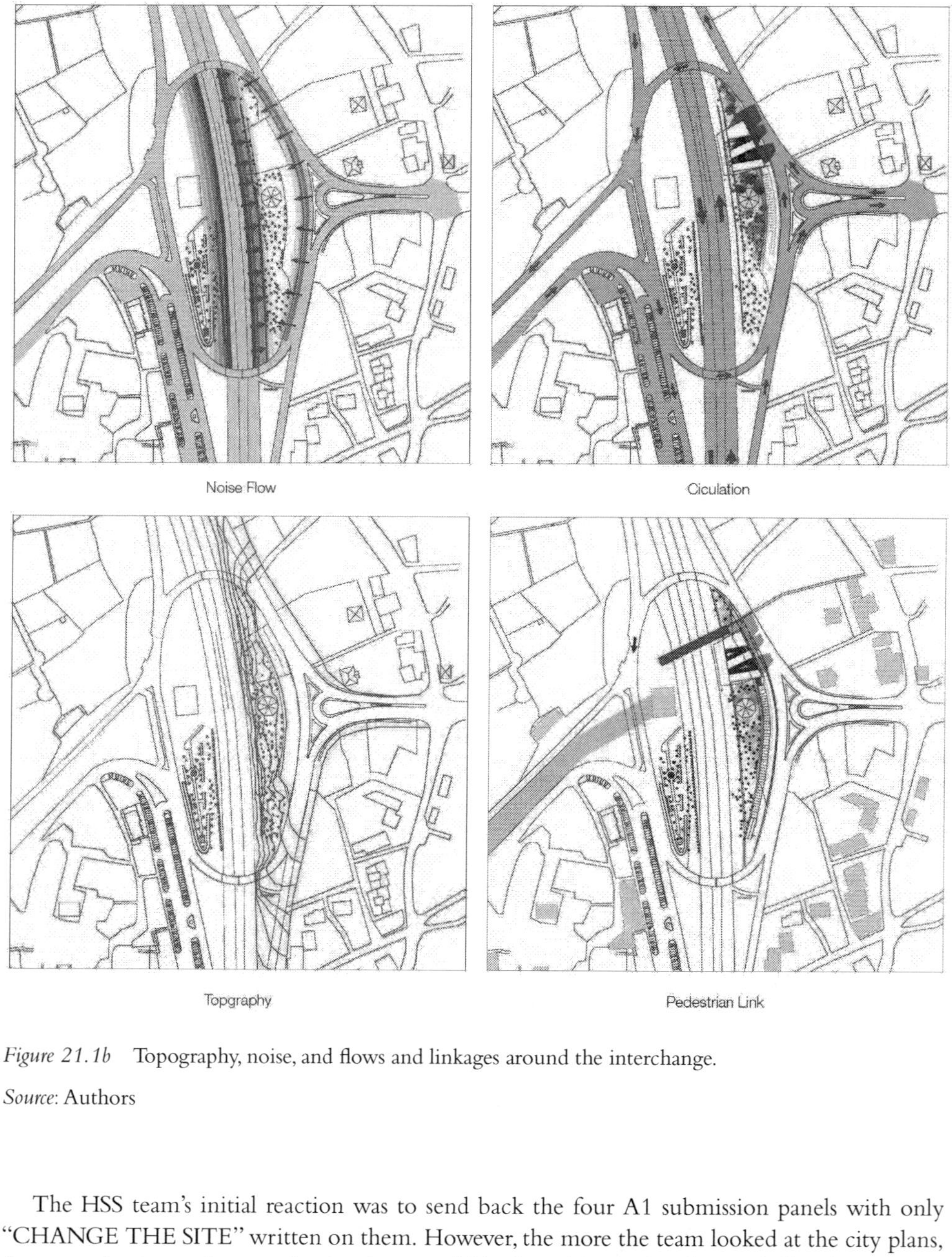

Figure 21.1b Topography, noise, and flows and linkages around the interchange.

Source: Authors

The HSS team's initial reaction was to send back the four A1 submission panels with only "CHANGE THE SITE" written on them. However, the more the team looked at the city plans, the more the site made sense despite the accessibility issue. By placing the city hall at the entrance of the city, it was made equally accessible to citizens and visitors alike, and by placing it on the highway interchange midway between the old and new town, the city seemed to acknowledge that the newer, eastern quarters of the city are as much part of Byblos as the historic core. By moving the city hall from the old quarters to the highway interchange, the municipality was aiming at re-centering a city that had lost its historic center to archaeological digs in the 1920s (Figure 21.2). Moreover, the old city residents who had moved out of the center to the adjacent fields and hills found themselves further removed from the historic core because of the new

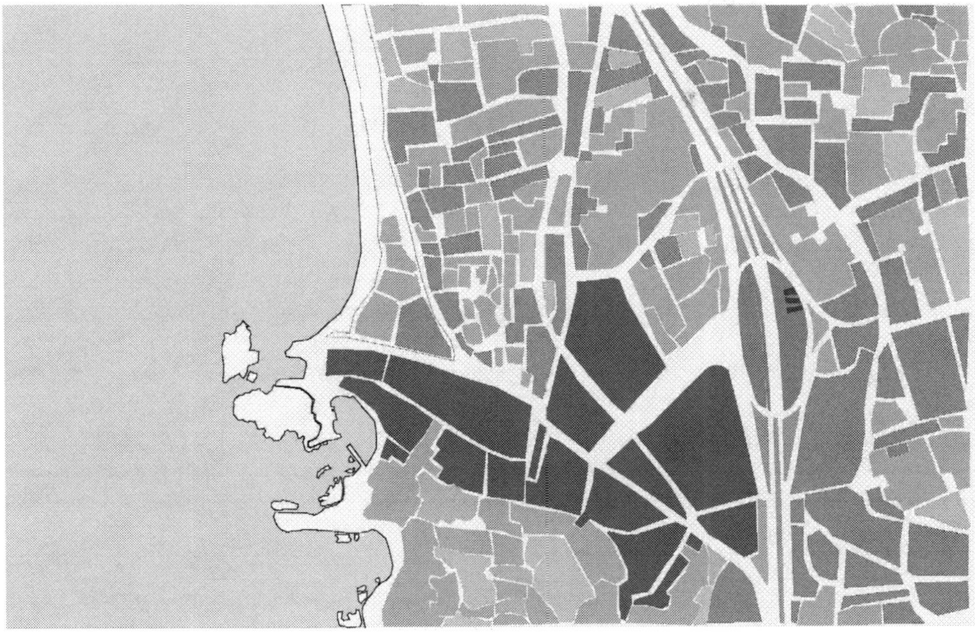

Figure 21.2 The municipality site as a linkage between the old and new town.

Source: Authors

highway that was implemented during the 1930s, creating an east-west divide between the town's expansion and its historic core. Since the competition proposal was not going to address the accessibility issue, the team focused instead on designing an architectural icon that captured the idea of re-centering in formal terms.

The HSS submission proposed a building that is broken down into three large building blocs, spanning over the park, and each housing a different component of the program, with the gaps between them bringing light into the park and the buildings. The park as well as the ground floor of the complex are maintained and open to the public. The three building blocs rest on circulation cores to the east and a long wall to the west that acts as a noise barrier with the highway. To protect the offices inside from noise and heat, the blocs are closed to the outside and clad with sandstone, the "official stone" of the city. The offices are open to the courtyards between the blocs from where they get their light and air.

Each of the three blocs contains a component of the program, namely the administration offices, the municipal council offices, and an interactive museum, that also provides a multi-purpose hall for the town. The volumes are at once separated from each other on the upper levels and connected at the ground level. Bridges between the municipal and administrative buildings provide a horizontal circulation that allows the two buildings to work in tandem on the programmatic and circulation levels. Terraces between the upper blocs provide communal spaces that overlook the city. The glass-enclosed ground-floor space combines the common elements of the program: entrances, information, city cashier, cafeteria, and exhibition space. Technical programs and the police department are located within the constructed base, at the level of the park.

It is not unusual for a public competition to rally a community and generate ideas that develop a collective imaginary around them. The jury converged on the HSS project because of the clarity with which it represented the program and created a public symbol out of it as well as public space around it. With hindsight, it is really this collective aspect of the project that ultimately expanded its impact from an architectural scale to an urban scale.

The building, the city and the Caza

Having won the competition, the HSS team held a set of meetings with the Municipality team to receive the commission and start working on the design. The tasks ahead were not clearly charted out because a budgetary uncertainty came with program crawl. The funding for the project was not yet fully secured, and there were more programmatic pieces that needed to be added to the building. One of the key proposed additions was the headquarters of the Union of Municipalities for the region or Caza of Byblos/Jbeil. The Union of Municipalities is a regional administrative body that governs the Caza with its 14 municipalities, of which historic Byblos is the center. The failure of the Union to provide funding for its piece of the building led to their withdrawal, but the process of negotiations did confirm the centrality of the building site not just to the city, but to the region at large. An additional auditorium underground and several above-ground changes were ultimately abandoned because of financial reasons. bringing the project back to the competition entry design several iterations and revisions later.

The negotiations and subsequent withdrawal of the Union may have been circumstantial and based primarily on lack of funds, but the incident does represent a tension that plays out in Lebanon and elsewhere between regions and their major cities. The main reason that unions among municipalities have been created is to facilitate the planning and management of shared infrastructure at an intermediary scale, between the national level and the local municipal level, especially in regions where some infrastructural projects, such as electricity or sewers, are autonomous. The Lebanese government often defers the responsibility of planning some infrastructural projects to these unions. However, with the recent devolution of powers and tax returns to local municipalities, the larger regional authorities do not hold much sway, or for that matter financial capabilities, as the larger cities. In the case of Byblos, the coordination between the city and the Union continued especially because many of the subsequent decisions, as the planning moved from the building outwards, extended into adjacent Union municipalities.

A stone of contention

The municipality's final scheme was built out of reinforced concrete. Each building bloc structure consisted of a circulation core, two columns to the west embedded in the noise barrier, and in the case of the central and south volumes, two additional columns in the ground-floor lobby. The spaces of the floor plans are open and clear for maximum flexibility that accommodates staff growth (Figure 21.3).

In order to avoid the unpredictable patterns of sandstone and express the monolithic nature of the stone wrapping the three buildings, the stones were cut into thin strips of 7 cm wide and a variety of lengths averaging about 40 cm. They were then separated into four shades and assembled on the façade following a pixelated pattern of striations that continue across the three buildings, derived from a sheet of yellow travertine (Figure 21.4). In contrast, the inner façades were clad in vertical aluminum louvers of varying widths spelling out the word "Byblos" in barcode, turning horizontally to become pergolas that weave between the blocks.

The lobby was treated with a monochromatic material, with counters and floors made of light gray concrete. On the inner face of the lobby, a mural representing a geometric abstraction of the Phoenician alphabet introduces one of different forms of expression between art and calligraphy that the building's architecture hosts.

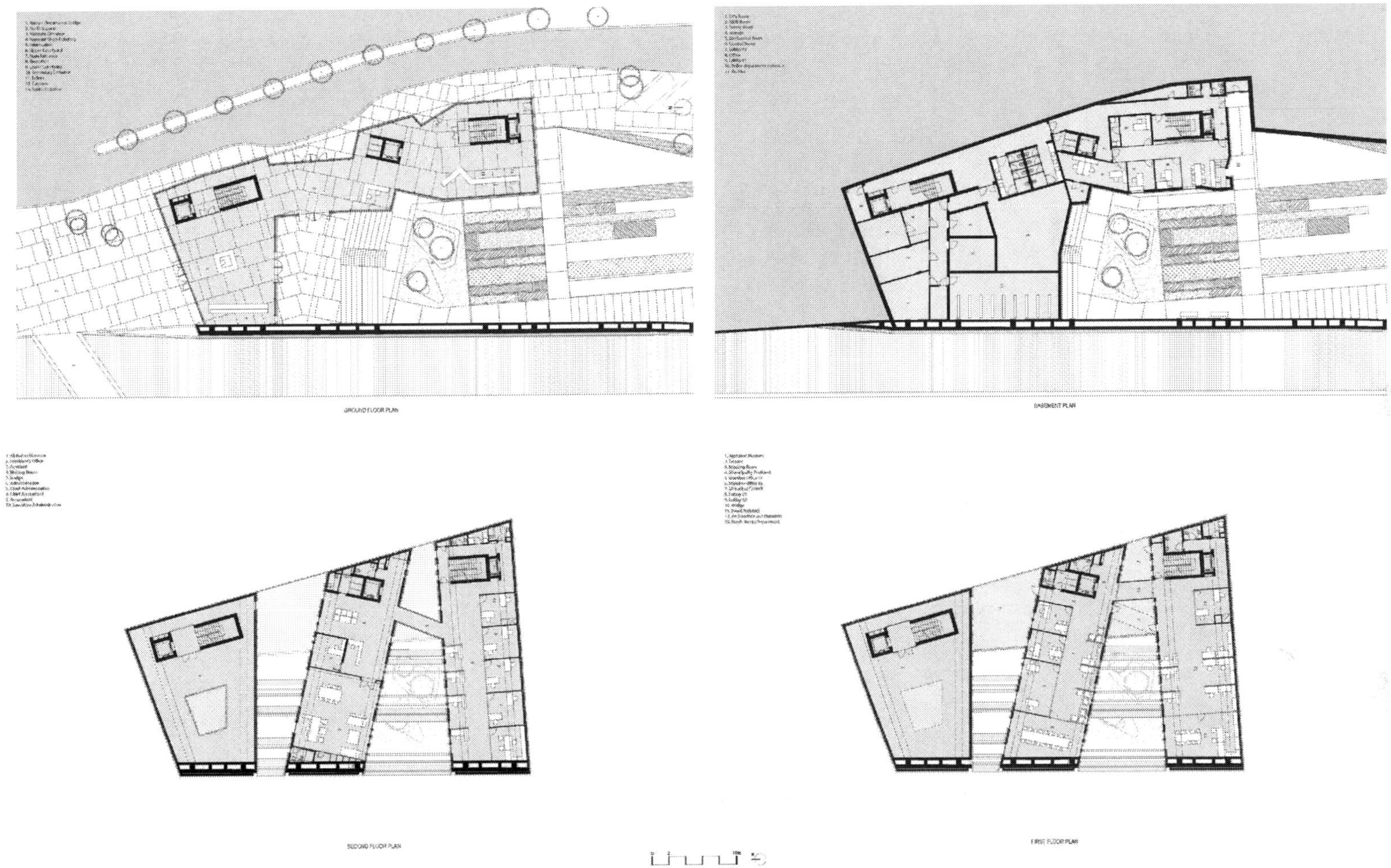

Figure 21.3 Byblos Town Hall floor plans.

Source: Authors

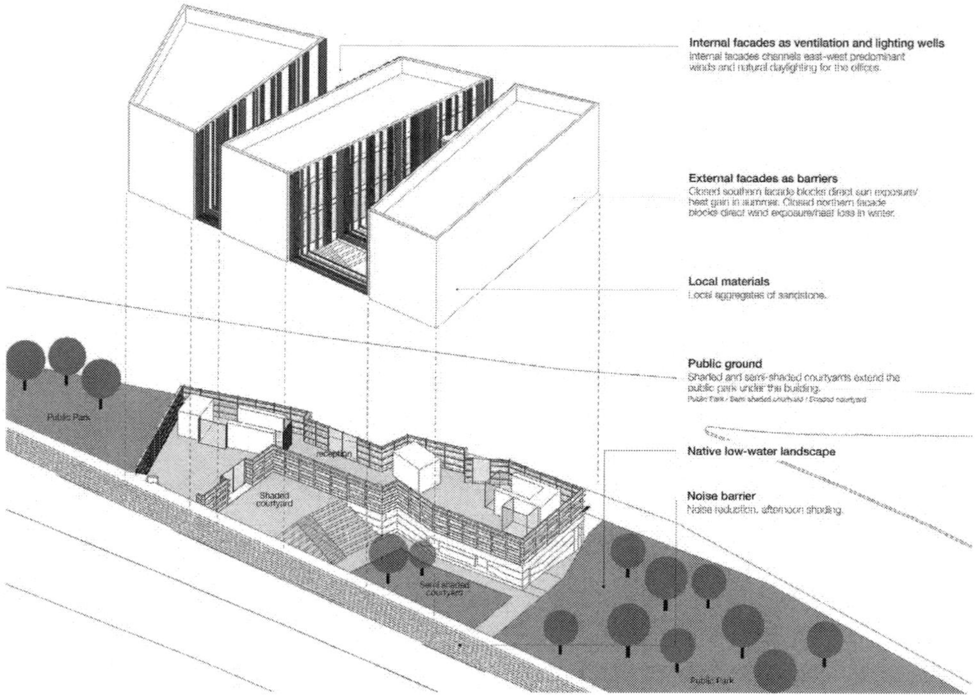

Figure 21.4 Byblos Town Hall exploded axonometric.

Source: Authors

From building to site

Halfway through the design development and the execution of the project, the city was able to consolidate the site and the property around it into one larger parcel, thus allowing connection of the grounds under the building to the existing park to the south and an envisioned plaza to the north (Figure 21.5). As such, HSS involvement expanded to include the design of the site, and to address the whole interchange. The team's urban design scheme proposed to increase pedestrian access from the city to this new urban core, by widening the bridges of the interchange to include generous sidewalks, and connecting the two parts of the interchange in a pedestrian bridge that extends the Roman Decumanus axis from the old city into the site and from the site to the modern/upper town.

The new pedestrian crossing lands in the project's northern square, connecting to the town's eastern neighborhoods, and allowing for the first time a safe pedestrian flow from the upper eastern neighborhoods to the historic core, through the municipality's northern square. The southern park provides a connection between the municipality's ground floor, passing by its courtyard, and the southern part of the interchange, with different possibilities of programming that vary from outdoor exhibitions to big events on its main lawn (Figure 21.6).

The proposed surrounding site strategies were immediately adopted and were the basis for two ongoing initiatives: The first initiative is the interchange's pedestrian connections that were adopted by the World Bank fund and considered part of its mission to showcase the World Heritage city. The bridge linking two parts of the interchange was implemented as a first phase, and the bridge reaching out to the eastern/upper parts of Byblos is now under development as

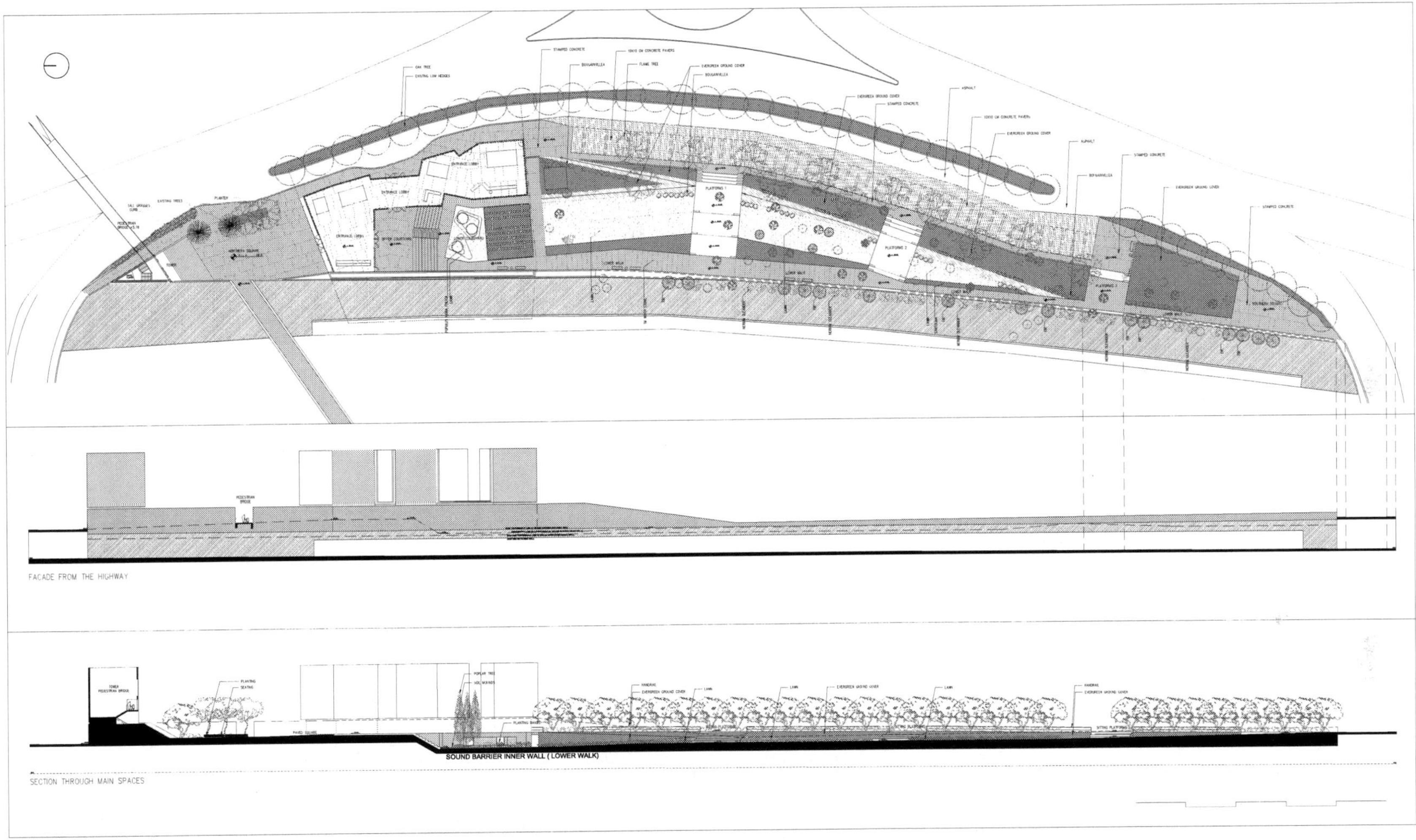

Figure 21.5 Public spaces south and north of the Byblos Town Hall is extended under the building.

Source: Authors

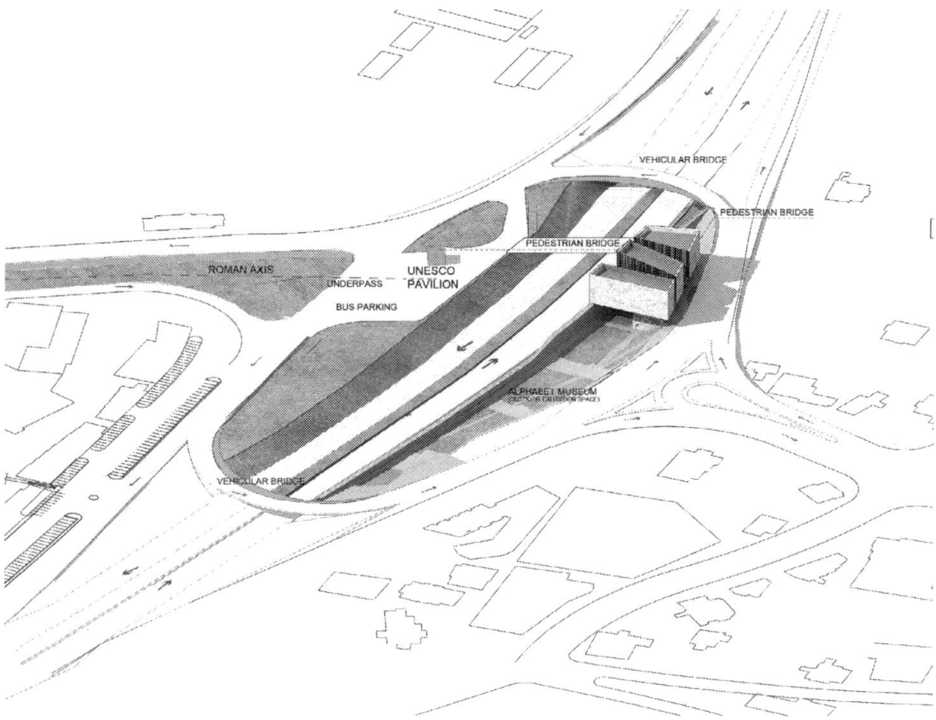

Figure 21.6 Byblos Town Hall in its extended site.

Source: Authors

a second phase. The second initiative is the southern park's landscape, which is expected to be implemented in phases from the municipality's own funds.

However, this scale change from the building to the surrounding site and the wider city brought up new discussions and old visions for the interchange, contending its role as a civic center for the city. One of the most controversial proposals was an earlier mayor's suggestion of roofing the whole interchange with the creation of underground parking on each side of the highway directly accessible from underneath, and extending the municipality's open space over it. This would have had to necessarily remove the parks on both sides of the interchange but, perhaps most seriously, would have created a bottleneck of traffic at the already-congested entrance to the city. The idea was championed by several members of the municipal board and was difficult to ignore. We had no tools other than precedents to explain the adverse effects that such a move would have on the city as a whole. Eventually, the idea was dropped in favor of the multi-entrance solution and the distribution of the parking along an inner public transport ring. This proposal is still to be negotiated through the Union of Municipalities because some of the more distant entrances are located in other municipalities.

From site to city: Byblos's urban strategies

In order to protect the Town Hall, HSS suggested ways to move away from the interchange's roofing plan by developing an urban design scheme that rethought access and mobility in the city, guided by the following ideas:

1 *Divert traffic pressure from the central interchange by clarifying and enhancing the north and south entrances of the city.* This will allow people living in peripheral parts of Byblos a decentralized access to their respective neighborhoods without having to go through the central interchange. Even though these two entrances are technically not within the municipality, the coordination through the Union of Municipalities is possible (Figure 21.7).
2 *Relieve congestion from the central interchange by promoting public transportation within Byblos.* This is achieved by introducing a shuttle bus route that overlays existing roads to form an intermediate ring between the central interchange and the city's edge entrances that connects the most active destinations in old and new parts of Byblos. In this case, the shuttle buses are those that are used during the festival season to bring visitors into the inner city but that are parked in the interchange and left idle the rest of the year.
3 *Reinforce the regional connection between Byblos and the district.* This strategy aims to reinforce Byblos's role as the center of the Caza by making it more accessible to the regional towns and villages. A network of regional buses is proposed to address the numerous commuters using Byblos as a destination or transit node. The development of this regional connection is now all the more urgent with the creation of a new sports center and a new mall to the east of the interchange. The connection will also link the city more effectively with the campus of the Lebanese American University and with several religious pilgrimage sites in the immediate outskirts of the city.
4 *Introduce a concentric network of parking spaces* to host incoming visitors' cars with connecting bus stops, that helps them navigate the use of the city's shuttle bus system, the regional connections to the Caza, and the coastal bus lines. The proposal here is to use strategically located parcels owned by the private sector and to provide development incentives that make parking more feasible financially. Another option is for the municipality to enter into public/private partnerships with the parcel owners to develop these sites into social infrastructure such as health clinics, schools, or social clubs with parking underneath.
5 *Deploy landscape as a unifying element for the highway façades* that currently host disparate edges and architectural styles. This project also relies on landscape elements to help unify and in some cases hide some of the development along the highway.

Working from this experience out, the municipality of Byblos and HSS developed a new overall framework for the city. The framework covers different sectors and enhances the city's ongoing initiatives by linking these sectors with each other and creating a coherent urban design for present and future action.

Drawing from the experience of the town hall site and central interchange urban design, HSS worked with the municipality to transfer the environmental strategies developed to other sites and to link this project with some of their other undertakings in different sectors. The main achievement of this collaboration was a new urban planning and design framework that gave coherence to ongoing and future projects in the city and addressed the physical disconnection of its different neighborhoods through multidimensional strategies. Aimed to serve both citizens and tourists, the framework consisted of: (1) a multimodal transportation network that improves public transport between city and outskirts and within the city; (2) new pedestrian linkages built along streams and old railroad tracks; (3) a network of key public/private development projects along the inner transportation ring that provide parking underground and open public spaces; and (4) a proposal to join agriculture and natural sites to protected areas, by promoting ecotourism for Byblos's northern district.

The first three elements of the framework build from existing isolated efforts to create a larger overarching coherence. However, the fourth point of the framework is by far the most

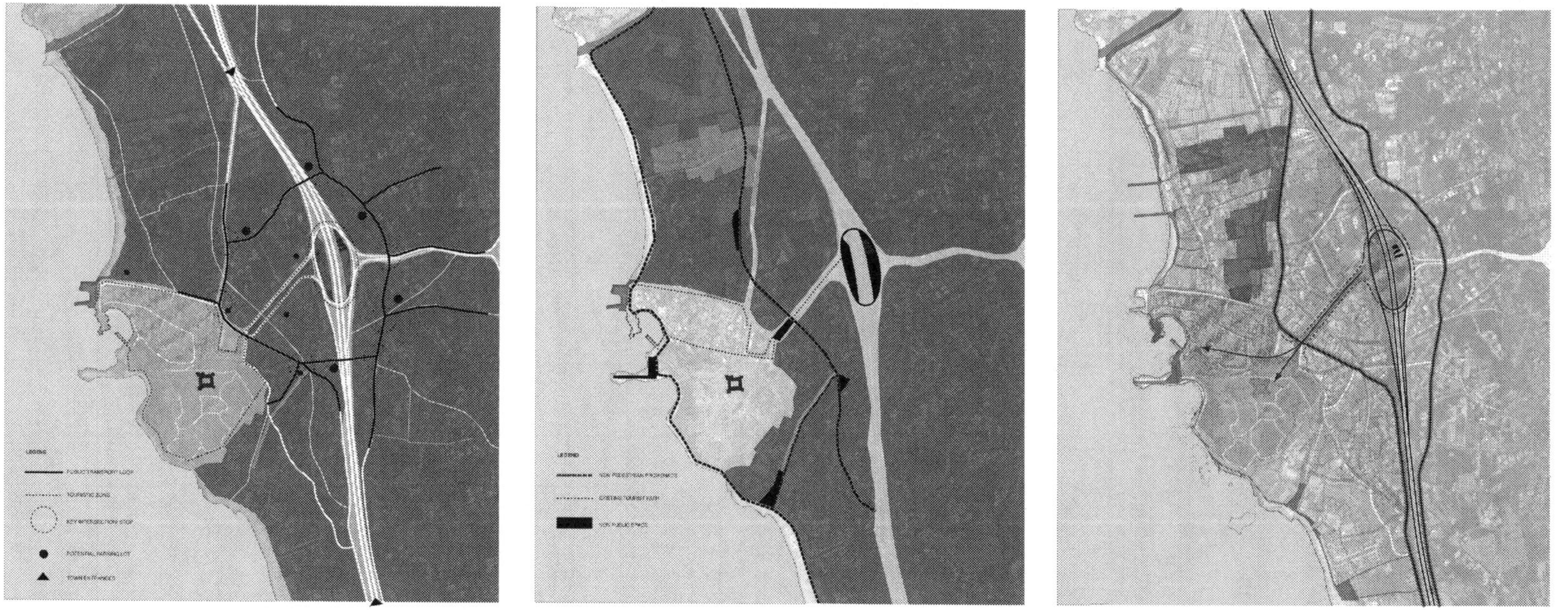

Figure 21.7 Urban strategy for Byblos: a multimodal network (left), pedestrian linkages, and public space network (center), promoting agrotourism as a way to protect agriculture.

Source: Authors

contentious because of the increasing pressure from some of the landowners to change the land use from agriculture to tourism and to allow for high-rise construction along the coast. HSS has been advising the mayor against such change, proposing instead a resort development model that could be integrated with the agricultural fields, allowing for a cohabitation between tourism and agriculture.

Resilient Byblos

Shortly after the submission of the Byblos Urban Strategies to the municipality, *100 Resilient Cities*, a program pioneered by the Rockefeller Foundation, selected Byblos among the first 33 cities that will receive assistance to develop strategies to become more physically, socially, and economically resilient. The concept of urban resilience promoted by the Rockefeller Foundation urged cities to consider a wide range of potential "shocks" and accumulating "stresses." As a result, Byblos was provided with financial support, logistical guidance, and expert advice for establishing an innovative resiliency strategy, in addition to access to solutions, service providers, and institutional partners.

The launching of the strategy in 2016, shortly after the inauguration of the municipal building, did not delineate physical strategies, but instead put forth broader "strategic pillars," general principles under which specific goals and projects would materialize. In the case of Byblos, the pillars were connectivity, resource efficiency, peacefulness, cultural identity, and prosperity. By doing so, the strategy also aimed to "scale up" existing projects and initiatives, to demonstrate that these projects can be more effective when evaluated under a multi-sectoral resiliency lens.

One of these projects was the Byblos Waterfront Boardwalk, a pedestrian pathway that was to link the four fragmented parts of Byblos's seafront in a continuous walkway. The municipality approached HSS to design the boardwalk in late 2014, when HSS was putting the final touches to Byblos's urban strategies. The boardwalk became one of the strategies as one of the key pedestrian linkages to be introduced in Byblos. Under the sponsorship of the 100 Resilient Cities initiative, the project became a part of the municipality's and 100 Resilient Cities' mutual top priorities.

Byblos waterfront boardwalk

At Byblos's western edge, the Waterfront is an astounding site that slowly grew disconnected from the city's social and physical fabric. In the past, Byblos's residents and visitors enjoyed an intimate connection to the water through a series of rituals: using the seaside as a pedestrian route to access different parts of the city, walking uninterruptedly along the waterfront and experiencing different aspects of the city's heritage, from the Armenian orphanage site, to the Phoenician harbor and site, to the festival grounds, all the way to the Crusader sea castle and the Adonis River. Today, such interactions are severed from the citizens for numerous reasons: poor accessibility, disconnection between northern and southern parts, private resorts that have exclusive access to their beach frontage, and remoteness of some parts due to the enclosed archaeological site, which has to be constantly circumnavigated.

In a country where the access to the seafront has become a privilege rather than a right, the boardwalk project's ultimate aim was to reclaim the waterfront as the natural public realm of the city. The overall project proposes various interactive nodes linked by a continuous boardwalk, to maximize the attractiveness, leisure potential, and multifunctionality of the waterfront, while increasing its socioeconomic value. This will be a truly unprecedented project, which, if implemented, will set a new standard for other Lebanese coastal cities to follow, especially at a time when many social activists in Beirut are fighting against the seafront encroachments of private developers.

Such nodes will be equipped with leisure infrastructures, such as open gyms, running tracks, overviewing steps to the water, seating, public art, public bathroom facilities, and locally owned/operated kiosks. In addition to these designated programs, the nodes will be collective open spaces that accommodate unstructured activities along the water, and will become a venue for various public, artistic, and cultural initiatives, allowing for temporal programs like festivals, markets, fireworks displays, concerts, spontaneous celebrations, and other high-energy gatherings.

With the support of the 100 Resilient Cities program funds of the Rockefeller Foundation, the project was advanced through stakeholders' meetings, integration of the archeological and festival components, social and environmental resilience into the project design, development of a price estimate, and appropriate write-ups to the attention of the Directorate General of Antiquities and potential donors. In early 2018, it was selected by the Agence Francaise de Development to receive a full grant for its implementation, through the Council for Development and Reconstruction, a central government organization, in charge of the planning and execution of large national projects.

Epilogue

While typical urban development works from the large scale at the planning level, zooming into urban design and then architecture, the Byblos experience proved that the opposite order was a plausible scenario, and that, in some cases, could be more feasible for municipalities with no history in planning, and whose timeline and phasing are more adaptive to a context of political uncertainty. However, it is important to highlight that much of the bottom-up approach already had larger urban implications embedded in the architectural and urban design moves themselves: a site that marks the entrance of the city and the point of connection or division between its historic center and its modern extension, a landmark building of special public programs, pedestrian and vehicular access, and circulation issues that radiate from this symbolic center far out enough to impact the larger territory governed by the Union of Municipalities.

As the municipality of Byblos searched for visible improvements in its urban environment, our role expanded from designing the center piece of the Town Hall to choreographing a succession of strategic interventions – quick-win projects that build on each other's positive synergies – as a bottom-up approach leading from architecture to long-term planning.

Clearly, the approach was bolstered by many projects already underway at the municipal and regional levels and by serious steps to improve the participatory planning process in the city. What the town hall project provided was an urgency to execute some of the projects, bring forth new ones, and give morphological coherence, based on this new centrality, to an otherwise disparate set of urban proposals.

The ripple effect from this center to the edges continues. The new town hall and the web of bridges and sidewalks has positively impacted the accessibility and legibility of the city's entrances. Whether the second ring of decisions is implemented or not depends on the upcoming elections and the priorities of the new municipal board and mayor. If the past changes in leadership have shown anything different in Byblos, it is the ability of the new mayor to carry forward the agenda of the previous mayor and not negate it. This has only strengthened the city's resilience and gained it an exemplary, albeit fragile, status among Lebanese cities.

The uniqueness of Byblos as a historic city and region with an excellent potential cannot be underemphasized. However, the implications and applications of this design/planning proposal could be extended to other, similar settings in Lebanon and elsewhere. For one, the possibility of reconstituting a new whole, a new gestalt, out of historic towns and their haphazard extensions is a shared aspiration among many of the smaller cities, especially those that lie beyond the metropolitan area of Beirut. For another, the scale of an architectural project could still be impactful in such settings. A project could still generate a plan. This approach amplifies limited resources but also encourages cities to attend to the urban implications of architectural projects. Third, the decentralization of administration and resources towards secondary cities has given these localities the possibility of taking matters in their hands and creating their own processes of participation and inclusion. The Byblos experiment shows that architectural competitions are one way of generating civic engagement. Last, the collectivity that urban planning necessitates and generates could be drawn in part from the collective imagination that architecture continues to exude.

Further reading

Appleyard, D. (1979). *The Conservation of European Cities.* Cambridge, MA: MIT Press. Multiple case studies of conservation in European cities – challenges of change and continuity in historic cities – based on a conference held at the American Academy of Rome.

Lynch, K. (1974). *What Time Is This Place?* Cambridge, MA: MIT Press. Critical reflections on time and place, change and continuity in the urban form, and the challenges for urban design.

PART II.V

Gentrification and displacement

Introduction

Gentrification and displacement are often byproducts of development in cities both of the Global North and the Global South, where market-driven urbanism is at play. The term "gentrification," first coined by British sociologist Ruth Glass (1964) in the 1960s, indicates neighborhood transformation and upscaling via flows of capital and people (Zuk et al. 2015). Glass was referring to the invasion of self-selected "urban pioneers" into some gritty, working-class neighborhoods of London's central city. This phenomenon, which would pre-occupy sociologists, urban geographers, and urban planners from both sides of the Atlantic over the following decades, was spurred by a back-to-the city movement by some segments of the public. The purchasing and upgrading of residential and commercial properties by higher-income residents moving into working-class neighborhoods would ultimately alter their socio-demographics, transform their built environment, and push land values and rents up.

Gentrification was originally associated with low-income, and often minority, neighborhoods of the central city, and referred to the transformation of their existing built environment by private market forces. However, state-led gentrification also became prominent, as urban renewal in the United States in the 1960s and 1970s and a series of contemporary urban regeneration projects in European and East Asian cities have poignantly shown. And it is no longer only the central city neighborhoods that are experiencing the forces of gentrification. As Suzanne Lanyi Charles and Richard B. Peiser describe in their chapter, suburban retrofitting and the densification of some suburban neighborhoods (often in first-ring suburbs) may also bring about the gentrification effect. This type of gentrification is mostly "new-build," "made up of large, newly constructed apartment complexes and luxury residential estates" and characterized by an infusion of private capital and also substantial public interventions. It is not the "urban pioneer," but the private developer facilitated by the municipal planner, who is the major actor in new-build gentrification. While Charles and Peiser do not find negative gentrification impacts (higher rents or change in population) in the adjacent neighborhoods of the Belmar redevelopment project in Lakewood, Colorado, they remain nevertheless cognizant of the possible adverse effects that new-build gentrification may have on vulnerable population groups there.

The most significant negative effect of gentrification is, of course, displacement. Displacement happens when forces outside of people's control make them move out of their house, and often out of their neighborhood. These forces can be instigated by the private sector, when private landlords initiate upgrades in their properties with the goal of extracting higher prices and rents, or when developers purchase properties with the purpose of "flipping" them and reselling them at higher prices. Such actions can lead to the displacement of the existing occupants, who cannot afford the increased rent.

But displacement can also be instigated by the public sector, as the chapter by Darshini Mahadevia discusses. She argues that in India, "the state's reliance on market principles and mechanisms to increase supply of low-income housing using public policy and finance, has resulted in peripheralization, marginalization, and displacement of the urban low-income households." Examining the city of Ahmedabad, she finds that the households that have been displaced by new and "modern" housing developments are worse off after their move to the periphery as their living conditions (access to jobs, services, water, etc.) have deteriorated. Ironically, such displacement happens under the rubric of Smart Cities Mission, which calls for infrastructure provision that will offer a better quality of life for citizens. This leads Mahadevia to note that "the process of seeking to make the city 'smart,' leads to 'unsmart' outcomes for households such as social exclusion, peripheralization and marginalization."

What is urban design's role in gentrification and displacement? There is a certain ambivalence among urban designers regarding their power to intervene in a phenomenon that is largely market-driven. And to some extent, urban design can be inadvertently responsible for gentrification as urban designers undertake retrofit and regeneration projects to upgrade declining neighborhoods and districts and to create more walkable and aesthetically pleasing places, which may lead to gentrification and displacement of original residents. But rather than being mute to these issues, it is important that urban designers situate their work in the broader political and socio-economic framework of the city, understand the consequences of proposed spatial interventions, anticipate gentrifying trends, and challenge and counteract displacement outcomes by working with communities to create more inclusive rather than exclusive spaces.

References

Glass, R. (1964). *London: Aspects of Change.* London: MacGibbon and Kee.

Zuk, M., Bierbaum, A.H., Chapple, K., Gorska, K., Loukaitou-Sideris, A., Ong, P., and Thomas, T. (2015). *Gentrification, Displacement and the Role of Public Investment: A Literature Review.* San Francisco: Federal Reserve Bank of San Francisco.

22

INNER-RING SUBURBAN RETROFIT AND NEIGHBORHOOD CHANGE IN THE POST-SUBURBAN ERA

Suzanne Lanyi Charles and Richard B. Peiser

Introduction

American metropolitan regions are in a "post-suburban era" characterized by continued development on the fringe and increased suburban employment growth (Lucy & Phillips 1997). Post-war inner-ring suburbs are particularly vulnerable to decline in the post-suburban era. As new suburban forms arise on the edge of metropolitan regions, many older inner-ring suburbs face declines in population, relative household incomes, and property values, and increases in poverty and fiscal distress (Anacker & Morrow-Jones 2008; Bier 2001; Hanlon 2008). Within the highly fragmented suburban landscape, mature suburban municipalities actively pursue efforts to increase their individual competitiveness in order to attract private-sector investment and thereby resist the forces of decline (Sweeney & Hanlon 2017). Suburban retrofitting – the redevelopment of defunct suburban shopping malls, strip retail centers, and parking lots into pedestrian-oriented mixed-use neighborhoods – has emerged as a possible solution to the car-oriented, low-density, sprawled landscape of many postwar inner-ring suburbs as well as the forces of post-suburban decline.

Urban planners and urban designers argue that American inner-ring suburbs are in urgent need of a retrofit to repair the environmentally unsustainable and unappealing suburban landscape (Dunham-Jones & Williamson 2009). In recent years, there has been much discussion of the physical form and development potential of suburban retrofits (Buntin & Pirie 2013; Calthorpe 1993; Dunham-Jones & Williamson 2009; Peiser & Schmitz 2007; Williamson 2013), as well as more recent discussion of the political motivation behind such suburban redevelopments (Sweeney & Hanlon 2017). Inner-ring suburban municipalities look to retrofit projects to differentiate themselves from their neighbors, to generate municipal revenue needed to provide public services, and to attract additional private-sector investment. But while pursuing these goals, inner-ring suburban retrofits may serve as a gentrifying force similar to "new-build gentrification" in central cities, leading to negative consequences for the original residents of the municipality. While urban redevelopment projects have been critically examined for their contribution to gentrification, suburban retrofit projects have largely escaped this scrutiny.

In this chapter we investigate the potential socioeconomic consequences of suburban retrofit redevelopments. We explore suburban retrofitting within the emerging literature on contemporary new-build gentrification, beginning with a discussion of post-suburbia and contemporary gentrification. Then, we examine the Belmar redevelopment in Lakewood, Colorado, an award-winning and well-publicized project that has been lauded as a model of a highly successful suburban retrofit.[1] In this case study, we discuss its history, development process, and urban design, and we probe the physical, social, and economic changes that accompanied it. We conclude with a discussion of whether the Belmar redevelopment is an example of new-build gentrification.

Post-suburbia

By the 1990s, many US metropolitan areas had entered a post-suburban era. The post-suburban era is characterized by the maturation of American suburbs, which involves the increasing location of employment in the suburbs and a consequent decrease in dependence on central city employment (Lucy & Phillips 1997). This results in a denser and more urbanized periphery taking the form of "edge cities," "stealth cities," or "technoburbs." The maturation of suburbia includes the aging of the older postwar housing stock and the diminution of its attractiveness for middle- and upper-income households, often accompanied by socioeconomic change. This blurs the once-distinct dichotomy between central city and suburb as inner-ring suburbs increasingly have more in common with the central city than with exurban places. Lucy and Phillips (1997: 273) predict, however, that while selective revival of central city neighborhoods through gentrification will occur, this type of neighborhood change is unlikely in suburban neighborhoods because of "insufficient neighborhood nonprofit institutions and unaesthetic, inconvenient, and unwalkable community designs." They argue that with suburban maturity comes accelerated decline and disinvestment. Scholars contend, however, that disinvestment may set the stage for subsequent reinvestment, followed by gentrification (Smith 1979, 1996).

Contemporary gentrification

Gentrification, a term first coined by Ruth Glass (1964), broadly refers to residential development in disinvested *urban* neighborhoods that receive new private- or public-sector investment, leading to neighborhood upgrade and population shifts. Early scholarly conceptions of gentrification were faithful to this definition, restricting it to the rehabilitation of housing and the direct displacement of residents and making a theoretical distinction between gentrification and more benign redevelopment (Smith 1979). Over the course of the intervening decades, however, the definition of gentrification has expanded. In his later work, Smith (1996) argued that a distinction could no longer be made, and that gentrification refers to a much broader phenomenon. Contemporary gentrification defies a single definition because it occurs in a variety of locations and takes a myriad of forms (Davidson & Lees 2005, 2010), which leads some scholars to label it a "chaotic concept" (Beauregard 1986). Other scholars (e.g., Boddy 2007; Buzar et al. 2007), however, contend that such an expansion of gentrification diminishes its substantive meaning, nullifying its usefulness as a concept.

Hackworth (2002: 815) succinctly defines gentrification as "the production of space for progressively more affluent users." Lees et al. (2008) argue that any definition of gentrification must include the reinvestment of capital, landscape change, social upgrading by high-income groups, and the displacement of low-income groups. Central to the identification of gentrification is the concept of displacement. Slater (2006: 748) contends that "displacement is and always will be vital to an understanding of gentrification, in terms of retaining definitional coherence and of retaining a critical perspective on the process." Displacement can occur directly (e.g., through eviction), indirectly, or, as Marcuse (1985) identified, by exclusion, resulting from changes in the real estate market brought about by physical changes in the built environment (see Zuk et al. (2017) for a comprehensive discussion of the forms of displacement). "Exclusionary displacement" occurs when lower-income groups can no longer afford housing that has increased in price (Marcuse 1985). In addition to the physical dislocation associated with the various forms of displacement, Davidson and Lees (2010: 402), citing Tuan (1977), argue that displacement may also be rooted in a phenomenological understanding of place – that "displacement is much more than the moment of spatial dislocation." Even if residents are not physically displaced, they may experience the loss of a sense of place and connection with their neighborhood as it undergoes social and economic change.

New-build gentrification is a contemporary variant of the gentrification process that includes substantial public intervention and large-scale deployment of capital (Davidson & Lees 2005, 2010). Davidson and Lees (2005: 1169) argue that new-build gentrification differs from previous rounds because different landscapes are being produced and different actors are involved:

> New-build developments do not involve the restoration of old housing stock by a new middle-class resident rich in social and cultural but lacking in economic capital. . . . Rather, new-build gentrification is made up of large, newly constructed apartment complexes and luxury residential estates where the gentrification has been conducted by a developer.

New developments built on vacant or abandoned land may set off a process of exclusionary displacement of lower-income residents of nearby communities. New-build gentrification involves resettlement by wealthier residents and a substantial change in the socioeconomic status of the area. Other scholars, who prefer terms such as "residentialization" or "reurbanization," contend that including new-build development under the umbrella of gentrification stretches the term too far (Boddy 2007; Buzar et al. 2007). They argue that the impact of new-build developments has been to reverse population decline in urban neighborhoods (Boddy 2007). Davidson and Lees (2005, 2010) argue that the use of the term gentrification is necessary to continue to stress the contentious class-based nature of the process. But Boddy (2007: 103) suggests that gentrification is "almost too quaint and small scale of a concept to capture the process at work" and disputes the need to politicize the process.

A case study of new-build gentrification in Lakewood, Colorado

Lakewood is an inner-ring suburb of Denver, Colorado, and the fifth-largest city in the state of Colorado. It is located six miles west of downtown Denver, directly adjacent to Denver's western city boundary. In 1969, in response to the city of Denver's interest in expanding city limits westward through annexation, residents of then-unincorporated Jefferson County voted in favor of incorporating as a city. Lakewood's commercial and social center was the Villa Italia shopping mall, a 1.5-million-square-foot, 140-store enclosed regional shopping center. When it was built in 1966, the Villa Italia mall was the largest indoor shopping mall between Chicago and California.

As a bedroom community to Denver, Lakewood did not originally have a traditional historic commercial center, as do some nearby cities (e.g., Golden and Boulder, Colorado). The Villa Italia mall served this function. As a regional shopping center, it drew customers from throughout the Denver region but also served as a local hub of social activity. It was the site of community festivals, high school proms, walking groups, and memorable experiences, holding a prominent place in residents' memories of Lakewood (Buntin & Pirie 2013).

Through sales tax collections, the Villa Italia mall was a substantial revenue generator for Lakewood and thus lessened the municipality's reliance on residential property taxes and municipal fees and fines. At its peak in 1994, the Villa Italia mall generated $3.12 million in annual tax revenue to the city of Lakewood (Buntin & Pirie 2013). As the shopping mall entered a process of decline in the late 1990s, so too did Lakewood's municipal coffers. By the end of the decade, much of the retail space was vacant (Buntin & Pirie 2013). The first of the four major anchor stores closed in 2000, followed by two more in 2001, at which point sales tax revenue had declined to half of what it was at its peak. Redevelopment of the Villa Italia site held the potential to regain the sales tax revenue.

The Belmar redevelopment

We chose the Belmar redevelopment as the focus of this chapter because it is a highly regarded redevelopment project and is used as a model for suburban retrofits across the country. It has been the subject of numerous descriptive studies (e.g., Buntin & Pirie 2013: 79–96; Dunham-Jones & Williamson 2009: 154–171; Peiser et al. 2000: 2–11; Urban Land Institute 2012: 20–23). Belmar encompasses the entire 100-acre site of the former Villa Italia shopping mall. The Denver-based real estate development firm Continuum Partners acquired the financially struggling shopping mall and its land in 1999. Demolition and site remediation began in 2001, Phase 1 opened in 2003, and the redevelopment was fully complete in 2008. The Belmar redevelopment includes 3.3 million square feet of floor area – nearly three times that of the former shopping mall. The retail component (880,000 square feet) is smaller than that of the Villa Italia mall, but the redevelopment also includes 250,000 square feet of office space, 800 residential units (in the first phase), 5,000 parking spaces, two educational institutions, public plazas, and green space. In 2013, Belmar generated approximately $17 million in annual tax revenue for the city of Lakewood (Myers 2013). In many ways, the Belmar redevelopment has become the traditional mixed-use commercial and civic center – downtown – that Lakewood historically lacked.

Belmar urban design

The Belmar redevelopment adheres faithfully to New Urbanist design principles. The development is organized as a "traditional street grid that emphasizes pedestrian movement with shops oriented closely to the street" (Van Meter Williams Pollack 2017). Figure 22.1 presents an image of a commercial street in the Belmar redevelopment. Belmar provides the hallmarks of urban form that are believed to foster pedestrian use: sidewalks and a visually diverse and varied streetscape. However, the redevelopment is inextricably located within a larger postwar suburban context. It is bordered on the north and west sides by major arterial streets, and there are few pedestrian crossings, particularly from the Lakewood City Commons, which is located directly across Wadsworth Boulevard and houses the city administration, the public library, and the Lakewood Cultural Center. Belmar is essentially an island within Lakewood, distinguished by its physical form and mixed uses. The redevelopment itself cannot make up for the decades of distinctly car-dependent development that surrounds it. Like the 1960s shopping mall that preceded it, Belmar must accommodate significant numbers of cars as the primary means of accessing the redevelopment. The urban design of Belmar incorporates metered street parking in some areas, but the majority of the parking in the core of the redevelopment is concealed in parking structures on the interior of parcels.

Within the Belmar development, the land uses transition from mixed use to purely residential towards its south and east edges. Immediately east of the Belmar redevelopment is the Martindale

Figure 22.1 A street in the Belmar redevelopment. Small shops, street furniture, and metered street parking are intended to foster a pedestrian-friendly atmosphere.

Source: Photo by Suzanne Lanyi Charles

neighborhood. The well-maintained single-family houses are approximately 1,500-square-foot, one-story structures with attached garages built in the 1950s and 1960s. Figure 22.2 shows a typical street in the Martindale subdivision. The rectilinear street grid of the Belmar redevelopment aligns with the historic street grid present in Lakewood and Denver. However, the continuity of the street grid through Belmar is rather superficial on the ground. East-west streets into the adjacent Martindale neighborhood are not continuous. At the behest of residents of the Martindale neighborhood, some of whom feared increased commercial traffic through their neighborhood,

Figure 22.2 A street in the Martindale subdivision, directly adjacent to the Belmar redevelopment. The neighborhood is made up of single-family houses, the majority of which were built in the 1950s and 1960s.

Source: Photo by Richard B. Peiser

streets are terminated at the borders of the redevelopment, and the location of new condominium buildings in the direct path of the streets forestalls any future direct connection with the adjacent neighborhood (City of Lakewood Community Planning and Development Department, 2002; T. Parker, personal communication, October 13, 2017).

Immediately south of Belmar is the Riata neighborhood, a planned development of high-density residential condominium and apartment complexes, constructed in stages beginning in the late 1970s (City of Lakewood Community Planning and Development Department 2002). The rectilinear street grid of the Belmar redevelopment is terminated as it meets the curvilinear streets that form the superblocks of the Riata neighborhood. Pedestrian access between the Martindale and Riata neighborhoods and the Belmar redevelopment is created through footpaths; however, the discontinuity of the street grid reinforces the idea of Belmar as an island not directly connected to adjacent neighborhoods. Figure 22.3 presents figure-ground diagrams of the Villa Italia mall (top) and the Belmar redevelopment (bottom) and the surrounding neighborhoods.

Neighborhood change

The Belmar redevelopment has substantively changed Lakewood; a large and prominent site was transformed from a 1960s-era enclosed shopping mall to a mixed-use district. The quintessentially postwar physical form of the former enclosed shopping mall in the center of an expanse of parking is in sharp contrast to the rectilinear street grid and pedestrian focus of Belmar. In this section, we examine the changes that occurred in Lakewood alongside the Belmar redevelopment. The 2000 and 2010 decennial censuses and the 2015 five-year American Community

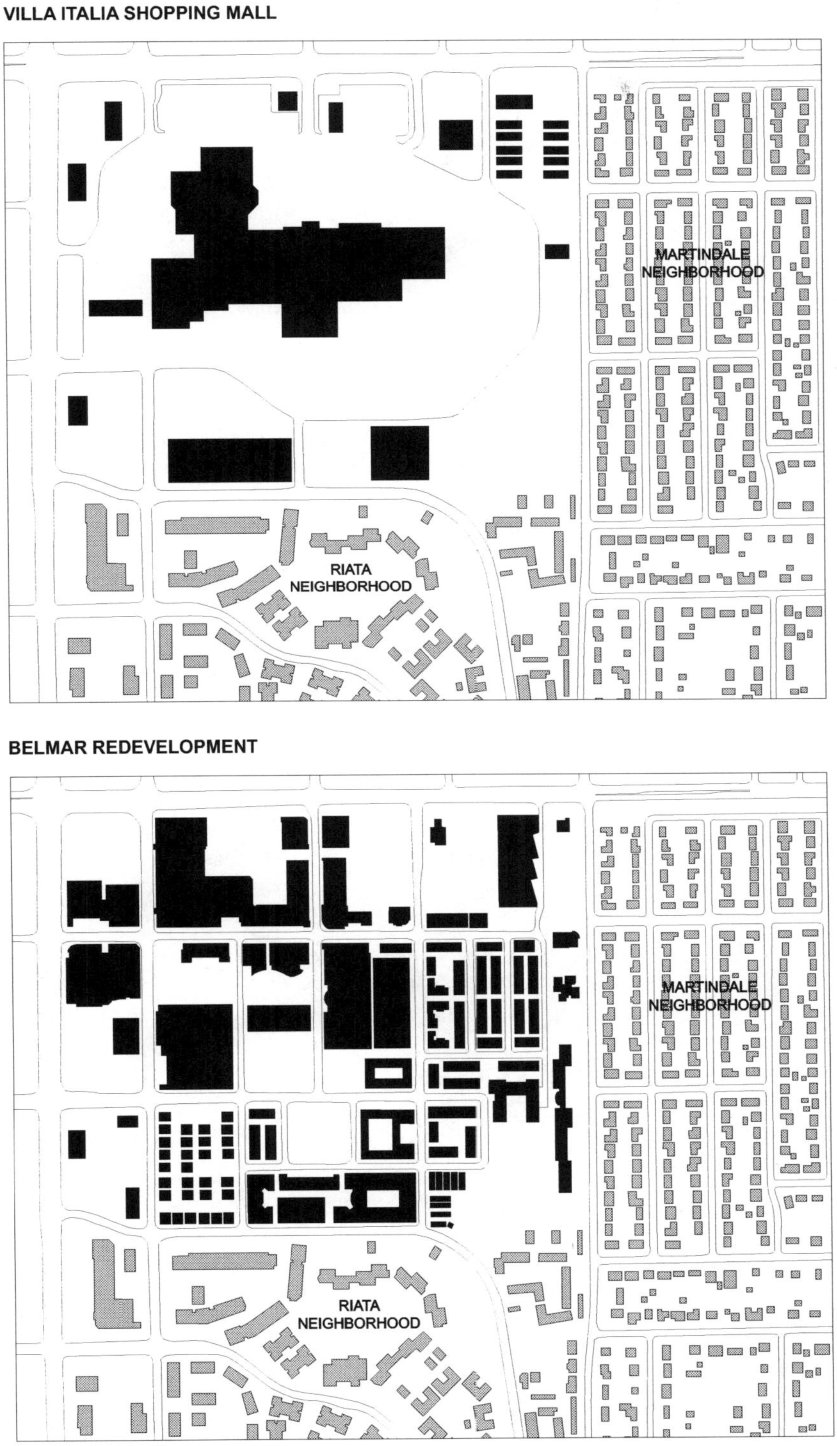

Figure 22.3 Figure-ground diagrams of the original Villa Italia shopping mall (top) and the Belmar redevelopment (bottom) and the surrounding Martindale and Riata neighborhoods.

Source: Suzanne Lanyi Charles

Survey (ACS) provide the socioeconomic characteristics of Lakewood before and after the Belmar redevelopment.[2] We discuss and compare the population and housing stock characteristics of the Belmar redevelopment with that of the neighborhoods adjacent to it and with the city of Lakewood. Table 22.1 presents these data.

The total populations of Lakewood and the neighborhoods immediately surrounding the Belmar redevelopment changed very little between 2000 and 2010. During that same period, the population living at the Belmar site increased from zero residents at the beginning of the decade to 2,000 in 2010, according to the decennial census. Similar to postwar suburbs across the country, Lakewood is experiencing the demographic shifts of the aging Baby Boom generation. In 2010, 20.8% of Lakewood residents were children (under 18), and 14.5% were seniors (older than 65). The population of Belmar, however, differs substantially. Only 15.9% of Belmar residents were under 18, and only 9.2% were over 65. The median age of residents at Belmar was 29.4 years in 2010, while it was approximately 39.2 in Lakewood overall.

The city of Lakewood overall became more racially and ethnically diverse between 2000 and 2015. In 2000, 78.9% of the population was non-Hispanic white, and this percentage decreased to 70.8% by 2015. While the percentage share of the non-Hispanic black and Asian populations remained constant, the Hispanic population increased from 14.5% to 22.1%. This trend is more pronounced in the neighborhoods directly adjacent to Belmar. The percentage of non-Hispanic white residents decreased from 66.2% in 2000 to 46.0% in 2015. The Hispanic population increased from 27.8% to 50.2% during that time period. According to the 2015 five-year ACS, the population of the Belmar redevelopment itself is more diverse than that of Lakewood: 47.8% of residents were non-Hispanic white, and 43.0% were Hispanic.

In 2000, the median household income in Lakewood was $68,500 (inflation adjusted to 2015 dollars); by 2015, it was $57,000. In the neighborhoods directly adjacent to Belmar, median household incomes decreased even further over the same time period to $41,400. The rate of families with incomes below the poverty level in Lakewood increased from 4.8% in 2000 to 8.6% in the 2011–2015 period. In the neighborhoods directly adjacent to Belmar, the poverty rate increased dramatically: it was on par with that of Lakewood overall in 2000, but then it increased more than fourfold to 16.3%. This is in sharp contrast to the Belmar development. At $81,543, the median income of households living at Belmar is substantially higher than that of adjacent neighborhoods, according to the 2015 five-year ACS, and the percentage of Belmar residents with a college degree or more is estimated at 40.5% compared to 15.8% in the adjacent neighborhoods.

The median owner-occupied house value in Lakewood was $254,700 in the 2011–2015 period, a 9.5% increase over the inflation-adjusted median house value in 2000, but lower than the median for the Denver-Aurora-Lakewood MSA ($267,000). The median owner-occupied house value is lower in the neighborhoods directly adjacent to Belmar, i.e., $214,600; the area has had an inflation-adjusted appreciation rate of 3.4% since 2000. The median house values are substantially lower than the $326,500 estimated for the Belmar redevelopment by the 2015 five-year ACS. It should be noted that much of the housing at Belmar is rental housing, so this house value reflects the value of a small portion of the housing. When we look at rents, though, a similar distinction between Belmar and the surrounding neighborhoods emerges. The median gross monthly rent at Belmar is estimated at $1,339, whereas it is estimated at $1,000 in Lakewood overall, and $774 in the directly adjacent neighborhoods. The house values and rents in Lakewood are on par with those of the region, while Belmar housing units are priced substantially higher.

Table 22.1 Socioeconomic characteristics of Belmar, Lakewood, and surrounding neighborhoods

	Belmar				*Adjacent Neighborhoods*[1]						*Lakewood*					
	2010 Census		*2015 5-year ACS*		*2000 Census*		*2010 Census*		*2015 5-year ACS*		*2000 Census*		*2010 Census*		*2015 5-year ACS*	
Population																
Total population	1,964		2,265		4,893		5,021		5,394		144,126		142,980		147,836	
Children (Population < 18)	312	15.9%	386	17.0%	1,208	24.7%	1,285	25.6%	1,380	25.6%	32,042	22.2%	29,734	20.8%	29,802	20.2%
Seniors (Population 65+)	180	9.2%	147	6.5%	735	15.0%	701	14.0%	892	16.5%	17,410	12.1%	20,749	14.5%	22,465	15.2%
Median age	29.4		30.7		36.1		35.8		37.9		36.5		39.2		38.1	
Race and ethnicity																
Non-Hispanic white	1,244	63.3%	1,083	47.8%	3,237	66.2%	2,565	51.1%	2,480	46.0%	113,755	78.9%	101,504	71.0%	104,619	70.8%
Non-Hispanic black	35	1.8%	54	2.4%	49	1.0%	48	1.0%	27	0.5%	1,910	1.3%	1,924	1.3%	1,732	1.2%
Asian	52	2.6%	89	3.9%	118	2.4%	102	2.0%	102	1.9%	3,854	2.7%	4,347	3.0%	4,195	2.8%
Hispanic	569	29.0%	974	43.0%	1,358	27.8%	2,147	42.8%	2,709	50.2%	20,949	14.5%	31,467	22.0%	32,680	22.1%
Economic status																
Median household income[2]	n.a.		$81,543		$63,767		n.a.		$41,393	-35.1%	$68,458		n.a.		$56,954	
Bachelor's degree or more	n.a.		697	40.5%	516	15.3%	n.a.		558	15.8%	32,221	32.9%	n.a.		38,006	36.5%
Poverty rate	n.a.		0	0.0%	196	4.0%	n.a.		224	16.3%	1,776	4.8%	n.a.		3,097	8.6%
Housing																
Median house value[2]	n.a.		$326,500		$207,502		n.a.		$214,619	3.4%	$232,684		n.a.		$254,700	9.5%
Median gross monthly rent[2]	n.a.		$1,339		$971		n.a.		$774	-20.3%	$1,051		n.a.		$1,000	-4.9%
Owner-occupied housing	239	23.5%	219	20.1%	1,311	66.8%	1,218	62.6%	1,249	59.3%	36,851	60.9%	36,528	58.9%	36,673	58.1%

1 The adjacent neighborhoods include three census block groups (including the census block group that encompasses the Martindale subdivision) that extend one-half mile east and north of the Belmar redevelopment.

2 Inflation adjusted to 2015 dollars.

Conclusion

Dunham-Jones and Williamson (2009) argue that individual suburban redevelopment projects that embody the important urban design features initially missing from postwar suburbia, such as walkable pedestrian-scaled streets and mixed-use "downtown" districts, are part of a larger project of retrofitting suburbia itself. Lakewood is an older inner-ring suburb that is vulnerable to socioeconomic decline in the post-suburban era; however, it has largely been able to fend off decline. It has managed to retain households as well as attract private investment. However, as suburban municipalities undertake retrofit projects to remake inner-ring suburbs into denser, more pedestrian-friendly, aesthetically pleasing places, they may begin to foster gentrification of middle-class suburban neighborhoods akin to that previously seen only in urban areas.

Belmar has the potential to incite new-build gentrification similar to that caused by the central city redevelopments identified by Davidson and Lees (2005, 2010), potentially resulting in exclusionary displacement (Marcuse 1985). The residents who now inhabit the residential units at Belmar are much wealthier, younger, and more highly educated than the residents of the surrounding neighborhoods and Lakewood overall. The median household income of Belmar residents is nearly twice that of households living in the neighborhoods adjacent to it. However, neighborhood socioeconomic change associated with gentrification is not presently evident in the residential neighborhoods surrounding Belmar or in Lakewood overall. While median house values increased between 2000 and 2015 in the neighborhoods adjacent to Belmar, the rate of increase was on par with that of the region overall, and the rate of increase in the neighborhoods nearest to the Belmar redevelopment was in fact lower than that of Lakewood. Rather than gentrification-induced increases in the socioeconomic status of neighborhoods surrounding Belmar, the reverse appears to be occurring. Census data gives some evidence of socioeconomic decline in neighborhoods surrounding Belmar, as indicated by decreasing household incomes and increasing rates of poverty.

Mixed-use multifamily retrofits in postwar suburban neighborhoods are often touted as diversifying the housing stock in suburbs in which the majority of the original housing stock is single-family homes. A benefit of housing-type diversification is that original, long-term residents of the neighborhood now, as empty-nesters, can downsize their housing without moving away from the municipality. Belmar offers a substantial number of multifamily housing units in a pedestrian-oriented neighborhood with commercial and civic amenities nearby. However, the price of housing units at Belmar is substantially more expensive than that of Lakewood generally. Thus, original residents may not have the means to live in the new units as they downsize. Belmar has attracted residents that are much wealthier than those who live in the neighborhoods around the redevelopment.

Although Belmar has many of the hallmarks of new-build gentrification, as yet it does not appear to have produced many negative effects associated with the phenomenon. Its effect on the surrounding residential neighborhoods has been, by all accounts, benign. Although the Belmar redevelopment is organized by a street grid aligned with that of the directly adjacent Martindale neighborhood, the two neighborhoods are not directly connected. While the impetus of this disconnection might have been to separate commercial traffic from the residential neighborhood, it results in reinforcing Belmar as a highly affluent island, physically separate from the residential neighborhood adjacent to it. The original residents of Lakewood were predominantly white, middle-class homeowners – although not wealthy, they are not typically the population particularly vulnerable to displacement associated with urban gentrification. The high rate of homeownership and the lack of preceding widespread disinvestment in the surrounding neighborhoods, combined with the effects of the Great Recession, likely slow socioeconomic changes in the neighborhoods surrounding Belmar. Moreover, the Belmar redevelopment replaces a regional shopping center with a new regional shopping center, albeit one that also includes residential, civic, and office uses; therefore, changes in the phenomenological sense of place often associated with new-build gentrification are also likely muted. The former shopping mall served an important commercial function and formed a central part of Lakewood's identity: Belmar has assumed both of those roles in much the same way.

Scholars contend that new-build gentrification results in decreased racial and ethnic diversity in the neighborhoods surrounding a redevelopment (Davidson & Lees 2005, 2010). That has not occurred in the neighborhoods surrounding the Belmar redevelopment. They became more ethnically diverse after Belmar opened. The percentage of non-Hispanic white residents decreased, while the percentage of Hispanic residents increased. This is similar to changes occurring in the Denver-Aurora-Lakewood MSA generally. Thus, the socioeconomic changes associated with new-build gentrification that might signal what Lees et al. (2016) describe as a homogenization of neighborhoods do not appear to be currently occurring in the neighborhoods surrounding Belmar.

Due to its adjacency to Belmar, the Martindale neighborhood is where one might expect to see the first physical and socioeconomic effects of new-build gentrification. Physical changes to the housing stock, such as the demolition of original houses and the rebuilding of much larger single-family houses, would signal significant capital reinvestment and a potential commensurate shift in the socioeconomic status of residents (Charles 2017). Although there is evidence of some new development and routine improvements such as roof replacements, we did not observe widespread reinvestment in the Martindale neighborhood. In the 2000s, new single-family housing was developed on vacant land in southern Lakewood, approximately 3.5 miles south of Belmar, as well as in recently incorporated municipalities south of Lakewood (W. Fleissig, personal communication, November 24, 2017). The Belmar redevelopment is situated a mile south of the new light-rail line that connects Lakewood to downtown Denver. Outside of Belmar itself, most new redevelopment of townhouses and multifamily housing is occurring within walking distance of the light-rail stations. The new housing developed elsewhere in Lakewood may have reduced the demand for reinvestment in established residential neighborhoods like the Martindale neighborhood. The Martindale neighborhood may be the canary in the coal mine of socioeconomic decline (and subsequent gentrification) in Lakewood. Between 2000 and 2010, annual household incomes and the rate of owner-occupied housing there declined. While median house values increased during the decade, they increased at a lower rate there than elsewhere in Lakewood. Belmar does not seem to have instigated the gentrification of adjacent neighborhoods; it may, however, have slowed the process of socioeconomic decline.

In the post-suburban era, inner-ring suburban municipalities have become entrepreneurial, seeking out private-sector investment in redevelopment projects so as to gain a competitive edge over their municipal neighbors in a fragmented suburban landscape. With limited ways of generating local municipal revenue, suburban municipalities with large developable sites have an advantage. Suburban retrofits like Belmar have the potential to generate municipal revenue and fend off the socioeconomic decline threatening some inner-ring suburbs, but they may also instigate suburban gentrification. The true test of whether Belmar is an example of new-build gentrification will come as the original residents of the neighborhoods surrounding it retire and move from their homes, and new households replace them. The change of population that comes with the aging of the baby boomers does not constitute direct displacement; however, if substantially more affluent households begin to move into the houses vacated by the retiring baby boomers, exclusionary displacement may appear. If middle-class households, similar to those that originally populated Lakewood, are increasingly outbid for housing by households similar in affluence to the residents of Belmar, gentrification-based exclusionary displacement will eventually result in significant socioeconomic change in Lakewood. Households like the ones that originally settled in Lakewood will instead have to settle in less well-located and less desirable suburbs. For now, however, the neighborhood change associated with the Belmar redevelopment has been benign, but future research should continue to monitor the socioeconomic changes in inner-ring suburbs like Lakewood.

Notes

1 ULI granted Belmar its 2006 Award for Excellence.
2 The 2000 and 2010 decennial censuses are used when data is available. The 2015 ACS five-year estimates are used when 2010 census data are not available, i.e., for educational attainment, median household income, race and ethnicity, and median house value.

References

Anacker, K. B. and Morrow-Jones, H. A. (2008). "Mature Suburbs, Property Values, and Decline in the Midwest? The Case of Cuyahoga County." *Housing Policy Debate, 19*(3): 519–552.

Beauregard, R. (1986). "The Chaos and Complexity of Gentrification." In Smith, N. and Williams, P. (eds.), *Gentrification of the City*. Boston, MA: Allen and Unwin.

Bier, T. (2001). *Moving Up, Filtering Down: Metropolitan Housing Dynamics and Public Policy*. Brookings: Washington, DC.

Boddy, M. (2007). "Designer Neighbourhoods: New-build Residential Development in Nonmetropolitan UK Cities: The Case of Bristol." *Environment and Planning A, 39*: 86–105.

Buntin, S. B. and Pirie, K. (eds.). (2013). *Unsprawl: Remixing Spaces as Places* (Second ed.). Los Angeles, CA: Planetizen Press.

Buzar, S., Hall, R., and Ogden, P. E. (2007). "Beyond Gentrification: The Demographic Reurbanisation of Bologna." *Environment and Planning A, 39*: 64–85.

Calthorpe, P. (1993). *The Next American Metropolis: Ecology, Community, and the American Dream*. New York: Princeton Architectural Press.

Charles, S. L. (2017). "Teardowns and Reinvestment in the Inner-ring Suburbs of Chicago." In Squires, G., Heurkens, E. and Peiser, R.B. (eds.), *Routledge Companion to Real Estate Development*. New York: Routledge.

City of Lakewood Community Planning and Development Department. (2002). *South Alameda Area Plan*. Lakewood, CO.

Davidson, M. and Lees, L. (2005). "New-build 'Gentrification' and London's Riverside Renaissance." *Environment and Planning A, 37*: 1165–1190.

Davidson, M. and Lees, L. (2010). "New-build Gentrification: Its Histories, Trajectories, and Critical Geographies." *Population, Space, and Place, 16*: 395–411.

Dunham-Jones, E. and Williamson, J. (2009). *Retrofitting Suburbia: Urban Design Solutions for Redesigning Suburbs*. Hoboken, NJ: John Wiley & Sons.

Glass, R. (1964). "Introduction." In Glass, R. (ed.), *London: Aspects of Change*. London: Center for Urban Studies.

Hackworth, J. (2002). "Postrecession Gentrification in New York City." *Urban Affairs Review, 37*(6): 815–843.

Hanlon, B. (2008). "The Decline of Older, Inner Suburbs in Metropolitan America." *Housing Policy Debate, 19*(3): 423–456.

Lees, L., Shin, H. B., and López Morales, E. (2016). *Planetary Gentrification*. Cambridge, UK & Malden, MA: Polity Press.

Lees, L., Slater, T., and Wyly, E. K. (2008). *Gentrification*. New York: Routledge.

Lucy, W. H., and Phillips, D. L. (1997). "The Post-suburban Era Comes to Richmond: City Decline, Suburban Transition, and Exurban Growth." *Landscape and Urban Planning, 36*(4): 259–275.

Marcuse, P. (1985). "Gentrification, Abandonment and Displacement: Connections, Causes and Policy Responses in New York City." *Journal of Urban and Contemporary Law, 28*: 195–240.

Myers, D. (2013). "Belmar: 'Urbanizing' a Suburban Colorado Mall. *Urban Land*. https://urbanland.uli.org/development-business/belmar-urbanizing-a-suburban-colorado-mall/

Peiser, R. B., Fleissig, W., and Zogran, M. (2000). *The Revitalization of Underperforming Shopping Centers into Mixed-Use Village Centers.* Cambridge, MA: Harvard Graduate School of Design.

Peiser, R. B., and Schmitz, A. (eds.). (2007). *Regenerating Older Suburbs.* Washington, DC: Urban Land Institute.

Slater, T. (2006). "The Eviction of Critical Perspectives from Gentrification Research." *International Journal of Urban and Regional Research, 30*(4): 737–757.

Smith, N. (1979). "Toward a Theory of Gentrification: A Back to the City Movement by Capital, Not People." *Journal of the American Planning Association, 45*(4): 538–548.

Smith, N. (1996). *The New Urban Frontier: Gentrification and the Revanchist City.* New York: Routledge.

Sweeney, G., and Hanlon, B. (2017). "From Old Suburb to Post-suburb: The Politics of Retrofit in the Inner Suburb of Upper Arlington, Ohio." *Journal of Urban Affairs, 39*(2): 241–259.

Tuan, Y. (1977). *Space and Place: The Perspective of Experience.* Minneapolis: University of Minnesota Press.

Urban Land Institute. (2012). *Shifting Suburbs: Reinventing Infrastructure for Compact Development.* Washington, DC: Urban Land Institute.

Van Meter Williams Pollack. (2017). *Multi-Purpose, Thoughtful Planning Creates Community.* Retrieved from www.vmwp.com/projects/belmar-master-plan/

Williamson, J. (2013). *Designing Suburban Futures: New Models from Build a Better Burb.* Washington, DC: Island Press.

Zuk, M., Bierbaum, A. H., Chapple, K., Gorska, K., & Loukaitou-Sideris, A. (2018). "Gentrification, Displacement, and the Role of Public Investment." *Journal of Planning Literature.* 33(1): 31-44.

Further reading

Boddy, M. (2007). "Designer Neighbourhoods: New-build Residential Development in Nonmetropolitan UK Cities: The Case of Bristol." *Environment and Planning A, 39*: 86–105. Drawing upon a critical reading of Davidson and Lees (2005), the article questions whether developer-led residential redevelopment should be viewed as a form of third-wave/postrecession gentrification using a case study of Bristol, UK.

Davidson, M. and Lees, L. (2005). "New-build 'Gentrification' and London's Riverside Renaissance." *Environment and Planning A, 37*, 1165–1190. Using a case study of Riverside redevelopment in London, the authors argue that new-build residential developments are examples of gentrification. They contend that new-build gentrification is a mature form of the process in the post-1990s recession era.

Davidson, M. and Lees, L. (2010). "New-build Gentrification: Its Histories, Trajectories, and Critical Geographies." *Population, Space, and Place, 16*: 395–411. The article addresses Boddy's (2007) contention that new-build development does not cause displacement and thus cannot be considered gentrification. It discusses displacement – in both its direct and indirect forms – associated with new-build developments in London.

Lees, L., Shin, H. B., and López Morales, E. (2016). *Planetary Gentrification.* Cambridge, UK & Malden, MA: Polity Press. This book presents a critical political economy approach to analyzing of gentrification in non-Western contexts, illustrating the differing trajectories and complexity of the phenomenon.

Lees, L., Slater, T., and Wyly, E. K. (2008). *Gentrification.* New York: Routledge. The book is a comprehensive presentation of the major concepts and ideas of gentrification.

23

UNSMART OUTCOMES OF THE SMART CITY INITIATIVES

Displacement and peripheralization in Indian cities

Darshini Mahadevia

Introduction

Smart Cities, the ascendant paradigm of urban development across the world envisioning new urban utopias (Datta 2015), adds technology to the early-20th-century utopian visions. The utopian and rational planning models that prevailed in many cities of the Global North (Taylor 1998) were also adopted by cities of the Global South, despite the criticism they had faced (Taylor 1998; Sandercock 1998, 1999; Friedmann 1987, 2011). By and large, rational planning has not worked in cities of the Global South due to the presence of fragmented social differentiation and multiculturalism presenting conflicting rationalities (Watson 2003), informality as integral to urbanization (Roy 2009), and a planning process that has been in a perpetual state of deregulation, ambiguity, and exception (Roy 2009). As also discussed in the chapters by Chalana (Chapter 13) and Ghertner (Chapter 27) in this volume, informality is quite widespread in India, as the ownership of land is not always clear and regulations cannot be applied in informal settlements (Roy 2009). Indeed, informal settlements have a complex web of ownership and occupancy rights, where planning legislation is hard to apply (Mahadevia 2015). Often rational planning exercises have led to exclusion and eviction of the urban poor (Watson 2003), resulting in protests, claims, and insurgency from the grassroots (Miraftab 2009).

Since India's independence, the country has practiced utopian as well as rational planning models. Hence, the highly technocratic Smart City concept, emerging from the Global North, has found ready acceptance by the Indian government and by local planners. The Smart City concept was formally introduced as one of the main urban programs in India in 2014, called the Smart City Mission (SCM). It promotes technological advances and their penetration to even the lower-income groups, for example, through smartphones, Internet, and other technologies. Projects under SCM attempt to address the question of housing for the low-income population, among others, calling for "housing and inclusiveness – expand housing opportunities for all" (Smart Cities website). But would these smart technologies address the housing challenges of the existing informal settlements, considered "unsmart"?[1] Have there been any lessons learned from earlier attempts at addressing such challenges? Taking example of the implementation of social housing projects in Ahmedabad city, this chapter attempts to answer the question as to whether the current "smart city" approach can truly address the concerns of the low-income housing of "unsmart" development, or whether it would lead to "unsmart" outcomes instead.

Social housing programs in Indian parlance are called slum redevelopment or housing for the Socially and Economically Weaker Sections (SEWSs). In recognition of the expanding informal housing, in 2006 the Government of India introduced the social housing program for the first time, called Basic Services for the Urban Poor (BSUP), under its largest urban development program called the Jawaharlal Nehru National Urban Renewal Program (JNNURM). Funds were allocated for the construction of about 12.5 million dwelling units (Mahadevia forthcoming). Recognizing the failure of this supply-driven housing program, in 2013, the Government of India shifted to a demand-driven and community participation-based program called Rajiv Awas Yojana (RAY). This program allowed community-led redevelopment of slums with options of in-situ upgrading and redevelopment, in contrast to the BSUP that focused on providing constructed housing.

The guidelines of RAY stated that the urban poor "live in inhuman conditions that deny them dignity, shelter, security and the right to basic civic amenities or social services in an environment in which crime, ill-health and diseases . . . [that draws] them deeper into vulnerability and poverty." A quarter of urban population living in slums "is an indication of the iniquitous and exclusionary urban planning system, urban land management practices and land legislation that have not been able to adapt themselves to the pace or profile of indigenous urban growth"; urban planning has not created spaces "within the formal system of planned living and working spaces to accommodate the informal working classes."[2]

While the national government was pursuing the new housing program, some states of India, namely Gujarat and Maharashtra, were pursuing slum redevelopment based on Public-Private Partnerships (PPP), wherein the private sector was given the sole responsibility of redeveloping the existing slum sites using neoliberal land management tools such as Transfer of Development Rights (TDR) and increase in permissible Floor Space Index (FSI) (Vidyarthi et al. 2017).[3] The national government changed in May 2014, and the right-wing political party (the Bharatiya Janata Party – BJP) came to power, replacing the former left-of-center party (the Indian National Congress – INC), which had introduced the demand-based RAY housing program. The BJP introduced two programs dealing with housing, the Prime Minister's Awas Yojana (PMAY – Urban) and Smart Cities Mission (SCM). As explained below, the PMAY fully adopts a neoliberal private sector-led housing supply, with a restricted government role.

This chapter uses Ahmedabad City as an example to examine the impact of the Smart Cities Mission (SCM) on housing for the poor. It finds that SCM continues to demonstrate imprudent decisions about the informal housing settlements and the needs of their low-income occupants in regard to links to employment. The new housing programs for the low-income households continue to produce unsmart and unjust settlement locations by consigning social housing to the periphery of the city, while seeking to formalize the informal.

The next section discusses the changing urban housing paradigm disguised as "housing reforms," presenting data about slums in Ahmedabad, their proposed development options, and ongoing construction of social housing projects in the city. It uses examples from one of the largest resettlement sites in Ahmedabad to show that the peripheral location of the urban poor results in their consequent marginalization and declining life quality. This case clearly shows what can be seen as the unsmart outcomes of the smart city initiatives. The concluding section summarizes an evaluation of the SCM.

The reforms in urban housing in India

The JNNURM initiative inaugurated in 2005, and its component BSUP, introduced urban housing in the national policy agenda. The BSUP was meant for the low-income urban households, offering a menu of housing options from slum improvement, slum upgrading, and slum redevelopment to new housing. Concerted lobbying by civil society organizations resulted in the BSUP component receiving 40% of the funds. In practice, however, the program largely became construction of small-sized housing. In all, 800,000[4] dwelling units were approved for construction, while construction of another 4.5 million units was approved under the Integrated Housing and Slum Development Programme (IHSDP).[5] Part of the BSUP funds came from the national government (50% for most cities and 80% for special cities such as in the hills and northeast states). The IHSDP also received 80% of its funds from the national government and produced highly diverse types of dwelling units.

The overall experience of the BSUP in particular has not been encouraging. Locations on the city's periphery dissuaded inner-city low-income households from applying for housing. The constructed units remained vacant for a long time. By March 2017, of the 687,231 BSUP dwelling units constructed nationwide, 15.5% were vacant (Mahadevia forthcoming). Some cities, such as Delhi and Jaipur, had very high numbers of unoccupied units; however, in Ahmedabad, only 3% of units have remained vacant. In some cities, the low-income households shunned the BSUP units for several reasons: peripheral location, poor quality of construction, faulty allotment process, socioculturally unacceptable design (for example, toilet adjoining the kitchen), and small unit size (Kumar and Singh 2014; Mahadevia and Datey 2015; Patel et al. 2015). Ahmedabad and other cities in Gujarat, however, witnessed higher occupancy rates because the state government decided to increase the size of the units, and, particularly in Ahmedabad, these went to households that had been displaced because of various infrastructure projects (such as the Bus Rapid Transit System, Kankaria Lake Development, various road widenings, and the Sabarmati Riverfront development).

BSUP's general lack of success in many cities was known early, and a new program called Rajiv Awas Yojana (RAY) was launched in 2013 as a demand-based housing program. It required community consent for deciding a development option for any slum. But RAY met its death in infancy, and only 21% of dwelling units had been constructed up to March 2017.[6] The state of Gujarat, instead of RAY, shifted to its own program for new housing called Mukhya Mantri Grup Yojana (MMGY). For existing slums, the Slum Redevelopment Scheme (SRS), an *in situ* slum redevelopment through PPP was proposed through a policy issued by the Government of Gujarat in 2010 (Mahadevia et al. 2018). The program provides incentives to developers through Floor Space Index (FSI)[7] increases and Transfer of Development Rights (TDR) to rehabilitate existing slum dwellers covered by the SRS. For slum dwellers or other low-income households not covered under the SRS, two private sector-initiated schemes, one for the economically weaker sections and another for the low-income groups, were also proposed under the MMGY program.

The Gujarat housing program's approach was introduced nationally when Gujarat's Chief Minister became India's Prime Minister. The new program, called Prime Minister Awas Yojana (PMAY) – Urban, is entirely private sector-led and modeled after the MMGY. The PMAY started with four elements:[8] (1) *in situ* rehabilitation of existing slum dwellers using land as a resource and through private-sector participation; (2) a credit-linked subsidy, which is providing interest subsidy to the Economically Weaker Sections (EWS)[9] and Low Income Groups (called LIG segment);[10] (3) affordable housing in partnership with the private sector, in which the national government provides a subsidy to the private developer for constructing housing projects in which 35% of the dwelling units are for the EWS section; and (4) a subsidy for beneficiary-led individual house construction or enhancement (extension), with the homeowner undertaking the construction either through "sweat equity" or with the help of a building contractor. To these four elements, six new PPP approaches have been added if the housing is to be constructed on public lands (Babar 2017). However, if the private sector is given public land for housing, this land is no longer available for housing low-income households, and the latter have to move to the city periphery if they desire formal housing.

Social housing in Ahmedabad – approaches

In 2012, Ahmedabad had 582 slum settlements with 145,455 dwelling units (Ahmedabad Municipal Corporation 2014: 46). The average dwelling unit size in the slums is 26.80 square meters (Ibid.: 43), which appears small. But all these dwelling units are on the ground and offer the dwellers the possibility of a vertical expansion or horizontal extension of their activities outside the dwelling unit (Figures 23.1 and 23.2). Only a quarter of the dwelling units are constructed with temporary materials (*kutcha* in local language); another 15% are constructed with permanent materials (*pucca*) and 60% are semi-*pucca*, with their walls out of brick but having a temporary roof, largely of corrugated iron or even asbestos sheets (Ahmedabad Municipal Corporation 2014: 42). Although asbestos is banned for use in India, slum dwellers continue to use it for roofing purposes.

Of these 582 slums, 15 slums have court litigation, or the slum dwellers were not willing to cooperate in their redevelopment. Hence, the Ahmedabad Municipal Corporation listed 567 slums with 139,584 households for redevelopment. After assessing the land conditions

Figure 23.1 Slum with spilling out activities in common spaces and vertical expansion, Ahmedabad.

Source: Shaurya Patel

Figure 23.2 Vertical extensions, roof use, and high density in a slum, Ahmedabad.

Source: Shaurya Patel

for each of the slums and their location, various strategies were proposed (Table 23.1). About 16% of dwellings – those on road margins,[11] near water bodies, near heritage monuments, or on railway lands – were proposed to be relocated. Another 28% could be redeveloped on site, using the PPP model (mentioned previously). In Ahmedabad, 30% of the low-income housing units constructed or under construction are through the Slum Redevelopment Scheme (SRS) approach, which is a component of the PMAY–Urban (Table 23.2). While the requirement was for construction of 77,757 dwelling units under the SRS, only 17,718 dwelling units have been constructed or proposed to be constructed (Figure 23.3). This approach has not attracted as many private developers as the city would have liked. Another 45% of slum dwellings could only be upgraded as they were on private lands and no redevelopment option could be chosen, since the Ahmedabad Municipal Corporation had no jurisdiction over them.

In parallel, through funds available from the national government under various programs, new low-income housing has been constructed or is under construction. Table 23.2 shows that a total of 40,786 (as of December 2017) new dwelling units have been constructed. The BSUP units (20,112 dwelling units) have been constructed under the former Jawaharlal Nehru National Urban Renewal Mission and fully used for resettlement of households affected by development projects. The RAY units include both *in situ* redevelopment as well as new construction for resettlement purposes. Both of these along with the *in situ* SRS have been for the existing slum dwellers. The dwelling units constructed under the MMGY and PMAY programs are for anyone who wished to purchase these units, through an application process when projects under these programs were announced. Since both these programs provided subsidies, they attracted more applicants than the number of dwelling units available, requiring a lottery to select beneficiary households.

Table 23.1 Development strategies for slums

	Strategy Suggested	*No. of Households*	*Distribution of Slums (%)*	*Distribution of Households (%)*
1	Relocation strategy for slums situated on road margin, railway margin, river bank, etc.	22,004	17.8	15.8
2	*In situ* redevelopment	77,757	18.4	28.0
	(a) Slums already taken for *in situ* redevelopment through PPP mode under RRRS-2010	*2,264*	*1.6*	*1.6*
	(b) Slums to be undertaken under PPP mode by open bidding (slums on public land with high land value)	*12,635*	*5.9*	*9.1*
	(c) *In situ* redevelopment on public land (slums on public lands)	*24,153*	*10.8*	*17.3*
3	*In situ* infrastructure upgrading for slums on private lands	62,858	58.9	45.0
4	Mixed strategy: combination of *in situ* upgrading + Redevelopment + Relocation	15,670	4.9	11.2
		139,584	100.0	100.0

Source: Government of Gujarat (2017) from AMC's SPCPoA prepared in 2012 (AMC 2014).

Table 23.2 Location of slum redevelopment and low-income housing projects, Ahmedabad

Scheme	*Total DUs*	*Distribution of DUs by Program*	*Distribution of DUs by Location*		
			Center	*Periphery*	*Total*
In situ SRS (completed, ongoing, and in process)	17,718	30.3	87.1	12.9	100.0
New construction under	40,786	69.7	29.7	70.3	100.0
–Prime Minister's Awas Yojana – EWS	*8,003*	*13.7*	*8.4*	*91.6*	*100.0*
–Rajiv Awas Yojana	*2,528*	*4.3*	–	*100.0*	*100.0*
–Mukhya Mantri Gruha Yojana	*10,143*	*17.3*	*41.0*	*59.0*	*100.0*
–Basic Services for the Urban Poor	*20,112*	*34.4*	*36.2*	*63.8*	*100.0*
Total	71,041	100.0	33.1	66.9	100.0

Source: Data collected from the Ahmedabad Municipal Corporation

The 20,000 BSUP units are constructed out of concrete (Figure 23.4). These two-room units of about 30–35 square meters are not suitable culturally or climatically and are not convenient for Indian joint family structures (households with more than one married couple). Further, the form does not allow for spillover of activities into semi-open spaces that is essential for households living in small dwelling units. The dwelling units radiate back heat inside in the evenings in the high summer temperatures that often exceed 40° C, making them very uncomfortable in the summer. There are no spaces for outdoor activities such as sleeping during summer months and socializing. Some of these units are in extremely poor condition, just three to four years after their construction (Figure 23.5).

Figure 23.3 *In situ* slum redevelopment.

Source: Shaurya Patel

Figure 23.4 BSUP dwelling units, constructed fully of concrete, Ahmedabad.

Source: Shaurya Patel

Figure 23.5 Poor condition of the BSUP units.

Source: Shaurya Patel

Peripheralization and marginalization

Peripheral locations

While public housing is found all around Ahmedabad (Figure 23.6), its larger concentrations are on the periphery. This will intensify in the coming years as more public housing will be built on the periphery in the proposed Residential Affordable Housing (RAH) Zone, along the outer ring road of the city named Sardar Patel Ring Road (Figure 23.7), following the market logic of low land prices in the urban periphery.

The new dwelling units were constructed under the prevailing programs of the time. The earlier program was BSUP, which was used for resettling project-affected families in Ahmedabad. In December 2016, of the 20,112 DUs constructed under the BSUP program, 2,197 (11%) were unoccupied (Government of Gujarat 2017: ix). The MMGY units are spread throughout the city and have partial occupancy. Much of PMAY-funded housing is expected to appear in the RAH Zone. The PMAY undoubtedly will place the low-income households at the city's periphery. About two thirds of the proposed or constructed public housing or SRS dwelling units are also built at the periphery (Table 23.2). Of the new dwelling units constructed, 70% are at the periphery, while 87% of the *in situ* SRS dwelling units are in the central parts of the city. More than 90% of the PMAY housing proposed is at the periphery, while 2/3 of the BSUP units are also at the periphery. As already mentioned, the BSUP program was used to house families displaced from the construction of various infrastructure projects. These low-income residents have been involuntarily forced by the state to move to the periphery. We turn to this issue next.

Figure 23.6 Location of public housing in Ahmedabad.

Source: From data collected from the Ahmedabad Municipal Corporation

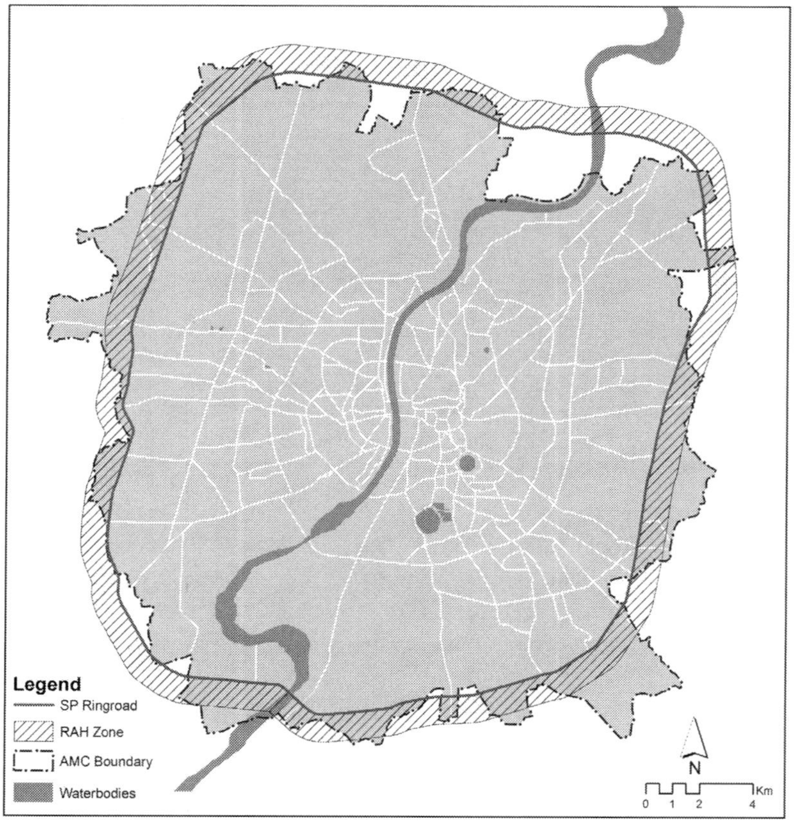

Figure 23.7 Peripheral Affordable Housing Zone, Ahmedabad.

Source: Extracted from Ahmedabad Urban Development Authority Development Plan

Everyday experiences

This section presents the living conditions in one resettlement site in the Vatwa area, on Ahmedabad's periphery (Figure 23.8). This site has 9,200 dwelling units distributed in seven locations.[12] The research was undertaken in three of the projects, Sadbhavnanagar (a mix of Hindu and Muslim households), Kusha Bhau Thakre (KBT) nagar (only Hindu households) and Vasant Gajendra Gadkar (VGG) nagar (only Muslim households). The resettlement is composed of units with electricity, running water, a toilet and bath in four-story buildings served by paved streets, streetlights, drainage, common open spaces, and social amenities such as *anganwadis* (government run/sponsored childcare and maternal care centers). For the relocated families, this was an improvement over their previous homes and neighborhoods.

However, the condition of the infrastructure was far from satisfactory. The residents had complaints about the hard drinking water, drawn from an underground well and leading to formation of kidney stones. On their former sites they had soft water from the AMC, which was sourced from the Sabarmati River. High salt content led to frequent choking of pipelines, and leakages reducing water availability. Lack of adequate water availability led to conflicts among residents and between residents and the water supply operator (Mahadevia et al. 2016a). The residents were forced to pump out water from the underground water storage tanks and wash their clothes in the open, dirtying the whole site (Figure 23.9). Mostly women were forced to carry water buckets to their residence, which for some meant carrying it to the fourth floor. The residents often had to depend on water tankers for their daily supply. After the resettlement, the

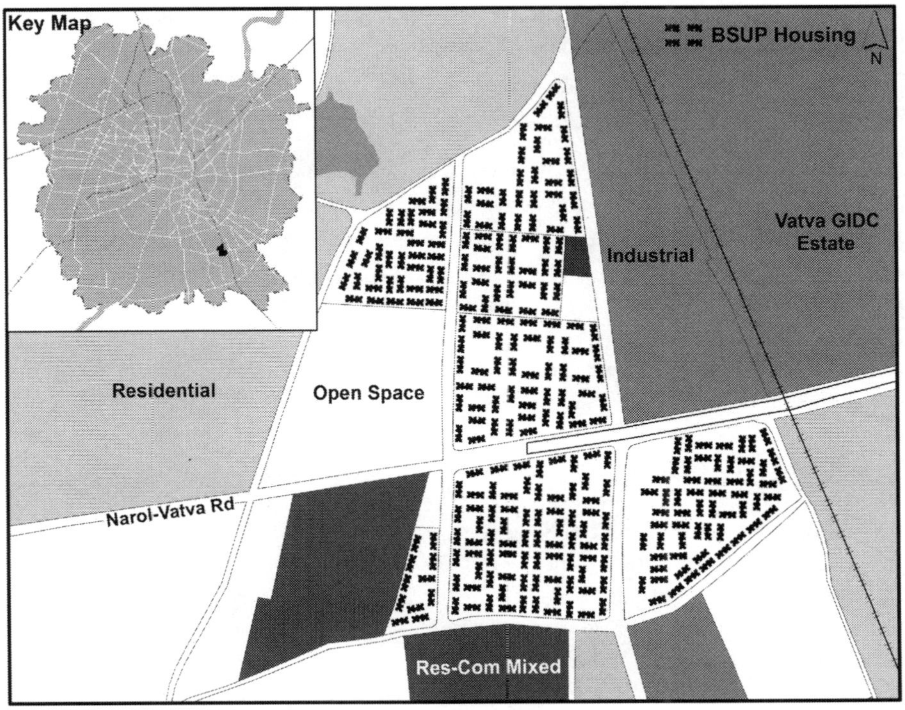

Figure 23.8 Vatwa rehabilitation site location and layout plan.

Source: Mahadevia et al. (2016a)

Figure 23.9 Washing around underground tank in KBT nagar, Ahmedabad.

Source: Mahadevia et al. (2016a)

governance of water supply was handed over to the residents, but this did not work out. The AMC had not considered or not cared to consider post-construction governance issues.

There were other concerns related to infrastructure. The cost of infrastructure went up for water supply system maintenance. The electricity bills also increased, whereas in the slums previously residents were entitled to subsidized electricity connection. All over the resettlement sites, one could see garbage strewn around and drainage pipes leaking. Streetlights were vandalized, and in many locations, they were not functioning. Residents revealed that local illicit trade operators deliberately broke streetlights so that they could carry on illicit trades of drugs and alcohol under darkness.[13] Some of the social amenities were also vandalized, and residents were not inclined to contribute funds for site maintenance, as they had made a one-time contribution already, and their living expenses had increased compared to their previous location.

Additionally, the residents' incomes had declined, due to three factors: (1) the additional distance from new sites to their work location meant higher transportation costs; (2) some women had to drop out of the labor market because of their inability to travel long distances to their former employment sites and also perceived insecurity on transit; and (3) lack of adequate work in the vicinity. Some residents reported that their post-relocation transportation expenditures were about a quarter of their monthly income (Mahadevia et al. 2016b). Those who made their livelihood by vending could previously walk to their vending locations and to wholesale markets to acquire goods, but now they could not carry out their trade from the new location or had to rent a spot to park their vending carts at a walking distance from the nearby market. Some women took up home-based work resulting in a loss of household income. Women, who were domestic workers, could not find work in the vicinity, as the entire neighborhood is low to middle income with few available work opportunities. The public transport available in the locality is very poor, and many residents were forced to use auto-rickshaws, which women did not find safe (Mahadevia et al. 2016c). Transport problems also resulted in decline in access to social amenities such as education and health facilities.

These relocation sites now host households displaced from different parts of the city and even from the different localities along the riverfront, and consequently lack social cohesion. Absence of decent jobs has led to a proliferation of illicit businesses, such as selling alcohol and drugs and operating gambling dens. Residents pointed out that men intoxicated on alcohol or drugs fight with each other in public spaces. There is also *goondagardi* (hooliganism) around the running of such illicit businesses, leading to physical threats and violence. Often, knives and swords are wielded (Mahadevia et al. 2016d).

Lack of livelihood opportunities and decline of family income has led young men, who are addicted to gambling and drugs, to resort to theft and robbery to finance their habits. Intoxicated men harass young women and girls. The parents fear leaving their children and teenage and unmarried girls at home alone. Also, some of the dwelling units were vacant at the time of our research and used by local goons for illicit activities. Lastly, men had turned to alcohol and gambling to deal with the stress of their destabilized livelihoods (Mahadevia et al. 2016d). All these lead to an atmosphere of insecurity. Increase in illicit activities and *goondagardi* leads to a loss of social control by community leaders, which was not the case in their previous housing sites. According to one resident: "in the earlier location no one would have dared to start an alcohol den, as the residents and the local leaders would shut it down."

Residents also reported increases in burglaries and small thefts, such as cash or bicycle thefts. Missing or vandalized streetlights have created a physical context amenable to robberies and thefts on the street, as well as harassment of women. Lastly, these sites have poor policing, which enhances the sense of insecurity. In the absence of police and law enforcement, local residents have taken to adjudicating justice themselves such as beating a thief, if caught (Mahadevia et al. 2016e).

Conclusion

The neo-modernist project of planning cities through infrastructure-led development and conversion of the informal into formal, that is from the supposedly "unsmart" to "smart," has unfortunately led to the creation of "unsmart" geographies and "unsmart" conditions for low-income households in Indian cities. The furthering of a neo-modernist agenda now takes place through the "Smart Cities" dogma, which in the Indian context means the provision of core infrastructure and a decent quality of life for the citizens. At the launch of the SCM program, the Prime Minister of India stated: "cities have to be strengthened through comprehensive and inter-connected approach combined with increased public participation . . . smart cities should not be seen as a fancy concept since the mission is meant for ensuring availability of basic services to people though infrastructure including houses for urban poor in a comprehensive manner."[14] Promoting construction of new houses for the urban poor under the SCM means continuing the same approach to housing that has not worked. Ahmedabad is among the 99 cities approved as a "Smart City." However, the city has proposed *in situ* slum redevelopment only in an area of 2.19 square kilometers, housing about 9,000 slum households living in the area and releasing land for other public amenities.[15] The Smart Cities Mission has two components: Area Based Development, focusing on investments in a small area; and Pan-City projects, which essentially are IT-based components (such as command and control centers giving real-time traffic condition information, closed circuit television cameras monitoring traffic junctions and crime, app-based heat wave warning, etc.).[16] While Ahmedabad has both components, the attention is on the tangible development of the 2.19 square kilometers, on about 0.5% of the city's total area of 450 square kilometers (Ahmedabad Urban Development Authority – AUDA n.d.). This is hardly making the whole city smart! For this to happen, a multi-pronged and bottom-up approach is necessary to addressing issues around informality.

Instead, the experience of Ahmedabad portrays that in the process of seeking to make the city "smart", the "unsmart" outcomes of social exclusion, peripheralization, and marginalization of some groups worsen. These urban dystopias will continue and probably expand, if large numbers of rural populations migrate to the cities. The "unsmart" would be the dominant paradigm under these circumstances. Over the past decades, many low-income households have survived and even progressed in Indian cities through some smart and indigenous methods such as incremental housing, negotiating tenure security through political processes, "occupancy urbanism" (Benjamin 2008), and above all showing tenacity to withstand hardships to gain a foothold in the city (Mahadevia 2015). But in the endeavor to make "smart" cities, there is a real danger of obtaining "unsmart" outcomes from seemingly "smart" initiatives.

Notes

1 Referring to the typical housing stock for the poor that existed prior to the "smart city" hype.
2 In the Preamble of the RAY guidelines (MoHUPA 2011), available at http://mhupa.gov.in/writereaddata/02-RAY-Guidelines.pdf (accessed on April 24, 2017).
3 The Floor Space Index is a ratio of built-up area to the land-plot area. Transfer of Development Rights gives developers of slum areas additional incentives in the form of extra Floor Space Index, which they can transfer to other plots where they would undertake development, if they are unable to use the Floor Space Index on the slum site they are redeveloping.
4 Data from the files were downloaded from the site http://moud.gov.in/cms/projects-under-jnnurm.php (site has removed the data). Instead, the data is now available on http://moud.gov.in/upload/uploadfiles/files/3City_wise_BSUP.pdf (accessed on January 8, 2017).
5 Data from the files downloaded from the site http://moud.gov.in/cms/projects-under-jnnurm.php (site has removed the data). Instead, the data is now available on http://moud.gov.in/upload/uploadfiles/files/4State_wise_IHSDP.pdf (accessed on January 8, 2017).
6 Data was available on http://mhupa.gov.in/writereaddata/Ray_all_India.pdf. It is now removed.
7 Also referred to as Floor Area Ratio (FAR).
8 Source: http://mohua.gov.in/cms/pradhan-mantri-awas-yojana.php (accessed on March 15, 2019).
9 Defined as household with annual income of Indian Rupees (INR) 300,000 (USD 5,000). Dwelling unit size should be up to 30 square meters.
10 Defined as household with annual income of INR 300,000 to 600,000 (USD 5,000 to 10,000). Dwelling unit size should be up to 60 square meters.
11 Unpaved road easements on both sides of the paved carriageway; in US terms can be considered as highway "shoulders."
12 This research was under the "Poverty, Inequality and Violence in Indian Cities: Towards Inclusive Policies and Planning," a three-year research project (2013–2016) undertaken by the Centre for Urban Equity (CUE), CEPT University in Ahmedabad and Guwahati, and Institute for Human Development in Delhi and Patna, and funded by the International Development Research Centre (IDRC), Canada and Department of International Development (DFID), UK, under the global program Safe and Inclusive Cities (SAIC). The research analyzes the pathways through which exclusionary urban planning and governance leads to different types of violence on the poor and by the poor in Indian cities. Other members of the research team were Renu Desai, Shachi Sanghvi, Suchita Vyas, Vaishali Parmar. In particular, I would like to acknowledge Renu's contribution to this research.
13 While the Gujarat State has had alcohol prohibition since 1927 and the selling of alcohol is illegal, in low-income neighborhoods, a health damaging type of local alcohol called "hooch" is brewed and consumed.
14 'PM Narendra Modi launches Smart City projects', The Economic Times, June 25, 2016. Source: https://economictimes.indiatimes.com/news/economy/infrastructure/pm-narendra-modi-launches-smart-city-projects/articleshow/52916581.cms (accessed on October 19, 2017).
15 Source: The proposal downloaded from: https://ahmedabadcity.gov.in/portal/jsp/Static_pages/pi_smartcity.jsp (accessed on June 2, 2018).
16 See http://smartcities.gov.in/content/innerpage/list-of-projects.php (accessed on June 1, 2018).

References

Ahmedabad Municipal Corporation (2014). *Ahmedabad Slum Free City Action Plan*. Ahmedabad: Ahmedabad Municipal Corporation. Available at: www.pas.org.in/Portal/document/PIP%20 Application/Ahmedabad%20Slum%20Free%20City%20Action%20Plan%20RAY.pdf [accessed on May 28, 2018].

AUDA (n.d.). *Revised Development Plan 2021*. Ahmedabad: Ahmedabad Urban Development Authority. Available at: http://deshgujarat.com/2013/02/07/audas-2nd-cty-draft-development-plan-2021-for-ahmedabad/ [accessed on June 24, 2014].

Babar, K. (2017). "Government Announces New PPP Policy for Private Investments in Affordable Housing." *The Economic Times*, September 21. https://economictimes.indiatimes.com/news/economy/policy/government-announces-new-ppp-policy-for-private-investments-in-affordable-housing/articleshow/60777583.cms [accessed on December 15, 2017].

Benjamin, S. (2008). "Occupancy Urbanism: Radicalizing Politics and Economy Beyond Policy and Programs." *International Journal of Urban and Regional Research*, 32(3): 719–729.

Datta, A. (2015). "New Urban Utopias of Postcolonial India: 'Entrepreneurial Urbanization' in Dholera Smart City, Gujarat." *Dialogues in Human Geography*, 5(1): 3–22.

Friedmann, J. (1987). *Planning in the Public Domain*. Princeton, NJ: Princeton University Press.

Friedmann, J. (2011). *Insurgencies: Essays on Planning Theory*. New York: Routledge.

Government of Gujarat (2017). *Report of the Comptroller and Auditor General of India on Local Bodies for the Year Ended 31 March 2016*, Report No. 2 of 2017. Available at: www.cag.gov.in [accessed on January 8, 2018].

Kumar, Y. and Singh, S. (2014). "Rehabilitation of Slum Dwellers: A Case Study in Bhopal." *Social Change*, 44(4): 567–578.

Mahadevia, D. (2015). "Informality and Shelter Security in India." In Mahadevia, D. (ed.), *Shelter Security in Urban India: Pathways, Barriers and Outcomes*. New Delhi: Concept, 1–31.

Mahadevia, D. (forthcoming). "Urbanization, Land and Housing in India: Options amidst the State and the Market Failures." In Wan, G. and Lu, M. (eds.), *Cities of Dragons and Elephants: Urbanization and Urban Development in the People's Republic of China and India*. Oxford: Oxford University Press.

Mahadevia, D., Bhatia, N., and Bhatt, B. (2018). "Private Sector in Affordable Housing? Case of Slum Rehabilitation Scheme in Ahmadabad, India." *Environment and Urbanisation, Asia*, 9(1): 1–17.

Mahadevia, D. and Datey, A. (2015). "From Land Tenure Regularization to Mass Housing in Bhopal: An Assessment." In Mahadevia, D. (ed.), *Shelter Security in Urban India: Pathways, Barriers and Outcomes*. New Delhi: Concept Publishers, 282–324.

Mahadevia, D., Desai, R., Sanghvi, S., and Vyas, S. (2016a). *Ahmedabad Policy Brief 2: Vatwa Resettlement Sites: Basic Services and Amenities: Deprivations and Infrastructural Conflicts, as a Part of Poverty, Inequality and Violence in Indian Cities: Towards Inclusive Polices and Planning*, Centre for Urban Equity, CEPT University, March. Available at: https://cept.ac.in/center-for-urban-equity-cue/policy-briefs

Mahadevia, D., Desai, R., Sanghvi, S., Vyas, S., and Parmar, V. (2016b). *Ahmedabad Policy Brief 1: Vatwa Resettlement Sites: Constrained Mobility and Stressed Livelihoods, as a Part of Poverty, Inequality and Violence in Indian Cities: Towards Inclusive Polices and Planning*, Centre for Urban Equity, CEPT University, March. Available at: https://cept.ac.in/center-for-urban-equity-cue/policy-briefs

Mahadevia, D., Desai, R., Sanghvi, S., Vyas, S., and Parmar, V. (2016c). *Ahmedabad Policy Brief 5: Vatwa Resettlement Sites: Gender Insecurity and Violence Against Women, as a Part of Poverty, Inequality and Violence in Indian Cities: Towards Inclusive Polices and Planning*, Centre for Urban Equity, CEPT University, March. Available at: https://cept.ac.in/center-for-urban-equity-cue/policy-briefs

Mahadevia, D., Desai, R., Sanghvi, S., and Vyas, S. (2016d). *Ahmedabad Policy Brief 4: Vatwa Resettlement Sites: Illicit Activities and Goondagardi, as a Part of Poverty, Inequality and Violence in Indian Cities: Towards Inclusive Polices and Planning*, Centre for Urban Equity, CEPT University, March. Available at: https://cept.ac.in/center-for-urban-equity-cue/policy-briefs

Mahadevia, D., Desai, R., Sanghvi, S., and Vyas, S. (2016e). *Ahmedabad Policy Brief 3: Vatwa Resettlement Sites: Thefts, Robberies and Burglaries, as a Part of Poverty, Inequality and Violence in Indian Cities: Towards Inclusive Polices and Planning*, Centre for Urban Equity, CEPT University, March. Available at: https://cept.ac.in/center-for-urban-equity-cue/policy-briefs

Miraftab, F. (2009). "Insurgent Planning: Situating Radical Planning in the Global South." *Planning Theory*, 8(1): 32–50.

Patel, S., Sluizas, R., and Mathur, N. (2015). "The Risk of Impoverishment in Urban Development-Induced Displacement and Resettlement in Ahmedabad." *Environment & Urbanization*, 27(1): 231–256.

Roy, A. (2009). "Why India Cannot Plan Its Cities: Informality, Insurgence and the Idiom of Urbanization." *Planning Theory*, 8(1): 76–87.

Sandercock, L. (1998). "Framing Insurgent Historiographies for Planning." In Sandercock, L. (ed.), *Making the Invisible Visible: A Multicultural Planning History*. Berkeley: University of California Press, 1–33.

Sandercock, L. (1999). "Expanding the 'Language' of Planning: A Mediation on Planning Education for the Twenty-First Century." *European Planning Studies*, 5(7): 533–544.

Smart Cities Website. Available at: http://smartcities.gov.in/content/innerpage/

Taylor, N. (1998). *Urban Planning Theory Since 1945*. London: Sage Publications Ltd.

Vidyarthi, S., Mathur, S., and Agrawal, S. (2017). *Understanding India's New Approach to Spatial Planning and Development: A Salient Shift?* New York: Oxford University Press.

Watson, V. (2003). "Conflicting Rationalities: Implications for Planning Theory and Ethics." *Planning Theory and Practice*, 4(4): 395–408.

Further reading

Bredenoord, J., van Lindert, P., and Smets, P. (eds.) (2014). *Affordable Housing in the Urban Global South*. London & New York: Earthscan from Routledge. An extensive overview of the facts and trends of current housing policies in Asia, Africa and Latin America.

Mathur, O. P. (Ed.) (2014). *Inclusive Urban Planning: State of the Urban Poor Report, 2013*. New Delhi: For Ministry of Housing and Urban Poverty Alleviation, Government of India, Oxford University Press. A compilation of articles on the state of urban poor in India and on issues around housing, land, basic services, and many others.

Rademacher, A. and Sivaramakrishnan, K. (eds.) (2013). *Ecologies of Urbanism in India: Metropolitan Civility and Sustainability*. Hong Kong: Hong Kong University Press. A compilation of articles that point towards ecological and housing conflicts in Indian cities, and against the urban planning practice in these cities.

Saglio-Yatzimirsky, M.-C. (2013). *Dharavi: From Mega-Slum to Urban Paradigm*. London, New York, and New Delhi: Routledge. This book presents a fascinating paradox: The convergence of stereotypes associated with the slum – poverty and misery – and an effervescent economic vitality, impelled by globalization and international capital flows.

Tripathi, D. (1998). *Alliance for Change: A Slum Upgrading Experiment in Ahmedabad*. New Delhi: Tata McGraw Hill Pub. Co. Ltd. Presents the story of partnerships in the first successful slum upgrading experience named Slum Networking Programme (SNP) in Ahmedabad city.

PART II.VI

Mimesis and simulacra

Introduction

The word mimesis is of mid-16th-century origin derived from the Greek word *μίμησις*, referring to imitation of real life and world in art and literature. Google Ngrams, the digital inventory of term mentions, shows a slow rise in the usage of the term beginning around 1950 and peaking in 2000. Simulacrum is a related concept referring to the representation of a person or an object. Thus, simulacra (plural of simulacrum) involve mimesis. Postmodern theorist Jean Baudrillard (1981), offers a more complex philosophical interpretation of these concepts, arguing that simulacra are essentially hyperreal, as effectively captured in Umberto Eco's (1990) *Travels in Hyperreality* or Orson Welles's (1973) film *F for Fake*, which challenges the viewer to differentiate between the real and the fake.

Lately, the question of what is authentic and what is fake has come to occupy scholars of urban design, as many recently built large-scale urban development projects utilize architectural forms that are essentially simulacra and outcomes of insouciant mimesis. The imaginative simulacra and hyperreality that Disneyland and its cohorts kept contained within their corporate theme park walls strictly for entertainment value have now seemingly escaped into the wider urban space, particularly and lately, in the Global South. The new towns built around Shanghai's urban core – each representing a particular European urban tradition and designed by urban designers from those countries – have become the icons of this extreme insouciance. Bianca Bosker (2013) and Tom Campanella (2008) have documented some of these examples in China; but the examples are not limited to China. They can be found in Bandung and greater Jakarta, in the outskirts of Mumbai, in the desert cities of UAE, Yamassoukro in Ivory Coast, and so on. Seemingly the epidemic of "disneyitis" has spread throughout the Global South. And as Eco's travels in the United States amply demonstrate, mimesis is not unique to the Global South.

But what is wrong with mimesis and simulacra? After all, "we all copy, don't we?" (Piazzoni and Banerjee 2018). Also, what's wrong with choosing a "good copy" over a "bad original"? In her recent book on the Shanghai new towns, urban designer Maria Francesca Piazzoni (2018) interrogates the notion of authenticity in Thames Town, which is designed to look like an English town, and wonders what can be inauthentic if the residents choose to live there with the expectation of achieving the life experiences of the West. Finally, and more significantly, what does it mean for urban design? Should urban designers encourage or oppose simulacra in their professional practice? Should urban designers find a common ground, developing policies and standards for managing mimesis and simulacra in city design? The four essays assembled in this section address these questions.

In his chapter, Scott A. Lukas addresses the issue of mimesis and sees it as an outcome of the "complex intersections of urbanism and consumer society." He argues that urban spaces are constantly being endowed with multiple narratives – stories, brands, concepts – in the remake of the identity of cities, presumably to be competitive in the global marketplace. He draws from the case studies of Berlin, New York, Los Angeles, Orlando, and Copenhagen, showing examples of the narratives of remake, branding, and identity. Mimesis becomes a part of the process and integral to the narrative overlay constructed to create a new city identity.

Li Xiangning continues the arguments of Lukas. In discussing theming of urban spaces throughout the world, in particular China, he notes that the mimesis and simulacra produced are beginning to show a tendency to integrate local cultural artifacts and practices. This is manifested in the construction of new urban spaces of consumerism, where aspects of local urban culture are increasingly integrated in the design of themed spaces

In her chapter, Bianca Bosker focuses on the multiple "true copies" of many iconic buildings and charters from the West. She refers to this phenomenon as "duplitecture" and asks why a country like China with a strong tradition of endogenous architecture would want to look for foreign templates. She wonders also if the tendency for duplitecture has reached its final phase. Bosker revisits some of these famous duplitecture and simulacrascapes to understand how they have evolved over the past two decades, and whether some of these structures have become part of China's ghost towns with unoccupied buildings and spaces. In concluding, Bosker argues that these trends in duplitecture and themed spaces may shed light on more fundamental cultural issues related to China's continuing development.

Finally, the chapter by D. Asher Ghertner reports on a different kind and purpose of mimesis, as is typical in unauthorized developments in Delhi's periphery. Here he discusses how planners create imaginary maps to accommodate buildings already occupied. To obviate demolition, the developers also fake occupancy by installing air-conditioning units, hanging laundry, and so on; thus appearance and aesthetics become a part of the acceptability and legitimization of many of these unauthorized buildings. Similarly, at the infrastructure level, deals are made with the suppliers of water. The nexus of these three aforementioned levels of mimicry undertaken by planners, developers, and water suppliers shapes the appearance and continuity of these fringe developments in Delhi.

References

Baudrillard, J. ([1981] 2017) (transl. by Sheila Faria Glaser). *Simulacra and Simulation.* Ann Arbor: The University of Michigan Press.

Bosker, B. (2013). *Original Copies: Architectural Mimicry in Contemporary China.* Honolulu: University of Hawai'i Press.

Campanella, T. (2008). *The Concrete Dragon.* New York: Princeton Architectural Press.

Eco, U. (1990). *Travels in Hyperreality: Essays.* Harcourt Brace Jovanovich.

Piazzoni, M.F. (2018). *The Real Fake: Authenticity and the Production of Space.* New York: Fordham University Press.

Piazzoni, M.F. and Banerjee, T. (2018). "Mimicry in Design: The Urban Form of Development." *Journal of Urban Design.* 23(4): 482–498.

Welles, O. (1973). *F for Fake.* Produced by François Reichenbach, Dominique Antoine, and Richard Drewitt.

24
BETWEEN SIMULATION AND AUTHENTICITY
The question of urban remaking

Scott A. Lukas

Introduction

In 2014, while conducting research in Berlin, Germany, I had the fortune of taking part in what was billed as a "subcultures tour" of the city.[1] For an urban researcher of contemporary cities, Berlin is a particularly notable site as its many spaces speak of the dialogues, debates, challenges, successes, and failures that are present in contemporary urban spaces and cityscapes. The subcultures tour represented an even more profound highlighting of these, and many other, cultural dynamics. Using descriptors that included "trends," "density," "open space," "underground and cosmopolitan," "urban renewal," "gentrification," "autonomous initiatives," "bizarre shops," and "street art," the tour articulated a view of the city that emphasized sensibilities of loss, challenge, even doubt. It illustrated the notable ways in which Berlin has relied on subcultural and avant-garde tendencies as a means of advertising its "city image" (Lukas 2018). Indeed, any visitor to the city will note that Berlin's status as a world city of culture, politics, and aesthetics is made possible due to the myriad spaces, architectural forms, and material contexts of the city that seem to hold together in uncanny senses. Yet, that same visitor, perhaps participating in a similar tour, might ask the question: to what extent are the cityscapes that I am experiencing moments of an authentic nature, as opposed to simulated unrealities? Walking in major world cities – New York City, Singapore, Shanghai, London, Tokyo – one is confronted with a multiplicity of images, memes, structures, architectural forms, performances, and contexts that, on their surfaces, defy explanation.

The question of what exactly constitutes the "urban" and the "city" is notably perplexing. Louis Wirth (1938), writing in the 1930s, warned of the presumed ease in which we define these terms. While factors that include population size, density, land uses, and infrastructure, among others, could all be said to define the urban and the city, overreliance on any one factor may result in myopic misrepresentation and essentialism in terms of meaning. Additionally, studying a given city space and its context, while analytically useful, may neglect other spaces, contexts, and issues. As Sassen (2013) has argued, a city is complex, diverse, often anarchic, and thus it is a considerable challenge to make the generalizations that are necessary – pedagogically – to an understanding of what constitutes a given city, and how its unique challenges and contradictions should be deconstructed. This difficulty of locating the urban and the city reminds us of the crisis of representation that is similarly at play (Banerjee and Verma 2010; Bender 2002). The city case studies considered in this chapter involve some level of contestation in terms of the urban visions of remaking and the related simulations in the city examples. On the surface, the debate that exists in the continuum of authenticity and simulation appears to be simplistic when, in fact, it complicates the matter of what constitutes a city.

The city as remake

Berlin, especially in the notable ways in which its cityscapes reflect the notion of *Vergangenheitsbewältigung*, or the sense of "struggling with the past," reminds us that it is a city of remaking (Lukas 2018). Sites like the Kaiser Wilhelm Gedächtniskirche, the historic church, which features both a modern spire and a historic one, show the ruinous effects of war[2] but also illustrate the city's ability to locate itself in conflicting temporal eras, and its notable emphasis on remaking. Remaking refers to the condition in which a person, thing, event, space, media form, brand, social institution, or feeling is transformed from one previous state into another state, in any combination of spatial, temporal, physical, technological, geographic, architectural, or existential dynamic. The prefix "re," present in the word, is particularly curious as it connotes "back" or "backwards" but, additionally, offers notions of a "starting point" and a "point reached" (*Oxford English Dictionary*). In the city, remaking may refer to any architectural or material project that creates a sense of the new while consciously, and sometimes unconsciously, referencing what existed before. It is an inherently hybrid, and often contradictory, form.

All cities and all spaces, of course, reflect dimensions of remaking. Yet, the architectural and spatial forms considered in this chapter are of a specific order of remaking that serve as heuristic or pedagogical devices, thus affirming principles or tendencies that may be imagined as "best practices" for future urban and city projects. Remaking, as an architectural or design practice, may take multiple forms. Narrativized space – that which, very generally, includes a foundational story – may be developed using *theming* (place, lifestyle, and other associations may be used to theme the space), *branding* (the brand is deployed as the primary signifier of the space), *heritage and history* (forms of heritage and history are used to orient people in the space), and *functionality* (the functions or utility of the space are the key definers). As a city, Berlin reflects layered narratives of space, often deploying multiple historic, thematic, and branded approaches in the same space. As such, it reflects a significant tendency within urbanism in terms of *eclecticism and pluralism*. Remaking, particularly in how it privileges pastiche, montage, and the palimpsest, reminds of the powerful potentials of density of information and representation, dialogue, and "thickness" in terms of any given space (or combinations of spaces) within an urban environment (Bender 2002: xii, 222; Jencks 1984: 111).

Two spaces of the remaking of Berlin's dark and disturbing past – the Topography of Terror and the DDR Museum – illustrate the curious and divergent courses that eclectic and pluralistic urban design may take. The Topography of Terror is a space that commemorates the war crimes committed under National Socialism, while the DDR Museum chronicles the brutal lives of Berliners during the time of the Stasi (the secret police of East Germany). Both spaces focus on memory and a disturbing past to orient the visitor in space, yet each diverges in terms of the material and theatrical ways in which it delivers context and information. For many critics, the Topography of Terror, in its sober, staid, and less immersive displays and material design forms, represents a more authentic detailing of the dark past; while the DDR Museum, with its over-the-top interactive, immersive, and even "fun" exhibits, is deemed to be inappropriate – an example of what some have called the Disneyization of the architectural and material world (Bryman 2004). Both spaces remake Berlin by retelling and representing a past, yet the disagreement about the appropriateness of the spaces – which reflects a perceived continuum running from the authentic to the simulated – illustrates the dramatic complexities evident in many urban architectural and material forms, regardless of the functions or content of the spaces in question.

The city as ideal

Among the major cities in the world, New York City remains the center of some of the most contentious debates about urban futures. The city is home to Frederick Law Olmsted's Central Park, one of the most significant urban landscape projects in history. It is the site that occupied the many urban musings of Jane Jacobs. It is the location of some of the most significant consumerist architectural projects, whether those of Times Square or of the many amusement parks of Coney Island; and is the location of the 9/11 terrorist attacks, which spawned fierce debates about public and commercial space and memorialization in their aftermath. Like many other cities of its size, New York City has also seen the development of significant forms of consumer and branded spaces, including those of Times Square. In the 1990s, Mayor Giuliani saw the renewal of Times Square as nothing short of a religious crusade that necessitated the "cleaning up" of seedy sex shops, the forcing out of criminal elements, and an overall sanitization of the façades and storefronts of this popular tourist center (Delany 1999).

The so-called revitalization of Times Square ushered in new debates about gentrification and, most notably, the role of branded, consumerist, and themed spaces in major tourist sectors of large cities. It also illustrated the trend of the "ideal city" – a utopian vision in which politicians, city planners, architects, and numerous others develop suggestions for the remaking of city space. Jane Jacobs (1961), while not writing of New York City in such consumerist times, was known for her own idealized vision of the city. Her ruminations on the size of city blocks, the age and height of buildings, and mixed-used spaces inspired many architects and city planners to consider urban spaces as organic entities and to privilege a "we make it what it is" humanism and optimism in spite of urbanism's challenges, controversies, and contradictions. We may deploy Jacobs's enthusiasm for the street life in New York City as a principle that may guide future urban endeavors. Namely, we should recognize the issue of *agency* and imagine the human actor's role – even her or his pleasure – in urban projects. At the same time, agency might be applied to *all* actors in urban space, including the buildings, streets, spaces, and contexts themselves. In contemporary social theory, Actor-Network Theory (ANT) has offered one of the most promising theoretical approaches to more complex understandings of agency and actors beyond human examples (Latour 2005). ANT's constructivist and relational approaches could have merit in terms of the future of urban studies, especially in cases in which certain categories – "authentic," "simulated" – are oversimplified or essentialized in the course of research. Social critics have argued that branded spaces, such as those at Times Square, have the unique potential to stifle the desires or agency of the individual who occupies and uses such spaces. While this is sometimes true, we must also address the "complicated agency" that is implicit in these venues (Lonsway 2016).

One clear concern noted in such criticism of contemporary urban simulacrascapes is the fact that these spaces rely on specific forms of space narrativization[3] (notably, those of a branded or consumer nature) that create an urban homogeneity, a "thinness" in Bender's terms (2002: 222). The other issue is the power and control that is vested in the political, corporate, and social institutions of the city. Giuliani's Times Square remake may thus be seen as the ushering in of both a homogenous branded-consumerist and a quasi-police state order (cf. Davis 1990). Combined, these two orders are countered by an overarching *social justice* focus of the contemporary city. The focus on social justice is one of the most significant trends of contemporary urbanism. Movements against gentrification, emphases on DIY (do-it-yourself) and open-source urbanism, and guerrilla action against surveillance and race- and class-based practices within cities, are all examples of this burgeoning trend (Douglas 2018).

The Museum of Reclaimed Urban Space, located in New York City's East Village, is one of the most curious records of New York City's numerous urban debates and its recent political foci in terms of social justice initiatives. The museum is the work of *Time's Up!*, an environmental activist collective that focuses on sustainable solutions for New York City. Some of the group's most popular initiatives have been a focus on transportation solutions for the city, notably biking, and public space campaigns such as community urban gardens. The museum chronicles the many collective actions undertaken in the city. Most notable in the museum and the group's suggestions for urban praxis is the emphasis on "city hacking."

The focus of *Time's Up!* on DIY or tactile urbanism reflects a significant trend in the remaking of cities worldwide. DIY urbanism locates a pathway through bureaucracy, hierarchical planning, and political and corporate power by vesting design solutions in the hands of everyday city dwellers. City hacking employs guerilla wayfinding, temporary bike lanes, pop-up parks, community gardening, flash mobs, neighborhood repair, or other such tactile urbanism[4] interventions to offer a form of immediacy that challenges the idea that a city is composed of passive inhabitants. DIY actions remind of the philosophical humanism of Jane Jacobs's views of the city and similarly suggest practical applications of the avant-garde principles of the Situationist International in terms of its members' suggestions for urban activism (Sadler 1999). In a praxis-oriented approach, both the DIY movement and the Situationist International orient our philosophical perspectives on urbanism to privilege remaking. What is key in such remaking is not so much that spaces are being remade, but rather the dynamics of their actors and their engagement in the process. The redevelopment of Times Square illustrates that many urban projects all too often eschew public participation of the actors who inhabit and come to enjoy such spaces.

New York City's many architectural and urban design controversies remind us of the need to imbue urbanism with forms of *dialogue*, *participation*, and *community ethos*. Perhaps in reworking the work of Jacobs, Bender, and others who have championed a pluralistic, on-the-ground approach to the city, we note significant challenges in creating community and inclusion in urban practices, big and small. Debates about the rebuilding and development of commercial and memorial spaces near the World Trade Center site, as profiled in the 2012 documentary *16 Acres* (by Richard Hankin), illustrate how challenging such calls for participation can be in terms of the end result of a project. Thus, we may question whether such dialogue and broad-based community participation in city projects come to appear more as ideal than real.

The city as brand

The Jerde Partnership, recognized for its significant consumer and themed projects that include the Bellagio and Fremont Street Experience in Las Vegas, Beursplein in Rotterdam, and World of Coca-Cola in Atlanta, is best known for its controversial project, Universal CityWalk, built in 1993 (Figure 24.1). CityWalk located in Universal City, a suburb of Los Angeles, in many ways, exemplifies the debates on urbanism that have taken place within the larger city. Like many contemporary city spaces, CityWalk uses a peculiar yet evocative approach to signage, lighting, storefront architecture, and theatricality that has led many to consider this entertainment/retail venue as a clear example of the movement of urbanism and city design into the realm of the theme park (Lukas 2008). Architectural critic Paul Goldberger (2001), echoing the previous work of Ada Louise Huxtable (1999), offered the following assessment of CityWalk:

> By now most of you have seen CityWalk, that pseudo-city in Los Angeles produced by Universal, which raises the curious question: is it a city street masquerading as a theme park, or a theme park masquerading as a city street? We are not quite sure, and that is

Figure 24.1 Universal CityWalk.

Source: Scott A. Lukas

> precisely how the designers of this place want it. CityWalk is a perfect metaphor for the city at this moment.

CityWalk resembles a theme park partly due to its proximity to Universal Studios Hollywood. As in the case of CityWalk in Orlando, the Los Angeles version utilizes its relationship to a theme park to govern the design and theatrical aspects of its own spaces. What is shocking to many is less this theming of the retail environment but more the exuberance that the CityWalk environment displays in terms of its celebration of the brand. Walking through Los Angeles's CityWalk, one is able to affirm the critiques of Goldberger, Huxtable, and many others, who have suggested that such garish consumerist and branded design leaves little possibility for authenticity – either in the material sense of the environments in which guests move, or in the cultural, psychological, and existential respects in which guests partake in behaviors that are deemed to be hedonistic and materialistic. At the same time, it is possible to imagine a view of the CityWalk project – and, by extension, similar urban projects that privilege the apotheosis of the brand and consumer orders – that considers these architectural forms and their corollary consumerist behaviors (which often induce pleasure) as "authentic" in their own right (Lukas 2016b, 2016c). While many architectural and urban critics argue that spaces like CityWalk are nothing more than material embodiments of consumer hegemony and examples of the inauthenticity of consumer society, numerous others – including designers and guests – experience these spaces in notably different, and, sometimes, authentic senses (Lonsway 2016; Lukas 2017).[5]

Clearly, the case of CityWalk necessitates rumination on the brand and its relationship to the debates over simulation and authenticity in the contemporary city. As a principle of future urbanism, we may create more dialogue in terms of the *reconciliation of the brand* and *consumerism* within the urban context (cf. Bender 2002: 227). Related to such dialogue, which may take place in formal or informal charrettes related to the spaces, uses, and contexts at hand, we may imagine new understandings of a conceptual nature that are noted in the work of architects like Bjarke Ingels, which will be discussed later in this chapter. One interesting positive conclusion that may be gleaned from urban projects like CityWalk is that the sense of design and context consistency are intrinsic to the holistic impulse common to themed and consumer space.[6]

The city as story

Disney Springs, located in Orlando, Florida, is the most curious of the spaces that are considered in this chapter (Figure 24.2). The project, which began in 2013 and was completed in 2016, represents the reconfiguration of a series of previous shopping and entertainment venues, including Pleasure Island, an unsuccessful adult club and entertainment district previously a part of the Downtown Disney shopping area. Disney Springs represents a form of entertainment remaking. It conflates, rearranges, and redefines four eras: the history of the Orlando land prior to Disney's acquisition, the entertainment history of the shopping area since the 1970s, the spaces of Disney Springs in the present, which connect to the fourth temporal era in which Disney creates a backstory that attempts to explain the evolution of the eponymous fictional town.

As imagined in the Disney backstory, the town includes a settler's cabin and stable of the mid-1800s (which have been transformed into a fast-food restaurant and Disney's guest services center in the present), a town center with Spanish Colonial Revival Style architecture of the 1920s, a marketplace that utilizes American Craftsman architecture of the 1930s, and an exposition center that imagines a previous expo within the town through reclaimed forms of 1950s Googie architecture. Dotting the various architectural and temporal zones of Disney Springs are other historical markers, including a bottling plant repurposed as a Masaharu Morimoto Asian restaurant, a STK Orlando steakhouse made to resemble a historic Orlando train station (complete with an

Figure 24.2 Disney Springs.

Source: Scott A. Lukas

accompanying faux train turnstile), and a created (and nonfunctioning) well machine connected to the springs. The namesake "springs" of the shopping venue is also fabricated – with colorful glass used to create a visual appealing effect. Local environmentalists took offense to Disney's riffing off of authentic (and threatened) Florida natural springs, though Disney retorted that its use of the springs as a metaphor was intended to evoke optimism and show respect for regional ecosystems.

The uniqueness of this shopping district is located not simply in its conflation of multiple historic periods, but in its willingness to remake what was previously a disparate arrangement of nightclubs, restaurants, and retail venues into a fictional town, complete with "historical" backstory (cf. Lukas 2016a). It should be noted that Disney Springs, while not situated within urban density and large-city life, reminds us of the processes of narrativization of space that will increasingly come to define larger cities and urban districts. The ultimate question of the narrativization of space, and specifically whether or not such fictional narratives should be seen as sound urban practices for the remaking of city spaces, is challenging to answer. One clear outcome of the Disney Springs development is its potential conceptual impact on urban design. In terms of future urban projects, it may be beneficial to reorient our focus from *functional* associations to *metaphorical* contexts of space (cf. Jencks 1984: 40). While it may be true that Disney's use of a fictional historic narrative and its metaphorical underpinnings (springs as optimism) was conditioned by the desire to remake and reconnect a struggling and disparate collection of retail locations in an "Disneyfied" approach, it may also be the case that the end result allows for new imagination of future urban spaces that, similarly struggling, make use of metaphor and storytelling as a means of revival.[7]

The city as concept

Copenhagen, Denmark, is one of the most remarkable cities in the world in terms of commitment to sustainability and eco-sensitive design. In 2014, the city was awarded the European Commission's European Green Capital Award, and it has set the goal of becoming zero carbon by 2025. Copenhagen is perhaps the premier example of an emphasis on urban remaking as a result not of desire but of necessity. The crisis of representation, evident in the architectural projects of the contemporary city, is often situated within debates about the material, formal, or aesthetic characteristics of buildings or spaces. The notable concerns of such debates often include contexts of mimesis, particularly as social critics argue (as in the previous case of CityWalk) that architects and designers are engaged in practices of excessive narrativization of space that are inconsistent with the lifestyles and needs of those who inhabit the space. In the case of theming, the narrative of space may be non-indigenous to the people involved (cf. Bosker 2013); while in the example of branded space, the narrative is shallow, hedonistic, and encouraging of consumption; and in the case of functional or abstract (postmodern) space, the narrative is either non-existent or too utilitarian in design and focused on the desires of the architect (or starchitect), not the experiences and perceptions of the people inhabiting the spaces. These concerns about the outcomes of the narrativization of space, while significant, overlook additional pressing issues that will inform the urbanism of the future – notably, the emphasis on a *crisis-adaptive foundation* of contemporary city management: Whether natural disasters, carbon impact, riots, terrorism, crime, gentrification, or social justice issues, the future urbanism may begin to focus on the ability to adaptively remake and create possibilities in the midst of crisis.

Danish architect Bjarke Ingels and his firm BIG have spent considerable time emphasizing a new direction for urban design – one that is consistent with both the crisis-adaptive emphasis and the reconciliation of the disconnect between the narratives of space presented by the architect, designer, or planner and what is experienced by the residents or end users of the space. According to Ingels, urbanism refers to a process of "creating a living city," (Ingels 2009: 151), and nowhere is this definition more evident than in his 2012 project Superkilen. The project is located in the Nørrebro district of Copenhagen and is an example of the potentials of architecture as a form of conceptual remaking.[8] Superkilen is a large multi-zone project that aims to provide revitalization to this district and also serve as a model of inspiration for future cities and urban zones around the world. It includes three zones: a red recreation zone (which is painted red), a black market square (designed to focus on social interaction), and a green zone (composed of hills and trees and focused on outdoor activities). Most importantly, the project was "conceived as a giant exhibition of urban best practices" and includes a collection of objects (exercise equipment from Los Angeles, Chinese palm trees, neon signs from Russia, a Moroccan fountain, and a Turkish bench, among others) from over 60 countries in the world, many of which are represented by the residents who live in this multiethnic community within Copenhagen (BIG 2017).[9] As BIG states, the project "serves as a surrealist collection of global urban diversity that in fact reflects the true nature of the local neighborhood – rather than perpetuating a petrified image of homogenous Denmark" (BIG 2017).

In the debate about the narrativization of space and the question of authenticity in light of contemporary practices of architectural simulation, Ingels's project suggests a considerable direction forward. It goes without question that Superkilen is highly narrativized – one could even say it is "themed." Yet what is so significant about how urban storytelling is undertaken in this Copenhagen example (and in notable contrast with CityWalk and Disney Springs) is how the project creates balance among the cultural and symbolic associations suggested by the many objects and urban features on display throughout its zones, and how the legitimacy of the objects and their inclusion is affirmed by the local residents, not by a particular politician, brand, or consumer order.[10] BIG's work – in this and in so many other spaces and projects – identifies a notable *conceptual play* of space within contemporary urbanism. Beyond the balance of form and function, practices of firms like BIG illustrate a performance, a theatricality of space and architecture, as well as important connections to nature and the environment and issues of diversity and social justice.

As Ingels has said, it is possible to imagine an architecture that is between "naively utopian [avant-garde architecture] or petrifyingly pragmatic [corporate architecture]" (Ingels 2009: 12), an architecture that maintains an emphasis on narrativization, while considering issues of authenticity, including those that have a direct social justice or environmental bearing on those who will interact with the spaces involved. This is perhaps one of the key lessons of future urban architecture and design in that those involved in the spaces will preserve the philosophical meanings inherent in remaking – the going back, the moving forward, the admixtures of forms, all the while emphasizing imminent concerns, sustainability, and agency and the immediate needs of residents.

With these contexts of remaking in mind, we may imagine more ways in which to create future urban spaces that emphasize the agency of the users and consumers who help make architecture and space manifest in each and every movement, use, and context within it. Politically viewing space through attitudes of the social contract – in which the individual gives up something of their claims/rights to the space, such as complete autonomy and agency, but receives something in return from it – suggests a new approach to urbanism that neither privileges the visions of the architect (or starchitect) behind the space nor the immediate uses, functions, or envisioned or predetermined context of the space. Instead, as a *process* (in the meaning of "limen"), the space is imagined as one in which the desires of the user (even as they change over time) are privileged (cf. Lukas 2015). In this sense, the etymology of "city," which has shifted from the person to the place itself, may begin to shift back to the person, her-or himself, and the collective identity (as in "civitas"; *Oxford English Dictionary*).

Social media, apps, augmented reality, and various forms of technology may all be envisioned as servants to the greater existential desires of the individual within the space at hand. In this regard we return, ironically enough, to a classic debate on the continuum of simulation and authenticity – that of the "guest," the anonymous yet significant individual who has been so labeled in the Disney, theme park, and service landscape discourse. In the theme park and themed space literature, the guest is the sine qua non – the archetypal figure – of the ills of consumer society. Critics have imagined that the mere mention of the guest in the design proposals of a space assures that the space will fall victim to the pitfalls of mimesis and narrativization that privilege inauthenticity, hedonism, and consumerism (Bryman 2004; Huxtable 1999). Yet, urban projects of the sort that imagine a meaningful role of the individual within the designed space offer opportunities to locate pathways between the presumed unreality of simulation and the ease of unexamined authenticity in contemporary cityscapes.

Notes

1 See Alternative Berlin Tours, http://alternativeberlin.com. All city case studies discussed in this chapter were developed through ethnographic research conducted in their various spaces.
2 The church was damaged in a bombing raid in 1943.
3 Narrativization refers to the ways in which spaces are imbued with stories, meaning, or symbolism that transcend abstraction. While all spaces have inherent meaning, certain contemporary spaces rely more heavily on meanings that convey plot, temporal or historic setting, and characterization that are typical to fiction and literary forms of artistic expression. This form of space reminds of the absolute versus program distinction in music in which the former is music that is abstract, nonrepresentational, and without story, while the latter is music that is representational and focused on story or forms of narrative expression.
4 Tactile urbanism refers to temporary, low-cost, pop-up approaches to urban design that often eschew planned, corporate, and hierarchical design tendencies of traditional urbanism. Tactile urbanism shares some of the foundations of the DIY and maker movements.
5 Such a move requires a philosophical reorientation of the concept of authenticity. Using the existentialist definition of the concept, we may imagine situations in which a guest's relationship with brands and consumer spaces is mediated by reflexive, self-conscious, even critical foundations.
6 Branded and consumer spaces like those of CityWalk suggest an order that is represented by the material, cognitive, and even existential consistency that is found in the overall site plan of the space. Of course, the nature of such consistency is part and parcel of the critiques of these same spaces offered by architectural and cultural critics.
7 This topic is especially relevant to the decline of brick-and-mortar retail in the United States. Storytelling, narrative development, and the use of experiences within such spaces could result in greater competitiveness with online retailers (Pine and Gilmore 1999).
8 The project included collaboration with Superflex and Topotek1.
9 A similar concept was attempted, in a much more subdued sense, at the Tribune Tower in Chicago, Illinois. Over 150 fragments of the world's great architecture are embedded in the walls of the Tribune Tower. Most people are unaware of the collection at the famous Chicago building, illustrating its purely aesthetic focus that contrasts with the multicultural emphasis of the collections at Superkilen.
10 The balance of the spatial zones is reminiscent of Ingels's notion that architecture should involve mutation and cross-breeding, and that it should evolve with life itself (Ingels 2009: 14).

References

Banerjee, T. and Verma, N. (2010). "Probing the Soft Metropolis: Third World Metaphors in the Los Angeles Context." *Planning Theory and Practice* 2(2): 133–148.
Bender, T. (2002). *The Unfinished City: New York and the Metropolitan Idea.* New York: The New Press.
BIG. (2017). SUK, Superkilen. Online source.
Bosker, B. (2013). *Original Copies: Architectural Mimicry in Contemporary China.* Honolulu: University of Hawai'i Press.

Bryman, A. (2004). *The Disneyization of Society*. London: Sage.
Davis, M. (1990). *City of Quartz: Excavating the Future in Los Angeles*. London: Verso.
Delany, S. (1999). *Times Square Red, Times Square Blue*. New York: New York University Press.
Douglas, G. (2018). *The Help-Yourself City: Legitimacy and Inequality in DIY Urbanism*. Oxford: Oxford University Press.
Goldberger, P. (2001). "Cities, Place and Cyberspace." Lecture, University of California, Berkeley, February 1, 2001. Retrieved online.
Huxtable, A. (1999). *The Unreal America: Architecture and Illusion*. New York: The New Press.
Ingels, B. (2009). *Yes Is More: An Archicomic on Architectural Evolution*. Cologne, Germany: Taschen.
Jacobs, J. ([1961] 1993). *The Death and Life of Great American Cities*. New York: Modern Library.
Jencks, C. (1984). *The Language of Post-Modern Architecture*. Fourth edition. New York: Rizzoli.
Latour, B. (2005). *Reassembling the Social: An Introduction to Actor-Network-Theory*. Oxford: Oxford University Press.
Lonsway, B. (2016). "Complicated Agency." In Lukas, S. (ed.) *A Reader in Themed and Immersive Spaces*. Pittsburgh: ETC. 239–248.
Lukas, S. (2019). "Postmodern Spaces of Berlin: Beyond the Confines of History, Self, and the Real." Forthcoming in Avitts, E., Noone, K. and Robinson, J. (eds.) *Thinking Through Things*.
Lukas, S. (2017). "Keeping It Real (on authenticity)." *Attractions Management* 22(1): 90–94.
Lukas, S. (2016a). "Time and Temporality in the Worlds of Theme Parks." In Carlà, F., Freitag, F., Mittermeier, S. and Schwarz, A. (eds.) *"Here You Leave Today": Time and Temporality in Theme Parks*. Hanover: Wehrhahn. 19–41.
Lukas, S. (2016b). "Should Architecture Be Entertaining?" *OAA Perspectives: The Journal of the Ontario Association of Architects* 24(3): 22–23.
Lukas, S. (2016c). "Judgments Passed: The Place of the Themed Space in the Contemporary World of Remaking." In Lukas, S. (ed.) *A Reader in Themed and Immersive Spaces*. Pittsburgh: ETC. 257–268.
Lukas, S. (2015). "Thresholds and Themed and Immersive Spaces." *OAA Perspectives: The Journal of the Ontario Association of Architects* 23(4): 28–29.
Lukas, S. (2008). *Theme Park*. London: Reaktion.
Oxford English Dictionary. Online. www.oed.com
Pine, B. and Gilmore, J. (1999). *The Experience Economy: Work Is Theater and Every Business a Stage*. Boston: Harvard Business School Press.
Sadler, S. (1999). *The Situationist City*. Cambridge: MIT Press.
Sassen, S. (2013). "Does the City Have Speech?" *Public Culture* 25(2): 209–221.
Wirth, L. (1938). "Urbanism as a Way of Life." *The American Journal of Sociology*, 44(1): 1–24.

Further reading

Ingels, B. (2009). *Yes Is More: An Archicomic on Architectural Evolution*. Cologne, Germany: Taschen. Ingels's "archicomic" represents a very curious musing on his firm BIG's many urban projects. Most notable is the architect's conceptual and pragmatic approach to urban design.

Sassen, S. (2013). "Does the City Have Speech?" *Public Culture* 25(2): 209–221. This short piece provides a very provocative assessment of the nature of urbanism. Sassen identifies the inherent complexities of the contemporary city.

Wirth, L. (1938). "Urbanism as a Way of Life." *The American Journal of Sociology* 44(1): 1–24. Louis Wirth's study of urbanism remains a classic. While written in the 1930s, Wirth considered the foundations of urbanism and city life that are still significant today.

25

GLOCALIZING THEMED SPACES

The creation of urban spaces in China, Asia, and other regions

Li Xiangning

Introduction: characteristics of consumer culture

> The whole world is passed through the filter of the culture industry. The familiar experience of the moviegoer, who perceives the street outside as a continuation of the film he has just left, because the film seeks strictly to reproduce the world of everyday perception, has become the guideline of production. The more densely and completely its techniques duplicate empirical objects, the more easily it creates the illusion that the world outside is a seamless extension of the one which has been revealed in the cinema.
>
> *Adorno and Horkheimer (2000: 95), "The Culture Industry"*

A century ago, sociologist and philosopher Georg Simmel astutely captured the state of mind of metropolitan residents transitioning to a modern way of life. Looking back at the evolution of urban societies around the world during the 20th century, Simmel essentially depicted the course of development that would take place in the century following his death: money, or a type of commodity-based economic model, would come to permeate every nook and cranny of urban life, altering, profoundly and utterly, the value orientations of modern metropolises and their residents.

The progression of contemporary consumer culture can be broken down into several aspects. First is the shift from the functional value of objects as consumer goods to their symbolic value as status symbols. Commodities, as cultural symbols, are bundled together with additional relevant information and serve to hint at this latent information. When we choose to purchase a specific commodity or service in a specific consumer venue, in a specific area, this decision is often made based upon a specific, definite value along the identity-status-class spectrum that is attributed to that particular area, venue, or commodity. In other words, consumers compare the cultural connotations of possible alternatives and select whichever they feel is most suitable to their own personal characteristics.

A second characteristic of consumer culture is its obsession with "images." Western society has had a tradition of ocularcentrism, privileging vision over other senses, since the days of Ancient Greece. But as explained by Walter Benjamin (2008 [1935]) in his 1935 essay, when the world entered the "Age of Mechanical Reproduction," the large-scale reproduction of cultural products started transforming objects and ideas into independent images. People no longer have the patience required to read written text, embracing instead the directness and clarity of visual media – but at the expense of depth. They are more focused on the images covering an object's faces than the actual function or content of the object itself. At the same time, postmodern architectural buildings are often reduced to the two-dimensional image of their façade.

A third characteristic of consumer culture is the revival of "aesthetics" and the process of "aestheticization." In the 1960s, the rise of structuralism undermined the position of aesthetics in the field of philosophy, but since the 1990s, postmodern culture has offered new support for the study of aesthetics, and "art" has become an ubiquitous concept. Since the time of Duchamp, modern art has come to incorporate every aspect of daily life into the realm of aesthetics, resulting in the phenomenon of "aestheticization" – the decoration and beautification of everyday environments, buildings, and even people themselves. For Guy Debord (1995: 26) images were the ultimate form of commoditization. As he reasoned:

> Here we have the principle of commodity fetishism, the domination of society by things whose qualities are "at the same time perceptible and imperceptible by the senses." This principle is absolutely fulfilled in the spectacle, where the perceptible world is replaced by a set of images that are superior to that world yet at the same time impose themselves as eminently perceptible.

These postmodern phenomena, which arise, fundamentally, from the logic of commodity society, are directly reflected in the design and creation of urban morphologies, urban spaces, and urban cultures. After all, urban construction is ultimately decided by economic factors and is thus fated to bear the brand of commodity and consumer cultures.

Theming and its evolution

The evolution of the trends described in the previous section has engendered an increasingly widespread cultural phenomenon: the theming of urban spaces. Themes and symbols have become important means for the promotion of commodities. Those of us living in metropolitan areas have become surrounded by themed restaurant chains, such as McDonald's, and theme parks, such as Disney World. Related to the concept of theming is the widespread practice of "theme-parkism," and the appearance in the English language of the terms "McDonaldization" and "Disneyfication" – corresponding to two of the most classic examples/settings of theming (Chaplin and Holding 2002: 188).

The culture and tourism industries have become increasingly influenced by the theming phenomenon. As the urban/rural boundary begins to blur, theming has seeped into all aspects of daily metropolitan living: on one hand, we exist amidst an ocean of advertisements, posters, symbols, themes, and motifs; on the other hand, the outer façades and inner spaces of many structures are constantly renovated to keep up with current fashions and changing consumer mentalities. The various media pervading our world – or, one might say, the culture industry as a whole – is steadily providing us with new cultural codes for commodities and consumption. Gourmet cuisine, bustling commercial districts, folk festivals, silhouettes of urban landmarks, even celebrities – all are enveloped in a rich cultural atmosphere that transforms them into codes to stimulate and attract tourism and consumption.

Since the 1990s, theme parks and other themed spaces have received increasing attention from urban researchers. Michael Sorkin (1992), Mark Gottdiener (2001), and other scholars have studied and criticized this phenomenon in great depth in their works. But it can be argued that themed spaces are not new but have been encountered throughout history – from ancient cities to modern urban spaces. However, globalization and global economic progress have resulted in the replication and proliferation of these spaces, making theming a universal phenomenon that can be observed anywhere in the world. Indeed, contemporary cities present us with a diverse, never-ending procession of themed spaces that bridge a vast range of scales, from display nooks in private residences and small themed restaurants, to themed urban commercial districts and even entire themed cities.

In fact, the various kinds of themed spaces seen around the world today have converted the spatial and cultural significance of the age of globalization and information into fragmented formal codes, which can be tweaked and implemented according to the logic of commerce and consumption and transplanted to any corner of the world. These codes might serve as a footnote to Foucault's "heterotopia" concept – in China, and other Asian countries, as well as in other locations around the world, themes originating from distant cultures ("the Other") have been applied indiscriminately, fusing with local cultures into a "glocal" cultural amalgam that drives the construction of a wide variety of themed spaces. This can be seen as the joint product of two influential factors: the creative reconstruction of culture by capital, and the resistant discourse of local everyday life.

One finds the predecessor of the modern theme park in the World Fairs and Expositions, hosted by different cities from the mid-19th century onward and designed to showcase national and city achievements. Before the appearance of modern recording technologies (photography, videography, etc.), experiencing a particular place or venue required observers to travel there personally and employ their own senses. Thus, knowledge of world affairs was inseparable from specific venue experiences. But modern society invented a brand new type of spatial experiences that revolutionized people's perception of distance. Starting from the 19th century, World Expositions that often included a simulacra of far-away places, have altered people's understanding of the world. All it now took to learn about something on the other side of the planet was a visit to a nearby Expo, instead of an arduous and expensive trek across the globe. This enormous conceptual change gave rise to a brand new industry: the "experiential industry" mentioned by Alvin Toffler (1980) in *The Third Wave*. World Expositions became instantly popular, attracting unprecedented numbers of visitors, a pattern that seems to continue: Expo 2010 Shanghai China set new records for both the number of participating countries and the number of visitors.

But Expos were not the only venues that sought to boost the visitors' experiences and fantasies. A superb example of success in the experiential industry, Disney World, employs a "film set" formula to take the stories we have read in books and seen in movies – together with travel experiences – and bring them to life in a theme park environment. This allows visitors to experience scenes

> that represent settings of the past, of a different land, and even of the future – all in a slightly edited, miniaturized, or exaggerated form. Disneyland, for example, offers a quaint but sanitized New Orleans square, a seveneighths scale rendition (to create a subtle "Gulliver" effect for the visitors) of a U.S. Main Street, and a Tomorrowland, a Frontierland, and a Fantasyland, playing upon our fascinations with technology, wilderness, and childhood romanticism. Even the everyday urban world becomes the object of emulation.
>
> *(Loukaitou-Sideris and Banerjee 1998: 256)*

At the same time, various aspects of everyday life have also become objects of imitation. For example, the Disney theme parks in Tokyo and Paris include models of local scenic sites within the park grounds. Shanghai Disney Resort, which opened for business in 2016, took this trend even further, incorporating traditional Chinese architectural styles into the park grounds and actively incorporating representative characters from China's cartoon culture. Similarly, Tobu World Square, a theme park in northwest Tokyo operated by the railway and department store giant, does an excellent job of integrating entertainment and experiences: the park contains scale models of famous architecture from around the world, and the selling point highlighted in the park's advertisements is that visitors can enjoy the experience of world travel and shopping without ever leaving Japan. An even more extreme example is Huis Ten Bosch, a "Netherlands village" located in Nagasaki, where the residents walking on the streets and working in the shops and restaurants are real-life Europeans. Japan also has similar "English villages" and "Austrian villages." An analogous situation can be seen at two theme parks in Shenzhen, Window of the World and Splendid China Folk Village, and at folk culture parks in other Chinese cities, where staff wear a variety of different ethnic costumes of indeterminate eras and seek to attract tourists with ethnic music and dance performances. Shanghai's Jin Jiang Hotel has begun to employ staff wearing red turbans reminiscent of Indian patrolmen, giving guests the illusion that they have walked back into the old Shanghai of the 1920s and 1930s. Theming can even be found in the Southeast Asian sex industry, where venues employ themed spaces, such as imperial bedchambers, and workers wear themed clothing, such as school uniforms or nurse outfits, to satisfy the fantasies of clients. But theme parks have been accused of blurring the lines between the real and the artificial and becoming synonymous with falsity, affectedness, and superficiality, which is why Sorkin (1992: 67) condemned them as an affront to taste: "Taste is the battleground of history, and theming tests the limits of taste."

Themed spaces

But Disneyland, Disney World, and the other theme parks that have popped up in different countries are not the only themed spaces. Another classic themed space is the shopping mall, which first appeared in its prototypical form – commercial structures known as galleries or arcades – in Europe during the 1830s and 1840s. Walter Benjamin (2002 [1927]) explored these structures in great depth in his work. Galleries were some of the most attractive commercial venues in consumer metropolises such as Paris, precisely because their function was not limited to consumption. More importantly, they provided an unrestricted location where urbanites wandering through the city could sightsee, experience, and observe.

Since the 1970s, shopping malls have served a similar function in our contemporary world, becoming a representative example of the consumer culture of modern metropolises. Shopping malls are large, complex venues offering a comprehensive set of services, including parking, shopping, dining, entertainment, and services. First originating in the United States, this commercial model has spread across the globe. Today, shopping malls have become a common fixture in many countries around the world. Shanghai has constructed the largest mall in East Asia, Super Brand Mall, and Beijing purportedly has plans to construct the world's largest mall in the near future. The architecture of the mall often assumes thematic qualities, seeking to appeal to, attract, and entertain consumers to come and spend their money on its premises.

The unchecked spread of the shopping mall model and the rapid development of the aviation industry has led to the emergence of a new type of themed space in airports around the world (as well as in many railway stations). A dramatic increase in the number of passengers and a greater incidence of layovers and delays has turned airports into new venues for encouraging consumption. As dining, shopping, entertainment, and other businesses have rushed to set up shop in these new venues, airports have become remarkably similar to yet another form of themed shopping center. The entire bottom level of Guangzhou Baiyun International Airport, for example, is one enormous supermarket. A narrow staircase and an inconspicuous sign reading "Departure Gates

Upstairs" are obscured by piles of snacks and fruit and display racks holding souvenirs, leaving travelers entirely puzzled as to whether or not they have actually entered the airport at all.

In much the same way, cultural institutions such as art galleries and museums have found it difficult to stave off the encroachment of theming. One classic example of this is the Guggenheim Museum in Bilbao, Spain, a stunning creation by superstar architect Frank Gehry, that helped develop the Guggenheim brand image and jump-start Bilbao's economy. Media packaging has allowed Gehry and the Guggenheim Foundation to transcend their roles as architect and public art institution and become powerful cultural resources, leading some media outlets to draw a comparison with McDonald's themed franchises, referring jokingly to the "McGuggenheim" brand. A recently announced project by the French Centre Georges Pompidou aims to create an outpost in Shanghai, using a local museum building designed by architect David Chipperfield.

Historic conservation districts have, unfortunately, become yet another form of themed space. Unprecedented levels of interest in history and historical awareness of the past have forced urban development and renovation projects to respect historical locations, and city conservation movements have become a panacea for urban transformation and the revival of city centers. However, historical heritage is often misused and reduced to the lowest common denominator, becoming a rallying cry and a catalyst for commercial consumption and losing sight of true historical facts. Perhaps overemphasis on heritage indicates a society's lack of confidence in the future and an inability to come up with more revolutionary solutions. While historians denounce theme parks for distorting history, how many of the countless historic conservation areas and conservation projects have truly preserved the original state and lively condition of their sites, and how many have deteriorated into "local history" theme parks? The tourism industry has become the greatest beneficiary of historical conservation projects. Today, when cities throughout China are scrambling to apply for world cultural heritage recognition, how many decision makers are concerned with the benefits of rediscovering history that such recognition would bring to local residents, and how many simply see this as a strategy to drive their city's economic development? Far too often, ancient cultural sites that successfully apply for world cultural heritage recognition or that become the target of conservation and renovation projects meet a swift end under the relentless footfalls of a massive influx of tourists.

Contemporary technological progress and the infatuation with "innovation" seen in many Asian countries, including China, have resulted in yet another type of themed space – a diverse collection of "incubation parks" for technology and innovation industries. From California's Silicon Valley to Shanghai's Zhangjiang Hi-Tech Park, multinational tech companies are setting up headquarters that could have come straight out of a Hollywood science fiction film, creating "futuristic technology" themed spaces with the use of flashing LED screens, VR equipment, and even robots that can lead tours, give introductions, and answer questions in the place of human guides (Figure 25.1). An even more ridiculous amalgamation can be found at Huawei's R&D base in Dongguan (Figure 25.2), where a Japanese planning and design company set up a space modeled after a quaint European town for the technology giant – who has declared that it will become the world leader for future communications. Such a seamless marriage of a historically European urban space with the theme of futuristic technology has produced a classic example of contemporary themed spaces that indiscriminately combine postcolonial cultural imagination, commercial identities, and the malleable cultural tastes of potential middle-class consumers.

Finally, we must turn back and consider the city itself. City spaces (or even cities themselves) may have already become their own theme parks – the largest of all theme parks. Los Angeles, the typical case study for postmodern urban research, is a city with an exceptional diversity of cultures and races. Each of the city's subcultures has formed its own communities and living spaces, where members can partake in shared cultural themes and symbols. The cultural and

Figure 25.1 Alibaba is testing its robots to facilitate restaurants and hotels with the theme of Artificial Intelligence, where robots will deliver most services.

Source: Author

Figure 25.2 The campus of Huawei's new headquarters under construction.

Source: Author

social phenomena of many other cities is reflected in Los Angeles. To urban researchers such as Michael Dear (2000) and Edward Soja (1992), Los Angeles is the aggregation of a set of theme parks, incorporating many different kinds of privatization and spatial division to adapt to all tastes, and the residents living in these embellished, fantasy-filled sites represent the world's ultimate consumers.

Sixty-five hundred miles west of Los Angeles, Shanghai began repositioning itself historically in the late 20th century. The city cast aside elements of traditional Chinese culture – of which

Longhua Temple and Yuyuan Garden are representative examples – and set its sights firmly on the heydays of the 1930s, when Shanghai was known as the "Paris of the East." This era became a code for historical nostalgia and consumer consumption, as Shanghai confidently authored a new cultural identity tinted with a fantasy aura spanning time and space.

As pointed out by Jane M. Jacobs (1996), the term "postcolonial" misleads one into thinking that the days of colonialism are long gone, but in reality we are still inextricably linked with colonial forms. Colonial constructions are not only part of a history that has been rejected today; they are also part of a history that is constantly looked back upon with nostalgia and employed for creative purposes today. Shanghai's semi-colonial history has become part of the city's proud heritage: structures built on the Bund by foreign banks, representing occupation by foreign forces, have become the most distinctive part of Shanghai, its "World Architecture Expo." The focus of Shanghai's architectural history research and city preservation efforts, consciously or not, all refer to a glorious period at the start of the previous century, when Shanghai was the largest metropolis in the Far East, with a prosperity on par with New York or Paris. As part of a new urban development plan, the Shanghai municipal government has come up with its "one city, nine towns" program, in which the urban space characteristics and architectural styles of various European nations will be used to "add flavor" to each of the nine towns. Today it would seem unthinkable to select a traditional Chinese roof style for an architectural structure in Shanghai, but there always seems to be justification from Shanghai's early modern history to support the burgeoning "European style" of architecture. Shanghai has chosen to emphasize one period of history and becloud others, so that it can restyle its cultural identity as the "international metropolis" of yore.

As the wave of globalization carries us forward, the scenic sites of metropolises have become increasingly dominated by the inexorable logic of commodity-based economies. Church steeples and the spires of municipal government buildings, once the principal features of urban scenery, have given way to the high-rise office buildings of international financial institutions. We can see an even more symbolic example in Shanghai: in the 1990s, the Eastern Orthodox church on Xiangyang Road was converted for use as a stock exchange; then in 2002 it was sold to a new owner and became a Hawaiian-themed restaurant, *Aloha*. President Putin's visit to China added another twist to the building's fate, as the restaurant was replaced by an art gallery displaying Russian paintings. If the original renovation in the 1990s transformed a site for religious ceremonies into an altar for mankind's current gods – money – then the 21st-century transition to an exotic foreign restaurant represents the domination of commodity-based economies over urban life. This domination is no longer the direct manipulation of finance and economics, but rather is achieved through culturally packaged consumption and aestheticized capital operation. This thematic progression from religion to economics to culture may echo the changing tune of Shanghai's development.

Conclusion

In conclusion, I will argue that perhaps architecture and urban design have not yet learned from postmodern thought how to truly understand and value culture, how to respect history, and how to serve the public interest. Instead, they have lost themselves under the swinging batons of market economics and the operation of capital. But the forms they create do not as much meet the consumer tastes of the general public, as they maximize benefits and profits to commercial and real estate interests; for it is within these commercial structures and urban spaces, prepared by architects and urban engineers, that the general public becomes entranced by an illusion born from commerce and culture, and begins purchasing and consuming fervently. Theme parks and other themed urban spaces provide the physical settings for this phenomenon.

Architecture and urban planning experts often believe that they are the actors altering or creating their city's physical spaces, but in reality, the global capital controls the design and construction of cities – even treating cities themselves as commodities that can be sold or exchanged, and using advertisements and media packaging to promote and market entire city spaces. Urban culture as we imagine it has, in fact, become merely a tool by which capital creates profits. The public administration of contemporary cities is becoming more and more akin to private economies, following operating models similar to those of businesses. Municipal government institutions have placed economic gains at the core of their policies, encouraging tourism and consumer spending through the conservation and renovation of historic districts – or, as in major Western cities, city center revival plans – or jumpstarting the city's economic development with entertainment and sightseeing activities, such as expositions and carnivals. Entire urban settings have become increasingly themed, as the dreams and demands of consumers regarding the culture of the metropolis are satisfied with visualized, symbolized, and superficial façades.

References

Adorno, T.W. and Horkheimer, M. (2000). "The Culture Industry: Enlightenment as Mass Deception," in Malcolm M., Hall,T. and Borden, I. (eds.), *The City Cultures Reader*, London and New York: Routledge,

Benjamin, W. ([1927–1940], 2002). *The Arcades Project* (translated by H. Eihlin H. and K. McLaughlin), Cambridge, MA: Harvard University Press.

Benjamin, W. ([1935] 2008). *The Work of Art in the Age of Mechanical Reproduction*, Penguin Great Ideas, Mass Market Paperback.

Chaplin, S. and Holding, E. (2002). "Addressing the Post-Urban: Los Angeles, Las Vegas, New York," in Leach, N. (ed.), *The Hieroglyphics of Space: Reading and Experiencing the Modern Metropolis*, London and New York: Routledge.

Dear, M.J. (2000). "Introduction," in *The Postmodern Urban Condition*, Oxford: Blackwell.

Debord, G. (1995). *The Society of the Spectacle* (translated by D. Nicholson-Smith), New York: Zone Books.

Gottdiener, M. (2001). *The Theming of America: Dreams, Media Fantasies, and Themed Environments*, 2nd Edition. Boulder: Westview Press.

Jacobs, J.M. (1996). *Edge of Empire: Postcolonialism and the City*, London: Routledge.

Loukaitou-Sideris, A. and Banerjee, T. (1998). *Urban Design Downtown: Poetics and Politics of Form*, Berkeley: University of California Press.

Soja, E. (1992). "Inside Exopolis: Scenes from Orange County," in Sorkin, M. (ed.), *Variations on a Theme Park: The New American City and the End of Public Space*, New York: Hill and Wang, 94–122.

Sorkin, M. (ed.) (1992). *Variations on a Theme Park: The New American City and the End of Public Space*, New York: Hill and Wang.

Toffler, A. (1980). *The Third Wave*, William Morrow edition.

Further reading

Lukas, S. (2016). *The Immersive Worlds Handbook: Designing Theme Parks and Consumer Spaces*, Burlington, MA: Focal Press. Focusing on the immersive experience and applied techniques for the creation of theme parks, this design handbook provides refreshing insights into the trends in consumer spaces.

Saunders, W. (ed.) (2006). *Commodification and Spectacle in Architecture*, Cambridge, MA: A Harvard Design Magazine Reader. This book is a theoretical reflection on the Bilbao effect in contemporary architectural culture, and it discusses various possible positions and alternatives related to consumer capitalism.

Sorkin, M. (Ed.) (1992). *Variations on a Theme Park: The New American City and the End of Public Space*, New York: Hill and Wang. Using case studies of various themed public spaces, this book is one of the landmark publications on the topic of theme parks and its application in American cities.

Li, X., Li, D., and Jiang, J. (2017). *Made in Shanghai*, Shanghai: Tongji University Press. This book showcases landmarks, infrastructure, everyday places and eccentric buildings of Shanghai, and tries to illustrate the hidden urban DNA of the legendary city of our time.

26

TRAVELS IN DUPLITECTURE

Bianca Bosker

Introduction[1]

Hangzhou's Paris was supposed to be empty. The development, Sky City, had been built in 2006 to house 10,000 people in a community modeled after France's capital, complete with its own Eiffel Tower, Champs-Élysées, and white Haussmann-style apartments (Figure 26.1). Yet word was that hardly anyone had moved in: It was too far, too inconvenient, too weird. In 2013, a videosurfaced showing Sky City's long boulevards empty of life and its Eiffel Tower choked with weeds, and news sites generated more than 60 copycat stories declaring Sky City a failure. They described the clip as a rare glimpse at China's "eerie" (Boissoneault 2013), "abandoned" (Miller 2013), and "post-apocalyptic" City of Light (Furdyk 2016). "Paris, now virtually a ghost town – streets empty, stores vacant," repeated a 2016 *Nightline* story, panning over bleak gray plazas populated only by fountains copied from Parisian gardens (Woodruff et al. 2018).

Figure 26.1 Sky City: A view of Hangzhou's Sky City, which boasts replicas of the Eiffel Tower, the Fountain of Apollo in Versailles, and the Fontaine de l'Observatoire in Paris, as well as carefully-manicured parterre gardens, all visible here.

For many, Sky City's demise was an "I told you so" moment. I have spent the past decade tracing China's "duplitecture" – the replica White Houses, Versailles Palaces, and even foreign cities, from Cambridge to Jackson Hole, Wyoming, that have multiplied throughout the country (Bosker 2013a, 2013b). From the very start of my research, planners and architecture critics have assured me the movement was on its last legs. "It's really just a trend and it's not sustainable," an architect at Ben Wood Studio Shanghai said in 2008. Soon after, a different Shanghai-based planner asserted duplitecture was "already outdated even within China."

Sky City became the poster child for other themed developments that had allegedly met the same fate: intended to house Chinese families in surroundings inspired by Orange County or Barcelona, these communities were said to have languished as ghost towns. An op-ed in the *Global Times* asserted, "These 'fake cities' are just so ridiculously similar to their Western originals that rather than anyone taking them seriously, they turned into residential amusement parks" – empty backdrops for wedding photos and tourist selfies" (Kibaroglu 2017).

Then again, reporting on Chinese culture has a tendency to turn into a game of telephone, with journalists too often regurgitating stories based off scraps of information obtained third-, fourth-, or fifthhand. That 2013 video of an allegedly abandoned Sky City, for example, was in fact filmed in 2008, as parts of the development were still under construction. Thus, when a documentary filmmaker who had read my book *Original Copies* (Bosker 2013a) invited me to join him in revisiting these duplitecture developments, some of which I hadn't seen in years, I leapt at the chance to check in on them firsthand. Had they been abandoned? Remodeled? Razed to the ground? After all, Liaoning's Holland Village –which installed windmills, canals, and a double of The Hague on an area three times the size of Brooklyn's Navy Yard – had been demolished ten years after its construction (Canaves 2009). Sky City had just celebrated its own tenth anniversary – it was now the same age Holland Village had been when it was razed. In May 2017, I ventured to the suburbs of China to see how its duplitecture had evolved.

La belle Chine

For nearly two decades, adventurous tourists in China have been able to complete a version of the classic Grand Tour without ever leaving the People's Republic. In Shanghai alone, you could start in Italian-inspired Città de Pujiang, continue to Holland Village and Spanish-style Fengcheng, then hit the "Scandinavia of Shanghai," Luodian Town, followed by British-themed Thames Town, a development where winding lanes open onto Eaton Square-style apartments patrolled by security guards in the red uniforms of the Buckingham Palace soldiers (Figure 26.2). Proceed south to Hangzhou, and you might have breakfast in Paris, then lunch in Water Town (Figure 26.3), where human-made canals flow beside pastel townhouses and a remake of the Piazza San Marco. Austria's Hallstatt, California's Orange County, and Manhattan's Midtown have all been doubled in China as part of the nation's duplitecture movement, which has seen the greatest hits of Western architecture replicated at a speed, scale, and scope that is unprecedented in history. "Replication, replication, replication" has supplemented "location, location, location" as the planning mantra of the day.

Developers have taken pains to ensure their replicas will appear realistic. Some expressly hire foreign architects whose nationality matches that of the architecture the development will emulate – Beijing's Orange County, for example, was designed by a team from Southern California. Other developers dispatch local architects to study the original firsthand, on location. Duplitecture creators might also import building materials from abroad (Palais de Fortune's

Figure 26.2 A couple poses for photographs in front of Thames Town's church, a brick-for-brick replica of the Christ Church in Bristol, England.

Figure 26.3 A view of Hangzhou's Venice Water Town, which was built on a series of manmade canals.

villas boast chandeliers from France), borrow foreign place names (Thames Town has an Oxford Street, plus a Gower Street), and recreate famous landmarks (Luoyang's Hallstatt has a picture postcard-perfect twin of the Austrian village's Central Square Marktplatz). Yet these duplitecture developments are not Las Vegas writ large. Though some do double-duty as tourist attractions, they are not theme parks. They were designed to create living, breathing communities in which thousands of Chinese citizens would raise families in surroundings mimicking the West's most iconic sites.

While these designs may be foreign, forces that are distinctly Chinese fueled duplitecture's rise. Where American culture, for example, fetishizes originals and rejects copies as illegitimate "fakes," traditional Chinese belief systems have encouraged a more nuanced and permissive stance. Replication is not necessarily a sign of failure, but can be a marker of skill, talent, and achievement. The American view casts copycats as unimaginative thieves; the Chinese perspective allows that a gifted replicator can be worthy of celebration and prestige. This cultural backdrop, percolating through abundant global capital, has arguably created an environment more permissive of replication. Not long ago, the director of China's National Copyright Administration praised copies as a manifestation of "cultural creativity" (Schmidle 2010).

More practical motivations have also underpinned the duplitecture boom. Imitating can be faster and cheaper than innovating, and many developers, racing to keep pace with China's breakneck urbanization, found it more efficient to adopt ready-made models – ones that had a built-in brand to boot. Still others saw an opportunity to learn from the West's expertise: in explaining its 2001 decision to commission a British firm to build Thames Town, the Shanghai Urban Planning Bureau cited the opportunity to "draw on the successful expertise of foreign nations" (Bosker 2013a). These Western-style communities got an additional boost as real estate firms found that giving their communities a distinctive – and specifically European – theme helped spur sales. A statistical analysis of housing prices in China, published in a 2017 working paperfrom the Lincoln Institute of Land Policy, found that merely incorporating "Napa," "Rhine," or "Hawaii" into a development's name made a difference: the researchers concluded that "most properties with foreign names have higher sales prices than those with traditional Chinese names" (Sui et al. 2017).

The trappings of the Old World elite appealed to a generation of upwardly mobile homebuyers who, emerging from the austerity of Communism, sought fresh templates through which to show off their success. In my conversations with homebuyers, many associated traditional Chinese architecture with the deprivations of the past, considering it ill-suited to convey their new affluence, and saw Western styles as more synonymous with luxury and achievement. Thus, the Baroque façades and Rococo flourishes on their homes effectively function as the residential equivalent of Chanel logos on a purse – status symbols connoting wealth and sophistication. For potential homeowners who might not reach this conclusion on their own, developers' glossy billboards, elaborate brochures, and chandeliered sales offices relentlessly link foreign surroundings with "the good life." When it first opened, Sky City assured buyers that its "European-style living space" was "one of the necessities of aristocratic life"; Beverly Hills promised to deliver a "land of courtly enjoyment"; Palais de Fortune guaranteed its residents would "steer the world's power." This idea seems to have taken hold. I spoke with more than one homeowner who told me that the ideal life is "to eat Chinese food, drive an American car and live in a British house."

But China's private real estate developers have not always spearheaded the Western-themed developments in response to market demand. The local state often championed, supported, and sponsored numerous such developments. Thus local officials in Chengdu, for example, built a replica of Dorchester, England, to house 200,000 people after reportedly falling in love with a picture of the town on a Christmas card they received. There are countless government buildings, from offices to courthouses that replicate the US Capitol, the White House, the Versailles Palace, or some combination of the three. Shanghai's archipelago of foreign towns was the brainchild of its former mayor, who decided to build ten satellite cities, each in the architectural style of a different European country.

China's leaders have long been keenly aware of the propaganda value of architecture, and present-day duplitecture appears to be no exception. The tradition of architectural mimicry dates back over 2,000 years, and, even in pre-modern times, Chinese rulers used such practice to declare their superiority and strength. In the third century BC, for instance, the First Emperor Qin Shi Huang celebrated his victory over rival kingdoms by building replicas of their palaces in his capital city (Bosker 2013a). Other rulers asserted themselves by building elaborate gardens that were pre-modern versions of today's theme parks. That past is key for understanding the present and suggests that the replication of foreign landmarks stems not from a desire to pay homage to the West, but as a testament to China's power and status. Consider that the White House is one of the most-copied buildings in China, as if to declare that it can figuratively – and, in a sense, literally – possess the seat of power for the leader of the free world.

Duplitecture looks different in this light – Sky City, for example, appears more a landmark to *la belle Chine* than *la belle France*. But these developments ultimately meant to be fully inhabited. And the question stirring in my mind as I arrived in Shanghai was whether anyone had decided to take up residence in these outsized monuments to China's rise – and whether this appropriated architecture had seen its own stature fall.

The Chinese dream

Once in China, I did not have to go hunting for duplitecture. I caught my first glimpse at the airport, while in line for customs: The flat-screen TV mounted overhead played footage of starched Chinese soldiers saluting a government building that was the lovechild of the US Capitol and the White House. En route to Hangzhou, I spotted sprawling Italian palazzos the color of Easter eggs; a British hamlet; a red-domed structure that could have passed for the Duomo in Florence; another US Capitol; and, in line for taxis at the Hangzhou train station, an airbrushed ad for the be-fountained villas of Cam-Town Riviera, which apparently offered a heady cocktail of British intellect and French *joie de vivre.*

A taxi driver who had last traveled to Sky City two years ago – and never with a foreigner – took me out of the dense thicket of Hangzhou's skyscrapers into the gangly, mismatched landscape of its suburbs, past a ball-bearing factory, a pink house frosted by white balustrades, and, at last, to a four-lane boulevard, at the end of which sat Sky City's Eiffel Tower, stretching 35 stories into the air.

Both sides of the road were flanked by row after row of high-rise apartments girded by scaffolding and waving cranes from their roofs. The walls bordering the buildings advertised homes in the shadow of the Eiffel Tower ("Don't wait!") and reminded passersby to pursue the "Chinese dream." For a deserted ghost town, construction was booming.

Expecting to find Sky City an empty shell, I had stocked up on water and peanuts at the train station. Instead, the streets hummed with the mosquito-whine of scooters and bustled with pedestrians: Parents pushed strollers, young couples queued for Pocky, teenage boys lounged on shady benches, and elderly women shuffled under their neon umbrellas. I snuck into the back entrance of what I thought was an abandoned hotel, only to discover myself in the chandeliered consulting room of a plastic surgery clinic. It advertised a procedure of "exquisite carvings" that would give patients a "U.S.-nose."

Long populated by Juliet balconies and Art Nouveau streetlights, Sky City had added other sights common to French towns. I passed groceries, barbershops, day care establishments, cafés, a cosmetics store, a "Baby Bilingual Education Center," and a boutique with a mannequin dressed in chic black shorts and a Yves Saint Laurent purse. Just like France's Paris, Hangzhou's Paris was also filled with Chinese tourists snapping photos.

I learned it had been two years since a new management company had taken over the town. Where an earlier breed of "build-it-and-they'll-come" developer had measured success in concrete poured, this more enlightened manager had recognized the importance of luring services and stores that would attract residents. The company's chairman promised he would bring Sky City a Montessori school, "French research institutes," and spas offering the "world's most authentic and advanced beauty treatments" (Hangzhou News 2016); a year later, he pegged the town's population at nearly 40,000[2] people – though an employee at Madenjoy Real Estate told me that between 14,000 and 18,000 residents had moved in. Still, it appeared something was working: According to *Hangzhou Daily*, when 663 new units went on sale in August, they sold out in less than four minutes for an average of 14,000 yuan per square meter – about $200 per square foot (Zhang 2017), slightly more than the average price for an apartment in the center of Houston (Numbeo website "Price Rankings. . . ."). The average price-per-square-foot for apartments in downtown Hangzhou, two hours away by public transportation, is about triple that, which might explain the high proportion of young families in Sky City (Numbeo website "Property Prices. . . "). The developers behind the Hangzhou Paris did not consider it an "eerily depressing ghost town." They described it as the foundation for a new satellite city (Finberg 2013).

In its early years, Sky City had, like other themed communities, pushed a European lifestyle to match its European surroundings. Thames Town courted English pubs, German-themed Anting Town served bratwurst, and Sky City hosted crash courses on Gallic customs that covered everything, from the time of day French diners take their meals (according to the organizers' website: "Most French restaurants offer lunch between 12:00 and 14:00") to how they savor caviar ("use the tip of the tongue to slowly crush each individual grain one by one").

However, tastes had evolved since then. I stopped into a bakery just a few steps away from the Eiffel Tower, imagining I might pick up a baguette or brioche to pair with my Parisian stroll. But not a single thing in its glass cases could reasonably be described as "French" – not the rolls of taro-infused "Purple Cake," not the triangles of Barbie-doll pink cream, not the fluffy mounds of dough uniting mayonnaise and hot dog in a perplexing number of permutations. These were hybrid pastries with a Chinese sensibility. Like Sky City itself, this *pâtisserie* had taken a European classic, then reconceived it to suit Chinese tastes.

Duplitecture, developed

Other duplitecture developments had undergone similar transformations. The ghost towns had both filled out – I hit traffic getting into Shanghai's Thames Town – and mellowed out, the Western surroundings giving way to local habits. These communities had once emphasized their foreign themes by courting businesses with ties to Western culture or adopting strict design covenants meant to preserve their foreign look and feel. Planting vegetable gardens, hanging laundry outside to dry, or enclosing balconies – all common sights in typical Chinese neighborhoods – were usually prohibited. Now, at Hangzhou's Venice Water Town, nearly every Moorish window had a view of glass-encased porches or underwear fluttering in the breeze. One middle-aged woman had planted yuzu, pomelo, peaches, mint, chives, and squash in a yard no bigger than a bus stop.

In Thames Town, where British pubs and wine stores had once far outnumbered Chinese restaurants, couples and families now perused pu'er tea shops, slurped down noodles, and waited in line for bubble tea, or Baskin-Robbins. The 27-year-old owner of a boutique selling clothes by up-and-coming Chinese designers told me Thames Town had grown busier since 2014, thanks in part to the expansion of the subway system, and in part to the swelling population of Shanghai proper. Indeed, between 2000 and 2016, the city had grown by the population of New York City, pushing the city limits closer to Thames Town. When I asked whether Thames Town had tried to court British restaurants to match its English architecture, he choked, mid-bite, on his sticky rice. Why would anyone want that? "But the English don't have anything good to eat!" protested the owner's friend. "All they have is fried fish and French fries! British food is *disgusting*."

Not every former ghost town has come to life. In Shanghai's Holland Village (no relation to Liaoning's), most storefronts along the main street stood empty or deserted, their dusty concrete floors littered with desiccated bouquets or curled posters. Like something out of a fairy-tale fever dream, I met an elderly woman who lived inside the town's wooden windmill – the previous tenant, a wedding photography studio, had left it in her care after its business closed. Several buildings, including replicas of Amsterdam's Maritime Museum and De Bijenkorf department store, were under construction – just as they had been during a previous visit I made in 2008. Since then, the developers had successfully completed a stone cathedral, which they had outfitted with crucifixes, a crèche, and a wooden altarpiece, then rented to local businesses for use as office space.

A 36-year-old entrepreneur, one of the few shop owners in Holland Village, had chosen the ground floor of a brick townhouse for the headquarters of her wine-importing firm. Since wine was European, she explained, it was appropriate to bring clients to drink in a European setting. And besides, she thought Holland Village was beautiful – a glimpse at a continent she had never experienced firsthand. "I think there's a lot of foreign architecture in China because people who can't afford to see the world can see what it's like overseas," she said. "It's a great thing that we now have all these different styles of buildings in China. It's not just architecture. It's also a cultural exchange in a way."

Weird hysteria

Anthony Mackay, a British architect and urban planner who worked on Thames Town, is less enthusiastic about the "cultural exchange" duplitecture represents. I visited Thames Town with him one afternoon, and we followed the curving cobblestone streets past columned apartments drawn from London's Belgravia, past black-and-white Tudor-style inns, and past a replica of the Christ Church in Bristol, England. (The original church is a 20-minute drive from the offices of the firm, Atkins, that planned Thames Town, and it seems the designers conveniently opted to copy from their own backyard.)

"When I discovered that the architecture of Thames Town was a pure imitation of buildings around Bristol and England, I felt ashamed. I felt ashamed that we – I mean Atkins – had succumbed to the client and built such a place," said Mackay, stressing that he did not design Thames Town's buildings but consulted on the urban plan for Thames Town and the larger suburban district, Songjiang, to which it belongs.

To Mackay, while the original, British buildings were authentic to the time and place from which they emerged, this Chinese interpretation was copy-and-paste architecture – a rootless imitation. "These buildings have no history. They are pure theater. They are pure replica," he said. "Tourists in Paris and Venice know that beyond the façade is a genuine history. They are trodding on the ground that Henry the Eighth trod on. Here, you're trodding on the ground that was a duck farm."

Mackay has a powerful ally in his disdain for duplitecture: the Chinese government, which, after years of spearheading ambitious and eye-catching European-themed projects, recently turned its back on these undertakings. In 2016, following in the spirit of President Xi Jinping's celebration of frugality and embrace of traditional Chinese culture, China's cabinet and the Central Committee of the Communist Party issued a ban on "weird" architecture, specifically prohibiting buildings that are "oversized," "xenocentric," and lacking "cultural tradition," while calling for structures to be "suitable, economic, green, and pleasing to the eye" (Bosker 2016). Though not specifically named, Thames Town – along with other duplitecture – would, according to their definition, no doubt be considered "weird." Given the Chinese government's history of using architecture to serve political ends, it seems unlikely that the ban is merely an aesthetic choice, and more probable that it represents an effort to extend national interests and political priorities to the built landscape.

The prohibition recalls long-standing criticisms of duplitecture – that it is tacky, kitschy, unimaginative, and in bad taste. I considered the antagonism towards these developments as I crisscrossed Thames Town by foot. To be sure, the landscape housed more than its fair share of (weird) absurdities: there were the many sculptures honoring British greats, including statues of Princess Diana, Winston Churchill, and Harry Potter; there was the limestone façade of an English school that commanded students to "Conquer English to Make China Stronger"; the dozens of brides dressed like princesses and oversize Tinker Bells, feigning surprise as photographers instructed grooms to proffer fake-flower bouquets. Like "uncanny-valley" robots that unsettle us by falling just short of being human, Thames Town's buildings represent "uncanny architecture," arousing suspicion because they are nearly identical to the originals, yet a smidge too big, or too new.

But then we forget the originals were once off-putting and conspicuous themselves. We expect columns and crenellations to have the respectable patina of age. They didn't always. Remarking on the flashy newness of J. Paul Getty's Malibu villa, a copy of an ancient Roman country house, Joan observed that the ornate surroundings sparkled a little too much with the shine of wealth and status – just as the original would have. "Ancient marbles once appeared just as they appear here: as strident, opulent evidence of imperial power and acquisition," she wrote (Didion 1979: 76).

Over time, the glitter and strangeness of duplitecture will fade, until its inhabitants all but forget the replica ever belonged anywhere except its adopted milieu. Back in the United States, on my drive home from the airport, I passed New York's own duplitecture – Joseph Pulitzer's copy of a Venetian palazzo on East 73rd Street, and Gertrude Rhinelander Waldo's Loire Valley *chateau* on Madison Avenue. (Of course, there is a double standard: while the Western creators of these buildings are often said to have been "inspired by" European landmarks, China's developers get accused of "knocking off" the greats.) Europe's aristocrats might never have trod in these duplicated structures. But it doesn't matter: They've acquired a new history and transformed into Manhattan's treasures, rather than Europe's.

China's duplitecture will also grow old. Already, its Paris, its Holland, its Venice, and its England have become Chinese in spirit and acquired a Chinese soul, if not in appearance.

Notes

1 This essay, which draws on material from *Original Copies*, was adapted from a piece, "We'll Always Have Sky City," that originally appeared in the website *Roads & Kingdoms*. https://roadsandkingdoms.com/author/bbosker/.

2 "《乐居会客厅》王欣：天都城是"城"，不是"镇"." 搜狐. http://www.sohu.com/a/209917391_692704.

References

Boissoneault, L. (2013, September 25). "Tianducheng: An Eerie Paris Ghost Town in China" (PHOTOS). *Weather.com*. Retrieved January 31, 2018, from https://weather.com/travel/news/tianducheng-eerie-paris-ghost-town-china-photos-20130925

Bosker, B. (2013a). *Original Copies: Architectural Mimicry in Contemporary China*. Honolulu: University of Hawai'i Press.

Bosker, B. (2013b, February 08). "In Chinese Buildings, a Copycat Craze." *The Wall Street Journal*. Retrieved January 31, 2018, from www.wsj.com/articles/SB10001424127887323951904578290520134759306

Bosker, B. (2016, February 29). "China's Ban on 'Weird; Architecture Is a Global Power Play." *The New Republic*. Retrieved January 31, 2018, from https://newrepublic.com/article/130589/chinas-ban-weird-architecture-global-power-play

Canaves, S. (2009, April 7). "Goodbye Holland Village: A Development Mired in Corruption Goes Down." *The Wall Street Journal*, China Real Time Report. Retrieved March 26, 2018, from https://blogs.wsj.com/chinarealtime/2009/04/07/goodbye-holland-village-a-development-mired-in-corruption-goes-down/

Didion, J. (1979). *The White Album*. New York: Farrar, Straus and Giroux.

Feinberg, A. (2013, August 9). "China's Replica of Paris Is Now an Eerily Depressing Ghost Town." *Gizmodo Australia*. Retrieved January 31, 2018, from www.gizmodo.com.au/2013/08/chinas-replica-of-paris-is-now-an-eerily depressing-ghost-town/

Furdyk, B. (2016, April 8). *The Fake Paris of China Sits Empty and Abandoned*. HGTV Canada. Retrieved January 31, 2018, from www.hgtv.ca/real-estate/photos/the-fake-paris-of-china-sits-empty-and-abandoned-1052678/#

Hangzhou News. (2016, September 6). 浙江天都实业有限公司老总王欣:我们要为杭州造一座理想的国际小镇. Retrieved January 31, 2018, from http://hznews.hangzhou.com.cn/chengshi/content/2016-09/06/content_6288571.htm

Kibaroglu, O. (2017, September 4). "Westerners Oughtn't to Criticize China for Its 'Fake Cities.'" *Global Times*. Retrieved January 31, 2018, from www.globaltimes.cn/content/1041487.shtml

Miller, L. (2013, August 7). "Paris in China: Tianducheng Is an Eerie, Abandoned City of Lights Clone." *HuffPost*. Retrieved January 31, 2018, from www.huffingtonpost.com/2013/08/07/paris-china-tianducheng_n_3714385.html

Numbeo website (n.d.). "Price Rankings by City of Price per Square Meter to Buy Apartment in City Centre (Buy Apartment Price)." http://www.numbeo.com/cost-of-living/city_price_rankings?itemId=100.

Numbeo website (n.d.). "Property Prices in Hangzhou…." http://www.numbeo.com/property-investment/in/Hangzhou.

Schmidle, N. (2010, August 19). "Inside the Knockoff-Tennis-Shoe Factory." *The New York Times*. Retrieved June 15, 2018, from www.nytimes.com/2010/08/22/magazine/22fake-t.html

Sui, D., Zhao, B., and Kong, H. (2017, October). *The Development of Copycat Towns in China: An Analysis of Their Economic, Social, and Environmental Implications (Working Paper WP17DS1)*. Lincoln Institute of Land Policy. Retrieved January 31, 2018, from www.lincolninst.edu/sites/default/files/pubfiles/sui_wp17ds1.pdf

Woodruff, B., Yiu, K., and Valiente, A. (2018). "China's 'Fake' Cities Are Eerie Replicas of Paris, London." *ABC News*. Retrieved January 31, 2018, from http://abcnews.go.com/International/chinas-fake-cities-eerie-replicas-paris-london-jackson/story?id=36525453

Zhang, L. (2017, September 8). 天都城：大城已现 蝶变中焕新启程. *Hangzhou Daily*. Retrieved January 31, 2018, from http://hzdaily.hangzhou.com.cn/hzrb/2017/09/08/article_detail_1_20170908A295.html

Further reading

Bosker, B. (2013). *Original Copies: Architectural Mimicry in Contemporary China.* Honolulu: University of Hawai'i Press. The first definitive account of China's duplitecture movement and an in-depth exploration of copy culture drawing on original research in the field.

Eco, U. (1986). *Travels in Hyperreality.* Translated by William Weaver. San Diego: Harcourt. In the title essay that lends its name to the volume, Eco describes his journey through the United States in search of the "absolute fake."

Fong, W. (1962). "The Problem of Forgeries in Chinese Painting, Part One." *Artibus Asiae* 2/3: 95–119. An exploration of Chinese attitudes toward copying and replication.

Piazzoni, M. F., & Banerjee, T. (2018). Mimicry in Design: The Urban Form of Development. *Journal of Urban Design* 23 (4): 482–498. A discussion of the diffusion of Western-like built environments in rapidly urbanizing regions. It draws from Marshall Berman's seminal treatise on modernity, which uses Goethe's Faust as a metaphor. In the Chinese context, Berman's analogy suggests Faust as the consumer who craves the Western mode of consumption.

27

IMPROVISED URBANISM IN THE DESIGN OF INDIA'S UNAUTHORIZED COLONIES

D. Asher Ghertner

Introduction: India's unauthorized colonies

This chapter explores mimicry as a design practice central to the planning, development, and occupation of what is rapidly becoming the most popular neighborhood form in Delhi, and indeed, much of metropolitan India – the unauthorized colony. Built on land beyond the urbanizable area of the master plan in place at the time of their formation, these peri-urban settlements house as much as 30% of Delhi's population, or around six million people (Sheikh and Banda 2014). They form as developers consolidate rural farmland and cut plots for private sale. Varying drastically in their size and income level, the category "unauthorized colony" encompasses everything from peripheral neighborhoods with services worse than those found in slums, to vast manicured enclaves composed of elite "farmhouses" – off-grid manors where the rich and famous go to escape the city's bustle, building codes, and property taxes. Although landowners in unauthorized colonies possess formal documents showing detailed payments for their homes – thereby making them more secure than squatter settlements – these transactions cannot be registered with the state because unauthorized colonies fall outside the development area of the master plan or exist on land not zoned for residential use (Bhan 2013).

Unauthorized colonies hence blur the line between the *de jure* and the *de facto*, representing a preeminent form of improvisational urbanism to which increasing numbers of people – lower-income and middle-class alike – have had to turn to find affordable shelter.[1] They are settled on rural land and with densities far exceeding those that can be supported by the village infrastructure to which they only sometimes have access. Everything from the plotting of individual housing lots, to the paving of roads, to the installation of water lines, to the disposal of household sewage must be built up incrementally – a pattern of self-organization characteristic of the "auto-constructed" peripheries prominent across the Global South (Caldeira 2016; Silver 2014). This range of improvised neighborhood-making practices represents a multifaceted and decentralized urban design process oriented towards balancing a range of competing interests. The developers who have historically initiated unauthorized development – often called land colonizers, given their role in aggregating village lands into the residential "colony" model typical of Delhi (Tiwari and Rao 2017) – often hire architects to design layout plans that superficially resemble those in planned colonies. In this sense, Delhi's more than 1,600 unauthorized colonies typically have a more "planned" look than other informal settlement types, such as squatter settlements, urban villages, or slum resettlement colonies (Ghertner 2015). However, key requirements of neighborhood design are often left out of the unauthorized colony plans so as to maximize the area of private, sellable lots. Space for schools, parks, and other social infrastructure thus tends to be limited, and developers often squeeze road widths and setbacks, thereby constraining community livability while also potentially hindering sales to wealthier households who want to preserve spatial separation from neighbors and nonresidential land uses.

From their earliest beginnings, unauthorized colonies mimic the appearance of planned neighborhoods through the use of the same hierarchy of spatial units (individual streets form pockets or blocks, which fit into sectors or phases of a wider colony), as well as in the materials and design elements that make up what has been their principal architectural form: the bungalow (Figure 27.1), including its more vertical version prominent in the smaller lots found in lower-income settlements. Developers mimic in this way to appeal to middle-class aesthetic standards, but also with a partial eye towards the possibility of future "regularization." This is a contested process by which town planners – often at the behest of politicians – retroactively approve the layout plans for unauthorized colonies. The idea behind regularization is to attempt to align colonies and buildings with layout norms and building codes to which they have never formally been subjected. Unauthorized colonies are "unauthorized" not necessarily because they violate code (although they often do), but rather because they exist in areas not subject to plan oversight or code enforcement. Mimicking code in anticipation of regularization is thus a prefigurative act of producing the appearance of being "planned." The regularization process allows unauthorized colonies to be registered with the revenue department, thereby making their residents' property titles legal and eligible for state service provision such as water, sanitation, road paving, and street lighting – which must be privately arranged prior to regularization (Zimmer 2012). So-called "regularized unauthorized colonies" remain officially unplanned based on the fact that they remain outside the city's development area (Bhan 2013: 61) but benefit from improved neighborhood stability and services, as well as associated property price increases.

This chapter builds on four months of fieldwork carried out in 2013–2015 in the Chattarpur area of South Delhi, where the most rapidly growing middle-class unauthorized colonies exist today. It considers three different spaces of unauthorized colony design to show how the making of this popular housing form functions neither as a "state of exception" that reinforces state power through the suspension of law (Roy 2005) nor as a subversion of that power (Miraftab 2009) – perhaps the two most influential frameworks for understanding urban informality today. Rather, informality in Delhi operates as a practice of neighborhood design and building that emerges through the improvisational practices by which diverse actors – developers, homeowners, town planners – mimic an original plan that exists only through its many scattered imitations. As I will argue, unauthorized colonies in Delhi are designed outside of planning norms, but, by selectively borrowing from those norms, they reveal them to be unstable elements of design and interpretation, not fixed rules or a rigid code (the formal) from which informal practices necessarily deviate or spill over. Building on Ananya Roy's (2005, 2009) key insight that informality emerges with and through state practices, Delhi's unauthorized colonies show the imagination and making of the city to be founded on what Walter Benjamin (1999) calls "the mimetic faculty," or the historically and geographically specific capacity to copy, draw on the power of the other, and become the other. As I hope to show, this mimetic faculty is central to urban design in Delhi, informing professional standards as much as community interpretations of "authentic" form, and thereby showing improvised urbanism to be neither necessarily marginal to the state nor exclusively driven by poverty, as is typically presumed.

The first space I examine is the town planning department, where planners compelled by pressures to regularize unauthorized colonies create maps depicting those colonies in conformance

Figure 27.1 A bungalow in a middle-class unauthorized colony in South Delhi, borrowing architectural and design elements (tiered iron gates, Grecian balustrades, single-person balconies) common in planned Delhi colonies.

Source: Photograph by author, June 8, 2013

with planning norms, while knowing that these "imaginary maps" will never be implemented. Drawing hypothetical roads, parks, and dispensaries, state planning becomes a design practice centered on the reproduction of an open fiction. The second space is the site of the unauthorized building, where multi-story apartments on designated agricultural lands are demolished, not because they violate building code and zoning rules – which all such structures do – but because they are as yet unoccupied. Mimicking occupancy, developers thus install air-conditioning units, hang laundry, and add religious iconography on vacant or incomplete buildings to avoid site demolition. Surface embellishments and signs of occupation thus become design features central to successful project completion, showing aesthetic appearance to be potentially as significant as calculative protocols of building inspection – such as the measurement of setback or the determination of floor area ratios – for the adjudication of legality. The third space consists of the improvised water systems of unauthorized colonies, which mimic state systems and develop through collaboration with state agents, even though such systems are illegal in unauthorized colonies. The post-regularization ability to network this privately built infrastructure into state water mains raises the question of when the copy becomes the original. Taken together, these three spaces show the design and planning of unauthorized colonies to mock power as much as they centralize or appropriate it.

Space 1 – town planning

In the summer of 2012, the beleaguered chief minister of Delhi, Sheila Dikshit, ordered the urban development department to issue a plan for regularizing Delhi's unauthorized colonies. This was not the first time that a chief minister pandered to the unauthorized colony vote through the offer of regularization, but it was a uniquely ambitious one. Whereas the town planning department in the municipal corporation had previously been charged with finalizing regularization lists on the basis of existing regularization guidelines, in this case, the chief minister took the lead, issuing her own list of 895 "regularizable" unauthorized colonies without any consideration of their relationship with regularization guidelines.[2] The town planning department was subsequently handed this list and given the new task of evaluating the layout plans for colonies already deemed "regularizable." While some unauthorized colonies already had layout plans, most did not, and once an unauthorized colony was added to the regularization list, its residents' welfare associations – informal governance bodies that typically organize local service delivery and interface with elected officials and bureaucrats – were given a mere 15 days to submit layout plans for approval.

When I visited the town planning department in the summer of 2013, planners were seeking ways of depicting unplanned neighborhoods, with faulty and hastily assembled layout plans, that would align with town planning requirements. As an engineer in the department told me, "We haven't found a single layout plan done properly. Most of them have basic errors of scale, North/South, colony boundaries, and the size of plots." The whole process, he said, "is entirely against planning principles" (personal communication, June 5, 2013).

Each time new regularization guidelines are established, a new date is set by when colonies have to be built up to qualify for regularization. In 1977, when Delhi's first wave of regularizations took place, this cutoff date was set at 1975. It was later advanced to 1992, then 1997, and then 2002. To be eligible for regularization, colonies must have at least 50% of their area built up by the cutoff date, as confirmed by aerial photographs. There are three main problems with the notion of 50% built-up area, however, that make unauthorized colony design a process of simultaneous mimicry and mockery. First, aerial photographs are not always accurate or comprehensive. Second, the definition of "built-up area" changes over time. Even though most unauthorized colonies are subdivided and sold off at a very early stage, they often remain vacant until land prices rise and residents' welfare associations form and organize the provision of basic infrastructure. In 2008, as part of the chief minister's effort to win the unauthorized colony vote, the definition of "built-up area" was relaxed. Accordingly, if somebody owned a large plot, it could be included within the colony's built-up area measure so long as they had constructed even a small shed on it. Consequently, colonies or sectors of colonies with minimal development saw their property prices skyrocket with the promise of state services and title registration.

The third problem with the cutoff date is that politicians regularly declare that they will eventually advance it. The new government formed out of the state assembly elections in 2013 advanced the cutoff date up to 2008. Even if a colony were to miss this updated 2008 deadline, though its residents could rest assured that they would make the next regularization list. This condition, what urbanist Amita Bhide (2015) calls "the regularizing state," allows landowners to avoid regulatory oversight, taxation, and building codes in the early stages of development of a colony, only to later be regularized. Town planners are forced to somehow reconcile unauthorized developments with a planning and building regime that these colonies continue to violate.

Can we describe this as a "state of exception," as much of the urban informality literature might urge us to do? Is this process best described as a condition under which the state suspends the rule of law to build up sovereign command and retain discretionary authority? Perhaps, but from the perspective of state officials, regularization appears anything but a coherent strategy to build sovereign authority. Consider the perspective of the chief town planner. Expressing deep frustration with the regularization process, the planner half-jokingly told me of his plans to write a book upon retirement called "How to Unplan a City" (personal communication, July 3, 2013).

At the time of our meeting, the layout plans he had received from unauthorized colonies on the regularization list all had major problems, including basic errors of cardinal direction and colony boundary identification. He was nonetheless expected to give these colonies provisional approval. How did he and his staff of senior planners proceed? How did they reconcile hundreds of plan-violating settlements with a plan they were schooled and hired to uphold?

They began by overlaying layout plans of unauthorized colonies upon government-supplied aerial photographs and land use data. The problem is that government records at the time were based on a 2007 aerial survey, which often indicated adequate space for the social infrastructure – such as parking and green space – available to complete the regularization process. However, when the planners (or the universities to whom they contracted this work) visited the colonies to confirm the ground reality, they almost universally found that the areas they had assumed to be vacant were already built up.[3] Instead of trying to reconcile such discrepancies and being dragged into an extensive "ground-truthing" process, these planners proceeded by acting as if the reality in the present was the same as was displayed in the 2007 photograph. In other words, they made plans that resembled not reality, but a photograph of the recent past. This took place through the simple design practice of drawing traces on digitized versions of the layout plans, producing hypothetical maps for these colonies that mimicked master plan norms for such things as road width, green space, and dispensaries, knowing full well that these colonies already violated these maps. This was done by marking with dashed lines "proposed roads" wide enough to meet code – the irony being that such roads would have required the demolition of the hundreds of structures already built up to the edges of the existing, narrow roads.

Consider, for example, Shiv Ram Park Extension, an unauthorized colony located in West Delhi. The heavily dashed lines going from the bottom-left to the upper-center of the map (Figure 27.2), indicated as a "proposed master planned road" on the legend, shows that the colony encroaches approximately 10 meters upon what is supposed to be a 36-meter-wide main trunk road. The thin, dashed lines on the inner-colony roads show where existing 6- and

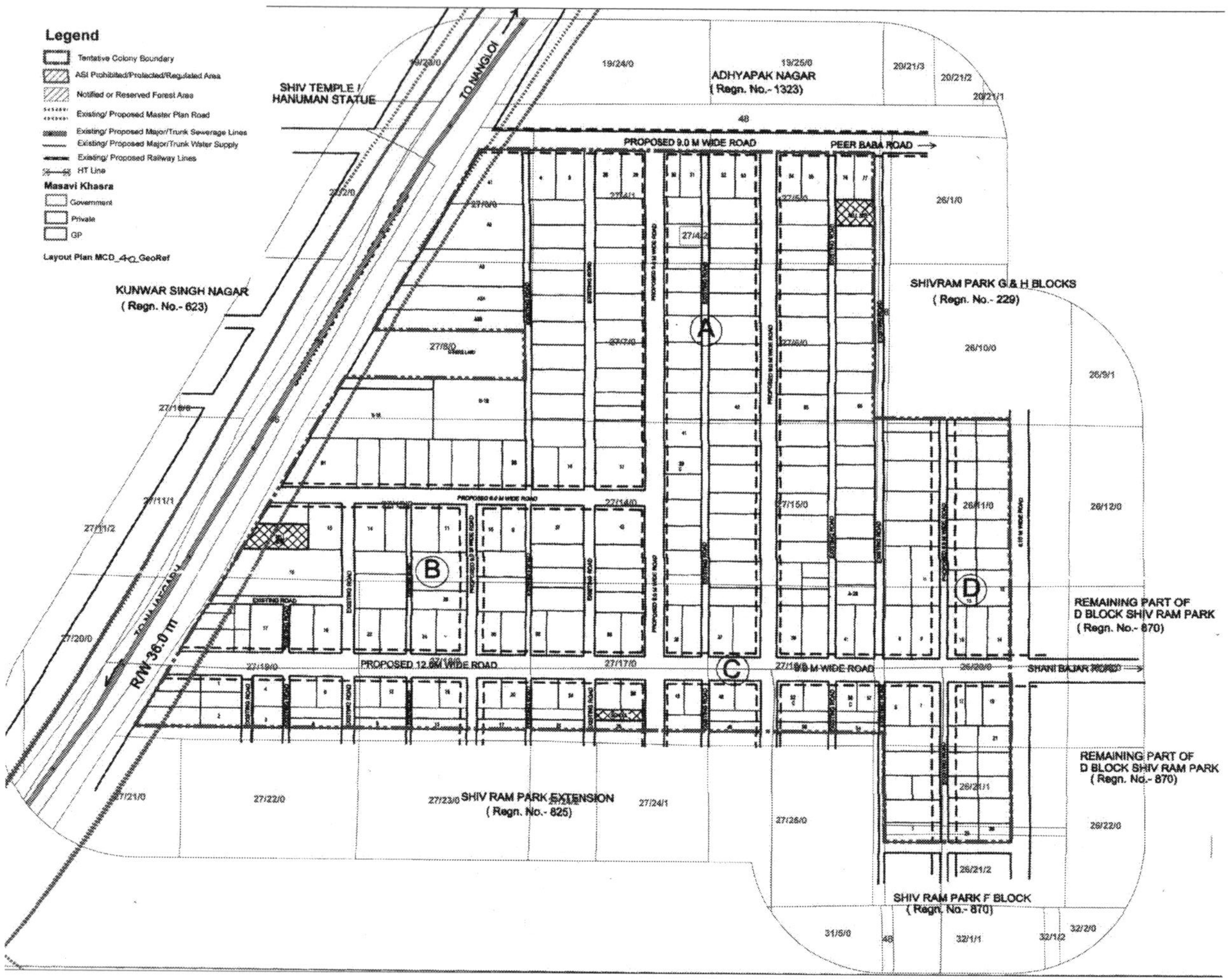

Figure 27.2 Section of "Scrutinized Layout Plan for Shiv Ram Park Extension," an unauthorized colony in West Delhi. The dashed lines shown inside the colony boundaries indicate where proposed, code-conforming roads would be located.

Source: Unauthorized Colonies Cell, Department of Urban Development, Government of the National Capital Territory of Delhi

9-meter-wide roads are "proposed" to be widened into 9- and 12-meter-wide roads, respectively. Were these "proposals" to ever be implemented, every structure in the colony would have to be at least partially demolished, since nearly all structures, as shown in the aerial photographs that accompany the layout plan, were built up to the very edge of the existing, narrower roads.

The planners next added to their layout plans necessary social infrastructure – such as parks and medical dispensaries – to what appeared as vacant or unused lands on the layout plan. However, these vacant lands are almost always privately owned and were mostly already built up by the time the planners made their amendments. As the chief planner told me about this process,

> On paper we do it, for the sake of planning. [The planning] authority will say "ensure all these things [like roads and parks] are there [on the layout plan]." If we ever try to implement the layout plan, the politician [who represents the people] will come and do *dharna* [sit-in]. So we are left making this layout plan knowing it will never become reality. Ours is a paper game.

And so, town planning in Delhi operates through the production of fake maps, which mimic planning norms but bear little resemblance to reality, showing how one key aspect of urban design – the location of social infrastructures and the functional organization of neighborhood space – operates not as a monumental state performance, but rather as a tentative representation of an ideal state that is simultaneously disavowed.

I am here drawing on Homi Bhabha (1994: 122), according to whom "mimicry emerges as the representation of a difference that is itself a process of disavowal." In Delhi, the representation of difference emerges as the exemplary form of the unauthorized colony, whose very condition of being is its definition as "unplanned," and yet whose surface resemblance to planned neighborhoods and "regularizability" call into question what, indeed, is so different about it. "In order to be effective," Bhabha writes, "mimicry must always produce its slippage, its excess, its difference." This slippage gives the mimicking subject – the one who copies – an incomplete and virtual presence. Bhabha writes about mimicry as a British colonial discourse through which Indian colonial subjects were brought into the sphere of liberal law, but never fully integrated. The virtual presence of the mimicking subject, in other words, is defined by his position both inside and outside the law, by the condition of being *similar*, but never same. In Delhi's urban design space, this virtual presence is visible in the dashed lines outlining the virtual road that upholds planning code but that will never quite meet it. For Bhabha, mimicry is an "ironic compromise" that conflicts with the assumptions that authorize it. We see this in the town planning department, where the planner himself mimics the plan. Urban mimicry in unauthorized Delhi seeks resemblance, but, in failing to achieve this perfect condition, also mocks that original. Or, as the anthropologist Michael Taussig (1993: xiii) puts it, "The wonder of mimesis lies in the copy drawing on the character and power of the original, to the point whereby the representation may even assume that character and that power." As the unauthorized colony "passes," via regularization, the power of the original diminishes, as does the system of rules upon which it depends. Informality here is not a deviation from the formal, but rather its inappropriate performance.

For Bhabha, mimicry's success depends precisely on the proliferation of what he calls *inappropriate objects*, which ultimately also ensure its failure. Bhabha's point is that efforts at appropriation, whether by those in power or by those marginal to it, cannot evade this menace – mimicry always produces a mocking double. In Delhi, the unauthorized yet "regularizable" colony is the plan's double – a figure that has a virtual presence within planning but whose inherent unplannedness destabilizes and mocks planning power. This is evident in the fact that, while unauthorized colonies are unplanned in the sense that they do not conform with code, they are very much an expected outcome of planning. Delhi's planning authority, the Delhi Development Authority, after all, has approved historically far fewer planned housing units than the master plan itself projected to be necessary, leaving a large mass of the population to rely necessarily on unauthorized development (Ghertner 2005; Verma 2002). Unauthorized colonies, in other words, are an internal product of the planning process, what Bhan (2013) calls a "planned illegality." The regularization of unauthorized colonies confirms that planning depends on state performances that merely imitate but do not replicate the plan, and that the viability of the plan itself rests on ongoing inappropriate performances, another version of which I now discuss through a second design space, the unauthorized building.

Space 2 – the unauthorized building

The rise in property prices in unauthorized colonies added to the regularization list led to construction boom and sale of "builder flats" – apartments in 8-, 12-, or 16-unit, multi-story high-rises, a building form that emerged only after the 2002 cutoff date and that contributed to a massive increase in their density. Responding to this escalation of illegal construction, in 2011 the lieutenant governor of Delhi – an official appointed by the central government to an oversight role independent of the chief minister (the head of the state government) – constituted special task forces for each city district with the charge of "acting against unauthorized/unsafe constructions whether on public or private lands."[4] Over the following two years, the special task force in South Delhi ordered the demolition of hundreds of buildings under construction in unauthorized colonies (see Figure 27.3) (Yardley 2013). In response to complaints of homeowners,

Figure 27.3 A typical mix of vacant lots (foreground), finished homes (middle ground, right), and partially demolished "builder flats" (background) in the outskirts of a middle-class, South Delhi unauthorized colony during the 2013 summer of demolitions.

Source: Photograph by author, June 19, 2013

threatened with demolition but claiming to have unknowingly bought illegal flats from small-scale and often difficult-to-trace developers, the task force conceded that once a structure was built, sold off, and occupied, it could not be demolished on humanitarian grounds. The Delhi High Court similarly granted relief to individuals who had purchased finished flats from builders, arguing that liability for plan violation lay with the builder, not the purchaser.

Developers, recognizing this occupancy loophole, began faking occupation. They did so by transforming their construction approach where possible. After the basic shell was completed, they would quickly finish the ground floors so that they could be sold, sometimes at a discount to add a building occupant, while the upper floors were being finalized over subsequent months. Finished ground floors, with marble cladding or faux brick – preferred middle-class design elements – or religious markings, such as the Hindu swastika, emblematic of house inauguration ceremonies (*griha pravesh puja*), thus became visible in buildings even before elevators were installed or upper-floor windows and balconies were added. In a rapidly developing, middle-class unauthorized colony, security guards took advantage of developers' offers of free housing and moved their families into ground-floor flats or parking spaces, usually before running water or electricity was connected or a bathroom had been built.

When such efforts to secure ground-floor occupation were not enough to deter the task-force-initiated demolition crews from attacking the upper floors with sledgehammers, developers began a practice of affixing satellite television dishes and air conditioners on unfinished or unsold upper-floor units. In some cases, they hung laundry over balcony rails to produce the appearance of occupancy (Figure 27.4).

Developers hence found a way to complete building projects through a simple semiotic shift: embellishing the façades of their structures with the resemblance of occupation. Installing family names on doors and including religious iconography in entranceways or other design techniques that signaled that the basic threshold of occupancy had been met. Hanging strings of green chilies and lemons or displaying masks of the goddess Kali – common Hindu practices to ward off the evil eye (*buri nazar*) – proved to be materially effective, as if here warding off building inspectors and charges of unauthorized construction. Urban design of the most elementary sort, hence, emerged as a most effective tool of illegal building and speculative real estate.

Ironically, these more superficial elements of urban design became more crucial to project success than building code conformance. Whereas regularization is supposed to bring unplanned areas into alignment with code, the elimination of the field survey during the regularization process, described earlier, also meant that the evaluation of building and design standards – the job of the municipal corporation – was mostly eliminated from regularization protocols. Task force oversight of illegal construction in unauthorized colonies thus had to do less with design rules, floor area ratio, or building codes and more with building completion and occupancy status, although particularly egregious projects that, for example, pursued floor area ratios far greater than even existing "builder flats," do appear to have been particularly vulnerable to sealing or demolition.

Space 3 – incremental infrastructure

One of the greatest benefits of regularization for unauthorized colony households is the promise of future state services, especially water and sanitation. In the years before regularization, though, households must organize and plan for the delivery of these basic services on their own. For lower-income unauthorized colonies, this poses challenges similar to those found in slums: household budgets are stretched to pay for private water delivery, and communicable diseases are higher than they should be due to neighborhood reliance on open drains (Zimmer 2012). Armed with

Figure 27.4 A vacant building's "occupied flats," as semiotically evidenced by the presence of air-conditioning units and laundry on balcony railings.

Source: Photograph by author, December 28, 2014

resources far greater than their poorer counterparts, better-off unauthorized colonies, such as the ones shown in Figures 27.1 and 27.4, improvise their own infrastructures. Delhi has a rapidly declining water table, and to discourage the formation of new unauthorized colonies, in 2010 the state government passed an order effectively banning the drilling of bore wells there. Yet, in wealthier unauthorized colonies, new wells were being dug regularly in 2013 and 2014. In the booming Chattarpur area of South Delhi, I counted fifteen bore wells, which I was told were six-hundred-feet-deep, pumping water 24 hours a day. Many of these illegal wells were overseen by water board officials, who trained local security guards to operate a series of "gate walls" that switched the flow of water from one lane to another according to need. How did water board officials come to operate bore wells in neighborhoods barred from bore well construction?

The officials had what they consider a well-reasoned answer to this question, one in which they – the face of the state – were simultaneously inside and outside of the state (Ghertner 2017). The local representative of Delhi state's legislative assembly paid the officials using his discretionary fund, a non-water board source. As water board employees, they operated the bore wells legitimately; yet the water board could claim noninvolvement by noting that none of its funds were allocated to unauthorized colony wells. As state officials who first exited the formal employment of the state but then reconstituted a cloak of state authority, these officials improvised a vast, informal infrastructure that provided reliable groundwater to thousands of households.

Regularization brings with it the promise that these colonies would become eligible for municipal water supply, but earlier, privately built water lines do not always become redundant when this occurs. When time comes to lay new pipes, state water mains are linked to existing neighborhood pipes, which are almost always designed to code, often with the oversight of water board engineers. Many households in planned Delhi neighborhoods use bore wells as informal/illegal backups to the irregular and intermittent municipal water that is their formal/legal source of water. Unauthorized water systems have an inverse relationship with the formal/informal: An illegal system first mimics state systems and then becomes redefined/repurposed as planned urban development, serving as the very vehicle for municipal water delivery, a perfect example of the mimetic capacity of the copy not just to draw on the power of the other, but also to become the other (Benjamin 1999).

Conclusion

The vast terrains of informal urbanization in India, like in much of the Global South, lack a regulated system of urban design. In this chapter, I have thus treated the range of improvised practices of neighborhood layout, infrastructure organization and provision, building design, and public administration as a collective design practice that at once orders the functional uses of the neighborhood and governs the aesthetic and organizational character of space, typically through a close, even copied, relationship with officially planned neighborhoods that do have a more formal system of urban design, planning, and oversight. Focusing on three distinct urban design spaces, I have sought to show informality to be something practiced by a whole range of actors, from the chief minister to the informal settler, from the water engineer to senior planners. This conforms with the treatment of informality not as a survival strategy of the poor and marginalized or "a sphere of unregulated, even illegal, activity outside the scope of the state," as it is often depicted; but rather as a form of spatial governance, as Ananya Roy (2008) argues, that maintains the constructed distinction between planned and unplanned upon which land price differentials and unequal service provision depend. At the same time, however, Delhi's unauthorized colonies make it difficult to describe informality as a coherent modality of planning that solidifies the discretionary power of a sovereign body. Informality in unauthorized Delhi appears more as an improvisational mechanism for pulling a diverse field of practices into the field of urban planning and design. Be it the production of virtual regularization maps, the installation of an air conditioner on an unauthorized building, or the private employment of state workers, Delhi's unauthorized colonies mimic the formal city, but in doing so, simultaneously make a mockery of it.

Even imperfect copies, as Taussig (1993: 17) asserts, can be effective in acquiring the power of the original. Contemporary Delhi – where the central, formally planned city has become largely stagnant, and the political energies of the unauthorized peripheries absorb increasing state bureaucratic attention – reveals that urban design might be best understood as an ongoing mimetic drama in which dominant rules of neighborhood form and layout are inappropriately followed, as a form of "symbolic piracy" (see Romberg 2005) – a borrowing of signs and rituals of legitimacy. In other words, through the vernacular forms of borrowing evident in unauthorized colonies, rules of neighborhood form and organization intended to exclude the informal and the "unplanned" are rechanneled to serve an unintended settlement form that also happens to be essential to the reproduction of the city. (Where else would the six million inhabitants of unauthorized colonies go?) Rather than denying or overturning the original, such performances reveal the extent to which the original depends on the copy in the end.

Notes

1 Unauthorized colonies are one of the main "solutions" to the shortfall of planned housing in Delhi, which the 2021 Delhi Master Plan (Delhi Development Authority 2007) estimates at 400,000 units. As Bhan (2013: 60) observes, no new land was notified as an urban development area between 1962 and 1990, despite a massive increase in need. Four thousand hectares were added to the city's development area in 1990, but this was a mere 4.5% expansion, despite the city's population growing by 3.5 million in the preceding period. An additional 20,000 hectares was added in 2007, by which time the population grew by another 4 million. In 1990, when the updated Delhi Master Plan 2001 was issued, and in 2007, when the current Delhi Master Plan 2021 was issued, "areas far beyond the notified area in the master plan were already built up. Both [plans] chose not to notify already built-up areas as development areas within the plan" (Ibid.). These un-notified but built-up areas are the principle spaces where unauthorized colonies exist today. They mostly appear as unshaded spaces on the master plan's land use map, although many such colonies also encroach on forest and other protected lands.

2 A total of 567 previously regularized unauthorized colonies also exist in Delhi (Sheikh and Banda 2014), which house just shy of 13% of the city's population (GNCTD 2009). Dikshit's initial list included 1,208 unauthorized colonies, but those that obviously encroached on green spaces, institutional lands, or other protected areas were excluded.

3 Official requirements for regularization, besides the preparation of a layout plan, include the assembly of land records for the properties in question, and a statement by the residents' welfare association that they shall abide by the layout plans and that they shall transfer any available land to the relevant municipal body free of cost so that the government might provide social infrastructure (Sheikh and Banda 2014: 4).

4 For the order constituting the special task forces, see Government of the National Capital Territory of Delhi, order No. F-27/SDM/KJ/2010/96, March 30, 2011, http://delhi.gov.in/wps/wcm/connect/e9b5290047d5252b93d8df0fed934187/stf+order.pdf?MOD=AJPERES&lmod=-1825763891.

References

Benjamin, W. (1999). "On the Mimetic Faculty." In Jennings, M.W., Eiland, H. and Smith, G. (eds.), *Selected Writings, 1926–1934*. Cambridge, MA: Belknap Press.

Bhabha, H. (1994). *The Location of Culture*. London: Routledge.

Bhan, G. (2013). "Planned Illegalities: Housing and the 'Failure' of Planning in Delhi: 1947–2010." *Economic and Political Weekly* 48 (24): 58–70.

Bhide, A. (2015). "The Regularising State." *Economic and Political Weekly* 49 (22): 92–100.

Caldeira, T. (2016). "Peripheral Urbanization: Autoconstruction, Transversal Logics, and Politics in Cities of the Global South." *Environment and Planning D: Society and Space* 35 (1): 3–20.

Delhi Development Authority. (2007). *Delhi Master Plan 2021: People-Friendly Version.* New Delhi: Rupa.

Ghertner, D. A. (2005). "Purani Yojana ki Kabr par, Nayi Yojana ki Buniyad: *Dilli Master Plan 2021* ki Chunauti aur Sambhavnae [Building the New Plan on the Grave of the Old: The Challenges and Possibilities of the *Delhi Master Plan 2021*]." *Yojana* 24 (7): 14–20.

Ghertner, D. A. (2015). *Rule by Aesthetics: World-Class City Making in Delhi.* New York: Oxford University Press.

Ghertner, D. A. (2017). "When Is the State? Topology, Temporality, and the Navigation of Everyday State Space in Delhi." *Annals of the American Association of Geographers* 107 (3): 731–750.

GNCTD. (2009). *Delhi Economic Survey 2008–09.* New Delhi: Department of Economic Affairs, Government of the National Capital Territory of Delhi.

Miraftab, F. (2009). "Insurgent Planning: Situating Radical Planning in the Global South." *Planning Theory* 8 (1): 32–50.

Romberg, R. (2005). "Ritual Piracy: Or Creolization with an Attitude." *New West Indian Guide* 79 (3 & 4): 175–218.

Roy, A. (2005). "Urban Informality: Toward an Epistemology of Planning." *Journal of the American Planning Association* 71 (2): 147–158.

Roy, A. (2009). "Why India Cannot Plan Its Cities: Informality, Insurgence and the Idiom of Urbanization." *Planning Theory* 8 (1): 76–87.

Sheikh, S., and Banda, S. (2014). "The Thin Line Between Legitimate and Illegal: Regularising Unauthorised Colonies in Delhi." In *A Report of the Cities of Delhi Project.* New Delhi: Centre for Policy Research.

Silver, J. (2014). "Incremental Infrastructures: Material Improvisation and Social Collaboration across Post-Colonial Accra." *Urban Geography* 35 (6): 788–804.

Taussig, M. (1993). *Mimesis and Alterity: A Particular History of the Senses.* New York: Routledge.

Tiwari, P., and Rao, J. (2017). *Delhi's Changing Built Environment.* New York: Routledge.

Verma, G. D. (2002). *Slumming India: A Chronicle of Slums and Their Saviours.* New Delhi: Penguin Books.

Yardley, J. (2013). "Illegal Districts Dot New Delhi as City Swells." *The New York Times*, April 27.

Zimmer, A. (2012). "Enumerating the Semi-Visible: The Politics of Regularizing Delhi's Unauthorised Colonies." *Economic and Political Weekly* 47(30): 89–97.

Further reading

Benjamin, S. 2008. "Occupancy Urbanism: Radicalizing Politics and Economy beyond Policy and Programs." *International Journal of Urban and Regional Research* 32 (3): 719–729. This foundational analysis of low-income neighborhoods in India shows how diverse tenure systems and mixed economic landscapes that appear disorderly and inefficient nonetheless maintain diverse economies and pro-poor agglomerations that divert real estate surpluses toward the urban majority.

Ghertner, D. A. 2015. *Rule by Aesthetics: World-Class City Making in Delhi.* New York: Oxford University Press. This ethnographic study of slum demolition and urban beautification in India's capital city shows how globally circulating aesthetic norms became the dominant means of adjudicating the legality of land uses in Delhi, allowing slums to be removed simply because they looked illegal, and plan-violating malls and other such "world-class" developments to be approved because they looked planned.

Roy, A. 2009. "Why India Cannot Plan Its Cities: Informality, Insurgence and the Idiom of Urbanization." *Planning Theory* 8 (1):76–87. This article explores how the appearance of a lack of planning implementation and coherent design standards is not the outcome of state failure or bureaucratic neglect, but rather a planned product of "unmapping," a strategic deployment of exceptions and ambiguous rules that allows state agents to flexibly enforce rules to sustain patron-client relations.

PART III
Aspirations

PART III.I

Resilience and sustainability

Introduction

Resilience and sustainability are relatively new terms in the lexicon of urban and environmental studies. Nevertheless, few concepts in the last decade have attracted as wide attention in these fields as these two. The term "sustainable" first appeared in the midst of the international energy crisis of the early 1970s[1] and initially referred almost exclusively to the preservation and enhancement of natural resources. Eventually, sustainability came to be defined as the balance of three E's – environment, economy, and equity – even though the challenges of reconciling the E's have been well documented, with equity often the loser among the three (Campbell 1996). As Google's Ngram Viewer shows, the term "Resilience" – which implies a system's capacity to recover quickly from difficulties, drawbacks, or catastrophes – has a longer history of use, as it first appeared in the early 19th century. But the term was used very sparingly until the 1940s, and it became increasingly popular and widely used only after the 1980s, as many cities became concerned about their risks from and responses to natural and human disasters, including prominently, climate change.

Resilience and sustainability represent aspirations for urban design, and the four chapters in this section seek to address some of the following questions: How is resilient urban design defined, and how can it instill resilience in cities? How can urban design help the groups that are the most vulnerable? And what tools do cities have available to instill and foster sustainability and resilience principles in development?

For Zachary Lamb and Lawrence Vale, *resilient urban design* is "the merging of environmentally protective infrastructure and a quality built environment that promotes equitable habitation." They bring attention to the fact that, all too often, efforts to promote sustainability and resilience in cities are not necessarily beneficial to the most vulnerable groups and instead lead to their displacement or further economic marginalization. Drawing examples from New Orleans, Rotterdam, and Dhaka, they argue persuasively that traditional design approaches seeking to achieve resilience have not given proper emphasis to the needs of vulnerable social groups. They call for an urban design that gives at least equal attention to social and spatial resilience, is sensitive and adaptive to the relative power of different actors, integrates protective infrastructure that can adapt to the city fabric, and balances the need for control and flexibility in urban landscapes.

Along a similar line of thought, in the following chapter, Diane E. Davis argues that it is the least privileged and most vulnerable social groups living in settings characterized by violence and insecurity that need resilient urban design the most. She defines resilience as "the establishment of social, spatial, and political conditions at the level of the community, which could allow citizens a semblance of normalcy and a daily life without extreme fear." Drawing lessons from the positive impacts of social urbanism in Medellín, Colombia, and from her own inquiries about urban design and resilience strategies in African and Latin American cities with chronic violence, she concludes that Global North strategies such as Crime Prevention Through Environmental Design (CPTED) cannot be successfully exported to the Global South. Instead, urban design strategies that can most effectively reduce risk in the Global South are those that are cognizant of the particularities of each context, reduce conditions of socio-spatial exclusion, and help promote collaboration among citizens and between citizens and the state.

A challenge for urban design is to help cities respond to existing and emerging conditions of risk. In the following chapter, Yosef Jabareen theorizes the concept of risk, arguing that it is interpreted differently by different actors based on their interests. In particular, politicians and decision makers follow political, economic, and social considerations in manipulating the concept of risk and promoting antidotes in different forms: "sustainable city," "ecological city," "zero-carbon city," "resilient city," "green city," "healthy city," and the like. Jabareen argues for the necessity of "spatializing contemporary risks and situating them at the city level." He describes how New York City is achieving all this in its sustainability plan, *PlaNYC 2030*, by incorporating visioning, adaptive spatial practices, energy-saving practices, and planning practices that promote justice.

How can cities foster sustainability principles in development? Since the mid-1990s, the LEED green building certification system has helped promote energy efficiency and sustainability principles in buildings worldwide. Frederick Steiner and Rebecca S. Popowsky discuss the recent development of a parallel system – the Sustainable Sites Initiative (SITES) – that focuses on outdoor environments. In their words, SITES "provides a comprehensive road map for defining and enhancing sustainability in site development, measuring performance, and ultimately assessing the environmental, economic, and social benefits of projects." They present two case study projects – the Washington Canal Park, in Washington, D.C. and Navy Pier in Chicago – to illustrate how SITES helped encourage an urban design that demonstrated sustainable practices, innovation, community engagement, and performance monitoring in development.

Note

1 The term "sustainable" was first used in the book *The Limits to Growth* (Meadows et al. 1972), in which the authors warned that the levels of resource consumption would lead to a catastrophic depletion and collapse of global systems and urged for "a condition of ecological and economic stability that is sustainable far into the future."

References

Campbell, S. 1996. "Green Cities, Growing Cities, Just Cities? Urban Planning and the Contradictions of Sustainable Development." *Journal of the American Planning Association* 62 (3): 296–312.

Meadows, D.H., Meadows, D.L. Randers, J., and Behrens, W.W. (1972). *The Limits to Growth*. New York: Universe Books.

28
PURSUING RESILIENT URBAN DESIGN
Equitably merging green and gray strategies

Zachary Lamb and Lawrence J. Vale

Introduction

The terms *urban design* and *resilience* each gained traction in the mid-20th century and initially inhabited almost entirely separate realms of theory and practice. Resilience emerged from the fields of engineering, social psychology, and ecology and focused on weakened infrastructure, traumatized individuals, and threatened ecosystems (Vale 2014). Urban design, as a profession, grew out of the ambitious projects of Modernist architects and planners for urban renewal and functional city design. In spite of the broad ambitions of early urban designers and a proliferation of normative aims (Loukaitou-Sideris 2012), urban design practice has frequently been limited to improving the aesthetic experience of the public realm between buildings.

In more recent decades, however, the expansion of urban settlements into more hazard-prone territories and the amplification of risk through environmental and technological change have made risk and resilience increasingly central concerns for practitioners and scholars of urban design. The impacts of recent urban disasters have revealed the dire consequences of building cities that are profoundly at odds with the biophysical processes of their landscapes. The need for adaptive responses to and preparation for urban disasters, including those linked with climate change, have encouraged city leaders to proactively consider altering the form of cities to reduce vulnerability.

As acute shocks become more widespread and more frequent, they reveal and exacerbate the chronic socioeconomic stresses that have long characterized cities. As starkly illustrated by the levee failures and flooding brought on by Hurricane Katrina in New Orleans, preexisting disparities among urban residents too often shape the uneven distributions of disaster impacts and recovery resources (Vale and Campanella 2005; Ehrenfeucht and Nelson 2011). As risks grow in cities across the Global North and South, vulnerability is distributed radically unequally both within and between countries and cities; those with the least economic and political resources frequently face the greatest harm from hazards (Wisner et al. 2004).

As city leaders around the globe seek to address these uneven vulnerabilities, urban designers (along with allied professionals) are increasingly involved in ambitious projects to restructure urban infrastructure and land use in the name of hazard resilience. Scholars and practitioners of pre-disaster planning and post-disaster recovery increasingly draw upon notions of both resilience and urban design (Rebuild By Design n.d.; National Disaster Resilience - HUD Exchange n.d.). Yet, both of these concepts have been faulted for lacking clarity and for not seriously considering issues of social equity.

In light of the widespread embrace of urban design as a tool for improving hazard resilience and concerns about the specificity and utility of the underlying concepts, the following sections address two linked questions: (1) can urban design be a domain for concrete projects, processes, and practices that ground resilience, creating places that are simultaneously more just and environmentally sound? And relatedly, (2) can a refined focus on resilience enable scholars and practitioners of urban design to synthesize social, ecological, and aesthetic analyses to make places that are simultaneously more flexible to changing environmental and social conditions and more attuned to the concerns of the most vulnerable members of society?

In addressing these questions, the chapter recounts the emergence and development of both "resilience" and "urban design" as analytical and practical constructs. We outline challenges and opportunities in developing a social-equity-focused practice of urban design, including the threat of opportunistic "disaster capitalism" (Klein 2008) and displacement via resilience-driven green gentrification (Hodson and Marvin 2009; Gould and Lewis 2017).

For the sake of clarity, we propose a working definition of *resilient urban design* as the merging of environmentally protective infrastructure and a quality built environment that promotes equitable habitation. We expand upon this definition by outlining a series of principles intended to simultaneously ground resilience and refocus urban design on the creation of just and environmentally sound places. To illustrate these principles, and the promise and pitfalls of resilient urban design, we draw from emblematic projects in cities of the Global North and Global South, with particular attention to three river delta cities, New Orleans, Rotterdam, and Dhaka.

Resilience and urban design: concepts in need of one another

The challenges of rapid urbanization and climate change-linked hazard amplification have motivated recent interest in applying the principles of "resilience" to "urban design." Yet for much of the time since the independent popularization of each term in the mid-20th century, resilience and urban design occupied distinctly different professional and theoretical spheres. Even so, both resilience planning and urban design have something important in common: both practices share a long-standing uncertainty about how best to engage with the natural systems that accommodate urban settlements. Proponents of resilience are torn between engineering ("bounce back") definitions of resilience that focus on hardened "gray" infrastructure, and definitions rooted in ecology that prioritize working more holistically with (rather than against) natural systems ("green" infrastructure). Similarly, urban designers may also pursue "green" approaches that emphasize the role of natural systems in shaping uses of urban space or may treat the abstract arrangement of buildings as the central determinant, relying on "gray" approaches to protect the built environment from natural hazards.

Gray and green resilience

In structural engineering and materials science, the concept of resilience describes the capacity of a material or system to return to a predefined steady state after some perturbation. This engineering resilience presupposes the existence of one desirable equilibrium state to which a system should return (Gallopín 2006). In the context of managing landscape risks, engineered resilience strategies tend to deploy infrastructure as a means of reducing the frequency of environmental disturbances in order to protect people and property. We term this *gray resilience*.

After a heyday of structural intervention in the mid-20th century, such gray strategies have been heavily criticized in recent decades. Critics hold that structural approaches frequently lead to a host of unintended consequences, including disturbing underlying ecological processes, unevenly distributing the benefits and burdens of infrastructural intervention, and even causing more damage, rather than less, in the event of major environmental disturbance (Burton, Kates, and White 1978). Flood control levees present a clear example of the critiques of gray resilience. Through a mechanism sometimes labeled the *levee effect*, natural hazards researchers have argued that levees meant to contain floodwaters can encourage more development and less well-adapted building patterns in hazardous areas. When levees are inevitably overwhelmed, as famously happened in New Orleans after Hurricane Katrina, these poorly adapted settlement patterns actually place more people and property at risk than would have suffered had the structures not been built (Burby 1998).

These engineering-centered structural strategies of gray resilience aim to keep hazards away from people, but there is also a countervailing approach that might be termed *green resilience*. Green approaches to hazard mitigation are resonant with the conceptualization of resilience that emerged from social psychology and the ecological sciences in the 1960s and 1970s (Holling 1973). This socio-ecological framing positions resilience as a relational concept, rejects the notion of a single homeostatic "balance of nature," and assumes that many socio-ecological systems, including cities, can have multiple stable states (Walker and Cooper 2011; Davoudi 2012). This suggests a more flexible coexistence between people and the natural systems that may come to imperil them under certain extreme circumstances. For those places subject to water inundation, green strategies tend to retain, manage, and drain water through mechanisms such as floodable public squares, permeable paving, and sacrificial floodplains that double as parkland with recreational and ecological benefits. In many cases, these green flood mitigation strategies mark a shift from "vertical" methods (e.g., levees and floodwalls) to "horizontal" methods that allow floodwaters to spread over larger areas (Warner 2008).

Gray and green urban design

The distinction between gray and green resilience has a parallel in the field of urban design.

Gray urban design: organizing buildings and the space between them

By the late 19th century, industrialization had driven the expansion of cities to unprecedented extents in Europe and North America. Urbanization in this era was largely driven by what Maria Kaika (2004) calls the "Promethean project" of "infrastructural modernization," through which states mobilized science and engineering for massive infrastructure projects meant to control the once fickle landscape forces on which cities rely.

The aggressive intervention of infrastructural modernization allowed the formal geometric vision of the Beaux Arts to be applied across large territories, enabling the classically inspired symmetrical grandeur of broad avenues, sprawling plazas, and precisely arranged civic buildings and monuments of the City Beautiful era from Chicago to St. Louis to Washington, D.C. (Wilson 1989). The predominant attitude of City Beautiful practitioners towards natural landscapes was exemplified in the 1901–1902 McMillan Commission proposal for extending the monumental axis of Washington, D.C. The plan called for the restoration and extension of L'Enfant's original grand east-west axis through the existing wetlands (Savage 2009).

In the years between the First and Second World Wars, this "gray" approach to urban design was adopted and stylistically renovated by Le Corbusier and his fellow CIAM Modernists. Many iconic projects of Modernist urban design from the 1930s through the 1960s – such as Ville Radieuse, Chandigarh, and Brasília – feature precisely arranged buildings on monumental and rigorously landscape-agnostic axes (Evenson 1969; Vale 2008).

Although prominent figures in early Modernist urban design, including Le Corbusier (Evenson 1969), Sert (1942), and Hilberseimer (1949), all cited defense from aerial bombardment (and later nuclear attack) as a rationale for radical spatial restructuring (Lamb and Vale 2017), they rarely discussed similar measures to address the threat of natural hazards. This lineage of geometrically abstract, gray urban design has continued across a range of settings, employing a variety of stylistic forms, including neo-traditional new town centers in North America (e.g., Seaside, Florida), grand civic plazas (e.g., Shenzhen's City Center complex (Cartier 2002)) and sprawling peripheral housing developments (Buckley, Kallergis, and Wainer 2016) in rapidly urbanizing regions.

Designers in the "gray urban design" lineage tend to place little emphasis on the forces and functions of landscapes, including those that yield natural hazards. Their schemes frequently apply highly abstract formal geometries in the name of aesthetic purity or narrow functionalism.

Green urban design: landscape as structure

In the mid-19th century, at the same time that infrastructural modernization and "gray urban design" were shaping the expansion of industrial cities, an alternative "green urban design" lineage was also taking root. Projects by early American landscape architects like Olmsted and Vaux created natural-looking landscapes to perform a range of functions, from floodwater retention to improving the physical and moral health of booming urban populations (Fisher 1986). Ebenezer Howard and later acolytes of the Garden City movement similarly argued for the functional, aesthetic, and spiritual value of structuring new cities according to landscape functions (Howard 1902).

While Howard's conception of landscape function remained largely diagrammatic and abstract, Patrick Geddes and those who followed in his path (from Lewis Mumford to Team 10) advocated integrated planning according to place-specific biophysical and social conditions (Geddes 1915). By the middle of the 20th century, landscape architects led by Ian McHarg advocated the idea that landscapes should provide the structuring framework for settlements through a "design with nature" approach (McHarg 1969). For practitioners of this "green urban design" lineage, urban design should be driven by a complex overlay of site-specific biophysical conditions, including existing soil types, hydrology, and ecological communities (Spirn 1985).

In recent years, branches of urban design research and practice known by such labels as "landscape urbanism" (Waldheim 2012) and "ecological urbanism" (Mostafavi and Doherty 2010; Spirn 2014) have made the ecological functioning of urban landscapes a central concern. They continue the legacy of "green urban design," calling for projects to create flexible urban spaces that adapt to natural processes.

These "green urban design" models have been widely criticized because their great expense and reliance on powerful central planning and control makes them incompatible with the constraints found in the vast majority of global urban settings, especially in the rapidly urbanizing Global South (Mukhija 2011). Landscape urbanism and its stylistic cousins have been criticized for paying insufficient attention to the human experience of urban space (Duany and Talen 2013). While "green urban design" is highly responsive to the geophysical dimensions of natural hazards, these approaches are largely blind to the social dimensions of uneven vulnerability.

Concerns regarding the relative inattention to the social dimensions of vulnerability can apply equally to green and gray urban design. These critiques suggest that, before resilience and urban design can be productively united, further attention is needed to how urban design shapes inequity in urban vulnerability.

Resilience and urban design: the dual challenge of safety and equity

While the proliferation of research studies about "resilience" in planning and urban design is a relatively recent phenomenon, the dual framing of "gray resilience" and "green resilience" has its roots in a broader debate about how hazard vulnerability is produced and how it is best mitigated. Scholars from a range of perspectives, including natural hazards research (White 1945), vulnerability research (Wisner et al. 2004), and political ecology (Collins 2008), have all made the case that uneven hazard vulnerability is socially produced. If social equity is to be an intrinsic part of the definition of resilient urban design, it is important to examine how the benefits of resilience are shared. In other words, achieving resilient urban design means more than constructing attractive places that reduce exposure to hazards; it also entails efforts to ensure equitable access to such places.

Achieving resilient urban design therefore entails a dual challenge. While it is certainly sensible to promote efforts to keep people away from hazards, all too often the efforts to design less vulnerable places have been accompanied by parallel processes of socioeconomic sorting and displacement (Anguelovski et al. 2016). Scholars and practitioners devoted to merging resilience and urban design have tended to focus on creating safer and more attractive public spaces in climate-vulnerable cities. Given the troubling patterns of inequality in hazard vulnerability and in the distribution of costs and benefits of urban resilience projects, we suggest that there must be a parallel focus on ensuring equity and access for all. In this reading, urban design that excludes or displaces parts of the population from areas rendered safer is, by definition, not resilient. If equity and access is part of the definition of resilient urban design, it is important to ask: resilient for whom?

Addressing this equity dimension of resilience entails looking carefully at how the benefits of urban design interventions are distributed. Behavioral approaches to hazards research came under heavy critique in the 1980s by scholars identified with the "vulnerability approach," who argued that such strategies focused too much on the role of individual choice and not enough on the ways that broader structures of political economy shape vulnerability (Wisner et al. 2004). In a similar vein, ecological framings of resilience – what we are calling green resilience – have been criticized for over-emphasizing the responsibility of individual actors in shaping their own vulnerability, ignoring power relations and naturalizing radical inequality by fetishizing "self-organizing" complex socio-ecological systems (Watts 2011).

These critiques of green resilience resonate with recent concerns raised by urban researchers. Some have raised alarms that opportunistic actors frequently use extreme events to restructure so-called "crisis cities" (Gotham and Greenberg 2014) and to advance agendas of privatization and accumulation through "disaster capitalism" (Klein 2008). Others argue that a focus on preparing for and recovering from "natural disasters" can obscure debates over the deeper political and economic roots of uneven vulnerability (Marcuse 2013; Fainstein 2015). Other critics have worried that the concept of "resilience," like "sustainability" before it, has been so liberally applied as to have lost any analytical utility (Vale 2014). In response to both the critique that resilience is too readily used to defend regressive status quo conditions and that the concept is hopelessly fuzzy, planning scholars have sought to reframe resilience with an integral normative orientation. They argue for the concept of "progressive" (Vale 2014), "evolutionary" (Davoudi 2012), or "critical" (Vale 2014) resilience, which enables urban communities to "bounce forward" towards greater equity following major shocks.

Navigating between gray and green resilience and between gray and green urban design

Informed by these critiques, we maintain that the concept of resilience continues to hold both practical and analytical potential for practitioners and scholars of urban design. Given the dialectical tensions between the gray and green framings of resilience, scholars, disciplines, and institutions tend to gravitate towards one side or the other of this spectrum. However, the complexity and dynamism of contemporary urban environments require that urban design practitioners and scholars continuously navigate the terrain between these poles. The legacy of previous generations of structural intervention and the mounting pressures placed on previously "safe" urban areas by climate change require that resilient urban design include consideration of collective gray hazard mitigation infrastructure. At the same time, the negative impacts of previous generations of gray protections, the dynamism of contemporary urbanization, and the uncertainties of climate change argue for green strategies that acknowledge socio-ecological complexity and aspire to greater flexibility.

In response to recent urban disasters and dire projections of climate change-related vulnerability, many cities are increasingly enrolling urban designers and allied professionals as central players in envisioning climate adaptive futures. Just as "resilience" must transcend the binary division between gray and green to be useful for urban design scholarship and practice, urban design must also move beyond the narrow tribalism that has often defined its debates in order to meaningfully contribute to the demands of adapting settlements to climate change. Neither "gray urban design," with its focus on the abstract composition of buildings for symbolic and narrow functional ends, nor "green urban design," with its fetishizing of geophysical landscape functions and relative inattention to social dimensions of vulnerability, will be adequate to create a robust model of resilient urban design.

Negotiating the tensions between gray and green resilience and between gray and green urban design is not a matter of splitting the difference or triangulation. Rather, in each case, the choice to move towards one pole or another carries distributional consequences, bestowing costs and benefits unevenly across space, across time, and across social groups. For instance, the choice to build a levee to protect a settlement in a floodplain – an option that fits squarely within the gray resilience and gray urban design frameworks – might reduce the incidences of flooding and enable expanded settlement in that area. However, such a project might also increase flooding elsewhere and may ultimately increase the vulnerability within the "protected" zone by encouraging risk-blind development. On the other hand, the choice to adopt landscape-based flood mitigation strategies would likely be cheered by advocates of ecological (green) resilience frameworks and green urban design. Yet proposals for such projects, shifting from "vertical" gray flood mitigation strategies to "horizontal" green strategies, have encountered fierce resistance in recent years from the Netherlands to New Orleans, as concerned residents fear losing the implicit or explicit guarantee of security that comes with engineered protections.

While recent frameworks for urban resilience identify dozens of normative goals and scores of indicators (Rockefeller Foundation and Arup 2015), we argue that forging new, equity-centered models for resilient urban design requires a dynamic and context-specific navigation of the gray-versus-green dialectical tensions of resilience that are paralleled by the gray-versus-green approaches to urban design. We argue that resilient urban design requires attention to four principles: (1) design for geophysical and social resilience simultaneously; (2) integrate adaptive infrastructure with the inhabited fabric of cities; (3) adapt to the relative strength of state versus non-state actors; and (4) balance the need for control and flexibility in urban environments.

The following sections expand on these four principles, drawing on the examples of recent and ongoing projects from three particularly flood-prone and infrastructure dependent cities: Dhaka in Bangladesh, Rotterdam in the Netherlands, and New Orleans in the United States. In all of their complexities, successes, and failures, the examples illustrate both the opportunities and the challenges of mobilizing urban design to advance social equity and environmental sustainability agendas in cities of the Global North and South.

Designing for geophysical and social resilience simultaneously

Projects pursuing resilient urban design, from both the gray and green perspectives, too often focus on the geophysical roots of hazard vulnerability, underappreciating the well-documented social dimensions of uneven vulnerability. Any attempt at resilient urban design must take these social functions seriously and treat them on an equal footing with geophysical factors. Unfortunately, many proposals that purport to unite urban design and resilience have actually exacerbated uneven social vulnerability by displacing or otherwise harming low-income residents.

The Hatirjheel project in central Dhaka was intended to simultaneously address several critical infrastructural deficits in the city, providing for stormwater retention while creating much-needed public space and improving road connectivity. Since its completion in 2013, the project has been celebrated as a rare instance of complex urban infrastructure improvement in the rapidly growing city. However, the Hatirjheel project's costs and benefits have been highly uneven. While thousands of low-income residents of informal settlements were displaced by the project, the powerful Bangladesh Garment Manufacturers and Exporters Association (BGMEA) has been allowed to continue occupying the illegal site of their headquarters building in the center of the project (see Figure 28.1). While project promoters claimed that displaced residents would be rehoused, interviews with participants indicate that resettlement has lagged and that the most vulnerable displaced residents, those with no documented tenure rights, received no compensation at all (Bangladesh University of Engineering and Technology official 2017). Given such equity concerns, this project cannot be considered as meeting all criteria for resilient urban design.

A few examples from post-Katrina New Orleans indicate promising directions for linking social and geophysical resilience in urban design projects. The Adaptation Support Tool, a spatial

Figure 28.1 Hatirjheel project, Dhaka.

Source: Satellite imagery from Google Earth (2017), modified by authors.

planning tool developed by Deltares with the city of New Orleans, employs touch screens and mapping software to provide instant estimates of the effectiveness of various flood resilience-related project options, including bio-swales, green roofs, and porous pavement. While the tool is primarily designed to allow planners and decision makers to assess the relative value of projects in improving geophysical resilience, the project team has also included some elements related to social equity and social resilience. The tool includes social equity criteria to encourage fair distribution of projects and includes options for green infrastructure projects like water squares that might support social capital formation through creating new inhabitable gathering spaces (Van de Ven 2017). In another example of integrating social and geophysical resilience, a nonprofit organization called The Ripple Effect has created both inhabitable water processing landscape projects and water education curriculum materials for urban public schools in New Orleans (Chang 2017).

Integrating adaptive infrastructure with the inhabited fabric of cities

Conventional gray approaches to resilience and urban design tend either to treat environmental infrastructure as completely separate from urban fabric (as in the case of the US Army Corps of Engineers' engineered levee landscapes in New Orleans that are not intended to be an inhabited part of the public realm), or to render these infrastructures invisible (as in the case of buried storm sewer infrastructure). Green strategies, on the other hand, tend to reveal these infrastructural functions and to incorporate them into urban life through multifunctional design strategies. In the words of one Dutch urban designer, "a sustainable plan" for urban water management is about "making every day better," not just those days when extreme events strike (Thomaes 2017).

The integration of infrastructural protections into the fabric of urban life can take a range of forms across scales, from the multifunctional infrastructural corridors proposed in the regional "delta planning" exercises in Bangladesh (e.g., Bangladesh Delta Plan 2100) and in the lower Mississippi delta (e.g., Changing Course competition), to proposals for small-scale stormwater retaining green infrastructure to be integrated into the urban fabric of New Orleans. As one component of the larger Gentilly Resilience District project, the city of New Orleans is planning a series of "blue and green streets," in which water retention will be incorporated into the "neutral ground" or median strips on several broad avenues in the Gentilly neighborhood (see Figure 28.2). The project will both reduce pressure on the city's aging stormwater systems and visually reinforce the relationship between New Orleans and its watery surroundings. The proposal for blue and green streets was included in the Greater New Orleans Water Plan and the federally funded Gentilly Resilience District and signals a return to a historical arrangement wherein many of New Orleans's broad streets were built straddling drainage canals (Waggonner and Ball 2013).

Incorporating infrastructural functions into the urban fabric can have a number of benefits, including creating redundant and distributed systems, adding recreational and ecological benefits, and re-visualizing previously invisible landscape processes. However, the specific configurations of these urban fabric-integrated infrastructures can have important equity implications. Because distributed green infrastructure can create new urban amenities, planners and urban designers must ensure that those amenities are distributed equitably and do not lead to "green gentrification" (Hodson and Marvin 2009; Gould and Lewis 2017). While the Hatirjheel and other "water urbanism" projects in Dhaka have combined water management infrastructure with recreational amenity, these projects have overwhelmingly benefited high-income residential areas. The projects have threatened low-income groups both with direct displacement through project-based

Figure 28.2 New Orleans's green and blue streets.

Source: Waggonner and Ball

Figure 28.3 Rotterdam Water Square.

Source: Photo by the Authors

clearance and with indirect displacement through gentrification of surrounding areas following the installation of the new amenities.

To insure that fabric-integrated infrastructure projects benefited existing residents and users of urban space, the "water square" project in Rotterdam's economically diverse Benthemplein area was designed by De Urbanisten with input from students in adjacent schools (see Figure 28.3) (De Urbanisten 2013). As an example of green approaches to resilience and to urban design that support the needs of less advantaged populations, this project meets all criteria for resilient urban design.

Adapting to the relative strength of state versus non-state actors

Both green and gray approaches to "resilient urban design" are typically premised on the presence of capable and stable state actors who can coordinate activities in the interest of a commonly accepted public good. Gray infrastructure strategies require major up-front financial investments as well as ongoing costs of maintenance and operations. Gray infrastructure strategies also generally rely on state regulation for some of their protective value. For instance, to reduce vulnerability, flood control levees depend on the capacity of state actors to regulate land use and building, prohibiting or discouraging vulnerable construction outside of protections. Because green infrastructure strategies typically require larger areas (e.g., to accommodate, store, and infiltrate floodwaters), such strategies are even more dependent on the capacity of state actors to regulate land use and building practices.

Unfortunately, this muscular, reliable, and public-spirited vision of state actors does not describe governance conditions in the majority of urban areas around the world. While there are many positive stories of neighborhood-based recovery and redevelopment in post-Katrina New Orleans, ambitious efforts to remake the city's relationship with water have been starkly limited by the fragmentation and limited capacity of state actors. In Dhaka, state actors not only do not have the capacity to regulate the city's land use, but they are, in many cases, among the most egregious violators of the city's plans and regulations (Sabet and Tazreen 2015).

While resilient urban design strategies developed in places like the Netherlands – with well-coordinated, well-financed, and accountable state actors – may provide some useful models, the translation of such models to radically different contexts like New Orleans and Dhaka can be deeply problematic. Rather than transplanting infrastructural strategies that require high levels of administrative and governance capacity, urban design strategies must be conceived to operate with the governance constraints and urbanization dynamics that actually exist on the ground in a given context. Models like the self-built water and sewer infrastructure of the Orangi Pilot Project in Karachi, Pakistan (Ahmed and Sohail 2003), incremental infrastructure upgrading in Shenzhen's urban villages (Kochan 2015), and the Baan Mankong community-led effort to restructure canal-side settlements in Bangkok to reduce flood vulnerability and water pollution (Boonyabancha 2005) suggest some promising models for resilient urban design strategies that are not premised on the involvement of strong state actors.

Balancing the need for control and flexibility in urban environments

While conventional gray infrastructure approaches promise to control natural hazards, green approaches tend to prioritize the need for flexibility in the face of uncertainty, including uncertainty regarding future urbanization and climate conditions. While events like the failure of New Orleans's levees from Hurricane Katrina demonstrate the limits of gray infrastructure, attempts to shift towards a paradigm based on the flexibility of green strategies can be fraught. In some cases, the development of flexible green strategies to benefit certain areas or populations can come at the expense of the perceived or real security of other residents.

Dutch landscape architects who worked on the removal of flood protection dikes, or "depoldering," of the 2000-hectare Noordwaard area described the deep resistance and resentment of rural residents who saw their levee protection sacrificed to create "room for the river" to protect nearby urban areas (de Konning 2017). Similarly, residents whose neighborhoods were threatened by the so-called "green dot plan" in post-Katrina New Orleans resisted the idea that they should give up their perceived security in the name of greater flexibility and resilience for the city at-large (Gotham and Greenberg 2014).

While recent failures of conventional gray infrastructure clearly demonstrate the need to develop alternative strategies, any shift towards more flexible green strategies will require transparent consideration of tradeoffs to ensure that this new flexibility does not disproportionately burden a city's most economically and politically vulnerable residents.

Conclusion: who benefits from resilient urban design?

While resilience and urban design emerged over the mid-to-late 20th century as distinct arenas of theory and practice, in recent years, urban design practice and scholarship have focused increasingly on mitigating environmental risk. Both resilience frameworks and the treatment of landscape in urban design get deployed across a spectrum between gray and green. The contemporary embrace of green models of resilience and urban design holds considerable promise for adapting and creating urban environments that are better suited to changing landscape conditions. However, there is no inherent correlation between advancing green models of resilience and urban design and improving social equity. In fact, the least advantaged urban citizens are often those most at risk of displacement and other forms of harm from projects undertaken to enhance landscape amenities. The flexibility demanded by many models of green resilience and urban design does not always appear so flexible to the urban poor. Too often, seemingly appealing strategies for creating recreational opportunities that double as buffer space or catchment areas during extreme weather events can also yield negative impacts on some members of the public. Many projects, such as Dhaka's Hatirjheel initiative, are far better at retaining water than they are at retaining low-income people. In such instances, placemaking gets paired with displacement with unease. In these cases, we argue, the intervention cannot meet the threshold of truly resilient urban design. For urban design to be resilient, the resulting interventions must do more than enhance the public realm and adapt to environmental challenges; they must also support social equity, rather than undermine it (see Figure 28.4).

While projects and programs meant to improve urban resilience can combine a range of approaches across the gray-to-green spectrums of resilience and urban design, to qualify as "resilient urban design," they must place consideration of social equity at their center. We propose four principles for social-equity-centered resilient urban design.

To achieve equitable outcomes when navigating between gray and green extremes, a project must be informed by a clear understanding of site-specific conditions that shape how the costs and benefits of interventions will be distributed. To guide this navigation, we have proposed a series of four principles that seek to reconcile four key tensions: (1) between geophysical and social drivers of vulnerability; (2) between spatial strategies centered on separate infrastructure and those focused on inhabited urban fabric; (3) between institutional strategies centered on state and non-state actors; and (4) between strategies whose normative goals are to control environmental forces and those focused on improving the flexibility of the built environment to adjust to environmental forces. If these tensions are addressed, resilient urban design can deliver well-designed, environmentally adaptive public space that serves the full spectrum of the public.

To advance this equity-centered model of resilient urban design, it is essential that scholars and practitioners of urban design critically engage with the distributive implications of their

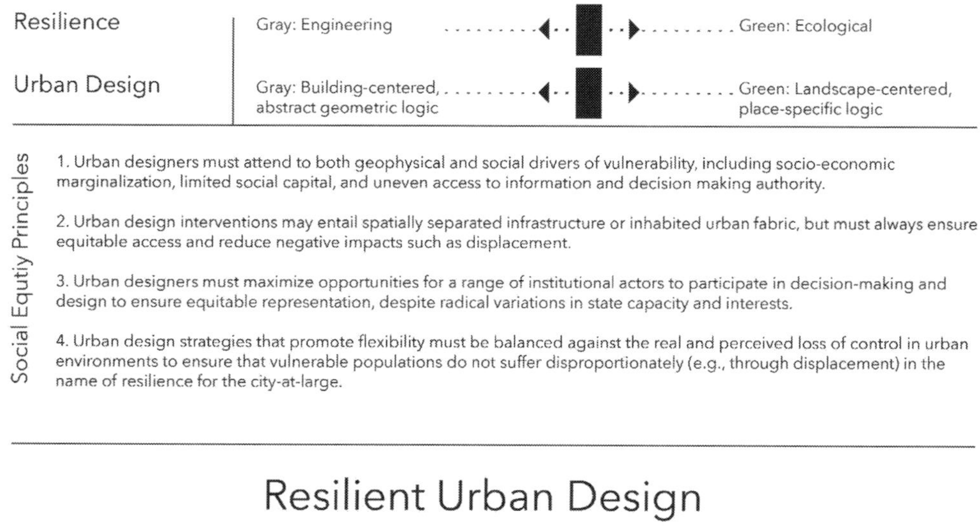

Figure 28.4 Resilient urban design as the equitable merging of green and gray strategies.

Source: Authors

interventions. In short, urban design cannot be resilient, and resilience cannot be fully urban, unless projects and practices are in service of social justice.

References

Ahmed, N., and Sohail, M. (2003). "Alternate Water Supply Arrangements in Peri-Urban Localities: Awami (People's) Tanks in Orangi Township, Karachi." *Environment and Urbanization* 15 (2): 33–42.

Anguelovski, I., Shi, L., Chu, E., Gallagher, D., Goh, K., Lamb, Z., Reeve, K., and Teicher, H. (2016). "Equity Impacts of Urban Land Use Planning for Climate Adaptation: Critical Perspectives from the Global North and South." *Journal of Planning Education and Research* 36 (3): 333–348.

Boonyabancha, S. (2005). "Baan Mankong: Going to Scale with 'Slum' and Squatter Upgrading in Thailand." *Environment and Urbanization* 17 (1): 21–46.

Buckley, R. M., Kallergis, A., and Wainer, L. (2016). "Addressing the Housing Challenge: Avoiding the Ozymandias Syndrome." *Environment and Urbanization* 28 (1): 119–138.

Burby, R.J. (1998). *Cooperating with Nature Confronting Natural Hazards with Land Use Planning for Sustainable Communities*. Washington, DC: Joseph Henry Press.

Burton, I., Kates, R.W., and White, G.F. (1978). *The Environment as Hazard*. New York: Oxford University Press.

Cartier, C. (2002). "Transnational Urbanism in the Reform-Era Chinese City: Landscapes from Shenzhen." *Urban Studies* 39 (9): 1513–1532.

Chang, A. (2017). Interview with the authors, January 19.

Collins, T.W. (2008). "The Political Ecology of Hazard Vulnerability: Marginalization, Facilitation and the Production of Differential Risk to Urban Wildfires in Arizona's White Mountains." *Journal of Political Ecology* 15 (1): 21–43.

Davoudi, S. (2012). "Resilience: A Bridging Concept or a Dead End?" *Planning Theory & Practice* 13 (2): 299–307.

de Konning, R. (2017). Interview with the authors, May 4.

De Urbanisten. (2013). "Water Square Benthemplein in Rotterdam, the Netherlands." *Landscape Architecture Frontiers* 1 (4): 136–143.

Duany, A. and Talen, E. (eds.) (2013). *Landscape Urbanism and Its Discontents: Dissimulating the Sustainable City*. Gabriola Island, BC: New Society Publishers.

Ehrenfeucht, R. and Nelson., M. (2011). "Planning, Population Loss and Equity in New Orleans after Hurricane Katrina." *Planning Practice & Research* 26 (2): 129–146. https://doi.org/10.1080/02697459.2011.560457.

Evenson, N. (1969). *Le Corbusier: The Machine and the Grand Design*. London: George Braziller.

Fainstein, S. (2015). "Resilience and Justice." *International Journal of Urban and Regional Research* 39 (1): 157–167.

Fisher, I.D. (1986). *Frederick Law Olmsted and the City Planning Movement in the United States*. Architecture and Urban Design: No. 15. Ann Arbor, MI: UMI Research Press, c1986.

Gallopín, G.C. (2006). "Linkages Between Vulnerability, Resilience, and Adaptive Capacity." *Global Environmental Change*, Resilience, Vulnerability, and Adaptation: A Cross-Cutting Theme of the International Human Dimensions Programme on Global Environmental Change, 16 (3): 293–303. https://doi.org/10.1016/j.gloenvcha.2006.02.004.

Geddes, P. (1915). *Cities in Evolution*. Williams & Norgate.

Gotham, K.F., and Greenberg, M. (2014). *Crisis Cities: Disaster and Redevelopment in New York and New Orleans*. Oxford: Oxford University Press.

Gould, K.A., and Lewis, T.L. (2017). *Green Gentrification: Urban Sustainability and the Struggle for Environmental Justice*. 1st edition. Abingdon, UK: Routledge.

Hilberseimer, L. (1949). *New Regional Pattern Industries & Garden*. Paul Theobald.

Hodson, M. and Simon M. (2009). "'Urban Ecological Security': A New Urban Paradigm?" *International Journal of Urban and Regional Research* 33 (1): 193–215.

Holling, Crawford S. (1973). "Resilience and Stability of Ecological Systems." *Annual Review of Ecology and Systematics*, 1–23.

Howard, E. (1902). *Garden Cities of To-Morrow*. London: Swan Sonnenchein & Co., Ltd.

Kaika, M. (2004). *City of Flows: Modernity, Nature, and the City*. New York: Routledge.

Klein, N. (2008). *The Shock Doctrine: The Rise of Disaster Capitalism*. 1st edition. New York: Picador.

Kochan, D. (2015). "Placing the Urban Village: A Spatial Perspective on the Development Process of Urban Villages in Contemporary China." *International Journal of Urban and Regional Research*. http://onlinelibrary.wiley.com/doi/10.1111/1468-2427.12287/full.

Lamb, Z. and Vale, L.J. (2017). "From the Cold War to the Warmed Globe: Planning, Design-Policy Entrepreneurism, and the Crises of Nuclear Weapons and Climate Change." *Planning Perspectives*, 1–33.

Loukaitou-Sideris, A. (2012). "Addressing the Challenges of Urban Landscapes: Normative Goals for Urban Design." *Journal of Urban Design* 17 (4): 467–484. https://doi.org/10.1080/13574809.2012.706601.

Marcuse, P. (2013). *Un-Natural Disasters, Recursive Resilience, Unjust Compensation, Visionless Planning*. www.plannersnetwork.org/wp-content/uploads/2013/09/PPM_Marcuse.pdf.

McHarg, I.L. (1969). *Design with Nature*. New York: American Museum of Natural History.

Mostafavi, M., and Doherty, G. (2010). *Ecological Urbanism*. Baden: Lars Müller Publishers.

Mukhija, V. (2011). "Urban Design for a Planet of Informal Cities." In Banerjee, T. and Loukaitou-Sideris, A. (eds.). *Companion to Urban Design*, New York: Routledge, 574–584.

"National Disaster Resilience – HUD Exchange." (n.d.). Accessed November 4, 2017. www.hudexchange.info/programs/cdbg-dr/resilient-recovery/.

"Rebuild by Design." (n.d.). Accessed November 4, 2017. www.rebuildbydesign.org/.

Rockefeller Foundation, and Arup. 2015. "City Resilience Framework." Accessed March 14, 2019. https://assets.rockefellerfoundation.org/app/uploads/20140410162455/City-Resilience-Framework-2015.pdf

Sabet, D.M., and Tazreen, A. (2015). "Governing Growth Understanding Problems in Real Estate Development in Dhaka, Bangladesh." *Journal of South Asian Development* 10 (1): 1–21. https://doi.org/10.1177/0973174114567367.

Savage, K. (2009). *Monument Wars: Washington, D.C., the National Mall, and the Transformation of the Memorial Landscape*. Berkeley: University of California Press.

Sert, J.L. (1942). *Can Our Cities Survive?* Cambridge, MA: Harvard University Press.

Spirn, Whiston, A. (1985). *The Granite Garden: Urban Nature and Human Design*. Basic Books.

Spirn, Whiston, A. (2014). "Ecological Urbanism: A Framework for the Design of Resilient Cities (2014)." *The Ecological Design and Planning Reader*, 557–571. Springer.

Thomaes, S. (2017). Interview with the authors, April 26.

Vale, L.J. (2008). *Architecture, Power and National Identity*. 2nd edition. New York: Routledge.

Vale, L.J. (2014). "The Politics of Resilient Cities: Whose Resilience and Whose City?" *Building Research & Information* 42 (2): 191–201.

Vale, L.J. and Campanella, T.J. (2005). *The Resilient City: How Modern Cities Recover from Disaster.* New York: Oxford University Press.
Ven, F. (2017). Interview with the authors, May 5.
Waggonner and Ball. (2013). *Greater New Orleans Urban Water Plan: Vision.* New Orleans: Greater New Orleans Urban Water Plan.
Waldheim, C. (2012). *The Landscape Urbanism Reader.* Chronicle Books.
Walker, J. and Cooper, M. (2011). "Genealogies of Resilience from Systems Ecology to the Political Economy of Crisis Adaptation." *Security Dialogue* 42 (2): 143–160.
Warner, J.F. (2008). "The Politics of Flood Insecurity: Framing Contested River Management Projects." Unpublished PhD dissertation, Wageningen University.
Watts, M. (2011). "Ecologies of Rule: African Environments and the Climate of Neoliberalism." *The Deepening Crisis: Governance Challenges after Neoliberalism.* New York: New York University Press, 67–92.
White, G.F. (1945). *Human Adjustment to Floods: A Geographical Approach to the Flood Problem in the United States.* Chicago, IL: University of Chicago.
Wilson, W.H. (1989). *The City Beautiful Movement.* Creating the North American Landscape. Baltimore: Johns Hopkins University Press.
Wisner, B., Blaikie, P., Davis, I., and Cannon, T. (2004). *At Risk: Natural Hazards, People's Vulnerability, and Disasters.* London; New York: Routledge.

Further reading

Gotham, K.F. and Greenberg, M. (2014). *Crisis Cities: Disaster and Redevelopment in New York and New Orleans.* New York: Oxford University Press. A critical account of "crisis cities" and "crisis-driven urbanization" processes in post-Katrina New Orleans and post-Sandy New York City, highlighting the spatial politics of post-disaster planning and urban restructuring.
Klein, N. (2008). *The Shock Doctrine: The Rise of Disaster Capitalism.* 1st edition. New York: Picador. A critical account of opportunistic "disaster capitalism," advancing agendas of privatization in the wake of natural disasters and war.
Meyer, H. and Nijhuis, S. (2013). "Delta Urbanism: Planning and Design in Urbanized Deltas: Comparing the Dutch Delta with the Mississippi River Delta." *Journal of Urbanism: International Research on Place Making and Urban Sustainability* 6 (2): 160–191. An analysis of the urbanized delta regions in the Netherlands and the lower Mississippi, rooted in McHargian "green" urbanism of layered landscape and human systems understood through the lens of complex systems theory.
Pickett, S.T., Cadenasso, M.L., and McGrath, B. (2013). *Resilience in Ecology and Urban Design: Linking Theory and Practice for Sustainable Cities.* New York: Springer. Edited volume bringing together scholars from ecological and social sciences with design scholars and practitioners. Firmly rooted in the "green urban design" lineage.
Watts, M. J. (2015). "The Origins of Political Ecology and the Rebirth of Adaptation as a Form of Thought." *The Routledge Handbook of Political Ecology.* New York: Routledge, 19–50. A critical analysis of the rise of resilience and adaptation discourses, particularly as applied to development and the Global South.

29

REDUCING VULNERABILITIES THROUGH URBAN DESIGN

Interventions for resilience in communities at risk[1]

Diane E. Davis

Introduction

Urban designers frequently conceive of their practice through the lens of aesthetics and with a commitment to making the built environment more visually appealing. This is not to say that urban design is not also a tool to foster sociability or to generate positive social and economic conditions (Madanipour 2006; Punter 2007; Gunder 2011). Yet the widespread adoption of urban design tactics in entrepreneurial cities motivated by global competitiveness aims reinforces the view that urban design interventions are associated with a degree of privilege and are more likely to be deployed in prime spaces in the city. Sometimes they are even seen as a luxury indulgence that is deemed appropriate only after basic livability priorities are met. Indeed, scholars have suggested that this is more likely to happen when the urban design skill set is treated separately from urban planning, allowing its tools to be co-opted by other actors without clear social aims (Cuthbert 2007). In recognition of these exclusionary tendencies, a recent publication by Armborst et al. (2017) argues for the importance of crafting urban design interventions that facilitate social inclusion. This chapter builds on that sentiment but takes it in a slightly different direction. It starts from the assumption that the poorest, least privileged, and most highly vulnerable communities with unmet basic needs like safety and security are most in need of urban design ingenuity and innovation. To the extent that the field of urban design is indeed capable of focusing on more than just attractiveness, it holds the potential to positively enhance human experience as well as the material properties of a place, even in those settings where vulnerabilities are extreme.

In what follows, we examine recent urban design interventions undertaken to reduce risk and generate resilience in cities of Latin America and Africa experiencing chronic violence. Although urban violence is not unique to these regions, it has become an everyday fact of life in cities of the Global South where the roots of violence are structural – related to poverty, income inequality, informality, and the crushing coercive power of the perpetrators of violence, whether criminal mafias or the police themselves. This chapter reviews three distinct approaches: (1) those associated with Crime Prevention through Environmental Design (CPTED), an urban design paradigm which originated several decades ago but is being adopted in cities across Africa and other parts of the Global South; (2) the program of social urbanism, an innovation emanating from Medellín, Colombia; and (3) a body of spatial strategies and urban design tactics revealed through research on urban resilience in situations of chronic violence.[2] The aim here is to lay bare the assumptions underlying the deployment of these various urban design approaches, to assess their demonstrated capacities to reduce violence, and to ask whether they have made a difference in improving quality of life and urban conditions in communities at risk of crime and other related forms of insecurity. We conclude with a discussion of which spatial strategies and urban design tactics have been most effective in reducing risk and generating resilience, and in what combination with other common urban planning or policy measures, how, and why.

Preventing violence through design: CPTED and VPUU

The urban design guidelines known as CPTED focus on the modification of urban infrastructures so as to reduce opportunistic criminal activity and foster citizen security. The principles that embody CPTED first evolved from Oscar Newmans's *Defensible Space* (1972) and his work in American cities (specifically New York's social housing projects),[3] but they also trace to writings by Jane Jacobs (1961) and C. Ray Jeffrey (1969, 1971). Newman emphasized the potential of building and neighborhood design as a basis for reducing crime (Newman 1972: 255). He conceived of four main strategies to produce defensible, or secure, spaces: (1) establish a sense of territoriality in the physical environment using both symbolic and physical barriers; (2) design opportunities for residents to perform natural surveillance; (3) engage aesthetics to influence positive behaviors; and (4) juxtapose safe, adjoining areas that foster neighborhood continuity. Together, these strategies were intended to dissuade potential offenders from committing opportunistic crimes and also to ensure that perpetrators would be easily detected (Owusu et al. 2015). Urban design's potential to foster a sense of community was key, since a breakdown of the collective social fabric was thought to cause neighborhood vulnerability to crime; and with a focus on community residents, Newman (1972: 3) also imagined a return to "policing" in the oldest Western tradition: "the responsibility of each citizen to ensure the functioning of the polis."

Kelling and Wilson's (1982) "Broken Windows" theory later evolved from these CPTED principles.[4] Broken Windows was premised on the assumption that small signs of disorder lead to larger deterioration, and thus efforts to control disorderly behavior by police should be direct and unambiguous. Though Broken Windows marked a fundamental turn away from design towards a singular and explicit focus on citizen behaviors (Cassidy et al. 2014), planning and design sensibilities did not entirely disappear. Amongst others, Wekerle and Whitman (1995) continued to massage CPTED's preoccupation with the "defensible space" of architecture and physical security into an urban planning framework, focusing more broadly on characteristics of the environment and the social – not merely physical – modifications that might enhance security. This included "situational" elements, such as lighting, sightlines, land use mix, activity generators, and informative signage, as well as measures to foster a sense of "ownership" and reduce entrapment spots.

Despite the fact that CPTED and Broken Windows theory originated in the Global North, these techniques have become quite influential in Latin America and elsewhere in the developing world. One of the region's most important scholars of urban violence, Caroline Moser (2004), considered CPTED to be valuable as a cross-sectoral strategy for linking citizen security aims to urban renovation, thus addressing two of the principal challenges facing authorities in poor cities of the Global South. Urban upgrading programs of the 1980s and 1990s had long focused on bettering slum housing conditions – on the assumption that changes in the physical built environment would translate into social benefits that might generate security, albeit primarily economic.[5]

Despite the fact that most of these programs were not originally developed with a stated intent to eliminate crime and violence, they were soon recast along these lines. Samper (2016) argues that CPTED principles translated relatively easily into the goals of many slum improvement programs, and that similar infrastructure interventions adopted in the Favela-Bairro upgrading program, such as the opening of pedestrian-only streets and public space, contributed positively to a sense of security amongst *favela* residents, despite the fact that citywide violence persists.

One of the more widely celebrated versions of CPTED adopted in the Global South was formulated for poor, crime-ridden areas of Cape Town, South Africa, in 2005. Called Violence Prevention through Urban Upgrading (VPUU), this initiative embraced the physical dimensions of crime prevention more explicitly than most Latin American upgrading programs, primarily by focusing less on housing and more on public space (Sanya 2012; Ugur 2016). Although advocates of this strategy were also committed to human development and social sustainability (Barolsky 2016), VPUU identified its primary aim as reducing opportunities for crime in public space by improving the usability of existing space and promoting citizen-led surveillance (Krause et al. 2014). Other aims included: a sense of community ownership over space; well-defined access and safer movement; improved image and aesthetics; elimination of physical barriers impeding movement; reconfigured logistics of operation; maintenance and management; and inclusive design practices via participatory processes (Ibid.: 37).[6]

The VPUU placed itself on the global stage as a model for the application of physical techniques towards violence reduction aims, and since its development, the program has been awarded numerous accolades. Both VPUU and Favela-Bairro have been widely celebrated for their immediate physical improvements in housing and surrounding neighborhood conditions, as well as for the community engagement processes that marked their implementation. Yet both have been assessed as falling short in terms of their identifiable impacts on security. Some scholars blame the failures of design-based crime prevention strategies like VPUU to a lack of transparency and insufficient communication with local residents (Brown-Luthango et al. 2016).[7] Of course, some positive effects did accrue, such as the increased perception of safety following electricity provision and targeted public space interventions for children. Police also report decreased levels of mob justice and a better relationship with community members, who are now more likely to report crime (Ibid.). Yet, most observers still note, the program produced very little measurable impact on violence.

Some of this may have to do with the flaws in the strategy. Even within the US and European contexts, CPTED and Broken Windows have been criticized as policy failures (Harcourt 2001) or assessed as "inconclusive," at best, in reducing violence (Cozens et al. 2005). But an equally significant barrier to success is the context, primarily the socioeconomic and spatial conditions in cities of the Global South, that limit the potential of US-derived urban design interventions to address violence. In Latin America and many parts of Africa, urban violence derives from complex political, social, and economic inequalities that are deeply ingrained in institutions and people's mentalities. In these settings, violence is usually far from opportunistic, but rather is motivated by larger aims of controlling territory and/or establishing robust criminal economies built on illicit practices. As such, intervening in the built environment in an effective manner will entail producing policy priorities and governance conditions that strengthen community autonomy and challenge socio-spatial exclusion – both of which will require a capacity to work across multiple sectors and scales simultaneously. Rather than focusing on "violence prevention" solely through CPTED, VPUU, and other such situationally oriented violence reduction programs, urban designers and planners must also identify policy measures that allow communities to marginalize armed actors and to foster quotidian security in situations of deeply ingrained, structurally produced violence.

Social urbanism

One city known for its successes in these regards is Medellín, Colombia, which has developed a spatially integrated package of multi-scalar interventions built around strategies and tactics drawn from the urban design toolkit. Medellín's problems with violence owe to years of armed struggle between competing armed forces seeking sovereignty and political power across much of the Colombian national territory. At one point in the late 1990s, it was considered one of the most dangerous cities in the world, with paramilitary and drug-trafficking violence made even more dangerous by extreme socio-spatial inequality in the city. But today, Medellín has both challenged its patterns of socio-spatial exclusion and made headway in dealing with chronic violence, primarily through the implementation of urban design strategies classified under the rubric of "social urbanism."

Medellín's spatial and design interventions are in fact dramatically different from those deployed in the cities noted earlier. First, the city committed itself to a high level of service provision across all its informal areas, with a central utility company providing water, sewer systems, and electricity to almost 99% of the city. Second, "targeted areas" were selected to host *Integrated Urban Projects* (PUI, or Proyectos Urbanos Integrales). These sites were chosen primarily on the basis of low human development index, and each PUI directly or indirectly benefited between 150,000 and 300,000 inhabitants (Osorio 2011). These programs were often connected with emblematic public architecture projects and accompanied by community engagement measures, including participatory budgeting, so as to bring citizens into conversation with local authorities (Davis 2012). Planning, design interventions, and governance priorities were central to the success of PUIs, with all three formulated around a commitment to robust community participation (Samper 2012). The community's overall mobility was also improved by the addition of Metrocable services[8] in several of these PUI areas, and new public spaces were strategically connected through the upgrading and regularizing of streets, and through the addition of lighting and walkways in the city's long-neglected peripheral neighborhoods.[9] Third, and perhaps most significantly, all these initiatives were seen as a means for connecting the urban poor with the state in both material and symbolic ways. PUIs allowed communities to see the "state as a non-repressive actor," and "offered legitimacy to informal communities in which the projects [were] executed, even when land titles themselves were not granted" (Samper 2012: 18).

Far from being understood as a series of individual, targeted interventions, then, the strategy of social urbanism must be understood as a set of principles for urban governance that were jointly crafted and carried out by both local authorities and citizens. In this sense, the Medellín case suggests a new way of thinking about violence reduction built on a threefold effort to: (1) reduce conditions of neglect and vulnerability at the scale of the neighborhood and district, not merely the house or the street; (2) empower citizens in the process of demanding services from the state, and by so doing create new forms of vertical political connections and legitimate governance; and (3) inculcate a sense that the city as a whole is a site for upgrading, thus introducing a much larger scalar context for targeted interventions, and by so doing, strengthen horizontal relationships that tie citizens together behind a larger urban project around which city leaders would be held accountable.[10]

Medellín's authorities were very skilled in introducing what we term *kinetic* and not merely static interventions. By kinetic, we mean interventions that generate movement and constant dynamism within and across space, conceiving of the city as a metabolic organism, where the parts and the whole are connected to each other. Transport interventions such as the Metrocable were intended to push back against spatial exclusion and ghettoization of the city's most vulnerable residents in ways that helped deal with the looming threat of violence. This meant freer socio-spatial integration intended to enable movement across "invisible borders" established by armed groups without fear (Heinrichs and Bernet 2014: 65), as well as symbolic integration to help tear down "imaginary walls" between the formal and informal city (Pérez, quoted in Coupé et al. 2013: 61). Following Medellín's publicized successes, numerous other cities began experimenting with cable cars, though not always with the same degrees of success in terms of violence reduction (Freire-Medeiros and Name 2017), thus suggesting that fostering mobility is not enough unless it complements other strategic spatial and planning interventions. As Rivadulla and Bocarejo (2014: 2034) argue, the Metrocable's embeddedness in a whole package of procedural and very costly material interventions in Medellín, much more "than the cable cars themselves," was responsible for community transformation. Similarly, Milan and Creutzig (2017: 130) suggest that Medellín's urban design authorities produced "synergies between participatory urban planning and the development of new public spaces and transport infrastructure, both of which helped promote ownership of the environment and close bonds of trust within and between communities and authorities."

However, one cannot deny that Medellín's program of social urbanism – and any claims about its positive impact on the security situation – came in the context of national-level peace negotiations that led to demobilization of armed groups immediately prior to the formulation of the strategies of social urbanism. Medellín had long pursued a multitude of security strategies, including repressive militarist policies, and ongoing negotiations with different armed actors (Samper 2012). As such, the city's progress in fostering security was the outcome of more than just the local authorities' urban design strategies, despite the clear importance they played. Equally important was the historical weight of chronic violence, which brought citizens, local elites, and local authorities together in ways rarely seen across the Global South, thus allowing for many ground-breaking programs and public-private partnerships to emerge. This in turn raises questions about the transferability of social urbanism. In the absence of peace negotiations and wide stakeholder collaboration, could urban design approaches still make a difference?

Urban resilience

Evidence suggests that the answer is yes, or so that is the conclusion of a report titled *Urban Resilience in Situations of Chronic Violence* (Davis 2012). Using a focus on four cities in Latin America and three in Africa, researchers sought to move beyond CPTED's focus on situational crime and/or the redesign of spaces, and to assess the impacts of local power structures, community relations, support from the state, and the role of security forces (i.e. police) in mediating community responses to design interventions intended to increase urban quality of life and/or reduce violence. Rather than asking if urban design policies could address the root causes of violence, then, the project sought to understand whether spatial interventions could enhance urban resilience, which itself was understood as a bulwark against violence. Resilience was defined as the establishment of social, spatial, and political conditions at the level of the community, which could allow citizens a semblance of normalcy and a daily life without extreme fear. In looking for evidence of resilience, researchers found that when formulated with larger socio-political conditions in mind, and when jointly involving citizens and public authorities in implementation, urban design can play the role of uniting citizens to each other and the state in ways that foster resilience.[11]

Two conditioning factors explained whether spatial strategies and designed tactics were likely to be successful in reducing risk and generating pathways towards community security: the *location* and the *scale* of the design intervention. Only in some parts of the city – primarily in locations with mixed land use and vibrant community activities – would urban design interventions hold the potential to engage a diversity of stakeholders – whether citizens, the private sector, or the state – in ways that strengthened community connection and autonomy. Further, the researchers found that site-specific, grand redevelopment projects often fell short of intended aims by disrupting local social relationships or displacing residents. A third main takeaway from the report was that urban design strategies were often most transformative when they extended across territorial scales, because such projects involved multiple stakeholders and widened the shared commitment to security. The report concluded that in order to build resilience through their actions, urban designers must consider the interrelationship between intervention strategies and the strengthening of horizontal and vertical relationships within and between citizens and local authorities. Stated differently, cities with the best examples of positive resilience not only counted on strong community organizations capable of pushback against violence, but their capacities to do so derived partly from the fact that the programs they pursued (urban renovation, participatory budgeting, community-led reconstruction, collectively administered water and sanitation projects, etc.) fostered strong horizontal linkages among a number of constituencies extended in physical space (Davis 2012).

Nowhere was this clearer than in Mexico City, where the redesign of downtown pedestrian shopping streets (Regina and Alhondiga) and new investments in the historical mariachi neighborhood of Plaza Garibaldi brought new visitors from across the city and thus reversed these neighborhoods' decline as a result of violence (Figure 29.1). Both sites experienced a resurgence of popularity as well as restored perceptions of security as evidenced by the increasing number of people on these streets both day and night (Davis 2012). Interventions were multifaceted and multi-scalar, built around pedestrianization, upgrading existing markets, adding trees and

Figure 29.1 Plaza Garibaldi, Mexico City.

Source: By EpMe – Own work, CC BY-SA 3.0, https://commons.wikimedia.org/w/index.php?curid=17143375

benches, and upgrading building façades. Design was used as a tool to reestablish "normal" use of these spaces. For Plaza Garibaldi, this also included controlling access through the construction and placement of a new museum (Broid and de la O 2012). Tellingly, these projects involved collaboration with the Secretary of Urban Development and Housing, Secretary of Public Safety, the prosecuting police, and the residents.

The results in Mexico City contrast dramatically with those revealed in São Paulo, where urban design interventions were also undertaken but were modeled more around the Broken Windows paradigm. Unlike in Mexico City, which adopted a robust revitalization program intended to create more public space, housing, economic opportunities, and other amenities, São Paulo's Nova Luz Renewal Program was an example of single-use urban renovation focused on a single bloc. The efforts to transform the former cultural and commercial center focused primarily on the physical upgrading of a key symbolic building, based on the assumption that its degradation reflected and reinforced an environment of insecurity. But with a focus on mounting a new building project, planners and designers failed to address the main sources of insecurity and violence on that particular bloc: the presence of local crack addicts. To the extent that the presence of crack addicts could have been considered a public health matter as much as a law-and-order concern or physical design concern, the focus on renovating a single building complex was misguided. Critics in fact suggested that the urban renovation should have been linked more systematically with a wider range of social interventions to strengthen the social and economic conditions in the area. Instead, building demolition threatened the existence of local businesses on the streets surrounding the renovation site, thus undermining community support (Figure 29.2).

Figure 29.2 Closed businesses in downtown São Paulo.

Source: Diego Torres Silvestre, Flickr, Creative Commons

To be sure, the design renovation did have some value. The area became more "secure" during the day because of a heightened police presence, required by the new building owners; but at the same time, security remained difficult to sustain at night when drug addicts reclaimed the streets, and the state and private sector retreated, owing to the single-use commercial function of the newly renovated building. The limited successes of downtown must be contrasted to the periphery of São Paulo – which was an even more complicated site for urban design interventions. In contrast to downtown, where police were able to contribute to security gains by regulating the spaces surrounding the Nova Luz redevelopment project, in the periphery drug mafias and other local strongmen governed community spaces. The police rarely entered, thus leaving citizens needing to negotiate their own conditions of security vis-à-vis local strongmen engaged in violence (Denyer-Willis and Tierney 2012). In this context, the state was absent in the realm of law and order, replaced by "informal" armed actors, who controlled the physical spaces of the community with the threat of violence. Citizen claims about state-financed infrastructure (utility access, paved main streets, stoplights, buses that travel to the end of the metro line, and so on) were as likely to be directed towards these strongmen as to local planning authorities, thus reinforcing a vicious cycle of state illegitimacy and criminal hegemony in which the latter negotiated directly with the city for services. Thus, the community had a public school and a public health center as well as other important infrastructural projects, but their presence did very little to reduce the looming threat of violence and insecurity.

The latter point is a critical one because it underscores the fact that new urban projects alone will not turn around a neighborhood when it comes to security concerns. In the São Paulo periphery, continued criminal presence may have reinforced a sense of stability, at least when mafia aims were not challenged by local police, but the community lived in fear because the perpetrators of violence remained unchallenged (Denyer-Willis and Tierney 2012). This situation also reinforced the stigmatization of the entire neighborhood as a site of insecurity, a reputation that reduced private interest in investing or remaining in this area. In this context, it was difficult for local planning authorities to find partners for new redevelopment projects, thus further undermining capacities of either local or multilateral authorities to carry out their desired urban upgrading and public space programs (Denyer-Willis and Tierney 2012). All this suggests that certain sites may be more amenable to urban spatial interventions intended to reduce violence. But it further shows that the very same policy priorities or active commitments on the part of urban planning and design experts can produce very different outcomes, depending on the socio-spatial context and precise urban location in which they are deployed.

Conclusion

The divergent paradigms and programs discussed in this chapter show that urban design has a role to play in enhancing security and fostering resilience, although not all urban design strategies will produce the same outcomes in all times and places. Socio-spatial context, community power structures, and stakeholder assemblages matter, as does the nature and source of violence. Urban design sensibilities are most likely to produce positive results when they reinforce urban planning principles of community engagement, when they are scaled beyond the street and towards the entire neighborhood, and when they take historically specific urban conditions into account – not just the differences between cities of the Global North and the Global South, but also the spatial and social peculiarities and how they lay the foundations for chronic violence. In such settings, the mere redesign of street lighting, commitment to better garbage collection, or construction of straighter pathways may be welcome, but they may be unable to fundamentally challenge the illicit networks and local political conditions that empower the perpetrators of violence. For this reason, the deployment of urban design strategies in situations of vulnerability and risk must be framed less as a technical fix and more as a means for realizing socially engaged projects that produce strong horizontal and vertical connections within and between citizens and authorities, in turn laying the foundation for collective pushback against the perpetrators of violence. In identifying potential projects that achieve such aims, urban designers must cultivate a good understanding of how spaces are used by violent actors, how citizens' capacities to enter into or move through these spaces are a reflection of the intensity of violence, and how design can facilitate freedom of movement and reduce fear in everyday urbanism. This will be a much larger project than merely addressing "opportunistic" crime.

With the specificities of city life in the Global South in mind, urban designers must also be wary of "fast policy transfer" solutions that extract generalizable strategies from brand-name models that may not be universally replicable. This is not just because a given policy or design strategy may morph depending on which elements are most easily deployed, but also because imported strategies often fail to draw on local knowledge and thus may produce outcomes that do more harm than good. If spatial strategies addressing violence and insecurity are understood not merely through the lens of design, but also as modes of urban planning and community governance, their deployment will inevitably involve local residents and other key stakeholders. In that sense, designing more secure and resilient spaces may be a formidable task, but it could be precisely when a city embraces this challenge that it will collectively set itself on a more positive, constructive, and sustainable urban development pathway.

Notes

1 Special thanks to Taylor Davey for her collaboration and research assistance with this project.

2 I served as the Principal Investigator for this project, titled Urban Resilience in Situations of Chronic Violence (URCV) and funded by USAID's Research Division. For more on the overall findings, see: https://cis.mit.edu/programs/urban-resilience-situations-chronic-violence.

3 In its early years, CPTED theory was just as popular in the United Kingdom as in the United States. See Coleman (1985), Poyner and Webb (1991), Crowe (1991), and (Cozens and Love 2015).

4 Ideas about defensible space were readily adopted by criminologists (Brantingham and Brantingham, 1975) and police officials (Clarke and Mayhew 1980; Poyner 1983), perhaps even more so than urban designers, at least in the United States and Britain.

5 Though upgrading programs often began with shelter, later iterations embraced larger renovation schemas that acted as the basis for establishing community empowerment and good governance at the local level, building on cooperative partnerships between public, civic, and private sectors (Van Lindert 2016). The 2000 *World Bank Cities in Transition* report marked this paradigm shift.

6 These objectives were achieved by the deployment of certain tools associated with the profession of urban design, including: assemblage of activities; integration of uses; site layout; active frontages and landscaping; visual connections; signage; and movement networks (Krause et al. 2014: 45).

7 Ugur (2016) explains VPUU failure by arguing that institutional barriers to scaling up pilot programs limited the extent to which the entire urban environment becomes crime free.

8 Metrocable is a gondola-style lift system that allows better transit access for Medellín's most impoverished residents, many of whom live in hills surrounding the city that are not easily served by traditional mass transit operations such as buses. By providing this cable car system, city authorities sought to connect citizens who have been socially and spatially excluded (Davila 2011).

9 Scholars have noted, however, that interventions complementing the Metrocable in the west were not as successful: the "topography was more difficult, complementary interventions were spatially disconnected from each other, there were lower levels of community participation, and the J Line to a large extent sought to leap-frog neighborhoods closer to the Metro station in a bid to connect as efficiently as possible the newly developed area of *Ciudadela Nuevo Occidente*" (Brand and Dávila 2011: 53).

10 One might think of Medellín's multifaceted strategy as accommodating both *physical* public space and *procedural* public space (Iveson 2007: 11, 17).

11 Some urban design interventions empowered rather than disabled the perpetrators of violence, leading to what the report called negative resilience, which can reinforce an environment of insecurity. By contrast, positive resilience was understood to be embodied in policy and urban design interventions that established a longer-term pathway out of violence (Davis 2012: 35).

References

Armborst, T., D'Oca, D. and Theodore, G. (2017). *The Arsenal of Exclusion & Inclusion.* New York: Actar Publishing.

Barolsky, V. (2016). "Is Social Cohesion Relevant to a City in the Global South? A Case Study of Khayelitsha Township." *SA Crime Quarterly* 55:17–30.

Brand, P. and Dávila, J.D. (2011). "Mobility Innovations at the Urban Margins." *City* 15(6):647–661.

Brantingham, P.J. and Brantingham, P.L. (1975). "Residential Burglary and Urban Form." *Urban Studies* 12:273–284.

Broid, D. and de la O, M. (2012). *Urban Resilience in Situations of Chronic Violence: Case Study of Mexico City, Mexico.* Cambridge, MA: Massachusetts Institute of Technology, Department of Urban Studies.

Brown-Luthango, M., Reyes, E., and Gubevu, M. (2016). "Informal Settlement Upgrading and Safety: Experiences from Cape Town, South Africa." *Journal of Housing and the Built Environment* 32:471–493.

Cassidy, T., Inglis, G., Wiysonge, C., and Matzopoulos, R. (2014). "A Systematic Review of the Effects of Poverty Deconcentration and Urban Upgrading on Youth Violence." *Health & Place* 26:78–87.

Clarke, R. and Mayhew, P., eds. (1980). *Designing Out Crime.* London, UK: HMSO.

Coleman, A. (1985). *Utopia on Trial: Vision and Reality in Planned Housing.* London, UK: Hilary Shipman.

Coupé, F., Brand, P., and Dávila, J.D. (2013). "Medellín: Institutional Context and Urban Paradigm Change." In: *Urban Mobility & Poverty: Lessons from Medellín and Soacha, Colombia.* Ed. J.D. Dávila. Medellín, Colombia: Development Planning Unit, UCL & Faculty of Architecture, Universidad Nacional de Colombia.

Cozens, P. and Love, T. (2015). "A Review and Current Status of Crime Prevention Through Environmental Design (CPTED)." *Journal of Planning Literature* 30(4):1–20.

Cozens, P., Saville, G., and Hillier, D. (2005). "Crime Prevention Through Environmental Design (CPTED): A Review and Modern Bibliography." *Journal of Property Management* 23(5):328–356.

Crowe, T. (1991). *Crime Prevention Through Environmental Design: Applications of Architectural Design and Space Management Concepts.* Boston, MA: Mutterworth-Heinemann.

Cuthbert, A. (2007). "Urban Design: Requiem for an Era: Review and Critique of the Last 50 Years." *Urban Design International* 12:177–223.

Davila, J. (2011). "Mobility Innovation at the Urban Margins, Medellin's Metrocables." *City* 15(6):647–661.

Davis, D. (2012). *Urban Resilience in Situations of Chronic Violence.* Cambridge, MA: Massachusetts Institute of Technology, Department of Urban Studies. Available from: https://cis.mit.edu/programs/urban-resilience-situations-chronic-violence.

Denyer-Willis, G. and Tierney, J. (2012). *Urban Resilience in Situations of Chronic Violence: Case study of São Paulo, Brazil.* Cambridge, MA: Massachusetts Institute of Technology, Department of Urban Studies.

Freire-Medeiros, B. and Name, L. (2017). "Does the Future of the Favela Fit in an Aerial Cable Car? Examining Tourism Mobilities and Urban Inequalities through a Decolonial Lens." *Canadian Journal of Latin American and Caribbean Studies* 42(1):1–16.

Gunder, M. (2011). "Commentary: Is Urban Design Still Urban Planning? An Exploration and Response." *Journal of Planning Education and Research* 31(2):184–195.

Harcourt, B. (2001). *Illusion of Order: The False Promise of Broken Windows Policing.* Cambridge, MA: Harvard University Press.

Heinrichs, D. and Bernet, J.S. (2014). "Public Transport and Accessibility in Informal Settlements: Aerial Cable Cars in Medellín, Colombia." *Transportation Research Procedia* 4:55–67.

Iveson, K. (2007). *Publics and the City.* Oxford, UK: Blackwell.

Jacobs, J. (1961). *The Death and Life of Great American Cities.* London, UK: Jonathon Cope.

Jeffrey, C.R. (1969). "Crime Prevention and Control Through Environmental Engineering." *Criminologica* 7:35–58.

Jeffrey, C.R. (1971). *Crime Prevention Through Environmental Design.* Beverly Hills, CA: Sage.

Kelling, G. and Wilson, J. (1982). "Broken Windows: The Police and Neighborhood Safety." *Atlantic Monthly* 249(3):29–38.

Krause, M., Giles, C., Shay, D., Cooke, J., Smith, E., Taani, I., and Lange, U. (2014). *Violence Prevention through Urban Upgrading: A Manual for Safety as a Public Good.* Available at: www.saferspaces.org.za/uploads/files/VPUU_a_manual_for_safety_as_a_public_good.pdf.

Madanipour, A. (2006). "Roles and Challenges of Urban Design." *Journal of Urban Design* 11(2):173–193.

Milan, B.F. and Creutzig, F. (2017). "Lifting Peripheral Fortunes: Upgrading Transit Improves Spatial, Income, and Gender Equity in Medellín." *Cities* 70:122–134.

Moser, C. (2004). "Urban Violence and Insecurity: An Introductory Roadmap." *Environment & Urbanization* 16(2):3–16.

Newman, O. (1972). *Defensible Space: Crime Prevention Through Urban Design.* New York: Macmillan.

Osorio, D.P. (2011). "Integral Urban Projects." In: *Medellín Laboratory: An Exhibit of Ten Ongoing Practices.* Trans. G. Meija. Medellín: Mesa Editores. 70–83.

Owusu, G., Wrigley-Asante, C., Oteng-Ababio, M. and Owusu, A.Y. (2015). "Crime Prevention through Environmental Design (CPTED) and Built Environmental Manifestations in Accra and Kumasi, Ghana." *Crime Prevention and Community Safety* 17(4):249–269.

Poyner, B. (1983). *Designing against Crime: Beyond Defensible Space.* London, UK: Butterworth-Heinemann.

Poyner, B. and Webb, B. (1991). *Crime Free Housing.* Oxford, UK: Butterworth Architecture.

Punter, J. (2007). "Developing Urban Design as Public Policy: Best Practice Principles for Design Review and Development Management." *Journal of Urban Design* 12(2):167–202.

Rivadulla, M.J. and Bocarejo, D. (2014). "Beautifying the Slum: Cable Car Fetishism in Cazucá, Colombia." *International Journal of Urban and Regional Research* 38(6):2025–2041.

Samper, J. (2012). *Urban Resilience in Situations of Chronic Violence: Case Study of Medellín, Colombia.* Cambridge, MA: Massachusetts Institute of Technology, Department of Urban Studies.

Samper, J. (2016). "Urban Upgrading in a Context of Violence: Perceptions of Security and Physical Space in the Case of the Favela-Bairro in Rio de Janeiro." *International Relations and Diplomacy* 4(12):760–778.

Sanya, T. (2012). "Participatory Design: An Intersubjective Schema for Decision Making." *International Journal of Architectural Research* 10(1): 62–74.

Ugur, L. (2016). "Managing Complexity: A Case for Adaptive Management in Tackling Urban Violence in South Africa." *Emergence: Complexity and Organization* (1). Doi: 10.emerg/10.17357.83b679e69ba1e0ab7e49964b8883ea2c.

Van Lindert, P. (2016). "Rethinking Urban Development in Latin America: A Review of Changing Paradigms and Policies." *Habitat International* 54:253–264.

Wekerle, G. and Whitman, C. (1995). *Safe Cities: Guidelines for Planning, Design and Management.* New York: Van Nostrand Reinhold.

Further reading

Cozens, P. and Love, T. (2015). "A Review and Current Status of Crime Prevention Through Environmental Design (CPTED)." *Journal of Planning Literature* 30(4):1–20. Periodizes the various eras of CPTED and carries forward into the present recent developments in the field of urban design, security, and crime prevention, situating these themes in the context of planning theory and action. It also evaluates evidence associated with CPTED and provides a detailed overview of its main criticisms.

Gaffkin, F. and Morrissey, M. (2011). *Planning in Divided Cities: Collaborative Shaping of Contested Space.* Oxford: Wiley-Blackwell. Examines urban planning and policy in the context of deeply contested spaces where place identity and cultural affinities shape the urban experience. It offers a conceptual and theoretical discussion of various types of conflict, and it discusses possible policy actions through a focus on case studies.

Davis, D.E. and Tirman, J. (2012). *A Toolkit for Urban Resilience in Situations of Chronic Violence.* USAID and MIT. Available at: www.wilsoncenter.org/sites/default/files/USAID%20Urban%20Resilience%20Toolkit.pdf. Identifies the spatial correlates of resilience, seeking to determine whether and how physical conditions in a neighborhood will affect the nature, degrees, and likelihood of resilience.

de Medellín, A. (2011). *Medellín Laboratory: An Exhibit of Ongoing Practices.* Edited by B.A. Sanin. Translated by G. Meija. Medellín, Colombia: Mesa Editores. Available at: www.scribd.com/document/79852057/Medellin-Laboratory-An-exhibit-of-ten-ongoing-practices. This catalog accompanied a 2011 exhibition that summarized the various spatial and institutional practices undertaken under the rubric of Social Urbanism (SU) in Medellin. It presents contributions from UN-Habitat writers and members of the municipal administration, amongst others.

Krause, M., Giles, C., Shay, D., Cooke, J., Krause, M., Smith, E., Taani, I., and Lange, U. (2014). *Violence Prevention Through Urban Upgrading: A Manual for Safety as a Public Good.* Available at: http://vpuu.org.za/wp-content/uploads/2017/05/VPUU_a-manual-for-safety-as-a-public-good.pdf. Documents the technical and conceptual strategies associated with the Violence Prevention through Urban Upgrading (VPUU) program. It includes case studies of particular site interventions.

30
PLANNING THE RISK CITY
Emerging practices

Yosef Jabareen

Introduction

Cities are facing mounting levels of risk and vulnerability, stemming from social polarization, poverty proliferation, violence, and, most recently, climate change. Thus, a great challenge for urban design is how to plan contemporary cities under the new emerging conditions of risk. The literature suggests that climate change poses new risks and uncertainties, which often lie outside our range of experiences and have the potential to affect the social, economic, and physical systems of any given city, impacting urban security and threatening the safety, well-being, and even the very existence of residents (Barnett and Adger 2005; IPCC 2017). Thus, the phenomenal climate change risk poses a great challenge for planning contemporary cities under the new circumstances of anticipated risk and associated uncertainties, and leads cities to develop plans to address it.

Strikingly, there is a lack of theorization in the planning profession that could help identify, characterize, and conceptualize the emerging risk-oriented practices. Thus, this chapter aims to build a framework for understanding the risk-oriented practices and applies this framework to the case of one of the most ambitious recent planning efforts: New York City's *PlaNYC 2030*. More particularly, the chapter examines the decisive role that risk from *climate change* has played in framing and forging the recent planning practices of New York City, and it identifies its major elements in coping with climate change and its anticipated threats. The next section presents the conceptual framework of the chapter, which help us understand how risk affects planning practices. It then presents the case study of New York City planning. The final section presents conclusions for planners and urban designers.

The framework of the risk city and its practices

This section presents the logics that articulate the framework of the *risk city* and risk-oriented planning practices. In my conceptualization of the *risk city*, I build on Giddens's and Beck's concepts of the *risk society* (Jabareen 2015). Giddens (1999) and Beck (1992, 2005, 2006) conceptualize modernity and modern societies in terms of risk. Giddens (1999) views risk as inseparable from modernity and as the mobilizing dynamic of societies that are bent on change and determined to control their own destiny rather than leaving it to religion, tradition, or the whims of

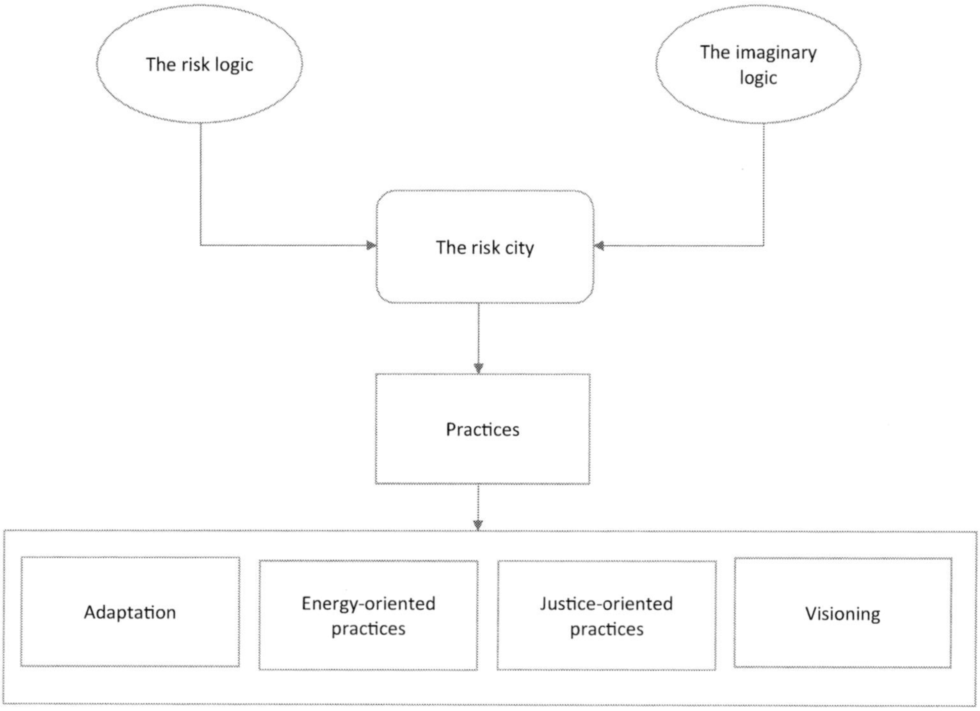

Figure 30.1 The logic of the risk city and its climate change-risk practices.

Source: Author

nature. Beck (1992) defines the *risk society* in terms of those threats that emerged in the 1960s, such as the development of nuclear and chemical weapons. Modern society, he maintains, "has become a risk society in the sense that it is increasingly occupied with debating, preventing and managing risks that it itself has produced" (Beck 2005: 332). From his perspective, this was "an inescapable structural condition of advanced industrialization." He further maintains that modern societies face new types of risk, and that their foundations are shaken by the worldwide anticipation of global catastrophes.

Yet, in contrast to Beck's and Giddens's lack of focus on the spatial aspects of the risk society, I argue for the necessity of "spatializing" contemporary risks and situating them at the city level. Therefore, my aim in this chapter is to develop the theoretical framework of the *risk city* with the primary goal of filling a gap in the literature with a framework that not only theorizes urban risk and its uncertainty, but also contributes to our understanding of the nature of risk-oriented planning practices.

The framework, or what Giddens (1976) called a "frame of meaning," for risk city is a construct of two main logics: the *logic of risk* and the *imaginary logic*, which articulate the concept of the *risk city*, and which provide us with a framework for understanding its practices (see Figure 30.1).

The logic of risk

Risk is "a virtual threat," and "is not an objective condition, but a social construction of reality, which starts with the question of how people explain misfortune" (Hoogenboom and Ossewaarde

2005: 606). Following Mary Douglas's and Aaron Wildavsky's (1982) pioneering work on risk perception, social scientists have argued that risk behaviors and perceptions can neither be understood nor analyzed outside the social and cultural contexts in which they evolve (Sommerfield et al. 2002). Accordingly, some argue that understanding a person's interpretation of risk requires attention to the broader social, cultural, and historical contexts within which such interpretation occurs (Beamish 2001; Erikson 1994). Thus, risk perception varies according to historical traditions and cultural beliefs, as well as political and administrative structures (Jasanoff 1986; Rohrmann 2006; Jabareen 2015). Consequently, each society has its own conceptions of the risk city based on its own understanding and interpretation of uncertainties, knowledge, political organization and values, political and market powers, and resources.

The logic of risk entails conflict over the meanings and interpretations of the risk that a city may face. For the most part, knowledge regarding risk is questioned and challenged, not only by the public but also by the experts themselves, which means that the risk city exists in the shadow of unstable, challenged, and incomplete knowledge. Furthermore, this logic suggests that risk is interpreted and manipulated differently by people with different interests and backgrounds. Since risk means different things to different people depending on their social, economic, and political capacities and conditions, we need to understand different approaches to plan and cope with different concepts of risk in cities. For example, for the underprivileged masses in cities of the Global South, "global warming" is not a major perceived risk, even though they might suffer more than others from the impact of climate change. For them, the most prevalent vocabulary for expressing risk (or lack thereof) pertains to access to food, clean water, employment, and urban hygiene.

Furthermore, some risks are conveniently ignored while others are emphasized, since risk relates to power and resource allocation, and risk conception can be a tool of political and social power in cities. Without a doubt, decision makers and politicians prioritize risk based on political, economic, and social considerations. It would be naive to suggest that politicians and decision makers consider scientific facts alone in their dealings with risk and the risk city. Convincingly, Beck (2006: 333) argues, "even the most restrained and moderate objectivist account of risk implications involves a hidden politics, ethics and morality."

The imaginary logic

The conceptualization of the imaginary logic is built on Lacan's (1988, 2006) concepts of lack, desire, and fantasy. Lacan suggests that at the "mirror stage," a child in his early months "primordially identifies with the visual gestalt of his own body. In comparison with the still very profound lack of coordination in his own motor functioning, that gestalt is an ideal unity, a salutary imago" (Lacan 2006: 113). The difference between the whole, total, and ideal image and the fragmented experience of the infant constitutes "lack," which must be understood as "the lack of being" (Lacan 1988: 223). Notably, "lack" causes a desire to arise (Lacan 1988). Thus, the desire of being is to "fill" the "lack" and become "whole" and "complete."

According to the logic of risk, cities live under the assumption of existing threats and risk that face them. Thus, we can think of the city as a "lacking subject"; it lacks the sense of full protection and complete security. Thus, based on the Lacanian framework, a "fantasy" of seeking "complete" and "whole" security emerges. While the logic of risk grasps with the existing and anticipated risk conditions of the city, the imaginary logic captures the motives behind the desire to change the current insecure conditions and to imagine the more peaceful and harmonious future for the city and its society. This logic deciphers the visions of decision makers and practitioners for the city's future, as they seek to bridge the current and anticipated risk conditions. The imaginary mode transcends the undesirable conditions of the current risk city through a provision of urban scenarios of a secured city, such as the "sustainable city," "ecological city," "zero-carbon city," "resilient city," "green city," and "healthy city." Therefore, this logic grasps the imagined scenarios that seek to fill the "lack" of security that is related to risk. Thus, the imaginary logic offers the means to understand why specific spatial and social practices are constructed, maintained, and transformed. The function of the fantasy is to address the insecurity of the risk city by providing scenarios for a better city and a sustainable life, following urban plans and practices oriented to risk management.

The logic of risk and the imaginary logic, which articulate the risk city, provide the motives and justifications for planning the risk city, as well as reveal the nature of planning practices, which are applied to cope with threats that face the risk city, as Figure 30.1 demonstrates. The risk city is always attempting to overcome its perceived risk through spatial, social, and political practices, through representations and discourse, and through maintaining a certain imaginary. John Friedmann (2002: 3) posits that utopian thinking is about imagining a future that is radically different from what we know as the prevailing order of things. Thus, the risk city seeks to plan for certainty, aiming at achieving a more secure city. Therefore, the planning of risk city is related to the dreams, desires, and visions of how to contend, and ultimately eliminate the currently insecure city conditions. The "fantasy" of achieving "fullness," "harmony," a "sustainable" and "resilient" city in the future, transcend the insecurity gaps of cities at present.

In recent years, under the assumption that climate change poses serious risk to cities, many local authorities around the world have been grappling with climate change using master, strategic, and action plans aimed at mitigating greenhouse gas emissions and adapting to the anticipated, albeit uncertain, impacts of climate change. This next section examines how New York City has addressed risk resulting from climate change in framing and forging its planning practices.

Planning the risk city: the case of New York City

New York City: a city at risk

New York City is a megacity of approximately eight million people, with a target population of nine million people by 2030 (OneNYC 2017). Our discussion draws from a review of the master plan of New York City, *PlaNYC 2030*, which was launched on Earth Day 2007, and many other relevant planning documents. A review of various recent plans prepared by the city (especially *PlaNYC 2030*, OneNYC (2015, 2017, 2018), and NYC Panel on Climate Change) shows that city decision makers and planners portray New York City as a city at risk (Jabareen 2013, 2014, 2015a).

A fundamental assumption of *PlaNYC 2030*, which is the first master plan that addresses risk, is that "climate change poses real and significant risks to New York City" (PlaNYC: Progress Report 2009: 39); it poses "a range of hazards to New York City and its infrastructure" and suggests "a need for the City to rethink the way it operates and adapts to its evolving environment" (NPCC 2009: 3). The planning documents suggest that climate change is likely to bring warmer temperatures to New York City, as the mean annual temperatures projected by global climate models are expected to increase by 1.5–3 degrees (Fahrenheit) by the 2020s, 3–5 degrees by the 2050s, and 4–7.5 degrees by the 2080s (NPCC 2009). The city will also experience more intense rainstorms, while heat waves are expected to become more frequent, intense, and longer in duration, and sea levels are likely to rise, with an increase of 2–5 inches by the 2020s, 7–12 inches by the 2050s, and 12–23 inches by the 2080s. Climate change will prompt increased summertime strain in the form of: higher peak electricity loads; voltage fluctuations, equipment damage, and service interruptions; transportation service disruption; increased street, basement, and sewer flooding; loss of water quality; inundation of low-lying areas and wetlands; increased structural damage; and increased need for emergency management procedures.

New York City has almost 578 miles of coastline and over half a million residents living within the current floodplain. This poses a particularly dangerous risk. NPCC holds that the city already faces the probability of a "hundred-year flood" once every 80 years, which could increase to once every 43 years by the 2020s, and to once every 19 years by the 2050s. As a coastal city, *PlaNYC* concludes, "we are vulnerable to the most dramatic effects of global warming: rising sea levels and intensifying storms" (PlaNYC 2007: 133). At the state level, the State Commission 2100 of New York also summarizes the demographic, poverty, and climate change-related risks that face the state and New York City as follows:

> There are significant climate change risks including sea level rise, changing patterns of precipitation, temperature change and increasingly frequent extreme weather events. There are demographic pressures, with significant population growth predicted for New York state, and structural changes within the population, including further urbanization, the growth of suburban poverty, as well as the continuing needs of those living below the poverty level and a growing aging population.
>
> *(NYS 2100 Commission 2013: 10)*

In sum, *PlaNYC* acknowledges that the shifting climate patterns will have a wide range of effects on the city, taking lives, posing "major public health dangers," and impacting the property and livelihood of many (PlaNYC 2007: 138).

Risk-induced planning practices

These planning documents, mainly *PlaNYC* and OneNYC, identify the main planning objectives. They are ambitious in their goals for sustainability and resiliency, aiming at charting the city's future for the coming decades and addressing the challenges of climate change-related risk (Rosenzweig et al. 2011; Jabareen 2015). Four major categories of risk-induced planning practices are evident in these plans: (1) visioning; (2) just planning practices; (3) adaptive spatial practices; and (4) energy-saving spatial practices. As we will see, urban design strategies play an important role for the latter two categories.

Visioning

Countering the climate change-related risk is central to the visioning of New York City and its planning. *PlaNYC* begins with a diagnosis of the local and global climate change crisis as problematic and critical for the city and the world as a whole. *PlaNYC*'s vision generates a sense of local and global urgency: "we will face an increasingly precarious environment and the growing danger of climate change that imperils not just our city, but the planet" (PlaNYC 2007: 141). Thus, *PlaNYC* promises a better future: It is a vision of providing New Yorkers with the cleanest air of any big city in the nation; of producing more energy more cleanly and more reliably and offering more choices on how to travel quickly and efficiently across our city. Interestingly, the 2017 plan *OneNYC* (OneNYC 2017: 5) is organized around four primary visions, which are related to the emerging and anticipated climate change risk: economic growth, equity, sustainability, and resiliency. More specifically, the plan articulates the following visions: (1) *growth*; (2) *equity*; (3) *sustainability*; and (4) *resiliency*.

Just planning practices

Equity, justice, equality, fairness, and urban rights are crucial moral terms for planning the risk city, and for addressing the injustices and inequalities of contemporary cities (Eizenberg, Jabareen, 2017). The literature suggests that the more just the city, the more efficiently it will cope with climate change (Jabareen 2015). While inequality leads to greater environmental degradation, a more equitable distribution of power and resources would result in improved environmental quality (Agyeman et al. 2002). Furthermore, climate change injustice occurs along ethnic, gender, class, and racial lines involving neighborhoods and communities (Mohai et al. 2009). Since all societies have individuals and groups who are more vulnerable than others and lack the capacity to adapt to climate change, planning practices must address this issue and promote just outcomes.

New York is a diverse city with five boroughs, 59 community districts, and hundreds of neighborhoods. Moreover, all five boroughs have a vulnerable coastline. The massive growth proposed by *PlaNYC* will certainly affect these communities and may even "erase the character of communities across the city" (PlaNYC 2007: 18). Yet, despite the significant planning it embodies and the crucial dilemmas it raises, the mechanism and procedures proposed for facilitating citizen participation in the planning process was inadequate in *PlaNYC* and insufficient for meeting the planning challenges stemming from climate change for one of the world's most socially and culturally diverse cities (Angotti 2008, 2010; Brian 2011; Finn and McCormick 2011; Rosan 2012). *PlaNYC* identified important issues but did little to address them through real community participation. Brian (2011: 2–3) argues that: "*PlaNYC* 2030 is a top-down bureaucratic initiative with little community involvement and 'buy-in' and is not well-integrated with the rest of city policy making." Angotti (2008) notes that *PlaNYC 2030* is actually a strategic real estate growth plan, and its presentation of "sustainability" is primarily a public relations effort. Miner (2011: 1) suggests that "[t]he public wasn't involved in creating the plan, so no grassroots constituency or broad public support ever developed around it." And while *PlaNYC* notes the existence of environmental injustice in the city, it fails to address the issue in a serious manner and takes no practical measures to mitigate the phenomenon. Convincingly, Angotti (2010: 3) argues that the city's 59 community boards are "[s]till invisible in the 2030 plan barely mentioned in the scores of spreadsheets, maps and colorful images that herald the coming of the green city."

Significantly, some scholars show the persistence of unequal and unjust environmental conditions among various communities of the poor and people of color in New York (Rosan 2012; Sze 2007). *PlaNYC*, however, fails to address this issue. For example, planners acknowledge that most brownfields are concentrated in low-income communities, resulting in a case of severe environmental injustice (PlaNYC 2007: 41). The owners of such land "often find that their financial interests dictate development plans that minimize cleanup requirements" and "may choose new uses for the land" that "do not reflect community needs or desires" (PlaNYC 2007: 42). Moreover, "in some communities, the impacts of exposure to local air emissions have likely contributed to higher asthma rates and other diseases" (PlaNYC 2007: 119). But these clear cases of environmental injustices go unaddressed in the plan.

However, recently the term "justice" appeared in the *One New York: The Plan for a Strong and Just City* (OneNYC), published in 2015. And in its recent (2017) progress report, OneNYC (2017: 125) acknowledges that it "explicitly seeks to address the disparate environmental conditions that have historically resulted in diminished health outcomes, economic opportunities, and quality-of-life in some of our city's neighborhoods." These injustices are related to "public health impacts from poor air or water quality, disproportionate exposure to contamination and toxins, and increased risks related to chronic exposure to extreme heat and the urban heat island effect." According to the plan, the city has developed two bills, which "will provide the City and all New Yorkers the necessary information to enhance community participation and deliver environmental justice through the City's deliberations and actions" (OneNYC 2017: 125).

Adaptive spatial practices

The concept of adaptability pertains to the ability of a city to adapt to anticipated and uncertain threats. Adaptation means "controlling uncertainty – either by taking action now to secure the future or by preparing actions to be taken in case an event occurs" (Abbott 2009: 237). Understanding future vulnerabilities, their exposure, and the response capacity of interlinked human and natural systems is challenging, particularly because the interaction of various social, spatial, economic, and cultural factors have not been fully understood or considered to date (IPCC 2017).

Through its recent plans, New York City proposes various adaptation policies and spatial practices, including measures to strengthen coastal defenses, fortify the city's critical infrastructure, upgrade buildings, protect infrastructure and critical services, update the floodplains to better protect areas that are most vulnerable to flooding, and work with at-risk neighborhoods to develop site-specific plans.

Furthermore, the city developed preliminary climate resiliency guidelines, which aim to embed resiliency across its capital improvement programs by providing forward-looking climate data to be incorporated in all capital projects. This is a significant shift, which shows how "the City invests and operates to ensure that New York is prepared for the impacts of climate change including shocks like Hurricane Sandy and chronic stresses like sea level rise and rising temperatures" (OneNYC 2017: 183). Interestingly, the city launched a Community Emergency Planning Toolkit "to help communities become more resilient." It describes how New York City plans for hazards and offers guidance on how communities can create their own emergency plan.

Energy-saving spatial practices

Energy-saving practices refer to mitigation measures, which aim to reduce the sources of factors causing climate change, such as greenhouse gas emissions (CCC 2010). The recent plans of New York City promote the following design and planning initiatives to curtail energy, improve air quality, and reduce emissions by 30% by 2030.

Building code *PlaNYC* calls for focusing major efforts on buildings, the city's largest energy consumers. The plan asserts that over two thirds of the city's energy is consumed within buildings, compared to a national average of less than one third. By 2030, at least 85% of the city's energy will be used by buildings that already exist today. If no action is taken, New York City's carbon emissions will grow from 58.3 metric tons measured in 2005 to almost 74 million metric tons by 2030 (PlaNYC 2007: 9). Planners calculate that energy-saving measures in existing New York City buildings can result in a seven-million-ton reduction in global warming emissions. Approximately 50% of reductions in CO_2 emissions will come from increased energy efficiency in buildings, while 32% will result from improved power generation, and 18% from changes in transportation. *PlaNYC* also suggests "greening" the Building Code of New York and proposes focusing on the reduction of the amount of cement used in concrete, as cement production is an energy-intensive process that releases one ton of CO_2 for every ton of cement produced (PlaNYC 2007: 106–107).

Transportation The most effective strategy, the plan asserts, is to reduce the number of vehicles on the road and simultaneously expand the city transit system and implement congestion pricing (PlaNYC 2007: 136). It also calls for developing underutilized areas throughout the city that are well-served by public transportation and other infrastructure; for capturing the potential of transportation infrastructure investments; and for decking over rail yards, rail lines, and highways.

Urban design practices The planning approach to New York City uses different design strategies to help use energy more efficiently, mainly *compactness, rezoning, infill, mixed land uses, transit-oriented development*, and *greening* strategies. Today, less than 4% of the city's buildings account for roughly 50% of its built area (PlaNYC 2007: 102). *PlaNYC* proposes a number of strategies to increase compactness and density in the city. It suggests *infill* "everywhere it is possible" and development of spaces that "are now lightly used," such as parking lots in public housing areas that were developed in the 1930s (PlaNYC 2007: 23). Through *rezoning*, *PlaNYC* directs growth towards areas with strong transit access; reclaims underused or inaccessible waterfronts; and explores opportunities to spur growth through the addition of mass transit. *PlaNYC* fosters rezoning and redevelopment of brownfields, which according to the plan represents one of the city's greatest opportunities, since brownfields cover about 7,600 acres throughout the five boroughs. It also encourages *mixed land use* in future development. Furthermore, *PlaNYC* adopts *greening* as a major strategy and proposes three primary ways to ensure that by 2030, nearly every New Yorker will live no more than a 10-minute walk from a park: (1) by upgrading land already designated as play space or parkland and making it available to new populations; (2) by expanding usable hours at current, high-quality sites; and (3) by reconceptualizing streets and sidewalks as public spaces. The combined impact of these design strategies will be the creation of over 800 acres of upgraded parkland and open space across the city. *PlaNYC* also calls for beautifying the public realm and undertaking "an aggressive campaign to plant trees wherever possible, in order to fully capitalize on tree opportunities across the city" (PlaNYC 2007: 38). In addition, it proposes the expansion of Greenstreets Program, which transforms thousands of acres of unused road space into green space filled with trees, shrubs, and groundcover in an effort to capture stormwater. It also suggests offering incentives for green roofs, which can reduce runoff volume and aid other natural process by absorbing and/or storing water.

Conclusion

This chapter conceptualizes the contemporary city as a "risk city," which is a construct of two interrelated logics: (1) the logic of risk, which captures the perceptions of risks and the manipulation and reshaping of the public opinion regarding the nature and intensity of the risks faced by the city; and (2) the imaginary logic, which is based on the assumption that the city is lacking in terms of security, and which grasps the motives behind taking actions and deciphers the visions and the imagined scenarios that seek to fill this lack of security. From this perspective, planning and design practices are not mere objects, policies, or actions, but rational and functional actions that respond to targeted risk. They are aiming at reducing risk, filling security gaps, and achieving a more resilient and sustainable city.

New York City is a risk city. Climate change-related risk is a critical threat that the city strives to address through planning measures. Its recent ambitious plans imagine the city as sustainable, resilient, and adequately coping with anticipated high risk. The case of New York City reveals four categories of risk-oriented planning practices, as Figure 30.1 shows. These practices are illuminated through the two logics of risk, as Table 30.1 summarizes.

Table 30.1 The logics of risk-oriented practices

Category of Practices	*Risk Logic*	*Imaginary Logic*
Visioning	Climate change poses serious threats to the very existence of the city.	Visioning the city as resilient and sustainable is supposed to properly fill the "lacks" of the city and cope with future risk.
Adaptation	Risk reduction is achieved through adaptation measures.	Achieving future resiliency so that the city is better "fortified" and ready to withstand and "emerge stronger from the impacts of climate change and other 21st century threats." The idea is to ensure that the city's assets, services, and infrastructure continue to function appropriately in the face of climate change.
Energy-oriented	Energy-oriented and mitigation practices aim to reduce the effects of climate change by reducing GHG emissions.	Achieving a more sustainable city through energy planning practices.
Justice-oriented	The impact of climate change is unevenly distributed and "socially differentiated" in terms of communities' capacity to address the uncertainties. Risk is therefore a matter of distributional equity and justice.	Promoting resiliency and sustainability policies that also achieve a "just city."

This chapter concludes that emerging risk and its resulting uncertainties challenge the concepts, procedures, and scope of conventional approaches to city planning and design. Planning the risk city is an emerging approach in planning contemporary cities aiming at countering climate change impacts, adapting cities to future uncertainties, and protecting residents from environmental hazards and risk. Risk plays a central role in formulating the problem, visioning and goal setting, and the outcomes of such planning and urban design. Planning and design practices then seek to cope with issues of risk on the ground. These practices have a central function in serving the imaginary, or fantasy, regarding the future of cities.

References

Abbott, J. (2009). "Planning for Complex Metropolitan Regions: A Better Future or a More Certain One? *Journal of Planning Education and Research*, 28 (4): 503–517.

Agyeman, J., Bullard, R. D., and Evans, B. (2002). "Exploring the Nexus: Bringing Together Sustainability, Environmental Justice and Equity." *Space & Polity*, 6 (1): 77–90.

Angotti, T. (2008). "Is New York's Sustainability Plan Sustainable?" Paper presented to the Joint Conference of the Association of Collegiate Schools of Planning and Association of European Schools of Planning (ACSP/AESOP), Chicago.

Angotti, T. (2010). "PlaNYC at Three: Time to Include the Neighborhoods." *Gotham Gazette: The Place for New York Policy and Politics*. www.gothamgazette.com.

Barnett, J. and N. Adger. (2005). "Security and Climate Change: Towards an Improved Understanding." Paper presented at the Human Security and Climate Change Workshop, Oslo, June 21–23. www.gechs.org/downloads/holmen/Barnett_Adger.pdf

Beamish, T. (2001). "Environmental Hazard and Institutional Betrayal: Lay-public Perceptions of Risk in the San Luis Obispo County Oil Spill." *Organization & Environment*, 14 (1): 5–33.

Beck, U. (1992). *Risk Society: Towards a New Modernity.* London: Sage.

Beck, U. (2005). *Power in the Global Age.* Cambridge: Polity Press.

Beck, U. (2006). "Living in the World Risk Society." *Economy and Society*, 35 (3): 329–345.

Brian, P. (2011). *PlaNYC: A Model of Public Participation or Corporate Marketing?"* www.hunter.cuny.edu

CCC-Committee on Climate Change Adaptation. (2010). "*How Well Prepared is the UK for Climate Change?"* https://www.theccc.org.uk

Douglas, M.A. and A. Wildavsky. (1982). *Risk and Culture: An Essay on the Selection of Technological and Environmental Dangers.* Berkeley, CA: University of California Press.

Eizenberg, E. and Jabareen, Y. (2017). "Social Sustainability: A New Conceptual Framework." *Sustainability*, 9 (6): 1–16.

Erikson, K. (1994). *A New Species of Trouble: The Human Experience of Modern Disaster.* New York: Norton.

Finn, D. and McCormick, L. (2011). "Urban Climate Change Plans: How Holistic?" *Local Environment: The International Journal of Justice and Sustainability*, 16 (4): 397–416.

Friedmann, J. (2002). *The Prospect of Cities.* Minneapolis, MN: University of Minnesota Press.

Giddens, A. (1976). *The Rules of Sociological Method.* London: Hutchinson.

Giddens, A. (1999). *Runaway World: Risk.* Hong Kong: Reith Lectures.

Hoogenboom, M. and Ossewaarde, R. (2005). "From Iron Cage to Pigeon House: The Birth of Reflexive Authority." *Organizational Studies*, 26 (4): 601–619.

IPCC-The Intergovernmental Panel on Climate Change. (2017). *Decision and Outline of the Special Report on Climate Change, Desertification, Land Degradation, Sustainable Land Management, Food Security, and Greenhouse Gas Fluxes in Terrestrial Ecosystems.* Online: www.ipcc.ch/meetings/session45/Decision_Outline_SR_LandUse.pdf

Jabareen, Y. (2013). "Planning for Countering Climate Change: Lessons from the Recent Plan of New York City - PlaNYC 2030," *International Planning Studies*, 18 (2): 221–242. Jabareen, Y. (2014). "An Assessment Framework for Cities Coping with Climate Change: The Case of New York City and its PlaNYC 2030." *Sustainability*, 6 (9): 5898–5919.

Jabareen, Y. (2015a). "City Planning Deficiencies and Climate Change – The Situation in Developed and Developing Cities." *Geoforum* 63: 40–43.

Jabareen, Y. (2015). *The Risk City: Cities Countering Climate Change: Emerging Planning Theories and Practices around the World.* New York City: Springer.

Jasanoff, S. (1986). *Risk Management and Political Culture: A Comparative Study of Science in the Policy Context.* New York: Russell Sage Foundation.

Lacan, J. (1988). *The Seminar. Book II: The Ego in Freud's Theory and in the Technique of Psychoanalysis, 1954–55.* Trans. Sylvana Tomaselli. Cambridge: Cambridge University Press.

Lacan, J. (2006). *Ecrits.* Trans. Fink Bruce. W. W. Norton.

Miner, D. (2011). *How to Get Civic Groups More involved in PlaNYC: To Make City Efforts More Sustainable.* http://beyondoilnyc.blogspot.co.il/2011/07/planyc-30-laboratory-for-social.html

Mohai, P., Pellow, D. and Roberts, J.T. (2009). "Environmental Justice." *Annual Review of Environment and Resources* 34: 405–430.

NPCC (2009). *New York City Panel on Climate Change: Climate Risk Information.*http://www.nyc.gov/html/om/pdf/2009/NPCC_CRI.pdf.

NYS 2100 Commission. (2013). *Recommendations to Improve the Strength and Resilience of the Empire State's Infrastructure.* https://www.cakex.org/sites/default/files/documents/NYS2100.pdf

OneNYC. (2015). *One New York the Plan for a Strong and Just City.* http://www.nyc.gov/html/onenyc/downloads/pdf/publications/OneNYC.pdf

OneNYC. (2007). *A Greener, Greater New York.* http://www.nyc.gov/html/planyc/downloads/pdf/publications/full_report_2007.pdf

OneNYC. (2017). http://onenyc.cityofnewyork.us/wp-content/uploads/2017/05/OneNYC_Progress_Report_2017.pdf

OneNYC. (2018). https://onenyc.cityofnewyork.us/wp-content/uploads/2018/04/OneNYC_Progress_2018-2.pdf

PlaNYC 2030. (2007). *A Greener Greater New York.* New York: City of New York, Mayor's Office. www.nyc.gov/html/planyc/downloads/pdf/publications/planyc_2011_planyc_full_report.pdf

PlaNYC: Progress Report. (2009). New York: City of New York, Mayor's Office. www.nyc.gov/html/planyc/downloads/pdf/publications/planyc_progress_report_2009.pdf

Rohrmann, B. (2006). "Cross-cultural Comparison of Risk Perceptions: Research, Results, Relevance." Presented at the ACERA/SRA Conference. www.acera.unimelb.edu.au

Rosan, C.D. (2012). "Can PlaNYC Make New York City "Greener and Greater" for Everyone? Sustainability Planning and the Promise of Environmental Justice." *Local Environment*, 17 (9): 959–976.

Rosenzweig, C., Solecki, W.D., Hammer, S.A., and Mehrotra, S. (2011). *Climate Change and Cities: First Assessment Report of the Urban Climate Change Research Network.* Cambridge: Cambridge University Press.

Sommerfield, J., Kouyate, M.S. and Sauerborn, R. (2002). "Perceptions of Risk, Vulnerability, and Disease Prevention in Rural Burkina Faso: Implications for Community-based Health Care and Insurance." *Human Organization*, 2: 139–146.

Sze, J. (2007). *Noxious New York: The Racial Politics of Urban Health and Environmental Justice.* Cambridge, MA: MIT Press.

Further reading

Bulkeley, H. (2013). *Cities and Climate Change.* London and New York: Routledge. It gives an overview of the science and policy of climate change at the global level and the emergence of climate change as an urban policy issue. It includes case studies with examples from cities of the Global North and the Global South.

Jabareen, Y. (2013). "Planning the Resilient City: Concepts and Strategies for Coping with Climate Change and Environmental Risk." *Cities*, 31: 220–229. This article suggests a new innovative conceptual framework (the Resilient City Planning Framework or RCPF) that addresses the critical question of what cities and their urban communities should do in order to move towards a more resilient state in the future.

Jabareen, Y. (2015). *The Risk City: Cities Countering Climate Change: Emerging Planning Theories and Practices around the World.* New York City: Springer. It suggests that the uncertainties resulting from climate change challenge conventional planning approaches, creating a need to rethink and revise current planning methods.

31
THE POTENTIAL OF SITES FOR URBAN DESIGN

Frederick Steiner and Rebecca S. Popowsky

Introduction

The world has grown more urban, and urbanization is likely to increase as more people move to city regions. The planning and design of cities have always been fundamental human endeavors, but our current geological era, the Anthropocene, adds even greater urgency. Our activities are now influencing climatic, geological, and life processes at the global scale. How we plan and design our built environments will play a significant role in the health and welfare of humans into the future. For instance, in the United States, buildings account for about 50% of the energy used and of the greenhouse gas production with transportation accounting for about another 30%. As a result, green building practices have garnered increasing attention and popularity.

This chapter reviews the principles and purposes of the SITES system. A brief history of the system displays its utility. Two urban design projects in Washington, D.C., and Chicago illustrate the benefits, usefulness, and potential of the system.

The sustainable SITES initiative

Principles and purpose

The US Green Building Council's (USGBC) LEED (Leadership in Energy and Environmental Design) programs and a growing number of other systems are significantly changing how buildings are designed and built. LEED is now in use in more than 150 countries and territories. The Green Business Certification Inc. (GBCI) has certified LEED buildings representing more than 13.5 billion square feet (1,254.2 square kilometers) of space and has accredited more than 201,000 with LEED credentials. LEED for Neighborhood Design (LEED ND) advances more environmentally friendly and energy-conserving building communities. LEED ND reflects ideas about urban design advanced by the Congress for the New Urbanism. However, no rating system for outdoor environments existed. To fill this void, the Sustainable Sites Initiative was created for landscape and site scale design (Calkins 2008; Venhaus 2012; Windhager et al. 2010). This initiative resulted in the SITES Rating System (www.sustainablesites.org).

The SITES Rating System provides a comprehensive road map for defining and enhancing sustainability in site development, measuring performance, and ultimately assessing the environmental, economic, and social benefits of projects. SITES is available for use by landscape architects, urban designers, city planners, engineers, architects, developers, policymakers, and others that work on land design and development. SITES has special significance for the planning of green infrastructure and for urban design.

The SITES Rating System views ecosystem services as the foundation for design and development guidelines and metrics. Resource economists have adopted the term to describe those benefits that the natural environment provides to humans for free, and which humans would have to find ways to replace, if the natural environment ceased to provide them. These services include, but are not limited to: filtering air pollutants; controlling erosion; maintaining fishable, swimmable, and drinkable water; regulating climate; cycling nutrients and waste; and supporting tourism and recreation.

Process

In the early 2000s, US leaders in sustainable design, land use planning, landscape architecture, ecology, hydrology, soil, botany, public health, and other areas of expertise began discussing the need for and potential benefits of regionally appropriate landscape design based on ecological principles. This led to the Sustainable Sites Initiative. Founded and led by the Lady Bird Johnson Wildflower Center at the University of Texas at Austin, the US Botanic Garden, and the American Society of Landscape Architects, the leaders of the initiative acknowledged the tremendous impact of the LEED program in advancing green building. Although LEED included some resource-related strategies, such as use of native plants and water conserving technologies, it was apparent that more was needed and could be accomplished with a complementary system for landscapes that may or may not contain buildings.

Beginning in 2006, a committee was formed to guide the development of a rating system that would create clear and practical benchmarks for landscape performance. The resulting SITES v1 (2009) reflected public comments as well as extensive research from experts in a variety of disciplines.

The pilot program was launched in 2010 to test and refine SITES v1, by assessing the level of difficulty of the credits and prerequisites for certification, and the fairness and appropriateness of the points assigned to each credit. The selected pilot projects were chosen to test whether SITES could be successfully applied to development projects in diverse climates, locations, and contexts. They also yielded information about SITES applicability for projects of different types, sizes, and budgets. By the end of the pilot phase in early 2015, 47 pilot projects had achieved certification by meeting all 15 of the prerequisites and accruing a minimum of 100 points from the 51 optional credits in the SITES 2009 Rating System.

The experience of the pilot projects contributed to the refinement of the program. *SITES v2 Rating System: For Sustainable Land Design and Development*, released in June 2014, represented the culmination of eight years of research and development, and it is the first rating system available for achieving landscape design certification. SITES v2 incorporates best practices in sustainable landscape architecture, ecological engineering, land planning, and ecological restoration, as well as knowledge gained through peer-reviewed literature, case study precedents, and projects registered in the SITES pilot program. Some new prerequisites and credits were also introduced (including a credit for promoting on-site food production), while others with similar intent were combined.

The organization of the rating system generally corresponds to the typical project design and development phases. The SITES v2 certification levels (Certified, Silver, Gold, and Platinum) mirror LEED certification nomenclature. The rating system includes 18 prerequisites and 48 optional credits in 10 sections that cover various stages of the design and development process.

Table 31.1 Goals for sustainable landscapes

1 Create regenerative systems and foster resiliency.
- Protect and restore natural resources such as soil, water, and vegetation.
- Encourage biodiversity.
- Enhance landscapes to provide multiple ecosystem services such as cleaning air and water, providing habitat, and storing carbon.
- Mitigate for evolving hazards and natural disasters.
- Plan for monitoring and adaptive management.

2 Ensure future resource supply and mitigate climate change.
- Minimize energy consumption and encourage use of low carbon and renewable energy sources.
- Minimize or eliminate greenhouse gas emissions, heavy metals, chemicals, and other pollutants.
- Reduce, reuse, recycle, and up-cycle[1] materials and resources.
- Conserve water.
- Increase the capacity of carbon sinks through re-vegetation.

3 Transform the market through design, development, and maintenance practices.
- Foster leadership in industry and professional practice.
- Use a systems-thinking, integrative, and collaborative design approach.
- Use lifecycle analyses to inform the design process.
- Support local economies and sustainable policies.

4 Enhance human well-being and strengthen community.
- Reconnect humans to nature.
- Improve human health (physical, mental, and spiritual).
- Foster stewardship by providing education that promotes the understanding of natural systems and recognizes the value of landscapes.
- Encourage cultural integrity and promote regional identity.
- Provide opportunities for community involvement and advocacy.

Projects receive SITES certification by fulfilling minimum requirements, as well as achieving specified numbers of points for different levels of performance. Bonus points are awarded for innovative and exemplary performance outside the 200-point system. The value or number of points assigned to each credit is based on its potential effectiveness in meeting the four goals outlined above (Table 31.1).

Establishing an integrative design process and conducting a thorough site assessment are critical first steps, yet are often not considered in projects. Measurable performance goals should be defined to guide the project team throughout the design process and subsequent decisions. Early engagement with the local community can enhance understanding of the project and its surrounding area, and can ensure that the undertaking meets community needs. The process begins with proper site selection, continues through design and construction, and includes planning for effective and appropriate operations and maintenance measures.

To better coordinate design with construction and maintenance, SITES requires project teams to communicate with contractors and maintenance professionals early in the process to inform the design of the project so that long-term sustainability goals can be met. It is important for the management to adapt to local conditions and performance monitoring to provide an active and valuable feedback loop for understanding how the site functions over time and as a means to collect data that strengthen the case for creating sustainable landscapes.

For project locations that include buildings, SITES v2 applies to the area from the building skin outward. It addresses the landscape attributes, hardscape, and site infrastructure and integration of buildings. For example, a vegetated roof or wall on a building can be considered as part of a project site. Strategic installation of appropriate vegetation near buildings is desirable in order to offset the energy load of a building's heating and cooling systems; providing unobstructed views of vegetation from common spaces in buildings is also favored because of the documented benefits such access provides to human health and well-being.[2]

SITES case studies

The two projects selected as case studies (see Francis 2001) for this chapter demonstrate three ways in which the Sustainable Sites Rating System may be transformational in the context of contemporary urban design practice. Each project demonstrates the capacity of SITES to *encourage* rigorous and measured practices in urban design, *facilitate* successful implementation of sustainable design intent, and *drive innovation* in sustainable practices through project work.

Pilot project case study: Washington Canal Park, Washington, D.C.

Washington Canal Park exemplifies the application of the landscape rating system in the context of urban design. One of the first parks built as part of the District of Columbia's Anacostia Waterfront Initiative, Washington Canal Park is a model of sustainability, attaining both LEED Gold certification and a three-star certification as a SITES pilot project (Figures 31.1 and 31.2). Located on three acres

Figure 31.1 Pre-design conditions: Washington Canal Park (2009).

Source: OLIN/Josh Leaskey

Figure 31.2 Washington Canal Parks after its design (2012).

Source: OLIN/Sahar Coston-Hardy

(1.21 hectares) of a former school bus parking lot, this three-block-long park lies along the historic former Washington Canal and has become a popular social gathering place and a catalyst for new economic activity in the area. Inspired by the area's waterfront, and designed by the Philadelphia landscape architecture firm OLIN and the Washington, D.C. office of Studios Architects, Canal Park's design evokes the history of its location, featuring a linear rain garden and three pavilions. A 9,000-square-foot pavilion that hosts a café and dining area is reminiscent of the barges that once floated on the canal.

Underneath the site's ice rink, 28 geothermal wells provide a highly efficient energy supply for utilities in the park. The use of geothermal is expected to reduce Canal Park's overall energy use by 37%. Native and adapted plants were installed in the rain garden and throughout Canal Park. Other sustainable design elements included dark-sky lighting elements, high albedo paving, and site features that encourage sustainable practices: bicycle racks, electric car charging stations, and recycling bins.

Overall, creating water resilience was a primary performance goal for Washington Canal Park and required addressing issues of scarcity, flooding, and quality. The park's focal point, the linear rain garden, functions as an integrated stormwater system that is estimated to save the District of Columbia 1.5 million gallons (5.7 million liters) of potable water each year. With the rain garden, low-impact development tree pits, and the capacity to store 80,000 gallons (302,833 liters) of water, almost all of the stormwater runoff generated by the park can be captured, treated, and reused to satisfy as much as 95% of the park's water needs for fountains, toilet flushing, irrigation, and an ice rink. But the project also contributes to neighborhood sustainability. Canal Park manages a community-scale program that reaches beyond its boundaries to capture, treat, and recycle roof and site runoff from nearby buildings. The Canal Park Inc. (the non-profit that manages the park) charges a fee to neighboring landowners for the use of this rainwater management service. This will contribute to the long-term economic viability of Canal Park.

Development of Canal Park involved collaboration with the Washington, D.C., Department of the Environment. Its success led to the department's adoption of more effective water management policies and water quality protections. The city agency considers Canal Park a demonstration project and a planning and design model for future projects with similar green infrastructure goals.

Encouraging sustainable practices in urban design: performance monitoring

Designers should view SITES not as an added constraint on project design, but as a tool that helps enable and support a sustainable design vision. In the case of Washington Canal Park, SITES added a level of rigor and measure to the sustainable principles that were already embedded in the early project conception.

As part of SITES certification, the Canal Park design team completed a formal post-occupancy evaluation (POE) of the project, following the opening of the park (OLIN, 2014). While post-construction evaluation of project performance is often a goal of landscape designers, such monitoring is often limited by budgetary and time constraints, as well as limited technical capacity within design firms to build meaningful evaluation metrics. SITES offers a clear framework within which such an evaluation can be implemented with a high level of scientific rigor and then documented and disseminated for maximum impact. Three credit options of SITES encourage the systematic evaluation of built work and the compilation and dissemination of knowledge gained from such analysis: Credit 9.1: "Promote sustainability awareness and education"; Credit 9.2: "Develop and communicate a case study"; and Credit 9.3: "Plan to monitor and report site performance" (SITES v2 Rating System, 113).

In the case of Canal Park, Credit 9.3 was particularly significant. The intent of this credit is to "Improve the body of knowledge on long-term site sustainability by monitoring and documenting sustainable design practices to evaluate their performance over time" (SITES v2 Rating System, 117). Strategies that SITES recommends to ensure monitoring include communicating early in the design process that monitoring is a project goal and setting aside funding for this purpose.

To measure the performance of Washington Canal Park, OLIN conducted an in-depth, year-long post-occupancy evaluation. A variety of methods were developed to address three SITES credits and topics of interest related to human health and well-being. The evaluation centered on the social and cultural performance of the project, including site accessibility, safety, wayfinding, mental restoration, social interaction, and noise and light levels. The post-occupancy evaluation found that one of the park's primary goals – that of attracting local residents as well as citywide visitors through year-round programming and events – had been met successfully. The park draws 28,000 visitors annually. Daily park usage peaked during summer months, centered on interactive water features and the event lawn. High site usage, associated with the ice-skating rink and holiday events, was observed during winter months, while the lowest level of site usage was during the fall. These findings encouraged park managers to consider more active spring and fall programming.

Another critical goal of the design team – that of creating an inviting public space that welcomes visitors and encourages social interaction – was monitored in the post-occupancy evaluation through visitor surveys. When asked whether they felt welcome in the park, 90% of respondents replied in the affirmative; over one third of respondents, when asked whom they perceived that the park was designed for, responded "everybody." Also, 27% of park visitors stated that they had made new acquaintances in the park. Surveys furthermore indicated that perceived safety within the park was very high.

In addition to meeting the SITES credit intent, the Canal Park POE supplied the design team with transferable insights for future design work, as well as recommendations for adaptive management. The structured method required to achieve SITES credit for performance monitoring substantially increased the rigor and intensity of observation and analysis performed, relative to the design team's typical post-occupancy practices.

Facilitating sustainable practices in urban design: community engagement

In the context of urban design, the ability to fully implement sustainable design principles often depends not only on the expertise of the design team, or on the vision of an individual client, but on the buy-in and cooperation of several distinct public and private entities and community members. Early adoption of the SITES Rating System in a project can facilitate the long and complex efforts involved in ensuring that environmental and social sustainability remain integral, from conceptual design through documentation, value engineering, construction, programming, and management. The pursuit of SITES certification ensures that community members play a substantial role in project ideation and creation, and performance goals are formalized and agreed upon by the design team and all project partners.

In order to achieve SITES certification, a design team must identify project sustainability principles and performance goals, incorporate them into a plan, identify all project partners and user groups, plan for construction oversight, and develop a strategy for preparing a site maintenance plan – all before the initiation of project design. To receive additional points, these efforts must be expanded to "identify specific, measurable, attainable, realistic, and timely project goals by engaging site users and stakeholders during the design process to supplement professional expertise with local knowledge" (SITES, Rating System).

Located in a particularly distressed and underserved neighborhood of Washington, D.C., one of the central objectives of Washington Canal Park was to support and enhance the economic, social, and physical health of local residents, including those living in the adjacent D.C. Housing Authority (DCHA) HOPE VI Arthur Capper and Carrollsburg public housing community. Canal Park was imagined not only as a citywide destination, but also as the future "backyard" for nearby residents, providing recreation, a farmer's market, cultural events, free entertainment programming and restorative outdoor space. DCHA became an early partner in project planning and allocated $13.5 million in New Markets Tax Credits to help fund the park in anticipation of positive quality-of-life impacts for residents. DCHA also worked with the Canal Park Development Association to put in place a Community Benefits Agreement (CBA) that would provide direct benefits to local residents. Importantly, the CBA guaranteed that 10% of the anticipated 150–160 construction jobs and six of the anticipated 30–40 post-construction jobs would go to low-income persons residing in the public housing developments. Currently, 47% of park positions are held by low-income residents and another 37% by local residents (LAF Performance Series).

In addition to pre-design outreach, the SITES system facilitates community engagement through the entire design process. A particular SITES credit requires engagement with potential site users and project partners during site assessment and program planning, schematic design, and design development, and at design completion. Site users are involved in the definition of project goals and identification of programmatic and functional needs of various user groups. Community members are provided with multiple design alternatives using visual presentations, and the final design is shared with the public in at least two formats (e.g., website, community meeting, newspapers, civic display).

Service to local communities is further facilitated after design and construction by another credit, which aims to "provide economic and social benefits to the local community by providing publicly available on-site events, facilities, amenities, or programming" (SITES v2, 84. In Washington Canal Park, in addition to the park tavern and ice-skating rink, the Capitol Riverfront Business Improvement District (BID) provides free programming year-round and attracts a wide range of local residents. A neighborhood study performed by the BID in 2014 showed that summer movies attracted 2,200 people, or 38% of the neighborhood. A weekly farmers' market takes place from May through October, and lunchtime concerts are scheduled from May through July.

Driving innovation in urban design: neighborhood-scale stormwater management and city-scale stormwater guidelines

As a tool for urban design professionals, the SITES rating system can be used to encourage rigor in the design process and facilitate engagement, cooperation, and buy-in from pre-design through post-occupancy. At a broader scale, the rating system aims to drive innovation in materials and technologies and in terms of market and policy transformation. The rating system offers up to nine "bonus" points for innovation or exemplary performance, which encourages project teams to drive forward sustainable strategies that fall outside of the predefined credits.

Some of the most impactful sustainable strategies implemented in Canal Park have to do with stormwater management and with the project team's partnership with the Washington, D.C., District Department of the Environment (DDOE). One such strategy is the use of the park as neighborhood-scale green infrastructure through the extension of rainwater best management practices (BMPs) beyond the site boundaries. Canal Park's rainwater collection infrastructure extends into adjacent properties to capture, treat, and recycle roof and site runoff from neighboring buildings. Canal Park, Inc. developed a plan to charge a fee to neighboring office building owners for the use of this rainwater management service. This fee will contribute to the long-term economic sustainability of Canal Park; at the same time, the availability of this infrastructural amenity will encourage urban infill development in the surrounding formerly industrial parcels. The DDOE sees the potential impact of this Shared BMP strategy as a forerunner to future "stormwater parks" that allow for neighborhood-scale stormwater management, rather than the less efficient citywide or individual parcel stormwater management models currently in use. In addition to these neighborhood-scale benefits, the use of "Shared BMPs" actually allows for a decrease in the size of the rainwater storage cisterns required to meet the water needs of Canal Park. Extending the watershed footprint to adjacent properties also increased the overall reliability of the system to provide water year-round.

The Canal Park project team also partnered with the District Department of the Environment to assess the benefits and risks of specific innovative rainwater management and reuse strategies that were not covered by prior city regulations. One such innovation was the recycling of rainwater from a highly contaminated site for use in interactive fountains and spray irrigation, both of which would entail a high level of human exposure. Since no comparable rainwater reuse system existed in the city, the design team worked with the DDOE to develop a risk assessment process that was incorporated into the agency's 2013 *Stormwater Management Guidebook*. This innovative approach to rainwater treatment and recycling, as well as Canal Park's neighborhood-scale stormwater management strategy, make the park a DDOE "demonstration project" to be used as a model for future urban developments with similar green infrastructure goals.

SITES v2 case study: Chicago's Navy Pier, Illinois

Phase One of Chicago's Navy Pier was the first project to receive Gold certification under SITES v2, in 2016. Originally built as part of Daniel Burnham's Chicago Master Plan, and opened to the public in 1916, the Navy Pier was designated as a Chicago landmark in 1977 and largely redeveloped in 1994. Despite its claim as the top tourist destination in the Midwest, the Pier was showing signs of deterioration when Navy Pier Incorporated (NPI) announced a design competition in 2012 to renovate the iconic structure. Landscape architecture firm James Corner Field Operations (JCFO) led the design team selected for the project, the first phase of which was completed in 2016 (Figure 31.3). The team addressed the sentiment among Chicago residents that the existing pier was overly tourist-oriented; it remained cluttered with stops, restaurants, and attractions and was confusing and increasingly disconnected from the rest of the city. The three primary strategies employed by JCFO in Phase One were: (1) strengthening connections between the city and lake; (2) refreshing and de-cluttering the pier to allow space for new green infrastructure; and (3) creating a place that reflects Chicago's identity. (www.fieldoperations.net)

Figure 31.3 Chicago Navy Pier (2018).

Source: Navy Pier, Inc.

The overall design concept for the mile-long pier was to create a "Pierscape" that aligned a series of thematic "social rooms" along a linear "green spine." At the west end of the project, Polk Brothers Park and Plaza act as the threshold between city and pier and include a dramatic series of earthworks that facilitate large events, festivals, and performances, and a new jet fountain that becomes an ice rink in the winter. The central Pier Park, with its dramatic Grand Stair and new Ferris wheel, will act as a primary space for entertainment and recreation. The east end of the pier, or the Lake Room, will focus on broad views of Lake Michigan, while supporting large events and performances. The key element that ties these spaces together is the linear, tree-lined South Dock Promenade, which acts as the park's "green spine."

One of the significant sustainability goals was that of controlling, filtering, and reusing stormwater, which had previously drained directly into Lake Michigan. On the pier portion of the site, stormwater was directed into constructed "tree tubs" which filtered water before releasing it into the lake. On the part of the site on firm ground, stormwater was collected in bio-filtration basins and cisterns prior to discharge. Through this system, harvested rainwater fulfills the total irrigation demand, and a substantial volume of sediment is kept from entering Lake Michigan each year (Re:Vision Architecture 2016).

High standards were implemented in terms of materials selection, energy efficiency, and long-term maintenance. The project reduced the pier's total energy consumption by an estimated 60% through the specification of efficient lighting, pumps, and other equipment. Approximately one third of materials used were made from recycled content, and practically 100% of structural, roadway, and infrastructural waste was diverted from landfills (Re:Vision Architecture 2016).

Encouraging sustainable practices in urban design: connectivity

As in the Washington Canal Park case study, the case of the Chicago Navy Pier offers a variety of insights into how the SITES Rating System can support and facilitate sustainable design practices and how it can drive innovation that extends beyond the boundaries of the project site. According to James Corner,

> SITES informed much of our design process, from access and circulation studies to plant and material specifications. It was an important tool that kept our entire team accountable to a high standard of best practices and resulted in an unprecedented project.
>
> *(Sustainable Sites Press Release 2016)*

Access and connectivity are central issues in any urban design project, but in the case of the Navy Pier, a general sense from the public that the pre-design condition of the site was cluttered, confusing, and disconnected pushed these challenges to the forefront. SITES supports strong urban connectivity in both the Site Context section of the rating system, as well as in its Human Health and Well-being section. In particular, one credit aims to increase site users' ability to understand, access, and feel safe in outdoor spaces.

In the case of the Navy Pier, the design team observed that the existing drop-off and pickup area, at the western entrance of the site, was dominated by vehicular traffic and significantly impeded pedestrian circulation, while the four existing traffic lanes made site access challenging and unsafe. To address these issues, the design reduced traffic lanes and relocated bus drop-off to a less busy site, and traffic-calming bump-outs at pedestrian crossings, so that pedestrians will have to cross a maximum of two lanes of traffic to enter the Gateway Park Plaza from the city.

Facilitating sustainable practices in urban design: institutional transformation

When asked to identify ways in which the pursuit of SITES certification had impacted project outcome, Sarah Weidner Astheime, the project manager for JCFO, referred to the power of the system to facilitate institutional and cultural change. Since the pier had existed for a century as a civic asset, for two decades in its current form, and since it would continue to be managed by an existing organization, the re-envisioning of the site required significant shifts in thinking, specifically around sustainability, among a large group of individuals. In other words, existing staff, managers, and other stakeholders had to redefine their image of the pier in order to ensure the project's long-term sustainability. According to Greenspan and Mason (2017: 4):

> Reinvesting in civic assets requires adequate funds and good design, but the challenge is not simply a matter of design and finance. Our research shows that building a robust network of civic assets requires deeper, longer-term, more abiding changes.

The SITES framework facilitates the types of "abiding changes" which include community engagement, institutional leadership, maintenance and programming, and measurement. In the case of the Navy Pier, shifts in the thinking and practices of the managing institution, facilitated by SITES, will have long-term effects on the space's social and environmental sustainability.

One of the key requirements of SITES, a prerequisite that has a synergistic relationship with many other aspects of the rating system, is a criterion asking for planning for sustainable site maintenance. This is a wide-ranging criterion, both in terms of subject matter (maintenance covers all areas of intervention) and in terms of time scale. Planning for the maintenance and operations strategies begins in pre-design, by including maintenance personnel in the integrative design team, and continues through adaptive management planning, which is reevaluated annually after project completion, to adapt to future conditions and unforeseen changes.

For a project like the Navy Pier, in which maintenance and management practices for the site were in place long before project initiation, it was particularly critical to engage individuals at all levels of site management from the earliest design phases, in order to facilitate mutual understanding and cooperation. Shared buy-in and early consideration of process and procedures allowed the project team to maximize the impact of the Rating System and to achieve Gold certification.

Driving innovation in urban design: market transformation

A compelling argument for the widespread acceptance of rating systems such as SITES in urban design and landscape architecture has to do with the potential of such systems to drive innovation and transparency in the building and materials industries. The Navy Pier case study illustrates this potential in the context of materials selection.

Much of the surface area of the Navy Pier is covered by a signature herringbone pattern of light-colored precast unit pavers. Because of the extensive use of this material across the site, in pursuing SITES certification, it was critical that the design team choose a paver that met very high standards in terms of material makeup (recycled content), provenance (regionally sourced), and color (high SRI value). In addition, the pavers had to support a variety of applications: there are areas where permeability was desired, including areas of heavy vehicular traffic and pedestrian areas. Within the range of regional suppliers, JCFO was not able to identify an available product that would meet such high standards, so the team was motivated to design one. Two paving manufacturers were selected as potential suppliers of a custom, highly sustainable unit paver and were given SITES materials standards as a baseline requirement for composition and manufacture. One of the suppliers was able to meet the standards and went on to create a custom paver that has subsequently been added to its standard product line (Kafka Granite n.d.).

Sarah Weidner Astheime, JCFO project manager, emphasized the significance of SITES as a tool to drive innovation in sustainability standards in this context. The design team was able to set up a scenario of competition between manufacturers on a high-visibility urban project that led to a novel product and material use. The project team was able to push their sustainability goals forward, facilitating cooperation and driving innovation, with the support of the framework provided by the SITES Rating System.

Conclusion

With climate change and increased urbanization, designers are seeking tools to help ensure that their work is sustainable and linked to community resilience (Ahern 2013, Steiner 2018). Grounded in ecosystem services, SITES presents such a system for landscape architects, architects, city planners, and urban designers. The Washington Canal Park and Chicago Navy Pier projects demonstrate the possibilities.

Today, a growing number of projects in the United States have been SITES certified. Most of these projects are located in urban places. International interest in SITES is also growing, principally in Australia and China (Pieranunzi et al. 2017). Meanwhile, designers, planners, environmental scientists, and others are taking the examination to become SITES Accredited Professionals. The continued evolution of the rating system will require sustained research and refinements of SITES v2. For instance, one research question is: if SITES is based on the ecosystem services concept, how well is it performing to recognize, protect, and enhance ecosystem services? With the increased number of projects receiving certification, there are more opportunities to assess and monitor performance. Such ongoing evaluation holds the potential to improve SITES and produce better urban spaces through design.

Notes

1 Up-cycling is the creative reuse of waste materials and unwanted products.
2 See Jenny Roe, "A New Health Urbanism," Chapter 33 in this volume.

References

Ahern, J. (2013). "Urban Landscape Sustainability and Resilience: The Promise and Challenges of Integrating Ecology with Urban Planning and Design." *Landscape Ecology* 28 (6): 1203–1212.

District Department of the Environment, Stormwater Management Guidebook. (2013). "Appendix M: Tiered Risk Assessment Management (TRAM): Water Quality End Use Standards for Harvested Stormwater for Non-Potable Uses," July. http://ddoe.dc.gov/swguidebook

Francis, M. (2001). "A Case Study Method for Landscape Architecture." *Landscape Journal* 20 (1): 15–29.

Greenspan, E. and Mason, R. (2017). *Civic Infrastructure: A Model for Civic Asset Reinvestment.* Philadelphia, PA: PennPraxis for the William Penn Foundation.

Kafka Granite. n.d. "Using Unilock at Navy Pier: A Kafka Granite Case Study." www.kafkagranite.com/using-unilock-at-navy-pier-a-kafka-granite-case-study

Lady Bird Johnson Wildflower Center, The University of Texas at Austin; United States Botanic Garden; and American Society of Landscape Architects. 2014. *SITES v2 Reference Guide: For Sustainable Land Design and Development.* Austin: Sustainable SITES Initiative.

Landscape Architecture Foundation. *Landscape Performance Series: Canal Park, Methodology for Landscape Performance Benefits.* https://landscapeperformance.org/case-study-briefs/canal-park

OLIN. (2014). "Canal Park Post-Occupancy Social Performance Evaluation," August, internal unpublished report.

Pieranunzi, D., Steiner, F.R., and Rieff, S. (2017) "Advancing Green Infrastructure and Ecosystem Services through SITES." *LA Frontiers* 5 (1): 22–39.

Re:Vision Architecture. 2016. "Chicago Navy Pier: Renovation, Sustainable SITES Case Study," August, internal unpublished report.

SITES Home. www.sustainablesites.org/

SITES Projects. www.sustainablesites.org/projects

SITES Rating System. www.sustainablesites.org/certification

Steiner, F.R. (2018). *Making Plans: How to Engage with Landscape, Design, and the Urban Environment.* Austin: University of Texas Press.

Steiner, F.R., Thompson, G.F., and Carbonell, A. (eds). (2016). *Nature and Cities: Urban Ecological Design and Planning.* Cambridge, MA: The Lincoln Institute of Land Policy.

Buente, S. (2016). "Navy Pier Awarded Prestigious SITES Gold Certification for Phase 1 Redevelopment," November 17. http://www.sustainablesites.org/navy-pier-awarded-prestigious-sites-gold-certification-phase-1-redevelopment

Venhaus, H.L. (2012). *Designing the Sustainable Site: Integrated Design Strategies for Small Scale Sites and Residential Landscapes.* Hoboken, NJ: John Wiley and Sons, Inc.

Windhager, S., Steiner, F.R., Simmons, M.T., and Heymann, D. (2010). "Toward Ecosystem Services as a Basis for Design." *Landscape Journal* 29 (2, Fall): 107–123.

Further reading

Calkins, M. (2008). *Materials for Sustainable Sites: A Complete Guide to the Evaluation, Selection, and Use of Sustainable Construction Materials.* Hoboken, NJ: John Wiley and Sons, Inc. Selection and evaluation guidance for landscape designers on a full range of traditional and innovative sustainable construction materials.

Calkins, M. (2012). *The Sustainable Sites Handbook: A Complete Guide to the Principles, Strategies, and Best Practices for Sustainable Landscapes.* Hoboken, NJ: John Wiley and Sons, Inc. This book builds upon information contained in SITES, offering a more in-depth and comprehensive look at related design principles, tools, and strategies.

Lady Bird Johnson Wildflower Center, The University of Texas at Austin; United States Botanic Garden; and American Society of Landscape Architects. 2014. *SITES v2 Reference Guide: For Sustainable Land Design and Development.* Austin: Sustainable SITES Initiative. Presents the basic framework for using the SITES Rating System.

Pieranunzi, D., Steiner, F.R., and Rieff, S. (2017). "Advancing Green Infrastructure and Ecosystem Services through SITES." *LA Frontiers* 5 (1): 22–39. This article lays out the case for employing SITES to help design green infrastructure and to create new ecosystem services.

Venhaus, H.L. (2012). *Designing the Sustainable Site: Integrated Design Strategies for Small Scale Sites and Residential Landscapes.* Hoboken, NJ: John Wiley and Sons, Inc. Focuses on the use of SITES for the residential and other smaller scale design projects.

Windhager, S., Steiner, F.R., Simmons, M.T., and Heymann, D. (2010). "Toward Ecosystem Services as a Basis for Design." *Landscape Journal* 29 (2, Fall): 107–123. This article illustrates why and how ecosystem services can be fundamental to design thinking.

PART III.II

Health

Introduction

Around the middle of the last century, public health researchers had a major influence in the planning of the residential environments in the United States. The concept of "neighborhood unit," initially advanced by Clarence Perry in the 1920s, was adopted in the late 1940s by the American Public Health Association (1960) for formulating "healthful and hygienic" standards for planning, design, and management of residential environments in America. Today, half a century later, public health scholars are trying to influence the design of the built environment with a more expanded mission. Over the last two decades, findings about the connection between health and the physical environment clearly point to the importance of urban design. Indeed, the design of built form, neighborhood layout, streets, and green spaces all have implications for healthy living (American Academy of Pediatrics 2009; Boarnet & Takahashi 2011). At the same time, researchers have established connections between environmental elements – namely the absence of amenities (such as parks, walking and biking trails, sidewalks, fresh food markets, etc.) and the presence of disamenities (such as high-traffic arterials and resulting air pollution) – and adverse health outcomes. Some have reasoned that ailments such as obesity, type II diabetes, and some cardiovascular and respiratory diseases are partly the result of neighborhoods that not only lack health-promoting features but also have an abundance or absence of certain land uses that may lead to health deterioration and unhealthy behaviors (for example, lack of greenery, predominance of concrete pavement, fast-food stores, etc.) (Jackson 2003).

The following three chapters articulate the clear links between specific features of the built environment and health. For Richard J. Jackson and Rebecca Ferdman, "urban design and planning strategies can act as effective and cost-efficient public health interventions to address both acute and chronic health crises." They emphasize that how we design and where we locate our housing, healthcare facilities, schools, transportation, and green infrastructures has significant impacts on health. Design interventions in the built environment can help tackle obesity-related illnesses, asthma, heat stress, and even mental health disorders. They present a proposed action agenda that, among other issues, suggests expanding the use of Health Impact Assessments that evaluate the health impacts of the built environment; using tax policies to privilege health outcomes; and developing an economic health metric.

Similarly, in her chapter Jenny Roe advocates a "new health urbanism" that "uses planning and design to foster fitness, happiness, better quality of life, longevity, better aging, and improved connectivity among people." Rather than single or isolated design interventions, she calls for a "systems approach" to health and for the creation of salutogenic environments, composed of certain interacting elements. These are "Living Streets" that are safe, comfortable, and walkable; "Urban Green Health" promoted by urban greenery; "Urban Blue Health" promoted by waterways that help reduce heat stress and heat island effects; "Compact Cities" that counteract sprawl and promote walking, biking, and social interaction; and "Enabling Environments" that foster experiences critical for the mental health and well-being of residents in both high-income and low-to-middle-income countries

In the last chapter of this section, Frank, Ulmer, Appleyard, and Bigazzi focus on one omnipresent element of the built environment: streets. They discuss how the various components of "Complete Streets" – streets designed to accommodate synergistically and seamlessly multiple transportation modes – have been found to promote health and safety. Drawing from a large body of recent literature, they show how even the micro-components of Complete Streets – sidewalks, crosswalks and intersections, lighting, street trees and street furniture, transit and cycling facilities, and adjacent buildings – have a role to play in creating safe environments that also encourage physical activity. But similar to other authors in this volume, they argue that design elements and characteristics to promote health should be context-sensitive and depend "on who are the current and expected future users, what are their needs, what health challenges exist."

References

American Academy of Pediatrics Committee on Environmental Health (2009). "The Built Environment: Designing Communities to Promote Physical Activity in Children." *Pediatrics*, 23(6): 1591–1598.

American Public Health Association, Committee on Hygiene of Housing (1960). rev. ed. *Planning the Neighborhood.* Chicago: Public Administration Service.

Boarnet, M. & Takahashi, L. (2011). "Interactions between Public Health and Urban Design." In Banerjee, T. & Loukaitou-Sideris, A. (eds.) *Companion to Urban Design*, pp. 198–207. London and New York: Routledge.

Jackson, R.J. (2003). "The Impact of the Built Environment on Health: An Emerging Field." *American Journal of Public Health*, 93(9): 1382–1384.

32

REFOCUSING PLANNING AND DESIGN TO MAXIMIZE PUBLIC HEALTH BENEFITS

Richard J. Jackson and Rebecca Ferdman

Introduction

Over the last two decades, there has been increased recognition in the United States of the importance of the built environment on human health. This has been triggered in part by the health community's documentation of staggering increases in the rates of chronic diseases, in particular obesity and diabetes. While the health sector has long recognized some of the health impacts of the built environment – for example, transportation infrastructure leading to car crashes as a cause of trauma – it has been slow to recognize the impacts the built environment can have on chronic diseases. For example, two thirds of US adults are now overweight or obese, and 9.4% of the US population has diabetes (Mitchell et al. 2011; Centers for Disease Control 2017). The skyrocketing rates of these disorders have been exacerbated by increasing rates of physical inactivity and sedentary lifestyles. This has both social and economic impacts. In the United States, close to 18% of the entire GDP is now spent on medical care, a large portion of which relates to chronic diseases (Centers for Medicare & Medicaid Services 2017). Also cogent in the shift in the incidence of human disease is the high rate of mental health disorders, in particular depression, which can be exacerbated by the urban form of contemporary cities that tends to isolate individuals from social contact and contact with nature (Maller et al. 2005).

A second driving force for the increased interest of the public health community in the interrelationship between built environment and health stems from the increased awareness of the importance of built environment on human health in other sectors of relevance. For example, the environmental health sector has raised the alarm that the burning of fossil fuels, the over-paving of forests and green spaces, and the degradation of air quality in cities are accelerating environmental threats to human health. The worldwide demographic shift from rural to urban areas, associated with sedentary lifestyles and reliance on personal cars, may provide further impetus to promote energy efficiency and human physical activity.

A third driving force is the proliferation of research literature documenting health impacts from the built environment. This research has accelerated, thanks to increased funding and investment by governments and private foundations, as well as by increasing recognition by public health and urban planning scholars of the importance of the design of the built environment on health. A good part of this research has been realized thanks to the availability and widespread uptake and use of personal monitors of movement (including mobile phones) and geographic information systems (GIS), along with a greatly increased computational ability related to large populations.

These three forces come together to provide the information and data necessary to demand new approaches to the issue of built environment and health. Changes must be implemented in the health sector, which includes public health, clinical sectors, and academic health researchers. However, this change must also include the planning and design sector – ranging from architecture to city planning, transportation, and infrastructure development. Armed with this new wave of knowledge from the world of technology and a growing body of experimental case studies and on-the-ground experiences of their own, planning and design professionals can translate health objectives and best practices into the physical planning and design interventions that will comprise the healthy built environments of the future.

The importance of planners and designers

Many architects, planners, builders, developers, real estate financers, transportation engineers, and those in similar disciplines are unaware of the importance of their role as it relates to the health of the overall population, especially for the people who live in the buildings they construct and the neighborhoods they develop. Because the metrics of success in these non-health disciplines focus on aesthetics, mobility, cost, return on investment, and similar measures, the impact on human health and well-being is often subordinated, if not completely overlooked. The focus on code compliance and cost reduction means that more difficult-to-measure outcomes such as social engagement and human vitality, though no less important, are sometimes lost. Forgotten is the need for the residents to get to work and their family members to get to schools, healthcare facilities, retail establishments, and more. These needs are particularly acute for the one third of the population who, because of age, disability, or income, is unable to drive.

Many transportation and design professionals may not see their roles as related to human health; they may understand healthcare as a service that is primarily clinical and medical. But in reality, such traditional interactions with the healthcare system often occur only at the end of a much longer "disease pipeline." For example, an individual who is facing a foot amputation due to complications from diabetes is not at the start of a healthcare intervention but rather at the conclusion of a series of missed health interventions. Many of these missed interventions are directly related to urban design decisions, such as access to grocery stores with affordable, nutritious foods; access to affordable housing located near jobs that do not require a lengthy car commute; access to parks and trails for exercise and play; and access to efficient and effective active transportation networks. Helping to prevent the need for acute medical care through these sorts of urban design interventions is not only better for the health of populations, but also often less costly in the long run. Further, planning and design professionals have access to strategies that can impact more than just physical health. Because health is not merely the absence of disease, but rather a state of physical, mental, and social well-being (see also Chapter 33 by Jenny Roe), many public health leaders have worked hard to "move upstream" in the development of disease process. Well-being cannot easily be found in the clinical setting, but it can be fostered through thoughtful planning and design decisions that promote social interactions, reduce isolation, connect us with nature, promote physical activity, and help us easily and reliably access our daily destinations. While all built environments have the potential to impact health and well-being, there are several aspects of the built environment that should receive priority from the planning and design professionals because of their potential importance for public health.

Understanding public health priorities for built environments

Housing

Housing is the most critical built environment for human health. Adequate housing is foundational; the unsanitary, unsafe, and undignified living conditions that prevail without it are antithetical to nearly all public health goals. The United Nations recognizes a human right to adequate housing to ensure "security, peace and dignity," but many countries, including the United States, do not yet officially recognize housing as a basic human right (Office of the United Nations High Commissioner for Human Rights). The UN definition of "adequate" housing is purposefully broad and includes several performance criteria that are basic components of public health such as "security of tenure," "availability of services, materials, facilities and infrastructure," and "habitability." Equally indispensable to human health, though, are the criteria that fall within the domain of urban planning, including "affordability" and "location." How much housing to build, what kind of housing to build, and where to build it are as much public health questions as urban planning ones. In many cities, much new housing is being created for those at the upper echelons of income, some of whom may purchase their residences as an economic investment rather than as primary dwellings. Middle- and low-income city residents are often driven to the outer fringes of the urban area, sometimes becoming "super commuters," who spend hours every day commuting to work (Moss & Qing 2012). This is particularly true in many cities of the Global North, from Vancouver to London and from New York to Los Angeles, where gentrification in many neighborhoods is leading to the displacement of low-income residents.

To meet the housing needs of a growing and increasingly urbanized population, cities should emphasize dense, high-quality multifamily housing developments that help put good housing within economic reach for all. Residential developments should be strategically located with respect to accessible pedestrian destinations, transportation networks, job centers, and community services such as schools, parks, and healthcare facilities.

As the 2017–2018 fires in Northern and Southern California have tragically shown, the layout of residential neighborhoods can have acute effects on human life, not just chronic ones. This is particularly true in the case of disasters. Imagine that a resident must evacuate her home, shown in Figure 32.1, due to a wildfire or mudslide, as experienced in 2018 in Montecito in California's

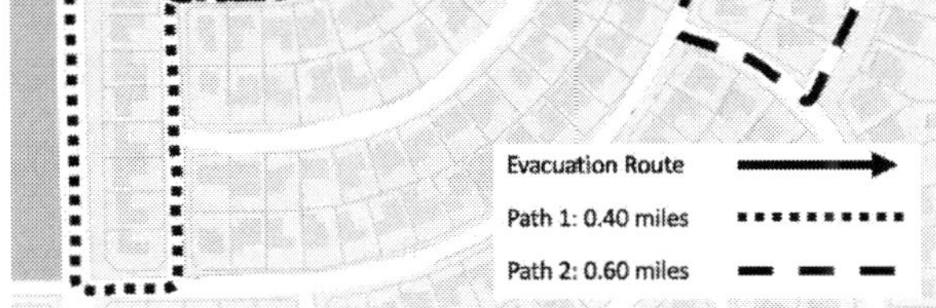

Figure 32.1 Suburban residential neighborhood design hinders efficient emergency evacuation.

Source: Authors

Santa Barbara County. The evacuation route is along the black arrow going east. Though the resident is less than a tenth of a mile from the evacuation route as the crow flies, the two shortest paths by car are 0.40 and 0.60 miles. In contrast, a grid-style development would have facilitated quicker evacuation in case of an emergency.

Hospitals and healthcare facilities

All of us interact with healthcare infrastructure throughout our lives, but the built environments of hospitals and healthcare facilities are most often experienced by populations that are acutely vulnerable from a public health perspective. These special populations include infants and children, seniors, people living with disabilities, people living with chronic illnesses, people facing acute health issues, pregnant women, and new mothers. While these populations visit healthcare facilities primarily to receive medical care, their interactions with the built environment of healthcare facilities can have important and often overlooked impacts on their health and well-being. For example, it has been shown that patients whose rooms have views of trees can recover from surgery more quickly than patients without views of nature (Ulrich 1984). The location and accessibility of healthcare facilities are also essential. Populations most likely to be unable to drive to a hospital, including seniors, people with disabilities, and economically disadvantaged people, are often the same populations with the greatest need for healthcare. Locating healthcare infrastructure near residential developments and planning for accessible pedestrian, bike, and transit routes can facilitate the delivery of medical care to the surrounding community.

Schools

All successful communities emphasize education for children, young adults, and people of all ages, more generally. Building a high-quality public education system requires investing in qualified educators, ensuring food security for students, and facilitating community engagement, to name just a few of many complex and interrelated factors. One investment that is deceptively simple is the built infrastructure of schools themselves. Designing schools that deliberately facilitate the learning process, while also promoting health and safety, takes relatively little extra time and money, but much more thought and care than designing "traditional" schools. For example, strategies such as ample natural lighting, attention to acoustic design, and green open spaces can all support health and learning for children. Equally important is school location and neighborhood context. Providing safe and accessible walking and biking routes to school, including through the use of programs such as "walking school buses," enables much-needed physical activity for both students and parents, cuts down on time spent commuting in cars, and reduces the risk of car crashes near schools (Mendoza et al. 2011). Due to their developmental stages, small body sizes, high breathing rates, and behavioral patterns, children are more vulnerable to poor air quality and other environmental health hazards than adults, so schools should be located away from environmental hazards such as freeways, major roadways, industrial sites, and contaminated brownfields. The walking-to-school experience for inner-city children is further aggravated by such social hazards as drug dealers, homeless, gang members, and bullies (Webb Jamme et al. 2018). Like much of the existing building stock, many existing school facilities were built using construction materials that are now considered hazardous, such as asbestos, lead paint, and lead plumbing. Because of the vulnerability of the children they serve, school facilities should undergo immediate remediation measures.

Transportation

The built environment is made up not only of buildings. One of the most impactful built environments for health is the transportation infrastructure. This includes automobile infrastructure such as highways and freeways; transit infrastructure such as trains, subway stations, and bus stops; and sidewalks, pathways, or other hardscapes where people walk, bike, skate, roll, or get around otherwise. While all forms of transportation infrastructure allow us to move from place to place, they are not all equal in their ability to help support public health goals. For example, while driving has many important benefits, it is often detrimental to health. Car crashes are one of the leading causes of injury and death for both drivers and pedestrians of all ages. Time spent commuting by car is time spent sitting rather than standing, walking, biking, or not commuting at all. Even the simple act of driving is physically stressful and has been shown to raise blood pressure (Maller et al. 2005). Driving also has significant externalities that affect the health of everyone, not just drivers, including noxious emissions, noise, and risk of crashes. Though a third of the US population does not drive due to age, ability, cost, or preference, much of our urban built environment is constructed primarily to accommodate driving, resulting in an automobile-oriented urban form. Houses and commercial buildings in the United States are designed around garages and large parking lots. They are built long distances apart and connected by roads and highways that are ideal for moving cars as quickly as possible; but they are largely remiss of the needs of pedestrians and are often inimical to people. In contrast, transportation infrastructure that supports diverse active modes of transportation can be supportive of and enable healthful lifestyles. People who have easy access to good walking and biking routes or efficient and reliable public transit to travel to and from work and basic services have the opportunity to easily incorporate more physical activity into their everyday lives and may find active transportation more enjoyable and less stressful than driving. For those who do not or cannot drive, the accessibility of these options may make the difference between making a trip and not traveling at all. This tradeoff is especially stark for older adults living in automobile-centric cities, who often must choose between continuing to drive when it is no longer safe for them to do so, or losing their mobility altogether.

Green infrastructure

While the term "green infrastructure" is often used to describe low-impact stormwater management strategies inspired by natural processes, the term can also refer more broadly to the sum of the constructed or managed parks, open spaces, landscapes, and recreational areas in a developed district. Rather than designing and managing these resources in isolation, considering green spaces together as a form of urban infrastructure is a systems approach that encourages efficiency and creative problem solving (see Chapter 33 by Roe). This can have important implications for public health, because exposure to green spaces has been shown to have beneficial health effects such as encouraging physical activity, decreasing stress, and facilitating social interaction (Maller et al. 2005). Thus, green infrastructure should be designed to maximize these benefits by considering the entire urban area. For example, rather than plan a few large parks, an urban development could plan an interconnected series of "green belts" spread throughout the project site. This will help equalize access to green spaces for all residents of the community regardless of their locations, and it will allow them to encounter and interact with natural areas as part of their daily lives rather than as a planned destination. Approaches like this have multiple benefits. In addition to maximizing exposure to green space, strategically linked green spaces can also act as low-impact stormwater management systems, flood mitigation measures, and urban wildlife habitat networks.

Targeting public health challenges of the future through urban planning and design

The public health challenges of the future, and increasingly of the present, are different from the challenges we have faced in the past in several important ways. First, with increasing advances in medical science and technology, acute medical issues such as infections, communicable diseases, and injuries can be prevented and treated with low cost and high reliability. Due to a constellation of factors, including but not limited to longer lifespans, changes in diet, and more sedentary lifestyles, chronic medical issues are on the rise. Chronic issues are often expensive to treat but cheap to prevent; once contracted, they often must be "managed" rather than cured. Second, climate change will be (and in many cases already is) a crucial driver of health issues of the future. It will lead to a tsunami of health effects. These include chronic effects – direct ones such as poor air quality and extreme weather, and indirect ones such as displacement – as well as acute effects such as changing exposures to both new and previously vanquished vectors and viruses. The effects of both the rise in chronic illnesses and of climate change will be most dire for populations that are already vulnerable. In our rapidly changing global society, there is much we do not know about how these changes will affect future public health challenges, and the solutions we should develop in the future. Despite this uncertainty, we can predict that some specific public health challenges will increase in importance in the future. In many cases, we can prepare to meet the challenges discussed here, at least in part through built environment solutions.

Obesity-related illnesses

One public health problem that countries of the Global North are facing today, and which will accelerate in the future, is the rise in obesity, especially childhood obesity, and the diseases that are associated with obesity including Type II diabetes, heart disease, stroke, and some types of cancer. Obesity can also lead to decreased ability to perform or enjoy physical activity, reinforcing a sedentary lifestyle and reducing overall quality of life. Obesity is complex and involves a variety of factors such as genetics, underlying health problems, and socioeconomic status. Two major factors contributing to obesity, diet and lifestyle, can both be approached from an urban planning perspective. For example, developing urban planning policies that require or encourage building grocery stores, farmers' markets, or other retail outlets that sell fresh fruit and vegetables can help alleviate "food deserts," or communities that lack easy access to fresh produce. Similarly, built environment policies can be put in place to limit or restrict the density of fast-food restaurants and liquor stores in high risk areas, such as near schools. Local initiatives may include creating community gardens where healthy produce can be harvested. The built environment can also be designed to facilitate active lifestyles by creating accessible, attractive, and safe open and green spaces in neighborhoods, efficient and reliable active and public transportation networks, and dense mixed-use residential and commercial developments.

Asthma

As populations continue to migrate towards urban centers with high levels of air pollution from cars and industrial uses, rates of asthma can be expected to rise. Studies show correlations between proximity to freeways or major roadways and rates of childhood asthma (McConnell et al. 2010). Many affordable housing options, especially in large cities, are located near high-traffic areas, forcing low-income families to make a tradeoff between affordable rent and acceptable air quality. Understanding this relationship can help us plan the urban form to reduce this conflict, minimizing both the health risks associated with poor air quality and those associated with housing insecurity. For example, urban planners can restrict residential and school developments in high risk zones such as within 500 feet of a freeway, and institute minimum distance from freeway requirements for housing developments. For developments that are in high risk areas, building codes can specify higher efficiency air filtration media. The built environment can also help contribute towards cleaner and healthier air in urban areas. For example, communities with high-quality public and active transportation networks may have less car traffic and therefore fewer automobile emissions. Densely developed communities with housing located near job centers and other daily destinations cut down on the need to commute, further reducing transportation emissions. Planners can also plan an electric vehicle charging network to encourage the convenient use of zero emissions vehicles. Abundant green and open spaces can filter and cool the air, improving its quality.

Heat stress

Increases in the frequency and severity of extreme weather events due to climate change, including extreme heat events, and the local warming caused by the urban heat island effect[1] will combine to give us more dangerously hot days in our cities each year. On days with high ambient heat, the rate of emergency room visits related to cardiovascular disease, asthma, dehydration, and other causes spikes (Basu et al. 2012). People living in areas without access to air conditioning and people living in conditions of social isolation, especially seniors, are at particular risk of heat stress and death from extreme heat. Climate adaptive urban design for these hotter cities of the future is an important way to mitigate these effects. For example, building codes can specify light-colored, reflective materials with high SRIs (solar reflective index)[2] for surfaces that commonly retain heat in urban environments, such as roofs and hardscapes. Green infrastructure is also an effective way to combat urban heat. Trees and other vegetation provide cooling and shade, while absorbing less heat than hardscaped areas. The built environment can also facilitate the types of community and social support that can make the difference between life and death for vulnerable individuals during an extreme heat event. For example, planners and urban designers can provide public community gathering spaces that can serve as "cooling centers," and they can use design techniques in residential developments to facilitate social interaction among neighbors rather than isolation.

Mental health

Many experts predict that, along with cardiovascular disease, mental health will be one of the biggest factors contributing to the global disease burden in the future (Maller et al. 2005). To help mitigate this health threat, the built environment can be designed in many ways to help promote mental health and well-being. Design strategies that stimulate social interactions, for example, can help combat feelings of anxiety and isolation. Open spaces like parks and plazas and vibrant walking places like dense, mixed-use commercial streets encourage people to gather and increase opportunities for socializing. Residential developments can also be designed with social interactions in mind, with residential units easily accessible to their neighbors, and residential streets that prioritize pedestrians over cars. Designing for diverse and accessible modes of transit can reduce isolation and loneliness for populations without access to cars such as seniors, teenagers, and people with disabilities, who are also often more vulnerable to mental health problems than the general population. Green spaces can also help prevent and alleviate mental health challenges such as stress and anxiety. Studies have shown that spending time in green, natural areas is associated with reductions in physical markers of stress and increased overall mental health (Maller et al. 2005).

Drawing from the previous observations, we turn to discuss an action agenda that urban designers, planners, and policymakers should follow to promote health.

An action agenda for refocusing planning and design to maximize health benefits

Action item: continued evaluation of health impacts of built environments

Just as a building is examined for its stability and robustness, energy efficiency, and utility during the design process, so too must the effectiveness of buildings and other built environments be assessed not merely in terms of sustainability and carbon footprint, but also in terms of health benefits. Collecting this information will help in the development of best practices for building for health that are based on real-world successes and failures.

Action item: communication training for researchers

Much of the data generated about the health and built environment interface is in the form of dense tables and graphs that are nearly incomprehensible for policymakers and the general public. Researchers must continue to develop a robust database on the interactions between built environment and health, but there is a need for accessible data, understandable information, and communication with and outreach to policymakers. Training researchers to find the most effective means to disseminate their findings, as well as to consider the modes of communication that have the greatest utility, is critical.

Action item: development of an economic metric for health

The development of economic metrics that capture the short- and long-term cost tradeoffs of various building styles in terms of human health is the "gold standard" for financiers and builders, as well as those working to market health-friendly communities. The problem has been that much of this economic evaluation has tended to be short term, while the benefits are likely to be slow to emerge. Measures of "quality-of-life" are often better surrogates for community health benefits than looking at simple metrics such as average weight or the Body Mass Index (BMI).

Action item: reevaluation of tax and other policy using the lens of health

Governmental economic policies such as subsidies and tax policies have long been drivers of both health and built environment decisions. For example, in terms of health protection and promotion, taxes on tobacco and alcohol have generated substantial improvements in human health. At the same time, subsidies for putatively unhealthy products such as sugar and corn have contributed to the obesity epidemic. The positive and negative impacts of policies that influence the built environment can be examined using this same health lens. For example, Proposition 13 in California under-taxes existing land ownership and tends to over-tax improvements such as private or commercial buildings. In addition, the under-taxation of land has pushed municipalities to raise sales tax rates and also compete for large sales tax-generating businesses such as automobile dealerships and "big-box" warehouse stores, which tend to occupy large amounts of suburban, rural, and agricultural land in order to meet their needs for parking lots and product displays. These land use changes set in motion by Proposition 13 have negative health effects by exacerbating sprawl, promoting sales tax-generating automobile-centric design, gutting funding for local governments, and discouraging the turnover and development of much-needed housing (Schwartz 1997). Examining the health impact of tax and subsidy policies on land use is a fruitful and potentially potent approach to making land use decisions more health-rational.

Action item: optimizing the use of health impact assessments

Environmental Impact Assessments (EIAs) of built environment projects do not adequately address human health impacts. Much greater use of a Health Impact Assessment (HIA) is needed, in terms of both scoping and evaluating the health impacts of projects during and after construction. The American Planning Association views HIAs as an effective process that "brings public health considerations into decision-making" and adds value across many stages of the planning process (American Planning Association August 2016). HIAs have been implemented on a wide variety of planning and policy projects, from the Beltline transit, trails, and parks project in Atlanta, Georgia, to San Francisco's living wage ordinance (Collins & Koplan 2009). However, many builders and developers are skeptical and cautious of HIA because of the challenges they experience during the EIA process. In reality, EIAs rarely capture the positive and negative short- and long-term health impacts of projects. Just as all clinical decisions convey health risks as well as benefits, and every patient and clinician must work to optimize decisions in challenging situations, so too should built environment professionals take into account the full array of options. For example, a subway line proposed in Los Angeles, California, will have the negative impacts of traffic disruption, noise and air pollution over the short term, risks to construction workers, and other risks. At the same time, over the long term, the subway systems will save travel time, help reduce congestion, fossil fuel use, and air pollution, and promote active transportation and other benefits. These evaluations should continue concurrently as well as in advance of the work.

Figure 32.2 below lists the process and components of a Health Impact Assessment (HIA). To be successful, HIAs require the expertise of both planning and health professionals. By combining the skills of these two disciplines, HIAs can help communities optimize their physical, mental, and social well-being.

1. Screening
2. Scoping
3. Assessment

Proximate Impacts ⇒
How will the built environment change? For example, things like:
- Walkability
- Air Quality
- Green Spaces

Key experts: *planning and design professionals*

Determinants of Health ⇒
How will this impact the health of the population? Consider all types of health:
- Physical wellbeing
- Mental wellbeing
- Social wellbeing

Key experts: *community residents and stakeholders*

Health Outcomes ⇒
What are the specific health outcomes we can expect? For example, consider:
- Obesity
- Heat Stress
- Depression

Key experts: *medical and health professionals*

4. Recommendations
5. Reporting
6. Evaluation

Figure 32.2 Process and components of a Health Impact Assessment.

Source: Authors

Concluding remark

The founders of the American Public Health Association included leaders who we would now call urban planners. At the same time, early urban planning laws and ordinances were motivated by public health concerns. Specialization has brought great scientific, medical, and engineering progress, and at the same time, balkanization, specialized language, ignorance of programs and roles, and attendant inefficiency and harm. Challenges and opportunities for the planning, urban design, and health professions lie ahead, and they will not be remedied by mere conversation. Society must commit to a new generation of leaders capable of working across these disciplines.

Notes

1 An urban heat island is an area in an urban setting that is significantly warmer than its surrounding areas because of human activity.

2 SRI is a measure indicating how hot a material can become under the sun.

References

American Planning Association. (2016, August). *Health Impact Assessment Can Inform Planning to Promote Public Health.* Retrieved from pewtrusts.org: https://www.pewtrusts.org/en/research-and-analysis/issue-briefs/2016/08/health-impact-assessment-can-inform-planning-to-promote-public-health

Basu, R., Pearson, D., Malig, B., Broadwin, R., & Green, R. (2012, November). The effect of high ambient temperature on emergency room visits. *Epidemiology*, *23*(6), 813–820.

Centers for Disease Control. (2017). *National Diabetes Statistics Report, 2017: Estimates of Diabetes and Its Burden in the United States.* Retrieved from cdc.gov: www.cdc.gov/diabetes/pdfs/data/statistics/national-diabetes-statistics-report.pdf

Centers for Medicare & Medicaid Services. (2017, December 7). cms.gov. Retrieved December 26, 2017, from National Health Expenditure 2016 Highlights: www.cms.gov/Research-Statistics-Data-and-Systems/Statistics-Trends-and-Reports/NationalHealthExpendData/Downloads/highlights.pdf

Collins, J., & Koplan, J. P. (2009). Health impact assessment: A step toward health in all policies. *JAMA*, 315–317.

Maller, C., Townsend, M., Pryor, A., Brown, P., & St. Leger, L. (2005, December 22). Healthy nature healthy people: "Contact with nature" as an upstream health promotion intervention for populations. *Health Promotion International*, *21*(1), 45–54.

McConnell, R., Islam, T., Shankardass, K., Jerret, M., Lurmann, F., Gilliland, F., . . . Berhane, K. (2010, July). Childhood incident asthma and traffic-related air pollution at home and school. *Environmental Health Perspectives*, *118*(7), 1021–1026.

Mendoza, J. A., Watson, K., Baranowski, T., Nicklas, T. A., Uscanga, D. K., & Hanfling, M. J. (2011, September). The walking school bus and children's physical activity: A pilot cluster randomized control trial. *Pediatrics*, *128*(3), e537–e544.

Mitchell, N., Catenacci, V., Wyatt, H. R., & Hill, J. O. (2011). Obesity: Overview of an Epidemic. *Psychiatric Clinics of North America*, *34*(4), 717–732.

Moss, M. L., & Qing, C. Y. (2012). *The Emergence of the Super-Commuter.* New York University, Wagner School of Public Service, Rudin Center for Transportation.

Office of the United Nations High Commissioner for Human Rights. (n.d.). *The Right to Adequate Housing: Fact Sheet No. 21 (Rev. 1).* United Nations: www.ohchr.org/Documents/Publications/FS21_rev_1_Housing_en.pdf

Schwartz, J. (1997). Prisoners of Proposition 13: Sales taxes, property taxes, and the fiscalization of municipal land use decisions. *Southern California Law Review*, 183–217.

Ulrich, R. S. (1984, April 27). View through a window may influence recovery from surgery. *American Association for the Advancement of Science*, *224*(4647), 420–421.

Webb Jamme, H.-T., Bahl, D., & Banerjee, T. (2018). Between "broken windows" and "eyes on the street": Walking to school in inner city San Diego. *Journal of Environmental Psychology*. doi:10.1016/j.jenvp.2018.01.004

33
A NEW HEALTH URBANISM

Jenny Roe

Introduction

Cities in high-income countries (HICs) are experiencing epidemics of non-communicable diseases (NCDs), including obesity and diabetes, widespread mental health problems, substance abuse, and traffic injuries. Low- to middle-income countries (LMICs) are also experiencing dramatic rises in NCDs while, at the same time, continuing to suffer from widespread infectious diseases often attributed to slums and informal settlements which lack basic services, including water and sanitation (Grant et al. 2017). Increasing air and noise pollution, reduced opportunities for physical activity, unhealthy diets, social isolation, and exposure to crime all play a role in a multiplicity of chronic health problems including stress, cardiovascular disease, obesity, and diabetes (Goenka and Andersen 2016). High levels of health disparities further worsen widening economic and racial inequity (Dickman et al. 2017). The effects of climate change, environmental degradation, and loss of biodiversity further exacerbate health outcomes in cities. The resultant health problems place increasing demands on public health services. And with increasing urbanization and a rising aging population, the pressure is set to increase.

Increasingly, global health agencies are recognizing a critical role for city planning in meeting public health objectives. Integrated multi-sector city planning, including urban design and transportation planning, are an important and underused force for sustaining and creating health within both HICs and LMICs (Giles-Corti et al. 2016). The World Health Organization (WHO), for example, recommends "placing health and health equity at the heart of [city] governance and planning" (WHO 2008), highlighting the need for integrated urban planning, transport, and housing policy. This builds on a 30-year old WHO Healthy Cities Program promoting comprehensive and systematic policy and planning for health (WHO 2014). Now established in over 1,000 cities worldwide (see de Leeuw and Simos 2017 for case studies), the healthy cities movement recently gained new momentum with all 53 member states of WHO Europe adopting the Copenhagen Consensus and committing to building healthier, happier cities (WHO 2018).

Presently, with a new global conversation on cities and health, comes the opportunity to capitalize on wider global strategies. In 2015, the United Nations drafted the first urban Sustainable Development Goals (SDGs). Goal 11, to "make cities inclusive, safe, resilient and sustainable by 2030," has sub-clauses on road safety, public transport, air quality, and safe, inclusive, and accessible green and public places (UN 2016a). Additionally, at the 2016 UN Habitat III Conference, a manifesto was issued embedding health within an ensuing urban agenda (UN 2016b). The 2030 UN Agenda for Sustainable Development aspires to "ensure healthy lives and promote wellbeing for all at all ages." These global policy initiatives offer a framework for forging a "new health urbanism."

The determinants of health

But what constitutes health? There are five major categories of health determinants: genetics, behavior and lifestyle, social circumstances, environmental influences, and medical care (McGovern et al. 2014). The contribution of the built environment to health is estimated to be approximately 10% (Minnesota Department of Health 2014). But this estimate overlooks the role the built environment plays in determining lifestyles and healthy behaviors, which contribute a further 20% to 50% towards health, depending on what model of determinants is applied.

Over the last two decades, the field of public health has moved towards a social determinant model of health (SDOH) exploring the political, social, economic, cultural, and environmental factors that shape health (WHO 2008). The environmental factors include neighborhood features (e.g. walkability, access to parks), availability of healthy food options, community cohesion, and social capital. Labeled the "causes of the causes" (Braveman and Gottlieb 2014), the social determinant model seeks to identify the underlying causes of health problems. For example, social inequality and poverty may explain why a child develops asthma due to poor housing conditions, lack of clean air, and lack of access to green space and trees. The unequal distribution of such social determinants contributes to health inequities; for example, in the United States, the risk of death from asthma is three times greater in African Americans than in white people (Partners in Healthcare 2010). The SDOH model, however, has led to singular intervention approaches to public health, trying to identify cause and effect, isolating, and testing one or two variables, rather than exploring more complex integrated models that might better explain health outcomes.

A salutogenic approach to health urbanism

The traditional focus in public health has been on ill-health, combatting disease, taking a "treatment" approach, and implementing interventions that reduce or minimize the risk of disease in human populations, working to achieve the WHO definition of health as an "absence of disease" and "complete health" (i.e. complete physical, mental, and social well-being). This approach views a community from the perspective of its deficiencies rather than its assets (Lindström and Eriksson 2005).

Increasingly, however, it is recognized that a "complete" state of health is difficult to achieve in an age of increasing NCDs, and that social and mental well-being can be maintained despite the presence of physical ill-health. This has led to a focus on health as an asset – rather than ill-health as a burden – and an interest in how urban design can achieve health and well-being. The original thinking lies in Antonovsky's (1979) "theory of salutogenesis" that focuses on the creation of health rather than the prevention of disease. This theory asks: how can people stay healthy despite life stressors? What creates health as compared to what causes disease? This assets approach to health focuses on individual and community capabilities and resources, which may actively help build health (Lindström and Eriksson 2005). An asset-based approach focuses on the protective factors that support health and well-being, and the environment is a significant contributing factor. A salutogenic environment is defined as the physical, social, and cultural contexts that support healthy behaviors and help build health. Salutogenic properties of the environment include access to space, daylight, parks, and gardens as well as the social dynamics of a place that support strong social networks and create social capital. This asset-based framework is evident in the UN's SDG framework (specifically Goal 11, identified earlier) and the new urban health agenda.

A systems approach to health

Recently, there have been calls for a more integrated approach to public health, using a systems-thinking perspective that conceptualizes health as a multitude of interdependent variables within a connected whole (Rutter et al. 2017). A systems approach to public health recognizes that complex health issues cannot be solved by a single intervention but requires the reshaping of the interacting factors within the system. Tackling asthma, for example, requires availability of healthy housing, free from mold, dust, and dangerous gases and other fumes; lifestyle change that removes the exposure of children to cigarette smoking in the home; reduced exposure to psychological stressors; and availability of clean air.

The DPSEEA model (Driving Force-Pressure-State-Exposure-Effect-Action) (Corvalán et al. 2000) provides a dynamic ecosystem framework to explore the complex interacting factors that contribute to health and well-being. It has evolved over time and embodies a set of interacting contextual factors that include demographics, socioeconomic, and behavioral influences (Morris et al. 2006; Reis et al. 2015). The modified ecosystem enriched model – the "eDPSEEA model" – is illustrated in Figure 33.1: Driving Forces (Box A) to Health and Well-Being (Box E) within the Ecosystem create Pressures (Box B), which change health-relevant characteristics

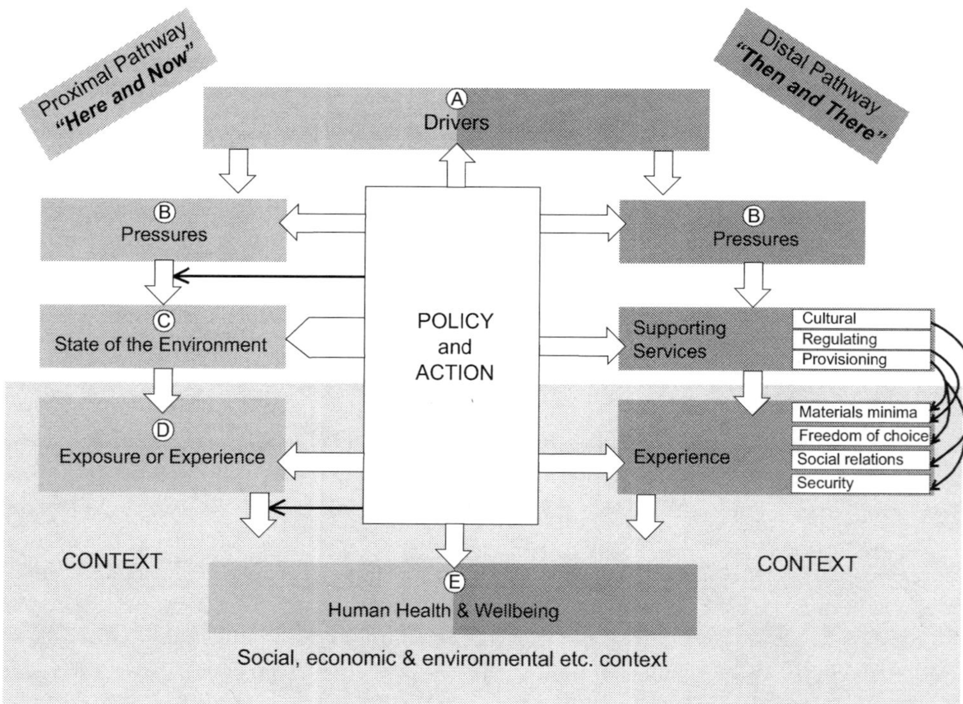

Figure 33.1 The Ecosystems Enriched DPSEEA Model [incorporating both social and ecosystem services dimensions].

Source: (Reis et al. 2015) as subsequently expanded (EEA 2015) (redrawn for publication)

of Environmental State (Box C) near, or "proximal" to, a community (the "here and now"). In addition, there is a "distal" pathway from Drivers to Health and Well-Being (the "then and there"), generating changes in Environmental State, which have environmental and health impacts far beyond the community, whose activities generate the harm (Reis et al. 2015).

Next, we consider some of the interacting environmental factors that create and support health, and explore how these operate synergistically and in tandem within complex urban systems.

Living streets

The configuration of city streets can promote health by minimizing noise and air pollution, providing opportunities for social interaction and promoting both mobility and physical activity, all in turn helping reduce the risk of obesity and cardiovascular disease. A living – or comfortable – street is a safe, enabling street for everyone, including vulnerable populations: an environment in which an older person, a child, or teenager can move with ease and freedom. A living street privileges pedestrian paths and non-motorized transport (e.g. cycling lanes) over cars and provides adequate separation of pedestrians and bikes from motorized transport. A living street is an attractive, desirable place to walk and is well connected to public transportation systems (buses, metro, tram). It engages curiosity and fascination (integral to our mental well-being) by offering visually stimulating architectural façades, shopfronts, and public art. Street trees and plants – when located alongside pedestrian paths and roadways – further promote walkability, provide shade, lower temperatures, mitigate noise, reduce air pollution, and increase comfort (see also the section on Urban Green Health in this chapter). A living street also nurtures social trust in the community and its visitors, who feel safe to linger, interact, and dwell in the space. There is now a significant body of evidence that shows how the design of streets – and their connectivity to the surrounding neighborhood – can support active living and health outcomes (Sallis et al. 2016).

For years many European countries have been creating safe and accessible pedestrian walkways, cycle paths, and public transport. In Holland, the "woonerf" concept (which translates as "living yard") is an early exemplar of safe, shared streets, conceptualized in the 1960s. Many other new exemplars are emerging: Oslo is repurposing parking spaces for bicycles and walkways and intends to make the city center car free by 2019, which, if achieved, will be the first total and permanent ban of private cars in a European city, and is part of a wider plan to cut greenhouse emissions in half over the next five years. In comparison, highly motorized countries, such as the United States and Australia, have been slower to integrate pedestrian-friendly and cycling-friendly streets in cities. Propelled by the Surgeon General's 2015 call to improve walkability in cities, the United States is increasingly embracing the living streets concept, driven by an active living research (ALR) initiative (see Sallis et al. 2016) and significant efforts by many city mayors (Designed to Move Active Cities 2015).

In LMICs, appropriate street design for health and well-being is an urgent priority. Rapid urbanization combined with lack of economic resources means that street design in these countries is often ill thought out and poorly planned. People in LMICs mostly walk and cycle because there are no other transport options, usually in intense heat, exposed to air and noise pollution, dust, road craters, and the risk of road traffic injuries. Improving urban mobility in these countries is an equity issue; poorer people in LMICs do not own cars and are often unable to afford public transport or bikes. At the same time, access to bikes is restricted for women and girls because of cultural or religious barriers. Some LMICs, however, are pioneering new approaches in urban design for health. Bogotá is one example of a LMIC city pursuing a new urban mobility via rapid implementation of bicycle infrastructure and a public transport system, repurposing road space from cars, for cycle pathways (see Sallis et al. 2016). The implementation of Ciclovía (a Spanish term that means "cycleway") – closing off streets on Sundays and public holidays – has been replicated worldwide and had been shown to improve quality of life for Bogotá residents (Sarmiento et al. 2010, cited in Sallis et al. 2016) and to significantly reduce air pollution. Other LMIC cities pioneering innovation such as motor-free streets and improving walking and biking infrastructure include Sorocaba, Brazil; Medellín, Colombia; and New Delhi, India (see Designed to Move Active Cities 2015).

Providing safe streets is integral to fulfilling the SDG 11 on inclusive cities, with an urgent need to accelerate the agenda in LMICs and highly motorized HICs. Design precedents are abundant, particularly from Europe, but policies are needed to better integrate public transport systems, reduce particulate emissions, encourage walking, cycling and public transport, discourage private driving, and build a new urban mobility. Exactly how smart technologies – driverless car technology, for instance – will impact the delicate balance of the street ecosystem is as yet unknown. Autonomous vehicles potentially offer isolated communities better opportunities to connect with the urban fabric of a city. There is a pressing need, therefore, to identify how such technology might improve urban mobility opportunities, particularly for marginalized, poorer communities.

Urban green health

Urban "green health" refers to the health benefits from contact with urban green space (e.g. parks, urban forests, green corridors), and there are now several systematic reviews documenting this evidence (Bowler et al. 2010; Kuo 2015; WHO 2016). The benefits of urban nature include increased opportunities for physical activity (Dadvand et al. 2016; Hartig et al. 2014; Nieuwenhuijsen et al. 2014), but by far the strongest benefits pertain to mental health including stress alleviation, reduced incidence of depression, improved sense of purpose in life, improved cognitive functioning, improved performance at work and school, and symptom alleviation in specific mental health illness, including schizophrenia, ADHD, and dementia (for evidence see Roe 2016). The evidence is stronger for deprived urban communities, with some researchers suggesting that green space may be "equigenic," i.e. that equitable access can help mitigate income-related health inequalities in relation to chronic disease and life expectancy (Mitchell and Popham 2008; Mitchell et al. 2015). Urban green space is also associated with increasing social capital, fostering greater place belonging and place attachment (WHO 2016), attributes that contribute towards community resilience and recovery from natural disaster (Marín et al. 2015), but the field of evidence is currently limited.

The literature posits four pathways by which green space benefits health and well-being: improving air quality, facilitating active living (e.g. walking or cycling), providing opportunities for social contact, and for psychological restoration (Hartig et al. 2014). A fifth biological pathway is that of enhanced immune functioning and the role of inflammatory cytokines known to impact stress and depression, arising from biodiversity and microbiota in natural settings (Kuo 2015). These pathways interact as part of a complex system: for example, the presence of street trees improves air quality in a neighborhood; this makes a street more walkable and promotes active travel, which in turn increases opportunities for social contact. This, combined with the aesthetic affordances of the trees themselves – termed "soft fascination" in the literature – in turn facilitates psychological and physiological restoration from stress (Ulrich 1983) and cognitive fatigue (Kaplan and Kaplan 1989).

Such is the strength of the evidence for the positive effects of urban green space on health, that several cities and towns around the world now promote "green prescriptions," i.e. the use of urban green space to improve physical and mental health outcomes. The best evidence to date comes from the New Zealand Ministry of Health's "Green Prescription program" which shows increases in physical activity levels and improved quality of life over 12 months, with increases in physical activity sustained in the long term over two to three years (NZ Ministry of Health 2013). Additional "green" or "park prescriptions" are now advocated in Japan, the United Kingdom, and the United States. These are complemented by "Green Care" programs – such as those currently offered in the Netherlands and the United Kingdom – that provide organized nature-based health interventions for people with specific mental health illnesses.

But how can green space be best integrated within the fabric of a city, and what are the active "green" ingredients that specifically contribute to well-being? This depends on contextual factors (e.g. city or neighborhood scale) and the culture, demographic characteristics, and health needs of the local community. While certain typologies of urban green space are known to bring health benefits – e.g. large-scale city parks (Larson et al. 2016), local neighborhood parks (CABE 2010), pocket parks (Nordh et al. 2009), and urban woodlands (Ward Thompson et al. 2013) – less is known about the specific physical attributes most likely to invoke health and well-being. The most conclusive evidence is for the prevalence of (1) street trees, which improve the walkability of a neighborhood (Sarkar et al. 2015), reduce depression (Taylor et al. 2015), reduce heat island effects (Shanahan et al. 2015), and increase life longevity (Takano et al. 2002); (2) hedges (Abhijith et al. 2017), which reduce the impact of air pollution (alongside trees); (3) front gardens, which foster social interactions and place belonging in a neighborhood (Roe and Ward Thompson 2011); and (4) urban water settings (referred to as blue space in the literature), which have an important role to play in urban health (see the next section). Recent research has begun to identify those design ingredients of an urban green space that are most likely to support psychological restoration, including structural properties (i.e. safety, openness, complexity, coherence) and physical attributes that relate to these structural properties (e.g. lighting, seating, animals, tree canopy) (Hunter and Askarinejad 2015). However, the empirical testing of this framework is still in its early infancy and to date has mostly been carried out in laboratory contexts.

Urban blue health

Urban waterways help buffer heat island effects and reduce heat stress. Their management is critical for climate adaptation; but their synergistic benefits to health and well-being are often overlooked. Sustainable urban drainage systems (SUDS), for example, if well designed and maintained, while managing surface water flow, can significantly contribute to the biodiversity of a city, and to its visual and sensory experience, including tranquility (e.g. as offered by bird sound) and screening noise (e.g. traffic).

A recent systematic review has concluded that the benefits of access to blue space are stronger for mental health (Gascon et al. 2017) than for physical health or social well-being. There is some limited evidence to show that proximity to inland waterways is associated with increased physical activity in adults. Living near coasts is associated with a reduced risk of obesity in children, and in teenagers and adults with reduced hypertension and cardiovascular risk (see Gascon et al. 2017 for a full review of the evidence). Owing to flood risk, living close to water – and attitudes to blue space – can also generate some negative health outcomes, if the risks are not well managed, particularly amongst vulnerable and marginalized populations (Miller et al. 2012).

To conclude, access to water is an underused health-promoting tool in the urban landscape. Many cities in the 1950s turned their backs to urban canals and rivers channeling waterways into culverts and channels paved in concrete; although some are being brought back to life now. Los

Figure 33.2 Sheffield's "Gold Route."

Source: © Jan Woudstra, University of Sheffield, Department of Landscape

Angeles, for example, is envisioning its river as a tool for urban regeneration, accompanied by a linear park to promote active living and systematically unify disparate city neighborhoods to downtown. Pittsburgh, Medellín, Chicago, West Palm Beach, New York, and many other cities around the world are also using their riverways and waterfronts to regenerate neglected parts of the city. Paris, a great innovator in urban parks, is turning to blue space to increase health and recreation opportunities, with free swimming at La Villette. Sheffield, in the United Kingdom, has sparked major regeneration of its city center using water features that form a "Gold Route," guiding visitors from the city station to the city's central square (Figure 33.2).

The compact city

Compact cities – where housing, jobs, shops, schools, healthcare, cultural, and leisure activities are concentrated within short distances – are posited to be more sustainable and health promoting by reducing the need for long commutes, and by promoting an urban form that facilitates active living, reduces traffic and noise pollution, and enhances social connectivity and life quality.

Some cities are putting strategic plans for the compact city to effect – Portland, Oregon, for example, has a "20-minute neighborhood" concept plan in place that aims to provide access to basic needs and services within a 20-minute walk from home (City of Portland 2010). But examples of compact urban design, together with longitudinal analysis of any health gain, are limited. In a recent hypothetical study of six cities (Melbourne, London, Boston, Copenhagen, São Paulo, and Delhi) estimating the effects of compact city scenarios (i.e. increasing land-use density by 30%, reducing average distance to public transport options by 30%, and increasing diversity of land use by 30%), the modeled scenarios generated health gains for cardiovascular disease, respiratory disease, and diabetes in all six cities (Stevenson et al. 2016). But it is currently difficult to substantiate these findings from real-world empirical studies. A recent synthesis review of the literature (Ahfeldt and Pietrostefani 2017) found compact urban form generally has positive effects, with the strongest effects on economic outcomes (e.g. productivity and innovation). The evidence also shows that compact urban form is associated with sustainable transport modes (i.e. non-automobile), lower crime rates, social equity, shorter trip lengths, lower energy consumption, and more efficient provision of local public services. However, negative health effects included lower subjective well-being, more traffic congestion, less open space, and mixed evidence on air pollution concentrations. While compact, mixed-use urban form appears to reduce car trips and car congestion, inferring improved health via increased walkability and cycling, higher residential density can have negative effects from increased noisy activities from neighbors, reduction in open space, lower sense of community cohesion, and reduction in subjective well-being. The longitudinal evidence on health outcomes, however, is currently limited.

It is also argued, but not substantiated by empirical evidence, that a compact city reduces health inequalities and increases social equity by reducing segregation and increasing job opportunities and accessibility to jobs, public transportation, healthcare, and cultural and recreational facilities. In summary, the compact city concept holds promise for promoting healthy, more sustainable lifestyles, but realizing these scenarios requires long-term strategic visioning and integrated regional and local planning and urban design.

Enabling environments

The demands that urban living places on mental health require new thinking about health urbanism. City dwellers are more likely to suffer from acute mental illness such as schizophrenia, depression, and anxiety (Meyer-Lindenberg 2013). For people with acute mental illness – or older people with cognitive difficulties – negotiating urban streets can be a terrifying and paralyzing experience resulting in social isolation, and increased health risks from sedentary lifestyles. Weight gain, obesity, heart disease, and diabetes – chronic illnesses associated with sedentary living – are higher for people with acute mental illness (Schizophrenia Commission 2012). While strides are being made in designing cities for active living, very few cities are spearheading attempts that put mental health at the core of urban planning. This is partly because convincing data on how cities aggravate or aid acute mental health problems is sparse. An Enabling Environments (EEs) framework offers one approach for building better mental health in cities (Johnson and Haigh 2011).

Enabling Environments are places that promote five experiences critical for mental well-being: a sense of belonging, safety, social interaction, inclusion, and agency. Failure to promote these qualities in an environment is not just unhealthy, but potentially toxic for mental well-being (Haigh 2013). The EE model has been successfully applied to improve well-being outcomes in UK residential care and prison settings, but the same approach could be applied more widely on a neighborhood scale. For instance, urban "green prescriptions" and "green care" (see the earlier section) are two examples of place-based interventions that promote better mental health. Improving access to good-quality city parks/gardens meets the criteria of the EE framework, potentially promoting social interaction, place attachment, a sense of belonging, and also stress regulation and anxiety reduction. However, the current EE framework could be better integrated within a whole systems approach to urban planning. This requires thinking about various aspects of the city: its soundscape (e.g. how noise pollution aggravates many mental health problems);

street connectivity and network design; cognitive wayfinding; and convenient, affordable, and easily accessible public transport, with friendly support systems at hand.

To conclude, an Enabling Environments framework offers a new approach to integrate mental health into the core of urban planning, placing the needs of the most vulnerable in our society at the core, and building healthier communities for all.

The future city and health impacts

Cities are our next greatest assets in achieving sustainable health and development (Kleinert and Horton 2016). Healthy placemaking – at a city and a community level – embracing biodiversity, inclusivity, mixed use, and compact urban form, and enabling those most vulnerable in our society to flourish – offers the prospect of sustaining not just our health now, but the long-term future of our planet. Many healthy city issues, touched upon in this chapter, need to be considered in greater depth, including opportunities for driverless cars to create more accessible cities; gender freedom; racial integration; improving urban mobility for elders; and intergenerational opportunities for play and living. While the future city is hard to predict, compact city modeling offers one method by which to predict how land use change can best support health and well-being. But it is urgent to act now, to retrofit our existing cities and accelerate health-oriented urban design, particularly in LMICs, maximizing access to open space, healthy housing, access to healthy food, urban surface water management, urban transit, and mixed-use development. The future also requires urban planning, transport planning, urban design, and public health to work together to address the global challenges ahead, including climate change, environmental degradation, health disparities, migration, and the fast-changing demographics of our cities.

Acknowledgment

Thank you to George Morris, Honorary Visiting Professor, European Centre for Environment and Human Health, University of Exeter, for advising on the eDPSEEA model.

References

Abhijith, K.V., Kumar, P., Gallagher, J., McNabola, A., Baldauf, R., Pilla, F., Broderick, B., Di Sabatino, S,. and Pulvirenti, B. (2017) "Air Pollution Abatement Performances of Green Infrastructure in Open Road and Built-up Street Canyon Environments: A Review." *Atmospheric Environment* 162(August): 71–86.

Ahfeldt, G.M., and Pietrostefani, E. (2017) "The Compact City in Empirical Research: A Quantitative Literature Review." *ERC Discussion Papers, SERCDP0215*. Spatial Economics Research Centre, London School of Economics and Political Science, London, UK.

Antonovsky, A. (1979) *Health, Stress, and Coping*. San Francisco: Jossey-Bass.

Bowler, D.E., Buyung-Ali, L.M., Knight, T.M., and Pullin, A.S. (2010) "A systematic Review of Evidence for the Added Benefits to Health of Exposure to Natural Environments." *BMC Public Health* 10(1): 456.

Braveman, P., and Gottlieb, L. (2014) "The Social Determinants of Health: It's Time to Consider the Causes of the Causes." *Public Health Reports*. 129 (January–February, Suppl 2): 19–31. doi: 10.1177/00333549141291S206

CABE (2010) *Community Green Using Local Spaces to Tackle Inequality and Improve Health*. London, UK: Commission for Architecture and the Built Environment (CABE).

City of Portland (2010) www.portlandonline.com/portlandplan/index.cfm?c=52256&a=288547

Corvalán, C., Briggs, D., and Zielhuis, G. (2000) *Decision-making in Environmental Health: From Evidence to Action*. London, UK: E & FN Spon.

Dadvand, P., et al. (2016) "Green Spaces and General Health: Roles of Mental Health Status, Social Support, and Physical Activity." *Environment International* 91: 161–167.

de Leeuw, E. and Simos, J. (Eds.) (2017) *Healthy Cities: The Theory, Policy, and Practice of Value-based Urban Planning*. Springer.

Designed to Move Active Cities (2015) http://e13c7a4144957cea5013f2f5ab26d5e83af3ea377013dd602911.r77.cf5.rackcdn.com/resources/pdf/en/active-cities-full-report.pdf

Dickman, S.L., Himmelstein, D.U., and Woolhandler, S. (2017) "Inequality and the Health-care System in the USA." *The Lancet* 389(10077): 1431–1441. doi: 10.1016/S0140-6736(17)30398-7

Gascon, M., Zijlema, W., Vert. C., White, M.P., and Nieuwenhuijsen, M.J. (2017) "Outdoor Blue Spaces, Human Health and Well-being: A Systematic Review of Quantitative Studies." *International Journal of Hygiene and Environmental Health*, 20 (8): 1207–1221 https://doi.org/10.1016/j.ijheh.2017.08.004

Giles-Corti, B., Vernez-Moudon, A., Reis, R., Turrell, G., Dannenberg, A.L., Badland, H., Foster, S., Lowe, M., Sallis, J.F., Stevenson, M., and Owen, N. (2016) "City Planning and Population Health: A Global Challenge. *The Lancet* 388(10062).

Goenka, S., and Andersen, L.B. (2016) "Urban Design and Transport to Promote Healthy Lives." *The Lancet* 388(10062): 2851–2853. doi: 10.1016/S0140-6736(16)31580-X

Grant, M., Brown, C., Waleska, C.T., Capon, A., Corburn, J., Coutts, C., Crespo, C.J., Ellis, G., Ferguson, G., Fudge, C., Hancock, T., Lawrence, R.J., Nieuwenhuijsen, M.J., Oni, T., Thompson, S., Wagenaar, C., and Ward Thompson, C. (2017) "Cities and Health: An Evolving Global Conversation." *Cities & Health* 1(1): 1–9. doi: 10.1080/23748834.2017.1316025

Haigh, R. (2013) "The Quintessence of a Therapeutic Environment, Therapeutic Communities." *The International Journal of Therapeutic Communities* 34(1): 6–15. http://growingbetterlives.org/wp-content/uploads/2014/06/Haigh-2013-Quintessence.pdf

Hartig, T., Mitchell, R., de Vries, S., and Frumkin, H. (2014) "Nature and Health." *Annual Review of Public Health* 35: 207–228.

Hunter, M.R., and Askarinejad, A. (2015) "Designer's Approach for Scene Selection in Tests of Preference and Restoration along a Continuum of Natural to Manmade Environments." *Frontiers in Psychology* 6: 1228. doi: 10.3389/fpsyg.2015.01228

Johnson, R. and Haigh, R. (2011) "Social Psychiatry and Social Policy for the 21st Century: New Concepts for New Needs: The 'Enabling Environments' Initiative." *Mental Health and Social Inclusion* 15(1): 17–23.

Kaplan, R. and Kaplan, S. (1989) *The Experience of Nature: A Psychological Perspective.* New York: Cambridge University Press.

Kleinert, S., and Horton, R. (2016) "Urban Design: An Important Future Force for Health and Wellbeing." *The Lancet* 388(10062): 2848–2850. http://dx.doi.org/10.1016/S0140-6736(16)31578-1

Kuo, M. (2015) "How Might Contact with Nature Promote Human Health? Promising Mechanisms and a Possible Central Pathway." *Frontiers in Psychology*, August 25. https://doi.org/10.3389/fpsyg.2015.01093

Larson, L.R., Jennings, V., and Cloutier, S.A. (2016) "Public Parks and Wellbeing in Urban Areas of the United States." *PLoS One* 11(4): e0153211.

Lindström, B., and Eriksson, M. (2005) "Salutogenesis." *Journal of Epidemiology and Community Health* 59(6): 440–442.

Marín, A., Bodina, O., Gelcichc, S., and Crona, B. (2015) "Social Capital in Post-disaster Recovery Trajectories: Insights from a Longitudinal Study of Tsunami-impacted Small-scale Fisher Organizations in Chile." *Global Environmental Change* 35: 450–462.

McGovern, L., Miller, G., and Hughes-Cromwick, P. (2014) "Health Policy Brief: The Relative Contribution of Multiple Determinants to Health Outcomes." *Health Affairs*, August 21. http://healthaffairs.org/healthpolicybriefs/brief_pdfs/healthpolicybrief_123.pdf (last accessed October 26, 2017).

Meyer-Lindenberg, A. (2013) "Big City Blues." *Scientific American Mind* 59–61, March/April.

Miller, D., Roe, J., Brown, C., Morris, S., Morrice, J., and Ward Thompson, C. (2012) *Blue Health: Water, Health and Wellbeing.* Aberdeen: Centre of Expertise for Waters, James Hutton Institute. www.crew.ac.uk/publicationswww.hutton.ac.uk/sites/default/files/files/snc/CREW%20Blue%20Health%20project%20FINAL.pdf

Minnesota Department of Health (2014) *Advancing Health Equity in Minnesota: Report to the Legislature.* www.health.state.mn.us/divs/chs/healthequity/ahe_leg_report_020414.pdf

Mitchell, R.J., and Popham, F. (2008) "Effect of Exposure to Natural Environment on Health Inequalities: An Observational Population Study." *The Lancet* 372(9650): 1655–1660. doi: 10.1016/S0140-6736(08)61689-X

Mitchell, R.J., Richardson, E.A., Shortt, N.K., and Pearce, J.R. (2015) "Neighborhood Environments and Socioeconomic Inequalities in Mental Well-being." *American Journal of Preventive Medicine* 49(1): 80–84. doi: 10.1016/j.amepre.2015.01.017. PMID:25911270.

Morris, G.P., Beck, S.A., Hanlon, P., and Robertson, R. (2006) "Getting Strategic About the Environment and Health." *Public Health* 120: 889–907.

Nieuwenhuijsen, M.J., Kruize, H., Gidlow, C., Andrusaityte, S., Antó, J.M., Basagaña, X., Cirach, M., Dadvand, P., Danileviciute, A., Donaire-Gonzalez, D., Garcia, J., Jerrett, M., Jones, M., Julvez, J., van Kempen, E., van Kamp, I., Maas, J., Seto, E., Smith, G., Triguero, M., Wendel-Vos, W., Wright, J., Zufferey, J., van den Hazel, P.J., Lawrence, R., and Grazuleviciene, R. (2014) "Positive Health Effects of the Natural Outdoor Environment in Typical Populations in Different Regions in Europe (PHENOTYPE): A Study Programme Protocol." *BMJ Open* 4: e004951. doi: 10.1136/bmjopen-2014-004951

Nordh, H., Hartig, T., Hagerhall, C. M., and Fry, G. (2009) "Components of Small Urban Parks that Predict the Possibility for Restoration." *Urban Forestry & Urban Greening* 8: 225–235. doi: 10.1016/j.ufug.2009.06.003

NZ Ministry of Health (2013) *Green Prescription Patient Survey*. www.health.govt.nz/our-work/preventative-health-wellness/physical-activity/green-prescriptions/green-prescription-research/green-prescription-patient-survey

Partners in Healthcare (2010) *Breath of Fresh Air: Chapter 15, Poverty and Asthma, Partners Asthma Center*. www.asthma.partners.org/NewFiles/BoFAChapter15.html (last accessed October 26, 2017).

Reis, S., Morris, G.P., Fleming, L.E., Beck, S., Taylor, T., White, M., Depledge, M.H., Steinle, S, Sabel, C.E., Cowie, H., Hurley, F., Dick, J.M., Smith, R.I., and Austen, M. (2015) "Integrating Health and Environmental Impact Analysis." *Public Health* 129(10), October: 1383–1389. www.publichealthjrnl.com/article/S0033-3506(13)00242-4/fulltext (last accessed October 27, 2017).

Roe, J. (2016) "Cities, Green Space, and Mental Wellbeing." *The Oxford Research Encyclopedia of Environmental Science*, November. http://environmentalscience.oxfordre.com/view/10.1093/acrefore/9780199389414.001.0001/acrefore-9780199389414-e-93 Evidence (last accessed October 26, 2017).

Roe, J., and Ward Thompson, C. (2011) "The Impact of Urban Gardens and Street Trees on People's Health and Wellbeing." In Bomans, K., Dewaelheyns, V., and Gulinck, H. (eds.), *The* Powerful *Garden*. Flemish Supporting Centre of Housing and Space.

Rutter, H., Savona, N., Glonti, K., Bibby, J., Cummins, S., Finegood, D.T., Greaves, F., Harper, L., Hawe, P., Moore, L., Petticrew, M., Rehfuess, E., Shiell, A., Thomas, J., and White, M. (2017) "The Need for a Complex Systems Model of Evidence for Public Health." *The Lancet*, published June 13. http://dx.doi.org/10.1016/S0140-6736(17)31267-9

Sallis, J.F., Bull, F., Burdett, R., Frank, L.D., Griffiths, P., Giles-Corti, B., and Stevenson, M. (2016) "Use of Science to Guide City Planning Policy and Practice: How to Achieve Healthy and Sustainable Future Cities." *The Lancet* 388: 2936–2947.

Sarkar, C., Webster, C., Pryor, M., Tang, D., Melbourne, S., Zhang, X., and Jianzheng, L. (2015) "Exploring Associations between Urban Green, Street Design and Walking: Results from the Greater London Boroughs." *Landscape and Urban Planning* 143: 112–125.

Sarmiento, O., Torres, A., Jacoby, E., Pratt, M., Schmid, T.L., and Stierling, G. (2010) "The Ciclovía-Recreativa: A Mass-recreational Program with Public Health Potential." *Journal of Phyisical Activity and Health* 7(Suppl 2), July: S163–S180.

Schizophrenia Commission (2012) *Schizophrenia: The Abandoned Illness*, November.

Shanahan, D.F., Lin, B.B., Bush, R., Gaston, K.J., Dean, J.H., Barber, E., and Fuller, R.A. (2015) "Toward Improved Public Health Outcomes from Urban Nature." *American Journal of Public Health*: e1–e8.

Stevenson, M., Thompson, J., de Sá, T.H., et al. (2016) "Land Use, Transport, and Population Health: Estimating the Health Benefits of Compact Cities." *The Lancet* 388(10062). http://dx.doi.org/10.1016/S0140-6736(16)30067-8

Takano, T., Nakamura, K., and Watanabe, M. (2002) "Urban Residential Environments and Senior Citizens' Longevity in Mega-city Areas: The Importance of Walk-able Green Space." *Journal of. Epidemiology and Community Health* 56(12): 913–916.

Taylor, M.S., Wheeler, B.W., White, M.P., Economou, T., and Osborne, N.J. (2015) "Research Note: Urban Street Tree Density and Antidepressant Prescription Rates: A Cross-sectional Study in London, UK." *Landscape and Urban Planning* 136: 174–179.

Ulrich, R. (1983) "Aesthetic and Affective Response to Natural Environment." In Altman, I. and Wohlwill, J.F. (eds.), *Human Behaviour and Environment: Advances in Theory and Research*. New York: Plenum Press, 85–125.

UN (2016a) *Sustainable Development Goal 11*. https://sustainabledevelopment.un.org/sdg11 (last accessed October 26, 2017).

UN (2016b) Habitat III. *New Urban Agenda*. http://habitat3.org/wp-content/uploads/Habitat-III-New-Urban-Agenda-10-September-2016.pdf (last accessed October 26, 2017).

Ward Thompson, C., Roe, J., and Aspinall, P.A. (2013) "Woodland Improvements in Deprived Urban Communities: What Impact Do They Have on People's Activities and Quality of Life?" *Landscape and Urban Planning* 118, October 1: 79–89.

WHO (2008) *Closing the Gap in a Generation: Health Equity Through Action on the Social Determinants of Health.* www.who.int/social_determinants/final_report/csdh_finalreport_2008_execsumm.pdf (last accessed October 26, 2017).

WHO (2014) *Healthy Cities for Health.* Kobe: World Health Organization Centre for Health Development. www.who.int/kobe_centre/publications/cities_for_health_final.pdf?ua=1 (last accessed October 26, 2017).

WHO (2016) "Urban Green Spaces and Health: A Review of the Evidence." www.euro.who.int/__data/assets/pdf_file/0005/321971/Urban-green-spaces-and-health-review-evidence.pdf?ua=1 (last accessed October 26, 2017).

WHO (2018) *Copenhagen Consensus of Mayors: Healthier and Happier Cities for All,* 13 February. www.euro.who.int/__data/assets/pdf_file/0003/361434/consensus-eng.pdf?ua=1 (last accessed March 1, 2018).

Further reading

Barton, H., Thompson, S., Burgess, S., and Grant, M. (eds.) (2015) *Routledge Handbook of Planning for Health and Well-being.* Abingdon, UK: Routledge. This book includes over 40 contributions from academics and practitioners addressing healthy city planning, includes a section on healthy planning in global contexts, and serves as a useful reference point and introduction to the subject.

De Leeuw, E. and Simos, J. (eds.) (2017) *Healthy Cities, the Theory, Policy, and Practice of Value-based Urban Planning.* Springer. This book addresses the social, economic, and political determinants of healthy cities, drawing on case studies from high and middle to low income countries and includes over 20 contributions from experts in the field.

The Lancet (2016) *Urban Design, Transport and Health Series,* Published: September 23. www.thelancet.com/series/urban-design This three-part series in the *Lancet,* a leading global medical journal, addresses how urban design can act as a future force to drive health and well-being.

34
COMPLETE AND HEALTHY STREETS

Lawrence Frank, Jared Ulmer, Bruce Appleyard, and Alexander Bigazzi

Introduction

How we design our streets – including sidewalks, furniture, trees, and lighting – influences human health. In the last decade, many North American cities have adopted ordinances that require making streets "Complete" when they are built or redesigned. "Complete Streets" are "designed for all ages, abilities, and modes of travel" so as to meet the needs of pedestrians, cyclists, transit users, and other vulnerable users better than traditional automobile-oriented streets (Lynott et al 2009). Complete Streets policies are one component of a truly multimodal environment. Other core features include compact, mixed-use development, cycling infrastructure, a connected street network, and convenient access to regional destinations through transit.

Studies have found associations between Complete Streets and more walking and bicycling, and increased levels of physical activity (Cain et al 2014; Boarnet et al 2011). One recent study documented a causal increase in physical activity and reduction in sedentary time from an urban greenway retrofit. Residents within 300 meters of the intervention were twice as likely than others to get recommended physical activity levels (Frank et al 2019). Some researchers have also found lower rates of obesity (Sallis et al. 2012; Frank et al. 2004), type II diabetes and cardiovascular disease (Ewing et al. 2014; Glazier et al. 2014) and increased a sense of community (French et al. 2014) after the implementation of Complete Streets and enhancement of overall walkability. Complete Streets policies may lead to reduced vehicle traffic and less exposure to air pollution (Bigazzi & Rouleau 2017). Evidence shows that many aspects of such policies help to calm traffic and reduce the per capita risk of injury (Ewing & Dumbaugh 2009; Lee & Maheswaran 2011). Safety from collisions and security from crime influence travel decisions and health outcomes; they are both significantly influenced by the street environment (Appleyard & Ferrell 2015; MacDonald 2015).

This chapter reviews the literature on health-related impacts of Complete Streets components. Most studies focus on the physical activity and traffic safety of Complete Streets. Fewer studies address more distal health outcomes such as body weight, chronic disease, and mental and social health. Most studies focus on adults or the general public rather than seniors, children, and youth, or persons with disabilities.

This chapter is organized by Complete Streets design elements: street networks, pedestrian spaces, transit facilities, cycling facilities, roadways, and adjacent buildings and land uses.

Street networks

Higher street connectivity (providing shorter, more direct paths) is widely recognized as one of the most important features of an urban environment for promoting physical activity (Appleyard 2012, 2015; Berrigan et al. 2010; McCormack & Shiell 2011). It includes density of intersections, short street blocks between intersections, and frequency of four-way intersections. In a meta-analysis of studies linking built environment characteristics to travel behavior, intersection density was the strongest predictor of more walking and a strong predictor of more transit use and fewer vehicle miles of travel (Ewing & Cervero 2010). While street connectivity has consistently been associated with active modes of transportation, its influence on leisure physical activity is less clear.

Two studies have examined if frequently spaced intersections encourage traffic calming and reduce collision severity (Clifton et al. 2009; Marshall & Garrick 2011), but they had conflicting results over the relationship between intersection density and frequency of injury.

Pedestrian space

The following sections consider elements typically found between buildings and the curb, including sidewalks, buffers, lighting, furniture, trees, transit stops, and similar amenities. Proper design of these features helps ensure a safe and inviting space for pedestrians.

Sidewalk presence and width

While the presence of sidewalks has been shown to have a positive impact on pedestrian and physical activity, there are inconsistencies due to different research methods and approaches. Some studies of neighborhood sidewalk availability, completeness, and width have found significant positive correlations with walking and overall physical activity, while others have not (Boarnet et al. 2011; Forsyth et al. 2008; Pikora et al. 2006). Sidewalk width is a core aspect of a complete street as shown in Figure 34.1.

In a meta-analysis of self-reported neighborhood conditions, people reported to be more physically active in the presence of sidewalks (Duncan et al. 2005). Another review logically found sidewalks to be associated with physical activity from walking but not with the all-inclusive category of general physical activity (Wendel-Vos et al. 2007). Another review reported no association between sidewalks and physical activity (Ewing & Cervero 2010). One synthesis of studies linking sidewalk availability with child and teen physical activity found mixed results (Ding et al. 2011): While younger children's physical activity was correlated with traffic speed/volume, access/proximity to recreation facilities, mixed land use, residential density, and walkability (including sidewalks); physical activity in adolescents was only correlated with land use mix and residential density.

The positive influence of sidewalks on pedestrian safety is clear. Two reviews noted that the presence of sidewalks helps to reduce vehicle-pedestrian collisions along street segments (Ewing & Dumbaugh 2009; Retting et al. 2003). The condition of sidewalks can also affect people's choice to walk. An exploratory study in Ottawa logically reported that environmental hazards related to the risk of falls can be significant barriers to walking for seniors (Lockett et al. 2005). A Northern California study of people 45 and older also logically found walking to be the most common activity related to falls, highlighting the importance of the design of sidewalks, curbs, and streets (Li et al. 2006).

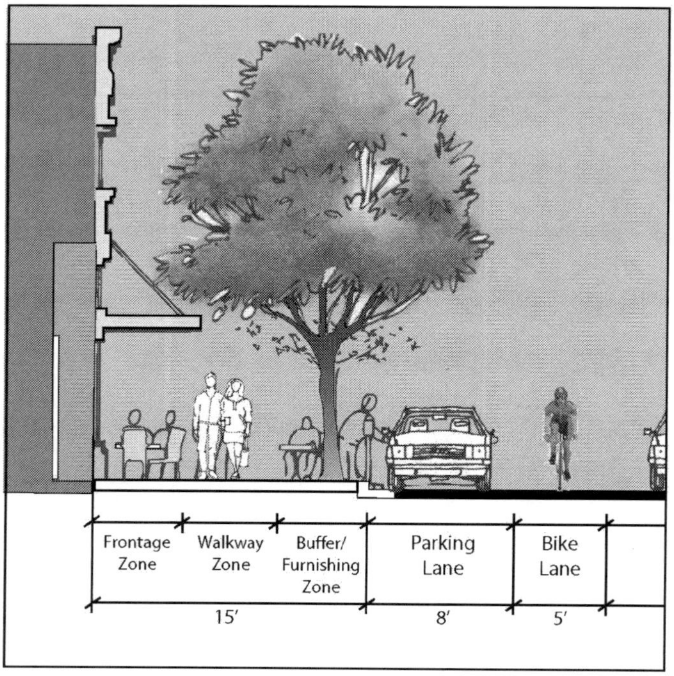

Figure 34.1 Sidewalks can encourage social activity by being wide enough for two people to walk side by side and a third to pass comfortably. Sidewalks along commercial streets should also allow space for seating, displays, and traffic buffers/furniture zones.

Source: Bruce Appleyard

Buffers

Little research has investigated the impact of a buffer between the sidewalk and the curb, or the size of this buffer, on physical activity and health. The presence of at least five feet of separation between the sidewalk and the curb was identified as an important predictor of walking attractiveness in a Portland, Oregon, study (Adkins et al. 2012). While one study found buffers to be associated with more transportation walking, buffers were not associated with recreational walking or overall moderate/vigorous physical activity (Boarnet et al. 2011).

Lighting

In terms of safety, a review of multiple studies found street lighting to be associated with a reduction in pedestrian-vehicle collisions at crosswalks (Retting et al. 2003). Two studies found lack of street lighting associated with increased collision frequency for cyclists (Kim et al. 2007; Romanow et al. 2012). Presence of lighting is also known to improve perceptions of safety and deter criminal activity (Foster & Giles-Corti 2008; Painter & Farrington 2001). Studies of the relationship between physical activity and lighting have found mixed results. In Minneapolis and St. Paul, Minnesota, density of streetlights near home was associated with more walking but not related to leisure or objectively measured moderate to vigorous physical activity (Forsyth et al. 2008). In a longitudinal study of older adults in Alameda County, California, "inadequate lighting" was identified as the strongest self-reported neighborhood-level determinant of the inability to complete physical tasks over time (Balfour & Kaplan 2002).

Street furniture

Research has shown the presence of street furniture is associated with higher pedestrian counts (Ewing et al. 2016), but it is unclear the extent to which this is a function of increased walkability and transit access in these locations. One group of researchers developed the "Microscale Audit of Pedestrian Spaces" (MAPS) audit tool and found several streetscape features to be positively related with physical activity (Cain et al. 2017; Kurka et al. 2016). Researchers have shown the presence of seating to be associated with more transportation-related but not leisure-related physical activity (Cain et al. 2014). Two other studies found no association between street furniture and physical activity (Boarnet et al. 2011; Werner et al. 2010).

Overall, most researchers confirm what intuition would suggest: provide amenities and lighting, and more people will participate. In a Boston study, seating was identified as an important predictor of social activity (Mehta 2007). In studies in Savannah, Georgia, and St. Louis, Missouri, trees, benches, or other comfort amenities on more than 25% of street segments were associated with more transportation-related physical activity among adults (Hoehner et al. 2005).

Street trees and vegetation

The health benefits of nature have been suggested by a wide variety of theories such as the biophilia hypothesis that humans have an instinctive need for a connection with the natural world (Wilson 1984). Attention restoration theory posits that exposure to nature can improve directed attention and reduce stress (Kaplan & Kaplan 1989), while nature deficit disorder or a lack of exposure to nature is associated with behavioral, developmental, and physical disorders (Louv 2012). Evidence to date suggests that better access to nature and green space is associated with more physical activity and better physical, mental, and social health (Bowler et al. 2010; Bratman et al. 2012; Lee & Maheswaran 2011).

Contrary to conventional engineering theory, suggesting the presence of fixed objects (such as trees) contributes to increased vehicular collisions, accumulating evidence indicates that tree-lined streets may actually have lower collision rates (Ewing & Dumbaugh 2009). One possible reason is that trees provide a visual constraint that signals drivers to slow down, whereas a wide open streetview may encourage higher speeds. A second possibility is that trees convey a calming effect and add interest, encouraging lower speeds. These findings are supported by Rapoport's (1991) "numbers of noticeable differences" hypothesis, whereby a corridor's interest is a function of its level of detail and the rate at which it changes as one moves through it. In sum, scale influences speed; a lack of detail and a coarser design scale encourage higher speeds.

Trees also affect exposure to sunlight and wind, which in turn affects thermal comfort and the decision to walk or cycle. Trees mitigate the vulnerability to extreme "heat island" effects. Research shows that neighborhoods with more pavement and less vegetative coverage have higher surface temperatures (Harlan et al. 2006; Rinner & Hussain 2011). Furthermore, street trees have been found to reduce cooling costs for adjacent buildings (Bowler et al. 2010). Evidence also suggests greater tree coverage reduces exposure to ultraviolet radiation, which causes skin cancer (Saraiya et al. 2004).

Quantifying the connection between trees, walking, health, and physical activity has led to promising, but mixed results. A recent study in Sacramento, California, used high-resolution multi-spectral LIDAR data to remotely assess vegetative cover and linked it with address-level health outcome data from a large sample of participants in the California Health Interview Survey. This study reported several health benefits of increased tree canopy: a 10% increase in tree canopy was associated with an 18% reduction in obesity (Ulmer et al. 2016). A study in Lille, France, found higher pedestrian counts on street segments with trees (Foltête & Piombini 2007). A study of adults in New York City found tree density to be associated with lower body mass index, but not with demonstrably longer walking or biking trips (Lovasi et al. 2012, 2013). In Minneapolis and St. Paul, Minnesota, neighborhood street trees were associated with greater leisure walking, but not transportation walking or total moderate/vigorous physical activity (Forsyth et al. 2008). In a longitudinal study in Tokyo, older adults living near more parks and tree-lined streets had a higher (five-year) survival rate (Takano et al. 2002). However, several studies found no relationship between trees and various measures of physical activity (Hoehner et al. 2005; Pikora et al. 2006). In a study of four Dutch cities, streetscape greenery was associated with better self-reported health conditions, including better mental health status, as mediated by stress reduction and higher social cohesion (de Vries et al. 2013). In a San Francisco study of street livability, residents on tree-lined boulevards indicated less annoyance and better social cohesion, despite heavier traffic (Bosselmann et al. 1999).

In a review of child and teen studies, positive associations were found between streetscape vegetation and physical activity in 36% of the children, and 12% of teens studied (Ding et al. 2011). One of the critical limitations of studies linking green space and vegetation with health outcomes is that wealthier people often choose to live in closer proximity to nature and where tree canopy is greater. Most studies "control" for income through a categorical measure of income that fails to capture the variation that exists. More importantly, income and wealth are different measures. Wealth determines where people can choose to live and impacts access to healthcare and other services, facts that are seldom, if ever, captured.

Public transit facilities

Recent research shows that transit users engage in more physical activity than non-transit users (Freeland et al. 2013; Lachapelle & Frank 2009; Lachapelle & Noland 2012). Several studies found that higher density of transit stops/stations leading to shorter distances from home to transit to be associated with more physical activity and lower body weight (Forsyth et al. 2008; Hoehner et al. 2005; Rundle et al. 2007). A review of several studies noted a reduction of collisions after relocating bus stops to the far side of intersections, keeping pedestrians from having to cross in front of the bus and being hidden from passing drivers (Campbell et al. 2004).

Cycling facilities

Many cities are developing their cycling networks, and while dedicated cycling facilities are not the only factor that facilitates higher rates of cycling, most cyclists express a preference for bike lanes or other designated cycling facilities, with stronger preferences for facilities with greater separation from motor vehicles. Cycling preferences vary, with those less experienced willing to divert from the shortest path to use protected cycling facilities. Figure 34.2 is of the Comox Helmken Greenway in Downtown Vancouver. Bicycle parking facilities also affect the perceived convenience and likelihood of cycling; there is a broad preference for sheltered and secured bicycle parking (Buehler & Dill 2016; Pucher et al. 2010; Winters & Teschke 2010).

Definitions of cycling facility types vary, but the National Association of City Transportation Officials (NACTO) Urban Bikeway Design Guide defines bike lanes as a portion of the roadway that has been designated by striping, signage, and pavement markings for the preferential

Figure 34.2 Bicycle boulevards or greenways use traffic speed and volume calming treatments (medians, diverters, chokers, traffic circles, stop signs, and ped/cyclist-activated signals) to make streets more safe, comfortable, and convenient for cyclists.

Source: Dylan Passmore, City of Vancouver, Engineering

or exclusive use of cycle tracks as "an exclusive bike facility that combines the user experience of a separated path with the on-street infrastructure of a conventional bike lane"; and bike boulevards as "streets with low motorized traffic volumes and speeds, designated and designed to give bicycle travel priority" (National Association of City Transportation Officials 2014). The design guide also includes various intersection treatments for cycling networks such as bike boxes, bike signals, and median refuge islands.

Safety impacts of bicycle facilities vary in the literature, partly due to methodological inconsistencies. Bike lanes are often associated with lower collision frequency than similar roads without bicycle facilities. On-street motor vehicle parking directly adjacent to bike lanes is perceived as less safe by cyclists, particularly due to the risk of "dooring" incidents (when parked motorists open their doors into the paths of oncoming cyclists). A popular countermeasure to this problem is providing a one-meter buffer (using painted stripes) between parked cars and cycle lanes. Cyclists generally feel safer in cycle tracks than in other types of facilities. One-way cycle tracks are generally safer and more preferred than two-way cycle tracks. Bike boxes at intersections are another strategy to promote safety as shown in Figure 34.3. The safety effects of cycle tracks at intersections are less clear, but intersection treatments such as pavement markings, warning signs, raised crossings, and merging cyclists onto an on-street (non-buffered) bike lane in advance of

Figure 34.3 Bike Box in Vancouver, Canada.

Source: Dylan Passmore, City of Vancouver, Engineering

an intersection have been shown to improve safety (Buehler & Dill 2016; DiGioia et al. 2017; Reynolds et al. 2009; Thomas & DeRobertis 2013).

Exposure to and inhalation of traffic-related air pollution is a risk for urban cyclists (Bigazzi & Figliozzi 2014), which is partially mitigated when cyclists choose routes away from motor vehicles when available and practical (Broach & Bigazzi 2017). Despite the safety and air pollution risks, there is a large and growing body of evidence that cycling overall increases public health due to significant health benefits from physical activity (Mueller et al. 2015).

Roadways

Design considerations for motor vehicle travel include the overall roadway width, number of vehicle lanes, vehicle lane width, medians, and traffic calming features. These factors have an influence on traffic volume and speed, which in turn influence safety and the attractiveness and likelihood of walking and cycling.

Narrower roadways tend to be safer, even after adjusting for traffic volume (Ukkusuri et al. 2012). Minimizing crossing distances for pedestrians is especially important for seniors and pedestrians with disabilities (Lynott et al. 2009). Road diets provide strong evidence of the effects of road width on safety, by reducing vehicle speeds, improving mobility and access, reducing collisions and injuries, and improving overall livability (Chen et al. 2013; Noland et al. 2015). A typical road diet consists of removing one lane in each direction of a four-lane street section, adding a center turn lane and/or landscaped median, and repurposing the additional road space for landscaped buffers, bike lanes, and sidewalks.

Narrow lane widths can lead to reduced motor vehicle travel speeds, as the visual constraint likely causes drivers to operate more cautiously. Although narrow lanes can provide safety benefits in some urban contexts, they are not appropriate or beneficial on all roadways (Manuel et al. 2014; Mecheri et al. 2017). It should be noted that narrow street designs can make travel and especially turning movements more challenging for emergency vehicles, buses, and other large vehicles, but this can be accommodated through thoughtful design.

There are conflicting impacts of on-street parking on safety and active travelers (Klassen et al. 2014). While cyclists prefer routes without on-street parking, others argue that parking buffers moving vehicles and pedestrians and increases pedestrian comfort and security (Buehler & Dill 2016; de Vries et al. 2007). Raised medians can improve safety by enhancing vehicle separation, limiting turning movements, and providing pedestrian refuges (National Association of City Transportation Officials 2014). Access management, such as limiting driveways and left turns, can also provide safety benefits (Ewing & Dumbaugh 2009).

Traffic calming (speed humps, warning signs and lighting, narrowed lanes, speed tables, curb extensions, etc.) is designed to reduce speeds and volumes to improve street safety and attractiveness for walking and cycling. However, the empirical evidence for resulting increases in cycling, walking, and physical activity is still limited (Brown et al. 2017).

Intersections and crossings

Strategies to make crosswalks more effective include advance stop lines, flashing warning lights, and high-visibility treatments (Chen et al. 2013). Pedestrian hybrid beacons can be used to stop motor vehicle traffic with a red indication while pedestrians cross (Pulugurtha & Self 2015). Street geometry at intersections and crosswalks also influences safety. Large corner radii are associated with higher vehicle turning speeds and longer crossing distances, which in turn increase collision rates (Fitzpatrick et al. 2006). Curb extensions can be used to improve pedestrian safety and comfort and reduce pedestrian delay at intersections and crosswalks (Bella & Silvestri 2015). Finally, leading pedestrian intervals (pedestrians given a walk indication before the signal turns green), exclusive phases for pedestrians, increasing crossing times, and devices that prompt pedestrians to check for vehicles before crossing improve the level of service for pedestrians (Chen et al. 2013; Saneinejad & Lo 2015).

Strategies to improve intersection safety for cyclists, such as protected intersections (separating them from motor vehicle traffic) and bike boxes (allowing cyclists to queue in front of cars), show promise but need further research. Roundabouts can reduce vehicle speeds but can increase conflict points between vehicles and cyclists and pedestrians. As expected, evidence on safety effects of roundabouts for cyclists and pedestrians is inconsistent (DiGioia et al. 2017; Pucher et al. 2010).

Adjacent buildings and land uses

Land uses and buildings can have implications for a range of street-related health outcomes. A fairly consistent base of evidence suggests that the presence of retail uses encourages transportation-related physical activity (Frank et al. 2004). Studies in Boston, New York, and Hong Kong found that pedestrian commuters were more likely to choose a route with a high proportion of pedestrian-oriented retail frontage, even after adjusting for route distance (Guo & Loo 2013). In mode choice studies, transit users were more likely to walk or bike to transit when their route contained more community oriented, human-scale retail uses (coffee shops, bookstores, and restaurants) (Appleyard 2012; Werner et al. 2010). One Boston study found the presence of retail to be an important predictor of social activity, including conversing, eating, sitting, or walking with another person (Mehta 2007). A few studies have investigated the influence of small, human-scale retail (sidewalk café, dining, and vending space) on pedestrian and physical activity. In two studies in New York City, the presence of a sidewalk café within one kilometer of home was associated with a higher likelihood of walking or biking for at least 10 blocks and lower rates of body mass (Lovasi et al. 2012, 2013). The distinction between what is locally versus regionally serving – whether these are retail services or parks and open space – is critical when considering the impact that a particular facility is likely to have on a given community.

Urban open space (e.g. parks, plazas) has a fairly well-established positive influence on physical activity. One study found open space to be associated with walking and objectively measured physical activity (Boarnet et al. 2011). Another study found open space associated with recreational physical activity for children, but not for other age groups or other types of physical activity (Cain et al. 2014). One earlier study found no association between open space and physical activity (Pikora et al. 2006). A Portland, Oregon study found the presence of an adjacent park to be an important predictor of walking attractiveness (Adkins et al. 2012). In New York and Hong Kong, commuters were more likely to choose a route with a high proportion of adjacent open space, even after adjusting for route distance (Guo & Loo 2013).

Although building characteristics (e.g. scale, massing, height/right-of-way ratio, façade design) are commonly theorized to promote pedestrian activity, little supporting evidence directly connects building design to transportation or physical activity. One New York City study showed pedestrian counts to be higher on streets with "continuous building façades forming a street wall" and "windows overlooking the street" than those without these characteristics (Ewing et al. 2016). In another study, higher building enclosure (the ratio of building height to the width between buildings on opposite sides of the street, creating an outdoor room) was associated with more walking and biking for transportation purposes for all age groups (Cain et al. 2014). In the same study, perceived attractiveness of building façades was associated with recreational physical activity for seniors. Another study found an association between walking and bicycling and proxies for human-scale street enclosure (small parcels, narrow streets, and lower levels of parking) (Appleyard 2012 & 15). Figure 34.4

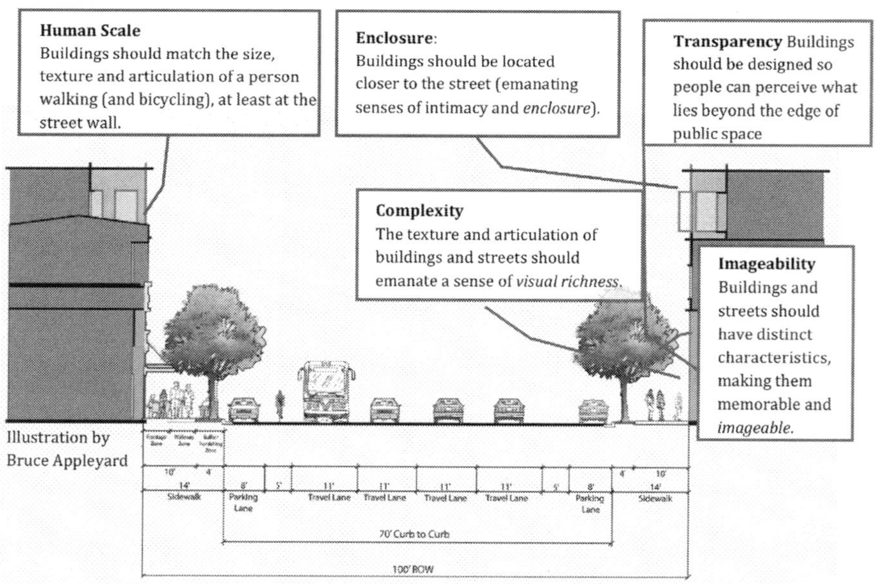

Figure 34.4 Illustrating findings from Appleyard (2012, 2015)

Source: Bruce Appleyard

illustrates findings from Appleyard (2012, 2015) in support of conclusions drawn from a survey of urban design experts by Ewing and Handy (2009) regarding key urban design qualities of a street environment that encourage walking, including (1) provide a "human scale"; (2) provide a stronger sense of enclosure; (3) emanate imageability; and (4) have windows on the ground floor (transparency) and detailed fenestration, emanating a visually rich sense of complexity.

Still, other studies found no association between building design and physical activity (Boarnet et al. 2011; Hoehner et al. 2005; Pikora et al. 2006).

A synthesis

The aforementioned discussion of the evidence of the influence of Complete Streets on safety, physical activity, and social activity outcomes is summarized in Table 34.1. Positive associations with health outcomes are indicated by "+" and negative associations by "-."

Table 34.1 Associations of Complete Streets elements with safety, physical activity, and social activity

Category	*Design Element*	*Safety*	*Physical Activity*	*Social Activity*
Street & network	Street connectivity	+	+	NK
Pedestrian category	Sidewalk presence & width	+	+	+
	Buffer zone	+	+	NK
	Lighting	+	+	NK
	Furnishings	NK	+	+
	Trees & vegetation	+	+	NK
	Public transit facilities	NK	+	NK
	On-street parking	+/−	+	NK
Cycling category	Bike lanes	+	+	NK
	Cycle tracks	+	+	NK
	Off-street bike paths/trails	−	+	NK
	Bike boulevards	NK	+	NK
	Bicycle parking	NK	+	NK
	On-street parking	−	−	NK
Roadway	Minimize street width/# lanes	+	+	+
	Narrow lane width	+	+	NK
	Median inclusion	+	+	NK
	Traffic calming features	+	+	NK
Intersections & crossings	Intersection control	+	+	NK
	Midblock control	+/-	+	NK
	Corner radii & curb treatments	+	+	NK
Adjacent buildings and land uses	Retail uses	+	+	+
	Open space uses	NK	+	NK
	Building enclosure & façade	+	+	NK
	Café/vending space	+	+	+

Source: Urban Design 4 Health, Inc. (2014). Healthy Streets: Evidence Review. *Toronto Public Health: City of Toronto.*

NK = not known
+/− = both positive and negative impacts were found, conflicting evidence

Some elements can have both positive and negative impacts. For example, on-street parking can provide safety benefits for adult pedestrians but have negative impacts on children, bicyclists, and drivers. Midblock intersection controls (crosswalks) tend to have positive impacts for pedestrians on low-volume streets but negative impacts on higher-volume, multi-lane streets.

Conclusion

People's decision to be physically active through walking and cycling can be influenced by providing streets and other facilities that allow for convenient access to destinations in a safe and enjoyable way. In response to a large and growing body of research and public interest, urban planners and public health professionals have been working together to find ways to redesign our streets and cities to facilitate increased physical activity and improve traffic safety. Complete Streets typically include sidewalks, bike facilities, public transit access, and a motor vehicle right-of-way that encourages safe automobile use at an appropriate speed. Creating policies, plans, and design guidelines that are supportive of Complete Streets can lead to a transportation system designed for people of all ages, abilities, and modes of travel.

Indeed, a supportive pedestrian environment with complete street features can reduce car dependence and greenhouse gas (GHG) emissions – particularly where macro-scale walkability factors including density, mixed use, and street connectivity, along with regional accessibility, are high. A case in point is the city of Vancouver, which has adopted a greenway corridor program that entails retrofitting existing roadways with seating, lighting, trees, and other features along with re-appropriating road space to support walking and cycling. A pre-post assessment of the Comox Corridor Greenway in Downtown Vancouver found a 21% reduction in GHG emissions for those living within 300 meters of the corridor (Ngo et al. 2018). That same study found significant increases in active transportation for those within 300 meters of the corridor reported in the beginning of this chapter.

Streetscape features are one of the most modifiable and inexpensive aspects of urban environments and likely among the most cost-effective approaches for promoting physical activity and active travel in urban settings. This is especially the case in locations that have good transit access and locally serving destinations. The best mix of Complete Street design elements and characteristics to promote health is largely dependent on context. It also depends on who are the current and expected future users, what their needs are, what health challenges exist, and what opportunities can be leveraged to encourage healthy living. Urban designers should consider the following actions for prioritizing the implementation of critical Complete Street elements:

- *Provide accessibility* via street connectivity (short blocks and four-way intersections or permeability through existing long blocks), continuity of pedestrian and cycling facilities throughout the network, and places to walk to and from along pedestrian and cycling networks, including residences, jobs, retail destinations, transit facilities, and recreational amenities.
- *Ensure safety* via continuous sidewalk and bicycle facilities, adequate lighting, safe intersection crossings, medians with pedestrian refuges, enhanced crosswalk and signalization treatments, curb extensions, and other traffic calming features.
- *Enhance the experience* through the provision of wide sidewalks, commercial seating and vending on the sidewalk, public seating, trees and landscaping, human-scaled buildings, and attractive, transparent façade design.

Acknowledgments

This chapter is based on "The Healthy Streets Evidence Review," conducted in 2014, written by Urban Design 4 Health, Inc., and funded by Toronto Public Health. Jared Ulmer (co-author) is acknowledged for his leadership of the original study. Jim Chapman with Urban Design 4 Health contributed considerably to the original study and is acknowledged for his diligence in ensuring a high-quality final product. Dr. Monica Campbell with Toronto Public Health provided considerable guidance and leadership for the original study.

References

Adkins, A., Dill, J., Luhr, G., and Neal, M. (2012). "Unpacking Walkability: Testing the Influence of Urban Design Features on Perceptions of Walking Environment Attractiveness." *Journal of Urban Design*, *17*(4): 499–510. https://doi.org/10.1080/13574809.2012.706365

Appleyard, B. (2012). "Sustainable and Healthy Travel Choices and the Built Environment: Analyses of Green and Active Access to Rail Transit Stations along Individual Corridors." *Transportation Research Record*, *2303*(1): 38–45. https://doi.org/10.3141/2303-05

Appleyard, B. (2015). "New Methods to Measure the Built Environment for Human-Scale Travel Research: Individual Access Corridor (IAC) Analytics to Better Understand Sustainable Active Travel Choices." *Journal of Transport and Land Use*, *9*(2): https://doi.org/10.5198/jtlu.2015.786

Appleyard, B., and Ferrell, C. E. (2015). *The Meaning of Mean Streets: Associations Between Crime, Casualties and Sustainable & Active Travel Choices.* Presented at the Transportation Research Board 94th Annual Meeting. Retrieved from https://trid.trb.org/view.aspx?id=1336985

Balfour, J. L. and Kaplan, G. A. (2002). "Neighborhood Environment and Loss of Physical Function in Older Adults: Evidence from the Alameda County Study." *American Journal of Epidemiology*, *155*(6): 507–515.

Bella, F. and Silvestri, M. (2015). "Effects of Safety Measures on Driver's Speed Behavior at Pedestrian Crossings." *Accident Analysis & Prevention*, *83*: 111–124. https://doi.org/10.1016/j.aap.2015.07.016

Berrigan, D., Pickle, L. W., and Dill, J. (2010). "Associations Between Street Connectivity and Active Transportation." *International Journal of Health Geographics*, *9*: 20. https://doi.org/10.1186/1476-072X-9-20

Bigazzi, A. Y., and Figliozzi, M. A. (2014). "Review of Urban Bicyclists' Intake and Uptake of Traffic-Related Air Pollution." *Transport Reviews*, *34*(2): 221–245. https://doi.org/10.1080/01441647.2014.897772

Bigazzi, A. Y. and Rouleau, M. (2017). "Can Traffic Management Strategies Improve Urban Air Quality? A Review of the Evidence." *Journal of Transport & Health*, 7(Part B): 111–124. https://doi.org/10.1016/j.jth.2017.08.001

Boarnet, M. G., Forsyth, A., Day, K., and Oakes, J. M. (2011). "The Street Level Built Environment and Physical Activity and Walking: Results of a Predictive Validity Study for the Irvine Minnesota Inventory." *Environment and Behavior.* https://doi.org/10.1177/0013916510379760

Bosselmann, P., Macdonald, E., and Kronemeyer, T. (1999). "Livable Streets Revisited." *Journal of the American Planning Association*, *65*(2): 168–180. https://doi.org/10.1080/01944369908976045

Bowler, D. E., Buyung-Ali, L., Knight, T. M., and Pullin, A. S. (2010). "Urban Greening to Cool Towns and Cities: A Systematic Review of the Empirical Evidence." *Landscape and Urban Planning*, *97*(3): 147–155. https://doi.org/10.1016/j.landurbplan.2010.05.006

Bratman, G. N., Hamilton, J. P., and Daily, G. C. (2012). "The Impacts of Nature Experience on Human Cognitive Function and Mental Health." *Annals of the New York Academy of Sciences*, *1249*(1): 118–136. https://doi.org/10.1111/j.1749-6632.2011.06400.x

Broach, J. and Bigazzi, A. Y. (2017). "Existence and Use of Low-Pollution Route Options for Observed Bicycling Trips." *Transportation Research Record: Journal of the Transportation Research Board*, *2662*: 152–159. https://doi.org/10.3141/2662-17

Brown, V., Moodie, M., and Carter, R. (2017). "Evidence for Associations Between Traffic Calming and Safety and Active Transport or Obesity: A Scoping Review." *Journal of Transport & Health*, 7: 23–37. https://doi.org/10.1016/j.jth.2017.02.011

Buehler, R., and Dill, J. (2016). "Bikeway Networks: A Review of Effects on Cycling." *Transport Reviews*, *36*(1): 9–27. https://doi.org/10.1080/01441647.2015.1069908

Cain, K. L., Gavand, K. A., Conway, T. L., Geremia, C. M., Millstein, R. A., Frank, L. D., . . . Sallis, J. F. (2017). "Developing and Validating an Abbreviated Version of the Microscale Audit for Pedestrian Streetscapes (MAPS-Abbreviated)." *Journal of Transport & Health, 5*: 84–96. https://doi.org/10.1016/j.jth.2017.05.004

Cain, K. L., Millstein, R. A., Sallis, J. F., Conway, T. L., Gavand, K. A., Frank, L. D., King, A. C. (2014). "Contribution of Streetscape Audits to Explanation of Physical Activity in Four Age Groups Based on the Microscale Audit of Pedestrian Streetscapes (MAPS)." *Social Science & Medicine, 116*: 82–92. https://doi.org/10.1016/j.socscimed.2014.06.042

Campbell, B. J., Zegeer, C. V., Huang, H. H., and Cynecki, M. J. (2004). *A Review of Pedestrian Safety in the United States and Abroad (FHWA-RD-03-042).* Retrieved from https://trid.trb.org/view/1158297

Chen, L., Chen, C., Ewing, R., McKnight, C. E., Srinivasan, R., and Roe, M. (2013). "Safety Countermeasures and Crash Reduction in New York City: Experience and Lessons Learned." *Accident Analysis and Prevention, 50*: 312–322. https://doi.org/10.1016/j.aap.2012.05.009

Clifton, K. J., Burnier, C. V., and Akar, G. (2009). "Severity of Injury Resulting from Pedestrian-Vehicle Crashes: What Can We Learn from Examining the Built Environment?" *Transportation Research Part D: Transport and Environment, 14*(6) Retrieved from https://trid.trb.org/view/897740

de Vries, S. I., Bakker, I., van Mechelen, W., and Hopman-Rock, M. (2007). "Determinants of Activity-Friendly Neighborhoods for Children: Results from the SPACE Study." *American Journal of Health Promotion: AJHP, 21*(4 Suppl): 312–316. https://doi.org/10.4278/0890-1171-21.4s.312

de Vries, S. I., van Dillen, S. M. E., Groenewegen, P. P., and Spreeuwenberg, P. (2013). "Streetscape Greenery and Health: Stress, Social Cohesion and Physical Activity as Mediators." *Social Science & Medicine (1982), 94*: 26–33. https://doi.org/10.1016/j.socscimed.2013.06.030

DiGioia, J., Watkins, K. E., Xu, Y., Rodgers, M., & Guensler, R. (2017). "Safety Impacts of Bicycle Infrastructure: A Critical Review." *Journal of Safety Research, 61*: 105–119. https://doi.org/10.1016/j.jsr.2017.02.015

Ding, D., Sallis, J. F., Kerr, J., Lee, S., and Rosenberg, D. E. (2011). "Neighborhood Environment and Physical Activity among Youth a Review." *American Journal of Preventive Medicine, 41*(4): 442–455. https://doi.org/10.1016/j.amepre.2011.06.036

Duncan, M. J., Spence, J. C., & Mummery, W. K. (2005). "Perceived Environment and Physical Activity: A Meta-Analysis of Selected Environmental Characteristics." *The International Journal of Behavioral Nutrition and Physical Activity, 2*, 11. https://doi.org/10.1186/1479-5868-2-11

Ewing, R., & Cervero, R. (2010). "Travel and the Built Environment." *Journal of the American Planning Association, 76*(3): 265–294. https://doi.org/10.1080/01944361003766766

Ewing, R., and Dumbaugh, E. (2009). "The Built Environment and Traffic Safety: A Review of Empirical Evidence." *Journal of Planning Literature, 23*(4): 347–367. https://doi.org/10.1177/0885412209335553

Ewing, R., Hajrasouliha, A., Neckerman, K. M., Purciel-Hill, M., and Greene, W. (2016). "Streetscape Features Related to Pedestrian Activity." *Journal of Planning Education and Research, 36*(1): 5–15. https://doi.org/10.1177/0739456X15591585

Ewing, R. and Handy, S. (2009). "Measuring the Unmeasurable: Urban Design Qualities Related to Walkability." *Journal of Urban Design, 14*(1): 65–84.

Ewing, R., Meakins, G., Hamidi, S., and Nelson, A. C. (2014). "Relationship Between Urban Sprawl and Physical Activity, Obesity, and Morbidity: Update and Refinement." *Health & Place, 26*: 118–126. https://doi.org/10.1016/j.healthplace.2013.12.008

Fitzpatrick, K., Schneider, W., and Park, E. (2006). "Operation and Safety of Right-Turn Lane Designs." *Transportation Research Record: Journal of the Transportation Research Board, 1961*: 55–64. https://doi.org/10.3141/1961-07

Foltête, J.-C., and Piombini, A. (2007). "Urban Layout, Landscape Features and Pedestrian Usage." *Landscape and Urban Planning, 81*(3): 225–234. https://doi.org/10.1016/j.landurbplan.2006.12.001

Forsyth, A., Hearst, M., Oakes, J. M., and Schmitz, K. H. (2008). "Design and Destinations: Factors Influencing Walking and Total Physical Activity." *Urban Studies, 45*(9): 1973–1996. https://doi.org/10.1177/0042098008093386

Foster, S., and Giles-Corti, B. (2008). "The Built Environment, Neighborhood Crime and Constrained Physical Activity: An Exploration of Inconsistent Findings." *Preventive Medicine, 47*(3): 241–251. https://doi.org/10.1016/j.ypmed.2008.03.017

Frank, L. D., Andresen, M. A., and Schmid, T. L. (2004). "Obesity Relationships with Community Design, Physical Activity, and Time Spent in Cars." *American Journal of Preventive Medicine, 27*(2): 87–96. https://doi.org/10.1016/j.amepre.2004.04.011

Frank, L.D., Hong, A., and Ngo, V. D. (2019). "Causal Evaluation of Urban Greenway Retrofit: A Longitudinal Study on Physical Activity and Sedentary Behavior." *Preventive Medicine, 123*: 109–116. https://doi.org/10.1016/j.ypmed.2019.01.011

Freeland, A. L., Banerjee, S. N., Dannenberg, A. L., and Wendel, A. M. (2013). "Walking Associated with Public Transit: Moving toward Increased Physical Activity in the United States." *American Journal of Public Health, 103*(3): 536–542. https://doi.org/10.2105/AJPH.2012.300912

French, S., Wood, L., Foster, S. A., Giles-Corti, B., Frank, L., and Learnihan, V. (2014). "Sense of Community and Its Association with the Neighborhood Built Environment." *Environment and Behavior, 46*(6): 677–697. https://doi.org/10.1177/0013916512469098

Glazier, R. H., Creatore, M. I., Weyman, J. T., Fazli, G., Matheson, F. I., Gozdyra, P., Booth, G. L. (2014). "Density, Destinations or Both? A Comparison of Measures of Walkability in Relation to Transportation Behaviors, Obesity and Diabetes in Toronto, Canada." *PLoS ONE, 9*(1): e85295. https://doi.org/10.1371/journal.pone.0085295

Guo, Z., and Loo, B. P. Y. (2013). "Pedestrian Environment and Route Choice: Evidence from New York City and Hong Kong." *Journal of Transport Geography, 28*: 124–136. https://doi.org/10.1016/j.jtrangeo.2012.11.013

Harlan, S. L., Brazel, A. J., Prashad, L., Stefanov, W. L., and Larsen, L. (2006). "Neighborhood Microclimates and Vulnerability to Heat Stress." *Social Science & Medicine (1982), 63*(11): 2847–2863. https://doi.org/10.1016/j.socscimed.2006.07.030

Hoehner, C. M., Ramirez, L. K. B., Elliott, M. B., Handy, S. L., and Brownson, R. C. (2005). "Perceived and Objective Environmental Measures and Physical Activity Among Urban Adults." *American Journal of Preventive Medicine, 28*(2): 105–116. https://doi.org/10.1016/j.amepre.2004.10.023

Kaplan, R., and Kaplan, S. (1989). *The Experience of Nature: A Psychological Perspective.* New York, NY: Cambridge University Press.

Kim, J. K., Kim, S., Ulfarsson, G. F., and Porrello, L. A. (2007). "Bicyclist Injury Severities in Bicycle: Motor Vehicle Accidents." *Accident Analysis & Prevention, 39*(2): 238–251. https://doi.org/10.1016/j.aap.2006.07.002

Klassen, J., El-Basyouny, K., and Islam, M. T. (2014). "Analyzing the Severity of Bicycle-Motor Vehicle Collision Using Spatial Mixed Logit Models: A City of Edmonton Case Study." *Safety Science, 62*: 295–304. https://doi.org/10.1016/j.ssci.2013.09.007

Kurka, J. M., Adams, M. A., Geremia, C., Zhu, W., Cain, K. L., Conway, T. L., and Sallis, J. F. (2016). "Comparison of Field and Online Observations for Measuring Land Uses Using the Microscale Audit of Pedestrian Streetscapes (MAPS)." *Journal of Transport & Health, 3*(3): 278–286. https://doi.org/10.1016/j.jth.2016.05.001

Lachapelle, U., and Frank, L. D. (2009). "Transit and Health: Mode of Transport, Employer-Sponsored Public Transit Pass Programs, and Physical Activity." *Journal of Public Health Policy, 30*: S73–S94.

Lachapelle, U., and Noland, R. B. (2012). "Does the Commute Mode Affect the Frequency of Walking Behavior? The Public Transit Link." *Transport Policy, 21*: 26–36. https://doi.org/10.1016/j.tranpol.2012.01.008

Lee, A. C. K., and Maheswaran, R. (2011). "The Health Benefits of Urban Green Spaces: A Review of the Evidence." *Journal of Public Health (Oxford, England), 33*(2): 212–222. https://doi.org/10.1093/pubmed/fdq068

Li, W., Keegan, T. H. M., Sternfeld, B., Sidney, S., Quesenberry, C. P., and Kelsey, J. L. (2006). "Outdoor Falls Among Middle-Aged and Older Adults: A Neglected Public Health Problem." *American Journal of Public Health, 96*(7): 1192–1200. https://doi.org/10.2105/AJPH.2005.083055

Lockett, D., Willis, A., and Edwards, N. (2005). "Through Seniors' Eyes: An Exploratory Qualitative Study to Identify Environmental Barriers to and Facilitators of Walking." *The Canadian Journal of Nursing Research = Revue Canadienne De Recherche En Sciences Infirmieres, 37*(3): 48–65.

Loon, J. van, and Frank, L. (2011). "Urban form Relationships with Youth Physical Activity Urban Form Relationships with Youth Physical Activity: Implications for Research and Practice, Implications for Research and Practice." *Journal of Planning Literature, 26*(3): 280–308. https://doi.org/10.1177/0885412211400978

Louv, R. (2012). *The Nature Principle: Reconnecting with Life in a Virtual Age* (Reprint edition). Chapel Hill, NC: Algonquin Books.

Lovasi, G. S., Bader, M. D. M., Quinn, J., Neckerman, K., Weiss, C., and Rundle, A. (2012). Body Mass Index, Safety Hazards, and Neighborhood Attractiveness. *American Journal of Preventive Medicine, 43*(4): 378–384. https://doi.org/10.1016/j.amepre.2012.06.018

Lovasi, G. S., Schwartz-Soicher, O., Neckerman, K. M., Konty, K., Kerker, B., Quinn, J., and Rundle, A. (2013). "Aesthetic Amenities and Safety Hazards Associated with Walking and Bicycling for Transportation in New York City." *Annals of Behavioral Medicine: A Publication of the Society of Behavioral Medicine*, *45*(Suppl 1): S76–S85. https://doi.org/10.1007/s12160-012-9416-z

Lynott, J., Haase, J., Nelson, K., Taylor, A., Twaddell, H., Ulmer, J., and Stollof, E. R. (2009). *Planning Complete Streets for an Aging America.* Retrieved from https://trid.trb.org/view/889160

MacDonald, J. (2015). "Community Design and Crime: The Impact of Housing and the Built Environment." *Crime and Justice*, *44*(1): 333–383. https://doi.org/10.1086/681558

Manuel, A., El-Basyouny, K., and Islam, M. T. (2014). "Investigating the Safety Effects of Road Width on Urban Collector Roadways." *Safety Science*, *62*: 305–311. https://doi.org/10.1016/j.ssci.2013.09.005

Marshall, W. E., and Garrick, N. W. (2011). "Does Street Network Design Affect Traffic Safety?" *Accident; Analysis and Prevention*, *43*(3): 769–781. https://doi.org/10.1016/j.aap.2010.10.024

McCormack, G. R., and Shiell, A. (2011). "In Search of Causality: A Systematic Review of the Relationship Between the Built Environment and Physical Activity among Adults." *International Journal of Behavioral Nutrition and Physical Activity*, *8*: 125. https://doi.org/10.1186/1479-5868-8-125

Mecheri, S., Rosey, F., and Lobjois, R. (2017). "The Effects of Lane Width, Shoulder Width, and Road Cross-Sectional Reallocation on Drivers' Behavioral Adaptations." *Accident Analysis & Prevention*, *104*: 65–73. https://doi.org/10.1016/j.aap.2017.04.019

Mehta, V. (2007). "Lively Streets: Determining Environmental Characteristics to Support Social Behavior." *Journal of Planning Education and Research*, *27*(2): 165–187. https://doi.org/10.1177/0739456X07307947

Mueller, N., Rojas-Rueda, D., Cole-Hunter, T., de Nazelle, A., Dons, E., Gerike, R., . . . Nieuwenhuijsen, M. (2015). "Health Impact Assessment of Active Transportation: A Systematic Review." *Preventive Medicine*, *76*: 103–114. https://doi.org/10.1016/j.ypmed.2015.04.010

National Association of City Transportation Officials (NACTO) (2014). *Urban Bikeway Design Guide.* New York: NACTO. Retrieved from https://nacto.org/publication/urban-bikeway-design-guide/

Ngo, V. D., Frank, L. D., and Bigazzi, A. Y. (2018). "Effects of New Urban Greenways on Transportation Energy Use and Greenhouse Gas Emissions: A Longitudinal Study from Vancouver, Canada." *Transportation Research Part D: Transport and Environment*, *62*: 715–725. https://doi.org/10.1016/j.trd.2018.04.013

Noland, R. B., Gao, D., Gonzales, E. J., and Brown, C. (2015). "Costs and Benefits of a Road Diet Conversion." *Case Studies on Transport Policy*, *3*(4): 449–458. https://doi.org/10.1016/j.cstp.2015.09.002

Painter, K. A., & Farrington, D. P. (2001). "The Financial Benefits of Improved Street Lighting, Based on Crime Reduction." *Transactions of the Illuminating Engineering Society*, *33*(1): 3–10. https://doi.org/10.1177/136578280103300102

Pikora, T. J., Giles-Corti, B., Knuiman, M. W., Bull, F. C., Jamrozik, K., and Donovan, R. J. (2006). "Neighborhood Environmental Factors Correlated with Walking Near Home: Using SPACES." *Medicine and Science in Sports and Exercise*, *38*(4): 708–714. https://doi.org/10.1249/01.mss.0000210189.64458.f3

Pucher, J., Dill, J., and Handy, S. (2010). "Infrastructure, Programs, and Policies to Increase Bicycling: An International Review." *Preventive Medicine*, *50*(Suppl): S106–S125. https://doi.org/10.1016/j.ypmed.2009.07.028

Pulugurtha, S. S., and Self, D. R. (2015). "Pedestrian and Motorists' Actions at Pedestrian Hybrid Beacon Sites: Findings from a Pilot Study." *International Journal of Injury Control and Safety Promotion*, *22*(2): 143–152. https://doi.org/10.1080/17457300.2013.857694

Rapoport, A. (1991). "Pedestrian Street Use: Culture and Perception." In Vernez Moudon, A. (ed.), *Public Streets for Public Use.* New York, NY: Columbia University Press.

Retting, R. A., Ferguson, S. A., and McCartt, A. T. (2003). "A Review of Evidence-Based Traffic Engineering Measures Designed to Reduce Pedestrian-Motor Vehicle Crashes." *American Journal of Public Health*, *93*(9): 1456–1463.

Reynolds, C. C., Harris, M. A., Teschke, K., Cripton, P. A., and Winters, M. (2009). "The Impact of Transportation Infrastructure on Bicycling Injuries and Crashes: A Review of the Literature." *Environmental Health*, *8*(1): 47–65.

Rinner, C., and Hussain, M. (2011). "Toronto's Urban Heat Island: Exploring the Relationship Between Land Use and Surface Temperature." *Remote Sensing*, *3*(6): 1251–1265. https://doi.org/10.3390/rs3061251

Romanow, N. T. R., Couperthwaite, A. B., McCormack, G. R., Nettel-Aguirre, A., Rowe, B. H., and Hagel, B. E. (2012). "Environmental Determinants of Bicycling Injuries in Alberta, Canada." *Journal of Environmental and Public Health* Vol. 2012, Article ID 487681, 12 http://dx.doi.org/10.1155/2012/487681

Rundle, A., Diez Roux, A. V., Free, L. M., Miller, D., Neckerman, K. M., and Weiss, C. C. (2007). "The Urban Built Environment and Obesity in New York City: A Multilevel Analysis." *American Journal of Health Promotion: AJHP*, *21*(4 Suppl): 326–334. https://doi.org/10.4278/0890-1171-21.4s.326

Sallis, J. F., Floyd, M. F., Rodriguez, D. A., and Saelens, B. E. (2012). "The Role of Built Environments in Physical Activity, Obesity, and CVD." *Circulation, 125*: 729–737. https://doi.org/10.1161/CIRCULATIONAHA.110.969022

Saneinejad, S., and Lo, J. (2015). "Leading Pedestrian Interval: Assessment and Implementation Guidelines." *Transportation Research Record: Journal of the Transportation Research Board, 2519*: 85–94. https://doi.org/10.3141/2519-10

Saraiya, M., Glanz, K., Briss, P. A., Nichols, P., White, C., Das, D., . . . Rochester, P. (2004). "Interventions to Prevent Skin Cancer by Reducing Exposure to Ultraviolet Radiation: A Systematic Review." *American Journal of Preventive Medicine, 27*(5): 422–466. https://doi.org/10.1016/j.amepre.2004.08.009

Takano, T., Nakamura, K., and Watanabe, M. (2002). "Urban Residential Environments and Senior Citizens' Longevity in Megacity Areas: The Importance of Walkable Green Spaces." *Journal of Epidemiology and Community Health, 56*(12): 913–918.

Thomas, B., and DeRobertis, M. (2013). "The Safety of Urban Cycle Tracks: A Review of the Literature." *Accident; Analysis and Prevention, 52*: 219–227. https://doi.org/10.1016/j.aap.2012.12.017

Ukkusuri, S., Miranda-Moreno, L. F., Ramadurai, G., and Isa-Tavarez, J. (2012). "The Role of Built Environment on Pedestrian Crash Frequency." *Safety Science, 50*(4): 1141–1151. https://doi.org/10.1016/j.ssci.2011.09.012

Ulmer, J. M., Wolf, K. L., Backman, D. R., Tretheway, R. L., Blain, C. J., O'Neil-Dunne, J. P., and Frank, L. D. (2016, November). "Multiple Health Benefits of Urban Tree Canopy: The Mounting Evidence for a Green Prescription." *Health and Place, 42*: 54–62.

Wendel-Vos, W., Droomers, M., Kremers, S., Brug, J., and Lenthe, F. V. (2007). "Potential Environmental Determinants of Physical Activity in Adults: A Systematic Review." *Obesity Reviews, 8*(5): 425–440. https://doi.org/10.1111/j.1467-789X.2007.00370.x

Werner, C. M., Brown, B. B., and Gallimore, J. (2010). "Light Rail Use Is More Likely on "Walkable" Blocks: Further Support for Using Micro-Level Environmental Audit Measures." *Journal of Environmental Psychology, 30*(2): 206–214. https://doi.org/10.1016/j.jenvp.2009.11.003

Wilson, E. O. (1984). *Biophilia.* Cambridge, MA: Harvard University Press.

Winters, M., and Teschke, K. (2010). "Route Preferences Among Adults in the Near Market for Bicycling: Findings of the Cycling in Cities Study." *American Journal of Health Promotion, 25*(1): 40–47. https://doi.org/10.4278/ajhp.081006-QUAN-236

Further reading

McCann, B. and Rynne, S. *Complete Streets: Best Policy and Implementation Practices.* PAS Report 559. Chicago, IL: American Planning Association. Retrieved from www.planning.org/publications/report/9026883/ Drawing from the experiences of 30 US communities, this report outlines successful policies and implementation practices for Complete Streets.

National Association of City Transportation Officials (NACTO). *Urban Street Design Guide.* New York: NACTO. Retrieved from https://nacto.org/publication/urban-street-design-guide/. Outlines the principles and current practices for the development of Complete Streets.

National Association of City Transportation Officials (NACTO) (2014). *Urban Bikeway Design Guide.* New York: NACTO. Retrieved from https://nacto.org/publication/urban-bikeway-design-guide/. Drawing from the experiences of the best cycling cities in the world, this booklet gives useful guidance for the design of urban bikeways.

PART III.III

Conservation/restoration

Introduction

A major challenge of rapid urbanization and urban development is managing change and continuity in the extant urban form, in cases where there is significant historical and cultural heritage. In *What Time Is This Place?* Kevin Lynch (1974) addresses the question of historic preservation and wonders if all of our history and the inherited urban form are worth saving. What should be saved and what shouldn't? This is a difficult question for urban design, as it is not clear that the urban designer necessarily has the authority to affect that decision. Although one could argue that in some urban transformations that have required demolition of the old and construction of the new, urban designers have certainly played an active role.

But when the walls of historic Beijing were taken down, we were told that the decision came down from the very top, that is Mao Zedong himself, to enable better circulation, create space for housing, and so on. Almost 50 years later, the historic Ming-Qing dynasty city of Beijing is pretty much scraped clean of its past, which has been replaced by office and residential towers, shopping malls, and freeway development, save the Forbidden City, some important temples and gardens, and a few historic streets with traditional hutongs. The denouement of the historic city of Beijing – considered the epitome of the canonical Chinese city – can be seen as a case of Faustian obsession with growth and change, to draw from Marshall Berman (1982).

We don't know how Lynch would have responded to the rapid transformation of historic places in the Global South today, but a narrative of loss often accompanies writings of contemporary scholars in urban design. Yet, urban design is seen as a means for managing the transition from tradition to modernity. Also, there is considerable interest today in the bottom-up resistance to the forces of change, especially when these lead to displacement and disruption of the social life of communities. Similarly, there is interest today in reconceptualizing the challenges of conservation and restoration in the context of the goals of sustainability, resiliency, ecology – reflecting the moral, rather than the technical order in development (Peattie 1981).

The essays assembled in this section variously discuss the urban design challenges for managing change and continuity, and the passage from tradition to modernity. The chapter by Vinayak Bharne considers three rather unique circumstances of cultural history in different parts of the world, and how their continuity may require a different perspective of resiliency, ecology, and social mediation. In the case of the Ise Shrines of Japan, the rituals of building and rebuilding are intrinsically linked to the visual and aesthetic design of the landscape. The historic *qanat* (subterranean water supply) system in Yazd in Iran presents opportunities for urban design to link sustainability issues with ecology and technology. Finally, he presents the case of the Taj Mahal in India and the social ecology of the settlements surrounding this important tourist attraction as a problem that can be addressed in the urban design plans and proposals.

In her chapter, Mary Ann O'Donnell presents an enlightening case study about how the civil society – albeit fledgling in China until recently – can play a role in establishing the continuity of a historic village as the cultural core of the sprawling metropolis of Shenzhen. A group of local intellectuals – academics, architects, planners, public intellectuals – were able to begin a civil and rational dialogue with the local state authorities through exhibitions, seminars, and public meetings. Ultimately, they were able to convince the municipal authority to direct the developers to prepare a new redevelopment plan that will preserve the "heart" – the historic core – of the city. Here the role of urban design was more in the form of educating the local public and the authorities through images, ideas, and historical narratives.

The third chapter in this section, by Pietro Calogero, is also from the Asian Global South but with a somewhat different focus. Here, the story is about how the royal and political leadership of the country had wanted to transform the historic city of Kabul to a modern city. Urban design was the principal mechanism for this transition. This process has been complicated as the history of Kabul's development went through different regimes of control and conflict since the colonial era. Calogero points to the institutions of planning and the ideological differences about such mechanisms of planning among various regimes that guided the development of Kabul. In this case, the story is more about urban design managing change than continuity.

References

Berman, M. (1982). *All That Is Solid Melts into Air.* New York: Simon and Schuster.
Lynch, K. (1974). *What Time Is This Place?* Cambridge, MA: MIT Press.
Peattie, L. (1981). *Thinking About Development.* New York: Plenum.

35

BROADENING HERITAGE CONSERVATION THROUGH URBAN DESIGN

Perspectives from Asia

Vinayak Bharne

Introduction

Heritage conservation is often seen by the architecture *avant garde* as a challenge or even obstacle to urban design: the perception of heritage instigates populist reactions, lengthening community outreach requirements and entitlement processes. It often calls for intervention of local, state, and national agencies, demanding negotiation and monetary investment. The presence of historic buildings within or around a site also raises difficult questions about design appropriateness of proposed schemes, and their compatibility and place character that may limit design freedom and innovation.

There are, however, alternative perspectives on the relationship between heritage conservation and urban design. One of them stems from the overarching moral and ethical imperative of sustainability and environmental protection. Going beyond extant conflicts, bureaucracies, and technocracies, this perspective seeks to identify and prioritize deeper ecological concerns as the driver for broadening the scope of heritage conservation through urban design. Prioritizing the social and ecological setting over a single object as the focus of engagement, it blurs the boundaries between heritage conservation and urban design through their disciplinary overlaps and interdependencies.

This inclusive and interdependent relationship is in part a response to the debacles of urban renewal and sprawl. Over the past decades, the disciplinary boundaries of urban design have expanded through the influences of social, geographical, and natural sciences, eschewing its exclusive co-option by the architecture and urban planning professions (Banerjee & Loukaitou-Sideris 2011). Scholarship on "information society and globalization" (Castells 2010), "third space" (Soja 1996), "postcolonialism" (Spivak 1999), and environmentalism (Hawken 2010) has deepened urban design's scope to embrace numerous urgent concerns such as ethnic tensions, informality, poverty, and climate change. At the same time, new ideas such as ecological urbanism, new urbanism, and everyday urbanism are representative of the multiplicity through which architects and designers have come to redefine urban design from various standpoints. Urban design has evolved into a formal, social, or ecological remodeling of the city – an effort of negotiating what must be shed, obliterated, or replaced with new products, versus retained, revived, and reused through conservation and heritage. The act of identifying, protecting, restoring, and reusing built "heritage" – be it buildings, districts, landscapes, or entire towns and regions of historic or cultural significance – is now increasingly recognized as a far more strategic act of urban and ecological transformation (Bharne & Sandmeier 2019).

The goal of this chapter is to explore this symbiosis between urban design and heritage conservation towards generating a more holistic reading of a place and mediating the degree to which experts intervene. How can a broader perspective on "heritage" through the lens of urban design serve to expand a conservation effort beyond the singular focus on the artifact as monument? How can the interface of urban design and heritage conservation expand their perceived boundaries of practice and offer deeper reflections about the past, present, and future of a setting?

In what follows, I will discuss three different places in Asia, each responding to these questions in various ways: (1) the Ise Shrines in Mie, Japan; (2) the *qanats* (vernacular hydro-infrastructure) in Yazd, Iran; and (3) the Taj Mahal in Agra, India. These three narratives will offer perspectives on the pluralism of urban design as a discipline. In the case of Ise, I will examine the architectural potential of urban design as a formal, visual, and even aesthetic engagement with the built environment. In the case of the *qanat*, I will speculate on how urban design might enmesh with sustainability prerogatives through engineering and technology. In the case of the Taj Mahal, I will explore urban design as mediation between complex social and ecological contentions and power structures.

The "complete" Ise Shrine, Mie, Japan

The Ise Jingu (or Ise shrine) is not only one of Japan's earliest recorded architectural masterworks, but also one of its unique places. Its *shikinen sengu*, the 1,300-year-old ritual of dismantling and replacing the wooden structures with new ones every 20 years, symbolizes a reverence to the deities by cleansing their household, while also responding to the pragmatic need to maintain and renew parts of a timber building. There are no precise dates of the shrines' origin; records suggest that the first *sengu* took place in the fourth year of empress Jito (approximately 686–697 CE), and some speculate that the first shrine goes back as early as 4th century BC.

For most, the Ise shrine refers to a modest cluster of impeccably designed thatched wooden buildings within a pine forest setting. In reality, however, the term Ise shrine refers to two autonomous precincts, the geku (outer shrine) and the naiku (inner shrine), located around five kilometers apart. Centered around a main shrine, each precinct contains other ancillary objects, from storehouses to bridges, totaling more than 60 structures, all rebuilt every two decades. A pilgrimage path connects the precincts, with 125 other related shrines located within and outside the municipal boundaries of Ise city. Thus, the original perceived sacred territory of the Ise shrines and its *sengu*'s ceremonies has four scales: (1) the regional geography, commemorating the surrounding natural elements – mountain, forest, river, and agricultural fields; (2) the land bounded by the 100-odd tutelary shrines; (3) the pilgrimage path and the town surrounding the two precincts; and (4) the two precincts containing the main buildings, in which the deities are housed (Bharne 2014).

It is the current condition of the Ise shrines that make them a relevant case study for this discussion. The shrine precincts are meticulously preserved, attesting to the attention and cultural importance they continue to garner. The *sengu* ceremonies are carried out with the same dedication and ritual rigor as ages ago. Millions of tourists visit the shrines each year, and the larger public attraction to this place, local and national, remains strong. However, there is a gap between the physical continuity of the shrine precincts versus their broader environs that form part of the Ise sacred territory. This gap is evident at a number of scales:

The first is the urban space immediately outside the shrine precincts. The entry to the Ise Geku is a "non-place" public realm with wide asphalted thoroughfares and painted crosswalks to direct pedestrian activity. There is no sense of any conscious urban design in the streets or open space to create a transition worthy enough of one of Japan's most sacred places. At the Ise Naiku, the land between the precinct and the Isuzu River was designed in 1993 as a figment of an Edo period (1603–1868) street called Okage Yokocho, catering to tourists. But here again, this invented street is not the mandatory entrance into the shrine precinct; a visitor's transition into the sacred precinct occurs through a tourist bus stop and parking lot, devoid of any design sensibility or sense of reverence to a special place.

The second scale is the one perceived in the areas between the shrines. The pilgrimage path between the Ise Geku and Ise Naiku has several historic sacred markers including stone lanterns, giant torii (gateways), and bridges (Figure 35.1). Yet, this former spine is little more than an undulating asphalted thoroughfare with a haphazard mix of historic and modern buildings with no semblance of any visual or aesthetic unity. Likewise, the transition from the Ise train station to the first shrine, the Ise Geku, is an un-designed townscape of ad-hoc buildings and streets. It neither announces the importance of arriving at one of the nation's most sacred places nor augments the sense of anticipation of walking or driving towards the sacred grounds (Figure 35.2). In essence, some of the most significant transitions towards the shrines appear devoid of any conscious design attention.

The third scale is evident in the sprawl within the towns that surround the two shrines. These towns were formerly an integral part of the sacred territory and housed most of the clergy that participated in their daily workings. Today, the one- and two-story traditional fabric of timber houses is randomly interspersed by modern mid- and high-rise buildings along major avenues. Within the neighborhoods, several older homes have been demolished and replaced by two- and three-story dwellings that are set back from the street to accommodate carports and parking lots, negating street definition and place character. Historic canals winding through the towns are embanked as concrete channels with barbed wire fences, which render them visually unappealing and socially and ecologically dysfunctional. My discussions with Ise city officials (in 2013 and 2017) revealed that this loss of character has not gone unnoticed. Yet, no effort has been made to date to create guidelines or planning mechanisms to control such ad-hoc development.

The heritage rubric of the Ise shrines is limited to their two precincts, with no concerted effort to recognize numerous other elements and areas that are an integral part of their history and evolution. While such recognition would require significant design effort and economic investment, one must consider the massive budget that the *sengu* requires every two decades. The 62nd *sengu* that took place in 2013 had an estimated cost of approximately US $0.5 billion, more than the US $0.3 billion-dollar investment spent on the *sengu* 20 years earlier, in 1993.

This investment, however impressive, raises difficult questions when examined through an environmental lens. Historically, trees needed for the shrines' reconstruction were available within their surrounding forests. Today, the 4,400 hectares of this 5,500-hectare national state forest, which is reserved exclusively for the shrines' reconstruction, suffers from lowered soil quality resulting from bombing, wars, and industrialization. Consequently, the tree yield is far lower in both quantity and quality than in earlier eras. Because of this, much of the timber supply for the *sengu* today – some 14,000 pieces of timber and 25,000 sheaves of miscanthus reeds – now comes from distant forests, transported by trucks, requiring significant infrastructure and cost. Even as discussion about the shikinen *sengu*'s future continues, other *shinto* shrines have already responded to such environmental pressures: the Izumo shrine's reconstruction ritual, *shikinen senza*, is performed once every 60 years, and is limited only to a few shrines rather than the entire

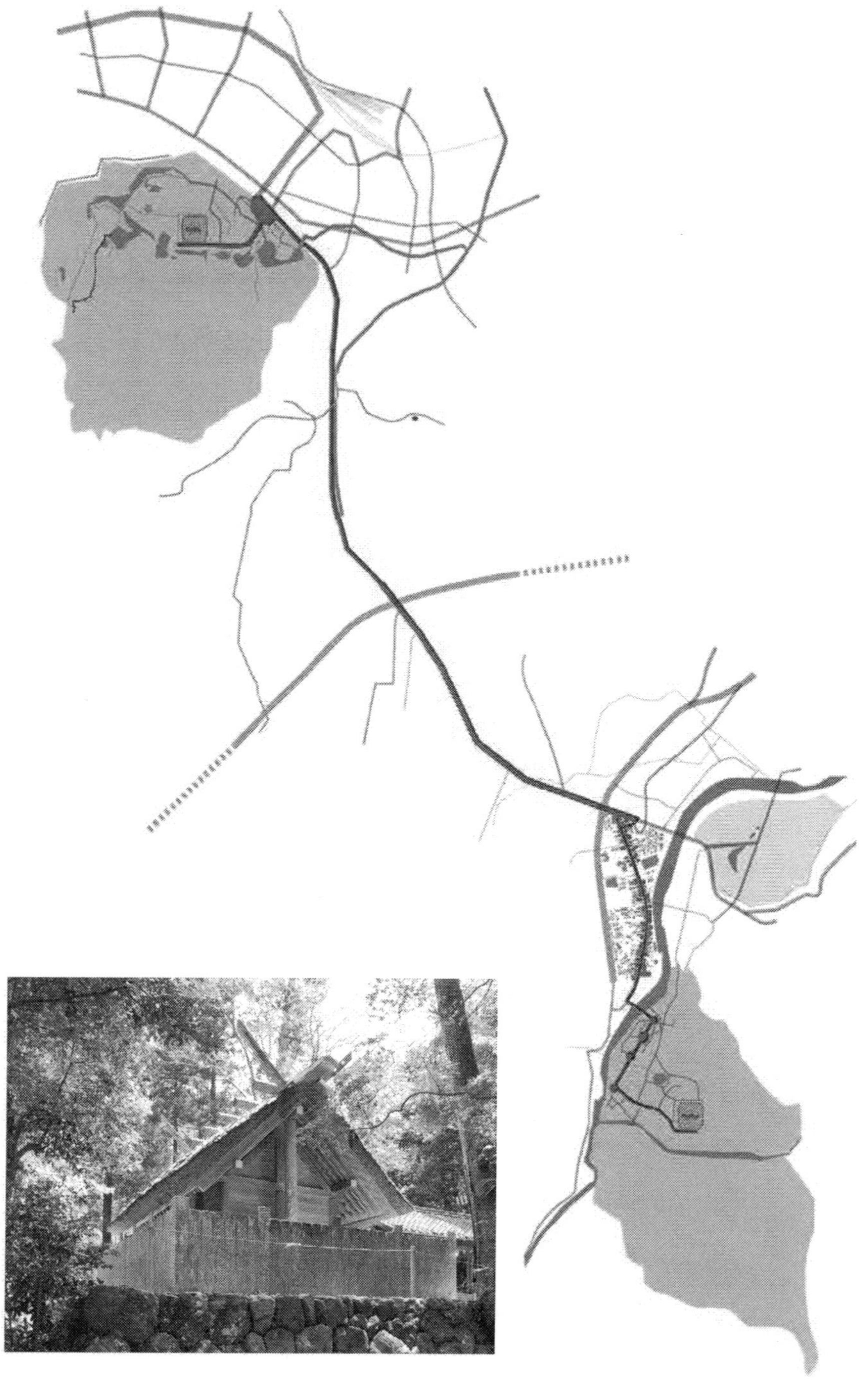

Figure 35.1 Diagram showing the separate sacred precincts of the Ise Geku and Ise Naiku with the pilgrimage path between them. The inset shows one of the Ise Naiku tutelary shrines.

Source: Diagrams and photo by author

Figure 35.2 (Top) The mundane path from the Ise train station to the Ise Geku belies its sacred presence. (Bottom) Typical insensitive infill in the neighborhood surrounding the Ise Geku.

Source: Photos by author

complex. At the Osaki shrine, the structural material has been changed to new kinds of timbers, with proximate land reserved to grow these trees.

An urban design perspective can serve to broaden the conservation rhetoric that currently operates at the Ise shrines. This perspective can highlight the importance of celebrating the two shrines as a single experiential whole, using urban design as a spatial and aesthetic tool. In Japan, there is a long and rich tradition of designing the paths to sacred precincts as sequential experiences that prepare a visitor for entering the special place. The use of trees, shrubs, winding paths, stepping stones, and lanterns, and the conscious manipulation of the terrain and landscape, help visually cut off the mundane world, creating the impression of walking through a dark forest towards a place with special significance (Nitschke 2007). Likewise, the path between the Ise shrines, the sequence from the station to the Ise Geku, and the commercial foregrounds to the precincts are areas that deserve conscious design deliberation as parts of a single sacred territory. Simultaneously, guidelines and regulations could control the character, size, and scale of new infill development within the towns, while façade improvement strategies such as fences and other public-private interface elements could mitigate the gaps and disjoints between traditional and modern buildings. Celebrating the broader environs of one of Japan's most revered ancient places remains an urgent issue of national cultural significance.

Reinventing the *qanat*, Yazd, Iran

The city of Yazd is situated 1,215 meters above sea level, at the confluence of the Dasht-e-Kavir and the Dasht-e-Lut deserts in Iran. Beneath and around this city lays an ancient network of subterranean water channels called *qanats*, stretching for more than 10 miles from the urban core, that nay have been created as early as the first millennium BC. In 2017, Yazd became the 22nd site in Iran to be included in UNESCO's World Heritage list, representing a long overdue recognition of this indigenous heritage. Transporting water over long distances is a significant challenge in Yazd's climate, where diurnal temperatures fluctuate from 50 to -20 degrees Celsius within 24 hours, and seasons vary from a long hot summer (mid-March to mid-September) to a cold winter (October to February). The city gets about 55 mm of precipitation annually, significantly less than Iran's average annual rainfall of approximately 242 mm, which in itself is less than one third of the global average annual rainfall of approximately 860 mm.

The origin of *qanats* is in the Zagros mountains crossing the Yazd province that gather winter snow and trap water in their crevices forming a subterranean aquifer. Experts have surveyed the mountain soil deposits mapping a potential source well. Vertical shafts of successively increasing depths were once dug at 50-meter intervals and connected horizontally underground by subterranean tunnels, approximately one meter wide and two meters high, gently sloped to ease gravity flow (Figure 35.3). Water was directed first into the surrounding agricultural fields and the remainder towards the city. The *qanat*'s entry into the urban context was marked by domed reservoirs called *ab anbars* (Figure 35.4). These reservoirs were designed to stabilize a low water temperature, withstand water pressure, and resist earthquakes. These subterranean reservoirs were capped above ground by a semi-circular brick-lined dome with central escape vents that cooled the water, while protecting it from dust and pollution. Wind catchers framing the dome maintained fresh air circulation and prevented water deterioration.

Ab anbars also played a pivotal role in the urban structure of Yazd. Distant *qanats* split into a smaller canal network called *karez*, bringing water to a hierarchy of city-center and neighborhood-specific reservoirs, and each reservoir provided water to a limited number of streets and houses, defining a distinct community shed. With each dwelling located within easy reach of their only water source, the formal complexity of Yazd's historic communities was thus ordered around existing *qanats* and *ab anbars*, as much as mosques.

The past six decades have transformed the role of Yazd's *qanats* and *ab anbars*. Modern piping systems are today considered far more reliable to supply the quantity and quality of water needed for the growing city, over an indigenous gravity-fed system dependent on a source aquifer susceptible to contamination and needing recurrent maintenance. *Qanat* use consequently

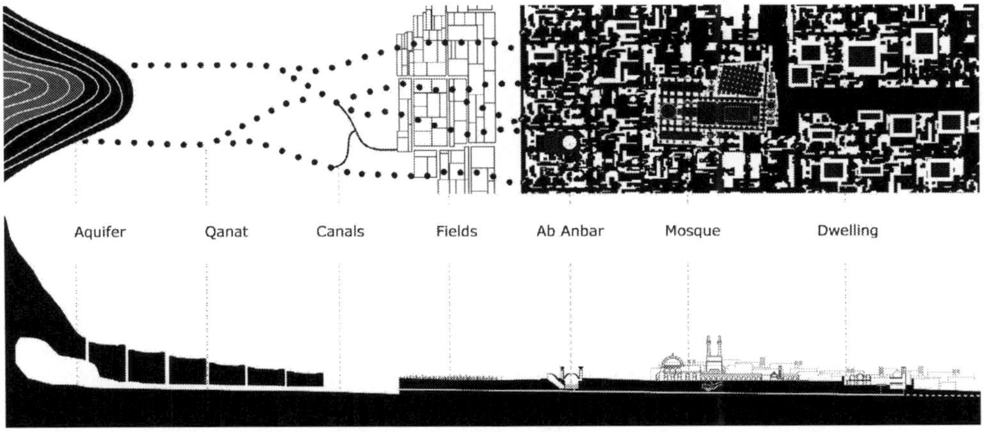

Figure 35.3 Anatomy of the *qanat*: the mountain aquifer and source wells are to the left, the domed reservoirs in the center, and the mosques and dwellings within the city to the right. The horizontal line to the bottom of the diagram represents the subterranean *qanat*.

Source: Diagram by Vinayak Bharne and Biayna Bogosian

Figure 35.4 Rostam Giv *ab anbar*, one of the most well-preserved historic reservoirs in Yazd today.

Source: Photo by Brian Mcmorrow

decreased "from 70% prior to 1950, to 50% around 1950 and to 10% in the year 2000" (Haeri 2003). Today, Yazd's principal water supply comes from the Zayandeh River in the city of Isfahan, more than 200 km away. No new *qanats* have been built using traditional methods since 1963, and of the 3,300 *qanats* within the Yazd province, around 3,000 are increasingly polluted from industrial discharge.

An obvious reason for the *qanat*'s decline was Yazd's rapid urbanization. Until circa 1925, Yazd grew at a steady pace, but subsequently, in as little as three decades, it exploded into a sprawling city of 500,000 residents. The *qanat*'s diminishing appeal was also in part a political phenomenon. With the advent of the white revolution in 1963 and its land reform program against feudalism, the government purchased agricultural fields from landowners and sold them to peasants at a much lower price. Large traditional land holdings were broken down, leading to confusion about subterranean channel ownership rights and maintenance responsibility. Water distribution, traditionally controlled by a few, now came under a larger purview, and subsequent government intervention had to be geared towards giving this larger demographic their fair share. Simultaneously, increased agricultural production necessitated the drilling of source aquifers, and many *qanats* began to dry out or became seasonal, even as the drilling industry attracted more people into the city.

Is there a future for the *qanat* beyond an abandoned hydro-infrastructure and an embalmed "heritage"? Can the conservation of *qanats* and *ab anbars* rise above its strictly nostalgic appeal to serve contemporary urban concerns of sustainability? Can this indigenous infrastructure be repurposed for a new use in the contemporary city? Recent concerns over dwindling groundwater reserves and costly sewage treatment have generated increasing global interest in the reuse or recycling of gray water for domestic use and commercial irrigation. Gray water is the wastewater generated from domestic activities such as dish washing, laundry, and bathing and is significantly less contaminated than black water that contains fecal matter and urine. In Yazd, such gray water is available from a number of sources. Besides dwellings and rainwater collection from roofs, ablution water from numerous mosques also offers a unique opportunity to integrate sustainability with cultural practices.

Could *qanats* become a parallel infrastructure for gray water harvesting and distribution? Abandoned *qanat* networks could be reactivated, saving significant expenditure on what would otherwise be an infrastructural installation from scratch. They could carry gray water for adequate processing to existing *ab anbars*. These reservoirs could be redesigned as local gray water treatment plants. Other *qanats* could return the water to tanks below buildings to be used for flushing toilets, or watering gardens, or direct it to larger destinations such as urban parks. Treating *qanats* and *ab anbars* as gray water infrastructure could retain their original forms for a contemporary cause, incentivizing their reuse and maintenance. New *qanats* and *ab anbars* could be built to localize water collection on a neighborhood scale, with their capacities and performance monitored with modern technology. This could significantly reduce the load on the city's current water supply through decentralized water harvesting, treatment, recycling, and management (Bharne 2011).

There are challenges to this proposition. As a gravity dependent system, *qanats* are always sloped towards a destination such as a reservoir or dwelling. Thus, within an existing *qanat* network, gray water recycling from a dwelling to a reservoir would require a reverse water flow against gravity, needing electrical pumps. Additionally, gray water recycling through passive means is most efficient when done on a smaller scale; that is, when the water is treated within or near the site and returned within short distances. Thus, *qanats* and *ab anbars* as gray water systems are justifiable on a lot or even block scale, but their large-scale application as citywide gray water systems remains technically dubious. The greatest challenge to this proposition is that a single *qanat* is situated under multiple properties and will require collective participation of households along a street or neighborhood to activate and maintain it.

The idea of *qanats* and *ab anbars* as living, repurposed contemporary infrastructure networks expands their heritage discourse into complex dialogues on urbanism, sustainability, and participatory planning. Their reuse as a decentralized gray water system is an engineering project with lasting urban design opportunities: streets and open spaces with repurposed *qanats* beneath them could be designed as special places for public use. The conversion and rebuilding of *ab anbars* as gray water treatment plants could add new visual elements within the contemporary urban maze creating civic art and navigation markers. Finally, new or old *ab anbars* could also be designed as multifunctional structures that combine their technical aspects with uses such as Internet cafes and restaurants, making them iconic places for public life. The combination of heritage conservation and urban design together can offer a new approach to recreating Yazd's urban future.

Re-reading the Taj Mahal, Agra, India

The famous Taj Mahal has been a UNESCO World Heritage Site since 1983. To safeguard the marble exterior of its main mausoleum building, 300 coal-based regional industries have switched to natural gas. No transport is allowed within 500 meters (1/3 mile) of the monument, and an air monitoring system continuously measures air quality. While such efforts are laudable, the conservation rhetoric surrounding the Taj Mahal remains largely limited to its buildings and gardens, when in fact its physical stability, as well as historic authenticity, relies on factors far bigger than the confines of its walled complex.

In September 2010, the Yamuna River (Figure 35.5) touched the base of Taj Mahal's 300-meter-long terrace for the first time in more than two decades. Heavy rains in northern India had raised the water level of this semi-dry riverbed fronting the monument (Figure 35.6). This is one of the most significant events in the Taj Mahal's recent history, as several studies have now concluded that the strength of the wooden shafts holding together the Taj Mahal's foundations depends on

Figure 35.5 Villagers appropriate the dry Yamuna riverbed fronting the Taj Mahal as a cricket field.

Source: Photo by author, circa 2013

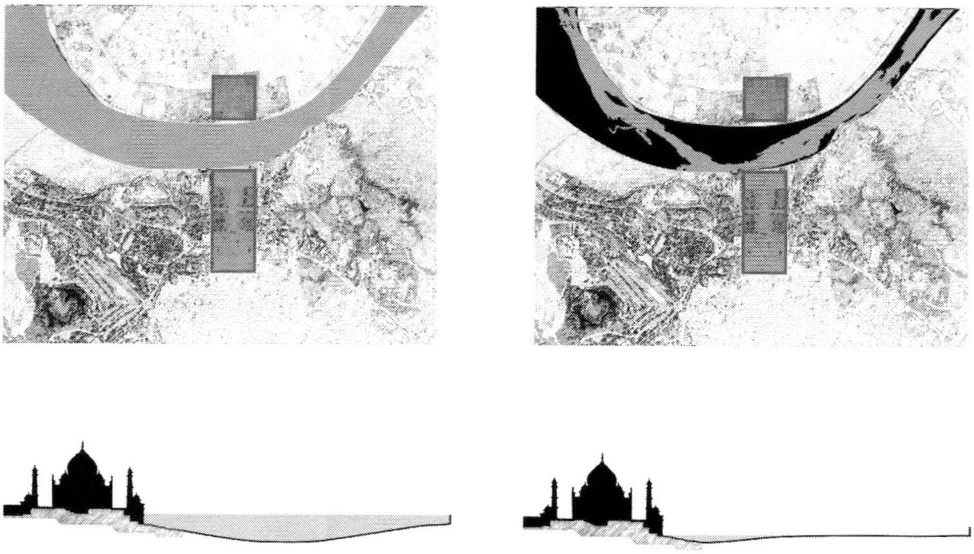

Figure 35.6 Yamuna River water levels fronting the Taj Mahal: (left) circa 1630; (right) circa 2018.

Source: Diagram by author

being constantly moistened by the river's water. The water's presence is also important considering the fact that the Taj Mahal was conceived as a riverfront monument within the larger urban landscape of the city of Agra, thereby highlighting the Taj Mahal's urban design dimensions (Bharne 2013).

The flourishing riverfront capital of Agra had grown over four generations around the Agra fort, the emperor's red sandstone palace on the western bank. Its entire riverfront was lined with opulent villas, mansions, and mausoleums, creating a five-kilometer-long public promenade as the city's dominant open space. This was the physical and aesthetic context within which the Taj Mahal was built in 1631, following the fifth Mughal emperor Shah Jahan's decision to memorialize his deceased queen. Unlike the other riverfront buildings that were generously set back from the water's edge, the Taj Mahal was built close to the bank, with an 8.7-meter-high and 300-meter-long river-facing terrace with steps at its base leading to the water. The river was thus the regal entrance and public foreground of the Taj Mahal as travelers entered the city via ship from the north.

The Yamuna River's dismal state today is, thus, a degradation of the Taj Mahal itself. While the river's restoration is intrinsic to the monument's physical stability, as noted earlier, its significant water shrinkage and contamination and the ruin and decay in many of the historic mansions pose significant challenges to its implementation. Adding to these challenges are events such as the 2002 "Taj heritage corridor" scandal: the kilometer-long stretch of the Yamuna bed between the Agra fort and the Taj Mahal was filled up with four million cubic meters of soil. The stretch was supposed to accommodate the "heritage corridor" beautification plan with shopping malls, parks, restaurants, and entertainment centers, undertaken by the National Project Construction Corporation, New Delhi, without clearance from the archaeological survey of India and other authorities. There was fear that construction on the riverbed might alter the river's course and lead to ecological problems. This drew community opposition, and the project was stopped by the central government on the grounds of violating the World Heritage Monuments conservation norms. Ongoing riverfront restoration efforts continue to face an uphill climb. Yet the potential of the river as a renewed historic setting and as public amenity is an acknowledgment that the Taj Mahal is larger than the limits of its marble-clad structure.

The Taj Mahal was originally conceived as a large 22.4-hectare campus. Closest to the water was the mausoleum framed symmetrically by two buildings: a mosque on the west and a twin assembly hall to the east. Just south was the 300 x 300 meter square *Char Bagh* (four quadrant) garden, planned along two perpendicular walkways. Next came the *jilaukhana*, the transitional forecourt between the funerary and commercial realms of the complex, where royal guests were received. And eventually closest to the city was the Caravanserai, a two-story arcaded complex also planned on a quadrangular theme, with its intersecting streets forming a commercial bazaar (market). By the 1640s, the area surrounding the Taj Mahal had in fact become an identifiable district, with the bazaars contributing financially to the complex's maintenance. Thus, the symbolic and commercial dimensions of the Taj Mahal were conflated from very early on.

Following the capital's shift to Delhi in 1648, the market declined, and the Caravanserai was gradually transformed into what is now known as Taj Gunj. Over four centuries, it was incrementally possessed and appropriated by the local merchants to meet their demands. By the early 20th century, nothing but the four-gated walls once framing the central market square remained identifiable. Its four quads had been in-filled with a haphazard development of labyrinthine walls, reducing the street space to tenuous lanes. Today, despite its continuing commercial activity, there is little to reveal the Caravanserai's original form.

But even so, as part of a Caravanserai, the Taj Mahal was always meant to be part of the plebeian experience of Agra. In this sense, the Gunj can be read as an extension of Taj's "commonplace" dimension; an energized world that is different from the desolate riverfront, invigorating the monument through a critical counter-image. As part of a tourism economy, however, the need for hygienic conditions within this informal habitat remains the dominant pretext for intervention. Conservationists remain bent on vacating the illegal settlement even as its residents argue for the Taj as their rightful legacy, claiming their ancestors as its builders and subjects of the emperor. If the restored riverfront can become a new socioeconomic engine for the Taj Mahal's experience, why can't a strategic nurturing of the Gunj – its skilled labor force, stone-inlay crafts and micro-commerce – play an essential role in the monument's recurrent upkeep, reviving a historic relationship with it? Arguably, the Taj Mahal today seems all the more meaningful this way: as a nexus between a potentially restored riverfront and the plebeian realities of the Gunj; a paradisiacal realm embracing the disparate social aspirations of two concomitant worlds, both of which want a stake in it.

Unfortunately, the Taj Mahal remains a de-contextualized cultural icon whose urban design and landscape dimensions are significantly undermined. It attests to myopic heritage conservation approaches as well as to the cautions against doctrinaire and linear attitudes to contemporary understandings of monumentality. The intention behind highlighting the Taj Mahal's broader dimensions should therefore be understood as an urgent advocacy to expand the simplistic dialogues that surround its future.

Conclusion

This chapter sought to examine how a heritage conservation scope can expand or shift using an urban design lens. An evident expansion occurs through an emphasis on ecology. Today, in the wake of climate change and global warming, this should be an increasingly dominant aspect of both urban design and heritage conservation efforts. The issues of deforestation at Ise, drought and water scarcity in Yazd, and river depletion in the case of the Taj Mahal are emblematic of a larger scenario that also encompasses the thermal expansion of the oceans and melting glaciers that loom on the horizon. This calls for a re-prioritization of the ecological issues that frame a project over its architectural and socioeconomic aspects. The ecological layers of a setting are as profoundly significant as its social layers, even though they both remain underestimated in mainstream urban design and conservation practices.

An ecological lens also serves to expand a project's perceived physical boundaries. Such an expansion emerges from a repositioning of the focus from monument-centric to place-centric. This emphasizes the evolving relationships between buildings, their social context, open spaces, and natural features that have shaped the larger contemporary context over a singular entity or object. This suggests that the study boundaries of any conservation project should not emerge from bureaucratic or economic limitations, but rather be determined by holistic means, where multiple actors might be able to deliberate about them from different standpoints. In this sense, any heritage conservation project must always begin as a larger planning effort and end with decisions about what to conserve and when, depending on the socio-political conditions at the time.

This discussion also reveals the multiple ways in which urban design intersects with heritage conservation. The case of the Ise shrines highlights connections of religion and ritual. Here, the overarching goal of both disciplines is the acknowledgment and commemoration of the intangible aspects of a setting – people's beliefs, cultural traits, and knowledge structures – and their eventual translation as a tangible place. Urban design and heritage conservation together seek to augment cultural identity by negotiating complex issues of authenticity, continuity, and change at multiple scales. The case of the *qanats* highlights the intersection of indigene and modernity, indicating the strategic need for pragmatic rather than nostalgic approaches to fragile, aging built elements. It speculates on how the technical dimensions of urban design might serve to repurpose heritage towards new uses and meanings amidst rapidly changing demographics and economies. The case of the Taj Mahal highlights how urban design and heritage conservation intersect through tourism capital. The evident irony here is that tourism industries, as local economic stimulators, incentivize communities to cater to obvious tourist expectations, even as the economic benefits of tourism often tend to overshadow this connection.

Collectively, the studies remind us that origins and deep pasts are not the only ways of understanding a place; shifting histories can mutate into new and unforeseen guises – at once an architectural and anthropological issue. The environmental challenges facing the Ise shrines today, the abandoning of the *qanats*, and the Taj Mahal's deteriorated surroundings suggest that the idea of heritage is not an embalmed but dynamic entity. Heritage must transcend its historic and monumental values to consider how contemporary places, forms, practices, and community interactions can contribute to its formation. Heritage can be a catalyst to improve the livability of historic areas, through a unified policy and practice framework, without approaching conservation and development as separate tasks – and it is here that urban design becomes such a crucial part of the conservation discussion.

As a final note to this chapter, I wish to highlight the "non-western" context of these three studies. Asia and the Global South today not only contain some of the oldest and most fragile examples of cultural heritage, but also offer an urban design and heritage conservation discourse that is a counterpoint to the formal and authorized one of the West. While an elaboration on this aspect is beyond the scope of this chapter, it is nonetheless important to emphasize that the efficacy of the urban design and heritage conservation interface, discussed in this chapter, will eventually emerge from the specific political and administrative realities of these places. Japan became a democracy in 1947 following the American occupation, and India also the same year after gaining sovereignty from British rule; but even so, their regulatory stringency and political structures are significantly different from those of the West. Iran by contrast is a theocratic government. Such differences will have a direct impact on the process and products shaping their built environment. The specific players that engage in acts of urbanism and conservation, and the actual processes and products that bring about bigger and deeper changes, will themselves be significantly different.

Such differences are shifting the field of urban design and heritage conservation globally, moving them from the realm of the super-professional to the territory of the locality. The role of the contemporary conservationist or urban designer as a professional expert is being superseded by new processes and initiatives, wherein conservation practice is now becoming the domain of ordinary citizens and activists, as well as numerous other actors that did not traditionally play a role on the heritage conservation stage (Manzini 2015). Thus, while the aspirations outlined in this discussion might seem universal, their implementation may occur through numerous alternative approaches, agencies, and processes, that could serve to both challenge long-held practice stereotypes and expand the urban design-heritage conservation discourse in even more expansive and inclusive territories.

References

Banerjee, T. and Loukaitou-Sideris, A. (Eds.) (2011). *Companion to Urban Design*, Milton Park, Abingdon: Routledge.

Bharne, V. (2011). "The Qanats in Yazd: The Dilemmas of Sustainability and Conservation," In Sang, L. (ed.) *Aesthetics of Sustainable Architecture*, Rotterdam: 010 Publishers.

Bharne, V. (2013). "The Paradise Between Two Worlds: Rereading Taj Mahal and Its Environs," In Bharne, V. (ed.) *The Emerging Asian City: Concomitant Urbanities & Urbanisms*, Milton Park, Abingdon: Routledge, 36–45.

Bharne, V. (2014). *Zen Spaces & Neon Places: Reflections on Japanese Architecture and Urbanism*, San Francisco: AR+D.

Bharne, V. and Sandmeier, T. (eds.) (2019). *Companion to Global Heritage Conservation*, Abingdon, UK: Routledge.

Castells, M. (2010). *The Rise of the Network Society* (2nd ed.), Chichester, West Sussex: Wiley-Blackwell.

Haeri, M.R. (2003). *Kariz (Qanat): An Eternal Friendly System for Harvesting Groundwater*, http://unfccc.int/files/meetings/workshops/other_meetings/application/pdf/121103_iran.pdf; accessed January 2011.

Hawken, P. (2010). *The Ecology of Commerce: A Declaration of Sustainability* (revised ed.), New York: Harper Business.

Manzini, E. (2015). *Design, When Everybody Designs: An Introduction to Design for Social Innovation.* Translated by Rachel Coad, Cambridge, MA: MIT Press.

Nitschke, G. (2007). *Japanese Gardens: Right Angle & Natural Form*, Köln: Taschen.

Soja, E. (1996). *Thirdspace: Journeys to Los Angeles and Other Real-and-Imagined Places*, Cambridge, MA: Blackwell Publishers.

Spivak, G. (1999). *A Critique of Postcolonial Reason: Toward a History of the Vanishing Present*, Cambridge, MA: Harvard University Press.

Further reading

Bharne, V. (2013). *The Emerging Asian City: Concomitant Urbanities & Urbanisms*, London: Routledge. An edited volume examining the multifarious forces, social, political, cultural, and historic, that shape the traditional and contemporary urban environments of Asia today.

Bharne, V. (2014). *Zen Spaces & Neon Places: Reflections on Japanese Architecture and Urbanism*, San Francisco: AR+D. A critical examination of the Japanese built environment from the historic to the present day – including an analysis of the Ise Shrines and the dilemmas surrounding their present and future.

Koch, E. (2006). *The Complete Taj Mahal and the Riverfront Gardens of Agra*, London: Thames & Hudson. A comprehensive examination of the Taj Mahal as an architectural monument and urban place across space and time.

36

HEART OF SHENZHEN

The movement to preserve "Ancient" Hubei Village

Mary Ann O'Donnell

When I want hometown taste, I go to Dongmen.

Lyrics from "Chao-Shan Boy in Shenzhen" by Pan Qionglin

Introduction

In the spring of 2016, a group of urban planners, architects, and public intellectuals started to discuss what they could do to preserve a section of the historical Shenzhen Market as a heritage site. The preservationists called themselves "Hubei 120." The name referred to Hubei Village, the surviving section of the historical Shenzhen Market, and to Shenzhen's emergency phone number, 120. Hubei 120 aimed to do something that many thought impossible – convince the Luohu Department of Urban Planning that it should require the developer, China Resources Real Estate, to submit a new redevelopment plan, one that would preserve the city's "heart (*xin wozi*)." The preservationists argued that given its role as the epicenter of early Reform and Opening, the historical Shenzhen Market was a public resource and should be commemorated as such (Figure 36.1).

Before Hubei 120 sounded the call to preserve Hubei, many Shenzheners, including the Luohu government and the developer, assumed that the historical Shenzhen Market had already been commemorated as the "Dongmen Commercial Area," which had been designated a historical area in preparation for the Return of Hong Kong to China (*huigui*).[1] In fact, when Hubei 120 organized in 2016, Shenzhen Market had already been part of the city's official identity for two decades. At stake in the preservationist movement wasn't whether or not the historical Shenzhen Market was the city's heart, but rather how this heart should be represented. Hubei 120 claimed that although the Dongmen restorations constituted accurate representations of preservation knowledge circa 1997, without actual historical buildings the commemoration of Shenzhen's history would be incomplete. They called to preserve a minimum area of 15,662.3 square meters, which included the layout and 78 historic buildings. However, they suggested that if the preservation area was expanded to 18,797.2 square meters, then sites built in the 1980s and 1990s could also be included. This larger area included the preservation of Luohu Cultural Park, one of the largest green spaces in the area. The inclusion of a park as a "historic site" was in keeping with Hubei 120's

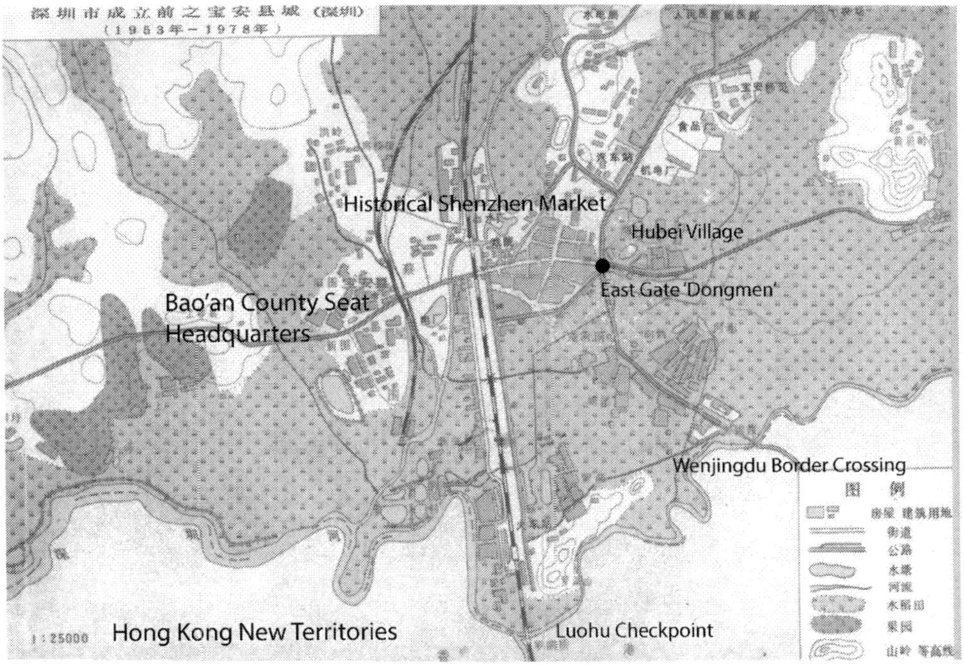

Figure 36.1 The "Map of Bao'an County Seat before the Establishment of Shenzhen City, 1953–1978" shows the location of the key sites that comprised "the Special Zone" during the 1980s and into the new millennium. Public source, Shenzhen Special Economic Zone 1985.

Source: Image adapted by the author

claim that historic areas were public resources and should be accessible by all Shenzheners (Figure 36.2, Figure 36.3).

The Hubei 120 movement revealed the fault line between locals and migrants that has not only informed the urban form of Shenzhen, but also the sense of belonging to the city and, in turn, the development of civil society. The original redevelopment plan had been developed in accordance with China's national heritage standards and traditional regional identities, recognizing the Hubei Zhang's claims by preserving and renovating the one superior (*youxiu*) building in the area – the Zhang Family Ancestral Hall. For the Hubei Zhangs, their identity as Shenzhen locals (*bendi ren*) was an ascribed status that they had been born into; the ancestral hall was the architectural representation of the family genealogy and the generations who had lived in the area for over 500 years. In contrast, for preservationists, their identity as Shenzheners was an achieved status that they had created by migrating to the Special Economic Zone (SEZ) – urban villages such as Old Hubei were important gateways (Bach 2017).

Ostensibly, the debate between preserving the entire village or a single ancestral hall was over migrant versus native identity as the truly "local" Shenzhen. However as we will see in this chapter, the Hubei 120 movement was part of a larger historical process that has dispossessed local villagers of their traditional identities and claims to the land; the Hubei Zhangs had little influence over the process. Moreover, even though migrant experience was foregrounded in the Hubei 120 articulation of Shenzhen history, migrant workers did not directly participate in Hubei 120 actions, except as audience for the actions of intellectuals. Consequently, at stake in the Hubei

Figure 36.2 Historical Shenzhen Market, circa 1990. This photograph hints at the original layout of market streets as well as the pace of urbanization.

Source: Photograph taken by the author at the Dongmen Pedestrian Street History Exhibition

Figure 36.3 Dongmen Pedestrian Street, 2010, one of Shenzhen's most popular tourist destinations.

Source: Photograph by the author

120 preservationist movement was not simply the idea of "come and you are a Shenzhener," but rather ongoing attempts to bring educated and rational voices to public conversations about the urban form of Shenzhen and the well-being of the city's 20 million residents.[2]

Hubei 120: the second generation of Shenzhen intellectuals

Not precisely top down, but not exactly bottom up, Hubei 120 represents what many Shenzhen intellectuals see as the future of civil society – an impartial, professional third party that speaks for the common good in dialogue with the government and business interests. Their activism can be understood within and against the changing roles of China's intelligentsia within a shifting public sphere.

In the post-Mao era, the marketization and professionalization of Chinese intellectuals created a lively, if relatively weak, public sphere. In turn, the plurality of ideas led to changing understandings of what an intellectual is and does, most famously culminating in the Tiananmen protest movement (Gu and Goldman 2004). The importance of Shenzhen in the post-Mao reconstitution of the public role of Chinese intellectuals has been twofold. On the one hand, the lesser-known Shekou Storm marked an important victory for young intellectuals and was a precursor to the 1989 protests (Xu 1995). On the other hand, many mark the start of post-Tiananmen intellectual activism with Deng Xiaoping's 1992 Southern Tour, when he visited and praised the city's level of market liberalization and openness to the outside world (Wright 2004). In 2016 the emergence of publicly active intellectuals within and against the ideological tightening under the leadership of Xi Jinping was significant, precisely because it set out to redefine the public sphere and create a viable public space for a third voice.

The first generation of Shenzhen's migrant intellectuals came to Shenzhen as part of the early movement to build the SEZ. They understood their role as intellectuals with respect to a national project to reform and open the Maoist state apparatus (Mason 2016). In addition to technical intellectuals, there were also sociologists and economists interested in economic reform, as well as journalists and writers, architects, and urban planners. This generation actively explored how privatizing sections of the state apparatus (such as media and cultural production) would, in turn, create conditions for a different kind of civil society. At Shenzhen University, for example, the Institute for the Study of SEZs, Hong Kong, Macao, and Taiwan hosted national conferences such as "What's Still Special about the Special Zone," while some of the most influential articles published by the *Shekou Bulletin* and the short-lived *Street Magazine* were exposés of conditions in factories and calls for transparency in business operations (Chen *et al* 1991). Shenzhen intellectuals encouraged migrant workers to write about their experiences, which they published in company newsletters and the state media. In each case, these interventions aimed to "oversee" the processes through which the Chinese state was being selectively dismantled and restructured in Shenzhen, bringing a third voice to the attention of state actors.

The first generation received local *hukou* when joining work units such as Shenzhen University or the Bureau of Urban Planning. Thus, unlike rural and unofficial migrants, many first-generation migrant intellectuals had access to subsidized housing, medical care, and school for their children. Nevertheless, the first generation did not identify emotionally with the city; they continued to identify with their childhood hometowns, joking that their children – the so-called "Shenzhen 2s" – were the "first" Shenzheners (O'Donnell 2006; forthcoming). In contrast, the second generation of Shenzhen intellectuals identified emotionally with the city. Crudely put, Hubei 120 intellectuals acted to preserve Old Hubei not only because it exemplified Chinese culture, but also because they loved the city; this "*xiangchou* (hometown nostalgia)" informed their sense of intellectual obligation and purpose.

As with the first generation of Shenzhen intellectuals, the decision of the Hubei 120 intellectuals to act on what they saw as the public good simultaneously articulated Shenzhen civil society and their right to speak *for* it. In this sense, the "public sphere" referred to a series of actions, in which intellectuals brought their interests and the interests of society to the attention of government officials and its appointed developers, the state-owned enterprise China Resources. The Shenzhen public sphere that Hubei 120 invoked was ironically predicated on the very inequality between "officialdom (*guanfang*)" and "common folk (*minjian*)" that had excluded many Hubei 120 intellectuals from participating in the drafting of the Shenzhen master plan. Habermas (1989) has emphasized that family institutions and property ownership comprised the threshold for participating in the early public sphere. In Shenzhen, intellectuals allied themselves to the state project of deepening reforms through education and the traditional role of Chinese intellectuals, justifying their actions because they "served the People (*wei Renmin fuwu*)."

Hubei 120 intellectuals aimed to widen public discourse about Shenzhen's urban form beyond the voices of the government and developers. With respect to urban planning in general and the redevelopment of Old Hubei specifically, their actions constituted a public sphere comprised of the government, the developer, and intellectuals. These three actors were in conversation to determine the public good, even as each actor represented a different aspect of it: The government was responsible for social order, realized as the Master Plan. The developer was responsible for material resources to realize the plan as a specific place – a mega multiplex comprising a shopping mall, office buildings, and residential towers. Hubei 120 asserted a "watchdog" responsibility, reminding both the government and the developer about the weaknesses of the plan and its implementation in order to make Shenzhen even better. In this sense, it is notable that the Hubei 120 intellectuals did not see themselves as opposing the government and developer, but rather working with them to craft better plans and build a better city.

Pedagogical activism: extending the Shenzhen public sphere

Hubei 120 participants were all college educated, many had advanced degrees from prestigious universities both at home and abroad, and some key figures were professors at Shenzhen University's School of Architecture. They had connections to the municipal and district governments, allowing them to have their opinions and suggestions heard by sympathetic government actors. They believed it was necessary to leave their classrooms, studios, and offices in order to educate both government officials and the public about the history of Shenzhen and how urban planning ought to serve a more comprehensive understanding of the city, its responsibilities, and potentials. Their strategy might be called "pedagogical activism" because, through the strategic use of "classrooms," Hubei 120 aimed to influence Shenzhen's urban plan.

The primary "classroom" was the Position (*Youfang*) Studio, where Hubei 120 organized its first public event, a two-day workshop held in July 2016. Position had opened an office in Shenzhen in 2016 with the aim to "rebuild the city and the public sphere." During the first morning of the workshop, over 100 citizens learned about Hubei's 500-year history, its connection to the historical Shenzhen Market, and the importance of its living traditions. To connect present-day Hubei to China's urban traditions, Hubei 120 pointed to the narrow alleys of Old Hubei, its intense arrangement of shrines and paper blessings, and its bustling commercial corners.

Architecture professor Yang Xiaochun presented the results of an architectural survey suggesting that the movement to preserve Hubei had been simmering for several years. She argued that Hubei was important not only because it contained 78 historic and traditional buildings, but also because development had not distorted its layout of "three vertical streets and eight horizontal roads (*san zong ba heng*) (Figure 36.4)." This layout, she maintained, was unique to Guangdong

Figure 36.4 Map of Old Hubei Village and Location of Historic Buildings. This rendering shows the location of the 78 historic buildings as well as the "Three Verticals, Eight Horizontals" layout (solid brown line). Building #16 is the Zhang Family Ancestral Hall.

Source: Image used courtesy of Professor Yang Xiaochun

villages, and thus the entire footprint of Ancient Hubei needed to be preserved. Some of Shenzhen's most famous architects also led working groups in generating alternative development plans for the Hubei area, and encouraging guided discussion about urban planning, history, and spatial thinking. Additionally, eight urban planners and architects worked with workshop participants to develop alternative urban plans for the Hubei area that were discussed during the final workshop.

On July 25, 2016, Hubei 120 submitted *Hubei Calls for Rescue: "The Hubei 120 Public Urban Plan" Petition* to the Shenzhen Central Arts Commission and the Luohu District Urban Renewal Office, calling on the government to stop China Resources from demolishing Hubei Ancient Village and replacing it with high-rise residences, malls, and office space. The petition was the outcome of ideas generated during the workshop and included an alternative development plan, an analysis of the historic buildings that were located within the village, and images of everyday life.

Sham Chun: from walled cities to markets

In 1996, I interviewed a sociology professor from Shenzhen University. He maintained that whatever culture or history Shenzhen claimed was not representative of "authentic" Chinese culture and history. According to him, "culture (*wenhua*)" glossed imperial history and concomitant rural feudalism in China's central plains. He suggested visiting Xi'an, Beijing, and Kaifeng to get a sense of Chinese culture, historically speaking, Shenzhen was a small market town of no significance. This 1996 interview was conducted against the backdrop of preparations for the 1997 Return of Hong Kong. The British had carved the colony of Hong Kong out of Xin'an County, the territorial precursor to both Shenzhen and Hong Kong. As part of those preparations, Nantou was being restored as a walking museum that celebrated the reunification of the historic Xin'an County, even though the majority of Shenzhen intellectuals thought that Nantou was an inferior example of Chinese culture. Unsurprisingly, that iteration of "local" Shenzhen identity failed to resonate with migrants (O'Donnell 2001, 1999).

Hubei 120 preservationists argued for the primacy of Shenzhen Market – rather than Nantou – as the city's heart. First mentioned in Chinese documents in 1688, Shenzhen Market was large enough to constitute a landmark on the map that was included in the 1819 *Xin'an County Gazetteer* (Figure 36.5). The market was a walled compound named for the river that flowed just beyond its western gate. The name of the river used "*chun*," a Hakka term for "the ditch between fields," suggesting that the market arose when Hakka families entered Xin'an County after the Coastal Evacuation Edict was rescinded in 1669. According to missionary reports, there had once been several markets in the Shenzhen River valley, but by the 19th century, "Sham Chun" Market dominated the region (Hase 1990). During his lecture at the Hubei 120 workshop, local historian Liao Honglei emphasized that Hubei had a direct genealogical connection to Sham Chun. The historical Shenzhen Market was a compound situated in land controlled by Huangbeiling Zhangs. Their relatives, the Hubei Zhangs, held measurement rights over transactions conducted in the market, arbitrating disputes over weight and price with a designated scale.

In 1866, an unnamed Western cartographer produced a *Map of Sun-On-District* (Figure 36.6). The first comprehensive map of the region after the colonialization of Hong Kong (1842) and Kowloon (1860), this map provided accurate information about relative distances between towns and villages, the topography of the territory, and the location of navigable rivers. Place names were provided in Chinese characters and Western romanization of local pronunciations. Shenzhen appears as "Sham Chun," the pronunciation of Shenzhen in local dialect (Ng 1983). Like the 1819 *Gazetteer* map of Xin'an County, the 1866 *Map of Sun-On-District* simplified the complexities of Xin'an County society in order to facilitate the actions of non-locals – imperial

Figure 36.5 Gazetteer map of Xin'an County, 1819. The largest walled city on the map is "Xin'an County Seat" on the Nantou Peninsula.

Source: Public document copied from Xin'an County Gazetteer

Figure 36.6 Everyday life in Old Hubei – children playing in an alleyway.

Source: Image used courtesy of Peng Xin

magistrates and colonial merchants, respectively. Comparison of the two maps also shows that the colonial incursion of the British exacerbated on-the-ground tensions between western (*Punti*) and eastern (Hakka) towns and villages, which have continued to inform Shenzhen's vernacular geography. This history of successive occupations has been a constitutive tension in the region, making the construction of the "local" a continuously vexed project (Siu and Faure 1995).

Hubei 120 appropriated aspects of this history, not simply to claim that Dongmen was the Heart of Shenzhen, but to specify its physical location. During the Direction workshop, Professor Yang used an annotated version of the 1866 *Map of Sun-On-District* to achieve several aims. First, she accepted that scientific accuracy was the necessary precondition for civic discourse. Her own work mapping the historic buildings and footprint of Ancient Hubei, like the 1866 *Map of Sun-On-District*, organized spatial information according to cartography. Second, she anachronistically included the Shenzhen-Hong Kong border, which was not drawn until 1898 or 32 years after the *Map of Sun-On-District* was drawn. Third, she highlighted the locations of the two walled cities, Nantou and Dapeng, as well as four market towns – Shajing, Guanlan, Shatoujiao, and "Dongmen," equating the cultural value of imperial cities and local markets. The appearance of these four markets on Yang's annotated version of the 1866 *Map of Sun-On-District* suggests the extent to which the contemporary celebration of Shenzhen's informal market culture informed Hubei 120's historical understanding. During the 1990s, Shenzhen intellectuals had dismissed local market culture with respect to China's imperial past. Two decades later, Shenzhen intellectuals accepted the area's markets as a cultural root. What's more, these four markets were famously peripheral to the imperial culture of Nantou and Dapeng. During the 19th and early 20th centuries, Shajing was part of a stretch of land infamous as a haven for pirates (Lai 2004), Guanlan was known as an independent market that was essentially ungoverned and overrun by bandits (Krone 1967), Shatoujiao was a Hakka market (Hase 1995), and "Sham Chun" was a borderland upstart (O'Donnell 1999; 2001).

In retrospect, it is clear that Shenzhen Market's regional importance grew with Hong Kong's and the increasing importance of the Sino-British border to Chinese domestic politics (O'Donnell and Wan 2016). Local markets, Chinese heritage, and Western hegemony comprised historic strands that could be rewoven into new ways of belonging to the area (Siu and Faure 1995). By making this version of history available to officials and their appointed developer as well as the common folk, Hubei 120 aimed to specify what made Shenzhen "special," claiming that because markets and entrepreneurs were the origins of "local" identity, this identity should be commemorated through the preservation of Old Hubei.

Old Hubei: rediscovering the heart of Shenzhen

When Hubei 120 began organizing, there were two Hubei Village settlements: Old (*Jiu*) and New (*Xin*). Built during the 1990s, New Hubei was an orderly neighborhood comprising "handshake" and office buildings, a school, a clinic, and shops. The expression "handshake building" is a colloquialism used to describe six to eight story tenement buildings that have been built so closely together that it is possible to reach through one's window and "shake hands" with the neighbor. Handshakes are the distinguishing architectural typology of Shenzhen "urban villages," and best understood as the city's informal, working-class neighborhoods. In contrast, Old Hubei was a decrepit cluster of crowded and dank row houses, some of which dated from the late Qing and Republican periods. The walls were moldy, and the open gutters needed to be cleaned at least twice a day; during heavy rains, the area was subject to flooding. Migrants opportunistically retrofitted these residences with electrical connections and modern plumbing. They not only lived in the buildings but also used them as shops, warehouses, and workshops. Thus, when Hubei

120 claimed "ancient" status for "Old" Hubei Village, they also needed to transvalue the cultural importance of urban villages to Shenzhen identity.

In addition to holding "classes" and publishing articles about the history of Shenzhen Market and the importance of Old Hubei to that earlier settlement, Hubei 120 encouraged young Shenzhen residents, especially young intellectuals, to go to Hubei Old Village and discover the city for themselves. This (re)discovery of Shenzhen's history was pedagogical in both form and ethics. On July 2, 2016, for example, preservationist Meng Yan took architect Ruan Yisan on a tour of Old Hubei. Local media reported on the visit, publishing images of Shenzhen's most iconic architect telling China's most well-known preservationist about the late Qing and Nationalist features that made Old Hubei architecture distinctive. Throughout the tour, Meng Yan respectfully referred to Ruan Yisan as "teacher" and Ruan Yisan diligently listened to the story of Shenzhen's "heart" and why it needed to be protected. Indeed, these images emphasized that "lifelong" learning and teaching were simultaneously the proper roles of an intellectual and the form of an ethical life; Ruan Yisan's tour of Old Hubei not only gave legitimacy to preservationists' claims, but also modeled a proper intellectual life for younger generations.

The Hubei 120 workshop and concomitant petition to preserve Ancient Hubei Village ignited a powerful discussion throughout Shenzhen, including the "Everyone's Hubei" series of public art interventions. Over the second half of 2016, 46 individuals presented spatial analyses of the area, created site-specific artworks, and organized public interventions to pressure the Shenzhen government to preserve the area. The underlying logic of these interventions was an extension of Hubei 120's pedagogical activism. Architect Feng Guochuan brought middle-school students into Old Hubei and had them build models of the area, discussing the city's complicated needs both to develop its downtown area and preserve its history. Performance artist Shen Piji gave a classical zither concert in the ancestral hall, ending it with a "rock and roll" style destruction of his instrument. The transition from harmonious music to gratuitous destruction of the instrument called attention to the need for material sites to house or shelter the city's "heart."

Beyond the architecture of Old Hubei, however, the Shenzhen intellectuals did not emphasize the indigenous village life, but rather the informal entrepreneurialism of Chao-Shan migrants. Better known as Teochew people, outside China, the Chao-Shan diaspora transformed Southeast Asia, their entrepreneurial activities becoming synonymous with Overseas Chinese capitalism, especially in Hong Kong and Singapore (Li 2016). When the Special Zone was first opened to foreign capital, many of the city's earliest investors were overseas Chao-Shan. Locally, Chao-Shan entrepreneurs controlled Shenzhen's produce, seafood, electronics, smuggled goods, construction, and unofficial currency markets; they have been the demographically dominant residents of Old Hubei since the first years of Reform. Many of Shenzhen's most famous private companies grew out of this informal, but highly organized diaspora. As one Hubei 120 preservationist remarked on Chao-Shan success, "They grabbed both ends of the spectrum," not only confirming that there was no level of Shenzhen society that was not influenced by Chao-Shan people, but also acknowledging that influence as a source of contemporary Shenzhen's culture.

The influence of Chao-Shan entrepreneurialism on the "local" culture that Hubei 120 aimed to commemorate resulted from the demographic structure of early Shenzhen. In 1980, the population of Bao'an County was roughly 358,000, but the indigenous population of pre-reform Shenzhen market area was only about 45,000 (SHPO 2001). During the first few years of the Special Economic Zone's (SEZ) existence, government leaders mobilized the state apparatus in order to bring administrators, engineers, investors, and labor to the newly established SEZ. The most significant number of official migrants were the 20,000 members of the Engineering Corps, who were deployed from Liaoning, Tianjin, and Shanghai to build the city. Official or "employed" migrants numbered about 50,000 (SSY 2016). In contrast, during the roughly nine months (between 1980 and 1981) that Shenzhen's second Party Secretary, Wu Nansheng, was in office, he mobilized over 100,000 Chao-Shan compatriots to settle in the early SEZ with an eye to jumpstarting administration, manufacturing, and commerce (Wu 2015). Dongmen and Old Hubei were only two of the areas settled by Chao-Shan migrants. Other key areas included the electronics markets of Huaqiangbei, many of the township and village industrial parks in Longhua and Longgang Districts, and the fishing villages of Nan'ao in Dapeng District.[3]

As artists and intellectuals continued to explore Old Hubei, their lessons increasingly departed from the themes of "architectural preservation" and "local tradition," focusing instead on the role of urban villages and informal entrepreneurialism in the ongoing reproduction of Shenzhen's social ecosystem. Artists Kong Xiaoya and Da Yu, for example, held a "Fisherman's Banquet" on the raised floor of a demolished house. The awkwardness of this piece drew attention to the contradictions that have been built into the Shenzhen cityscape. On the one hand, the artists wore formal attire and ate their meal at a well-set table. Their clothing and the starched white tablecloth set off the uniforms and working-class clothing of the residents of Old Hubei, calling attention to the class differences between the Hubei 120 intellectuals and the urban village residents. On the other hand, in addition to housing downtown Shenzhen's service workers and manual laborers, many Chao-Shan migrants had transformed Old Hubei row houses into warehouses for nearby hotels, supplying the area with food via commercial networks that extended throughout the province. Consequently, "Fisherman's Banquet" also highlighted the interdependence of Shenzhen's high-end formal areas and its informal, urban village settlements.

Others, such as photographer Peng Xin, went into Hubei to capture images of everyday life. Her simple but lush images include scenes of a mother taking care of an infant, students wearing the Shenzhen school uniform, meat venders at an informal shop, and a grandfather burning incense at a small shrine. Most of these "traditional" activities were in fact local to the Chao-Shan area. Other, equally moving images, included a garbage collector with a cigarette dangling from his mouth, a young girl doing homework outside her parents' shop, and an elderly couple holding hands as they walk through the street. Overwhelmingly, images of children growing up in Old Hubei Village emphasized Hubei 120's assertion that demolishing Shenzhen's past was tantamount to razing its future, and if that happened, where would the city be? (Figure 36.6).

Heart of Shenzhen: Shenzhen identity, urban villages, and limits to the public sphere

Hubei 120 introduced the idea of historic preservation as the city was preparing to celebrate its 40th anniversary. In newspapers and on the radio, in bookstores and lecture halls, people learned about the city's history, archaeology, and different migrations to the area. The emphasis on history allowed Hubei intellectuals to expand the question of urban villages from "affordable housing" to "urban identity." In turn, throughout the second half of 2016, Shenzhen newspapers, television stations, and independent news media used the preservation of Old Hubei and ongoing interventions to discuss the importance of urban villages as sites of Shenzhen's past, present, and possible future. After this flurry of activity passed, however, it was possible to evaluate the possibilities of and limits to the Shenzhen public sphere, circa 2016.

Urban villages have been an important site of informal articulations of Shenzhen identity and history. Hubei 120, for example, emerged out of ongoing public debate over the role of urban villages to Shenzhen's prosperity. Since the demolition of Gangxia (2009–2011), Shenzhen intellectuals have been increasingly aware of the value of urban villages for redevelopment, informal housing, and entrepreneurialism. What's more, as Shenzhen's downtown urban villages have been demolished, newspapers and social media have increasingly included accounts that emphasize the social value of urban villages. In China, the media is known as the "throat and tongue of the Communist Party (*Dang de houshe*)." Thus, media coverage about Old Hubei was understood to reflect state support for the issue specifically, and awareness of the importance of urban villages more generally. The clearest indication that there was government support for rethinking the meaning of urban villages came in February 2016, when the governing body of the Shenzhen-Hong Kong Bi-City Biennale of Architecture \ Urbanism (UABB) announced that the next Biennale would be held in an urban village.

In addition, many public and private media outlets worked with Hubei 120 to publicize the intellectuals' claims. Through their efforts to educate government officials and the general public, Hubei 120 used Shenzhen identity to momentarily detach the government's interest in redevelopment from those of the developers and village property owners, creating a space for preservationists and other intellectuals to act. Many read this flourishing of civic engagement as a sign that the Shenzhen government agreed with the preservationists' position that a third voice would enhance urban planning in the city, defining Shenzhen identity through a willingness to contribute to public debate. By December 2016, hundreds of ordinary residents had participated in events like the Hubei 120 workshop, and thousands had signed an online petition to save Ancient Hubei, with one notable exception – the Zhangs of Hubei Village. Their absence from this conversation was a visible reminder that the Shenzhen public sphere was predicated on the separation of officials from common folk. In comparison to Hubei's Chao-Shan migrant population, the Hubei Zhangs were relatively privileged. Nevertheless, in the articulation of the Shenzhen public sphere, the Hubei Zhangs were also spectators to the actions of intellectuals.

Moreover, the question of what will happen to Old Hubei's residents and shopkeepers once the plan has been finalized continues to haunt the project. After all, when artists went into Old Hubei, their work highlighted the city's need for affordable and dignified housing options; renovating the village row houses will result in forced evictions. This process is graphically captured in the Chinese expression, "love the cage and change the birds (*teng long huan niao*)," which is often used to criticize preservationist projects that merely raise property values for the city's elite. As Hubei 120 participants were well aware, the extent that Shenzhen can remain an immigrant city, where it is possible to proactively become a Shenzhener through one's own labor, hinges on how this broader question is answered. Indeed, many who participated in Hubei 120 were arguably less interested in preserving the old row houses than in setting a precedent for bringing independent voices into the development of future urban plans. More immediately, many hoped that the Hubei 120 model could be redeployed to save other urban villages from demolition, most specifically Baishizhou, the largest urban village in downtown Shenzhen.

These problems notwithstanding, the preservationists' success shocked many in Shenzhen and elsewhere. No one expected that they could push back against the Luohu government and China Resources. Consequently, Hubei 120's tactics for pushing forward their agenda and the extent of their success tells much about the norms and forms of civil society in Shenzhen, circa 2017. Hubei 120 members were architects, engineers, urban planners, and academics. Their vision of civil society was based on professional standing and technocratic proficiency, and many understood their role to be an impartial "third party," which could evaluate proposals for urban change from the disinterested perspective of knowledge, providing a counterweight to the biased perspectives of political expediency and economic necessity. Their strategy of pedagogical activism assumed the need for informed public discourse to move the city forward. Shenzhen's civic intellectuals reworked the city's history of professional and entrepreneurial migration to create a role for public intellectuals. Importantly, this understanding of the public sphere also found support in government halls and popular forums. That said, even as the successful preservation of Hubei village revealed the intersection of civic engagement and Shenzhen identity, it also suggested current limits on a more inclusive public sphere.

Acknowledgments

I would like to thank volume editors Anastasia Loukaitou-Sideris and Tridib Banerjee for their careful reading of this chapter and helpful comments. Jesse Fields and Nick Smith both provided insightful comments that have made this a better chapter. I am also indebted to members of Hubei 120 who shared their time and passion with me, especially Professor Yang Xiaochun and her hardworking and generous graduate students.

Notes

1 In English, the transfer of Hong Kong from British to Chinese authority is known as the "Handover," while in Mandarin it is known as the "Return."

2 Population figure announced by then Shenzhen Party Secretary Ma Xingrui during his state-of-the-city address, 2016. This is the "administration population," which estimates all residents. In contrast, the "official population" of roughly 15 million people refers only to people with legal registration to be in the city.

3 These three sites – township and village enterprise (TVE) industrial parks, the Huaqiangbei electronics market, and the deep fishing industry centered on Nan'ao in eastern Shenzhen – have been central to the formation of Shenzhen identity. The TVE industrial parks made Shenzhen "the factory of the world" during the 1990s and 2000's; Huaqiangbei pirated or "*shanzhai*" cell phones have made Shenzhen the "Silicon Valley of hardware," and the presence of working fishing villages on Shenzhen's eastern coastline continues to support the net-to-riches myth that "Shenzhen used to be a fishing village." Chao-Shan migrants have dominated the informal development of all three areas.

References

Bach, J. (2017) "'They Come in Peasants and Leave Citizens': Urban Villages and the Making of Shenzhen," in O'Donnell, M.A., Wong, W. and Bach, J. (eds.), *Learning from Shenzhen: China's Post-Mao Experiment from Special Zone to Model City*, pp. 138–170, Chicago: University of Chicago Press.

Chen, B. A., Hu, G. and Liang, Z. S. (1991) *Shenzhen de Sifenkesi zhi Mi [The Secret of the Shenzhen Sphinx]*, Shenzhen: Haitian Publishing.

Gu, E. and Goldman, M. (2004) "Introduction: The Transformation of the Relationship Between Chinese Intellectuals and the State," in Gu, E. and Goldman, M. (eds.), *Chinese Intellectuals Between State and Market*, pp. 1–18, London: RoutledgeCurzon.

Habermas, J. (1989) *The Structural Transformation of the Public Sphere: An Inquiry into a Category of Bourgeois Society*, Cambridge, MA: MIT Press.

Hase, P. H. (1990) "A Village War in Sham Chun," *Journal of the Royal Asiatic Society Hong Branch* 30: 265–281.

Hase, P. H. (1995) "The Alliance of Ten: Settlement and Politics in the Sha Tau Kok Area," in Siu, H.F. and Faure, D. (eds.), *Down to Earth: The Territorial Bond in South China*, pp. 161–187, Stanford: Stanford University Press.

Krone (1967) "A Notice of the Sanon District 新安县, Read before the Royal Asiatic Society Hong Kong Branch, February 24, 1858," *Journal of the Royal Asiatic Society Hong Branch* 7: 104–137.

Lai, W. J. (2004) *Shajing Jiyi [Shajing Memory]*, Hong Kong: China Review Academic Publishers.

Li, H. X. (2016) *Chaoshan Huaqiao Shi [A History of Chao-Shan Overseas Chinese]*, Guangzhou: Jinan University Press.

Mason, K. A. (2016) *Infectious Change: Reinventing Chinese Public Health after an Epidemic*, Redwood City: Stanford University Press.

Ng, P. Y. L. (1983) *New Peace Gazetteer: A Chinese Gazetteer of the Hong Kong Region*, Hong Kong: Hong Kong University Press.

O'Donnell, M. A. (1999) "Path Breaking: Constructing Gendered Nationalism in the Shenzhen Special Economic Zone," *Positions: East Asian Cultures Critique* 7(2): 343–375.

O'Donnell, M. A. (2001) "Becoming Hong Kong, Razing Baoan, Preserving Xin'an: An Ethnographic Account of Urbanization in the Shenzhen Special Economic Zone," *Cultural Studies* 15(3/4): 419–443.

O'Donnell, M. A. (2006) "What the Fox Might Have Said About Inhabiting Shenzhen: The Ambiguous Possibilities of Social- and Self-Transformation in Late Socialist Worlds," *TDR: The Drama Review* 50(4) (T192): 96–119.

O'Donnell, M. A. (forthcoming) "The Handshake 302 Village Hack Residency: Shenzhen, Chicago and the Experience of Assimilation."

O'Donnell, M. A. and Wan, Y. (2016) "Shen Kong: Cui Bono?" in Bolchover, J. and Hasdell, P. (eds.), *Border Ecologies: Hong Kong's Mainland Frontier*, pp. 35–47, Basel: Birkhauser.

Shenzhen History Publication Office [SHPO] (2001) *Shenzhenshi Dashi Ji 1979–2000 [Shenzhen Municipality Great Events 1979–2000]*. Shenzhen: Haitian Publishing.

Shenzhen Statistics Bureau & NBS Survey Office in Shenzhen (2016) *Shenzhen Statistical Yearbook [SSY]*, Beijing: China Statistics Press.

Siu, H. F. and Faure, D., (eds.) (1995) *Down to Earth: The Territorial Bond in South China*, Redwood City: Stanford University Press.

Wright, T. (2004) "Intellectuals and the Politics of Protest," in Gu, E. and Goldman, M. (eds.), *Chinese Intellectuals between State and Market*, pp. 158–180, London: RoutledgeCurzon.

Wu, Y. Q. (2015) "Chaoshan Ren Zai Shenzhen: Shanyu Baowo Shangye Faze, Geng Jingwei Shangye Faze [Chao-Shan People in Shenzhen: Talented at Knowing Commercial Law, Even Better at Respecting Commercial Law]," *chaoren.com*. www.chaoren.com/news/people/news/2015-12-21/104045.html.

Xu, L. (1995) "The Shekou Storm: Changes in the Mentality of Chinese Youth Prior to Tiananmen," *The China Quarterly* 142: 541–572.

37

PLANNING MODERNITY IN KABUL

Urban design as developmental politics

Pietro Calogero

Introduction

In this chapter I argue that Afghan leaders have used urban design in Kabul to express specific conceptions of modernity, as a way to promote the legitimacy of their rule. Previously, rulers from Alexander the Great to Ziaruddin Babur had used urban design to express their power. However, in the 20th century, Afghan leaders began to use formal elements that did not simply represent grandeur or authority; but also a very specific project of modernization. This project has been especially difficult for Afghanistan, because the national government has been violently overthrown eight times over the last century, often due to geopolitical interference. Thus, the shifting expressions of modernity in the built landscape of Afghanistan's capital represent a record of global urban modernity over the last 100 years. This chapter describes these urban design interventions in the city chronologically.

Urban design as modernization (jadidism), 1919–1929

My discussion begins with King Amanullah's efforts at urban modernization. In 1919, Amanullah fought the Third Anglo-Afghan War with British India and gained full independence. In 1921, formalized relations with the USSR, who saw him as the "Red Prince" – a monarch, who nevertheless was bringing his people towards the modern condition in which communism would ultimately prevail (Shah 1939). Amanullah also closely followed events in Turkey after the collapse of the Ottoman Empire. His wife, Queen Soraya, was the daughter of Mahmud Tarzi, the leading Afghan advocate for modernization at that time. Tarzi grew up in Istanbul in the late 1800s and brought back a substantial understanding of the West from the perspective of the late-Ottoman society.

The influence of Tarzi and Turkish culture on Amanullah helps explain a contradiction. Although Amanullah resisted European colonial domination, he regarded the West as the model for modernization and used Beaux-Arts design as a symbolic vocabulary to declare that modernization was underway. One example was the reconstruction of the Mosque of the King of the Two Flaming Swords, from 1922 to 1925 (Figure 37.1). Its new design was an emulation of the Dolmabahçe Mesjid of Istanbul, expressing Afghanistan's solidarity with the Young Turk's conception of a non-Christian form of modernity.

Figure 37.1 Mosque of the King of the Two Flaming Swords. Rebuilt 1922–1925.

Source: Photo by the author

Amanullah's main project was to build a new, modern capital outside of the dense, mud-walled streets of the existing city. He commissioned the French urban designer André Godard to design grand, tree-lined boulevards emulating Haussmann's boulevards in Paris, and German architect Albert Harten to design a new palace of government (Dar ul-Aman Palace) that deliberately looked like the Reichstag in Berlin (Figure 37.2).

The new capital was to be called Kabul Jadid, which means "modern Kabul." Jadid became an important word for Muslims grappling with the sudden dominant power of the West. Jadidism came to mean modernization in the broader political, cultural, and techno-economic sense (Khalid 1998).[1] Most importantly, Amanullah used the urban space of Kabul to push modernity on the Afghan people, by requiring Western behavior in his new capital. Afghans were not allowed to wear customary clothes in public: men had to wear shirts, trousers, boots, and European hats. They also were required to behave in Western ways while in public. Author Roland Wild recounted an incident in which he and a fellow Westerner were gently reprimanded by a police officer for sitting on the grass in a new park; all people in the new public parks were now required to sit on the park benches (Wild 1932). The police officers explained that the Westerners were setting a poor example of Western behavior. The purpose of these behavioral enforcements was to modernize the Afghan people through behavioral norms and the experience

Figure 37.2 Dar ul-Aman Palace at the end of Dar ul-Aman Avenue, 2012.

Source: Photo by the author

of a modern urban space. Therefore, the design of urban space was viewed as the medium for Afghans to adopt and assimilate modernity.

The design of Dar ul-Aman Boulevard, which linked the new government center back to the old urban core, was simultaneously a major piece of infrastructure and a major political act. The boulevard was designed like a Parisian boulevard, with wide, tree-lined sidewalks. When contrasted with the narrow streets in the old city, this 60-meter-wide (197 feet) street was clearly designed for more than circulation, but also for public strolling and shopping in a distinctly modern way. It was designed to support what Walter Benjamin (1999 [1935]) described as the Parisian flânerie in new spaces of modern consumerism. The boulevard became an iconic act in the king's claim to modernity through urban design.

When viewed from both the central roadway and the original sidewalks, Dar ul-Aman Boulevard creates a long, one-point perspective vista of the grand government palace at the south end. The word "goal" in English, the Dari word hadaf, means both a physical destination and a policy objective; the Palace was positioned symbolically as the hadaf-e jadid, the purpose and justification of Amanullah's political agenda (Figure 37.3).

Although Amanullah equated modernity with Europe, modernity did not simply originate in Europe (Chakrabarty 2000). The axial boulevard-and-palace design resembled Edward Lutyens's 1913 design of the Central Vista and King's Way for New Delhi; but two of the greatest inspirations for its urban design were actually local and non-Western: the Maidan-e Shah in Esfahan and the Mughal designs in Delhi, Agra, and Fatehpur Sikri (Kostof 1991). However, in 1923, Amanullah's commissioning of a French and a German to design this specific form of a modern Kabul was a message that modernization, Westernization, and strong political power were intertwined (Wild 1932). This was the urban spatial expression of an aspiration, but not a reality: Amanullah overextended his treasury, and in the era of global financial destabilization in the late 1920s, his regime collapsed (Shah 1939).

Figure 37.3 Left: Dar ul-Aman Palace, originally designed as the Palace of Government by Albert Harten, 1923. Dar ul-Aman Avenue is the major street extending to the right (northeast) across the image.

Source: Google Earth/Digital Globe

The Musahiban monarchy that re-stabilized the country in 1930 officially abandoned Amanullah's aggressive modernization project but quietly continued it with expansion of education, the conversion of royal gardens into public parks, and royal land into government property. Officially, the Kabul Jadid plan was abandoned, but in fact, the land use controls remained in effect and, years later, enabled the Karzai government to start relocating national ministries to the area around the Dar ul-Aman Palace after 2005. More importantly, from that time onward, Afghan urban planning was aware of, and linked to, transnational urban planning concepts and practices.[2]

Urban design as developmentalism: 1949–1962

From 1949 to 1953, Kabul's Mayor, Ghulam Farhad, used urban renewal to modernize the old core city of Kabul (Nemat 1976). He cut two avenues through the old city to provide vehicular access, promote modern commercial development, and prevent the old city from becoming a slum (Figure 37.4). Urban renewal in Kabul was used during the same and for the same reasons as in American cities; modernization was generally considered an improvement, and a way to promote economic development (AlSayyad 1992). Westerners tend to react with horror when they hear that the creation of these two avenues demolished about one fifth of the old city; but Afghans still regard much of the remaining old city as a slum. In his description of the creation of these avenues, Abdul Khaliq Nemat expressed the city government's concern with center-city urban decline (Nemat 1976). Like the Napoleon-Haussmann strategy in Paris, the creation of the new avenues in old Kabul included expropriation, demolition, and construction of new commercial buildings (Ibid.). The strategy succeeded in retaining core distribution and retail activities in the Old City. However, the residential neighborhoods behind these new frontages continued to become poorer. Starting in the 1930s, most families with any significant income moved out of the Old City. By the 1970s, the original urban core was referred to as shahr-e kona. Shahr means "city," but kona means "old" in a specifically negative way. Kona is used to refer to a "dog-eared book" or a used car; in urban terms, it marks out an area as decrepit and pre-modern.

Figure 37.4 Old Kabul (shahr-e kona) in 2009. Maiwand Avenue runs east-west through the long axis of the urban core, and Nader Pashtun Avenue runs north-south, intersecting at the circular plaza at the center of the image. Almost the entire northwestern quadrant of the old city was rebuilt as a modern commercial quarter after 1949.

Source: Google Earth/Digital Globe

The city of socialist man: comprehensive master plans, 1962–1992

The linkage between urban design and development became even more explicit after Afghanistan signed a technical cooperation agreement with Eastern Bloc countries in 1962. Escobar (1995) argues that the political project of developmentalism was driven by the US/USSR global competition for dominance. This was keenly felt in Afghanistan, where the king's cousin, Daoud Khan, cultivated relationships with both superpowers through development projects. In the 1950s and 1960s, the United States built the Kajaki Dam and developed Lashkar Gah as part of the Helmand Valley Authority project, which was named and modeled after the New-Deal era Tennessee Valley Authority (Ansary 2002).

While the United States focused on Afghan rural development, the Eastern Bloc countries focused on urban planning. Eastern Bloc planners developed a two-level urban planning system, with the whole urban area as a regional plan (Rayon) comprised of superblocks called microregions (mikrorayon). The design of microregions was a synthesis of the idea of integrated planning from Ebenezer Howard – who lectured in the USSR in the 1920s (French 1994) – and the car-free zielenbau plans developed in Germany and the Netherlands in the 1920s (Bauer 1934). Microregions were planned neighborhoods with shops, playgrounds, clinics, and district heating, where local streets only extended into the neighborhood to access parking areas adjacent to the apartment buildings. Soviet planners developed their prototype superblock design in 1957. Once deemed successful, the

Rayon/Mikrorayon system was incorporated into their national urban development program for the Seventh Soviet Five-Year Plan of 1959–1965 (French and Hamilton 1979). Soviets shared this system with Afghans, building the first microregion in Kabul in 1964. Microregion-1 has three- and four-story brick walk-up apartment buildings built in bars running east-west. Microregions 2, 3, and 4 (1974–1992) are also superblocks comprised of bar-shaped buildings, but these are made from precast concrete panels and range from four to six stories.

These microregion apartments in Kabul remain some of the most sought-after housing in the city because they are affordable and heated in winter. American planners reject superblock planning because in the United States, it is associated with public housing, top-down design, and crime. Jane Jacobs's critique of modernist planning (1961) is often misinterpreted as a categorical rejection of this design style. In fact, Jacobs's primary concern was with the paternalistic disregard of white planners for the needs of nonwhite residents (Jacobs 1961).

But the Afghan experience contrasts sharply with the US one. The Soviet-era president Dr. Muhammad Najibullah was one of the most effective presidents of Afghanistan and is fondly remembered in Kabul.[3] American planners may reject the paternalistic, autocratic approach of Robert Moses, but Afghans experienced far more stability and development under autocratic "Dr. Najib." For Kabulis in particular, the microregion apartment blocks stand as a reminder of better days in the past.

Market-led reconstruction: Kabul under the Bush/Karzai regime, 2002–2014

Soviet urban technical cooperation with Afghans continued up through the end of the Soviet occupation of the country in 1989. The Najibullah government continued to implement the Third Master Plan (Pekrosfki et al. 1978) including the construction of Microregion 4, up until the surrender of President Najib to US-backed mujahideen forces in April 1992 (Calogero 2011a). Once the government collapsed, construction was immediately abandoned. Because of the severe housing shortage in Kabul, families have occupied these incomplete buildings as well (Calogero 2011a).

Within one month of taking power, the mujahideen turned their weapons on each other within Kabul, inflicting serious harm upon the civilian population and the city's built environment. The Taliban restored order when they occupied Kabul in September 1996 (Rashid 2000). However, the collapse of the national economy, the Taliban's relative lack of interest in Kabul, and their ideological opposition to atheistic Soviet socialism meant the Taliban did not support the Soviet-trained planners in Kabul.

Once Americans overthrew the Taliban, Afghan planners thought they would finally be able to resume implementation of the Third Master Plan of 1978. To their dismay, they discovered that the Americans had the same anti-Soviet attitudes as the Taliban. When American forces occupied the city in December 2001, the United States began a "de-Sovietization" program. In Kabul, this was most clearly expressed in the USAID "Land Titling and Economic Restructuring Activity" (USAID 2009), and the overt declaration of Afghanistan as a market society in the US-supported Constitution of 2004 (Constitutional Loya Jirga 2004). As in the 1950s, US aid did not focus on urban development and planning but was overwhelmingly military. Non-military aid focused heavily on the rural sector and refugee repatriation back to rural areas. The

Provincial Reconstruction Team (PRT) program focused on rural schools, bridges, wells, and other rural infrastructure projects that were built by inter-agency teams led by the US Department of Defense (First Army Public Affairs 2007).

From 2002 to 2014, the United States essentially did not have an urban policy in Afghanistan other than the privatization of land. Implicitly, the US policy under Presidents Bush and Karzai was for "market-led reconstruction." Central planning was discouraged, including the 2005 dissolution of the Ministry of Planning – the national development ministry that had overseen the Five-Year Plans since the 1960s – and the liquidation of state-owned enterprises. For more than a decade, Soviet-educated planning engineers in the municipal government clashed with Western-educated planners in the Ministry of Urban Development and Housing (MUDH). MUDH planners worked closely with UN-Habitat, who promoted participatory planning and tried to block the implementation of the 1978 Master Plan (Calogero 2011a). In practice, participatory planning is a slow process, and that proved disastrous in Afghan cities in the first years of the 21st century

The decade-long stalemate over the resumption of master planning in Kabul coincided with a period of rapid urbanization. Though Americans and the Karzai regime wanted Afghan refugees in Pakistan and Iran to repatriate and return "home," in many parts of Afghanistan there was no home to return to. The Soviets had waged a scorched-earth campaign in the 1980s that left villages and irrigation systems destroyed and landmines in the fields. Meanwhile, families had grown and shifted, and many returnees had lost claim to their land (Turton and Marsden 2002). Furthermore, a severe four-year drought (1997–2001) had rendered some parts of rural Afghanistan uninhabitable. Therefore, "repatriation" meant only that families had reentered the political boundaries of Afghanistan and had therefore lost their internationally protected status as refugees. More than half of the returnees could not actually resettle in their ancestral towns and villages. The lack of space, security, and opportunity in rural areas has been an exceptionally strong "push" factor for Afghan urbanization since 2001.

A powerful "pull" factor was created inadvertently when Coalition forces made Kabul one of the safest spaces in Afghanistan. UK prime minister Tony Blair created an International Security Assistance Force (ISAF) to protect the new Afghan government and the returning diplomatic community. That security was not meant to benefit common Afghans, but it turned out to be a non-excludable public good for any household that could find a place to live within Kabul. Between December 2001 and May 2002, rents in the capital decupled because of the demand from both returnees and Western agencies seeking locations within the capital (Bearak 2002). Aid agencies and embassies rented houses in the center of the city for at least USD 5,000 cash per month, and in some cases multiple times that rent. Meanwhile, Afghan returnees lost their war refugee status and became merely internally displaced persons (IDPs) within their own country. While war refugees have many rights and protections under international treaties, there are no rights-claims associated with IDP status (Turton and Marsden 2002).

More than two million repatriated Afghan refugees fled the violence, poverty, and insecurity of rural Afghanistan and settled in Kabul. They did so in violation of the 1978 Master Plan, by building dense, irregular neighborhoods of clay-brick houses on the hillsides that had been designated as open spaces, and the fields that had been designated as agricultural reserve land. And since Kabul had no successor Master Plan, the city government had no guidance for in-migrating families. By 2004, about 75% of the population of Kabul was living in informal settlements. The extensive informal settlements of Kabul and other Afghan cities were the spatial product of a laissez-faire, "market-led reconstruction" policy.

Rights-based urban design: 2014–present

Since 2014, the influence of market-fundamentalist policy advocates has faded in Kabul. This seems to be due to several factors: (1) the idea of "market-led reconstruction" failed in practice; (2) the US-backed government turned out to be one of the most corrupt governments both in Afghanistan and the world (Transparency International 2017); and (3) the withdrawal of American military control was accompanied by the withdrawal of hundreds of aid programs. Fourth, the new president, Ashraf Ghani, seems to be more committed to rights-based policy rather than market-fundamentalist ideology. Unofficially, support for at least some socialist governance among the population has been indicated by the widespread popular display of the portrait of President Najibullah. Officially, the Ghani Administration has revived the reputation of Amanullah Shah, which was marginalized by the Musahiban family that succeeded him from 1930 to 1978. Today, portraits of Amanullah and Queen Soaraya are exhibited on billboards in Kabul.

A significant shift has taken place in the urban sector, as planners have acknowledged the value of continuity from previous planning practices. The Japan International Cooperation Agency (JICA) developed a Fourth Master Plan for Kabul in 2012, which acknowledges and builds on the previous three Master Plans. In March 2018, the urban design firm SASAKI completed an urban design framework for Kabul that builds on the previous four Master Plans (including the 2012 plan) as well as Amanullah's Kabul Jadid plan of 1923. The firm calls their plan an urban design framework – which is what President Ghani requested – but it embeds policy ideas ranging from regional planning to sustainability to the promotion of women's rights (SASAKI 2018). SASAKI was able to embed substantial policy within the plan because it was expressed in the visual-spatial language of physical planning.

SASAKI's strategy of policy-through-urban-design has important implications for Asian urban planning as a whole.[4] Furthermore, in Afghanistan, engineering is a widely recognized profession. Detail Plans (plan-e tafsili) were and are still understood as the civil-engineered implementation of Master Plans. Policy is certainly implemented through these plans, but it is implicit in the width of roads, the design of commercial areas, and distribution of schools, clinics, and mosques.

However, the politics of urban planning is now also becoming more publicly understood in Kabul. Under the new President, Ashraf Ghani, and the new Minister of Urban Development and Housing, Sadat Mansoori Noorzad, the Ministry was renamed back to Ministry of Urban Development and Housing in 2014. In 2015, Ministry staff expressed a revival of commitment to social housing through the design of a Detail Plan for a new Microregion 5 (MUDH 2016), which literally picks up where the construction of Microregion 4 halted in April 1992. Staff-level planners in the Ministry regard the Microregion 5 Detail Plan as their best recent work, as a superblock design with integrated social services and generous green space. As a Detail Plan, it is a declaration of national housing policy through urban design. However, this plan has not been adopted due to the rise of another way of understanding housing: as a right to be protected, rather than a technical problem to be solved. The site of the proposed Microregion 5 has been densely and informally settled over the past 20 years, and the construction of the microregion would require the eviction of more than a thousand very poor households. A more socially responsive approach to housing policy continues to be expressed through urban design: the upgrading of existing informal settlements.

The adoption of an upgrading policy is the clearest expression of rights-based planning through urban design in Afghanistan. In 2013, the Ministry and UN-Habitat started working on an Urban Upgrading Policy, to establish a regular set of procedures for improving the conditions of the informal settlements that now constitute the majority of the Afghan urban landscape. This policy was being developed in an institutional environment in which there were only two legal planning instruments: the comprehensive Master Plan, and the highly engineered Detail Plan. The Upgrading Directorate of the Ministry has been using a modified version of the Detail Plan to design the infrastructure, urban services, and land uses within informal areas. This has been problematic because Detail Plans for upgrading usually violate standards, especially the minimum street width of 12 meters (39 feet). But now, for informal areas, they are specifying roads as narrow as three meters (10 feet) and concretized paths and stairways as narrow as one meter (3.2 feet). In 2003, this formal approach to urban upgrading would have been unthinkable among the Ministry planners and engineers.

The most striking aspect of the Upgrading Policy is that it is a rights-based approach to urban design – legitimizing the rights and claims of households living in informal settlements. President Ghani has been issuing Occupancy Certificates to these households since 2016. The Upgrading Policy is designed to extend roads, drainage, water supply, and other urban services into these areas. Since some informal settlements are dense neighborhoods on extremely steep hillsides, upgrading might involve only structural stabilization and the construction of concrete stairways. The policy – and current practice of the Upgrading Directorate – is to design infrastructural extensions that balance improvement of living conditions with a minimum amount of demolition and resettlement, usually below 10%. Through physical planning and the creation of upgrading engineering plans, the Ministry is also formalizing the neighborhoods.

In essence, the physical extension of infrastructure is also a political extension of recognition. Urban design is the method by which the right to the city for half of the Afghan population is being secured (MUDH 2018). This is a vitally important aspect of urban design in the Global South: the very act of upgrading means that the urban regime of the city is recognizing the existence of that community. The creation of maps for the design of infrastructure also means that the settlement is included in the formal mapping and planning process of the city. In Afghanistan, the most consistent term for "informal settlement" is ghair-e plan, or "outside the plan." The process of surveying, designing streets and drains, negotiating with community leaders, and conducting social-demographic surveys are all ostensibly technical, but the outcome is that the status of the settlements shifts to being recognized as "within the plan."

Next order of business: Afghan sidewalks

Within the formally planned parts of Afghan cities, the market-led reconstruction policies of 2002–2014 resulted in substantial reconstruction and new construction of private houses and businesses, and the repaving of roads to serve those private investments. What has suffered is the space between: the sidewalks. Starting with Amanullah's Kabul Jadid plan in 1923, Afghan planners have allocated ample space for sidewalks in formally planned areas; often five meters including the planting-strip adjacent to the roadways. However, during the Karzai regime, this space was not protected and was encroached upon. Private landowners have been permitted to build driveways across sidewalks that occasionally drop 10 to 30 centimeters (4–12 inches), but more often are 30 to 70 centimeters (1–2.3 feet) above the grade of the sidewalk. Since many houses were shelled flat by the mujahideen from 1992 to 1996, new houses were often built on the rubble of the old, raising the base elevation within private lots by more than half a meter (1.6 feet). New driveways were simply built to link the new higher internal yards with the existing streets, with no regard to what that might do to the continuity of the sidewalk they crossed (Figure 37.5).

This is a class conflict. Driveways serve the wealthier minority of Afghans who own cars; sidewalks serve the pedestrians, disabled wheelchair-users, and informal merchants, who constitute

Figure 37.5 Sidewalks disrupted by private development in Kabul.

Source: Photo by the author

the majority of the urban population. Ria Hutabarat Lo (2010) and Annette Kim (2015) noticed similar class-based conflicts in Jakarta and Ho Chi Minh City, respectively. While wealthier classes try to enforce an idea of the "proper," segregated-use, modern city, most people use a variety of spaces in a contingent, diverse way. Kim (2012) coined the term "mixed-use sidewalk" to emphasize that the more informal, contingent ways of using sidewalks align with recent Western planning ideas of New Urbanism, transit-oriented development, and walkable cities.

Increasingly, planners within the Ministry and Kabul Municipality share this concern about the neglected condition of urban sidewalks. The head of the Housing Directorate, Sayedd Osman Sadat, argues that new urban development and regeneration of existing Afghan urban neighborhoods should follow the Compact City model, with smaller lot sizes and a focus on walkability. Whether this concern gains traction depends upon how the needs of poorer Afghans might get heard by elite policymakers. As Anastasia Loukaitou-Sideris and Renia Ehrenfeucht argue (2009), this is a core conflict over the meaning and use of public space. It is not clear whether progress will be made in this urban design arena, but the current shift towards rights-based urban design makes this much more likely in the near future.

Conclusion: Kabul as ordinary city, Kabul as extraordinary teacher

This chapter showed how the aspirations and claims of different actors and regimes in Kabul were expressed and materialized through urban planning and design. Indeed, the history of urban planning and urban design in the Afghan capital since 1919 bears witness to the long history of a global set of linkages. Because of its density, political fragmentation, and environmental stress, Kabul is dissimilar to North American cities, but actually quite representative of urban conditions for most cities in the world today. From a sustainability perspective, it is North American cities which are extremely low density and resource-intensive, whereas Kabul is in many ways normal; or, as Jennifer Robinson (2006) would argue, an Ordinary City.

The urban design lessons that can be learned from Kabul are therefore more applicable for the Global South and even the Global North than many lessons learned in North America. Ananya Roy (2005) has questioned the appropriateness of shifting from a "Chicago School" to a "Los Angeles School" of urban theory, because neither city is an appropriate representative of urbanism as a whole. I agree that no single city should be used as the basis for a generalized urban model because each has unique (and wonderful) idiosyncrasies; but Kabul is instructive of early-21st-century urbanism and urbanization of the Global South more than any city in the Global North.

The ordinariness of Kabul is also worth emphasizing, precisely because the city is presumed to be so abnormal. Kabul is officially a war zone; it is treated as a frontier in the fight against politicized Islam and terrorism; a site of failure in the American War on Terror; a place where some Americans accepted the idea that the United States is an empire (Roy 2006; Calogero 2011b); and a city where 15 years of concerted foreign aid efforts have failed to achieve peace or prosperity. Most of these feelings are valid concerns about contemporary American politics, but they run strong interference on any attempt to describe the mundane, sometimes common, and occasionally really distinctive traits of Kabul as an "actually existing" city. The general lessons of urban design we can all learn from this city are globally relevant. This is a core argument for this entire collection of essays.

Notes

1 Amanullah undertook many forms of modernization that not only focused on the built environment, but also included the adoption of a national constitution, the creation of a parliament, and the creation of an independent municipal government for Kabul (Gregorian 1969).
2 The current Director of Housing has reduced the typical lot size in new Detail Plans from a range of 300–450 square meters down to 200–300 square meters in order to promote compact, walkable cities (Sadat 2017).
3 Even in 2003, I noticed many merchants in central Kabul had put his picture up rather than the then president (Karzai).
4 Recently (2015–2017), I taught planning in Suzhou, China, and found that planning is thought of almost exclusively as physical planning, sitting directly between civil engineering and architecture. Nevertheless, implementation of socially beneficial urban policy *through* urban design is welcome and encouraged.

References

AlSayyad, N. (1992). *Forms of Dominance: On the Architecture and Urbanism of the Colonial Enterprise.* Aldershot; Brookfield, VT: Avebury.

Ansary, T. (2002). *West of Kabul, East of New York: An Afghan American Story.* New York: Farrar, Straus and Giroux.

Bauer, C. (1934). *Modern Housing.* Boston: Houghton Mifflin Company.

Bearak, B. (2002, May 14). "Kabul Journal: In the Afghan Capital, Rents Go through the Roof." *New York Times*.
Benjamin, W. and Tiedemann, R. (1999). "Paris, the Capital of the Nineteenth Century." In *The Arcades Project*. Cambridge, MA: Belknap Press., 3-13.
Calogero, P. (2011a). *Planning Kabul: The Politics of Urbanization in Afghanistan* (Ph.D. Thesis). University of California, Berkeley, CA.
Calogero, P. (2011b). "Kabul Cosmopolitan: Geopolitical Empire from the Planner's Viewpoint." *Planning Theory*, 10(1), 66–78. https://doi.org/10.1177/1473095210386105
Chakrabarty, D. (2000). *Provincializing Europe: Postcolonial Thought and Historical Difference*. Princeton: Princeton University Press.
Constitutional Loya Jirga. (2004). *Constitution of the Islamic Republic of Afghanistan*. Kabul: Government of Afghanistan.
Escobar, A. (1995). *Encountering Development: The Making and Unmaking of the Third World*. Princeton, NJ: Princeton University Press.
First Army Public Affairs. (2007, March 21). "First Army's 189th Infantry Brigade Trains Provincial Construction Teams for Afghanistan." *Army News Service*.
French, R. A. (1994). *Plans, Pragmatism and People: The Legacy of Soviet Planning for Today's Cities*. Pittsburgh, PA: University of Pittsburgh Press.
French, R. A. and Hamilton, F. E. I. (1979). *The Socialist City: Spatial Structure and Urban Policy*. New York: Wiley.
Gregorian, V. (1969). *The Emergence of Modern Afghanistan: Politics of Reform and Modernization, 1880–1946*. Redwood City, CA: Stanford University Press.
Hutabarat Lo, R. (2010). "The City as a Mirror: Transport, Land Use and Social Change in Jakarta. "*Urban Studies*, 47(3), 529–555. https://doi.org/10.1177/0042098009348557
Jacobs, J. (1961). *The Death and Life of Great American Cities*. New York: Vintage Books.
Khalid, A. (1998). *The Politics of Muslim Cultural Reform: Jadidism in Central Asia*. Berkeley: University of California Press.
Kim, A. M. (2012). "The Mixed-Use Sidewalk." *Journal of the American Planning Association*, 78(3), 225–238. https://doi.org/10.1080/01944363.2012.715504
Kim, A. M. (2015). *Sidewalk City*. Chicago: University of Chicago Press.
Kostof, S. (1991). *The City Shaped: Urban Patterns and Meanings through History*. Boston, MA: Little Brown and Co.
Loukaitou-Sideris, A. and Ehrenfeucht, R. (2009). *Sidewalks: Conflict and Negotiation over Public Space*. Cambridge, MA: MIT Press.
MUDH (Ministry of Urban Development and Housing). (2016). *Detailed Plan of Fifth Microregion (from microregion four to the airport)*. Unpublished report, Kabul, Afghanistan.
MUDH (Ministry of Urban Development and Housing). (2018). *Urban Upgrading and Renewal Policy for Unplanned Settlements*. Unpublished draft policy under review for adoption. Kabul, Afghanistan.
Nemat, A. K. (1976). *A Strategy for Restructuring of the Central Business District of Kabul* (Master's Thesis). School of Planning and Architecture, New Delhi.
Pekrosfki, E., Pazanov, R., Pekroskaya, V., Pasanova, E., Kartisva, M., Shujaee, G. H., and Asad, S. (1978). *The 1978 Master plan of Kabul City*. Kabul: Design and Implementation Department.
Rashid, A. (2000). *Taliban: Militant Islam, oil, and fundamentalism in Central Asia*. New Haven: Yale University Press.
Robinson, J. (2006). *Ordinary Cities: Between Modernity and Development*. London and New York: Routledge.
Roy, A. (2005). "Urban Informality: Toward an Epistemology of Planning." *Journal of the American Planning Association*, 71(2):147–158.
Roy, A. (2006). "Praxis in the Time of Empire." *Planning Theory*, 5: 7–29.
Sadat, S. O. (2017). Interview with the author.
SASAKI Associates. (2018). *Kabul City Urban Design Framework*. Boston, MA: Unpublished.
Shah, I. A. (1939). *Modern Afghanistan*. London: S. Low, Marston & Co., Ltd.
Transparency International. (2017). *Corruption Perceptions Index 2017*. Retrieved from www.transparency.org/news/feature/corruption_perceptions_index_2017
Turton, D. and Marsden, P. (2002). *Taking Refugees for a Ride? The Politics of Refugee Return to Afghanistan*. Kabul: AREU.
USAID. (2009). *Land Titling and Economic Restructuring Activity (LTERA)*. Retrieved December 29, 2017, from www.usaid.gov/node/52186
Wild, R. (1932). *Amanullah: Ex-King of Afghanistan* (1st ed.). London: Hurst & Blackett, Ltd.

Further reading

Hosseini, K. (2003). *The Kite Runner.* New York: Riverhead Books. This book was extremely popular in the United States a decade ago for the powerful story Hosseini writes. However, what most readers are not aware of is that it is a vivid portrayal of the topos of Kabul in the 1990s. In some cases, I could identify the part of the block in which events took place.

Wild, R. (1932). *Amanullah: Ex-King of Afghanistan* (First ed.). London: Hurst & Blackett, Ltd. Roland Wild was a journalist of his time: rather bigoted against "Oriental" peoples. However, his bias should be apparent to contemporary readers, which gives us a glimpse into Amanullah's modernizing project in spite of Wild's own cynicism about it.

Forbes, A. (1892). *The Afghan Wars, 1839–42 and 1878–80.* New York: C. Scribner's Sons. Curiously unlike Wild, Archibald Forbes was much less patronizing towards Afghans in his analysis of the first two Anglo-Afghan Wars. While describing the first conflict, in particular, he describes the spaces of the Afghan capital as well.

PART III.IV

Justice

Introduction

The idea of justice in urban design was formally introduced in the early 1980s by Kevin Lynch (1981) as part of his classic treatise, *A Theory of Good City Form*. In proposing the five essential performance characteristics of good city form – vitality, sense, fit, access, and control – he also suggested that efficiency and justice should be seen as the meta-criteria for achieving such performance outcomes. In explaining the application of justice as a meta-criterion towards achieving the five basic characteristics of good city form, Lynch drew from John Rawls's (1971) *A Theory of Justice*, especially in discussing the distributive inequality of all five performance criteria, including redressing such inequality by applying the "difference principle" argued by Rawls.

Lynch wasn't around to consider Rawls's (2001) subsequent concept of "justice as fairness," a more political reformulation of justice. While Lynch's discussion of justice in achieving good urban form remains somewhat in the normative realm, subsequent discussions focused on the processes and outcomes of obtaining a "just" city. Susan Fainstein (2010) emphasized justice as a political outcome, where community participation, and empowerment through participation, can help achieve it. Although Fainstein did not address the role of urban design specifically, this notion of bottom-up approach in the production of urban form is being emphasized in contemporary design literature, including in some of the chapters in this book and in this section in particular.

Of the two chapters included in this section, Bryce C. Lowery and Lisa A. Schweitzer expresses similar themes. Here the authors argue against the expertise-based, top-down urban design that is structurally removed from ground level experiences and deprivations of the minority and low-income groups. Instead, they call for participatory design, and for reconciliation of the market choice-based outcomes with the efforts of the more communal and collective enterprises. They argue that "co-production of the built environment" through participatory design would provide opportunities for promoting just and equitable communities.

The following chapter in this section by Kian Goh presents a novel conceptual approach to achieving justice through design. She considers justice through design in the context of large-scale urban development, especially as required to address resiliency and sustainability for a larger environmental outcome. Drawing from examples in Jakarta, Indonesia – particularly recent initiatives for greening the city and protecting future environmental threats of global warming and rising sea water requiring major infrastructure intervention – Goh finds opportunities for subaltern group participation and establishing claims on the new spatial arrangements. In this scenario, urban design works in tandem with political organization, advocacy, and engagement.

References

Fainstein, S. (2010). *Just City*. Ithaca, NY: Cornell University Press.
Lynch, K. (1981). *A Theory of Good City Form.* Cambridge, MA: MIT Press.
Rawls, J. (1971). *A Theory of Justice.* Cambridge, MA: Harvard University Press.
Rawls, J. (2001). *Justice as Fairness: A Restatement.* Cambridge, MA: The Belknap Press of Harvard University Press.

38
JUSTICE AND URBAN DESIGN

Bryce C. Lowery and Lisa A. Schweitzer

Introduction

Justice is a desirable goal for urban design, both as an outcome of its efforts as well as a source of its inspiration. Focusing design practice on justice raises difficult questions about who determines the form of the physical landscape, particularly as undesirable social values such as racism, income inequality, sexism, and elitism continue to influence the physical milieu as well as social life. In this chapter, we provide multiple perspectives on justice and relate these ideas to the conceptualization and act of design. The intent is to encourage continued vigilance of the ways that planners, landscape architects, and architects influence the opportunities afforded to users of space to engage meaningfully in the process of determining the future appearance, structure, and function of the landscape. Engaging communities to help shape the physical environment requires recognizing the ways expertise and power can serve to foster injustice and meaningfully engaging with users of space to foster shared learning about the role of design in alleviating past and future injustice.

We start by presenting a number of theoretical perspectives to illuminate the interactive relationship between social values, the built environment, and justice. The physical and spatial structure of the landscape is the result of contextual and global factors such as markets, the hopes and beliefs of individuals, the ideals and imagination of designers, and our collective beliefs about how communities should function and what neighborhoods should look like. Designers are confronted with important choices about how to design based on beliefs about two competing visions: (1) a laissez-faire approach that allows markets and individual choice to determine the structure of the landscape; and (2) engaged, participatory approaches that suggest a collaborative and didactic design focused on consultation, collaboration, and partnership with users of space.

We then present an historical account of design theory that suggests these ideas coincide with a movement away from expertise-driven visions of the "good city" and towards increasingly democratic ideals of how the built environment should appear and function. This transformation challenges design professionals to balance expertise-driven design strategies with locally derived social and cultural values that may or may not match perfectly with innovative techniques employed to generate more equitable, sustainable, and efficient places. Co-creation of the built environment may require recognition that the vision held by residents and the local community may conflict directly with the goals established by designers. Earlier strategies frequently relied on the designer to guide the design and appearance of urban form; but today design professionals are tasked with encouraging individuals to help imagine, co-create, and steward built environments in order to ensure these places sustainably help residents achieve a self-determined quality of life.

Perspectives on justice and the built environment

One way to view the influence of design on justice is through the perspective of a free market. Market here captures the everyday economic activities where people trade goods, services, and capital. People trade their labor for a wage, their money for goods they need, or they may barter goods for services. Planning for markets focuses attention on technological innovation and capital redistribution rather than local, collective decision-making and land use regulation (Gordon & Richardson 1997). Markets differentiate among people and their skills with different and unequal wages, and consumers differentiate among products according to their individual needs, preferences, and budget. Urbanism created to achieve market outcomes has some important concepts to bear in mind during discussions about design and justice.

Market-based approaches to design are often seen as a way of liberating control of the physical environment from the state, emphasizing individual liberty over group actions or control. Liberty here means individual agency and choice, and libertarians emphasize voluntary and market relationships over government institutions because the former allow for individual choice and agency to govern the outcomes. Libertarian urbanists stress that urban design and architecture should be products primarily of market outcomes (Richardson & Gordon 1993). If urban design enhances the quality of specific places, such as walkability, individuals and households who value walkability can choose to pay more for housing in those locations or move to another location where the mix of cost and associated amenities are more desirable (Tiebout 1956). If governments mandate walkable designs, and those increase the cost of housing, this will force residents to pay extra for something they do not have enough to pay for themselves. Just about all theorists of justice treat respect for individual liberty as an important aspect of any just system. The difference is that libertarians, aptly named, emphasize liberty over other potential concerns about justice such as equality, equity, and subalternity resulting from historical roots in subjugation, discrimination, and exploitation.

Design in this context is driven solely by individual willingness to support a particular set of physical attributes within a community. Markets allow individuals to act on their own values and preferences, and to the degree that they wish to have a designed neighborhood, they can create voluntary neighborhood associations to create one. Fundamentally, the libertarian critique of state planning and design comes down to information and coercion. Designers do not know as much about individual preferences as the individuals themselves do, and why, exactly, should groups use coercion to make individuals accept design or other goods they prefer less than other things? If societies let markets function and give individual liberty free reign, people will choose what matters to them, and however urban design gets distributed via those markets will be just because it has resulted from individuals selecting what they want.

An individualist perspective, however, may fail to recognize that coercion and restrictions on individual liberty come from many sources other than governments. In particular, poverty and economic deprivation can be coercive and self-reinforcing as well. Markets function in reality much differently from the ideal. Markets are frequently seen as generators of a particular type of landscape that promotes inequality and injustice rather than serving as a mechanism for ensuring fairness (Banerjee 1993). Some individuals have far more power than others, and they can readily use that power to coerce them. Information, as useful as it is, can also be severely limited or asymmetrical among people in markets, leading to power differences, limited choice, and perpetuating inequalities that inhibit justice. Design for the health, safety, and welfare of residents may be so fundamental to human life that societies may not want to leave some aspects of urban design to market distributions, even if that means impinging on individual choices. People who are impoverished deserve to have liberty and agency to choose, however limited their resources may be, rather than have governments or the market decide what they get.

The idea that at least some outcomes of urban design, such as pedestrian safety, are too important to leave to free-market distributions suggests that all neighborhoods and all places should have a baseline standard for design. Where to draw the line between what aspects of design are so important that they should be ubiquitous in the physical environment, and which ones might be left to individual choice, poses a thorny question. Philosopher John Rawls (1971) developed a theory of justice that offers a systematic way to parse what things may be fairly left to markets, and what goods should be provided for all. Rawls's entire theory is too big a topic to cover comprehensively here, but he contributed two key concepts that can help sort through how design might be more equitably distributed.

Rawls's (1971) perspective begins from what he calls "the original position" that seeks to change how we think about individuals' positions and "deservingness." He accomplishes this change through a thought experiment dubbed the "veil of ignorance." Rawls argues that rules for distributing goods like the physical environment would emerge via a deliberation where all of us must agree to the rules of distribution with no knowledge whatsoever about what personal characteristics we would have once subjected to those rules. Rawls argues that under such an arrangement, the consensus would promote rules that do the most for those with the least. In an urban design context, the veil of ignorance takes away – at least in theory – the idea that wealth will place one in an environmentally clean, aesthetically beautiful neighborhood, while the environmental and social burdens of supporting that lifestyle occur elsewhere, upon the impoverished. Random assignment for where one must live could, according to a Rawlsian approach, prompt societies to invest in a much higher standard for land use and urban design across urban geographies, so that nobody lives under conditions of material deprivation, poor environmental quality, and bad design.

Rawlsian standards for the minimum that society should provide create a potential baseline for sorting through which design "goods" may be distributed via the market and which should not. Rawls's standards of fairness potentially reinforce the necessity for an emphasis on public design and participatory urbanism. Markets may effectively respond to the needs of a wealthy family who want to own a private swimming pool and swing set for their own children to use, but all children, no matter what their social position or wealth, should have access to public places where they can exercise, play, experience nature, and develop in safety and security. Some people may be able to equip themselves with home air conditioning to ride out heat waves, where temperatures threaten health, but all people should have access to a physical environment that supports health and protection from disasters. Design is fundamental to achieving some minimum standards of city life.

Rawls's critics have pointed out some important limits to his approach. First, the approach is ahistorical. It is almost impossible to imagine any of us being able to neutrally evaluate the world given our existing experience. In some instances, it may require individuals to forgo certain privileges and comforts to achieve a minimum standard for others. His perspective also fails to recognize the important role of space, place, and geography in the manner that the physical environment is laid out. While equitable minimums might be a desirable goal, factors such as existing transportation and shipping corridors, entertainment and shopping amenities, and environmental conditions may make it challenging to achieve the same baseline across communities.

Amartya Sen provides another perspective on the possible conflicts that arise for designers seeking to address inequality through participatory design processes. In *Inequality Reexamined*, Sen (1992) asks the reader to consider two central questions about equality: Why equality? Equality of what? Sen suggests that the first question is inherent in the formulation of any normative beliefs about the outcomes of design and how social life should be arranged and organized. The second question, he suggests, must be defined by each individual based on that individual's own capabilities and dreams. Accordingly, we can argue that the built environment should offer each of us the freedom to decide our own future and maintain a reasonable level of well-being.

Martha Nussbaum (1988) elaborated on this concept, asserting that human dignity merits certain minimum capabilities that are necessary to ensure that every person flourishes. Germane to the design of the built environment, she asserts that in addition to health, integrity, thoughts, emotions, and reason, individuals should enjoy the capability to participate politically and control the environment (Nussbaum 2011). This approach suggests that designers focus on ways to integrate assets, amenities, and opportunities into the built environment. This can be challenging given the diverse nature of contemporary society with an equally diverse set of individual demands for control over the environment. The heterogeneity of demands suggests a complicated opportunity for design to work collaboratively with communities and users of space to collectively evaluate the demands for justice being placed upon the physical environment.

Edward Soja (2010) has argued that political theories of justice, from Plato onward to Rawls, do not help us fully understand cities or justice very well because they often emphasize institutions over geography and ignore spatial and social processes. Political theories treat spatial arrangements and the material form of the built environment as containers for people and social organization instead of active, structural components of human life, shaped by human action. Designers are encouraged to be sensitive to the ways their own work influences the arrangement of the physical environment and be supportive of community-level processes and efforts to address injustice (Soja 2010). Historical and contemporary site-specific physical and social processes can interact with design strategies to perpetuate locational disadvantage and foster spatial inequality.

Soja (2010) focuses on the social processes and power relationships that give rise to the form and arrangement of the urban landscape. Geography is not merely destiny, where institutions control and design the built environment and determine how humans live within neighborhoods. That is only part of the story. Communities facing domination from oppressive institutions that seek to control urban space find ways to build their resistance through self-expressive street art, modified streets and street use, and other practices that distinguish their places from the mainstream (Kim 2015). Institutions that direct and control the physical character of urban life contribute to a number of unplanned conditions that give rise to social, ecological, and economic vulnerability, necessitating new approaches to design that focus on redistributive justice, recognize the informality of urban life, and consider the social and cultural context (Roy 2005). What might be spatially just design is determined by who has the ability to make critical decisions about the spatial arrangement of neighborhoods and who does not; even then, communities will often find ways to design the environment with or without designers.

Soja (2010) does not suggest a pattern for just spatial outcomes other than to imply that justice can be obtained from local, empowered, community control over space and the design strategies that result from democratic participation. Soja's conception of space as an actor in shaping human life draws on and extends Henri Lefebvre's (1968) "right to city." Lefebvre's original conception of right to the city has undergone extensive discussion in urban theory, but its central tenet envisions a withering away of state institutions, including those that plan and design urban developments. In place of the state, Lefebvre envisioned a radical democratic equality among residents in the city, one that enables them to govern existing and new spaces together in an inclusive and just way. Designers participate in this vision by engaging in these discussions, allowing communities to also engage in the process of placemaking.

Local control over spaces, however, has recently clashed with contemporary controversies about justice and inclusion in cities. Urbanists arguing for building more affordable housing in urban neighborhoods face local residents, who often oppose such developments by asserting community control over the built environment. If community control is used to exclude other people from places that have good design, amenities, and public goods, then such a locally controlled "right" to the city simply favors neighborhood incumbents over newcomers. Some community advocates note that ignoring local preferences about new developments, even if done to favor new residents justified by sustainability or justice goals, simply empowers developers and city governments over residents. Favoring new development without consideration for the lifestyles, hopes, and dreams of existing residents may clear the way for gentrification that, in turn, erases the local norms and aesthetics.

Lefebvre's call for equality in governance suggests not a "new" right for local self-determination, but a radical equalizing of power to control neighborhood environments across geographies. Rights to control design and the built environment of neighborhoods, at least in a US context, are civil rights. Cultural theorist bell hooks (1990) describes "homeplace" and its importance to African American and other oppressed groups. For individuals in groups facing oppression and marginalization in city spaces, a homeplace provides respite from surveillance, violence, and stigmatization encountered in the rest of the city. Homeplaces, as hooks describes them, strongly suggest that justice in design supports individuals in creating, being in, and keeping places that express and reinforce identity through culturally relevant aesthetics and design.

A critical race perspective deepens other approaches to justice. Design occurs within structures of domination that need to be confronted and dismantled in order for designers to contribute to more spatially, socially, and economically just cities and neighborhoods. Abstract ideas about justice from political theory or philosophy that ignore white supremacy and colonization obscure how injustices structure social institutions and materially affect nonwhite people. Rights are always situated in histories and contexts. White supremacy and wealth have created systems of asymmetrical local control over neighborhoods in city politics, whereby affluent white neighborhoods can readily control their environments, keep out nuisances, and exclude others.

Places built and occupied by white and affluent residents have been celebrated for their appearance and used as the basis for safety and aesthetic standards. That ability to control neighborhoods, however, has not historically been extended to communities subjected historically to discriminatory redlining, destructive urban renewal policies, spatial segregation, and loss of access to opportunity. Residents of these communities have been denied residency in the neighborhoods of more powerful groups, and their contributions to street art and urban design have been variously denigrated and appropriated.

Charles Mills (1999) described a "racial contract" in political structures that illuminates the obstacles that white supremacy poses for promoting justice by design. Loosely argued, the racial contract mirrors and challenges the social contract tradition in political theories about justice and society. The social contract, most commonly associated with Thomas Hobbes, is a metaphorical description of the reasons why individuals allow some governmental control over their lives. Thinking about social contracts provides a metaphor for the agreements about conduct that govern the social life of a community or neighborhood. Mills challenged the idea that social contracts govern all social life. Instead, the metaphorical contract that governs social relations is a racial contract, one that hinges on governments systematically elevating and protecting white people and their interests at the expense of people of color. Carole Pateman (1988) presaged Mills's work with an exposition on how gender oppression factors into social contracts. The unspoken rules that govern daily life and everyday institutions create and maintain power and privilege for select groups and individuals over others. Awareness of those differences, and working to counter them, should influence designers both in their professional interactions with community members and with their design colleagues. Elevating the work of women and people of color in urban design is a necessary step towards justice.

One final approach to justice extends thinking about who controls our understanding of the role of design in promoting justice. Epistemic justice approaches hold that since knowledge and information have tremendous consequences for individuals and communities, they should be engaged in the framing and formulation of that knowledge (Fricker 2007). Design as a practice involves understanding, representing, and making new knowledge about places as well as critical judgments about what is good and bad aesthetics. Epistemic justice in design means that the design preferences and needs of individuals and community groups for their neighborhood take priority over those of outside authorities and professionals. Further, residents control crucial decisions about what knowledge they need in order to design, how they are going to obtain that information, and how they will use it, along with which priorities for design matter to them the most.

Epistemic justice aligns nicely with new approaches to design that de-center the professional designer as authority. Design recognizes historical and contemporary site-specific social processes that interact with the built environment to perpetuate locational disadvantage and foster inequity. Dori Tunstall (2013) suggests that creating physical environments that promote justice requires decolonizing design; liberating it from expert-driven, imperialist tendencies; and refocusing it to engage local inhabitants to co-design the built environment. Designers must be open to the possibilities offered by allowing residents of neighborhoods and cities to collectively determine the values that guide efforts to create a just city. Generating urban form requires thoughtful engagement of design professionals in the quotidian life of communities, where people want the opportunity to be happy, healthy, and safe.

Empowering communities to participate in design requires architects, landscape architects, and planners to be open to conflict between expertise and the values local community members wish to see reflected in the built environment. Professional designers enable epistemic justice in design by offering new ideas and perspectives that can support community members' own design knowledge as well as help them evaluate the consequences of their priorities and choices. There is still plenty of work for design professionals, even if design practice emphasizes sharing the power of creating the physical character of neighborhoods.

From "designing for" to "designing with" communities

These theories of justice inform a number of the historical perspectives on the practice of urban design. This history is marked by aspirations to move from expert-driven design practice towards increasing democratization of the processes and outcomes associated with neighborhood plans, design guidelines, and other comprehensive planning efforts. Urban design is traditionally viewed as a set of practices and strategies for creating physical environments that are aesthetically pleasing and functional. Often these perspectives share little in common; one focuses on notions of beauty and appearance imposed by designers, the other on co-creating along with the public spaces that promote social interaction and well-being. Over time, design imposed by architects, landscape architects, and urban planners has provoked calls for increasingly socially focused practices that require design expertise to cede some control of the creation of space to users, residents, and communities.

The first perspective finds inspiration in the work of Camillo Sitte (1945/1889), who in the late 19th century suggested that the creation of urban space is akin to making art for the masses. He criticized the technicality of design that led designers away from efforts to create humane public space, where individuals and communities could flourish. Sitte recognized the need to improve city life through design and suggested a set of normative goals derived from his personal beliefs about ways to cure the ill of contemporary urban life. Ebenezer Howard and later Le Corbusier shared these concerns. Both of them developed their own visions of the garden city and "towers in a park," respectively, as antidotes to the problems and disarray associated with the industrializing cities of the time. Howard developed a vision for a horizontal garden city that melded town and country (1902), while Le Corbusier suggested a vertical layout (1971/1946). Both schemes could be considered elegant and artistic visions, but ultimately they were derived from their own normative beliefs about the forms that facilitate good life.

In recent times, expert-driven perspectives have given way to visions of incorporating user experience and public participation in the design process. In the United States, this movement gained momentum following the urban renewal strategies of the mid-20th century that eradicated the built and social environments in a number of US cities (Hou 2011). Watching neighborhoods fall under bulldozers led many to question the role of design expertise in formulating a vision of the physical form of place. Kevin Lynch (1960) suggested that urban design is a temporal art; a process built upon the associations, memories, and experiences of residents and other users, whose mental maps provide valuable information about the way everyday experience should inform collective construction of the built environment. Individually derived urban design strategies provided only a small cross-section of the knowledge and expertise involved in constructing and maintaining the physical environment.

Participatory design strategies necessitated a shift away from purely scientific or artistic visions towards design informed by the values of the local community. The design profession responded with normative theories to help guide designers and communities through the analysis, design, and construction of the urban environment. Works addressing the relationship between locally held values and the built environment became increasingly common following the publication in 1981 of *A Theory of Good City Form*, in which Lynch challenges designers to help create cities based on "how people relate to their surroundings . . . rather than out of self-absorbed technical fantasy" (1981, 293). His normative theory of the built environment is structured around a set of professionally derived performance standards grounded in the values of local residents.

Lynch led the way to new conceptions about how designers can better balance rationality, artisanship, and justice. Donald Appleyard and Allan Jacobs (1987) directly challenged the utopian visions of the past, suggesting that expert guided designs too often fail to address injustices created by focusing attention on well-off communities at the expense of other neighborhoods. Their manifesto urges designers to create environments that promote social identity, increase access to opportunity, and enhance sense of shared understanding for everyone, not simply neighborhoods that can afford the physical features that make communities livable.

But engaging communities and incorporating the values of residents into the physical structure of the landscape remains largely aspirational for designers. Urban design plans in North American cities have increasingly focused on important quality-of-life goals like sustainability, walkability, and mobility but demonstrate less interest in participatory approaches, equity, and diversity (Linovski & Loukaitou-Sideris 2013). Manuel Castells (1983) suggests that the goals of design are often in conflict, built through both an expert-driven vision of function as well as a social construction derived from struggles between dominant forces and grassroots organization over the meaning of place. Urban design, from this perspective, physically manifests the beliefs and values of a particular locale as the result of historical and contemporary actions of both designers and inhabitants. Justice, alongside form and function, should serve to guide designers shaping the built environment.

Planning theorist Robert Lake provides an important way for contemporary planners and urban designers to think about the role of design as a contributor to justice in the city. Lake (2016) argues that justice might be the object, or goal, of urban design – that is, something that design is meant to accomplish as an outcome. Expertise plays a central role in determining desired outcomes. Prior generations of urban designers, like Howard and Le Corbusier, tended to epitomize this approach, where the professional designer creates urban form, and the consequences of this creation might be judged, post hoc, as accomplishing or embodying a just outcome or set of outcomes. Lake, however, suggests that justice must move from being simply an external, evaluative position to the center of design practice as a component of social learning. Such social learning relations among design professionals, community members, and elected leaders emphasize democratic design practices, where participants work in tandem and with mutual regard in what Lake calls an "anticipatory, generative" approach that connects social and spatial concerns across time, from a neighborhood's history, contemporary challenges, and future aims. Democratic design processes necessarily influence the desired outcomes of design, as the public engages, informs, and evaluates the built environment alongside physical interventions suggested by the designer (Lake 2016).

Challenges and opportunities ahead

Designing for justice necessitates reflecting on past and new perspectives for generating urban form. Participatory strategies challenge foundational theories of good design grounded in practice and guided by expertise. The trajectory outlined here suggests a growing emphasis on shared expertise and democratically inspired urban form. This presents architects, landscape architects, planners, and other designers of cities with challenges and opportunities.

Participatory approaches to design allow for increasingly tailored but potentially complex demands on the built environment and its role in perpetuating or alleviating disparities in the health, safety, and welfare of marginalized and under-resourced communities. Designers are often challenged to understand where to situate expertise in the diversity of these processes. Moving from historical models of nonparticipation and tokenism used to engage communities towards increasing levels of citizen control over the design process remains an important and necessary part of ending the victimization of local residents through design (Arnstein 1969). To fully embrace justice as a goal of design requires relinquishing some control over the design process as well as the physical and social outcomes that materialize. Such a perspective challenges notions of the primacy of the designer and creates the potential for disagreement between community determined goals and designs and broader societal goals based on promoting justice.

Designing for justice also encourages urban designers to create more equitable cities, where the physical landscape and the processes that shape it are available, accessible, and shared by everyone. Kevin Lynch (1981) suggested that justice is a meta-criterion for measuring the performance of form; justice as an outcome of design focuses on alleviating income inequality, helping individuals and families obtain a baseline quality of life, and empowering communities to determine a shared future. The theories provided here impel designers to recognize and address difference. Identifying instances of racism and sexism requires us to evaluate the participatory processes that confine innovation by ignoring advantage conveyed through the built environment. Design itself has been part of the racial and sexual contracts that determine who controls design as a process in cities, and to whom the benefits of urban design accrue. Awareness of how different individuals and communities engage with space and design from situations of difference in power and prestige, and broadening opportunities for them to determine the physical environment of their local neighborhood, can help ensure a more just urban form, in which benefits from the built environment are shared more equitably.

References

Arnstein, S. R. (1969). "A Ladder of Citizen Participation." *Journal of the American Institute of Planners, 35*(4): 216–224.

Banerjee, T. (1993). "Market Planning, Market Planners, and Planned Markets." *Journal of the American Planning Association, 59*(3): 353–360.

Castells, M. (1983). *The City and the Grassroots: A Cross-cultural Theory of Urban Social Movements*. Berkeley: University of California Press.

Fricker, M. (2007). *Epistemic Injustice: Power and the Ethics of Knowing*. Oxford, UK: Oxford University Press.

Gordon, P. and Richardson, H. W. (1997). "Are Compact Cities a Desirable Planning Goal?" *Journal of the American Planning Association, 63*(1): 95–106.

Hou, J. (2011). "Citizen Design: Participation and Beyond." In Banerjee, T. and Loukaitou-Sideris, A. (eds.) *Companion to Urban Design*, New York: Routledge. 329–340.

hooks, b. (1990). *Yearning: Race, Gender, and Cultural Politics*. Boston: South End Press.

Howard, E. (1902). *Garden Cities of To-Morrow*. London: Swan Sonnenschein & Co.

Jacobs, A and Appleyard, D. (1987). "Toward an Urban Design Manifesto." *Journal of the American Planning Association, 53*(1): 112–120.

Kim, A. M. (2015). *Sidewalk City: Remapping Public Space in Ho Chi Minh City*. Chicago: University of Chicago Press.

Lake, R. (2016). "Justice as Subject and Object of Planning." *International Journal of Urban and Regional Research, 40*(6): 1205–1220.

Le Corbusier. (1971/1946). *Looking at City Planning*, translated by E. Levieux. New York: Grossman. (Originally published as *Manière de penser l'urbanisme*).

Lefebvre, H. (1968). *Le droit à la ville*. Paris: Anthropo.

Linovski, O. and Loukaitou-Sideris, A. (2013). "Evolution of Urban Design Plans in the United States and Canada: What Do the Plans Tell us about Urban Design Practice?" *Journal of Planning Education and Research, 33*(1): 66–82.

Lynch, K. (1960). *The Image of the City*. Cambridge: MIT Press.

Lynch, K. (1981). *A Theory of Good City Form*. Cambridge: MIT Press.

Mills, C. W. (1999). *The Racial Contract*. Cornell, NY: Cornell University Press.

Nussbaum, M. (1988). "Nature, Function, and Capability: Aristotle on Political Distribution." In *Oxford Studies in Ancient Philosophy*. Oxford: Oxford University Press.

Nussbaum, M. (2011). *Creating Capabilities: The Human Development Approach*. Cambridge, MA: Harvard University Press.

Pateman, C. (1988). *The Sexual Contract*. Redwood City, CA: Stanford University Press.

Rawls, J. (1971). *A Theory of Justice*. Cambridge, MA: Harvard University Press.

Richardson, H. W. and Gordon, P. (1993). "Market Planning Oxymoron or Common Sense? *Journal of the American Planning Association, 59*(3): 347–352.

Roy, A. (2005). "Urban Informality: Toward an Epistemology of Planning." *Journal of the American Planning Association, 71*(2): 147–158.

Sen, A. (1992). *Inequality Reexamined*. New York: Russell Sage Foundation.

Sitte, C. (1945/1889). *The Art of Building Cities: City Building According to Artistic Fundamentals*, translated by C. T. Stewart. New York: Reinhold. (Originally published as *Der Städtebau Nach Seinen Künstlerishen Grundsetzen*).

Soja, E. (2010). *Seeking Spatial Justice*. Minneapolis, MN: University of Minnesota Press.

Tiebout, C. M. (1956). "A Pure Theory of Local Expenditures." *Journal of Political Economy, 64*(5): 416–424.

Tunstall, D. (2013). "Decolonizing Design Innovation: Design Anthropology, Critical Anthropology, and Indigenous Knowledge." In Gunn, W., Otto, T. and Smith, R.C. (eds.) *Design Anthropology: Theory and Practice*, Bloomsbury. 232–250.

Further reading

Agyeman, J., Bullard, R. D., and Evans, B. (eds.). (2003). *Just Sustainabilities: Development in an Unequal World.* Cambridge, MA: MIT Press. In this collection of essays, the authors use a set of global examples to illustrate how social and environmental justice must be a central component of economic physical development of human settlements.

Griffin, T. L., Cohen, A., and Maddox, D. (eds.). (2015). *The Just City Essays: 26 Visions for Urban Equity, Inclusion and Opportunity.* New York: J. Max Bond Center on Design. In this collection, a diverse set of authors, including designers, policymakers, and environmentalists, provide insight into the ways a more just city might manifest.

Harvey, D. (2012). *Rebel Cities: From the Right to the City to the Urban Revolution.* New York: Verso. The book builds on the author's previous work looking at global cities as sites of class conflict and social movements to suggest that the reorganization of physical urban environments offers an opportunity to combat both injustice and ecological decay.

Wilkins, C. L. (2007). *The Aesthetics of Equity: Notes on Space, Place, and Music.* Minneapolis: University of Minnesota Press. Suggesting that the history of architecture omits nonwhite voices, this book suggests a more critical perspective on the ways design and the built environment serve to marginalize and empower people of color.

39

TOWARD TRANSFORMATIVE URBAN SPATIAL CHANGE

Views from Jakarta

Kian Goh

Introduction

What is *just urban design*, in the context of burgeoning social and environmental challenges? Many cities around the world face threats from climate change, increasing inequality, and rapid urbanization. Urban responses to such challenges often involve large-scale urban design and development projects formulated through globally interconnected networks of practitioners and consultants, fueled by globalized flows of capital. Plans now frequently invoke concepts such as resilience and sustainability, promising safe and environmentally attuned urban futures.

Interrogations about justice and the city have rarely taken on the question of urban design directly. Design is often subsumed within the broader discourses on urban development. Theoretical concerns of urban design as concept or practice, or questions about the agency and authority of designers and firms, are regularly neglected. But urban design is often the medium through which contested ideas about the future of the city are visualized and justified. Not only does design dictate desired physical outcomes, but its invocation and implementation can expose power relationships in cities. Design practices generate claims on space, and on identity and belonging.

This article conceptualizes *just urban design* by tracing the possibilities of design practice through multiple scales and levels of urban spatial interventions, from large-scale urban environmental and infrastructural projects to more "local" scale, ground-up interventions by marginalized urban residents. It focuses on the ways through which design is used to envision alternative futures and to make claims on them. It looks at Jakarta, Indonesia, as a critical case, and two related kinds of design projects: initiatives to "green" the city, and to protect it from climate change and other environmental threats.

Just cities

Urban centers are seen alternately as sites of liberation and oppression. They are considered spaces of relative autonomy, where urban dwellers find individual freedom in anonymity, a line of thought present since the turn of the previous century (Simmel 1950). Cities have often been seen as places of refuge, congregation, tolerance, and support for immigrants, racial minorities, and gender and sexuality nonconforming people. At the same time, some have also considered cities as spaces of dispossession, exploitation, and other kinds of marginalization.

Urban researchers have explored justice and the city from myriad approaches. Without delving too deep into the debates (see e.g. Marcuse et al. 2009), it can be said that these have centered on issues of processes and outcomes, systemic factors, and worldviews. Scholars have asserted concepts of justice and injustice from the point of view of philosophical thinking (Young 1990; Fainstein 2010) and political-economic bases, often from a critical Marxist structuralist and spatial perspective (Harvey 1973; Smith 1996; Soja 2010). They have probed issues of identity around race and class, often intertwined with environmental disparities (Pulido 2000; Holifield 2001; Agyeman and Evans 2003; Swyngedouw and Heynen 2003). There has also been work spanning theory and practice through the "right to the city" concept (Lefebvre 1968/1996; Harvey 2008; Mayer 2009; Marcuse 2009).

Central to planning, Friedmann (1987) advocated for a normative basis for justice, and affirmation of utopian ideals of the good city (Friedmann 2000). Even though cities contain elements of injustice and oppression, planning theorists might – and ought to – provide "*guiding, normative images*" (Friedmann, 2000, italics in the original) of the "common good." Along these lines, Sandercock (2004) and Miraftab (2009) extended the calls for transformative, radical visions for how we see the city. These normative ideals of a just and "good city" persist even though debates continue around the possibilities and contradictions of such an ideal, with researchers also pointing out the systemic inequalities and oppressions evident in governing urban spaces (Yiftachel 1998; Amin 2006).

While not an exhaustive review, it appears that our major debates on justice neglect the possibility and potentiality of designers to engage, either formally or informally, in visionary and recursive processes of spatial change in cities.

Just design?

This neglect of design in discourses of justice in cities is unfortunate. Core concepts and discourses of what became a professionalized urban design stemmed directly from urban injustice. Conditions of overcrowding, poverty, and ill-health in European and American cities of the late 19th and early 20th centuries (Hall 2002) motivated efforts to envision better urban spatial form. Ebenezer Howard's "garden city" (1898/1965) took on the socio-physical conditions as well as the political-economic base of industrial cities of that era, and proposed a transformative, collective intervention in land use, housing, and urban economy as an interlinked reorganization of society and space. And Tony Garnier's *Cité Industrielle* plan (1899–1917) imagined a utopian industrial city, with segregated functions, open spaces, and modern materials and technology providing the ideal spatial form for a socialist reorganization of production and governance.

Similar broad considerations around societies and spaces drove the architects in and around *Congrès Internationaux d'Architecture Moderne* (CIAM). The somewhat abstract notions of justice suggested in the Athens Charter (1933/1946) underscored an almost self-evident belief that there could be a perfect functional relationship between the form of the city and who we are – ideal cities for ideal societies. Modernist urban ideas and projects attributed to CIAM were often maligned by its detractors as missing the human dimension, openly hierarchical in its conceptions of societal order, and myopic in its embrace of universal principles (see Birch 2011). But CIAM architects insisted that good design organizes social as well as spatial relationships in cities, and their idealized plans often advocated for housing and recreation for workers in the city, and minimum standards for space, light, and air.

The two driving forces of this movement, the rationalist-bureaucratic tendencies of planning and the ideal of the visionary designer, came to a halt in the 1960s, especially in the United States. That myopia of universalism, and changing political economies of cities and the nation state, caught up with urban design. Protests against urban renewal and infrastructural projects accompanied declining faith in the state. The demise of state-led, large-scale projects was epitomized by the spectacular demolition of Pruitt-Igoe in St. Louis in 1972. Driven by Jane Jacobs's (1961) narratives about the failure of city planning visionaries, and captivated by Kevin Lynch's (1960) ground-up public images of the city, many planners no longer believed that ideal cities could reflect and give form to ideal societies. But somewhat lost in this withdrawal from the power of visionary design was an intrinsic concept of justice in urban design.

Some critiques of modernist urban design, including Jacobs's (1961) and Scott's (1998), took issue with aspects of spatial form itself. However, many focused less on the concepts and practices of designing urban spatial interventions, and more on the universalist and exclusionary processes of top-down spatial plan-making. For example, when Holston (1999) attacked CIAM doctrines for not taking seriously existing social relationships in its imagined futures, he upheld the possibilities for insurgency in social practices. But he seemed to dismiss (or at least did not seriously engage) the possibility of justice by an alternative kind of spatial intervention.

So while urban design has been critiqued as being divorced from social realities – and thereby from dealing with issues of justice – few critiques have been leveled at the ways in which design concepts and practices are interlinked with specific processes of urban political economy that drive just or unjust outcomes. What we have, then, is a bifurcation, between those who challenge the asserted injustice of a particular kind of design, such as universalized modernist space planning, and those who push a vision of "good city form" in ways that often neglect foundational issues of justice itself.

This is reinforced by urban design thinking that presumes a kind of normative, just outcome, but does not probe the possibility of transformational change. Lynch's (1984) treatise on good urban form advocates for an approach that is socially responsive, yet largely apolitical, sidelining issues of inequality, power, and oppression. The Jacobs and Appleyard (1987) manifesto for urban design offers normative bases, including invocations of social justice. But it is divorced from urban political economy – the driver of uneven development and injustice in cities. In other works, design is embedded in the nexus between various profit-making imperatives of urban development and bureaucratic regulatory systems (see, e.g., Barnett 1982).

A number of recent initiatives, including contributions from landscape architects and architects involved in community-engaged practices, continue to push the boundaries of justice, spatially, politically, and ecologically (see, e.g., Spirn 2005; Hou 2010; Armborst et al. 2014; Orff 2016). But too often, design efforts that presume to challenge systems of exclusion and oppression fail to do that. Sometimes they result in depoliticized "place-making" or "tactical" urbanism (Brenner 2015). Missing in all this is a critique of the role of urban design in the political economy of cities, and a serious debate on what it would take to change this.

Toward transformative design

How do we conceptualize urban spatial, physical interventions in terms of justice? Many scholars have pointed out the somewhat awkward role that urban design occupies – disciplinarily positioned between planning and architecture, a mode of knowledge and action caught between epistemological frameworks of social science research and iterative methods and procedures of design practice (Loukaitou-Sideris 1996, 2012; Cuthbert 2007, 2011; Carmona 2014; Arabindoo

2014). These arguments, frequently aimed at a better formulation of the place of urban design in urban political-economic processes, are consequential for deliberations around justice.

While a full theoretical elaboration is beyond the scope of the present work, I will sketch out a broad framework that draws on three constitutive elements: (1) Lefebvre's conceptual triad of the social production of space – space as conceived, perceived, and lived – illustrating how space can be simultaneously material, political, and imagined (1991); (2) Cuthbert's (2007) notion of urban design's theoretical object as "civil society"; and (3) the general acceptance in planning scholarship and practice that the "public," in its many forms, is our purview and responsibility. Based on this framework, I contend that urban design justice is explicitly public and collective, a common "right" to change us and our city (cf. Harvey 2008) through urban spatial interventions, that transform unjust systems and power relationships. This is not a spatial version of Rawls's concept of individualized fairness (1971), but more one of acknowledging difference and being together (Sandercock 2006), beginning with a grounded view of community, but skeptical of community as a sole means to obtain broader urban societal justice (Young 1990).

Further, if empirical findings of justice so far have focused on distribution of good and bad, as in environmental justice (Bullard 1990; Holifield 2001), on the one hand, and on procedural and communicative processes (Forester 1989; Innes 1995) on the other, I look to a different approach. Scholars of urban political ecology, confronted with a parallel phenomenon (a tradition of environmental justice seen more as descriptive and practice-based, and conditions in cities seen as structural and systemic), have proposed research that engages explicitly with abstract concepts and concrete change, with what one might call a normative view of urban ecology (Swyngedouw and Heynen 2003; Heynen et al. 2006). This is a distinctly political research project calling for the democratic control of socio-ecological processes in cities.

I propose, then, design as the processes and practices through which control of intertwined urban social and spatial change is envisioned, contested, and conducted. And *just* design aims at the possibilities for spatial transformation towards more equitable outcomes in such change.

I now turn to examine how some of these issues played out in two kinds of urban design projects in Jakarta.

Design contestations in Jakarta

Jakarta exemplifies postcolonial cities with fragmented social institutions and spatial form accentuated by rapid urbanization. It is the quintessential megacity, with about 10 million in Jakarta DKI (*Daerah Khusus Ibukota*, or special capital region) and 30 million in the urban agglomeration. Low-lying and threaded by 13 rivers, Jakarta faces chronic flooding. It experienced massive inundations in 1996, 2002, 2007, and 2013. Severe land subsidence (caused primarily by groundwater over-pumping), decreasing permeability and infiltration (due to rampant urban development), and ongoing sea level rise threaten increasingly catastrophic flooding conditions (Abidin et al. 2011). Spatial and social disparities in the exposure to flooding have been heightened by privatized urban development (Bunnell and Miller 2011; Kusno 2013) that has displaced poorer residents in the informal *kampung* settlements to more vulnerable areas of the city.

While Jakarta exemplifies the tropes of megacities, it is also at the forefront of emerging global dynamics. Historically linked into cycles of global capital accumulation, in particular as the center of Dutch colonial expansionism in Southeast Asia from the early 17th through the mid-20th centuries, it is positioned prominently in new networks of global urban and environmental governance. It has a long history of World Bank-funded development projects, and is a member of C40 Cities and 100 Resilient Cities.[1] In part due to its historical interconnectedness with global capital flows and transnational institutions, and pressing environmental concerns, it has become a site of urban experiments to address these challenges, involving multinational infrastructure and development companies and public-private agencies, primarily from the Netherlands, initiating ambitious plans and projects (Goh 2019a). (Figure 39.1)

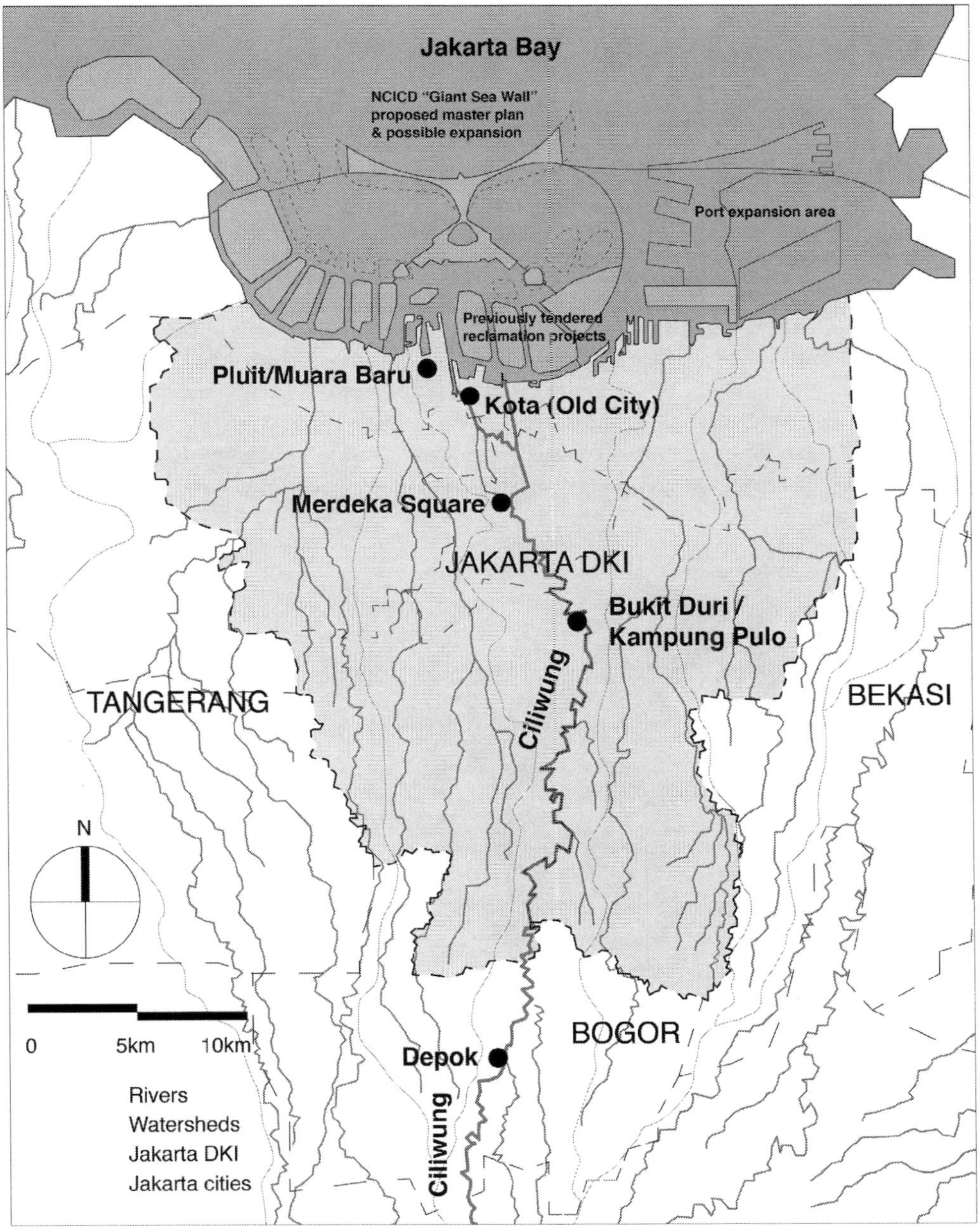

Figure 39.1 Map of Jakarta showing sites of concern, rivers and watersheds, administrative zones, and proposed NCICD master plan (Indonesia MENKO 2014).

Source: Map by author

Greening the city

Jakarta has embarked on urban design projects to green the city – to make it more environmentally sustainable as well as create more open spaces – and to protect it – to defend against or adapt to worsening environmental conditions.

Rampant urban development has reduced the amount of open space to less than 10% of the Jakarta capital region. This has contributed not only to loss of recreational and public spaces, but also to decreased ground permeability, worsening the flooding problems. The city government has a stated goal of increasing open, green spaces to 30% of its area, according to the Spatial Plan 2030 (Jakarta DKI 2012). Many recent efforts to increase green space have focused on areas of low-lying kampung settlements along the waterways and coastline.

Former Jakarta governor (and current Indonesian president) Joko Widodo (also known as Jokowi), on winning the capital region election in 2012, pledged improvements in government-civil society relationships and the daily lives of Jakarta residents. Jokowi made direct pitches to kampung residents, including promises to address the flooding problems, worsening housing conditions, and the lack of recreational and open spaces.

One "showcase" project involved the creation of a park along the Pluit Reservoir, a large catchment pond in North Jakarta. This entailed the eviction of thousands of residents and the demolition of a stretch of kampung in 2013. The cleared area was quickly turned into grassy fields and paths, with fledgling trees symbolizing the change. The park, officially Taman Kota Waduk Pluit (Pluit Lake Park), became popularly known as "Taman Jokowi" (Jokowi's Park) (Figure 39.2). The project, helped by accounts of Jokowi's moves towards government

Figure 39.2 "Taman Jokowi" overlooking Pluit Reservoir, July 2014.

Source: Photograph by author

transparency, signaled positive changes in the city. Evicted residents were transferred to newly built social housing, both close by and sometimes more than 20 km away at the eastern edge of the capital region.

Efforts to green the city have been criticized as part of a governmental strategy to manage "social imagination," while pursuing development goals (Kusno 2011). But they have also made headway in building relationships with pro-poor advocacy groups. An activist with the Urban Poor Consortium (UPC), a housing rights organization working with kampung residents around Pluit, expressed cautious optimism, pointing to the more participatory and open process (E. Saidi, personal communication, July 15, 2014).

Protecting the city

In Jakarta, efforts to green the city are invariably tied to efforts to protect it from environmental threats. Jokowi promised to take steps to mitigate the flooding, especially around the notorious Ciliwung River. He claimed that the city would engage the South Korean government to help restore the river and create a recreational riverfront. He promised to work with kampung activists and take seriously their proposals for rehousing-in-place alternatives to displacement. While Jokowi was making these promises, city agencies were proceeding with infrastructural projects to dredge and widen the rivers and canals (to "normalize" them, in officials' words), ostensibly to increase water flow. The World Bank-funded Jakarta Urgent Flood Mitigation Project (JUFMP)/Jakarta Emergency Dredging Initiative Project (JEDI) entailed the demolition of large areas of kampung settlements.

Jokowi won his bid for the Indonesian presidency in 2014. His successor as Jakarta governor, Basuki Tjahaja Purnama (also known as Ahok), continued the theme of government transparency but took a more hardline stance towards the urban poor. Ahok's government proceeded with a number of forced evictions, including, in 2015, demolishing parts of Kampung Bukit Duri and Kampung Pulo along a stretch of the Ciliwung in Central Jakarta (see Van Voorst and Padawangi 2015). In mid-2017, one can see the outcomes of the evictions and river normalization. The channelized, concretized river leaves gashes in the urban fabric where kampung houses were literally carved in half, with the blank backs of the high embankments just footsteps away from houses that were spared (Figure 39.3).

During this time, a more ambitious project was in the works. After the 2007 floods, teams of Dutch and Indonesian scientists discovered the severe land subsidence. Hydrologists and engineers began devising plans to protect against the deteriorating conditions. In 2014, the National Capital Integrated Coastal Development (NCICD) master plan, or, colloquially, the Giant Sea Wall, was announced. The master-planning process was funded by the Dutch government and led by a team of largely Dutch engineers and designers. It envisioned a new sea wall city for 1.5 million people to be built on reclaimed land in the Jakarta Bay.

The NCICD promised to deal with subsidence and flooding, ease crowding and traffic, stimulate economic growth, and build a modern, world-class waterfront (Indonesia MENKO 2014). But it also appealed to design symbolism and cultural history. The Giant Sea Wall was shaped like the Great Garuda, a mythical bird that is Indonesia's national symbol, heroically portrayed in graphic renderings (Figure 39.4). It married the hydrological feats of infrastructure projects such as the Singapore Marina Reservoir (which turned a river into a reservoir) with iconic reclamation developments such as the Palm Islands in Dubai, both projects featuring some of the same Dutch consultants involved in the NCICD. The designer of the Giant Sea Wall asserts that his role was to give form to the technical process in order to foster buy-in from the Indonesians (G. Van den Boomen, personal communication, October 1, 2014). According to a hydrologist involved with

Figure 39.3 Concrete embankments and partially demolished houses in Bukit Duri and Kampung Pulo, May 2017.

Source: Photograph by author

Figure 39.4 Rendering of NCICD Giant Sea Wall Master Plan.

Source: Urban design by KuiperCompagnons, from Indonesia MENKO (2014) the plan, the design turned what might have been perceived as a development aid project into "national ambition" (K. Bons, personal communication, July 11, 2014) (see Goh 2017).

The NCICD implementation is to be paid for by private investment on the reclaimed land, a strategy that has spurred recent coastal development in Jakarta. The plan has faced substantial opposition, with critics contending that it will harm fragile coastal ecologies and disrupt the livelihoods of fishing villages. Housing rights activists assert that the project uses the rhetoric of environmental catastrophe in order to promote privatized urban development, with little guarantee of social welfare (D.T. Irawaty, personal communication, July 27, 2014). The master plan has gone through iterations of scope, form, and stewardship. While one phase of the plan broke ground in late 2014, the figural "Garuda" element, a critical part of the symbolism of the project, appears to have been canceled as of late 2017 (see Anya and Wijaya 2017).

Counter-design

In these examples, often framed as for the common good, justice seems fleeting and elusive. On the one hand, improving civil society-state relations bodes well for greater participation, a key tenet of procedural justice. And increasing government transparency builds faith in democratic government. On the other hand, a more material and systemic form of justice – that is, increasing collective rights and transforming unjust power relationships – appears to be denied by outcomes that often devalue public control and historical social claims in favor of designed green spaces and increased real estate values.

A bright spot in Jakarta's urban spatial politics is the organization of counter-planning movements against state-corporate planning, including bottom-up, collaborative, place-based protest strategies (Padawangi et al. 2016; Goh 2017, 2019b; Leitner and Sheppard 2018). One notable aspect of these efforts is the prominence of "counter-design" strategies. Activists have formed coalitions with designers and architects to design and propose alternative visions, developing spatial ideas of insurgent planning.

In the Pluit area, UPC activists, with the Rujak Center for Urban Studies, architects from Arkomjogja, and students from University Indonesia developed a proposal for more socially attuned housing blocks. In Kampung Bukit Duri and Kampung Pulo, Ciliwung Merdeka, a community organizing group, worked with Kota Kita, a citizen participation and planning organization, and planning and landscape architecture students to conduct community mapping and propose new models of "humanitarian vertical kampungs." These proposals, merging the strategies of community organizing with the modes and methods of architecture and urban design, provided a platform on which to negotiate with city officials. According to a Ciliwung Merdeka activist, this process proved that kampung residents could give form to the ideals for their own kampung in concrete ways (S. Sumardi, personal communication, July 18, 2014).

In another example, residents and activists in Kampung Tongkol, along with UPC and Architecture Sans Frontières Indonesia, responded to plans for a riverfront access road, which would have entailed eviction and demolition of the kampung, by reorganizing the physical orientation of their houses to face the river and provide maintenance access, thus mitigating the eviction threat (Figure 39.5). Residents also initiated a solid waste management program, as well as local crafts and agriculture projects. This strategy builds on a previous project coordinated by UPC affiliates along a river in Surabaya, Indonesia (Some et al. 2009), a model that UPC hopes to build on in other kampungs throughout the country

These efforts in Jakarta have helped community organizations assert their agency in negotiations with political leaders. They have also captured the imagination of scholars and

Figure 39.5 Houses reoriented to face the river in Kampung Tongkol, May 2017.

Source: Photograph by author

designers from around the world, bringing research attention and volunteers. However, they have so far not succeeded on a systemic level. In Pluit, while some new government housing has been constructed close to the reservoir, the promised models of more "kampung-like" housing projects have yet to be built. Evictions and demolitions continue. Jokowi's Park, in the absence of more concerted rehousing-in-place efforts, stands to become an example of "green gentrification" (Gould and Lewis 2012). And while Ciliwung Merdeka's initiatives gained traction, residents could not prevent the eviction and demolition of parts of Bukit Duri and Kampung Pulo in 2015. The uncertainty revolving around the fate of the NCICD master plan can be seen as an example of continual struggle, although visionary alternatives have yet to be broached.

Indeed, one of the central challenges to and potential of counter-design movements is to conceive spatial visions beyond community-defined places. While the organizing strategies have crossed geographic scales (witness UPC's network and Rujak's advocacy), community-led plans, by nature, are not conducted at the scale of the city or region. Recalling the theoretical framework posed earlier in the chapter, these counter-design movements carve new ground in urban design justice struggles by asserting increasingly public, collective claims on space. The potential is for organizers and designers to develop design strategies that operate at the scale of social and spatial change critical for broader transformative outcomes. These strategies must still be ground up – reflecting the political struggles of marginalized people – yet interconnected across places and issues.

In Jakarta, such an approach would entail spatial-political organizing at an urban regional scale, extending Young's (1990, p. 251) ideas about "empowerment without autonomy," transposed to the strong state/fragmented governance context. And it would entail parallel, intertwined organizing on the scale of environmental exigencies, in particular the watershed (see Goh 2019b). Some community groups in Jakarta have begun nascent conversations to trace flood threats and protections across the Ciliwung watershed, trying to make clear the political and environmental links between new urban development in the upper reaches of the river and the place-specific displacement along the riversides in the urban center and waterfront. These efforts, necessarily collaborative and networked, offer crucial paths forward.

Terrains of contestation

In a previous work about urban climate change adaptation, my coauthors and I charged that unjust outcomes can come from both acts of commission and omission – land use actions that adversely impact marginalized communities, and those that leave them out (Anguelovski et al. 2016). The issues in Jakarta go beyond climate change adaptation strictly defined. They point to the interactions between claims on space, environmental change, and urban development. But the dynamics of action and exclusion play out in parallel ways. Here I have tried to illustrate how these "terrains of contestation" (Goh 2017) – spatial claims, with appeals to urban environmental futures – are made through design plans and counterplans.

In these examples, questions of justice appear on three levels. First, there is design justice as a *process*. This involves the deliberations between urban design, landscape architecture, and engineering consultants and city officials, as well as negotiations between city officials and community leaders. This level revolves around questions of participation and inclusion in the setting of priorities and constraints. The second level is design as *practice*. This involves the making of design visions within the formal and informal institutions of planning and design. Last, the third level is design as *outcome*, tracking and (re)evaluating the material struggles and repercussions of project iteration and implementation.

In places like Jakarta, the emerging model of combining organizing strategies, and theories of political change, with modes and methods of design practice, has repositioned marginalized peoples' struggles. It has literally pushed the physical boundaries of spatial claims and arguably has changed activists' relationships with government. This does not happen through commonly understood design practice alone. Rather, design works in conjunction with, and as a part of, social and spatial-political organizing. These initiatives – in part successful – broach further questions about space beyond the community-bound, and approaches that radically reconsider what a community-led design project might be. Such efforts to expand the agency of counter-design movements will be critical in order to gain broader, more enduring repercussions for material, political-economic relationships in the city.

Note

1 C40 Cities Climate Leadership Group is a transnational municipal network of 90 cities fostering knowledge sharing and action on climate change. The initiative 100 Resilient Cities, funded by the Rockefeller Foundation, promotes city institutions able to withstand social, economic, and environmental shocks and stresses.

References

Abidin, H. Z., Andreas, H., Gumilar, I., Fukuda, Y., Pohan, Y. E., and Deguchi, T. (2011). "Land Subsidence of Jakarta (Indonesia) and Its Relation with Urban Development." *Natural Hazards*, 59(3): 1753–1771.

Agyeman, J., and Evans, T. (2003). "Toward Just Sustainability in Urban Communities: Building Equity Rights with Sustainable Solutions." *The Annals of the American Academy of Political and Social Science*, 590(1): 35–53.

Amin, A. (2006). "The Good City." *Urban Studies*, 43(5–6): 1009–1023.

Anguelovski, I., Shi, L., Chu, E., Gallagher, D., Goh, K., Lamb, Z., . . . Teicher, H. (2016). "Equity Impacts of Urban Land Use Planning for Climate Adaptation: Critical Perspectives from the Global North and South." *Journal of Planning Education and Research*, 36(3): 333–348.

Anya, A., and Wijaya, C. A. (2017, December 11). "Govt Cancels Great Garuda Seawall." *The Jakarta Post*.

Arabindoo, P. (2014). "Urban Design in the Realm of Urban Studies." In Carmona, M. (ed.), *Explorations in Urban Design: An Urban Design Research Primer*. Farnham, Surrey, UK: Ashgate, 47–58.

Armborst, T., D'Oca, D., Theodore, G., and Gold, R. (2014). *The Arsenal of Exclusion & Inclusion*. Barcelona: Actar.

Barnett, J. (1982). *An Introduction to Urban Design*. New York: Harper & Row.

Birch, E. L. (2011). "From CIAM to CNU: The Roots and Thinkers of Modern Urban Design. In Banerjee, T. and Loukaitou-Sideris, A. (eds.), *Companion to Urban Design*. Milton Park, Abingdon, Oxon; New York: Routledge, 9-29.

Brenner, N. (2015, March 24). *Is "Tactical Urbanism" an Alternative to Neoliberal Urbanism?* Retrieved December 30, 2017, from http://post.at.moma.org/content_items/587-is-tactical-urbanism-an-alternative-to-neoliberal-urbanism

Bullard, R. D. (1990). *Dumping in Dixie: Race, Class, and Environmental Quality*. Boulder, CO: Westview Press.

Bunnell, T., and Miller, M. A. (2011). "Jakarta in Post-Suharto Indonesia: Decentralisation, Neo-liberalism and Global City Aspiration." *Space and Polity*, 15(1): 35–48.

Carmona, M. (2014). "The Place-shaping Continuum: A Theory of Urban Design Process. *Journal of Urban Design*, 19(1): 2–36.

Congress Internationaux d'Architecture moderne (CIAM). (1946). *La Charte d'Athenes or The Athens Charter, 1933*. (J. Tyrwhitt, Trans.). Paris, France: The Library of the Graduate School of Design, Harvard University. Retrieved from http://www.getty.edu/conservation/publications_resources/research_resources/charters/charter04.html

Cuthbert, A. R. (2007). "Urban Design: Requiem for an Era: Review and Critique of the Last 50 Years. *Urban Design International*, 12(4): 177.

Cuthbert, A. R. (2011). "Urban Design and Spatial Political Economy." In Banerjee, T. and Loukaitou-Sideris, A. (Eds.), *Companion to Urban Design*. Milton Park, Abingdon, Oxon; New York: Routledge, 84-96.

Fainstein, S. S. (2010). *The Just City*. Ithaca: Cornell University Press.

Forester, J. (1989). *Planning in the Face of Power*. Berkeley: University of California Press.

Friedmann, J. (1987). *Planning in the Public Domain: From Knowledge to Action*. Princeton, NJ: Princeton University Press.

Friedmann, J. (2000). "The Good City: In Defense of Utopian Thinking." *International Journal of Urban and Regional Research*, 24(2): 460–472.

Goh, K. (2017). "Terrains of Contestation: The Politics of Designing Urban Adaptation." In McAllister, M. and Sabbagh, M. (eds.), *Perspecta 50: Urban Divides the Yale Architectural Journal*, Cambridge, MA: MIT Press, 64-74.

Goh, K. (2019a). "Flows in Formation: The Global-urban Networks of Climate Change Adaptation." *Urban Studies*. https://doi.org/10.1177/0042098018807306

Goh, K. (2019b). "Urban Waterscapes: The Hydro-politics of Flooding in a Sinking City." *International Journal of Urban and Regional Research*, 43(2), 250–272.

Gould, K. A., & Lewis, T. L. (2012). "The Environmental Injustice of Green Gentrification: The Case of Brooklyn's Prospect Park." In DeSena, J.N. and Shortell, T. (eds.), *The World in Brooklyn: Gentrification, Immigration, and Ethnic Politics in a Global City*. Plymouth, UK: Lexington Books, 113–146.

Hall, P. (2002). *Cities of Tomorrow: An Intellectual History of Urban Planning and Design in the Twentieth Century*. Oxford, UK ; Malden, MA: Blackwell Pub.

Harvey, D. (1973). *Social Justice and the City*. Baltimore, MD: Johns Hopkins University Press.

Harvey, D. (2008). "The Right to the City." *New Left Review*, 53: 23–40.

Heynen, N., Kaika, M., and Swyngedouw, E. (eds.). (2006). *In the Nature of Cities: Urban Political Ecology and the Politics of Urban Metabolism*. London; New York: Routledge.

Holifield, R. (2001). "Defining Environmental Justice and Environmental Racism." *Urban Geography*, 22(1): 78–90.

Holston, J. (1999). "Spaces of Insurgent Citizenship." In Holston, J. (ed.), *Cities and Citizenship*. Durham, NC: Duke University Press, 155–174.

Hou, J. (Ed.). (2010). *Insurgent Public Space: Guerrilla Urbanism and the Remaking of Contemporary Cities*. New York: Routledge.

Howard, E. (1898/1965). *Garden Cities of To-morrow*. Cambridge, MA: MIT. Press.

Indonesia, Coordinating Ministry for Economic Affairs (MENKO). (2014). *National Capital Integrated Coastal Development Master Plan* (p. 124). Retrieved from http://ncicd.com/wp-content/uploads/2014/10/MP-final-NCICD-LR.pdf

Innes, J. E. (1995). "Planning Theory's Emerging Paradigm: Communicative Action and Interactive Practice." *Journal of Planning Education and Research*, 14(3): 183–189.

Jacobs, A. and Appleyard, D. (1987). "Toward an Urban Design Manifesto." *Journal of the American Planning Association*, 53(1): 112–120.

Jacobs, J. (1961). *The Death and Life of Great American Cities*. New York: Random House.

Jakarta DKI. (2012). *Rencana Tata Ruang Wilayah Jakarta 2030 (Jakarta Region Spatial Plan 2030)*. Jakarta, Indonesia: Jakarta BAPPEDA.

Kusno, A. (2011). "The Green Governmentality in an Indonesian Metropolis." *Singapore Journal of Tropical Geography*, 32(3): 314–331.

Kusno, A. (2013). *After the New Order: Space, Politics, and Jakarta*. Honolulu: University of Hawai'i Press.

Lefebvre, H. (1968/1996). "The Right to the City." In *Writings on Cities*. Oxford, UK; Cambridge, MA: Blackwell.

Lefebvre, H. (1991). The production of space. Oxford, UK ; Cambridge, Mass., USA: Blackwell.

Leitner, H. and Sheppard, E. (2018). "From Kampungs to Condos? Contested Accumulations Through Displacement in Jakarta." *Environment and Planning A: Economy and Space*, 50(2): 437–456.

Loukaitou-Sideris, A. (1996). "Cracks in the City: Addressing the Constraints and Potentials of Urban Design." *Journal of Urban Design*, 1(1): 91–103.

Loukaitou-Sideris, A. (2012). "Addressing the Challenges of Urban Landscapes: Normative Goals for Urban Design." *Journal of Urban Design*, 17(4): 467–484.

Lynch, K. (1960). *The Image of the City*. Cambridge, MA: MIT Press.

Lynch, K. (1984). *Good City Form*. Cambridge, MA: MIT Press.

Marcuse, P. (2009). "From Critical Urban Theory to the Right to the City." *City*, 13(2–3): 185–197.

Marcuse, P., Connolly, J., Novy, J., Olivo, I., Potter, C., and Steil, J. (eds.). (2009). *Searching for the Just City: Debates in Urban Theory and Practice*. London; New York: Routledge.

Mayer, M. (2009). "The 'Right to the City' in the Context of Shifting Mottos of Urban Social Movements." *City*, 13(2–3): 362–374.

Miraftab, F. (2009). "Insurgent Planning: Situating Radical Planning in the Global South." *Planning Theory*, 8(1): 32–50.

Orff, K. (2016). *Toward an Urban Ecology*. New York, NY: Monacelli Press.

Padawangi, R., Turpin, E., Herlily, Prescott, M. F., Lee, I., and Shepherd, A. (2016). "Mapping an Alternative Community River: The Case of the Ciliwung." *Sustainable Cities and Society*, 20: 147–157.

Pulido, L. (2000). "Rethinking Environmental Racism: White Privilege and Urban Development in Southern California." *Annals of the Association of American Geographers*, 90(1): 12–40.

Rawls, J. (1971). *A Theory of Justice*. Cambridge, MA: Belknap Press of Harvard University Press.

Sandercock, L. (2004). "Towards a Planning Imagination for the 21st Century." *Journal of the American Planning Association*, 70(2): 133–141.

Sandercock, L. (2006). "Cosmopolitan Urbanism: A Love Song to our Mongrel Cities. In Binnie, J. (ed.), *Cosmopolitan Urbanism*. London; New York: Routledge, 37–52.

Scott, J. C. (1998). *Seeing Like a State: How Certain Schemes to Improve the Human Condition Have Failed*. New Haven, CT: Yale University Press.

Simmel, G. (1950). "The Metropolis and Mental Life." In *The Sociology of Georg Simmel*. Glencoe, IL: Free Press.

Smith, N. (1996). *The New Urban Frontier: Gentrification and the Revanchist City*. London; New York: Routledge.

Soja, E. W. (2010). *Seeking Spatial Justice*. Minneapolis: University of Minnesota Press.

Some, W., Hafidz, W., and Sauter, G. (2009). "Renovation not Relocation: The Work of Paguyuban Warga Strenkali (PWS) in Indonesia." *Environment and Urbanization*, 21(2): 463–475.

Spirn, A. W. (2005). "Restoring Mill Creek: Landscape Literacy, Environmental Justice and City Planning and Design." *Landscape Research*, 30(3): 395–413.

Swyngedouw, E. and Heynen, N. C. (2003). "Urban Political Ecology, Justice and the Politics of Scale." *Antipode*, 35(5): 898–918.

Van Voorst, R. van, and Padawangi, R. (2015, August 21). *Floods and Forced Evictions in Jakarta.* Retrieved August 27, 2015, from http://asiapacific.anu.edu.au/newmandala/2015/08/21/floods-and-forced-evictions-in-jakarta/

Yiftachel, O. (1998). "Planning and Social Control: Exploring the Dark Side." *Journal of Planning Literature*, 12(4): 395–406.

Young, I. M. (1990). "City Life and Difference." In *Justice and the Politics of Difference.* Princeton, NJ: Princeton University Press, 226–256.

Further reading

Design Justice Network. Retrieved from http://designjusticenetwork.org/. Network of interdisciplinary scholars, artists, and activists working to create design processes and practices that center the most marginalized groups.

Goh, K. (2018). "Safe Cities and Queer Spaces: The Urban Politics of Radical LGBT Activism." *Annals of the American Association of Geographers*, 108(2): 463–477. Explores the ways in which marginalized urban groups organize to make spatial claims in cities.

Mock, B. (2015, May 14). *There Are No Urban Design Courses on Race and Justice, So We Made Our Own Syllabus.* Retrieved from www.citylab.com/design/2015/05/there-are-no-urban-design-courses-on-race-and-justice-so-we-made-our-own-syllabus/393335/. Reading list on race and justice in cities, arising from dialogues started by African American students at Harvard's Graduate School of Design.

Shi, L., Chu, E., Anguelovski, I., Aylett, A., Debats, J., Goh, K., . . . Vandeveer, S. (2016). "Roadmap towards Justice in Urban Climate Adaptation Research." *Nature Climate Change*, 6: 131–137. Outlines barriers to justice in urban climate change adaptation and provides paths forward, including on spatial planning and urban design.

Wachsmuth, D., Cohen, D. A., and Angelo, H. (2016). "Expand the Frontiers of Urban Sustainability." *Nature News*, 536(7617): 391. Critiques the prevailing view of urban sustainability initiatives and calls for better accountability around equity and ecological effectiveness.

PART III.V

Intelligence

Introduction

Today we have arrived at – indeed, crossed the threshold of – the brave new world of intelligence. This is the exciting new arena of information and communication technologies, sensors, big data, surveillance, artificial intelligence, automation, and robotics. In the late 1960s, Marshall McLuhan (1967) prophetically pointed out that "the medium is the message," discussing how technology shapes our everyday life. His anecdotes are cases in point. When the copy is perceived better than the original, or at least the same as the original, when the sensorium of our built environment challenges the veridicality of our senses – is the wood paneling real or Formica? Is that potted plant real or plastic? And, when the distinction between the fake and the real begins to dissolve (as we have discussed in our introduction to the section on mimesis and simulacra), we need to seriously interrogate the new imperatives of urban design, a practice deeply rooted in improving the sensory experiences, livability, social contact, and public life in cities.

More than 20 years ago, sociologist and communication theorist Manuel Castells (1996) wrote his classic book "The Rise of the Network Society," arguing that increasingly our lives will be consigned to spaces of flows rather than spaces of places. He urged urban designers to maintain, recover, and restore the spaces of places and even attempt to increase their supply. Today, we may ask the question: how does the brave new world of intelligence define the new imperatives and mandates for urban designers? The essays assembled in this section attempt to address this question.

Of course, there is quite a bit of enthusiasm and excitement about this new dawn of intelligent cities. The media, the blogs, and the podcasts are all abuzz with the idea of "smart cities," where the everyday drudgeries of life will be eased by autonomous cars, efficient utilization of the capacity of existing infrastructure, and where sensors will facilitate public safety through surveillance. Prime Minister Narendra Modi wants to have 500 smart cities in India, the Emir of Dubai wants Dubai to be the smartest city in the world, and so on. Yet doubts remain. The essay by Darshini Mahadevia in the section on Gentrification and Displacement, and the one by Balaji Parthasarathy and Brinda Sastry in this section present critical arguments about the "Smart Cities initiative" in India. In fact, we (the editors of this volume) have earlier expressed our own reservations about the surveillance regime of the intelligent city, and its possible violation of the Fourth Amendment rights in the United States (see Banerjee and Loukaitou-Sideris 2013).

Current news stories about identity thefts and loss of privacy in the social media, all underscore these concerns.

So, what are the possibilities for urban design in the ensuing "intelligence regime"? What are the challenges and mandates for urban designers? The essays presented here generally brim with optimism and enthusiasm, but with caveats where necessary.

In the first chapter of this section, titled "Tarzan vs. IBM: Value Paradigms of Urban Technology," Dietmar Offenhuber sets the framework for reviewing the promises and caveats of the new intelligence. He presents four value paradigms that help to capture the application and enthusiasm for various emerging technologies. Commonly used in what he calls "techno-urbanistic discourse," these value paradigms are "seamless efficiency," "responsive transparency," "empowering devolution," and "shared stewardship." He proceeds to discuss their similarities and differences and reviews their goals and critiques. These value paradigms are generally optimistic with a few cautionary notes and caveats.

In the second chapter of this section, Jennifer Stein and Scott Fisher focus on the enhanced capabilities and agency of lay citizens to collect, share, and represent in real time data about their cities, neighborhoods, local streets and the like from the new digital technology, social media, and other such "apps." The new digital technology thus allows for "crowd sourcing" of urban data, empowerment of citizens, and expansion of the possibilities for participatory design. Brettany Shannon's chapter in the ensuing Arts and Culture section offers an empirical example – a "digital fieldwork" involving Instagram – of how public perceptions and preferences for urban design and public art can be obtained through the "crowd sourcing" possibilities of the Internet. This is then good news for the Aristotelian model of pluralistic city design.

In the next chapter, Fábio Duarte and Carlos Ratti describe how digital technologies, geographic positioning systems, and sensor technology obtaining data about urban form and urbanism on a real-time basis can greatly facilitate the work of urban designer. They invoke Ildefonso Cerdà's penchant for quantitative data, and William H. Whyte's ethnographic field study of the uses of New York's public spaces as intellectual precedents of the use of empirical data in understanding urbanism and urban-activity patterns in the city. They discuss several of their projects on Senseable Cities, including the use of cell phone data to understand environmental health issues in the city. Using the metaphor that the built environment offers a "third skin," they argue that more and better data can make that "skin" more responsive to human needs.

The last chapter in this section by Balaji Parthasarathy and Binda Sastry is a thoughtful reflection on the current enthusiasm about India's Smart Cities Mission, promoted by the country's prime minister. They begin with a cautionary note about the history of technology transfer from the West, and how "techno-functional planning and design" of 20th-century modernism failed to capture the possibilities of human agency in transforming urban space in a socially inclusive manner. Similarly, the current Smart Cities Mission makes no promises of involvement of such agency and inclusion, while the promised infrastructure development might come at the cost of displacement of people in the margin, as Darshini Mahadevia has also discussed in an earlier chapter.

References

Banerjee, T. and Loukaitou-Sideris, A. (2013). "Suspicion, Surveillance, and Safety: A New Imperative for Public Space?" in Carmon, N. and Fainstein, S. (Eds.) *Policy, Planning, and People: Promoting Justice in Urban Development.* Philadelphia, PA: University of Pennsylvania Press. pp. 337–355.

Castells, M. (1996). *The Rise of the Network Society.* Oxford, UK: Blackwell Publishers.

McLuhan, M. ([1967] 2001). *The Medium Is the Message: An Inventory of Effects.* Gingko Press Inc.

40
TARZAN VS. IBM
Value paradigms of urban technologies

Dietmar Offenhuber

Introduction

The notion of the smart city remains stubbornly persistent in the urban planning and policy discourse. Its longevity is partly due to its ambiguity: the smart city is not a consistent concept, but a metaphor with many facets and variations. Far from a new idea, the city as an intelligent machine is a classic science fiction trope, treated in films such as Jean Luc Godard's 1965 film *Alphaville*, which the director provisionally entitled "Tarzan versus IBM." Yet, the words *intelligent* and *smart* have different meanings. While Godard's city exhibits human-like intelligence and the ability to engage in philosophical dialogue, technological smartness, as envisioned by architect Bill Mitchell, does not require such advanced mental capacities (Mitchell 1995, 2000). Perhaps Mitchell preferred the word "smart" because it means accomplishing much with little effort: being efficient by taking shortcuts, utilizing available resources and connections rather than mobilizing brute-force computational power.

After the 2008 recession in the United States, and in anticipation of infrastructure stimulus programs, IT companies such as IBM embraced the smart city concept, promoting technological solutions to manage urban systems – including transportation, energy, water, sanitation, and healthcare – more efficiently with the help of distributed sensor networks and predictive data analytics. As federal funds, however, failed to materialize and cities turned out to be complicated customers, IBM reduced its "Smarter City" activities. The term, however, did not disappear. The European Union continued to use the smart city label in its funding frameworks but framed it as a broader approach to regional development and innovation policy. In the Global South, smart city initiatives acquired yet another flavor: here, beyond their typical application to infrastructure modernization, the term is often used in the context of large-scale real estate development projects. Indicative of the most recent redefinition of the term, the Obama administration launched a nationwide smart city initiative in 2015 that aimed to combine civic media, transparency initiatives and neighborhood projects with technical infrastructure goals (Correa 2015). The initiative was based on the premise that smart urban technologies are not rocket science: many cities and civic initiatives have already developed an array of tools and practices to support planning, deliberation, and participation.

Value-oriented paradigms of urban technologies

The most tangible impacts of digital technologies on urban space are, however, rarely connected to smart city initiatives, but are the result of new services deployed by Silicon Valley giants such as Amazon, Uber, or Airbnb, often bypassing local administrations and regulations. Ride-sharing and social navigation apps such as Waze have changed traffic patterns and pose challenges to public transportation authorities (Alemi et al. 2017; Foderaro 2017). Dockless rental bikes and scooters appeared on sidewalks across the United States seemingly out of nowhere. The gig economy has changed the demographic and economic landscape of cities. The places where the new informalized workforce of Uber drivers, Task Rabbit[1] handymen, or Alfred[2] domestic workers prefer to operate are usually not the same where they can afford to live. Peer-to-peer rental services such as Airbnb are under scrutiny for their suspected impact on the availability of affordable housing in many cities. Social media spaces are fragmented into multiple parochial domains (aka filter bubbles), and as spatial behavior is increasingly shaped by social platforms, this fragmentation seeps into physical space (Ito Okabe and Matsuda 2006; de Waal 2014). Far from their alleged equalizing effect on geography, digital communication technologies have introduced new spatial inequalities.

These earlier examples illustrate that many traditional models about the spatiality of digital media technologies need to be revised. For a long time, many architects, urban designers, and planners were guided by an idea of cyberspace as an abstract domain of information that augments the physical space of the city. Augmentation, however, is no longer an apt description: the built environment and spatial practices are not just complemented, but increasingly regulated by media technologies. Digital networks and physical spaces become contingent upon each other; many physical infrastructures no longer function without digital networks (Kitchin and Dodge 2011). The idea of "cyberspace" itself, a consistent space that follows its own laws and principles, implies a determinism that is often found in technological narratives. In the words of futurologist Kevin Kelly (2010, 175), "by following what technology wants, we can be more ready to capture its full gifts." But not only techno-enthusiasts, but also critics frequently take a deterministic stance, understanding technology as something ontologically stable and inherently destructive.

On closer scrutiny, however, current examples of urban technologies do not support such sweeping characterizations. Each platform, tool, and project introduces its own spatialities, shaped by values, design decisions, and power dynamics rather than inherent technological principles. As critic Evgeni Morozov points out, the Internet is a heterogeneous assemblage of different protocols and infrastructures – some centralized, others distributed – rather than a consistent object with essential qualities. In other words, the Internet does not have an inherent logic that would determine its social effects, let alone possess agency and "wants" (Morozov 2014). This does, however, not imply a social determinism – that technologies are merely a representation of social forces, and their effects are only a matter of how they are used. The reality is more complicated: technologies such as ride-sharing apps, the Internet of Things, predictive analytics, or blockchains are deeply intertwined with social practices, power relationships, values, and ideologies. Their mutual relationships are rarely deterministic and could be better described through the metaphor of resonance. Emerging techno-cultures resonate with more established ideas and values of how cities and society should be organized and improved.

This chapter will attempt to disentangle urban technologies and techno-cultures and group them based on the goals and values articulated by their proponents. In this respect, techno-cultures cannot be treated as monolithic systems: the open-source movement, for example, encompasses a range of diverse and sometimes contradictory values and beliefs. In the remaining sections, I

Table 40.1 Urban technologies by value paradigm

Value Paradigm	*Examples*	*Goals*	*Critiques*	*Further Reading*
Seamless efficiency	Corporate smart city, data-driven urbanism, cybernetics, and systems theory	Using economies of scale, legibility, control (top-down), autonomy, convenience	Technocratic (not people centric), surveillance, not resilient, not attainable	Greenfield (2013); Mayer-Schönberger and Cukier (2013); Goodspeed (2014); Gordon and Walter (2016)
Responsive transparency	New Public Management, KPIs, Open Data Movement, civic technologies, accountability technologies	Freedom of information, control (bottom-up), outcomes over process, participation, generative value of data	Perverse incentives, political misuse, overestimating transparency	Lathrop and Ruma (2010); Goldsmith and Crawford (2014)
Empowering devolution	Maker culture and DIY, tactical urbanism, pop-up urbanism	Decentralization, participation, democratization of technology	Privatization, favoring the retreat of the state (big society)	Egyedi and Mehos (2012); Ratto, Boler, and Deibert (2014)
Shared stewardship	Digital communities of practice, cooperative movement, blockchain/smart contracts, inverse infrastructures	Good governance, inclusion	Gaps in capacity, access (digital divide), depoliticizing/trivializing politics (bitcoin example), hyperformalization of interactions	Greenfield (2017)

will identify four different value paradigms for urban technologies and discuss examples, their associated value propositions, and critiques. The value perspective reveals differences between apparently similar approaches, but also surprising alignments between seemingly opposed views. Table 40.1 provides an overview of the four value paradigms.

Seamless efficiency – the smart city and cybernetic feedback systems

Embodied in the image of the smart city, the paradigm of seamless efficiency is based on the notion that certain urban problems can be addressed only on the systemic scale. The paradigm is rooted in the urban cybernetics discourse of the 1960s, which conceptualized the city as a complex, dynamic, and self-regulating feedback system (Goodspeed 2014). As systems theorist Jay Forrester emphasizes, the behavior of such systems is often counterintuitive. Since the system dynamically reacts and adapts to changes, policy interventions can have the opposite of the intended outcomes: for example, attempts to reduce traffic congestion can lead to even more congestion. As Forrester (1969, 1970) argues, policymakers can succeed only if they consider the entire system and understand its dynamic behavior. Unsurprisingly, Forester's pilot projects attempting to provide objective and apolitical simulations for shaping low-income housing policy in Boston and New York led to intense public controversies (Alfeld 1995).

But in the paradigm of seamless efficiency, most decisions are not made by planners, urban managers, or policymakers, but by the system itself. Alluding to the cybernetic metaphor of the human brain as a feedback system, the concept of the smart city implies sentience and autonomy. To a certain extent, the smart city is meant to regulate itself; analyzing real-time data streams from the entire system and making autonomous decisions to optimize its performance. Such control systems are not new: since the 1980s, the city of Los Angeles has been operating the ATSAC system, a network of real-time traffic sensors connected to a computer system that controls the timing of traffic signals.

What has changed, however, is the central role of data. While the ATSAC system does not keep records of historic sensor readings, big data repositories have become central elements of smart city projects, used as a resource for predictive analytics as well as a monetizable product.

Implications for urban form

To achieve a real-time data representation of the whole system, the smart city requires standardized infrastructure, sensor specifications, and protocols. As a result, smart city thinking tends towards the monolithic rather than the heterogeneous. The most iconic examples of the first wave of smart cities are large-scale planned developments, including Masdar City in the United Arab Emirates or the Songdo International Business District in South Korea, both developed under the heavy involvement of IT companies (Halpern et al. 2013; Townsend 2013). Masdar City is a planned city project in Abu Dhabi adjacent to its international airport, designed by Foster and partners. The city's main theme is sustainability and energy efficiency, which is aimed for through both architectural and technical means. Its pilot stage banned cars and included a personal rapid transit system of pod-like electric vehicles. Masdar is monolithic not only in its shape – a compact block rising above the desert segmented by narrow streets – but also in its organization, infrastructure, and governance. Rather than a city in the traditional sense, it is in its current state more aptly described as a campus with residential facilities (Greenfield and Kim 2013; Vignal 2014). The Songdo International Business District is like Masdar located in close proximity to an international airport and targeted at a globally mobile elite. The city is built around the theme of ubiquitous computing, the integration of computation and broadband data networks into all aspects of the physical environment such as trash collection, traffic control, pedestrian information systems, and fully automated condos. As Halpern et al. (2013) note, Songdo is a conceptualized as a test bed for developing, perfecting, and showcasing urban technologies that can then be replicated elsewhere – a city as a physical representation of a consistent concept.

Besides large-scale deployments, IT companies offer smart city solutions also for infrastructure upgrades in cities characterized by a heterogeneous assemblage of old infrastructure systems. Under the premise that data-driven urban management can increase the efficiency of existing systems, it is argued that more costly infrastructure upgrades can be avoided. However, beyond applications in transportation and energy, ambitious smart city implementations are scarce in traditional cities. The complexities of public institutions, the messiness of infrastructure bound to historical legacies and contingencies, and the constraints of tight municipal budgets made smart city products eventually unattractive for many IT companies. After the partial retreat of private actors, the second wave of smart city projects is increasingly driven by cities. In public-private partnerships, municipalities act as landlords, offering companies market access and infrastructure as a conduit for data collection, for example in the form of public information kiosks, or wireless Internet on public transportation.

Smart city infrastructure is designed to be invisible, seamlessly connecting a multitude of networks in the background. Instead of exposing technical complexities, the system is expected to sense and adapt to the needs of the residents, for example in the case of on-demand public transport.

Critiques

It is fair to say that the smart city has a reputation problem – its critiques outnumber its praises (Greenfield and Kim 2013). Many critics consider its data-driven models as inappropriate technocratic abstractions that ignore social and political realities, sometimes with catastrophic results (Flood 2010). The smart city is often characterized as a project of top-down control and surveillance (Greenfield 2013). The underlying goals of efficiency are questioned as a technocratic fetish (Gordon and Walter 2016) but also as a potential weakness that makes cities less resilient; highly optimized systems tend to be fragile. The instrumentation of critical infrastructure with potentially hackable technologies is another risk (Kitchin and Dodge 2017). The design paradigm of seamlessness, the attempt to hide all technical complexities, can exacerbate these vulnerabilities, by making the system more rigid, opaque, and harder to repair.

Responsive transparency – e-government systems and civic technologies

While the hardware-centric smart city paradigm focuses on the technicalities of public management, the paradigm of responsive transparency deals with its procedural and political aspects. Nowadays, most US cities operate citizen feedback systems in the form of 311 hotlines and smartphone apps for submitting complaints or reporting infrastructure failures. Data collected by these systems are used to improve public services, shared on open data platforms, and repurposed for other civic applications. Based on freedom of information laws of the 1960s and 1970s, the open data movement goes beyond mere disclosure of public information; it now demands public datasets to be machine-readable and available in real time, so they can be better analyzed and used by citizens. Recently, land titles and public expenditures are recorded and opened for scrutiny on public blockchains – distributed databases that are considered especially transparent and tamper-proof. The paradigm of responsive transparency, however, goes beyond functions of public accountability. Civic technologies facilitate a two-directional flow of information, allowing the city to be responsive to the needs of its constituents.

Citizen feedback systems, participatory budgeting projects, and social accountability campaigns represent a new model of public accountability that favors outcomes over process. According to supporters of responsive transparency, this allows cities to experiment with unconventional and innovative ideas (Goldsmith and Crawford 2014). In contrast to the paradigm of seamless efficiency, this paradigm favors incremental changes rather than systemic interventions. The ecosystem of open-source development, with its fast and public cycles of prototyping, testing, and deployment, serves as an important inspiration. In many cases, tools developed for open-source development are repurposed for participatory urban planning and governance (Figure 40.1). This includes the use of version control systems – digital repositories that allow developers to keep track of code changes and to resolve conflicts – as tools for asynchronous and distributed decision-making, bug trackers for collecting and managing reports about infrastructure failures, or collaborative authoring tools for participatory planning and co-production projects.

Figure 40.1 g0v.asia ("gov zero") is a Taiwanese community dedicated to values of transparency, openness, and committed to building civic technologies. The group emerged from the 2014 protests of the Sunflower movement but has since entered a partnership with the Taiwanese government.

Source: g0v.asia

Implications for urban form

A review of reports submitted to citizen feedback systems[3] across many US cities shows that a significant portion of citizen reports are not complaints, but suggestions and ideas to improve urban space (Offenhuber 2014). A responsive planning paradigm encourages such citizen input and considers such proposals serious. Although the paradigm of responsive transparency focuses on process rather than questions of urban form, its policy of small steps leaves visible traces in the urban environment. The paradigm favors small interventions over large-scale, comprehensive projects; local solutions over system-wide changes.

But transparency is not merely a procedural principle; it can have concrete design implications. As an increasing number of invisible sensors and interfaces become part of the urban environment, the legibility of data collection infrastructures and their governance becomes an important issue. The concepts of ambient accountability and accountability-oriented design address the need for making governance in physical space legible (Zinnbauer 2012; Offenhuber 2017). This can include, for example, physical hints that alert to the presence of sensors and cameras, information about what kinds of data they collect, who has access, and how data are governed.

Critiques

Responsive transparency may seem like an unequivocally beneficial goal. However, the capacity to engage with the local government and participate in planning decisions is not equally distributed. Studies of citizen feedback systems have shown that rich neighborhoods with a high level of public services often submit the most complaints, while low-income areas with poor services complain less (Martínez, Pfeffer, and van Dijk 2009; Verplanke et al. 2010).

It is also easy to forget that transparency by itself does not actually solve anything, and relying too much on its positive effect can actually be detrimental, since transparency requirements can be easily circumvented by tactics of obfuscation. More importantly, transparency tends to call more attention to failures than to successes and can be easily politically misused through "gotcha" tactics (Lessig 2009). It is also worth noting that open-source projects are transparent, but not necessarily always democratic – the governance of software projects exhibits the full range from constitutional democracies to "benevolent" dictatorships. The premise of "outcomes over process" can be problematic since outcomes have to be measured, and measures such as Key Performance Indicators (KPI) introduce their own idiosyncratic processes that can be gamed and exploited. With regard to urban form, incrementalism can be a distraction from necessary systemic change. A constant stream of requests for minor fixes can tie up the resources of a municipal government, preventing more comprehensive measures to address underlying issues.

Empowering devolution – inverse infrastructures and platform urbanism

United by their large-scale system-building ambitions, the paradigms of seamless efficiency and responsive transparency complement each other; the former focusing on logistic, the latter on civic infrastructures. The third paradigm of empowering devolution is in many ways opposed to such ambitions. Like responsive transparency, this paradigm is rooted in the ethos of open-source development, but rather than focusing on the function of the finished system, it focuses on social practices of making and infrastructuring. While responsive transparency appropriates tools as means to an end based on their usefulness for civic purposes, this paradigm views the acts of making and maintaining as civic values in themselves. Often, new ideas emerge as side products of collaborative efforts. The "smart citizen" – a creative, techno-literate, and civically minded amateur expert – is heralded as a counter model to the smart city. Beyond the use of civic technologies, the collaborative production of such tools is considered at the core of democratic discourse. The paradigm questions the need for central planning and large-scale infrastructure interventions, since the democratization and proliferation of digital technologies make it possible for individuals to build and maintain self-organized infrastructures. Examples range from public wi-fi infrastructure and air quality-sensing networks to Twitter-operated food trucks or self-organized waste management systems (Figure 40.2).

Under various labels such as pop-up urbanism and tactical urbanism, empowering devolution promotes bottom-up organization as an alternative to public services and centralized urban design. Standardization is of no immediate concern; it is seen as an evolutionary outcome rather than a prerequisite for planning. If the paradigm of seamless efficiency strives for consistency and simplicity, empowering devolution embraces bricolage and experimentation. In many ways, devolution is a form of privatization; and commercial platforms, including ride-sharing services such as Uber or hospitality services such as Airbnb, can be counted in this category. Having emerged from outside the urban planning field, they have sometimes significant impact on the shape and socioeconomic structure of the city.

Implications for urban form

Tactical urbanism directly intervenes in urban form, sometimes with acts of anarchistic appropriation – painting bike lanes to instigate policy changes, asking for forgiveness rather than permission. Civic technologies and viral media are instrumentalized as means for "talking back" to

Figure 40.2 Digua Shequ (Sweet potato community) is a self-organized residential community squatting in the unused underground bunkers throughout Beijing. The Digua team organizes community improvement projects and connects residents above and underground.

Source: Digua Shequ

planning authorities (Sassen 2011). The paradigm draws inspiration from the organic structures of the informal city, informal transportation services and food distribution networks. Planning and doing often coincide, leading to systems of improvisational governance, in which multiple stakeholders negotiate policy through direct interventions in the urban environment (Offenhuber and Schechtner 2017). As a counter model to the traditional infrastructure paradigm of the large technological system (Hughes 1987), this paradigm embraces "inverse infrastructures," driven, developed, and maintained by their users (Egyedi and Mehos 2012).

Critiques

Similar to transparency, participation and self-organization are widely regarded as unquestioned positive values. However, it is doubtful whether user-driven models can replace traditional planning approaches. Self-organized infrastructures that rely entirely on a dedicated community of volunteers are difficult to maintain and scale; the results tend to be short-lived. Furthermore, praises for participation and volunteerism are often merely rhetorical maneuvers used, for example, to justify public spending cuts and privatization. Policy expert Sherry Arnstein identifies malicious forms of pseudo-participation used to placate, distract, or manipulate constituents (Arnstein 1969). Informality and bricolage are often romanticized but frequently insufficient means to address urban problems. Contrary to popular assumption, decentralized systems tend to be inefficient, fragmented, fragile, and opaque. Systemic planning and implementation guided by concerns for efficiency often lead to better and more sustainable results.

Similar to participation, empowerment is a frequently abused term, often condescendingly used by planners towards constituents. Especially in low-income areas, however, participation in public service provision is often more a burden rather than a form of empowerment, especially when this burden is unequally distributed and participants are expected to contribute significant resources and carry responsibilities without fully benefitting from them (Bovaird 2007).

Shared stewardship – digital cooperatives and blockchain governance

The paradigm of shared stewardship is more difficult to situate than the others in traditional dichotomies such as city vs. citizen, user vs. provider, top-down vs. bottom-up. Shared stewardship refers to technologically supported models for collaborative governance among diverse stakeholders, whose interests are not always aligned. Each stakeholder group has different roles and responsibilities in managing and governing a commons – a resource that is accessible to everyone, but at the same time limited and easily depleted. As a critique of the economic "tragedy of the commons" argument according to which common pool resources are doomed to fail (Hardin 1968), Elinor Ostrom (1990) famously showed how common pool resources such as user-managed water irrigation systems can be successfully managed by a collective that shares responsibility for governance. While the paradigm of responsive transparency assumes a traditional public authority as the governing entity, and the paradigm of empowering devolution tends to avoid questions of governance altogether, shared stewardship focuses on questions of consensus, conflict resolution, and alignment of interests.

Examples of shared stewardship include waste and recycling cooperatives in Latin America, Africa, and India, which use digital technology for planning their collection or interacting with customers, community members, and public authorities. They also include energy cooperatives such as the Brooklyn Microgrid initiative, which builds and maintains an infrastructure of photovoltaic panels mounted on the roofs in the neighborhood to provide local energy. The individual energy consumption and production is recorded in a dedicated blockchain ledger.[4] In models of participatory design and co-production, beneficiaries of public services are often involved in their development, maintenance, monitoring, and governance, ensuring that the services address their needs and maintain their quality. The digital systems are manifestations of the social contract underlying these initiatives. This social contract requires constant updates, renegotiations, and clarifications. The incentive structures for the participating actors require fine-tuning as the system evolves. In blockchain applications, where governance happens exclusively through algorithmic or highly formalized rules, seemingly miniscule technical details can shift the power relationships between the participating actors and become expressions of political interests within the system.

Implications for urban design

Cooperatives focusing on the collective use (sometimes temporary), design, and governance of residual spaces have blossomed in many cities. The city of Berlin houses a particularly active community of temporary use of residual or empty urban spaces, co-working initiatives, builder collectives, and co-housing initiatives (Ring 2013). Local administrations have experimented with models of co-production of housing projects and public services. They are often inspired by the practices of participatory design and co-design, which have originally emerged from Scandinavian labor unions during the first waves of digitization in the 1970s (Kensing and Blomberg 1998). The degree to which digital technologies shape models of shared stewardship ranges from simply facilitating coordination and communication to more central roles. Similar to empowering devolution, the paradigm of shared stewardship tends to emphasize the local context rather than the systemic scale, while also addressing overarching questions of governance.

Critiques

Traditionally, critiques of digital governance models have centered on questions of access, inclusion, and the digital divide. This concern no longer holds economically, as access to digital technology is cheaper than most other basic products. Age, language, cognitive abilities, and the digital divide still remain important issues.

Complex governance structures are difficult to represent in algorithmic approaches. Considering the wide range of shared governance projects, it is not always clear to what extent technology was critical for their success. Digital technologies thrive in situations requiring the execution of simple tasks at a massive scale. As issues of governance become more complex, the potential benefit of technology diminishes. In the best case, technology plays a facilitating role; in the worst case, it becomes an obstacle. The inherently informal and improvisational nature of governance negotiations also does not mix well with the rigid hyper-formalization of smart contracts and blockchain technologies, which execute formalized contracts without any space for interpretation and negotiation (Lehdonvirta 2016).

Conclusion

The four value paradigms for urban technologies discussed in this chapter are grounded in a system of practices and beliefs concerning the nature of urban change, its underlying problems, and appropriate courses of action. Urban planners have to ask themselves not only which values they subscribe to, but also which values a particular technology is aligned with in respect to its histories and practices, as well as which issues it can and cannot address. Comprehensive approaches that focus on system-wide efficiencies can be desirable in certain contexts (i.e. when resource consumption and infrastructure management are central challenges, which they often are) and inappropriate in others (e.g. when a concern for efficiency hinders civic discourse). Systemic interventions such as citywide networks of digital interfaces such as kiosks, responsive lampposts, or dynamic signage systems are aligned with the modernist planning tradition and emphasize a seamless experience of urban space. Most would agree that a paradigm driven by the principles of responsiveness to the needs of citizens and transparent conduct of government is wholly desirable. With regard to urban planning, the paradigm of responsive transparency promotes an incrementalism that is directly opposed to the systemic scope of seamless efficiency. Driven by civic technology tools such as citizen feedback systems, the approach is sensitive to the local context and to individual needs but can exhaust the capacity of government by focusing on short-term issues and distract from long-term systemic problems. In the paradigm of empowering devolution, communities co-produce public services and take over more responsibilities for shaping their environment. Decentralization and informalization as demonstrated by the open-source community, the maker movement, or pop-up urbanism can foster creativity and local solutions but require keeping an eye on larger questions of equity and public responsibility. Finally, paradigm of shared stewardship focuses on questions of community governance and the difficult process of reaching consensus about the use of scarce resources. Digital tools such as liquid democracy[5] are meant to facilitate this process, but it is not always clear to what extent such technologies can address the ambiguities and informal negotiations of governance in small groups.

The paradigms described come with a number of caveats. The paradigms are not mutually exclusive, and their real-world manifestations are mostly hybrid. Responding to criticisms of techno-centricity, current smart city proposals emphasize civic and participatory elements. The boundaries between the paradigms of responsive transparency, empowering devolution, and shared stewardship are fluid, with most projects combining elements from each paradigm in different proportions.

Furthermore, the list of four paradigms could be extended. Some might argue that the phenomenon of platform urbanism (the provision of urban services through match-making platforms and crowdsourced marketplaces) deserves its own paradigm due to the ubiquitous presence of platform companies as recently described by economists (Evans and Schmalensee 2016; Parker, Alstyne, and Choudary 2016). I have decided against this choice, because its scope would be narrower than the four others, and it could be seen as a special case of decentralization and devolution.

The four paradigms are grouped by normative ideas about how cities should be designed, or how societies should be governed. Real-world projects, however, are rarely ideologically pure and usually combine conflicting, even contradictory beliefs. Design intentions and goals may be explicitly articulated, but whether they match the actual outcomes is a different matter. Most urban design or infrastructure projects take shape in a complex network of many actors with different interests (Hughes 2000). As projects progress, goals and constituencies can morph and change shape.

Nevertheless, I believe it is important to shift the discourse around emerging urban technologies from a discourse of technical possibilities to a discourse of values and goals. Too often a technology that has proven successful in a particular situation is generalized and promoted as a universal solution.

An integrated approach to planning with technologies

The numerous interconnections between the paradigms show that simple conceptual models are not sufficient to guide the practice of planning and design. Such reductive models include the technological determinism of tech companies, which identifies technology as the agent of change and explains its impacts through its intrinsic qualities, as well as the social determinism of traditional social science, which considers technology neutral, a mere representation of larger socioeconomic forces that can be safely neglected from consideration.

Digital technologies require an integrated design approach, which can no longer be neatly constrained into isolated disciplines with distinct responsibilities and scales on which they operate. As crowdsourced navigation apps can disrupt entire transportation networks, technological design decisions on the micro scale have a noticeable effect at the urban scale.

A first challenge is to make these multi-scalar interconnections legible. Civic technologies have implications for how citizens read the city and its governance. The same technologies structure how planners, urban designers, and administrators read the city, as they try to extract knowledge from a multiplicity of heterogeneous data sources. This reading, however, is increasingly determined by algorithms that curate and filter the digital representation of the environment (Graham 2005; Hamilton et al. 2014). As light poles and park benches become conduits for data collection, design has implications for the enactment of governance, accountability, and the rights to privacy and anonymity in public space. Instead of tokenistic concepts such as the smart city, planners and urban designers should confront an assemblage of heterogeneous technologies, which are at the center of an equally diverse range of participatory cultures and practices.

Notes

1 Task Rabbit is online marketplace for freelance labor, mostly used for tasks such as assembling furniture; see www.taskrabbit.com.
2 Alfred is an online marketplace for domestic workers and their clients; see www.helloalfred.com.
3 In the United States, such systems are often described as 311 systems based on the telephone shortcode for reporting non-emergency issues.
4 See LO3 Energy. (2018). The Future of Energy. Retrieved March 16, 2019, from https://lo3energy.com
5 See, for example Liquid Democracy e.V. (2018). Liquid Democracy. Retrieved March 16, 2019, from https://liqd.net/en/

References

Alemi, F., Circella, G., Handy, S., and Mokhtarian, P. (2017). "What Influences Travelers to Use Uber? Exploring the Factors Affecting the Adoption of On-Demand Ride Services in California." *Travel Behaviour and Society*, 13: 88–104.

Alfeld, L.E. (1995). "Urban Dynamics: The First Fifty Years." *System Dynamics Review* 11 (3): 199–217.

Arnstein, S.R. (1969). "A Ladder of Citizen Participation." *Journal of the American Institute of Planners* 35 (4): 216–224.

Bovaird, T. (2007). "Beyond Engagement and Participation: User and Community Coproduction of Public Services." *Public Administration Review* 67 (5): 846–860.

Correa, D. (2015). "Launching a Smart Cities Initiative to Tackle City Challenges with Innovative Approaches." *Whitehouse.Gov* (blog). September 16. https://obamawhitehouse.archives.gov/blog/2015/09/16/launching-smart-cities-initiative-tackle-city-challenges-innovative-approaches.

Egyedi, T.M., and Mehos, D.C. (2012). *Inverse Infrastructures: Disrupting Networks from Below*. Cheltenham, UK: Edward Elgar Publishing.

Evans, D.S., and Schmalensee, R. (2016). *Matchmakers: The New Economics of Multisided Platforms*. Harvard Business Review Press.

Flood, J. (2010). *The Fires: How a Computer Formula, Big Ideas, and the Best of Intentions Burned Down New York City- and Determined the Future of Cities*. Penguin.

Foderaro, L.W. (2017). "Navigation Apps Are Turning Quiet Neighborhoods into Traffic Nightmares." *The New York Times*, December 24, 2017, sec. N.Y./Region. www.nytimes.com/2017/12/24/nyregion/traffic-apps-gps-neighborhoods.html.

Forrester, J.W. (1969). *Urban Dynamics*. Cambridge, MA: MIT Press.

Forrester, J.W. (1970). "Systems Analysis as a Tool for Urban Planning." *IEEE Transactions on Systems Science and Cybernetics* 6 (4): 258–265. https://doi.org/10.1109/TSSC.1970.300299.

Goldsmith, S., and Crawford, S. (2014). *The Responsive City: Engaging Communities Through Data-Smart Governance*. 1st edition. Jossey-Bass.

Goodspeed, R. (2014). "Smart Cities: Moving beyond Urban Cybernetics to Tackle Wicked Problems." *Cambridge Journal of Regions, Economy and Society*, August, rsu013. https://doi.org/10.1093/cjres/rsu013.

Gordon, E., and Walter, S. (2016). "Meaningful Inefficiencies: Resisting the Logic of Technological Efficiency in the Design of Civic Systems." In *Civic Media: Technology, Design, Practice*, 243. Cambridge, MA: MIT Press.

Graham, S. (2005). "Software-Sorted Geographies." *Progress in Human Geography* 29 (5): 562.

Greenfield, A. (2013). *Against the Smart City: A Pamphlet*. The City Is Here for You to Use 1. New York, NY: Do projects.

Greenfield, A. (2017). *Radical Technologies*. Verso. http://lib.myilibrary.com?id=1006751.

Greenfield, A., and Kim, N. (2013). *Against the Smart City (The City Is Here for You to Use)*. New York, NY: Do projects.

Halpern, O., LeCavalier, J., Calvillo, N., and Pietsch, W. (2013). "Test-Bed Urbanism." *Public Culture* 25 (2 70): 272–306.

Hamilton, K., Karahalios, K., Sandvig, C., and Eslami, M. (2014). "A Path to Understanding the Effects of Algorithm Awareness." In *CHI '14 Extended Abstracts on Human Factors in Computing Systems*, 631–642. CHI EA '14. New York, NY, USA: ACM. https://doi.org/10.1145/2559206.2578883.

Hardin, G. (1968). "The Tragedy of the Commons." *Science* 37 (162): 48–57.

Hughes, T.P. (1987). "The Evolution of Large Technological Systems." *The Political Economy of Science, Technology, and Innovation*: 51–82.

Hughes, T.P. (2000). *Rescuing Prometheus: Four Monumental Projects That Changed Our World.* Random House Digital, Inc.

Ito, M., Okabe, D., and Matsuda, M. (2006). *Personal, Portable, Pedestrian: Mobile Phones in Japanese Life.* Cambridge, MA: MIT Press.

Kelly, K. (2010). *What Technology Wants.* 1st Printing edition. New York: Viking.

Kensing, F., and Blomberg, J. (1998). "Participatory Design: Issues and Concerns." *Computer Supported Cooperative Work (CSCW)* 7 (3): 167–185.

Kitchin, R., and Dodge, M. (2011). *Code/Space: Software and Everyday Life.* Cambridge, MA: MIT Press.

Kitchin, R., and Dodge, M. (2017). "The (In)Security of Smart Cities: Vulnerabilities, Risks, Mitigation, and Prevention." *Journal of Urban Technology*, December, 1–19. https://doi.org/10.1080/10630732.2017.1408002.

Lathrop, D., and Ruma, L. (2010). *Open Government.* O'Reilly Media, Inc.

Lehdonvirta, V. (2016). "The Blockchain Paradox: Why Distributed Ledger Technologies May Do Little to Transform the Economy." *Internet Policy Blog* 21.

Lessig, L. (2009). "Against Transparency." *The New Republic*, October 8. https://newrepublic.com/article/70097/against-transparency

Martínez, J.A., Pfeffer, K.H., and van Dijk, T. (2009). "The Capacity of E-Government Tools: Claimed Potentials, Unnamed Limitations." In *Proceeding of the 10th N-AERUS Conference: Challenges to Open Cities in Africa, Asia, Latin America and the Middle East: Shared Spaces within and Beyond, IHS Rotterdam.*

Mayer-Schönberger, V., and Cukier, K. (2013). *Big Data: A Revolution That Will Transform How We Live, Work, and Think.* Houghton Mifflin Harcourt.

Mitchell, W.J. (1995). *City of Bits: Space, Place, and the Infobahn.* Cambridge, MA: MIT Press.

Mitchell, W.J. (2000). *E-Topia.* Cambridge, MA: MIT Press.

Morozov, E. (2014). *To Save Everything, Click Here: The Folly of Technological Solutionism.* PublicAffairs.

Offenhuber, D. (2014). "Infrastructure Legibility: A Comparative Analysis of Open311-Based Citizen Feedback Systems." *Cambridge Journal of Regions, Economy and Society*, March. https://doi.org/10.1093/cjres/rsu001.

Offenhuber, D. (2017). *Waste Is Information: Infrastructure Legibility and Governance.* Cambridge, MA: MIT Press.

Offenhuber, D., and Schechtner, K. (2017). "Improstructure: An Improvisational Perspective on Smart Infrastructure Governance." *Cities.* https://doi.org/10.1016/j.cities.2017.09.017.

Ostrom, E. (1990). *Governing the Commons: The Evolution of Institutions for Collective Action.* Cambridge: Cambridge University Press.

Parker, G.G., Van Alstyne, M.W., and Choudary, S.P. (2016). *Platform Revolution: How Networked Markets Are Transforming the Economy and How to Make Them Work for You.* 1st edition. W.W. Norton & Company.

Ratto, M., Boler, M., and Deibert, R. (2014). *DIY Citizenship: Critical Making and Social Media.* Cambridge, MA: MIT Press.

Ring, K.K. (Ed.). (2013). *Self-Made City: Self-Initiated Urban Living and Architectural Interventions.* Bilingual edition. Berlin: Jovis.

Sassen, S. (2011). "Open Source Urbanism." *Domus*, June 29. www.domusweb.it/en/op-ed/2011/06/29/open-source-urbanism.html.

Townsend, A.M. (2013). *Smart Cities: Big Data, Civic Hackers, and the Quest for a New Utopia.* New York, NY: W.W. Norton & Company.

Verplanke, J., Martinez, J., Miscione, G., Georgiadou, Y., Coleman, D., and Abdishakur, H. (2010). "Citizen Surveillance of the State: A Mirror for eGovernment?" In *What Kind of Information Society? Governance, Virtuality, Surveillance, Sustainability, Resilience*, 185–201. Berlin; Heidelberg New York, NY: Springer.

Vignal, Ms. L. (2014). *Under Construction: Logics of Urbanism in the Gulf Region.* Ashgate Publishing, Ltd.

Waal, M. de. (2014). *The City as Interface: How New Media Are Changing the City.* Rotterdam: Nai010 Publishers.

Zinnbauer, D. (2012). "'Ambient Accountability': Fighting Corruption When and Where It Happens." SSRN Scholarly Paper ID 2168063. Rochester, NY: Social Science Research Network. https://papers.ssrn.com/abstract=2168063.

Further reading

Gordon, E., and Mihailidis, P. (2016). *Civic Media: Technology, Design, Practice.* Cambridge, MA: MIT Press. An anthology focusing on the role of digital media in the civic realm, in areas of participation, community engagement, and governance.

Greenfield, A. (2017). *Radical Technologies.* Verso. http://lib.myilibrary.com?id=1006751. A critical discussion of emerging technologies and their not-so-obvious societal and political implications.

Halpern, O. (2015). *Beautiful Data: A History of Vision and Reason since 1945.* Durham, NC: Duke University Press. A historical analysis that traces the emergence of the smart city concept from the history of cybernetics and systems-oriented design since 1945.

Kitchin, R. (2014). *The Data Revolution: Big Data, Open Data, Data Infrastructures and Their Consequences.* Thousand Oaks, CA: Sage Publications Ltd. An introduction into the methodological toolbox of data science, their critiques, and implications for space and society.

Willis, K.S., and Aurigi, A. (2017). *Digital and Smart Cities.* London and New York: Routledge. A thorough discussion of smart city technologies at the intersection of urban planning and data studies.

41

SMART CITIZENS, PARTICIPATORY URBANISM, AND THE FUTURE OF CITY DESIGN

Jennifer Stein and Scott Fisher

Introduction

In 1966, architect Cedric Price offered the following provocation as the title of his lecture on the impact of technology on architecture: "Technology is the answer, but what was the question?"[1] The following year, in 1967, Henri Lefebvre called for radical city change in which citizens demand their "Right to the City." In 2007, Apple introduced the iPhone, the first smartphone to be widely adopted, transforming the ways in which people access and share information – including GPS, photos, tweets, the Internet – in real time and real space. Half a century after Price's and Lefebvre's provocations directed at architects, cities, and citizens, and just over a decade after the introduction of the iPhone, among other mobile smartphones, we are at a crossroads in urban design, as we must reconsider what the question is, and how technology is serving our hybrid physical and digital cities, and the people who make up their lively urban fabric.

Over the past decade, smart cities have been envisioned as powerful sensor-rich technology platforms that promise seamless data-informed city efficiencies, specifically for mobility, energy, buildings, and infrastructure. This vision has predominantly been led by large technology companies like Cisco, IBM, and, more recently, Sidewalk Labs,[2] all of which often imagine how new smart cities can be built from the ground up. The results of these top-down visions are often fairly generic "cities in a box," including New Songdo City in South Korea, Masdar City in Abu Dhabi, and future smart cities like Lagos Smart City in Nigeria, and Sidewalk Toronto. But these smart cities often fail to include in their visions the lived experiences of their residents, or their enterprises – the often inefficient and messy human bits that make cities thrive.

In response to these large infrastructural "smart" developments, we have seen a rise of bottom-up, citizen-led initiatives to reclaim the smart city for more meaningful, human-scale smart city interventions. Examples often include neighborhood-based projects that invite citizens to engage in participatory urbanism and citizen science, developing their own mobile smartphone applications or local sensor networks. But just as large, expensive smart city projects often prove to be too invisible or too inaccessible to citizens, many smart citizen projects go undiscovered or fail to gain traction with larger swaths of a city's population. We must therefore establish ways to meaningfully transition from one-off, citizen-led projects to long-term systemic and sustainable projects in the smart city.

Somewhere in between the top-down smart city and the bottom-up smart citizen exists an opportunity for urban intermediaries – universities, cultural institutions, city laboratories – to negotiate possibilities for more playful, participatory, and innovative smart cities with engaged citizens in mind. As noted by urbanist and designer Dan Hill (2013) in his website regarding smart cities and the need for a changing focus towards governments and citizens:

> But as well as a new urban hardware and software, it's in this interface between engaged citizens and engaged government that its real promise may lie, as evidenced by the way that citizens are racing ahead while the smart cities movement lags behind.

Going forward, urban designers and city planners can learn from a number of current efforts in both the Global North and the Global South, which will be described in more detail in the case studies in this chapter.

Smart cities then and now

Early smart city narratives expounded the values of sensor-rich infrastructure and buildings that could monitor energy usage, traffic signals and flows, air quality, and water usage across the city. The massive amounts of networked data produced by these sensors would be visualized through dashboards for cities and buildings, designed as a way to make visible the invisible systems and demonstrate the efficiency of smart cities. As noted by Hill (2013), "Partly due to privacy fears, but perhaps also due to a more general discomfort with engaging the messiness of humans, smart cities tend to sense objects not people, infrastructure not culture."[3] Existing smart city systems produce seamless feedback loops for city planners, for example real-time traffic management and air quality sensing, and lead to programmable, efficient cities. However, these systems are not necessarily accessible to citizens and are therefore often meaningless to the people who inhabit these smart cities.

The past decade has seen unprecedented technological advances that push us beyond the original vision for hyper-efficient, data-driven smart city infrastructure. Smartphones, GPS, and faster networks like 4G have put more power and information, including real-time maps, sensors, and services, into the hands of citizens. Platforms for the sharing economy,[4] like Lyft, Uber, Airbnb, and Postmates, have made cities more accessible and efficient for many citizens, redefining what a smart city might mean. Social media and real-time location-sharing by urban dwellers have mobilized with one tweet people to action, rapidly organizing protests and demonstrations, such as "Arab Spring" and the Occupy movements, and more recently Women's Marches across the United States. Efficiencies in cities have become more citizen-driven as people begin to engage more deeply and in new ways with their cities, in real time and real place.

Furthermore, great advances in the development of autonomous vehicles have been made in the past five years and are expected to hit the roads within the next two years. These vehicles are reshaping how we think about planning cities and streets of the near future, and could offer new possibilities for public and shared transportation. In addition, drones and other autonomous aerial devices have already started to be used for delivery, sensing, and surveillance in cities. Taking these recent developments as a whole, concepts of a smart city portend to rely less on the "city in a box" model and look more towards highly connected services and devices, many of which exist in the hands of citizens.

Smart citizens and participatory urbanism

In *The Right to the City*, Lefebvre (1996) imagines a radical reconceptualization of cities, based on the social needs of citizens beyond planning and consumption. As he explains, there are "specific needs which are not met by planners. This refers to the need for creative activity, for the oeuvre (not only of products and consumable material goods), of the need for information, symbolism, the imaginary and play" (147). Though Lefebvre did not live to see the smart city in its current instantiation, his recognition of the social needs for play and information was prescient, and most relevant to the notion of smart, engaged citizens in our current technological urban landscape.

Though technology has the potential to connect citizens to the city, it does not necessarily make the city smart. However, giving people access to the data and platforms that make up "smart city" infrastructure moves us closer to an engaged, smart citizen model, in which citizens can use data and information to help shape their cities into more livable, sustainable, and just places. As articulated by Hemmet and Townsend (2013: 2) in *Smart Citizens*:

> The notion of the Smart Citizen as a co-creator draws on a rich intellectual backdrop in both technology design and urban design. Digital culture has given rise to a collaborative code ethic, and there has been a trend towards applying thinking and methods from open source software development to other domains.

With this in mind, we argue that smart citizens have the power to help shape their cities environmentally, politically, economically, and culturally.

To explore this more deeply, we propose the following questions: *What do citizens want from their smart cities? How have readily available smart technologies – smartphones, social media, sensors – reshaped citizens' vision of the cities in which they live? How do city spaces become smart, while inviting citizens to actively participate in and shape new experiences in the city?* Going forward, we will not simply define smart cities as large top-down infrastructure projects; we will also consider how personal everyday technologies have played a part in making cities – and their citizens – smarter. We will look at specific instances of smart citizenship and participatory urbanism within more formalized examples than the many citizen-led smart projects that have been developed over the past decade.

As described by computer scientist and artist Eric Paulos:

> In the spirit of Urban Computing, Participatory Urbanism is the open authoring, sharing, and remixing of new or existing urban technologies marked by, requiring, or involving participation, especially affording the opportunity for individual citizen participation, sharing, and voice. Participatory Urbanism builds upon a large body of related projects where citizens act as agents of change.
>
> *(Paulos et al. n.d.)*

Communities in cities across the globe are appropriating smart city systems, big data, and social media to co-create participatory urban interventions, while others are inventing creative solutions for everyday problems that go unrecognized or ignored by city officials, planners, and urban designers.

Many local creative or arts-based community organizations, both formal and informal, have emerged to engage citizens in reimagining the cities they want in light of rapid "smart" development, while taking action to fulfill those visions. Participatory urbanism and civic engagement within the smart cities context often take the form of open data "hackathons," public art projects that aim to create meaningful and playful engagement with infrastructure, and digital platforms for communicating visualized city-specific information. Furthermore, with an estimated 2.4 billion people using smartphones worldwide in 2017, there is a growing opportunity to engage smart citizens worldwide in a massive effort to collect and share data about their cities and visualize such information.

When we think about the smart, participatory city made up of engaged citizens, we can also imagine the possibility of using smart city infrastructure, to create a digital public space comprising projects that engage citizens, and inspire them to further participate in digital placemaking and smart citizenship. We see many opportunities for meaningful and smart collaboration between cities and citizens leading to such outcomes as creating centers for urban innovation.

In addition to offering alternative approaches to what makes a city *smart*, this chapter will consider how much technological change we have seen in the last decade and what it might mean for the next decade. What will 5G connectivity and augmented reality mean for the future of smart cities, smart citizens, and urban planners and designers? How will machine learning, artificial intelligence, and automation contribute to how a smart city sees and thinks? And what is the role of the smart citizen in this future? Finally, if new smart city projects like Sidewalk Toronto continue to promise data-driven efficiency for cities and reflect citizen representation for future planning purposes, it is important to ask who is being represented within that data and if that representation is equitable and inclusive of all city inhabitants.

Case studies

The following four case studies offer examples of how smart city narratives can be both explored and juxtaposed as participatory, smart citizenship through playable interactions, social media, and citizen-based smart sensing. These examples, from both the Global North and Global South, highlight how citizens and cultural and arts organizations can make cities smarter and more accessible, and consider a middle ground between top-down technology solutions and bottom-up citizen-led initiatives. To do this, we will examine initiatives within academia and at the city government level, which often function as creative intermediaries in "smart," innovative projects. As noted by Catherine Mulligan (2013: 85), research fellow at Imperial College, "A robust debate between corporates, citizens, NGOs, academics, city leaders and technologists is required to push the smart city debate forward and fulfill its promise of a balance between environment, economy and the citizenry."

These case studies explore a number of different approaches to smart city interventions that consider different time horizons and city scales, from playful and playable, to participatory visual data collection, to sensor-based future visions. The role of the arts, interaction design, and hacking the city inform the choice of examples in what follows, arguing for a redefinition of smart cities and new approaches to smart citizenship. Rather than thinking of cities becoming smart merely through a layer of new technologies, we consider the role of innovation, imagination, and participation as paths to smarter cities. The following approaches for how technologies can organize and activate others in the city will be reflected in these case studies: City Infrastructure used Playfully; Innovation and Creative Urbanism; Social Media for Smart Citizen Engagement; and Speculative Smart Cities and Citizens.

City infrastructure used playfully: Playable City, Bristol, United Kingdom

Playable City, a Bristol, UK-based arts initiative developed in 2011 by Clare Reddington, Creative Director of the Watershed, demonstrates the power of playful experiences to connect people to their cities, and to think about how cities are designed. Developed as a way of rethinking the concept of the "smart city" from within a cultural institution, this program invites interactive art installation submissions that playfully use or misuse city infrastructure to create more playable, accessible, and equitable interactions with the city. At the heart of the Playable City initiative is the desire to think about technology as a way to connect people to each other and interact with the city, rather than technology as the thing that will make a city smart.

One of the first installations commissioned by Playable City, *Hello Lamp Post* by Pan Studio, encouraged citizens to interact with everyday city infrastructure in new ways. *Hello Lamp Post* made parts of the city playable that had become so engrained in everyday life that they had become invisible. These included urban objects such as mailboxes, lampposts, and bus stops, and they were made playable by asking citizens to "talk" to them via mobile text message. Citizens could awaken a mailbox by sending it a text message with a unique code located on the box. The mailbox would then ask the citizen questions via a text message, and a citizen's answers would inform the street objects' conversation with the next citizen to engage with the object. *Hello Lamp Post* plays with the notion of a citywide IOTs and, rather than thinking about smart city infrastructure, considers how everyday infrastructure could become smarter, friendlier, and playable.

Though Playable City initially commissioned projects in Bristol and London beginning in 2011, it has expanded to cities across the world, considering the cultural implications embedded within each city by exploring what "play" means in Seoul, Tokyo, Lagos, and Recife. The experiences, values, and lessons learned through the Playable City program led to the development of the Creative Producers International program in 2017. This program invites city change makers from around the world to participate in a three-week residency at the Bristol-based Watershed, provides small grants to each creative producer, and nurtures this community over three years to develop playful public art projects in their own cities. The first cohort of creative producers represent a diverse set of world cities, including Durban, Copenhagen, Tampa, Liverpool, Almatay, Tokyo, Sydney, and Mexico City.

This program provides a compelling example of how urban designers might think about the role of emerging technologies in cities, by supporting and empowering local creative producers to develop city-specific, playful interventions that are context- and culturally specific, while engaging with "smart city" narratives in new ways. Moving towards a model that engages bottom-up city change makers (activists, artists, and hackers passionate about their cities) in the process of top-down design from city planning offices offers a balance of voices towards smart citizenship within a smart city ecosystem.

Innovation and creative urbanism: Laboratorio para la Ciudad, Mexico City, Mexico

The Laboratorio para la Ciudad (Laboratory for the City) is a civic innovation and urban creativity lab embedded within Mexico City's government offices. The first of its kind in Latin America, "the Lab is a space for rethinking, reimagining, and reinventing the way citizens and government can work together towards a more open, more liveable and more imaginative city" (Laboratorio para la

Ciudad website).[5] As a different kind of smart city project initiated by Mayor Miguel Ángel Mancera in 2013, the Lab functions as a "skunkworks" space, or a space for unstructured radical innovation, for experimentation between the government and citizens of Mexico City. Laboratorio's director, visual artist and documentary filmmaker Gabriela Gomez-Mont, sees this initiative as an experimental think tank for the city, bringing together an interdisciplinary team of architects, technologists, editors, art historians, political scientists, journalists, urban planners, filmmakers, sociologists, designers, and urban psychologists, all of whom believe in change in the city through innovation and creativity.

Many of Laboratorio's projects bring citizens into discussions and decision-making from the very beginning and invite them to be active participants through provocations, prototyping experiments, and open data hackathons. For example, Mapatón CDMX[6] is a citizen-crowdsourcing exercise to create an open data map of all bus routes in Mexico City. Designed as a city-scale team-based game, Mapatón CDMX invited citizens to playfully and collaboratively track their public transit journeys – which include 30,000 minibuses, buses, and trolleys – across the city to fill in missing data about schedules, routes, and end-points at the outer edges of the city. Though the end goal was to develop a robust open-source database of transit routes, the use of game mechanics and playful citizen collaboration demonstrated the possibilities for everyday citizen-based activities to contribute to the creation of datasets to be used by citizens at large.

As a developing-world city, Laboratorio considers innovative solutions to citywide problems in Mexico City through not only technologies but also arts and citizen-centric innovation. With a clear understanding that the digital divide is a prevalent problem for the city, Laboratorio considers how to make the city smarter for everyone, not just those with access to technology. In thinking about innovation and creative urbanism, they underscore that:

> Many of our experiments take advantage of digital technology to design new participatory platforms and make government more open. . . . Still, we cannot rely exclusively on digital technologies in a city where only half of its inhabitants have access to the Internet on a regular basis. What other creative ways can we connect government and citizens?
>
> *(Laboratorio para la Ciudad website)*

When considering the role of creative and cultural intermediaries and how they can unite smart cities to smart citizens, government offices are often seen to be part of the problem – a clunky bureaucratic system that stifles innovation and creativity. However, Laboratorio makes it an explicit part of its mission to work against that sentiment. Rather, they ask: "How do we reconnect cities and citizens through government itself?" and explains that: "We believe that the growing disconnect between citizens and government, sometimes broadened by one-way conversations and lack of trust, can be addressed and actually reverted by reimagining roles and repurposing traditional mechanisms of engagement" (Laboratorio para la Ciudad website). This approach to innovation and creativity, which includes a highly interdisciplinary team of instigators and change makers, should be considered more broadly in future planning practices for engaging smart citizens in more participatory smart cities.

Social media for smart citizen engagement: Emerge LA, Los Angeles, California

Social media interventions have significantly reshaped citizen engagement in cities over the past decade. From the ousting of the Philippines dictator, Ferdinand Marcos, in the 1990s via a spontaneous protest coordinated by a single text message to many citizens,[7] to more recent riots and then post-riot cleanup efforts in East London in 2014, social media outreach and organization have transformed the ways in which citizens enact their right to the city. This next case study looks at how universities can act as smart city intermediaries by designing social media-based research tools for urban data collection at the street level. The *Emerge LA* project is a mobile application designed by the Emergent Cities Research Group, an interdisciplinary research group at the University of Southern California that demonstrates how urban designers, policymakers, arts researchers, and citizens can collaboratively map, visualize, and ask meaningful questions about their ever-changing cities and neighborhoods.

The *Emerge LA* app (Figure 41.1) spontaneously asks important questions about cities by presenting unique participatory opportunities for citizens to measure their cities through visual, qualitative data, such as photos, video, and audio recording using social media. These questions range from "What do you value most in your neighborhood?" to "What would you like to see change?" to "What building would you like to preserve?" – all of which could be answered through photos

Figure 41.1 *Emerge LA* mobile interface showing daily prompts to users (left) and images submitted to answer questions based on location (right).

Source: Authors

taken in the mobile application. With these kinds of questions, citizens have the opportunity to show researchers, city planners, and local governments real-time and location-specific examples of things they either love or would like to change about their city through emerging forms of crowdsourcing and participatory urbanism.

The kind of visual data generated by *Emerge LA* is easily uploaded and shared among participants with smart mobile devices, and it is also potentially more easily understandable to a critical mass of people. By designing playful interactions to collect data, the *Emerge LA* app sends daily requests for crowdsourced data, and is a way for citizens to actively participate by submitting their own questions. In addition, *Emerge LA* encourages citizens to make new discoveries in their neighborhood or the city at large, and to be able to see what others value, like public art or fresh produce, or things they would like to see changed, like access to more resources. This form of mobile phone-based participatory urbanism acts as a tool for generating meaningful stories within communities, where they can share their concerns about and visions of their neighborhoods and cities. Active engagement and sense of agency in data collection and visualization bestowed upon citizens have the potential to provide a much richer dataset than numbers alone.

Emerge LA introduces a different approach to the smart city, one that actively engages citizens in reporting back about their cities, goes beyond discovering patterns in qualitative data or making sense of complex problems through mobile data collection by giving agency to city dwellers to visually and sonically share their concerns through photos, videos, and audio recordings. Crowdsourced participatory urbanism can help city council members search for and view problems within their specific regions via geo-tagged media, help planners see what is working and what isn't through visual user feedback, and also help arts researchers connect citizens to other nearby public art or architecture through common tags and access to geodata.

Speculative smart cities and citizens: Seoul On-Air, Korea, and Hyper Reality, Medellín, Colombia

This final case study views speculative design practices as a means to imagine near future smart cities and smart citizenry. Each example demonstrates how artists, designers, and educators speculatively explore future city scenarios as both cautionary tales and as ways to encourage future urban designers and architects to imagine new interfaces for smart cities and smart citizens.

> *SEOUL ON-AIR: Augmented Environments for Urban Activism* (Figure 41.2) "focuses on the implementation of low-cost mobile urban sensing technologies and immersive environmental data exploration mechanisms with the ambition to enable citizens to reframe their role in the politics of urban air quality."[8] This project, designed by Maider Llaguno-Munitxa and Biayna Bogosian for the 2017 Seoul Biennale of Architecture and Urbanism, imagines a three-dimensional augmented reality interface for citizen activists to collect data about air quality and monitor it daily. Instead of developing the kind of two-dimensional data dashboard used to visualize smart city data in the past, the artists imagine an immersive environmental visualization interface to offer a more detailed understanding of pollutant concentrations and thermal comfort in their cities.

In a related speculative design project, Bogosian and Llaguno-Munitxa, who are both architects and educators, developed *Urban Environmental Sensing with Drones*[9] as part of the DigitalFUTURES2017

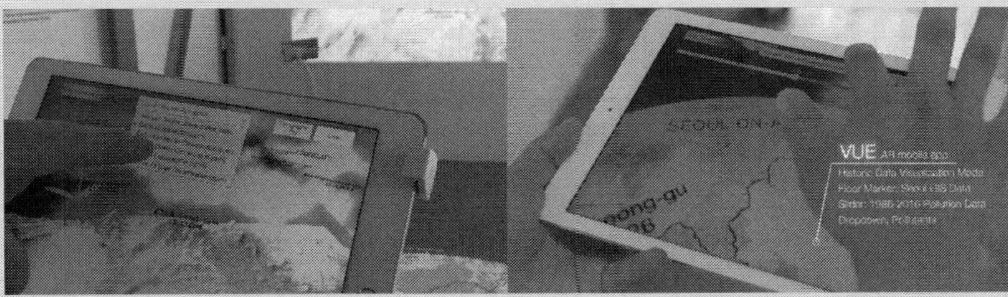

Figure 41.2 SEOUL ON-AIR Left figure, VUE app in Live mode. Right figure, VUE app in historic mode. Virtual Urban Environments (VUE) is a marker-based augmented reality mobile app that aims to enable off-site environmental data visualization and analysis.

Source: Photo by Biayna Bogosian

series organized by the College of Architecture and Urban Planning at Tongji University in Shanghai. Students imagined near future scenarios for a number of bicycle and drone-based environmental sensors that captured spatiotemporal aerial mapping of urban environments, and they developed a short film to tell a day-in-the-life story about how we might interact with our environments. By imagining alternative ways for cities and environments to be sensed, seen, and visualized, students considered how emerging technologies like drones and augmented reality glasses might become part of smart city and smart citizen narratives of the future.

In another near future vision of augmented reality in the urban setting, designer and filmmaker Keiichi Matsuda offers a provocative series of design fiction films titled *Hyper Reality*. Filmed in Medellín, Colombia, the *Hyper Reality* series imagines the experience of both driving in and walking through a hyper-mediated augmented reality layer over the urban landscape. Though *Hyper Reality* might present a hyperbolic vision of the future city, with its highly personalized and always-on augmented interface, it raises the concern of a smart city that might know too much about its citizens and asks the viewer to consider how much technology might be too much.

As a method for both critical thinking and playful world-building, speculative design prototyping and filmmaking afford artists, designers, and educators evocative ways to think about and visualize possible smart city and smart citizen futures.

Conclusion

Just as Lefebvre's *Right to the City* encouraged a radical reimagining of the role of city spaces in everyday life, so too did the smartphone bring about radical change to the ways in which citizens navigate, interact with, and capture their everyday lives in cities, in both the Global North and South. In the past decade, the scale has tipped so that more than half of the world's population now resides in urban areas. At the same time, technological innovation has brought about a sea change to our experience of cities and in our everyday lives. Faster networks, more connected objects within the Internet of Things, autonomous vehicles, augmented reality, machine learning, and artificial intelligence will continue to fundamentally transform how a growing number of engaged citizens and increasingly smart cities interact.

Emerging technologies will present many challenges and opportunities for city planners and urban designers. With this in mind, urban designers should look closely at the ways in which citizens use and misuse cities, how they access and use data, and how public spaces are impacted by an ever-growing number of connected devices and data-generating infrastructure. Planners and urban designers should further look at how social media is used to rapidly disseminate information, and how it has become more common across larger citizen populations as the first point of contact for news or updates related to the cities where people live. The evolving concepts of what makes a city smart, leading to new considerations in city planning, point to the importance of policymakers and intermediaries that work at the intersection of smart cities and engaged citizens.

At the same time, it is necessary for planners and urban designers to carefully consider who is and isn't being represented in data, and what that means for diversity and inclusivity in the kinds of urban services and experiences that are made possible by emerging technologies. Though the digital divide has decreased as mobile phones and Internet access have become more accessible and inexpensive to more people around the world, that is not to say that all are being represented equally in the kinds of projects and processes described in this chapter.

The case studies described in this chapter demonstrate this middle ground between smart cities and smart citizens, highlighting playful and innovative initiatives that take place at the intersection of cities and the people who inhabit them. Going forward, the role of active and engaged citizens needs to be considered in parallel to smart technology infrastructure, and needs to nurture spaces for creative and participatory interventions that bring together city governments and citizens to collaboratively imagine the future needs of cities. With Playable City and *Emerge LA*, playful social interactions are at the heart of a more participatory city experience, in which the city becomes smarter by engaging citizens in new ways. These two projects further demonstrate how cultural institutions and universities can assume the role of urban intermediary, while working closely with cities and citizens. The Laboratorio para la Cuidad, as an interdisciplinary skunkworks lab initiated by the city government, is nurturing both high-tech and low-tech participatory urbanism through civic innovation and urban creativity. Finally, artists, designers, and educators are using speculative design to both envision scenarios and explore the implications of the future of smart cities and smart citizens. Taken together, these approaches offer guidelines and opportunities for smarter and more participatory urbanism.

Though earlier smart city concepts consisted of city-in-a-box technologies and applications, which we argue often undervalued the role of smart citizens, a recalibration of this initial vision seems to be taking place. For example, the *Sidewalk Toronto Quayside Project* is moving closer to this vision for a more human-centric smart city. Though still ultimately led by a large technology company (Google/Alphabet), Sidewalk Labs has moved beyond the smart-city-in-a-box approach towards a more citizen-focused smart urbanism approach. Described as a "forward-thinking urban design and new digital technology to create people-centered neighborhoods that achieve precedent-setting levels of sustainability, affordability, mobility, and economic opportunity" (*Sidewalk Toronto* website), the vision for the *Sidewalk Toronto* quayside development includes the needs of citizens from the beginning, though it is still heavily focused on data collection as an integral part of the city.

The future of cities and citizens will continue to be enmeshed in technological change, as it has for hundreds if not thousands of years. With faster networks and more connected devices, along with rising concerns about data and privacy, urban design and city planning will encounter challenges to build not only smart, but equitable and inclusive cities for citizens, who have become increasingly more engaged and more participatory. Speculating about these futures through design prototypes and films offers a way to engage citizens, city governments, planners, urban designers, technologists, and educators in critical and playful dialogue about the kinds of cities we as citizens want to live in.

Notes

1 C. Price, "Technology is the answer, but what is the question?" quote taken from Cedric Price, London: Pidgeon Audio Visual/World Microfilms, 1979.
2 Alphabet (formally Google) Sidewalk Labs is a team of urbanists and technologists thinking about cities at the intersection of human-centric needs and emerging technologies. More is at: www.sidewalklabs.com/.
3 Hill, D (2013) "On the Smart City; Or, a 'Manifesto' for Smart Citizens Instead." From *City of Sound* blog: www.cityofsound.com/blog/2013/02/on-the-smart-city-a-call-for-smart-citizens-instead.html
4 It should be noted that these platforms are ripe with their own important issues and implications for cities that should not be overlooked, for example, contributing to housing crises (Airbnb) and unfair labor practices (Uber) in cities across the world.
5 https://labcd.mx/a-lab-for-the-city/.
6 More information about the game can be found at http://mapatoncd.mx/.
7 www.washingtonpost.com/wp-dyn/content/article/2006/08/24/AR2006082401379.html?noredirect=on
8 Seoul On-Air was exhibited at the 2017 Seoul Biennale of Architecture and Urbanism and is a collaborative project by Maider Llaguno-Munitxa and Biayna Bogosian.
9 Urban Environmental Sensing with Drones video: https://vimeo.com/224889539.

References

Hemmet, D. and Townsend, A. (eds.). (2013). "Here Come the Smart Citizens," in *Smart Citizens* (Vol. 4). FutureEverything Publications.

Hill, D. (2013). "On the Smart City: Or, a 'Manifesto' for Smart Citizens Instead." *City of Sound*, 1.

Jordan, M. (2006). *Going Mobile: Text Messages Guide Filipino Protesters*. Accessed from Washington Post Online, http://www.washingtonpost.com/wp-dyn/content/article/2006/08/24/AR2006082401379.html?noredirect=on

Lefebvre, H., Kofman, E., and Lebas, E. (1996). *Writings on Cities*. 63 (2):147. Oxford: Blackwell.

Mulligan, C. (2013). "Citizen Engagement in Smart Cities," in Hemmet, D. and Townsend, A. (eds.) *Smart Citizens* (Vol. 4), FutureEverything Publications.

Paulos, E., Smith, I., and Honicky, R.J. (n.d.). *Participatory Urbanism*. www.urban-atmospheres.net/ParticipatoryUrbanism/

Sidewalk Toronto. https://sidewalktoronto.ca

Case study references

Creative Producers International. www.playablecity.com/programmes/creative-producers-international/

EmergeLA. http://emerge-la.com/

Hyper Reality. http://hyper-reality.co/

Laboratorio para la Ciudad. http://labcd.mx/labforthecity/

Playable City. www.playablecity.com/

Seoul On-Air. http://seoulbiennale.org/en/exhibitions/thematic-exhibition-nine-commons/air/seoul-on-air%E2%80%8B

Urban Environmental Sensing with Drones. https://creators.vice.com/en_us/article/593xxa/drones-augmented-reality-cities-map-air-pollution

Further reading

Foth, M., Brynskov, M., and Ojala, T. (2016). *Citizen's Right to the Digital City: Urban Interfaces, Activism, and Placemaking*. Springer. ISBN 978-981-287-919-6 (eBook) Presents a number of ideas about how citizens have become activists and co-producers of cities, rather than passive citizens through new urban interfaces, citizen action, and new ways of city making.

Hemmet, D., and Townsend, A. (2014). *Smart Citizens*. Manchester, UK: FutureEverything Publications. Offers a number of insights and approaches in an edited volume of interdisciplinary thinking about cities and citizens from a global perspective.

Hill, D. (2013). "On the Smart City: Or, a 'Manifesto' for Smart Citizens Instead." www.cityofsound.com/blog/2013/02/on-the-smart-city-a-call-for-smart-citizens-instead.html (accessed from web on 10/10/17). A "manifesto" for smart citizens, it underscores the fundamental role of people in cities, the culture they create, and how new technologies shape the experience of cities for people.

42

DESIGNING CITIES WITHIN EMERGING GEOGRAPHIES

The work of Senseable City Lab

Fábio Duarte and Carlo Ratti

Introduction

Planners and urban designers have long dreamed of using urban data. Ildefonso Cerdà, the father of modern Barcelona, was one of many urbanists who dreamed of a more quantitative urbanism. Indeed, Cerdà grounded his proposals on quantitative descriptions, scientific studies, and statistics about the physical and social aspects of the city. In the early 20th century, Patrick Geddes based his pursuits to establish town and regional planning as a scientific discipline on data gathered at multiple scales. During the second half of the 20th century, urbanist William H. Whyte used on-site cameras to capture human flows at New York's public spaces. By extensively monitoring how people use urban plazas, Whyte wanted to understand whether and which physical elements and personal interactions induced different behaviors in public spaces. His work showed how quantitative data could reveal some of the intangible qualities of space and inform design. Whyte's methods were insightful but labor intensive. In the age of big data, this might change, opening up an unprecedented potential to urban planning and urban design.

More than a decade ago Howard Rheingold (2002, 86) noted that "[t]he kind of world we will inhabit for decades to come could depend on the technical architecture adopted for the emerging mobile and pervasive infrastructure over the next few years." Since Rheingold's remarks, global communication networks, the miniaturization of electronics, and the dramatic increase in data storage capability and processing speed have made digital technologies pervasive, productive, and powerful. The vital component that makes such technologies powerful is the generation and traffic of data from different sources that can be combined through multiple technologies. Indeed, the generation of huge amounts of data in real time and their fine-grained spatial and temporal resolution are beginning to play a major role in planning and urban design.

In 1995, Michael Batty (1995) proposed the vision of the "computable city." With the refinement of computer and digital technologies, this vision became what Rob Kitchin and Martin Dodge (2011) referred to as "programmable urbanism" – that is, our dynamic ability to sense a city's activity, understand its changes and fluctuations, and deliver tailored responses to meet the needs of its urban environment. This is partially due to a dimensional shift: we are now able to collect massive amounts of data of how urban activities change over time, whereas previously, space was generally perceived as a fixed or very slowly changing aspect of urban form. As William Mitchell (2001: 144) pointed out, "[b]y selectively loosening place-to-place contiguity requirements, wired networks produced fragmentation and recombination of familiar building types and urban pattern." Thus, we argue that the temporal dimension of urban life and urbanism is becoming a crucial outcome of contemporary urban design.

These changes have important consequences for urban planning and design. Until recently, urban planning was mostly concerned with large spatial and temporal scales. As Batty (2012) suggests, the little concern urban planning had for small-scale developments was partially due to the lack of available data. One might argue that the focus on comprehensive and long-term master plans, often based on coarse data updated every few years, which for long have characterized urban planning, was less an intrinsic characteristic of the field and more due to the lack of more detailed and dynamic data. The challenge urban scholars and designers face now is to find novel ways to integrate short-term big data with urban planning and design concerns and strategies.

Today, with the diffusion of handheld electronics, such as cell phones and smartphones, data collection is becoming effortless. With more than 7 billion mobile devices worldwide and 2.3 billion mobile-broadband subscriptions globally, cities in both the Global North and Global South are experiencing new forms of understanding urban phenomena, of creating social interactions, and proposing the future of cities. Moreover, global telecommunications and faster computational power are closing the temporal gap between data gathering, data processing and analysis, and actuation cycles.

In this chapter, we focus on one particular technology – movement and location sensing technologies embedded in personal mobile devices. We present a review of the use data collected by cell phones in the work of MIT's Senseable City Lab. We report some of our initial experiments that included using cell phone data to map people's quotidian flows in European cities, tracking discarded objects at a global scale, and predicting people's exposure to pollutants according to their daily movements in New York City. We group these cell phone-based movement and location sensing technologies into two categories: opportunistic sensors and purpose-built sensors. Sensors are embedded in an increasing number of devices we use daily, and many of these sensors collect data from the natural and built environment, and from human interactions without people's direct action. These are what we call *opportunistic* sensors. Another group of sensors are designed to collect data to respond to specific questions. These are the *purpose-built* sensors. Self-tracking sensors are an example of purpose-built sensors that have reached significant commercial success in recent years. Embedded in wearables, these devices register the user's location and speed, heartbeat, and other personal data that, in addition to giving back information to the user, at an aggregated level might be useful to help us better understand and design urban spaces.

The chapter is divided in three sections: *Real-Time Cities* describes how we used cell phone-usage data to map the complexities of the spatiotemporal use of cities; projects discussed in *Tracking Urban Flows* explore GPS trackers embedded in objects, such as trash, that are found in cities; and *Emerging Geographies* combines data from cell phones and self-tracking apps to propose new ways to understand urban phenomena, from comparing patterns in urban dynamics in cities around the world to having a better understanding of people's exposure to pollutants in urban areas.

Real-time cities

Ten years ago, the Senseable City Lab launched a research initiative to use cell phone data, aggregated at the cell phone tower coverage area, to understand the intensity in the use of different areas of the city by its inhabitants. Cell phone tower coverage area (cell) collects the location of cell phones based on the signals transmitted and received by these devices. Even when off, cell phones send signals to the transceiver station, and when in movement, they "hand over" signals between transceivers. Although not as precise as using the GPS location of each individual device, at the time we started this research the use of data from the cell phone towers presented the advantage of harnessing vast amounts of data from existing urban infrastructures.

Aggregated data of cell phone location at the tower level is anonymous – the log uses as unique identification the International Mobile Subscriber Identity (IMSI). An encrypted code that is renewed daily (which hinders long-term tracking), IMSI contains a random number instead of the mobile phone number, a code for the current service provider, the home country service provider (as roaming partner), the network area, and the network cell. Each time the cell phone sends and receives signals to/from the tower, an event with a timestamp is logged. This information permits to map the spatiotemporal distribution of the cell phones, which can be used as proxy for population distribution.

Mapping the movements of cell phones among the cells creates a fairly clear picture of human interactions in space and time. Indeed, not long after these initial explorations, researchers used cell phone data in Haiti, in the weeks following the 2010 earthquake, to quantify the number of Haitians (more than 600,000) who fled Port au Prince, the country's capital (Bengtsson et al. 2011), and track their movements within the country. Such dynamic maps based on the intensity of cell phone usage have also been used to understand the macro movements taking place in cities, and to inform urban and transportation planning (Calabrese et al. 2011; Çolak et al. 2015) or understand economic activities (Reades et al. 2009). Not surprisingly, IT companies such as Cisco and Siemens are investing in systems that can mine such data, not only to better understand but also to predict mobility patterns.

The Senseable City Lab's first initiative using cell phone data to understand human movements happened in Graz, Austria, in 2005. Customers of A1 Mobicom Austria network were invited to allow us to "ping" their cell phones, which generated data of their position and movements in the city. In 2007, we monitored cell phone usage combined with GPS location from volunteers, to have a more granular understanding of human movements during the Culture Night (Copenhagen's biggest annual one-day cultural event). The underlying research interest of these early projects was to use cell phone data to describe and represent urban dynamics at a moment when mobile devices had started to become ubiquitous. It was clear that a new tool of analytical inquiry for cities was emerging.

In 2006, the Senseable City Lab displayed *Real Time Rome* during the Venice Biennale. This project used aggregated data from cell phone towers and taxi and bus movements in Rome between July and September 2006. The dataset included more than 1 million TIM (Telecom Italia) subscribers and visitors to Rome, totaling 3.5 million observations. During this period, we focused on two special events: the FIFA World Cup final match, between Italy and France, and a concert by Madonna. Before the match, the location of the calls showed a regular pattern, representing a normal day in the city. During the match, though, it became clear how the events in the soccer field influenced the intensity in the number of calls – moments of silence when France scored, and peaks when Italy scored and, eventually, won the match and the World Cup. Peaks spread across Rome in celebration, culminating with the reception of the Italian team in the following day at the Circo Massimo. While the World Cup final match showed the city pulsating differently according to specific events, which varied in time and space, the Madonna concert showed a different spatiotemporal pattern. In this case, the intensity in cell phone use was focused on a single location and during a very specific time period.

Real Time Cities demonstrated how cities behave differently in the relative short timeframe of only a few months, reflecting how particular events influence urban dynamics. By revealing how the intensity of the calls varied spatially and temporally, we demonstrated novel ways to understand urban dynamics, also revealing the emotional landscape of Rome. In a more practical way, this knowledge can inform transportation management in a more responsive way. Indeed, in the following years, similar projects have emerged in different research centers, pointing to a tendency to create dynamic and collaborative mapping that can inform transportation management and urban design.

Tracking urban flows

Trackers used in the *Real Time Cities* projects were opportunistic sensors. The fact that these sensors collect data from the users – such as their location – is essential to the functioning of the technology. In this section we explore both opportunistic sensors and purpose-built sensors, which are designed and deployed with specific research questions in mind.

In 2009, the Senseable City Lab launched *Trash Track*. The basic idea behind the project was to understand the removal chain with the same accuracy that the contemporary society understands the supply chain. Virtually all components of consumables can be tracked back to their origin – in many cases, this is mandated by laws and regulations. However, once a product is discarded, it immediately disappears from the user's mind: you take the garbage to a trash bin, or put it at the curbside, and it magically disappears. We assume that someone is taking care of it, collecting our trash and bringing it to a recycling facility, an incinerator, or a dumpsite. But what actually happens with the things we discard? We decided to take a close look.

In Seattle, hundreds of people brought us items they were about to discard: sneakers, tires, canned food, computers. We attached trackers to 3,000 objects, and people returned home and discarded their respective objects using the usual channels. Each tracker was equipped with GPS, battery, and GSM/GPRS modems (Offenhuber 2017). Following and mapping the routes each object took to its final destination, we found that while perishables ended up close to the city, some types of objects (such as electronics) traveled a lot, ending up in different states, after months and thousands of kilometers far from the moment and location they had disappeared from the user's mind (Figure 42.1).

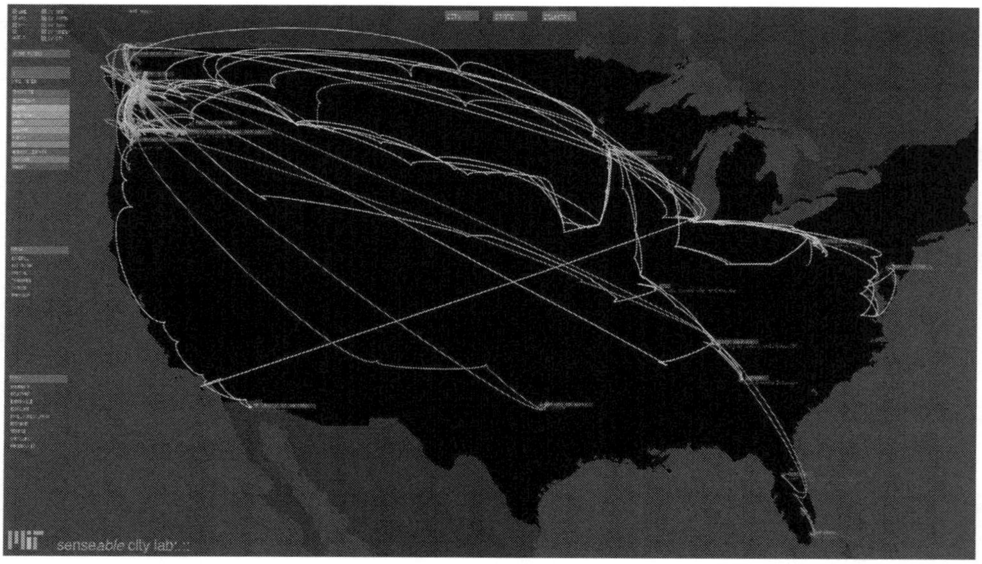

Figure 42.1 Mapping trash movement throughout the United States.

Source: MIT Senseable City Lab

Mapping trash movements throughout the United States made clear that the removal chain is overall inefficient. This in itself was an eye opener to many citizens, who could bring this information back to public authorities and demand more efficient ways to deal with discarded objects. Beyond this general picture, we found that electronic devices, such as CRT and LCD screens, and printers, showed a particularly awkward behavior: they traveled longer, and several of them ended up by the country's borders and in port cities. Regarding the country's borders, this was because for this project we had only domestic cell phone coverage: if a device had crossed the US borders, we would have lost track of it. But what about the port cities? Why were electronics ending up there?

In order to investigate this question, we partnered with the Non-Government Organization Basel Action Network (BAN), whose mission is to monitor international trade of toxic waste, including electronics. BAN deployed trackers in 205 printers and LCD and CRT monitors. BAN's researchers approached the owners of these devices before they donated this equipment to shops and recycling facilities certified by the e-Steward and R2 programs.[1] We were aware that 205 devices make a small sample and do not necessarily represent what happens in the country. Still, the experiment was important to test whether certified recycling programs were respecting the Basel convention, which controls transboundary movements of hazardous wastes, such as electronic devices.

However, when we mapped their traces, we found that after being delivered to centers where the owners imagined their equipment would be sent for recycling, electronic devices traveled on average 4,158 kilometers for about 100 days. Sixty-nine of the 205 devices ended up abroad, most of them in China (Lee et al. 2017). In the *Monitour* project, we used the deployment of purpose-built sensors as an investigative tool to understand a problem (export of electronic waste) that, because of its potential illegality, was hard to investigate using traditional means, and whose existence, amounts, and routes were largely based on proxies, such as the selling of electronics. Previously, only rough estimates were produced based on selling number leads (Lepawsky 2015). Using trackers, which basically employ simple off-the-shelf technologies, BAN was able to launch an investigation that shed light on an unglamorous side of the information society (Figure 42.2).

Cell phone-based movement and location sensing devices are carried by millions of people daily. As Gina Neff and Dawn Nafus (2016) argue, it is now impossible to live an untrackable life. Next, we discuss how health and wellness-related trackers have been used not only to mirror the individual's performance and health signals, but also as a social tool through which people interact and learn from each other.

The profitable niche of health trackers is composed by devices and applications. The former are gadgets that collect several signals from the users' activity, from the number of steps a person takes daily to how long a person sleeps, from the heartbeat rate to the approximate arousal based on skin conductance. The latter are applications that can be installed in smartphones. These apps use several features embedded in the smartphones, such as GPS and accelerometers to measure distances and speed, and social media apps that compare the user's performance with her peers.

Researchers and the app companies have been creating maps that show the most used routes, users' average speed, and the preferred areas of the cities to perform different outdoor activities. As a marketing strategy, apps are commonly designed for a specific public, such as cyclists, pedestrians, or runners, even though they use the same technologies embedded in most smartphones. In the *CityWays* project, we used data from a single app to understand and compare the behavior of cyclists, pedestrians, and runners in Boston and San Francisco. Based on the speed, the app defines which activity the user is performing – for instance, removing traces when the speed is

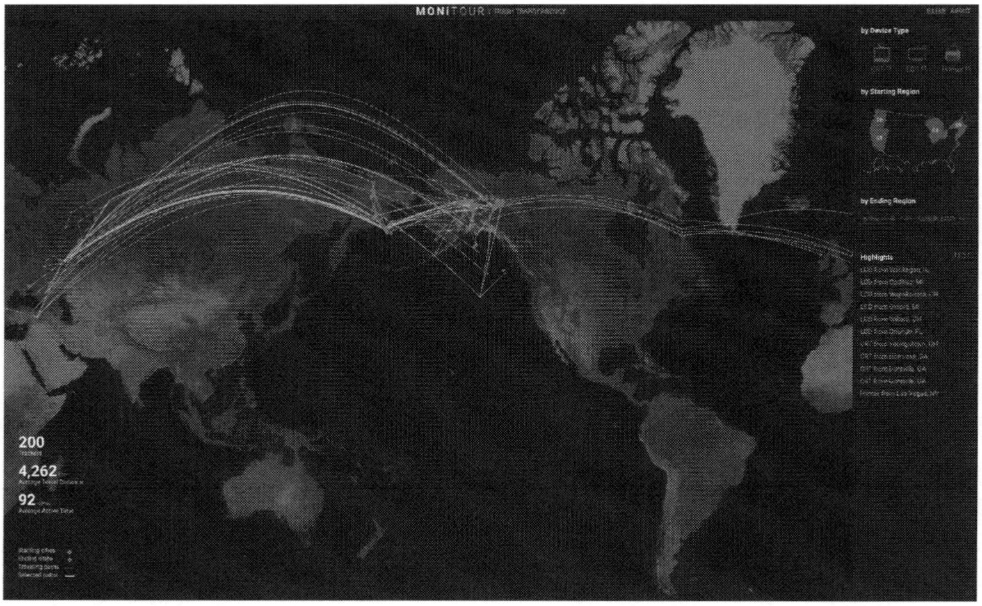

Figure 42.2 Mapping electronic waste.

Source: MIT Senseable City Lab

high, which would indicate the use of motorized modes. Rather than centering on the person's performance, our goal was to understand the city through self-tracking apps. Anthony Vanky (2017) used this data to understand whether and how weather, geography, street network, urban greenery, and the presence of amenities, such as shops or urban furniture, influence the location, frequency, and length of pedestrian, cyclist, and runner activities in Boston and San Francisco. Thus, such apps can provide pertinent information for urban designers.

However, in a time when we are not only self-tracking ourselves but making this data available – often times unaware of it – it is critical that we ask whether this phenomenon is empowering the users, or whether it is a disguised form of social control (Nafus 2016). Therefore, it is equally important to explore what benefits this data can bring to society. This was the focus of *CityWays*. Although designed to measure physical activities at the individual level, aggregated data from self-tracking devices can inform city officials and designers to propose public policies and design proposals that could trigger more physical activities and promote higher use of public spaces. Understanding the temporal use of cities by different people's profiles can make urban design more fitting to the city residents' habits.

Emerging geographies

The Senseable City Lab has demonstrated that the use of cell phone data can shed light onto urban phenomena that happen at different scales, from the national to the neighborhood level, which would be impossible to detect using traditional fixed- and long-term databases. For example, in the project *Many Cities*, we used the numbers of calls, SMS, and data requests, as well as the amount of data uploaded and downloaded by subscribers in four major global cities:

London, New York, Los Angeles, and Hong Kong. The goal was to detect repeating patterns or special events based on cell phone traffic data, and to compare how different cities behave using a common dataset. On the one hand, this project reveals that cities throughout the world share some activity patterns, such as the buzzing activities in the core financial centers, while other commercial and residential areas have more city-specific patterns. On the other hand, we also found unexpected urban patterns at the global scale, such as New York and Hong Kong sharing more similar urban-activity signatures than New York and London (for example, cell phone activity drops in London during the weekends much more than in New York and Hong Kong), even if the cultural and linguistic background of the latter are more closely related (Grauwin et al. 2015).

When we correlate data from cell phone usage with other datasets, we can understand urban phenomena in novel ways. In *Urban Exposures*, we faced one of the world's most pressing environmental health concerns: the increasing number of people contributing to and being affected by air pollution, leading to seven million early deaths each year. This problem is particularly relevant in cities, which usually have high rates of vehicle miles traveled (or VMT), making urban dwellers more exposed to pollutants. Measuring exposure to pollutants usually involves correlating pollution levels measured in fixed air quality network sites and static population distributions, such as home location.

However, correlating exposure to pollutants with home location misses a crucial aspect of urban life: people move, and move more frequently in urban areas, where pollution levels are also higher than in rural or suburban areas. In order to understand people's exposure to pollutants, we need to take into consideration how people move in space and time in cities. Researchers have used time-activity diaries and questionnaires to monitor a person's movement in space and time, but this method is laborious and requires the active participation of the subjects. Tied to GPS monitors, such time-activity diaries become less time-consuming and more efficient; however, so far, most of the studies focus on small samples (Steinle et al. 2013). The combination of large networks of air quality sensors and self-tracking devices is a fertile research field to correlate exposure to pollutants with people's daily movements. Nazelle et al. (2013) monitored 36 volunteers in Barcelona using CalFit-equipped smartphones (which uses the integrated GPS and accelerometer technologies embedded in smartphones to record a subject's time-location patterns and energy expenditure) to correlate their physical activity and geographic location with spatiotemporal air pollution mapping. Similarly, other research initiatives (Hasenfratz et al. 2012) have attached air quality sensors to smartphones to collect data at the individual level, with the constraint that this methodology requires finding volunteers, and data from low-cost monitors should be calibrated to static and professional ones.

Such studies use purpose-built sensors and are based on very limited samples. In *Urban Exposures*, we decided to take advantage of one of the most pervasive opportunistic sensors: cell phones, which constantly generate spatiotemporal logs of millions of users. Marguerite Nyhan et al. (2016) propose to integrate mobility patterns of millions of cell phone users with spatiotemporal PM2.5 concentration level estimates, collected at 155 locations over all seasons in two-week intervals in 71 districts in New York City. Mapping the movement patterns of cell phone users demonstrates that air pollution exposures vary temporally and spatially when we compare "home population exposure" with "active population exposure," which reflects population activity based on extensive mobile device usage records. This result might bring important contributions to public health policies, clearly showing the value of using anonymized and aggregated cell phone-usage data to understand urban phenomena in novel ways.

Conclusion

If the built environment is a kind of "third skin" – in addition to our biological and clothing – it has long been a rigid one. The understanding that the city exists through the information exchanges that happen among people, as well as between people and the natural and built environments, is not new. However, either such understanding was mostly argued conceptually or measured in time-consuming and very localized ways. The abundance of data we generate daily, and which structure our contemporary society, represents a "game-changer" in the way we make sense of urban phenomena. The widespread use of opportunistic sensors embedded in devices used daily, as well as purpose-built sensors more often deployed in buildings and infrastructures, opens the opportunity for gathering data from our daily activities with fine spatial and temporal granularity.

Gathering and analyzing such abundance of data is not an end in itself. Both opportunistic and purpose-built sensors using movement and location sensing technologies embedded in personal mobile devices can inform planning and urban design in novel ways. In a series of projects combined in *Real Time Cities*, we pioneered the use of cell phone data to understand the temporal variations in how people move in cities and use public spaces, which vary according to daily patterns and special events. In *Urban Exposures*, we used the same type of data to reveal a hidden side of public health: rather than quantifying the exposure to pollutants in urban areas based on people's residence, we mapped their daily movements throughout the city. This approach helps to understand the actual levels of exposure of people, while they are moving and using different parts of the city in their daily routines. Also using movement and location sensing technologies embedded in personal mobile devices, we proposed a series of research initiatives to follow trash disposal both at the local and global levels. This research demonstrated that contemporary consumption has spatial and temporal footprints far beyond what we usually imagine.

In practical ways, using movement and location sensing technologies embedded in personal mobile devices presents novel ways to understand urban dynamics that have the potential to improve traffic control and public transportation, public health, and waste management, and to inform urban interventions aimed at activating particular urban areas. On the other hand, sensors embedded in the built environment to measure and adjust humidity, temperature, vibration, lighting, and energy consumption are fostering responsive urban design, which responds to the surrounding environment as well as to the type and intensity of its use. Thus, what we argue here is that with better data, the built environment can be more responsive to human activities. The city can become a living, tailored environment that is molded on its inhabitants' behaviors.

Acknowledgments

The authors would like to acknowledge the many researchers at the Senseable City Lab who have worked on these projects – they were responsible for generating ideas, building and deploying sensors, visualizing the data, and writing scientific papers that discuss in depth concepts and analysis described in this chapter. Each of them are credited on the website of the respective projects. The authors also thank Dover Corporation, Teck, Lab Campus, Anas S.p.A., Cisco, SNCF Gares & Connexions, Brose, Allianz, UBER, Austrian Institute of Technology, Fraunhofer Institute, Kuwait-MIT Center for Natural Resources, SMART—Singapore-MIT Alliance for Research and Technology, AMS Institute, Shenzhen, Amsterdam, Victoria State Government, and all the members of the MIT Senseable City Lab Consortium for supporting this research.

Note

1 e-Stewards (Standard for Responsible Recycling and Reuse of Electronic Equipment®) and R2 (Responsible Recycling Practices) are the two standard certification programs for electronics recyclers in the United States.

References

Batty, M. (1995). "New Ways of Looking at Cities." *Nature* 377: 574. doi:10.1038/377574a0

Batty, M. (2012). "Editorial: Smart Cities, Big Data." *Environment and Planning B: Planning and Design* 39(2): 191–193. doi:10.1068/b3902ed

Bengtsson, L., Lu, X., Thorson, A., Garfield, R., & Schreeb, J. V. (2011, August). "Improved Response to Disasters and Outbreaks by Tracking Population Movements with Mobile Phone Network Data: A Post-Earthquake Geospatial Study in Haiti." *PLoS Medicine*, 8(8). doi:10.1371/journal.pmed.1001083

Calabrese, F., Colonna, M., Lovisolo, P., Parata, D., & Ratti, C. (2011, March). "Real-Time Urban Monitoring Using Cell Phones: A Case Study in Rome." *IEEE Transactions on Intelligent Transportation Systems*, 12(1): 141–151. doi:10.1109/tits.2010.2074196

Çolak, S., Alexander, L. P., Alvim, B. G., Mehndiratta, S. R., & González, M. C. (2015, January). "Analyzing Cell Phone Location Data for Urban Travel." *Transportation Research Record: Journal of the Transportation Research Board*, 2526: 126–135. doi:10.3141/2526-14

Grauwin, S., Sobolevsky, S., Moritz, S., Gódor, I., & Ratti, C. (2015). "Towards a Comparative Science of Cities: Using Mobile Traffic Records in New York, London, and Hong Kong." In: Helbich, M., Arsanjani, J. J., and Leitner, M. (eds.) *Computational Approaches for Urban Environments*. Springer Verlag, 363–387.

Hasenfratz, D., Saukh, O., Sturzenegger, S., and Thiele, L. (2012). "Participatory Air Pollution Monitoring Using Smartphones." 2nd International Workshop on Mobile Sensing, 1–5.

Kitchin, R., & Dodge, M. (2011). *Code/space: Software and Everyday Life*. Cambridge, MA: MIT Press.

Lee, D., Offenhuber, D., Duarte, F., Biderman, A., & Ratti, C. (2017). "Monitour: Tracking Global Routes of Electronic Waste." *Waste Management* (forthcoming).

Lepawsky, J. (2015). "The Changing Geography of Global Trade in Electronic Discards: Time to Rethink the E-Waste Problem." *The Geographical Journal*, 181(2): 147–159.

Mitchell, W. J. (2001). *Me++: The Cyborg Self and the Networked City*. Cambridge, MA: MIT Press.

Nafus, D. (ed.) (2016). *Quantified: Biosensing Technologies in Everyday Life*. Cambridge, MA: MIT Press.

Nazelle, A. D., Seto, E., Donaire-Gonzalez, D., Mendez, M., Matamala, J., Nieuwenhuijsen, M. J., & Jerrett, M. (2013, May). "Improving Estimates of Air Pollution Exposure Through Ubiquitous Sensing Technologies." *Environmental Pollution*, 176: 92–99. doi:10.1016/j.envpol.2012.12.032

Neff, G., & Nafus, D. (2016). *Self-tracking*. Cambridge, MA: MIT Press.

Nyhan, M., Grauwin, S., Britter, R., Misstear, B., Mcnabola, A., Laden, F., . . . Ratti, C. (2016, August). "'Exposure Track': The Impact of Mobile-Device-Based Mobility Patterns on Quantifying Population Exposure to Air Pollution." *Environmental Science & Technology*, 50(17): 9671–9681. doi:10.1021/acs.est.6b02385

Offenhuber, D. (2017). *Waste is Information*. Cambridge, MA: MIT Press.

Reades, J., Calabrese, F., & Ratti, C. (2009). "Eigenplaces: Analysing Cities Using the Space: Time Structure of the Mobile Phone Network." *Environment and Planning B: Planning and Design*, 36(5): 824–836. doi:10.1068/b34133t

Rheingold, H. (2002). *Smart Mobs: The Next Social Revolution*. New York: Perseus.

Steinle, S., Reis, S., & Sabel, C. E. (2013, January). "Quantifying Human Exposure to Air Pollution: Moving from Static Monitoring to Spatio-Temporally Resolved Personal Exposure Assessment." *Science of the Total Environment*, 443: 184–193. doi:10.1016/j.scitotenv.2012.10.098

Vanky, A. (2017). *To and Fro: Digital Data-driven Analyses of Pedestrian Mobility in Urban Spaces*. Massachusetts Institute of Technology. Department of Urban Studies and Planning (PhD Dissertation).

Further reading

Duarte, F., and Rodrigo, F. (2018). *Unplugging the City: The Urban Phenomenon and its Sociotechnical Controversies*. London: Routledge. The authors argue that modernity has entrusted technology with such power that it is treated as an autonomous entity, with its own manners and morals. Based on that premise, the authors discuss multiple sociotechnical complexities involved in the stabilization and disruption of urban technological arrangements.

Kitchin, R., and Dodge, M. (2011). *Code/space: Software and Everyday Life.* Cambridge, MA: MIT Press. This book proposes that software has extended their influence beyond computers to generate social behaviors and new spatialities, to the point that "living beyond the mediation of software means being apart from collective life."

McLuhan, M. (1964). *Understanding Media: The Extensions of Man.* New York: McGraw-Hill. This book is "old" only by its publication date; otherwise, it still keeps the vitality of analyzing how media has been constantly reshaping social relations and human interactions among ourselves and with nature and the built environment. The author analyzes works on digital media through critical eyes and opens up unexpected ways to understand contemporary technologies.

Mitchell, W. J. (1996). *City of Bits: Space, Place and the Infobahn.* Cambridge, MA: MIT Press. The author offers a pioneering discussion of how virtual spaces, interconnected by what at the time was called the information superhighway, and the miniaturization of electronics incorporated in our daily life would transform the practice of architecture and urbanism in the context of the digital telecommunications revolution.

Ratti, C. and Claudel, M. (2016). *The City of Tomorrow: Sensors, Networks, Hackers, and the Future of Urban Life.* New Haven: Yale University Press. The authors discuss how pervasive digital systems that layer our cities are transforming urban life, and propose the concept of *Futurecraft*, by which urban ideas will be generated by the symbiotic relation between designers and the public.

Stephen, G. and Simon, M. (2001). *Splintering Urbanism: Networked Infrastructures, Technological Mobilities and the Urban Condition.* London: Routledge. This book has become a key reference in the study of urban technologies. It discusses how information technologies embedded in personal devices or large infrastructures are reshaping the urban condition – from its physical organization to its social and political relations.

43

INTELLIGENCE FOR PLACE-MAKING AND SOCIAL INCLUSION

Critiques and alternatives to India's Smart Cities Mission

Balaji Parthasarathy and Brinda Sastry

Introduction: India's Smart Cities Mission

As the planet becomes increasingly urban, it is turning to advances in information/intelligent technologies (IT) to cope with the attendant challenges. Prominent in this turn is the creation of "smart cities," which "collect lots of data through *instrumentation*, bring these data together through *integration*, and then analyze the integrated data for *intelligence*" (WDR 2016: 240, original emphasis) to optimize data flows and use predictive analytics for evidence-based decision-making. The goals are to improve urban efficiency and city services, and to foster innovation and social inclusion to enhance quality of life.

India launched its Smart Cities Mission (SCM) in 2015 to promote 100 cities that "provide core infrastructure and give a decent quality of life to its citizens, a clean and sustainable environment and application of 'Smart' Solutions" by "developing the entire urban eco-system represented by the four pillars of comprehensive development – institutional, physical, social and economic infrastructure" (GoI 2015: 5). The rationale for the SCM is the extent of urbanization in the country: according to the 2011 Census, cities accounted for 31.16% of India's population and contributed to 63% of GDP. These figures are expected to touch 40% and 75%, respectively, by 2030 (Ibid.).

Stating that "there is no universally accepted definition of a Smart City," the SCM invites cities to participate in a "challenge," in a spirit of competitive federalism (*op. cit.*: 8). Each city has "to formulate its own concept, vision, mission and plan (proposal) for a Smart City that is appropriate to its local context, resources and levels of ambition" (Ibid.). Yet, the SCM goes on to claim that, "in the imagination of *any city dweller in India*, the picture of a Smart City contains a wish list of infrastructure and services" (Ibid., emphasis added) and expects all proposals to include infrastructure services and smart solutions, which are listed as being at the core of the mission – adequate water supply, assured electricity supply, sanitation and solid waste management, efficient urban mobility and public transport, affordable housing, robust IT connectivity and digitalization, e-governance and citizen participation, sustainable environment, safety and security, education, and health. Access to these infrastructural elements and services is to be improved, especially for the poor and the disadvantaged, using technology, information, and data, with the ultimate goal of creating inclusive cities.

Cities participating in the SCM can seek help from a preselected list of consulting firms or bilateral and multilateral institutions, such as the World Bank, to prepare proposals of four types: retrofitting,[1] redevelopment,[2] greenfield development[3] and pan-city development.[4] The proposals are first screened by the respective states, which then forward a short list for evaluation and final selection at the national level. While the SCM receives financial support from the Indian government, implementation in the selected cities is through a Special Purpose Vehicle (SPV) registered under the Companies Act 2013,[5] in which the respective state and city will have 50:50 equity. The SPV can also take the form of a public-private partnership (PPP) in which the state and city hold the major share. This has triggered interest among IT firms, including multinational corporations (MNCs) (Chengappa 2014), in a mission whose budget for the 2015–2020 period is USD 7.5 billion.

Despite the enthusiasm for smart cities, and the laudable goals of improving a broadly defined "quality of life" and social inclusion, the concept has proven controversial (WDR 2016). The next section examines critiques of smart cities in general, and those of the SCM in particular. While the critiques acknowledge that IT can help alleviate infrastructure deficiencies, there are concerns with socio-political exclusion, and how that can diminish the experiential richness of the city. The question of how the application of IT can improve the urban quality of life for a diverse citizenry, instead of treating technology deployment as a panacea, is explored in the following section. This section also examines cases from India that highlight how, with ubiquitous technology (mobile phones),[6] a decentralized and open-ended deployment of IT in cities can ensure human agency in placemaking and social vitality. The chapter concludes with lessons from the SCM for smart cities.

Interrogating the smart city controversy

"Thrilling" is how Sassen (2012: 13) describes the act of "installing, experimenting, testing, or discovering to generate innovations . . . as the city becomes a living laboratory for smart urban technologies that can handle all the major systems a city requires: water, transport, security, waste, green buildings." However, she hastens to add a concern with the "sensored" city becoming censored, i.e. the tendency to make technologies invisible, both physically and metaphorically, may put them in command rather than in dialogue with users. In a similar vein, Sennett (2012) notes that the development of new technologies has historically preceded our ability to apply them. Consequently, he argues that, unless we "learn" to apply IT, there is a risk of creating the "stupefying smart city" (16).

Pointing to the fate of two "stupefying" smart cities, Masdar in the United Arab Emirates and Songdo in South Korea, Sennett argues that incorporating "every bit of high-tech into your design," so as to monitor and regulate a city's functioning leaves little opportunity for an urban experience that is apparently disorderly but experientially rich. He describes a comprehensively planned smart city as a closed system where "unforeseen activity is either incorporated into existing rules – the algorithms – of the system or expelled as irrelevant 'noise'" (Ibid.). Sennett cautions against using our growing ability to centralize information gathering and processing to dictate choices that limit citizens' ability to make sense of the conditions in which they live. He says this is akin to conceiving the city in Fordist terms, which uses the metaphor of a machine and assumes that all activities have their spatio-temporal specificities.

Invoking a Fordist approach to planning and design is useful to understand the consequences of modernism's quest for a universal urban order for the machine age.[7] The quest ignored history, local social institutions and behavioral norms, and vernacular aesthetic sensibilities. Modernism envisioned a rationally planned city, characterized by hierarchy and standardization, geometrical order, and functional separation. Such an authoritarian conception, which assumed that the planner/designer was omniscient of the needs of the citizen, undermined plurality. For instance, design for urban circulation sharply segregated vehicular traffic from pedestrians for efficient flows, ignoring the many reasons why people might be on the street.

A mechanistic notion of an efficient city, and assumptions that any concession made to diversity will increase transaction costs, may ease the provision of certain services such as water supply or sewage lines. But, such a notion "not only fails to respect the autonomous purposes and subjectivity of those who live in it, but also fails to allow sufficiently for the contingency of the interaction between its inhabitants and what that produces" (Scott 1998: 144). As Jacobs (1992) pointed out, the city is a socio-spatial entity, whose visual or formal spatial order only weakly conveys a social order, which rarely has a regular pattern. She famously argued that public spaces are best secured when the efforts of formal institutions exclusively devoted to a cause, such as policing, are complemented by "eyes on the street." These "eyes" result from a rich street life generated by citizens during the course of their daily activities.

Although modernist city design principles emerged in the Global North, they found wide application in the Global South in part because of the promise of historical and institutional rupture. For instance, Chandigarh, designed by Le Corbusier, was born under the patronage of India's first Prime Minister Jawaharlal Nehru, who wanted a city "unfettered by the traditions of the past" to overcome the trauma of colonialism and India's bloody partition during independence (Crabtree 2015). The direct imprint of modernism is most evident in postcolonial capital cities such as Abuja, Brasilia, or Islamabad – all of which are essentially mono-functional administrative enclaves for relatively affluent bureaucrats. The quality of life created by public investments in physical infrastructure and social amenities in these icons of modernity was sustained, in part, through rigid land use and zoning regulations. However, this only served to reinforce the enclave and museum-like character of these cities. As a result, towns such as Mohali and Panchkula around Chandigarh became home to industrial and other non-administrative functions, including informal housing settlements (Prakash 2016). But the growing attraction of these towns, driven by their proximity to Chandigarh, is also leading to high real estate values and proving to be as high a barrier to surmount for the less affluent as the regulatory framework in the "old" city.

As with the modernist planning in Chandigarh, there are critiques of the fit between the SCM and the institutional context of an urban India characterized by "inadequate housing, absence of basic services, tenure insecurity and forced evictions, poor health and nutritional levels, unemployment, and stark levels of inequality" (HLRN 2017: 17). Mukhopadhyay (2015) bemoans the SCM as another instance of a top-down program that limits urban agency,[8] despite the mission acknowledging that a "one-size-fits-all" approach will not work. Although the SCM cities must formulate their own proposals, Mukhopadhyay maintains that the emphasis on providing infrastructure through SPVs and PPPs ignores that the significant economic contribution of Indian cities comes from the disproportionate provision by the informal sector[9] of what are typically considered public goods, or at least merit goods, such as water supply.[10] By reducing the demand for urban services to a matter of supplying physical infrastructure, the SCM overlooks the social fabric underpinning these informal networks (Mukhopadhyay 2015).

Once city proposals are selected, the SCM requires the local monitoring of projects by a Smart City Advisory Forum[11] comprising a member each from stakeholders such as resident welfare associations and local nongovernment organizations (NGOs). But there are fears that the coupling of PPPs, with the "slum-free" rhetoric in many proposals, will further undermine the limited safeguards against forced land acquisition and displacement of the poor (HLRN 2017). Further, the reliance on novel technologies, often supplied by foreign investment, raises concerns about data security, and the emergence of a technocracy that might replace the political process. Against this backdrop, Mukhopadhyay (2015: 79) argues that instead of "avoiding the messiness of political representation" by thinking of smart cities as symbols of "apps and wi-fi," and counting on data analytics and service level agreements, the SCM should focus on "governance, on giving citizens agency and resources." Otherwise, smart cities are likely to become "un-smart." Critics also question the SCM's claims to using technology for social inclusiveness by asking: "Smart for whom? Cities for whom?" (HLRN 2017: iii)

What emerges from the discussion is that the smart city concept is controversial, not because of an outright rejection of technology, but because of the worry about a deterministic conceptualization. Graham (1998: 167) refers to this as "*substitution* and *transcendence* – the idea that human territoriality, and the space and place-based dynamics of human life, can somehow be replaced using new technologies." Put another way, while technology to provide the missing infrastructure and amenities is welcome, the concern is that the mechanisms for technology provision may replace the possibilities for political negotiation and become so reified as to reinforce inequities. For the Global South, the danger can be twofold. First, this can lead to a city (whether newly built or otherwise) being treated as a *tabula rasa*, where technology is deployed with the arrogance of an imperial descent on a "people without history" (Wolf 1982). Second, technology can be used patronizingly, to nudge cities on to teleological trajectories, along predetermined "stages of development" experienced by the Global North (Rostow 1960).

Whatever their shortcomings, Le Corbusier and the modernists could at least claim to have a sense of urban form and city design. The same cannot be said about the images of smart cities in India. Some area-based and greenfield proposals call for walkable localities, preserving and developing open spaces to reduce congestion and air pollution, and promoting interactions in the public spaces to ensure security. The SCM also wants to improve urban mobility, not only for private vehicles and public transport, but also for pedestrians and cyclists. But such features have long been on the agenda of many cities not claiming to be "smart." Then there are other aspects of the envisioned urban form reminiscent of modernism (Figure 43.1). These visions have invited scathing criticism:

> [M]ost images related to "smart cities," . . . tend to consist of skyscrapers, multiple flyovers, glitzy streets, and bright lights. . . . Though the rhetoric of the Smart Cities Mission is one of resource efficiency and inclusion, none of the images seem to portray mixed-income neighbourhoods, social housing, street vendors, women's and children's security, and integrated development paradigms. If the predominant visual of the "smart city" is one in which there is no space for the poor . . . can this be a sustainable and equitable model to be pursued at this stage of India's development?
>
> *(HLRN 2017: iii)*

Thus, what is missing in the SCM is an articulation of how IT can improve urban placemaking or foster inclusion. Instead, there is an implicit assumption that technology deployment is a sufficient condition. Harrison and Dourish (1996) highlight the conceptual limits to thinking about city design from a cyber-spatial perspective. Arguing that we live and interact in *places* "rooted in sets of mutually-held, and mutually-available, cultural understandings about behavior and action," they proclaim that "space is the opportunity, place is the understood reality" (67).

Figure 43.1 Dholera smart city, Gujarat: Glitzy imagery, no people.

Source: www.dholera-smart-city-phase2.com_slider5.jpg

Harnessing IT for urban placemaking and inclusion

How can IT be harnessed not merely as enabling infrastructure and economic vitality but also as means for placemaking, informal social behavior, and inclusion? As an alternative to determinism, Graham (1998: 167) draws on actor-network theory to suggest *recombination*, which argues that:

> [a] relational view of the links between technology, time, space and social life is necessary. Such a perspective reveals how new technologies become enrolled into complex, contingent and subtle blendings of human actors and technical artifacts, to form actor-networks (which are sociotechnical "hybrids").

In similar terms, Sassen (2012: 14) calls for urbanizing technology and draws on the metaphor of open-source software to call for open-source urbanism, which permits the city to be partially made "through a myriad of interventions and little changes from the ground up." Chandavarkar (2008) uses the same metaphor to argue for the enabling of the self-organized complexity of cities. The complexity emerges by freely permitting and recognizing individual experiences and harnessing such experience patterns in a public domain in which all information and processes are transparent.

Sennett (2012: 6) also calls for understanding the city as an open, non-linear system, "programmed to evolve, being open to the unforeseen, changing its very structure as it absorbs new data," which values "noise" to provide cognitive simulation, and allows people to *make* their cities. He goes on to add that "informal social processes are the genius of cities – the source of innovation economically and the foundation of an arousing social life." In these processes, he sees machines/technology playing a role in coordinating human activity rather than being prescriptive. To gain an insight into how some of these approaches can be used to deploy technology in cities, the section turns to case studies drawn from India.

Wireless hotspots for urban places: tea stalls and railway stations

Coffee shops have long been information hotspots. For example, 17th-century European coffee-houses "formed the internet of the Enlightenment era," as they became outlets for newsletters, pamphlets, and other print media, where scientific discussion, literary and philosophical speculation, commercial innovation, and political ferment took place (*The Economist* 2003). Those coffee shops have their contemporary equivalents in outlets such as Starbucks, where patrons can be found discussing diverse topics over expensive cappuccinos or herbal teas in air-conditioned comfort. Wireless communication, instead of printed matter, is perhaps a crucial difference in the technological basis of information exchange.

Plush coffee-serving outlets also have their equivalents in the Global South, although personal income in a country like India sharply limits the "publicness" of these places. Further, while public wi-fi hotspots can play an important role in the last-mile delivery of broadband Internet access, India has only 31,000 of them, compared to 10 million in the United States, and 13 million in France (TRAI 2017: 1). To address this shortage, the Telecom Regulatory Authority of India is establishing public data offices (PDOs), with open wi-fi hotspots, in partnership with firms such as the Bangalore start-up Wifi Dabba.[12] The target locations for the PDOs include street-side tea (*chai*) stalls, which are ubiquitous and popular across urban India because they are affordable. By making available low-denomination wi-fi coupons, ranging from INR 2 (~0.03 USD) to INR 20 (~ 0.3 USD), to encourage use, the hope is that PDOs will "become bustling centers of economic activity, where consumption of data for the average Indian becomes as common as consuming a cup of hot chai" (Ibid.).

While the clientele of street-side tea stalls cuts (Figure 43.2) across class barriers, they are gendered spaces not frequented by women. Inspired by a desire to create safe and non-threatening

Figure 43.2 The *chai* stall as hotspot.

Source: www.wifidabba.com_7.jpg

meeting places that attract all genders and age groups, a woman in Udaipur opened a tea stall that offers not just tea, snacks, and free Internet access, but also books to read and board games to play (Abbas, 2016). Defying conventional coding of public space, such initiatives spearhead a movement where users reorganize and reinterpret space by spontaneous, often temporary, activities to incubate "third places" (Oldenburg 1989) using ubiquitous technologies.

A similar initiative is that of the Indian Railways which, in partnership with Google, is installing broadband wi-fi facilities at stations to provide free Internet access to anyone with a working mobile connection on a smartphone.[13] The typical Indian railway station is abuzz with a combination of formal and informal activity, as travelers and visitors, luggage porters, auto-rickshaw and taxi drivers, throng ticket counters, waiting rooms, and platforms, while hawkers peddle an array of reading material, food and beverages, and odds and ends to support a system that carries 23 million passengers every day (Baliga 2016). As railway stations become hotspots, and provide access to various structured services, such as online banking, without diminishing the importance of unstructured activities that already take place, Baliga (2016) argues that the 65,000-kilometer railway network will continue to enable public discourse, and bridge divides.

Intelligently enabling social networks for urban inclusion

In 2011–2012, 80% of employment in urban India was in the informal sector (Chen and Raveendran 2014). Inspired by research showing that job diversification is a route out of poverty, and that urban informal workers with strong social networks are more likely to succeed in creating job opportunities for themselves, an entrepreneur established a firm in Bangalore to aggregate and enlarge the information network for informal jobs.[14] The firm's portal connects employers – individuals and households, firms, and housekeeping services – seeking to cut the costs of recruiting workers – from chauffeurs and cooks, to domestic help and security guards – seeking to increase income and reduce commute time in Bangalore's notorious traffic.

Employers can use the web, mobile web, and SMS alerts to post their positions and access the database of job seekers. Those seeking jobs sign up for alerts on their mobile phones, provided through agreements between the firm and major telephone networks. The firm claims that those who have found jobs through its portal reported, on average, a salary increase of about 20% and a reduction in commute time by 14 minutes. By June 2017, the firm expanded to cover local labor markets in 20 cities, with 8.5 million registered job seekers and 500,000 employers. Facilitating the rapid growth was the availability of the firm's mobile app in various languages such as Hindi, Tamil, Kannada, Gujarati, and Marathi.

If IT can help include migrants into local labor markets, there is also a need to help them overcome the barriers to sharing the benefits of their urban labor. In 2011, only 41% of urban residents in India had access to financial services in the form of bank accounts (Demirguc-Kunt et al. 2013). Gopinath et al. (2010: 10) estimated that out of 100 million domestic migrants to cities, only 22% had access. For the nearly 12% of the urban migrant labor force in the construction industry, which moves from site to site and lacks documentary proof of a residential address, opening a bank account and accessing a branch is especially hard. Thus, to remit money home, many migrants prefer informal couriers over banks or the post office network, despite extra risk and cost involved.

In 2006, to widen financial inclusion, the Reserve Bank of India permitted micro-finance institutions, NGOs, and other civil society organizations to become Business Correspondents for banks.[15] A start-up firm providing value-added telecommunication services became a Business Correspondent and enrolled a network of *kirana* shops (a petty shop, usually informal) (Figure 43.3)[16] as Customer

Figure 43.3 *Kirana* shops offering phone recharge facilities.

Source: Brinda Sastry

Service Points (CSPs).[17] When a migrant approaches a CSP, the shopkeeper uses a mobile phone to dial the transaction for a given amount to a destination account. Once the details are verified, an instant credit appears on the account along with a confirmation message to both the shopkeeper and the customer. The recipient uses a similar procedure to withdraw money. In this manner, bank-less financial transactions are conducted using the existing urban retail, telecom, and banking infrastructure to help migrants.

Conclusion

Smart cities offer an alluring solution to the challenges of contemporary urbanization, especially in the Global South. The use of IT to gather and process information at decreasing marginal costs can especially help address those urban demands that are either characterized by routine transactions (such as the provision and operation of public infrastructural services), or require space-time optimization (such as mobility). But claims that smart cities will improve the "quality of life," or be socially inclusive, are unconvincing for at least two reasons. First, the history of modernism offers a cautionary tale about the impacts of techno-functional determinism. Second, even ignoring history, there is little articulation of the range of opportunities technologies can afford, and how those affordances can help placemaking to address diverse needs.

The case studies offer decentralized alternatives, in contrast to centralized IT interventions or the mere provision of amenities such as open spaces in technocratic master plans. The installation of hotspots at railway stations and tea stalls suggest how the urbanization of technology is possible without limiting the choices available to citizens and the range of individual experiences. The hotspots provide infrastructure for data access to complement the primary reason why a diverse cross-section of Indian society visits these public spaces. As theaters of everyday life, the tea stalls and railway stations are infused with new social meaning and associations. When existing spaces are unwelcoming, the technology can be appropriated, as the Udaipur example shows, to include women and other actors, and to create new socio-spatial experiences.

The second case exemplifies the blending of technologies to support contingent livelihood networks. In local labor markets, as employment is far from secure, technology not only can aggregate demand and supply for informal labor, but also allow workers (even those with inexpensive feature phones) to recombine space and time. Such recombination allows workers to broaden their social networks and improve their livelihood options for better integration with the urban economy – a goal that is missing in the SCM proposals. There is another kind of technology-enabled integration from which many migrants, who literally build the cities, benefit. When relying on informal channels to transfer their earnings to their dependents, they run the risk of undermining their very reason for migration. By contrast, a network, which integrates the ubiquitous *kirana* shops with the banking and telecommunications infrastructure for fund transfer gives migrant workers greater financial reassurance. The facilitation of urban-rural flows is also a reminder that smart cities are not enclaves, but places with ties to a hinterland.

Overall, this chapter argues that the idea of smart cities can be relevant when technology provides a backdrop for the performance of everyday urban life. For the cities of the Global South, such as Mumbai, which Correa (2010) pithily described as a "great city, terrible place," deploying IT can potentially arrest their physical decay and make them better places without trampling the greatness of their citizens' ingenuity. Instead of using technology to reinforce centralized and hierarchical plans, it is possible to create decentralized, yet large-scale, initiatives that are inclusive of diverse socioeconomic groups and support unstructured interactions. Thus, rather than being prescriptive about behavior and built form, the chapter has emphasized the use of technology as a resource affording human agency to shape experientially rich and socially vibrant cities using that powerful source of urban intelligence – the interactions of a plural citizenry.

Notes

1 Planning in an existing built-up area of at least 500 acres, to make the existing area more efficient and livable with intensive infrastructure and smart applications.

2 Replacement of existing built environment, in an area of at least 50 acres, to create a new layout with improved infrastructure using mixed land use and increased density.

3 Development of "smart" solutions for a vacant area of least 250 acres.

4 Citywide application of smart solutions to existing infrastructure.

5 For the role of SPVs in the SCM, see: http://smartcities.gov.in/upload/uploadfiles/files/SPVs.pdf (accessed July 7, 2018). For a sample draft of how to incorporate a SPV under the Companies Act, see: http://smartcities.gov.in/upload/uploadfiles/files/SPVs.pdf (accessed July 7, 2018).

6 By June 2017, urban India had a teledensity of 167.97 for wireless subscriptions, and 72.51 for Internet subscriptions. See: https://trai.gov.in/sites/default/files/Performance_Indicator_Reports_28092017.pdf (accessed July 7, 2018).

7 This paragraph draws from Scott's (1998) critique of modernist planning and design, especially the ideas of Le Corbusier and the CIAM Charter of 1929.

8 Mukhopadhay was reacting to a 2014 concept note on Smart Cities at http://c-smart.in/wp-content/uploads/2015/02/CONCEPT_NOTE_-3.12.2014__REVISED_AND_LATEST_.pdf (accessed January 3, 2018). But the broad scope of his argument applies to the final mission statement as well.

9 India's National Commission of Enterprises in the Unorganized Sector (NCEUS, 2009:3) uses the terms "informal" and "unorganized" interchangeably. "The unorganized sector consists of all unincorporated private enterprises owned by individuals or households engaged in the sale and production of goods and services operated on a proprietary or partnership basis and with less than ten total workers." The term "unorganized workers" refers to "those working in the unorganized sector or households, excluding regular workers with social security benefits provided by the employers and the workers in the formal sector without any employment and social security benefits provided by the employers."

10 Only 70.6% of the urban population has access to piped drinking water supply, typically for no more than one to six hours a day, of variable quality (Planning Commission, 2013: 321). Similar indicators for limited formal service provision for sanitation, public transport, and housing can also be found at the same source.

11 For details on Forum membership see www.smartcities.gov.in/content/innerpage/city-level.php# (accessed July 24, 2018).

12 For more on this start-up, see www.wifidabba.com/ (accessed February 4, 2018).

13 Details of the project are available at www.railtelindia.com/key-projects/station-wi-fi-project.html (accessed January 30, 2018).

14 This case draws primarily from an interview with a co-founder on June 12, 2013, in Bangalore, supplemented with data from the firm's website and media reports.

15 Business Correspondents offer financial services, such as the collection of small deposits, the receipt and delivery of remittances, and the disbursal of small value credit, at designated locations using mobile technology. See https://rbi.org.in/Scripts/NotificationUser.aspx?Id=2718&Mode=0 (accessed July 7, 2018).

16 Informality is particularly prevalent in the retail sector; of the more than 14 million retail outlets in the country, only 8% are in the formal sector (KPMG 2014). Barely 4% of the outlets are larger than 50 square meters, and most of them are mom-and-pop stores run by people with limited literacy and infrastructure.

17 This case draws primarily from an interview with the Vice President (Design) of the firm, on July 23, 2012, in Bangalore, supplemented with data from their website and media reports.

References

Abbas, Z. (2016). "Three Addictions, a 'Thadi Chai' for Girls by a Girl." https://udaipurtimes.com/three-addictions-a-chai-thadi-for-girls-by-a-girl/ (accessed July 7, 2018).

Baliga, V. (2016). "A Public Space: Reflections on the Indian Railways." www.addastories.org/a-public-space/ (accessed January 30, 2018).

Chandavarkar, P. (2008). "Rethinking the Social Contract: The Open Source City." Presented at the 5th Great Asian Street Symposium, National University of Singapore, December 5–7.

Chen, M. A. and Raveendran, G. (2014). *Urban Employment in India: Recent Trends and Patterns*. WEIGO Working Paper No. 7. http://wiego.org/sites/wiego.org/files/publications/files/Chen-Urban-Employment-India-WIEGO-WP7.pdf (accessed January 15, 2016).

Chengappa, S. (2014). "MNCs Line Up for a Slice of India's Smart-city Pie." www.thehindubusinessline.com/news/MNCs-line-up-for-a-slice-of-India's-smart-city-pie/article20879786.ece (accessed January 30, 2018).

Correa, C. M. (2010). *A Place in the Shade: The New Landscape & Other Essays*. Gurgaon, India: Penguin.

Crabtree, J. (2015). "Le Corbusier's Chandigarh: An Indian City Unlike any Other." www.ft.com/content/2a194cb4-1a8d-11e5-a130-2e7db721f996 (accessed January 28, 2018).

Demirguc-Kunt, A., Klapper, L. and Randall, D. (2013). "The Global Findex Database: Financial inclusion in India." www.worldbank.org/content/dam/Worldbank/Research/GlobalFindex/PDF/old%20pdfs/N8india6pg3.pdf (accessed January 16, 2018).

Economist, The. (2003). "Coffee Houses: The Internet in a Cup." www.economist.com/node/2281736 (accessed January 12, 2018).

GoI (Government of India). (2015). *Smart Cities: Mission Statements& Guidelines*. http://smartcities.gov.in/upload/uploadfiles/files/SmartCityGuidelines(1).pdf (accessed January 3, 2018).

Gopinath, S., Oliver, J., Tannirkulam, A., Bhattacharya, S. and Kulkarni, R. R. (2010). "Putting Money in Motion: How Much Do Migrants Pay for Domestic Transfers?" http://ifmrlead.org/wp-content/uploads/2015/OWC/42_Putting_Money_in_Motion.pdf (accessed January 12, 2018).

Graham, S. (1998). "The End of Geography or the Explosion of Place? Conceptualizing Place, Space and Information Technology." *Progress in Human Geography*, 22(2):165–185.

Harrison, S. and Dourish, P. (1996). "Re-placing Space: The Role of Place and Space in Collaborative Systems." Proceedings of the ACM Conference on Computer Supported Cooperative Work, Boston, MA, USA, November 16–20, 67–76.

HLRN (Housing and Land Rights Network). (2017). *India's Smart Cities Mission: Smart for Whom? Cities for Whom?* http://hlrn.org.in/documents/Smart_Cities_Report_2017.pdf (accessed January 13, 2018).

Jacobs, J. (1992). *The Death and Life of Great American Cities*. New York, NY: Vintage Books.

KPMG. (2014). *Indian Retail: The Next Growth Story*. www.kpmg.com/IN/en/IssuesAndInsights/Articles Publications/Documents/BBG-Retail.pdf (accessed January 23, 2018).

Mukhopadhyay, P. (2015). "The Un-smart City." *Seminar*, 665:75–79.

NCEUS (National Commission of Enterprises in the Unorganised Sector). (2009). *The Challenge of Employment in India: An Informal Economy Perspective, Volume I: Main Report*. http://nceuis.nic.in/The_Challenge_of_Employment_in_India.pdf (accessed January 30, 2018).

Oldenburg, R. (1989). *The Great Good Place: Cafes, Coffee Shops, Community Centers, Beauty Parlors, General Stores, Bars, Hangouts, and How They Get You through the Day*. New York, NY: Paragon House.

Planning Commission. (2013). *Twelfth Five Year Plan (2012–2017): Economic Sectors*. Volume 2. http://planning commission.gov.in/plans/planrel/12thplan/pdf/12fyp_vol2.pdf (accessed July 7, 2018).

Prakash, V. (2016). "Messy Work: Transnational Collaboration in Chandigarh." In Chalana, M. and Hou, J. (eds.) *Messy Urbanism: Understanding the "Other" Cities of Asia*. Hong Kong, HK: Hong Kong University Press. 175–192.

Rostow, W. W. (1960). *The Stages of Economic Growth: A Non-Communist Manifesto*. Cambridge, UK: Cambridge University Press.

Sassen, S. (2012). "Urbanising Technology." In Burdett, R. and Rode, P. (eds.) *The Electric City Newspaper*. 12–14. https://files.lsecities.net/files/2012/12/the-electric-city-newspaper.pdf (accessed January 13, 2018).

Scott, J. C. (1998). *Seeing Like a State: How Certain Schemes to Improve the Human Condition Have Failed*. New Haven, CT: Yale University Press.

Sennett, R. (2012). "The Stupefying Smart City." In Burdett, R. and Rode, P. (eds.) *The Electric City Newspaper*. 6. https://files.lsecities.net/files/2012/12/the-electric-city-newspaper.pdf (accessed January 13, 2018).

TRAI (Telecom Regulatory Authority of India). (2017). *Public Open Wi-Fi Pilot*. www.trai.gov.in/sites/default/files/TRAI_Public_Wifi_Pilot_070717_1.pdf (accessed February 1, 2018).

Wolf, E. (1982). *Europe and the People without History*. Berkeley and Los Angeles, CA: University of California Press.

World Bank. (2016). *World Development Report 2016: Digital Dividends*. Washington, DC: World Bank.

Further reading

Kitchin, R. and Dodge, M. (2014). *Code/Space: Software and Everyday Life*. Cambridge, MA: MIT Press. The book argues that the production of space is increasingly dependent on code and identifies the everyday relationship between software, space, and everyday life.

Townsend, A. M. (2013). *Smart Cities: Big Data, Civic Hackers, and the Quest for a New Utopia*. New York, NY: W.W. Norton. The book asks why and how cities across the world are deploying smart technologies and provides a look at the people behind these technopian visions.

Willis, K. S. and Aurigi, A. (2017). *Digital and Smart Cities*. Abingdon, UK: Routledge. Using a number of case studies, this book provides an overview of how technologies are shaping the urban built environment, and how design can react to technologies.

PART III.VI

Mobility and access

Introduction

At the most elemental level, the economy and spatial organization of cities require residents to be able to move to fulfill their daily commerce needs, maintain social contacts, and engage in public life. Mobility and access are necessary for people to physically interact with each other or reach the different spaces and amenities that a city has to offer. For this reason, enhancing mobility and accessibility are important goals for urban design. Since ancient times, the street has represented the fundamental physical space that is purposefully created to facilitate mobility and access. Indeed, the development of the street system is a necessary prerequisite for the development of the urban settlement. Historically, streets have served two distinct purposes: as *places* and as *routes*. As *places*, streets represent settings for social interaction, recreation, open-air commerce, information exchange, and even celebration and expression of dissent (Loukaitou-Sideris and Ehrenfeucht 2009). As *routes*, streets have to facilitate movement and the efficient reach of different destinations. Some of the most celebrated streets, such as Champs-Élysées in Paris or La Rambla in Barcelona, excel both as places and as routes.

With the proliferation of the automobile in the 20th century, however, cities started privileging the route rather than the place functions of their streets. As noted by Nicholas Blomley (2007), a "traffic logic" emerged that guided street design and consequently the layout of urban form, privileging the fast and unobstructed movement of motor vehicles. As a number of authors in this section mention, modernist planning and design that separated land uses into their own single-use functional cells also privileged vehicles over pedestrians and, in the process, destroyed the street as a multifunctional and social space in many cities of the Global North.

As many cities in the 21st century have come to realize, rampant automobility, however, has its banes: sprawl, traffic congestion, environmental pollution, even the destruction of nature and valued places. Therefore, some of the most promising urban design and transportation interventions in cities today are those that try to reinvent streets as both places and routes, without compromising mobility and accessibility, but recognizing that this should not be a privilege limited to travel by automobile.

Susan Handy in her chapter explains that a place's accessibility can be understood as two components: "impedance" – the cost (time and money) of reaching it – and "attractiveness" – the inherent value of the destination. She contemplates what it means to design for accessibility, particularly in ways that complementarily enhance local and regional accessibility; in short, designing "a community where residents can meet many daily needs within walking (or at least biking) distance and have good transit connections to regional activity centers." She demonstrates how this is possible, even in suburbia, by discussing the example of Davis, California, a city that has initiated design policies and measures that encourage and accommodate walking and biking to local destinations. Design guidelines that promote compactness by encouraging infill development, locating easily accessible commercial centers in each neighborhood, and building safe paths for walking and cycling are some of the design initiatives that Davis has taken to create a very high level of accessibility for its residents.

Yodan Y. Rofè details in his chapter the five tenets of modernism that have served to erode formerly vibrant urban spaces: (1) limits to development density; (2) separation of land uses through zoning; (3) artificial segmentation of the urban form into neighborhood units; (4) public open space standards that limit higher densities; and (5) planning for automobility. He focuses on the last tenet and discusses how minimum parking requirements in cities and the creation of hierarchical street networks that classify streets according to levels-of-service have led to cities designed for automobiles rather than people. He finds great promise, however, in the emerging paradigm for urban street design and transportation planning that reinvents streets as social spaces, mixing transportation modes, initiating traffic calming measures, increasing densities near transit stations through transit-oriented developments, and in general, "stressing accessibility rather than mobility, balancing efficiency and equity, changing priorities from automobile-oriented to people-oriented measures, and emphasizing safety and the protection of pedestrians."

Arguing that "street design is fundamental to urban design theory and practice," Paul Hess, Kelly Gregg, and Ryan Whitney review the North American pedestrian-oriented street design and its European influences. They describe how some current street design initiatives – namely Complete Streets, Parklets, and Pavement-to-Plaza projects – have been influenced by European street design precedents such as Woonerfs, Traffic Calming, and Shared Spaces. Ironically, while these new street design ideas and initiatives have appeared in the United States as a counterpoint to modernist street design paradigms and as efforts to reclaim streets from automobiles, Hess, Gregg, and Whitney find that they "continue to be deeply entangled with [street design] ideas about managing and accommodating vehicle traffic."

While the first three chapters examine mobility and accessibility in the context of surface travel, Max Hirsh examines the design implications of "the leap in aerial mobility" in East and Southeast Asian cities, or what he coins "airport urbanism." In recent decades, the rapid increase in air travel among residents of these countries has led their governments to develop massive airport infrastructures. Beyond enhancing air accessibility, such megaprojects have left major footprints on the ground as they are often accompanied by the building of master-planned new "airport cities." The newly built airports at the urban periphery are connected via high-speed transit lines to downtown check-in terminals. According to Hirsh, the success of the flashy airport-driven projects is often not as high as their ambitions. He contrasts them with smaller ad-hoc, mundane, and unpretentious terminals used by budget airlines that are often built by private entrepreneurs and lack in architectural quality, but serve rather well the needs of a larger flying public. The lessons that Hirsh draws from his review of Asian airport urbanism is the need for urban designers to recognize the shortcomings of airport megaprojects and develop design strategies that better fit their rapidly changing urban environments. This may mean more incremental, adaptive, and flexible design.

References

Blomley, N. (2007). "Civil Rights Meets Civil Engineering: Urban Public Space and Traffic Logic," *Canadian Journal of Law and Society*, 22(2): 55–72.

Loukaitou-Sideris, A. and Ehrenfeucht, R. (2009). *Sidewalks: Conflict and Negotiation Over Public Space.* Cambridge, MA: MIT Press.

44
ACCESSIBILITY-ORIENTED URBAN DESIGN

Susan Handy

Introduction

> The unique commodity that the city offers to location seekers is accessibility. . . . The history of city growth, in essence, is the story of man's eager search for ease of human interaction.
>
> *(Webber 1964, 85–86)*

Cities, small and large, exist because they facilitate human interaction, economic and social. The concentration of human activity in cities helps foster growth and innovation – new businesses, services, arts, and so on. At the same time, this concentration of activity benefits city residents in their daily lives: they have more opportunities for jobs, shopping, and social and cultural activities within a given distance. In short, cities provide a higher overall opportunity for interactions of all sorts, for both businesses and residents. This quality, known as "accessibility," fosters the human activity essential to modern life.

Accessibility is a characteristic of the built environment that is shaped over time by the countless actions, public and private, individual and collective. Sometimes these actions work to enhance accessibility, sometimes to hinder it. Of particular importance are the land use and transportation planning processes that establish a vision for a community and guide both public and private investments towards that vision. The intentional shaping of the built environment through these processes is urban design in its broadest sense, and it occurs at multiple scales – from the design of buildings and streets, to the layout of neighborhoods, to the structure of the metropolitan region.

If accessibility is a goal of design, the age-old dilemma of form versus function becomes moot: form and function both matter! To design for accessibility is first and foremost to design for function: how does a community function for its residents, and how does it help or hinder them as they go about their daily lives? But function is not independent of form: the aesthetic qualities of a place influence how accessible that place feels to residents, how well it functions for them beyond simply fulfilling their utilitarian needs. Just as urban design must cross multiple scales, it must span from the functional to the formal, from the objective to the subjective, from quantity to quality.

The goal of this chapter is to contemplate what it means to design for accessibility. I first define accessibility and its dimensions, then outline the ways in which urban design can contribute to accessibility, present an example of accessibility-oriented design, and consider what the future holds for accessibility.

Defining accessibility

> The quality of being able to be reached or entered.
>
> *(English Oxford Living Dictionaries 2017)*

Accessibility, in its simplest definition, is the ease of getting to places. It can be defined from different perspectives. Accessibility from the perspective of a place relates to the ease with which people can reach that place. This perspective is important in evaluating potential locations for a business, for example, or cultural centers, or public facilities. The accessibility of a place influences its market potential, since it affects the number of potential customers a business located there might attract. Accessibility from the perspective of a person relates to the ease with which that person can reach desired destinations. This perspective is important in evaluating how well an urban system serves its residents with respect to how easily they can get to jobs, schools, stores, services, recreational activities, and so on.

By most definitions, accessibility consists of two components: the cost of reaching destinations, and the value of the activity found at the destinations (Handy and Niemeier 1997). The cost of reaching destinations is also known as "impedance" and is usually measured in terms of travel distance or travel time, but monetary cost and other qualities of the trip can also factor into impedance. The value of the activity at the destination is usually referred to as the "attractiveness" of the destination. Attractiveness is most commonly measured by the size of destination; for a store, for example, this could be the number of employees working there or the square footage of the building. As with impedance, other qualities can also play a role in attractiveness, as discussed further in this chapter.

This two-component definition of accessibility suggests a tradeoff between impedance (measured as travel distance) and attractiveness (measured as destination size). Take the difference between a neighborhood corner store and a large supermarket. People will use the corner store if it is close by, but they won't go out of their way to get there; the corner store will draw customers from the local area. The supermarket, on the other hand, will draw customers from a much larger area, as people are willing to travel farther to shop there to take advantage of lower prices, a wider range of products, more specialized products, and/or other qualities the supermarket can offer that the corner store cannot. The supermarket offers greater attractiveness but also higher impedance on average; the advantage of the corner store is the low impedance rather than the attractiveness of the store itself.

This distinction between nearby-but-small and large-but-far – the difference between what might be called "local accessibility" and "regional accessibility," respectively – is an important one (Handy 1992). Both kinds of accessibility are necessary to satisfy residents' needs. Individuals do not have an unlimited ability to travel, so they depend on having opportunities within close proximity, particularly for their daily needs. Conversely, they generally cannot get everything they need nearby, at least not the more specialized goods and services, so they also need to reach opportunities found elsewhere in the region. They need, or at least benefit from having, both the local, proximity-dependent accessibility and the regional, mobility-dependent accessibility; they benefit from having both the corner store and the supermarket.

The challenge, then, is to design so as to enhance local and regional accessibility in complementary and sustainable ways. The tradeoff between impedance and attractiveness means that, for a given activity, a typical individual will sometimes choose the local option and other times the regional option (Handy 1996). The more attractive the regional options are relative to the local, the more likely that the individual will choose the regional. This is good for the individual: having both options increases the likelihood that the individual's needs will be satisfied. But this may not be good for society: the more that residents take advantage of regional opportunities, the more they will travel – that is, drive, at least in the US context. If the goal is to reduce car dependency while still meeting individual needs, then the goal must be a particular type of accessibility that emphasizes the local scale but provides for mode choices at the regional scale – in short, a community where residents can meet many daily needs within walking (or at least biking) distance and have good transit connections to regional activity centers. But how can we achieve this goal?

The role of urban design

> Thus, we have substantial information on which to base this particular dimension of performance. Even so, while many of the obvious measures [of accessibility] are well developed, there is a gap between them and some of those felt qualities of access which citizens prize. Systematic attention to the entire range of the dimension is lacking.
>
> *(Lynch 1981, 187–188)*

> Decisions at the city and site planning levels can establish the basis for the creation of well-functioning outdoor spaces. It is, however, only through careful consideration at the detail planning levels that the potential possibilities can come into their own. . . . Crucial to determining the acceptable distance in a given situation is not only the actual physical distance, but also to a great extent the experienced distance.
>
> *(Gehl 1986, 133, 139)*

Accessibility depends on urban design in its broadest sense. It is shaped by activities traditionally in the domain not just of professionals with the title of "urban designer," but also those with the titles of planner and engineer. Planners and engineers are largely responsible for the two basic functional dimensions of accessibility: attractiveness defined as the amount of activity at a destination, and impedance defined as travel distance or time. They are, in short, responsible for the quantitative aspects of accessibility. The important role of the design professions is to look out for the subjective aspects of accessibility, the qualitative components of both attractiveness and impedance. The process is not simple (Figure 44.1).

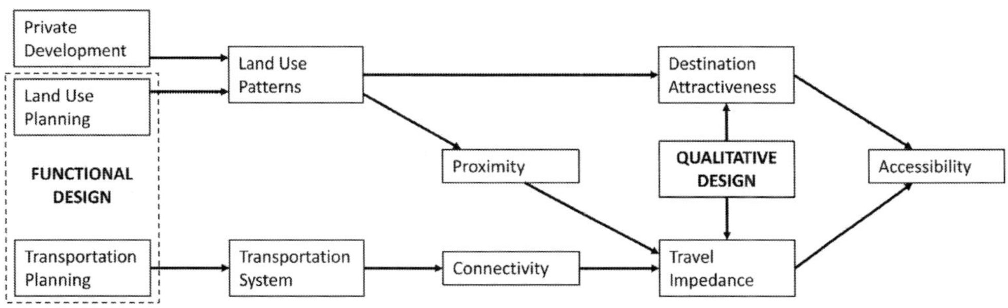

Figure 44.1 The Complex Accessibility Process.

Source: Author

Accessibility depends first and foremost on land use patterns, the spatial distribution of human activities; in other words, what kinds of activities are located where. Land use patterns determine proximity to potential destinations of different types and thus affect both the impedance and attractiveness components of accessibility. Land use patterns, in turn, are the product of the land development process as shaped by land use planning. In most of the United States, land use plans are under the purview of the planners working for local governments, who implement these plans through zoning codes and other mechanisms. But land use patterns largely depend on the decisions of private developers about the design of their projects given the rules set by the city; in many cases the role of the planner is more reactive than proactive. Land use planning at a regional scale is rare in the United States, so that land use patterns across the region are, in effect, the sum of countless decisions by local governments and private developers.

Accessibility also depends on the transportation system – the physical infrastructure and the services using that infrastructure that provide the spatial links, or "connectivity," between activities. The transportation system works in partnership with land use patterns to determine the travel distances between one activity and another. The more directly the transportation system links activities, the shorter the travel distances. The transportation system also determines the set of possible modes to get from one place to another. The system is the product of transportation planning and system design by multiple agencies: local governments for local roads and bicycle and pedestrian facilities, city-scale or regional-scale transit agencies for bus and rail services, regional transportation planning agencies for major roadway and transit investments, state departments of transportation for highways. Their decisions, guided by engineers as well as planners, about the location and design of links in the transportation system directly influence the impedance residents experience in reaching destinations.

But accessibility is also about the subjective qualities – Kevin Lynch's "felt qualities" – that are the traditional domain of architects and urban designers. Design in this narrower sense comprises the visual details of the built environment – the styles, textures, color, ornamentation, and other qualities of physical structures, both interior and exterior. It overlays both land use patterns and the transportation network and determines the felt qualities of destinations and of connections (Handy and Clifton 2001).

The qualitative design of transportation connections affects impedance. The design of streetscapes – the arrangement of activities and their nexus – are especially important. The better the street environment, the less the perceived impedance. A street that offers a safe, comfortable, pleasant environment for a pedestrian or bicyclist will lower the impedance for these modes and thus improve accessibility. The design of intersections, where pedestrians and bicyclists cross the paths of cars, can have a disproportionate effect on impedance. Streetscape design also matters for transit users, who often walk or bike to bus stops or rail stations. The design of stops and stations is equally important: facilities that offer safe, comfortable, and pleasant environments for transit users as they wait for the bus or train will lower the impedance for this mode and thus improve accessibility. The environment within the bus or the train matters as well.

The qualitative design of destinations affects their attractiveness beyond just the amount of activity found there. The attractiveness of a destination is greater if the experience of being there is better. Many things contribute to that experience, particularly the nature of the activity itself (e.g. products sold at what prices, types of services offered, etc.). But design of the physical environment, both inside and outside buildings, also substantially influences the experience. Included in the outside category are public spaces such as walkways, plazas, and gardens. The design of these areas makes a destination more or less appealing. Of underappreciated importance is the design of parking lots and garages (Ben-Joseph 2012), which serve as the connection point between the transportation system and the destination for drivers and which may serve to increase impedance for pedestrians, bicyclists, and transit users.

The multifaceted process that shapes the built environment and determines the quantity and quality of accessibility that residents experience has fallen short in the US context in several respects. One problem is that land use planning has focused on separating rather than connecting activities. Conventional zoning practices separate land uses, keeping residential areas away from commercial and industrial areas, thereby protecting residents from potentially noxious neighbors but reducing their proximity to destinations. They also tend to limit density by setting maximum building heights and floor-area ratios, thereby compounding the separation by dispersing activities across a larger area. A second problem is that transportation planning has focused on car mobility to the detriment of other modes. US streets have long been designed for the convenience of cars, to minimize their delays, an objective imbedded in the "level-of-service" standards widely used in transportation planning. The focus on level-of-service as it is conventionally measured equates to the minimization of just one aspect of impedance (time) for just one mode (cars). Users of other modes – walking, bicycling, transit – are also sensitive to travel time, but because they are not encased within a private vehicle while traveling, they are also more sensitive to the surrounding environment. That leads to a third problem, the lack of attention to the sensory qualities – qualitative rather than quantitative design of the physical environment. Design focused on felt qualities is often an afterthought, particularly in the design of the transportation system.

To improve accessibility in US cities, both land use and transportation must shift towards an accessibility framework. Land use planning in this framework focuses on connecting rather than separating land uses. The adoption of mixed-use zoning and form-based codes by a growing number of US cities suggests that such change may be underway. Transportation planning in an accessibility framework focuses on transportation as a means to an end, rather than an end in and of itself; it focuses on people and their needs, not the system and its needs. A movement to broaden the set of performance measures beyond level-of-service could open the door to a shift from mobility-oriented to accessibility-oriented transportation planning. Inspired by the "complete streets" movement (McCann 2013) as well as its predecessor concept "livable streets" (Appleyard et al. 1981), local as well as regional agencies are using new measures that consider the quality of the walking, biking, and transit experiences and aim for a balance between the modes. As a part of the adoption of an accessibility framework, planners and engineers need to accept that qualitative design is necessary rather than optional, and an integral element of accessibility. Conversely, designers need to understand that focusing on subjective qualities is not enough, that qualitative design is a complement to rather than a substitute for good functional design.

Accessibility beyond the city center

> Why should anyone have to take a car and drive a couple of miles to get a package of cigarettes or a loaf of bread, as one must often do in a suburb?
>
> *(Mumford 1955, 244)*

Although accessibility can be hard to define, you know it when you see it. European cities, for example, provide high levels of accessibility for residents as well as visitors at both local and regional scales. Commercial areas blend seamlessly with residential areas, putting shops and restaurants within walking distance in most neighborhoods. Integrated networks of bus, streetcar, and metro services link neighborhoods to the city center, and getting around by car is often more difficult than getting around by transit, walking, or biking. Although many Europeans choose to drive, they almost always have the option not to.

Accessibility is at the heart of the European concept of the "city of short distances" (or in the original German, "der Stadt de kürzen Wege"). Often used as a synonym for "compact city," the concept evokes a place where destinations are within a short distance from homes, making walking or biking a real possibility. In such cities, accessibility is achieved through proximity rather than mobility. Freiburg, Germany, epitomizes this idea and is often held up as a model of how to design a city so as to prevent dependence on cars. An integrated traffic management plan adopted in 1969 set the city on an alternative path by giving priority to more sustainable modes, including transit, walking, and biking (Making Cities Livable 2017).

Older city centers in the United States offer the "short distance" option, albeit with much less charm than in Europe. The escalation of housing prices in many city centers – New York, Seattle, San Francisco, and many others – reflects a growing appreciation for high-accessibility environments. Compare that to the situation in much of the rest of the United States, where residents have chosen (or, increasingly, been forced into) lower-density living in the suburbs over the good accessibility found in city centers. In these places, residential and commercial areas are separated by intention; streets prioritize cars over other modes by design; and relatively few residents ever consider getting around by transit, walking, or biking unless they have no choice. Americans have become inured to poor accessibility, rarely questioning the car-dependent lifestyle it engenders.

But can they have their cake and eat it too? Is it possible to create high levels of accessibility in suburban environments? The case of Davis, California, suggests that – with good urban design at all scales and attention to form as well as function – it is.

Davis, a municipality of about 67,000 residents and 10 square miles, lies 70 miles northeast of San Francisco and 15 miles to the west of the state capitol in Sacramento. It is home to the University of California, Davis (UC Davis), and many Davis residents work for the university, though many commute to Sacramento or the Bay Area, and many UC Davis employees commute to campus from surrounding cities. In this way it functions as both a suburb and a job center. With a large population of students, the share of multifamily housing is relatively high for a suburb, at 53%; even so, the community has a single-family, suburban feel. Yet it differs from typical American suburbs in important ways. Although its plans do not explicitly emphasize the concept of accessibility, Davis has in fact been implementing accessibility-oriented design for many decades, as is evident throughout its current (and soon to be updated) General Plan (City of Davis 2007).

Bicycling is the city's best-known accessibility quality. Although less than 1% of US residents bike to work, over 20% of Davis residents do so, according to the American Community Survey, and this number rises to over 50% for residents commuting to the UC Davis campus (Thigpen et al. 2015). Davis achieved these high levels of bicycling in part because of the luck of natural conditions, particularly flat topography and temperate weather for most of the year. But Davis also invested heavily in bicycling infrastructure beginning in the late 1960s, when just over 20,000 people lived in the city (Buehler and Handy 2008). Today, the city boasts over 50 miles of on-street bike lanes and 50 miles of off-street bike paths, with dozens of bicycle tunnels and several bicycle bridges over the two freeways that bisect the city (Figure 44.2). According to the Transportation Element of the General Plan, adopted in 2013, "Greenbelts and separated bike paths on arterials should serve as the backbone of much of [the bicycle] network" (City of Davis 2013, 40).

The city promotes a high level of connectivity for all modes through policies that state that development "shall incorporate a balanced circulation network that provides multi-route access for vehicles, bicycles and pedestrians to neighborhood centers, greenbelts, other parts of the neighborhood and adjacent districts and circulation routes" (City of Davis 2007, 152). But policies especially promote bicycle and pedestrian connectivity. The city allows cul-de-sacs,

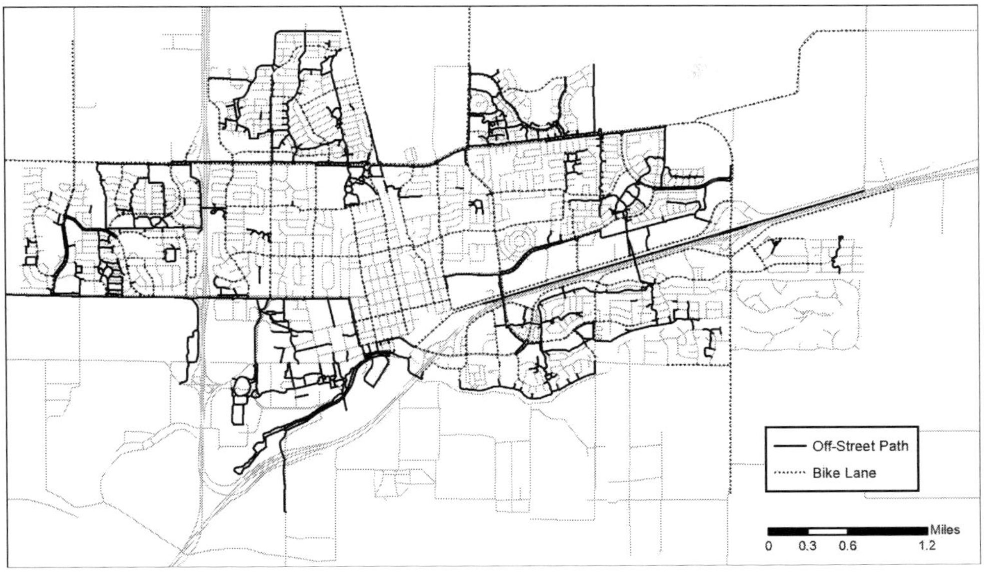

Figure 44.2 Map of bike facilities in Davis, California.

Source: City of Davis General Plan

provided they connect to bicycle/pedestrian corridors (City of Davis 2013, 29), and emphasizes the development of "destination-oriented trails" (City of Davis 2013, 42). As a result of such policies, the bicycle network has a higher level of connectivity than the street network, meaning that bicycling distances can be considerably shorter than driving distances (Tal and Handy 2010).

Bicycling works in Davis because of this network, but also because of a multifaceted approach to land use planning. Facing significant growth pressures in the early 1970s, Davis adopted a general plan designed to avert suburban sprawl and its environmental impacts. The city has partnered with the county and with other cities in the area to permanently preserve large tracts of agricultural land to ensure a permanent buffer between cities. Under current policies, development of agricultural land at the edge of the city must be approved by the voters, a majority of whom so far have consistently voted "no." Instead, the city actively encourages infill development and has identified enough developable parcels within its boundaries to accommodate enough new housing to meet its obligations under state policy. The first land use goal in the city's General Plan is to "Maintain Davis as a small, University-oriented city surrounded by and containing farmland, greenbelt, and natural habitats and reserves" (91) with "a distinct permanent urban edge" (City of Davis 2007, 94). The product of these policies is a relatively compact city, more akin to the European "city of short distances" than the typical US suburb.

Adding to this compactness is the city's long-standing policy that each neighborhood should have at least one commercial center within easy walking and biking distance of housing. According to the General Plan, "All neighborhoods, both new and existing, should include a centrally located hub or activity node within walking distance of housing in the neighborhood. . . . Transit stops, neighborhood commercial uses and activity centers should be in this hub" (City of Davis 2007, 57). The city defines "neighborhood shopping centers" as "shopping centers that serve the daily needs of the surrounding neighborhood for goods and services, such as groceries, pharmaceuticals, dry cleaning, and other uses" (City of Davis 2007, 79). Figure 44.3 shows the

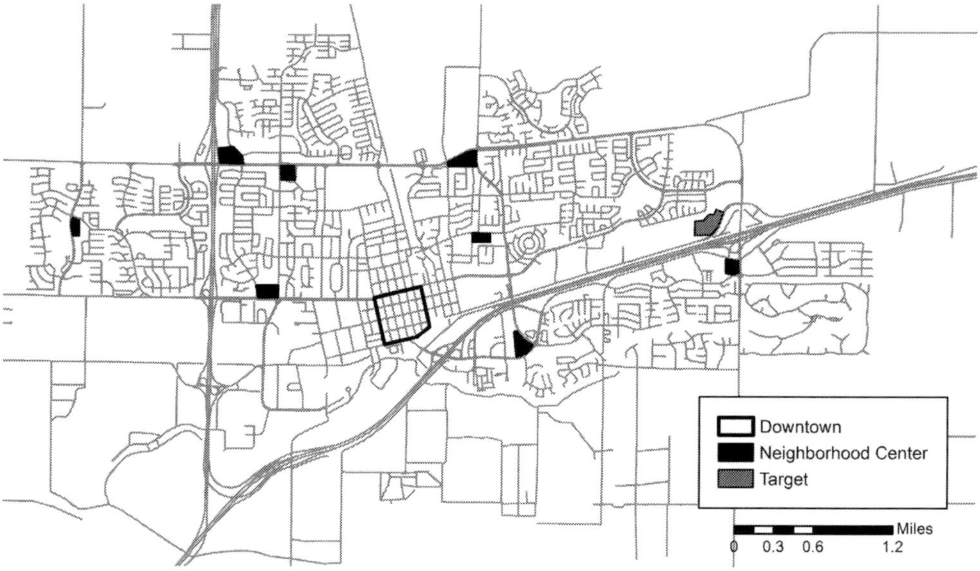

Figure 44.3 Neighborhood centers in Davis, California.

Source: Author

locations of these centers, numbering nine in total, throughout the city. The explicit goal of this policy is "to minimize the lengths of trips and to facilitate walking, bicycling and transit use as alternatives to auto use" (City of Davis 2007, 56).

Another key element of Davis's approach to accessibility is the preservation of its traditional downtown. A core part of the city's vision is to "Maintain a strong, vital, pedestrian-oriented and dynamic downtown area" (City of Davis 2007, 41). Downtown Davis, which dates back more than 100 years, has small blocks, a number of historic buildings, large street trees, small storefronts, and many sidewalk cafes. Much of the vitality of downtown Davis is rooted in activities other than retail, such as its bi-weekly farmers market, a monthly art open-house, public concerts, restaurants and bars, and movie theaters; services such as banks, medical offices, hair salons, and yoga studios add to the daily activity (Figure 44.4). A business improvement district actively promotes and advocates for downtown Davis as the town's primary business, social, and cultural center. The city contributes to downtown vitality through capital improvements, particularly pedestrian amenities such as "signage/wayfinding, street furniture, outdoor dining, crosswalks, drinking fountains, street lighting, street trees, and gathering areas" (City of Davis 2013, 27).

One of the policies that protected downtown for many years was a limit on the size of stores in the city that effectively prohibited big-box stores. In the 1990s, the city deemed "warehouse style retailers . . . [as] inappropriate given the nature and scale of the Davis market" (City of Davis 2006). Zoning codes restricted retail businesses outside of downtown to be sized to serve neighborhoods rather than the region, limiting them to 30,000 square feet. While this policy may have helped to protect downtown, it also restricted shopping opportunities within the city. In response, the city council and then the voters approved an exception to this limit in 2006 to allow a Target store, which opened in 2009. Opponents expressed concerns that Target would harm downtown businesses, but in fact the new store improved accessibility for residents by eliminating the need to drive to other cities to buy many products (Lovejoy et al. 2013).

Figure 44.4 Bicycle parking in front of Varsity Theater in Downtown Davis.

Source: Author

Given its relatively small size, Davis remains dependent on the rest of the Sacramento region for many services and other activities. Driving dominates travel to and from Davis, but alternatives are available. The city is linked to the region by an interstate freeway but also by express bus and commuter rail services. A bike path alongside the interstate links Davis to West Sacramento and enables bicycle commuting in both directions, while bike lanes on county roads enable bicycling between Davis and its other neighboring cities. Regional transportation planning is the responsibility of the Sacramento Area Council of Governments, which in 2004 adopted its Blueprint Plan that aimed to channel growth into existing areas to create a more compact region (SACOG 2017). The Blueprint Plan became the basis for the region's Sustainable Communities Strategy, required by the state to show how the region will reduce greenhouse gas emissions from transportation through both transportation and land use planning (SACOG 2016). Davis's local planning efforts contribute to this larger regional effort.

Qualitative aspects of design are important throughout the city's efforts. Davis has set a goal to "Encourage community design throughout the City that helps to build community, encourage human interaction and support non-automobile transportation" (City of Davis 2007, 152). Urban design standards include details about the appearance of houses and the siting of commercial buildings. Bicycle and pedestrian connectivity is an important theme in these standards, which specify, for example, pedestrian walkways separate from driveways and parking lots for commercial development. The city aims for "a sustainable community forest," important in the hot summer climate of Davis; the goal is to have 30% to 50% of the pavement area of streets covered in shade when street trees reach their maturity (City of Davis 2007, 157). The Transportation Element calls for shade trees in parking lots, along with other amenities (City of Davis 2013, 45). The strong growth boundary around the city is not just about preventing outward growth but also about creating a "visual edge" for the city (City of Davis 2007, 57).

The Davis example shows that, with a concerted effort, it is possible to design for accessibility in the suburban US context. But it takes a combination of land use and transportation planning within and beyond the city boundaries. It requires attention to functional considerations of proximity and connectivity but also to the felt qualities of pathways and destinations. The reward is a high quality of life with a lower dependence on driving (see the Appendix for a sample of city Davis urban design policy).

Accessibility in the future

> [P]eople and activities are connected regularly and at rapidly decreasing real cost through a variety of increasingly sophisticated telecommunications facilities. . . . Nevertheless, the [urban] cores remain highly significant for activities requiring face-to-face contact, and if anything, these are growing in importance.
>
> *(Hall 1996, 12)*

Places with high levels of accessibility tend to be either older places or newer places modeled on older places. The characteristics that contribute to accessibility – close proximity and intermixing of land uses, good connections for human-powered forms of transportation – are characteristics typical of cities prior to the proliferation and dominance of the automobile. In other words, accessibility can be achieved by looking to the past. But what about the future? As Webber (1964) recognized as early as the 1960s, when he articulated the idea of "communities without propinquity," information and communications technologies (ICT) raise the possibility of accessibility without either mobility or proximity. People can get many of the goods and services they need, connect with friends and family, and participate in the workforce without ever leaving home. But while this kind of accessibility has grown over the last five decades as technologies and the services that use them have improved, there is no sign that it is replacing the human desire – need, even – for physical accessibility. As Hall noted in 1996, in reflecting on Webber's ideas, the lack of evidence of a decline in travel as the use of ICT increases "is surely testimony to the value people put on face-to-face encounters" (Hall 1996, 14). Cities continue to be important to society because they provide a form of accessibility – a set of opportunities for interaction – that no other technology provides.

At the same time, design will be ever more important to cities, both to ensure that cities continue to provide a high level of accessibility and to guarantee that a high level of physical accessibility remains important to many people. The qualitative attractiveness of destinations becomes more important as goods and services are more easily ordered online; hence the emergence of "life-style centers," the growth of sidewalk cafes, and the renewal of public spaces. Reducing the unpleasantness and enhancing the safety of traveling, particularly by modes other than cars, also becomes more important so as not to deter residents from taking advantage of the opportunities the city has to offer. To avoid a vicious cycle, it is important to incorporate new forms of accessibility into the city in ways that do not dampen the old forms of accessibility – to avoid, for example, streets becoming increasingly crowded with delivery vehicles. Design in the broadest sense, encompassing land use planning, transportation planning, and attention to the qualitative aspects of places, from the site level to the neighborhood level to the regional level, is essential for preserving accessibility – and thus cities – in the future.

References

Appleyard, D., Gerson, M. S., and Lintell, M. (1981). *Livable Streets, Protected Neighborhoods.* Berkeley: University of California Press.

Ben-Joseph, E. (2012). *ReThinking a Lot: The Design and Culture of Parking* (Vol. 7). Cambridge, MA: MIT Press.

Buehler, T. and Handy, S. (2008). "Fifty Years of Bicycle Policy in Davis, CA." *Transportation Research Record,* 2074: 52–57.

City of Davis. (2006). Ordinance No. 2259. Davis, California.

City of Davis. (2007). *General Plan.* Available: http://cityofdavis.org/home/showdocument?id=3684 (accessed 12/27/17).

City of Davis. (2013). *General Plan Transportation Element.* Available: http://cityofdavis.org/city-hall/community-development-and-sustainability/planning-and-zoning/general-plan (accessed 12/27/17).

English Oxford Living Dictionaries. (2017). *Accessibility.* Available: https://en.oxforddictionaries.com/definition/accessibility (accessed 12/27/17).

Gehl, J. (1986). *Life Between Buildings: Using Public Space.* New York: Van Nostrand Reinhold Company.

Hall, P. (1996). "Revisiting the Nonplace Urban Realm: Have We Come Full Circle?" *International Planning Studies,* 1(1): 7–15.

Handy, S. L. (1992). "Regional Versus Local Accessibility: Neo-Traditional Development and Its Implications for Non-Work Travel." *Built Environment,* 18(4): 253–267.

Handy, S. L. (1996). "Understanding the Link Between Urban Form and Nonwork Travel Behavior." *Journal of Planning Education and Research,* 15(3): 183–198.

Handy, S. L. and Clifton, K. (2001). "Evaluating Neighborhood Accessibility: Possibilities and Practicalities." *Journal of Transportation and Statistics,* 4(2/3): 67–78.

Handy, S. L. and Niemeier, D. A. (1997). "Measuring Accessibility: An Exploration of Issues and Alternatives." *Environment and Planning A,* 29(7): 1175–1194.

Lovejoy, K., Sciara, G. C., Salon, D., Handy, S. and Mokhtarian, P. (2013). "Measuring the Impacts of Local Land-Use Policies on Vehicle Miles of Travel: The Case of the First Big-Box Store in Davis, California." *Journal of Transport and Land Use,* 6(1): 25–39.

Lynch, K. (1981). *A Theory of Good City Form.* Cambridge, MA: MIT Press.

Making Cities Livable. (2017). *Freiburg: City of Vision.* Available: www.livablecities.org/articles/freiburg-city-vision (accessed 12/27/17).

McCann, B. (2013). *Completing Our Streets: The Transition to Safe and Inclusive Transportation Networks.* Island Press.

Mumford, L. (1955). "The Roaring Traffic's Boom." *The New Yorker.* March 19, pp. 115–121.

Sacramento Area Council of Governments (SACOG). (2016). *Metropolitan Transportation Plan/Sustainable Communities Strategy.* Available: www.sacog.org/metropolitan-transportation-plansustainable-communities-strategy (accessed 12/27/17).

Sacramento Area Council of Governments (SACOG). (2017). *Sacramento Region Blueprint.* Available: www.sacog.org/sacramento-region-blueprint (accessed 12/27/17).

Tal, G. and Handy, S. (2010). "Travel Behavior of Immigrants: An Analysis of the 2001 National Household Transportation Survey." *Transport Policy,* 17: 85–93.

Thigpen, C. G., Driller, B. K. and Handy, S. (2015). "Using a Stages of Change Approach to Explore Opportunities for Increasing Bicycle Commuting." *Transportation Research Part D,* 39: 44–55.

Webber, M. B. (1964). "The Urban Place and the Nonplace Urban Realm." In M. B. Webber, et al. (eds.), *Explorations into Urban Structure.* Philadelphia: University of Pennsylvania Press.

Further reading

Brookings Institute. (2017). *Moving to Access.* Available: www.brookings.edu/interactives/moving-to-access/ (accessed 12/27/17). This report looks at policies, tools, and techniques that can help to ensure equitable access to the places people need to go across all income and demographic groups.

Handy, S. (1994). "Highway Blues: Nothing a Little Accessibility Can't Cure." *Access* 5: 3–7. Available: www.accessmagazine.org/fall-1994/highway-blues-nothing-a-little-accessibility-cant-cure/ (accessed 12/27/17). This essay defines the concepts of accessibility and mobility, explains the relationships between them, and discusses the potential for a shift from a mobility focus to an accessibility focus in transportation planning.

Levine, J., Grengs, J., Shen, Q., and Shen, Q. (2012). "Does Accessibility Require Density or Speed? A Comparison of Fast versus Close in Getting Where You Want to Go in US Metropolitan Regions." *Journal of the American Planning Association*, 78(2): 157–172. This study explores two contrasting ways of providing accessibility – by ensuring that destinations are close or by ensuring that travelers can get to destinations quickly – and concludes that proximity matters more than speed.

Levinson, D. M., Marshall, W., and Axhausen, K. W. (2017). *Elements of Access*. Available: https://gumroad.com/l/elements-of-access. This wide-ranging book starts from the premise that accessibility is the overarching purpose of the transportation system and provides a helpful introduction to transportation planning concepts for engineers and to transportation engineering concepts for planners.

APPENDIX

Sample of city of Davis urban design policies from the 2007 General Plan

City of Davis. (2007). *General Plan.* Available: http://cityofdavis.org/home/showdocument?id=3684 (accessed 12/27/17).

GOAL UD 1. Encourage community design throughout the City that helps to build community, encourage human interaction and support non-automobile transportation.

Policy UD 1.1 Promote urban/community design which is human-scaled, comfortable, safe, and conducive to pedestrian use.

Standards

a New neighborhoods shall be designed so that daily shopping errands and trips to community facilities can generally be completed within easy walking and biking distances.

b New development shall incorporate a balanced circulation network that provides multi-route access for vehicles, bicycles, and pedestrians to neighborhood centers, greenbelts, other parts of the neighborhood and adjacent districts and circulation routes.

c Garages on new single-family, duplex and split lot homes should be subordinate in visual importance to the house itself, especially the entry.

d Where possible, commercial buildings should abut the street or other public accessway, with parking located behind.

e Pedestrian walkways and/or building pass-throughs should be located to provide access from sidewalks into individual projects separate from major vehicular driveways and circulation.

f New cul-de-sacs shall have bicycle/pedestrian passages through them that connect to other streets, greenbelts or publicly accessible corridors.

GOAL UD 2. Maintain an aesthetically pleasing environment and manage a sustainable community forest to optimize environmental, aesthetic, social, and economic benefits.

Policy UD 2.1 Preserve and protect scenic resources and elements in and around Davis, including natural habitat and scenery and resources reflective of place and history.

Policy UD 2.2 Maintain and increase the amount of greenery, especially street trees, in Davis, both for aesthetic reasons and to provide shade, cooling, habitat, air quality benefits, and visual continuity.

Standards

c Streets that are planted in the future are expected to have wide canopies, sufficient to eventually provide, at maturity, at least 50 percent shade coverage of the pavement area of local streets and 30 percent shade coverage of the pavement area of collector and arterial streets.

Policy UD 2.3 Require an architectural "fit" with Davis' existing scale for new development projects.

Policy UD 2.4 Create affordable and multi-family residential areas that include innovative designs and on-site open space amenities that are linked with public bicycle/pedestrian ways to neighborhood centers and transit stops.

GOAL UD 6. Strengthen the city's neighborhoods to retain desirable characteristics while allowing for change and evolution, promoting public and private investments, and encouraging citizen involvement in neighborhood planning.

Policy UD 6.1 Recognize the existence of individual neighborhoods with general boundaries and facilitate the development of neighborhood strategies in partnership with residents and property owners. The strategies should recognize the unique characteristics of the individual neighborhood and the potential for change, within the context of a well-planned city.

45

MOBILITY, ACCESSIBILITY, AND URBAN FORM

Yodan Y. Rofè

Introduction

We are living today in the latter part of "the urban revolution." Starting at the beginning of the 19th century in the United Kingdom and Northern Europe, and gradually extending to the entire globe, urbanization has transformed human society, the economy, and the world itself. The human population has grown exponentially in size and has gained immensely in material wealth, health, and life expectancy. This process is due to be completed by the end of this century, as the world's population stabilizes and becomes roughly 80% urban (Angel 2012; United Nations, Department of Economic and Social Affairs 2017).

Cities are a rather late arrival in the human story. Throughout most of its life on earth, the human species has lived in small communities numbering at most several hundred people (Diamond 2012). Eventually, cities rose to dominance for a variety of reasons, not the least of which is that they increase the probability of fruitful contact among strangers, while controlling the dangers that such contact entails (Hillier 1996; Jacobs 1961). The network of streets, which constitutes most of urban public space and where movement occurs, allows mobility and access for everyone in relative safety. Yet, if we look at the 200 years of urban expansion, we see a marked difference between the shape of urban space as developed in cities of the 19th and early 20th centuries, and the urbanism of the postwar period (Dibble et al. 2017; Omer & Zafrir-Reuven 2015). This chapter explores this difference and its impact on human mobility, accessibility, and urban liveliness.

The chapter starts with a description of five elements of urban form that have evolved through history: the square, the street grid, the Baroque radial plan, the boulevard, and the urban park. Cities in the 19th century made use of all these elements to develop plans for large-scale expansions. It continues with a brief discussion of the internal forces and contradictions created by the economic success of cities, which led to the dissipation and decline of urban life. Modernist urban planning has not created the tools to counteract these forces, but rather exacerbated their impact. The chapter discusses the role of planning for automobility rather than accessibility, and how it contributed to the decline of urban space. In particular, it examines how hierarchical urban street planning and parking standards have compromised the most crucial element of cities: the urban street network that forms the majority of public space.

As the destruction of many loved urban spaces in cities (as a result of postwar policies) began to be visible, resistance began. Citizens revolted against urban renewal and freeway building, and urbanists and researchers started critiquing these policies. Cities began experimenting with human-centered street design. This resistance has gradually developed, over the last 50 years, as an alternative paradigm for building cities, returning in many ways to the 19th-century model. This chapter describes the major elements of this paradigm and concludes with a reflection on the reasons why – despite our better professional understanding of how to build good cities – its implementation is still rather rare, and automobile-oriented planning is still the norm.

The elements of urbanism

The square

Squares were the first to come about, perhaps even as a precursor of cities. Indeed, squares existed even in non-urban societies, often at important crossroads or necessary passages, providing a place for meeting, social activities, and trade. In biblical stories, a square was the place where the elders sat, "at the gate" of the city, and therefore at the interface between locals and strangers. The square was where disputes were arbitrated and judged, and important transactions occurred under the public eye. Often a square was associated with a religious shrine that brought together different people from afar. Such were the beginnings, for instance, of the Athenian agora, the Roman forum, or the Piazza del Campo in Siena, even before they were given shape by the growth of the city around them. However, it is only after a network of streets evolved around the square that a different kind of meeting place emerged – one that acquired importance at different scales: from the scale of the local neighborhood (for everyday life activities), to the scale of the metropolis embodied in main city squares, and even the scale of the nation (in squares used as stages for major demonstrations and public events).

The street network

The street is the basis of urban life (A. B. Jacobs 1994; J. Jacobs 1961; Marshall 2004). Streets are so commonplace to us as urban inhabitants that we forget that they are a human invention. For millennia, the main focus of city planning was the laying out of streets in a rectangular grid or other possible variations. The earliest street grids in Greek colonies were probably more concerned with the equal division of land than with creating urban space. However the street network, which facilitates movement and makes it more predictable, allows co-presence of people in space and generates encounters between city inhabitants and strangers; it is the very mechanism which enables social life in urban settlements. This mechanism sets the conditions for the city to become an economic generator.

The Baroque radial plan

Street grids may be a good way to organize a fairly small city, but over a large area they may become disorienting as they lack a natural centrality and orientation. Thus, in the late Renaissance and Baroque periods, a new element was added to the urban planning of cities – the radial plan, first used by Pope Sixtus V in his reorganization of Rome to facilitate the movement of pilgrims. The plan defined a set of important destinations in the city, connecting them visually and also with direct and wider streets, while allowing the rest of the city to remain untouched. In an existing city, such as medieval Rome, and later Paris during Hausmann's transformation of the city in the 19th century, the "rest" was the existing medieval maze of streets. In the case of a new city or part of a city, such as in Washington, D.C., or Barcelona under Cerdà's plan, the "rest" was an ordinary grid. The radial plan created a two-level hierarchy of movement and accessibility and two types of streets: at the citywide (or regional) scale, and at the local (neighborhood) scale. But while movement was hierarchically structured, accessibility and the interface between movement and access were similar for both types of streets – the streets for citywide movement were also part of the local interface and not separated from it.

The tree-lined boulevard

The tree-lined boulevard has its origins in various forms of tree-lined promenades, some of them imported from the architecture of palace landscapes at the end of the 17th century and the early 18th century (Jacobs, Macdonald, & Rofe 2003; Lawrence 1988). It completes the Baroque radial plan by providing an added uniqueness to the main avenues for movement; because of its multi-way form, it allows for a resolution of the conflict between movement and access (Jacobs et al. 2003).

The urban park

The urban park is a reflection of the effort to bring nature into urban public space. Thus, urban parks have their origin in two very different traditions. One is the village common or small common public garden, which existed in many small towns throughout Europe. The second is the tradition of opening up the gardens of large urban palaces to the public on particular days of the week. The growth of cities in the 19th century indicated the importance of providing large areas of natural settings and green open space for recreation and leisure activities. Park advocates touted the value of large urban parks, such as Central Park in New York City, to the areas surrounding them. Along with the tree-lined boulevards, parks often gave suburban development the promise of urban dignity and attraction (Macdonald 2005), seeming to combine the best of both worlds. In expanding urban regions, parks were viewed as a means to protect water and environmental resources, and as attractors to newly developing areas. The aim of both of these later additions to the urban design palette was to offer the urban public access to vegetation and nature – as this access had become remote as cities expanded.

Differences between 19th- and 20th-century urbanism

The 19th century saw the beginning of the urban revolution, with its associated phenomena of industrialization and immigration from the countryside to the city. In their expansion plans, cities made effective use of the elements of urbanism discussed earlier. They also developed the environmental infrastructure necessary for high concentrations of people: water provision systems, sewage and solid waste removal systems, large-scale paving of streets, and railway transportation. Railways, first used for interurban and freight transportation, allowed the first wave of suburbanization to nearby small towns surrounding the central city. Later, the omnibus (a horse-drawn carriage on rails), and at the turn of the century the electric tram and underground and elevated trains, eventually allowed the large-scale expansion of cities and the second wave of suburbanization. Another significant innovation of the late 19th century, the bicycle, allowed working people to extend the radius of their daily movements and thus take advantage of the growing opportunities created by larger cities.

The 19th-century city, however, was still a walking city, as most people walked to their destinations, with public transport used more by the middle class than the working class. Horse- or human-driven carts moved most freight within the city, while the speed of mechanized movement was still low. If we look at images of city streets from the late 19th and early 20th centuries, we are struck by the sheer number of pedestrians on the streets. People were not confined to the sidewalks, but were walking all over the street and crossing it at will.

While urbanization continued unabated in the 20th century, extending to other regions of the world and expanding cities in Europe and North America, it underwent a change in character along the way. This change had its origins in the ideas of the new profession of urban planning dating from the end of the 19th century: the Garden City Movement, the New York Regional Plan in the United States, and Le Corbusier and other modernist architects forming CIAM in Europe. It is also connected with the entry of mass-produced automobiles and trucks into the market during the 1920s. However, because the 1930s and 1940s were decades of economic depression and war, the impact of these ideas was only felt after World War II – in the suburban expansions of US cities, the postwar reconstruction plans of European cities, and planned urban expansions throughout the world. Before discussing the reasons for this different type of urbanism, it is worthwhile to visually contemplate its impacts by giving a specific example.

The difference between 19th- and 20th- urbanism is clearly visible in the city of Rishon Letzion in Israel (Figure 45.1). The areas east of Road no. 4 developed piecemeal, as the agricultural colony, established in 1882, urbanized as part of the growth of the Tel-Aviv metropolitan region. Possessing large reserves of land, the city continued, during the 1980s and 1990s, to grow west of Road 4, on land that had previously been uncultivated sand dunes. This growth took place

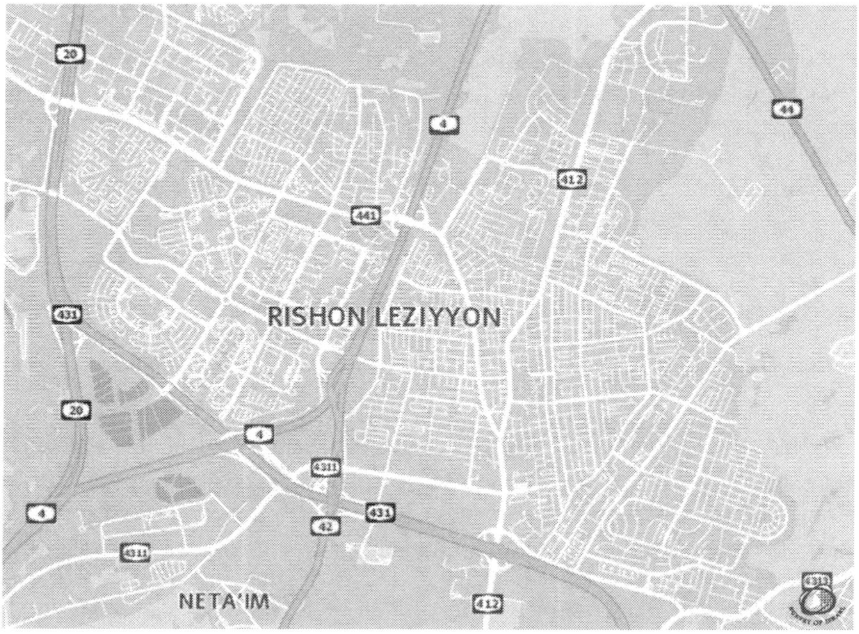

Figure 45.1 The city of Rishon Letzion in Israel showing the difference between 19th-century urbanism (east of Road 4) and 20th-century urbanism (west of Road 4).

Source: Survey of Israel, Geospatial Portal

on the basis of a Master Plan for the whole area. While the eastern part of the city is centered on a main road, which at some point splits into two branches and forms the main street of this part of town, the western part has no such central street. Neighborhoods are separated by arterial roads; they have some small shopping centers, but no main street. At the northwest corner of the city, there is a large commercial area, originally intended as an industrial area but which was transformed into what is primarily a series of enclosed shopping malls. More commercial and employment areas are currently developing on the southwest corner of the city, near the junction of three freeways. One can also see the impact of these freeways, bisecting the city, and cutting it off from neighboring towns, which have also become a part of the metropolitan area.

The erosion of the city as a generator of human contact

It is tempting to lay the responsibility for this change in the form of cities over the 20th century on the impact of urban planning doctrines (Jacobs 1961) or on street engineering standards (Hebbert 2005), but the reality is more complicated, as planning is often less influential than what planners would like it to be (Alfasi, Almagor, & Benenson 2012). Issues relating to accessibility and land availability, and the economics of land development, are, in the long run, much stronger forces in shaping city form. Over the last two centuries, many cities all over the world have expanded in order to accommodate population growth. But they have been decreasing in density at a mean rate of 1% to 2% per year (Angel 2012). Today, cities are much less compact (and therefore much less walkable) than they were at the beginning of the urban revolution.

Cities thrive on density and diversity, since these are necessary, if not sufficient, preconditions for fruitful face-to-face encounters among people. However, the very success of cities in creating wealth, and pulling people out of poverty, creates internal dynamics that also work to erode density and diversity. While I do not have space to discuss these dynamics in full here, it is important to mention them as a background against which to understand the failures of modernist urban planning. They are: (1) the success of the city as a mechanism for alleviating poverty and improving the economic well-being of its inhabitants, thus bringing about the rise of the middle class, which, with improved economic status, tends to seek more spacious places of residence, often at the city's outskirts; (2) economic mobility reducing familial, ethnic, and religious dependency and leading to a stratification along economic class, and thus a reduction in social and economic diversity within urban neighborhoods; (3) competition for central places and fashionable locations, which led to pricing out of less profitable uses and lower-income people, again reducing economic and social diversity within neighborhoods; (4) the economic allure of urban expansion due to the exceptional profitability of transforming non-urban to urban land; and (5) the freedom of location for households and businesses provided by mechanized transportation, and in particular automobiles, which allows urban expansion at an unprecedented scale.

These are powerful dynamics, which have led to the expansion of cities over the past century. The task of urban planning should have been to devise policies to counteract or manage the effects of these dynamics. Instead, operating from the perception that the city itself, with its density and diversity, is the problem, urban planning has adopted five basic tenets that have exacerbated the problems created by the dynamics eroding urbanism discussed earlier. At the basis of these tenets lies the apprehension of dynamic change characteristic of the city, and the wish to control and freeze city development.

The first two tenets defining 20th-century urban planning are limits to development density, and strict zoning controls separating primary land uses from each other, in particular housing from employment. Both tenets were denounced by Jane Jacobs (1961), but – despite her far-reaching influence – they remain present in many master plans. The third tenet is the "neighborhood unit," which broke down the continuous urban realm into artificial neighborhood areas that did not necessarily reflect people's use of space or sense of neighborhood (Mehaffy, Porta, Rofe, & Salingaros 2010; Mehaffy, Porta, & Romice 2014). The fourth tenet is public open space standards, which, because they are often defined on a per capita basis, have proven to be another limitation on achieving density. This was a misguided attempt to provide an alternative public space to the urban street network, which was relegated more and more to automobile traffic (Rofè, Zarchin, & Feierstein 2012; Rofè 2016). The fifth tenet, and what the rest of this chapter will discuss, is planning for cars and not for people. This aspect of modern city planning involves two separate issues that have had devastating results on cities in the 20th century: the design of street networks according to the functional hierarchy of streets, and the institution of minimum parking standards. Both resulted in the privileging of access for car drivers over all other users of urban streets.

Automobile-based urban planning

Perhaps the most damaging concept for cities has been the institution – sometime in the 1930s – of the functional hierarchy of urban streets, and its marriage with the idea of the neighborhood unit. The aim of this concept is twofold. On the one hand, it is designed to allow cars to travel at high speeds on major streets, without the friction caused by vehicle and pedestrian access to properties along the street, and by frequent intersections with side streets. On the other hand, the concept seeks to safeguard neighborhoods from the noise, pollution, and danger of fast-moving vehicular traffic. Thus, the functional hierarchy of urban streets dictates that the more a street is used for movement, the less it should allow access to properties fronting it, and the further apart the intersections should be (Figure 45.2). Thus, major urban streets that had previously given form to the city and served as the center of social and economic life were replaced by major arterials that separate neighborhoods and allow fast-moving traffic to flow through the city.

When compared to the overall number of streets that exist in a city, major urban streets constitute a minority. They make up about 15–20% of all street length in a city, and yet most of the city's economic and social life occurs within 200 meters of them (Carmona 2014; Rofè & Omer 2014). Major streets are not limited to the urban center but extend throughout the urbanized area (Carmona 2014; Vaughan & Griffiths 2013). Historic major urban streets defy the functional classification of streets (Figure 45.2). They account for most through movement (vehicular and pedestrian) in their urban context, and they provide relatively greater access to their surroundings. They typically have more intersections than their surrounding residential streets (Hillier 1996; Porta et al. 2009, 2012). Buildings fronting them usually have, aside from an entrance to their upper floors, several entrances to the shops lining the street.

It was the confluence of two of the major tenets of modern city planning – the functional classification of streets and the neighborhood unit – which caused the size of the area located between major streets to expand from a maximum of 400 meters by 400 meters (about a quarter mile squared), to at least double, if not triple, the distance between main streets (Dibble et al. 2017; Mehaffy et al. 2010). This extension of scale, coupled with the prohibition of direct access from arterial streets, resulted in practically eliminating street commerce in modern planned layouts. They have also caused severe separation between adjacent neighborhoods, as there are no easy walking paths, or safe crossing points on the arterials – making direct walking from one side of the arterial to the other difficult (Omer, Rofè, & Lerman 2015). Furthermore, they make public transport less viable as transportation planners have to choose between providing good accessibility to residences, causing bus lines to serpentine through the neighborhoods' disconnected

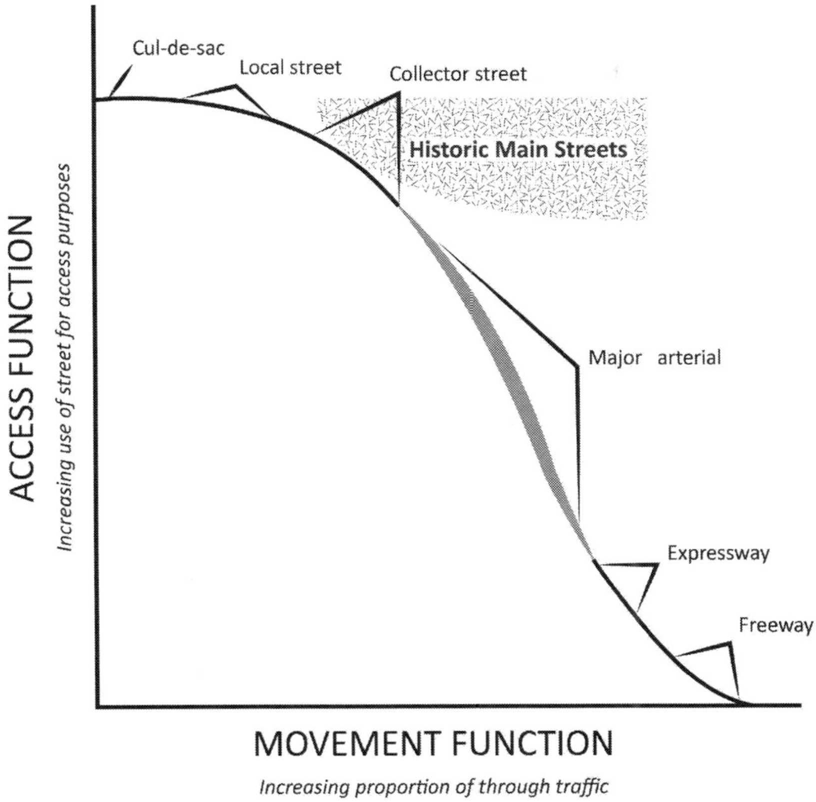

Figure 45.2 The functional hierarchy of streets and its contrast with historic main streets (hatched area).

Source: Author

grids, and having them run only along the arterials between neighborhoods, but not within easy and safe walking reach from most residential areas. The reduction of accessibility by walking and public transit and the elimination of street commerce result in an increased reliance on automobile-oriented shopping centers located near the major intersections of arterial streets, or near freeway entrances – thus further increasing automobile dependence.

Moreover, there is no evidence that this mode of planning actually reduces automobile traffic within the neighborhoods. Because there are very few destinations within the neighborhood, and almost all destinations of interest require a trip by automobile, this generates more travel by car within the neighborhood than in comparable nonhierarchical grid areas of the city, where more local trips can be on foot. Furthermore, because cars tend to drive faster on arterials, there is evidence that this way of planning street networks actually increases the risk of crashes, particularly those with pedestrian casualties (Dumbaugh, Li, & Joh 2013; Dumbaugh & Rae 2009; Marshall & Garrick 2011; Omer et al. 2017).

A second element of automobile-based planning that has caused tremendous damage to cities is allowing free parking for cars on urban streets, and the institution of minimum parking standards (Shoup 2017). The first has directly caused the deterioration of public space in historic central areas of cities and in residential neighborhoods built before the 1950s without provisions for car parking, as streets and urban plazas are overwhelmed with parked cars. The second has caused an oversupply of land for parking, which also limits high urban densities, or makes building at high densities more expensive as built or underground parking has to be provided.

The emergence of an old-new paradigm

It did not take long for criticism of the modernist city planning paradigm to emerge. As early as the 1960s, opposition to urban renewal, urban freeways, and the havoc they brought to cities had started in the United States, while in Europe people began opposing the destruction of historic city centers and the increased impact of automobile traffic in urban centers and residential neighborhoods. Some cities and countries, such as Copenhagen and the Netherlands, never abandoned the understanding that urban development should be based on fast and reliable public transport systems, and began in the 1960s to actively support these systems with compatible bicycle and pedestrian improvements to public spaces. Thus, the 1960s brought the beginning of urban conservation projects, pedestrian streets, and urban freeway revolts. In the 1970s, woonerfs and traffic calming were initiated as a way to limit the impact of cars on residential areas, while the 1980s saw the beginning of criticism of street design guidelines (Hebbert 2005). This in turn led to the development of new street design guidelines in the 1990s, which stressed the importance of the pedestrian realm and the need to design complete streets (Garrick & Wang 2005; NJDOT & PennDOT 2008). In the beginning of the 2000s, planners and urban designers started promoting the idea of shared space as a way of controlling the impact of cars on major urban streets and intersections (Hamilton-Baillie 2008), and even replacing urban freeways with boulevards (Mohl 2012).

During the last two decades, planners and urban designers formulated and refined the ideas of sustainable urban mobility, stressing accessibility rather than mobility, balancing efficiency and equity, changing priorities from automobile-oriented to people-oriented measures, and emphasizing safety and the protection of pedestrians. Simultaneously, a renewed interest in Transit-Oriented Development (TOD) brought a rethinking of the connection between transit and urban development, showing that in order to succeed in changing people's mode of travel, urban nodes have to be both well connected to each other in the urban region, and also designed in a pedestrian-friendly way (Ewing & Cervero 2010; Papa & Bertolini 2015). Furthermore, metropolitan governance and coordination of the transportation system emerged as a condition for creating a multimodal transportation network that has the ability to respond to people's accessibility needs (Kenworthy & Newman 1999; Sciara 2017; Vuchic 2017).

Thus, a new paradigm for urban street design and transportation planning has evolved for the creation of more livable and sustainable cities. At its basis is the realization that all streets should again be designed as places, and not only as movement channels. This is particularly important for major urban streets, which should be placed more densely than was the norm – at most at a distance of 400 meters apart (about a quarter mile) (Mehaffy et al. 2010; Porta, Romice, Maxwell, Russell, & Baird 2014). While most areas between these major streets are naturally traffic calmed (because of the narrowness of local streets and their indirect connections), the idea of traffic calming should extend to practically all the urban realm, wherever both pedestrians and vehicles are present. Designing urban streets to encourage travel at speeds lower than 30 kmh (20 mph) will not significantly reduce the volume of traffic moving through the city, as average speeds today are typically lower than this figure, but it will achieve a major improvement in pedestrian safety and sense of well-being of pedestrians (Muhlard 2007). In urban corridors, where there is much through traffic combined with access to adjacent properties, boulevards (instead of freeways) can provide a good solution for hosting multimodal movement, while privileging public transportation and maintaining pedestrian livability (Cervero, Kang, & Shively 2009; Rofè & Ishaq 2009). Parking should be managed and priced to at least equalize the accessibility benefits of non-automobile-dependent modes in dense urban areas. Transit-oriented development can be used to extend access to the whole metropolitan region by mandating it in new urban areas, and by transforming existing areas around important nodes of the public transportation system.

As an example of the transformative power of this strategy, one can refer to the possibility of redevelopment of a declining strip along the old road between Jaffa and Jerusalem. This road no longer serves as the main road between Tel-Aviv and Jerusalem, which now is a freeway several kilometers to the north. Nevertheless, it still serves major traffic, connecting the central city to its southeastern suburbs. Hemmed between this road and the freeway is the town of Azor, and to its south a declining industrial area in Holon, another suburban town. One of the future light-rail lines of the metropolitan transit system is supposed to run along the road, but because of the sparse development, it will not provide good accessibility to residents of Azor or to workers in the industrial and employment areas around the road. A design prepared by a student at Tel-Aviv University's Azrieli School of Architecture shows how, by rebuilding an urban grid based on a 400-meter distance between major streets, transforming the regional road into a boulevard as it passes through the urbanized area, and connecting the two towns by a pedestrian-oriented boulevard leading to the regional park to the northwest, a dense urban node can be created that can provide employment and services to the residents of the surrounding area and connect them to the regional transportation network (Figures 45.3 and 45.4).

Conclusion

Cities have transformed human life. They have allowed populations to flourish and created civilization as we know it. Over centuries of evolution, a coherent syntax of urban elements has evolved based on urban squares and streets combining into complex urban space networks. At the dawn of the enlightenment and in the 19th century, the Baroque radial plan, the boulevard, and the urban park were added to the urban design lexicon and used in the 19th century to design urban expansions and accommodate the burgeoning urban population. However, cities also contain, within their own dynamics, the seeds of their own erosion. Modern city planning, which began at the turn of the 19th century but became dominant after World War II, has exacerbated rather than abated this erosion process by privileging the automobile as a preferred mode of travel. If cities are to continue to flourish, and to become the basis for a human civilization that comes to terms with the limits of the earth's ecological system, then a new paradigm for mobility and accessibility must be adopted, one that returns to the lessons of urban history and reinterprets them in the light of current challenges.

This paradigm has gradually emerged in the closing decades of the 20th century and is now fully developed. It includes a revaluation of city streets as complex social spaces; traffic calming as a means of maintaining pedestrian safety and livability where different modes mix; boulevards designed as an alternative to freeways for major roads passing through urban areas; parking management that helps equalize automobile and public transit accessibility; and transit-oriented development that enables regional accessibility. There are now many built examples in the Global North and the Global South that show the viability and value of this paradigm. However, implementing it in practice is still difficult and has been met with opposition. Car owners expect their accessibility difficulties (traffic congestion and parking) to be resolved technically, without having to change their mobility patterns and travel behavior. Even environmentally concerned citizens do not necessarily connect their own preference

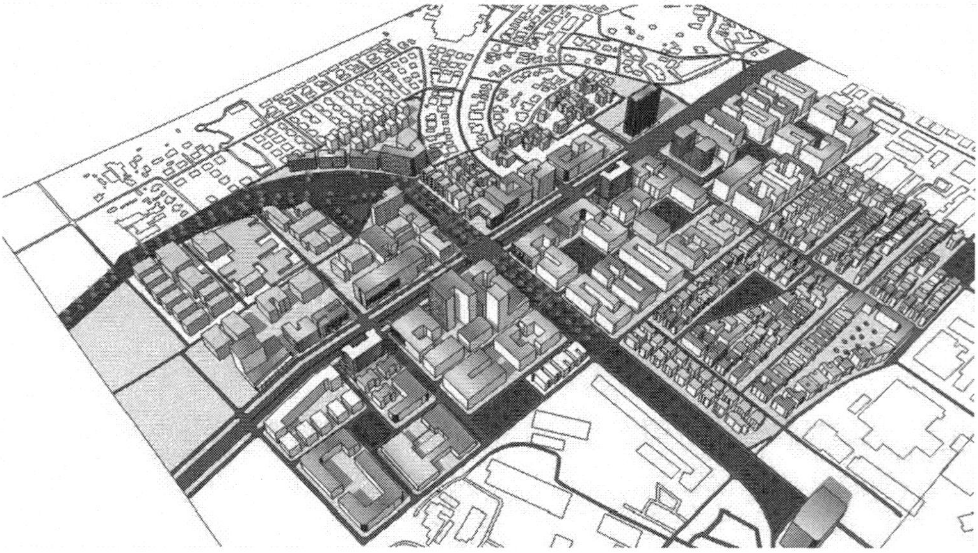

Figure 45.3 and Figure 45.4 Transformation of suburban regional road separating between two towns into an urban node that connects them, as well as connecting them to the regional transportation network.

Source: Google maps; Eyal Harari, Hillel Schocken, and Yodan Y. Rofè, advisors for low-density housing with ample open space to the prevalence of an automobile-centered culture. It is therefore unsurprising that many politicians balk at taking the lead in promoting this paradigm, while various panaceas, the most recent being shared and autonomous vehicles, are promoted as a way of maintaining an unsustainable suburban fabric.

The challenge for this century is not technical, but rather political and cultural. It is the challenge of reembracing urbanism as a way of life, which might be the only way of life that will allow us to survive this last century of the urban revolution and beyond.

References

Alfasi, N., Almagor, J., and Benenson, I. (2012). "The Actual Impact of Comprehensive Land-Use Plans: Insights from High Resolution Observations. *Land Use Policy*, *29*(4): 862–877. http://doi.org/10.1016/j.landusepol.2012.01.003

Angel, S. (2012). *Planet of Cities*. Cambridge, MA: Lincoln Institute of Land Policy.

Carmona, M. (2014). "London's Local High Streets: The Problems, Potential and Complexities of Mixed Street Corridors. *Progress in Planning*, *100*: 1–84. http://doi.org/10.1016/j.progress.2014.03.001

Cervero, R., Kang, J., and Shively, K. (2009). "From Elevated Freeways to Surface Boulevards: Neighborhood and Housing Price Impacts in San Francisco." *Journal of Urbanism: International Research on Placemaking and Urban Sustainability*, *2*(1): 31–50.

Diamond, J. M. (2012). *The World until Yesterday: What Can We Learn from Traditional Societies?* New York: Viking.

Dibble, J., Prelorendjos, A., Romice, O., Zanella, M., Strano, E., Pagel, M., and Porta, S. (2017). "On the Origin of Spaces: Morphometric Foundations of Urban Form Evolution." *Environment and Planning B: Urban Analytics and City Science*, 239980831772507. http://doi.org/10.1177/2399808317725075

Dumbaugh, E., Li, W., and Joh, K. (2013). "The Built Environment and the Incidence of Pedestrian and Cyclist Crashes. *Urban Design International*, *18*(3): 217–228. http://doi.org/10.1057/udi.2013.2

Dumbaugh, E., and Rae, R. (2009). "Safe Urban Form: Revisiting the Relationship Between Community Design and Traffic Safety." *Journal of the American Planning Association*, *75*(3): 309.

Ewing, R., and Cervero, R. (2010). "Travel and the Built Environment: A Meta-analysis." *Journal of the American Planning Association*, *76*(3), 265–294. http://doi.org/10.1080/01944361003766766

Garrick, N. W., and Wang, J. (2005). "New Concepts for Context-based Design of Streets and Highways." *Transportation Research Record: Journal of the Transportation Research Board*, *1912*(1): 57–64.

Hamilton-Baillie, B. (2008). "Towards Shared Space." *Urban Design International*, *13*(2): 130–138.

Hebbert, M. (2005). "Engineering, Urbanism and the Struggle for Street Design." *Journal of Urban Design*, *10*(1): 39–59.

Hillier, B. (1996). *Space Is the Machine*. Cambridge, UK: Cambridge University Press.

Jacobs, A. B. (1994). *Great Streets*. Cambridge, MA: MIT Press.

Jacobs, A. B., Macdonald, E., and Rofe, Y. Y. (2003). *The Boulevard Book: History, Evolution, Design of Multiway Boulevards*. MIT Press.

Jacobs, J. (1961). *The Death and Life of Great American Cities*. New York, NY: Random House Inc.

Kenworthy, J., and Newman, P. (1999). *Sustainability and Cities*. Washington: Island Press.

Lawrence, H. W. (1988). "Origins of the Tree-lined Boulevard. *Geographical Review*, *78*(4): 355–374.

Macdonald, E. (2005). "Suburban Vision to Urban Reality: The Evolution of Olmsted and Vaux's Brooklyn Parkway Neighborhoods." *Journal of Planning History*, *4*(4): 295–321. http://doi.org/10.1177/1538513205282143

Marshall, S. (2004). *Streets and Patterns: The Structure of Urban Geometry*. Taylor & Francis.

Marshall, W. E., and Garrick, N. W. (2011). "Does Street Network Design Affect Traffic Safety?" *Accident Analysis & Prevention*, *43*(3): 769–781. http://doi.org/10.1016/j.aap.2010.10.024

Mehaffy, M. W., Porta, S., Rofe, Y., and Salingaros, N. (2010). "Urban Nuclei and the Geometry of Sreets: The "Emergent Neighborhoods" Model." *Urban Design International*, *15*(1): 22–46.

Mehaffy, M. W., Porta, S., and Romice, O. (2014). "The "Neighborhood Unit" on Trial: A Case Study in the Impacts of Urban Morphology." *Journal of Urbanism: International Research on Placemaking and Urban Sustainability*, *8*(2): 199–217. http://doi.org/10.1080/17549175.2014.908786

Mohl, R. A. (2012). "The Expressway Teardown Movement in American Cities." *Journal of Planning History*, *11*(1): 89–103. http://doi.org/10.1177/1538513211426028

Muhlard, N. (2007). "A Short History of Pedestrian Safety Policies in Western Europe." In *ICTCT Special Workshop V (extra workshop)*. Beijing, China. https://www.ictct.net/workshops/beijing-extra-workshop/, 138–149.

NJDOT & PennDOT. (2008). *Smart Transportation Guidebook: Planning and Designing Highways and Streets that Support Sustainable and Livable Communities*. NJDOT and PennDOT.

Omer, I., Gitelman, V., Rofè, Y., Lerman, Y., Kaplan, N., and Doveh, E. (2017). "Evaluating Crash Risk in Urban Areas Based on Vehicle and Pedestrian Modeling." *Geographical Analysis*, *49*(4): 387–408. http://doi.org/10.1111/gean.12128

Omer, I., Rofè, Y., and Lerman, Y. (2015). "The Impact of Planning on Pedestrian Movement: Contrasting Pedestrian Movement Models in Pre-modern and Modern Neighborhoods in Israel." *International Journal of Geographical Information Science* (August), 1–22. http://doi.org/10.1080/13658816.2015.1063638

Omer, I., and Zafrir-Reuven, O. (2015). "The Development of Street Patterns in Israeli Cities." *Journal of Urban and Regional Analysis*, 7(2): 113–127.

Papa, E., and Bertolini, L. (2015). "Accessibility and Transit-oriented Development in European Metropolitan Areas." *Journal of Transport Geography*, *47*: 70–83. http://doi.org/10.1016/j.jtrangeo.2015.07.003

Porta, S., Latora, V., Wang, F., Rueda, S., Strano, E., Scellato, S., . . . Latora, L. (2012). "Street Centrality and the Location of Economic Activities in Barcelona." *Urban Studies*, *49*(7): 1471–1488. http://doi.org/10.1177/0042098011422570

Porta, S., Latora, V., Wang, F., Strano, E., Cardillo, A., Scellato, S., . . . Messora, R. (2009). "Street Centrality and Densities of Retail and Services in Bologna, Italy." *Environment and Planning B: Planning and Design*, *36*(3): 450–465. http://doi.org/10.1068/b34098

Porta, S., Romice, O., Maxwell, J. A., Russell, P., and Baird, D. (2014). "Alterations in Scale: Patterns of Change in Main Street Networks across Time and Space." *Urban Studies*, *51*(16): 3383–3400. http://doi.org/10.1177/0042098013519833

Rofè, Y. (2016). "Urbanism, Sustainability and Public Gardens: Towards a "New" Conception of Public Open Space Standards in the City." In Moccia, D. and Sepe, M. (eds.), *Reti e infrastrutture dei territori contemporanei: Networks and Infrastructures of Contemporary Territories*. Rome: INU Edizioni Srl, 52–62.

Rofè, Y., and Ishaq, R. (2009). "Building Liveable Major Urban Streets: Begin Road, Tel-Aviv." *Proceedings of the ICE: Urban Design and Planning*, *162*(4): 169–178.

Rofè, Y., & Omer, I. (2014). "How Urban Grids Generate Urbanism-Examples from Israel." *New Urban Configurations*. http://doi.org/10.3233/978-1-61499-365-0-678

Rofè, Y., Zarchin, I., and Feierstein, G. (2012). "Quantity and Quality of Public Open Spaces in Israel." *Proceedings of the ICE: Urban Design and Planning*, *165*(3): 177–187. http://doi.org/10.1680/udap.11.00021

Sciara, G. C. (2017). "Metropolitan Transportation Planning: Lessons from the Past, Institutions for the Future." *Journal of the American Planning Association*, *83*(3): 262–276. http://doi.org/10.1080/01944363.2017.1322526

Shoup, D. C. (2017). *The High Cost of Free Parking (Updated Edition)* (3rd ed.). Routledge.

United Nations, Department of Economic and Social Affairs, P. D. (2017). *World Population Prospects: The 2017 Revision, Key Findings and Advance Tables* (No. Working Paper No. ESA/P/WP/248).

Vaughan, L., and Griffiths, S. (2013). "Mixed Use Across the Street Network." *Urban Design Quarterly*, *2012*(1): 17–19.

Vuchic, V. (2017). *Transportation for Livable Cities*. Taylor & Francis.

Further reading

Hillier, B. and Hanson, J. (1984). *The Social Logic of Space*. Cambridge, UK: Cambridge University Press. Hillier, B. (1996). *Space Is the Machine*. Cambridge, UK: Cambridge University Press. These two texts are the foundation of the Space Syntax approach to urban space analysis, showing how the configuration of space fosters co-presence and movement, and thus generates the "movement economy," which in turn influences social ties, land use, crime, and other social phenomena in cities.

Jacobs, A. B. (1994). *Great Streets*. Cambridge, MA: MIT Press, and Jacobs, A. B., Macdonald, E., & Rofè, Y. Y. (2003). *The Boulevard Book: History, Evolution, Design of Multiway Boulevards*. Cambridge, MA: MIT Press. These books are a celebration of urban streets of all kinds and contexts, and of multiway boulevards in particular. Important both for their method of observation and comparative presentation as well as the knowledge they contain about streets, their evolution, and their role in cities.

Jacobs, J. (1961). *The Death and Life of Great American Cities.* New York, NY: Random House Inc. Jane Jacobs's classic is still a must read, and it has to be read carefully and attentively, as many of her insights on cities are still not very well understood, nor were her warnings heeded in time to save many American cities.

Kenworthy, J. and Newman, P. (1999). *Sustainability and Cities.* Washington: Island Press. This book was the first one to tie metropolitan urban form with transportation energy use. Its tying together of many different strands of public policy and urban planning make it still an important book to read in the context of the theme of this chapter.

46

MODERNISM, PEDESTRIANS, AND PUBLIC SPACE

A century of North American street design

Paul Hess, Kelly Gregg, and Ryan Whitney

Introduction

North American street design has prioritized moving motor vehicles for a century, but this is changing as cities implement policies and projects to create livable, walkable streets and vibrant public spaces. The celebrated pedestrianization of Times Square in 2009 is only one of New York's many recent pedestrian and public space projects. Transformations in other leading cities include San Francisco's Pavement to Parks program, and Mexico City's Avenida Madero, pedestrianized in 2010 and now attracting up to 200,000 pedestrians per hour (Kazis 2012). Important street design policies are also being adopted, including over 1,200 Complete Street policies across the United States, that declare streets must safely accommodate all users including pedestrians and cyclists (National Complete Street Coalition 2017). Together, these initiatives are redefining relationships between automobiles, pedestrians, and other street users. This chapter contextualizes these changes by examining pedestrian-oriented street design across the last century, and then briefly speculates about future change.

Street design is fundamental to urban design theory and practice. Debates over what makes good streets centrally involve the design of pedestrian facilities and public space. More broadly, walkability is related to network connectivity, land use mix, density, and proximity to destinations (Forsyth 2015), as well as perceptual characteristics such as "complexity," "enclosure," and "human scale" (Ewing et al. 2016). This chapter focuses more narrowly on the design of street facilities, and particularly on strategies that redistribute space devoted to motor vehicle movement or parking to active or passive pedestrian uses such as walking, sitting, or socializing. We do not examine broader policy approaches such as *Vision Zero* or *Safe Routes to School.*[1]

From a scoping review of academic indexing databases, we identified key articles focusing on basic definitions of design strategies, their goals, and design elements. We did not systematically review the huge literature of government, professional, and nongovernment reports, but we supplemented articles with published monographs. For current design approaches, we conducted interviews with lead organizations in the United States and Canada advocating for and/or implementing pedestrian-oriented design and, where appropriate, drew on interviews we conducted in Mexico City as part of a more focused project. Here, we focus on North America, particularly the United States, but also include European practices that were in dialogue with North American ideas.

Modernist streets and pedestrian malls

Modernist streets

Contemporary pedestrian-oriented street design evolved from and is also a reaction to modernist streets conceived before World War II. Futurist visions of urban streets were prevalent in proposals by CIAM architects, and in more populist forums such as magazines, films, and world fair exhibits. These visions rendered pedestrians and motor vehicles in separated spaces, many with elevated pedestrian walkways above motorized roadways. In practice, typically pedestrians and vehicles were segregated horizontally rather than vertically. Dividing streets between sidewalks and roadways predates automobiles (Loukaitou-Sideris & Ehrenfeucht 2009), but by the early 20th century, both roadways and sidewalks were regulated to prioritize circulation and movement over other functions. Thus, municipalities attempted to suppress social uses of sidewalks such as vending, political activities, and gathering in favor of pedestrian circulation (Ehrenfeucht 2012). Likewise, the roadway was increasingly defined as vehicle space, and the concept of the "jaywalker" was invented to denigrate pedestrians who interfered with traffic and to imply they bore responsibility for being killed if they entered roads (Norton 2007).

The separation between pedestrians and automobiles was established at larger scales with ideas about a hierarchy of street types. Hebbert (2005) notes that the concept of large streets as traffic conduits, devoid of buildings and pedestrians, inverted the tradition of urban streets as important public spaces lined with monumental buildings. At the other end of the hierarchy, small streets were to serve local traffic and pedestrian travel along with community and civic life, as articulated in Clarence Perry's (1929) influential Neighborhood Unit principle. The concept was modeled in the superblocks of Radburn, New Jersey, where pathways allow children to walk to school without crossing any streets, and in a cruder form was incorporated into development regulations and practices (Brody 2013). The idea of a street hierarchy, however, was fully institutionalized into traffic engineering, road standards, and road classification systems (Hess 2009; Southworth & Ben-Joseph 1997).

Pedestrian malls

During World War II, modernist ideas about pedestrianization were applied to both suburban shopping centers and downtown pedestrian malls. In both, pedestrian areas were separated from vehicles to create safe, comfortable shopping and civic spaces. The planned shopping center evolved in scale from initial visions of walkable neighborhood amenities, to regional destinations for shoppers arriving by automobile (Gillette 1985). After the war, suburban malls such as Northgate in Seattle, Washington, and Northland near Detroit, Michigan, were built across North America, demonstrating that separating pedestrians and automobiles was beneficial for retail sales. Jane Jacobs, as associate editor of *Architectural Forum*, featured Northland in the June 1954 issue, proclaiming it "a new yardstick for shopping center planning" and a "planning classic" like Radburn (103). Jacobs also promoted applying the pedestrian shopping center to downtowns, noting that Northland's "flexible market-town use of open spaces looks like a natural for coping with rehabilitation of blighted-spotted decaying shopping districts" (103).

Ideas of pedestrianization to support downtown revitalization circulated between Europe and North America after World War II. Essen, Germany, was first to pedestrianize downtown streets in 1930 as a strategy to manage traffic, followed by other European cities after the war (Brambilla & Longo 1977). As a result, many scholars attribute the North American pedestrian mall to European precedents, but the cross-Atlantic exchange of ideas was more complex

(Gregg 2018). The Lijnbaan opening in Rotterdam, the Netherlands, in 1953, for example, was influenced by American shopping center design and then became a prototype for other projects across both Europe and North America (Brambilla & Longo 1977; Gregg 2018; Wagner 2014). In Europe, however, cities continued to implement pedestrianization strategies as a part of their postwar rebuilding (Brambilla & Longo 1977), eventually expanding single pedestrianized streets into larger networks that continue to operate as central retail zones in many European cities.

In North America, where vehicle ownership rates were higher, the downtown pedestrian mall was employed as a strategy for countering suburban retail competition and as part of urban renewal programs (Gregg 2018). These malls were likened to suburban shopping malls but lacked their unified management, had limited parking, and were typically limited to a single street. Still, from 1959 through the mid-1980s, over 160 pedestrian malls, including variations of the concept such as transit- and semi-malls, were built across the United States and Canada (Robertson 1993). Few, however, were built after urban renewal funding ended, and over time, large numbers were removed as not living up to their promise of saving downtown retail districts and were even seen as promoting further decline (Knack 1982).

Humanist streets: from woonerfs to shared space

Woonerfs and home zones

By the 1960s, a backlash against modernist street design occurred in Europe and North America. In the United States, Jane Jacobs famously contrasted lifeless modernist environments with the social and economic benefits of the "sidewalk ballet" on traditional streets (Jacobs 1961: 50). In the early 1970s, Danish architect Jan Gehl similarly argued that streets should discourage traffic and support social life (Gehl 1987). In the United States, Donald Appleyard's 1981 *Livable Streets* became iconic for its graphic representation of traffic impeding neighborhood social life (Appleyard, Gerson, & Lintell 1981), and Allan Jacobs and Appleyard's 1987 "Urban Design Manifesto" placed pedestrians and public space at the center of good city design (Jacobs & Appleyard 1987).

US cities continued to remove pedestrian malls during this period, but new street designs were tested in Europe. In the Netherlands, the idea of the *woonerf* evolved in the 1970s, representing a new type of pedestrian-oriented street. "*Woonerf*" literally translates as "residential area" or "residential yard," and was defined as "an area in which the residential function quite clearly predominates over any provisions for traffic" (Royal Dutch Touring Club 1978: 418). The particular concept is traced to Niek De Boer of the University of Emmen, but the intellectual origins are often ascribed to that of "environmental areas" found in the 1963 UK Ministry of Transport report known as the Buchannan Report (Karndacharuk, Wilson, & Dunn 2014) – hardly an anti-modernist document. Indeed, *woonerfen*, restricted to small residential streets, fit comfortably with the modernist street hierarchy.

In contrast to earlier modernist designs, however, pedestrians and traffic share the same street surface, with Dutch regulations limiting vehicles to a "walking pace." *Woonerf* design includes entrance gateways, changes in paving textures, and the use of parking, landscaping, and street furniture to enforce abrupt changes in direction for moving vehicles and create opportunities for play and social interaction. Safety for pedestrians, particularly children, is emphasized to create residential, community-oriented spaces. Appleyard (1980: 113) argues that *woonerf* users must "take care of the others," and that "pedestrians and children have greater rights to residential street space than traffic." Likewise, Guttenberg (1982: 101) suggests that the *woonerf* is an attempt to save "valued practices and human relationships from forces that threaten them," principally the impact of traffic. Beyond, this, however, there is little discussion of equity, such as how preventing through traffic on some streets may impose higher traffic burdens on others.

Thousands of *woonerven* were built across the Netherlands. The idea also spread throughout Northern Europe, with many countries establishing their own guidelines and regulations. In the United Kingdom, the concept was branded as "home zones," but only a few early examples associated with New Town projects were initially developed. Starting in 1999, a number of pilot projects were implemented, leading to a general authorization by the national Transport Act of 2000 (Biddulph 2003). *Woonerven* are also found in North America but have not been systematically or widely adopted. Many projects that use the term are simply residential streets that include some form of traffic calming, rather than being faithful to the original concept.

Traffic calming

The concept of traffic calming is a broader, more diffuse idea than the *woonerf*. Lockwood defines traffic calming as "the combination of mainly physical measures that reduce the negative effects of motor vehicle use, alter driver behavior and improve conditions for non-motorized street users" (1997: 22). Thus, while the *woonerf* is residential and focused on public space, traffic calming is focused on traffic control, especially speed, irrespective of a specific urban setting.

Historically, the two concepts are interconnected. In the Netherlands, alternatives to full *woonerven* were initially explored with street projects limiting speeds to 30 kph, using speed humps, lane narrowings, and other design treatments less expensive than fully rebuilding streets (Clarke & Dornfeld 1994). In 1988, *woonerf* regulations were simplified, allowing treatments of low-volume, nonresidential streets such as local shopping streets (Pharaoh & Russell 1991). Traffic calming experiments were also tried in Danish village centers, culminating in the National Road Directorate's 1988 *Catalogue of Ideas* (Herrstedt 1992), as the national government sought alternatives to creating expensive bypass roads. Instead, village main streets were equipped with devices such as warning rumble strips, entrance gateways, shifting lane alignments, and roundabouts to improve pedestrian safety.

Like *woonerven*, residential slow zones and traffic calming on local commercial streets have become common practice in Europe (Ewing 2008). In the United States, the range and goals of traffic calming has been more limited. Appleyard's famous work on livable streets (Appleyard, Gerson, & Lintell 1981) shows experiments were occurring on residential streets in San Francisco and Berkeley relatively early. Yet, Hoyle, referencing her *Planning Advisory Service* report, notes that traffic calming was considered "a very radical concept" in 1995 (Hoyle 2016). In a series of surveys of US jurisdictions between 1997 and 2007, Ewing found that traffic calming beyond local and collector streets was rare, strategies were largely designed to improve safety and livability without reducing vehicle levels of service, and improving walking and cycling environments were only implicit goals (Ewing 2008). However, starting in 2006, a series of reports on "context sensitive solutions" sponsored by the Institute of Transportation Engineers (ITE) were a major step forward. They recommended reduced curb radii, midblock crossings, refuge islands, and other pedestrian-oriented strategies on "major urban thoroughfares" (ITE 2006: 3). This work has fed into the safety rubrics found *in Complete Streets*, *Vision Zero*, and *Safe Routes to School* policies.

Shared space

With street surfaces shared by both pedestrians and vehicles rather than being divided into different zones with sidewalks and roadways, the *woonerf* is also tied to current ideas of shared streets. However, Hamilton-Baillie (2008) identifies innovations by the Dutch Traffic Engineer Hans Monderman as key to the development of the concept. These include removal of signage in the town of Oudehaske in the mid-1980s, and removal of standard traffic signage, signals, and markings in the village of Makkinga in 1992. Unlike the *woonerf*, these projects were implemented in main village public spaces, rather than residential areas. Thus, they are more like Danish village traffic calming initiatives, but rely on psychological rather than engineered methods to control traffic.

In the United Kingdom, Hamilton-Baillie (2008) started advocating for shared space in the early 2000s, arguing their benefits for traffic circulation, walking and cycling, and enhancing the public realm. Some authors emphasize the economic benefits of shared space projects (Kaparias et al. 2013), but others have criticized them for reducing accessibility for the visually impaired who rely on curbs to safely navigate streets. Calls for a moratorium on new projects have generated proposed improvements such as changes in surface level and/or pavement texture to distinguish vehicle space from more protected pedestrian space that, to some degree, reintroduce spatial segregation between pedestrians and vehicle projects (Smithies 2015).

In North America, shared space projects are also being deployed, although documentation is limited. Such projects are usually implemented on low-volume streets and designate protected pedestrian space with differing paving. Bollards are often used as separators to mark clear zones, but they also suggest that pedestrians are invited to cross the roadway at any point. These types of designs common in Europe have also become standard for recent shared streets projects in Mexico City, even on some high-volume streets (Figure 46.1).

Figure 46.1 Calle 16 de Septiembre in the historic center of Mexico City after its conversion to a shared street in 2014. Vehicles and pedestrians are invited to use most of the street surface, with bollards creating protected pedestrian areas.

Source: Photo by Paul Hess

New street design initiatives in North America

Complete Streets

The idea of "complete streets" was branded in the United States in the early 2000s, although many core ideas are much older. Originally associated with efforts to include cycling facilities, the idea broadened to be defined as streets "designed and operated to enable safe access for all users, including pedestrians, bicyclists, motorists and transit riders of all ages and abilities" (National Complete Street Coalition n.d.). Thus, Complete Streets challenges design that prioritizes automobiles. Barbara McCann (2013: 21), a founder of the National Complete Street Coalition (NCSC), argues that the "movement is at its heart a policy initiative that seeks to change the way all roads are built."

Complete Streets policies rapidly spread across the United States following the national *Complete Streets Act of 2009*, with the NCSC cataloging more than 1,200 policies at all jurisdictional levels. Policies, guidelines, and projects have also been established in Canada and Mexico. Most policies do not, however, direct specific design strategies, although the provision of sidewalks, curb ramps, speed control devices, and other facilities are mentioned in some policies. Design strategies are more fully articulated in guidelines produced by large cities including Chicago, New York, and Toronto. As also described in Chapter 34 by Frank et al. in this book, the *Urban Street Design Guide* (2013), produced by the National Association of City Transportation Officials (NACTO), has become a significant source for design guidance (National Association of City Transportation Officials 2013). In interviews with advocacy, policy, and professional organizations, some participants characterized the NACTO guide as "a key linking document," indicating that NACTO plays a central role in facilitating knowledge sharing.

The NACTO guide lists public space as a basic principle for street design. Most municipal Complete Streets policies, however, treat streets as transportation infrastructure, with ideas of the street as public space missing or poorly articulated. The main goal is "safe accommodation," although health promotion is often articulated as well. Policies and guidelines rarely acknowledge that street design requires tradeoffs between users and functions, for example between the movement and safety requirements of different modes, or between movement, loading, and public space functions of the street. Almost all policies list automobiles, pedestrians, and cyclists as equal in their need for accommodation. In most guidelines, different modes are consigned to clearly defined zones, often replicating the basic modernist idea of specialized movement spaces, but with better facilities for pedestrians and bikes. Policies are also still evolving, but it is unclear how fundamentally they are challenging street building practices (Gregg & Hess 2018).

Some critics also question their equity impacts, suggesting Complete Streets have encouraged gentrification in marginalized and racialized neighborhoods (Zavestoski & Julian 2015). Awareness of equity issues is growing, and in its latest evaluation framework, NCSC is rewarding policies that prioritize marginalized neighborhoods and street users (Anzilotti 2017). The complex dynamics of equity may be difficult to assess, with street redesign promoting access and safety in some neighborhoods, but gentrification in others.

Parklets and pavement-to-plazas

Other North American street design strategies are more focused on creating new public space in roadway areas formerly devoted to automobiles (Figure 46.2). The most visible has been the pedestrianization of Times Square in New York City, with only Mexico City's Avenida Madero close in scale and impact. More influential have been smaller, "tactical" projects without full

Figure 46.2 A large, summertime parklet installation on Rue St-Denis, Montreal, originally developed in conjunction with the local commercial development corporation to help counter construction impacts, 2016.

Source: Photo by Paul Hess

pedestrianization, with San Francisco and New York leaders in doing things "lighter, cheaper, quicker" to affect change (Lydon & Garcia 2015: 234).

The concept most associated with San Francisco is "parklets," which effectively extend the sidewalk into vehicle parking spaces with seating areas, tables, landscaping, and other features, some of which can be quite fanciful. The origin of parklets is typically traced to the Rebar Group design studio, who in 2005 created a temporary one-parking space park with sod, a bench, and a potted tree after paying the meter fee (Birdsall 2013), although Lydon and Garcia (2015: 158) identify "Parking Meter Parties" in Hamilton, Ontario, as early as 2001. We suspect there are other antecedents as well. In San Francisco, the Rebar Group worked to replicate the strategy as an annual "PARK(ing) Day," and produced a how-to manual. PARK(ing) Days are now held in multiple cities around the world. In 2010 the strategy became part of the San Francisco's Pavement to Parks program, which allows merchants, neighborhood groups, and non-profits to apply to build parklets as a temporary, provisional use. There are now more than 50 parklets in San Francisco, and dozens of other cities across the United States and Canada have adopted similar programs (San Francisco Planning Department n.d.).

In San Francisco, parklets' stated purposes are to encourage non-motorized transportation, pedestrian safety, neighborhood interaction, and support of local businesses. They clearly serve public space more than transportation functions, but their publicness has been questioned. Parklets are encouraged and permitted by the city, but they are largely created by businesses, with permitting, construction, and insurance costs creating significant entry barriers (Mould 2014). About 80% are sponsored by restaurants, with most others sponsored by retail stores and business improvement districts (Littke 2016). San Francisco is careful to stipulate that parklets are open to the public; connecting a parklet to a restaurants' existing outdoor service space, providing table service, or posting advertising are not allowed. Yet some cities allow adjacent restaurants to restrict the use of parklets to customer use. Thus, while parklets can be seen as public space carved out of vehicular space, they also act as a kind of privately owned public space within the public right-of-way with the attendant problems of privatization.

Between 2007 and 2013, Janet Sadik-Khan, New York City Transportation Commissioner during the Bloomberg administration, held a powerful platform in promoting street redesign in New York. She has been influential through her central roles in NACTO, as part of Bloomberg Associates working with cities around the world, and through her book *Streetfight* (Sadik-Khan & Solomonow 2016). In our interviews with organizations working towards pedestrian-oriented design, New York was referred to as "the stand out model," and as "legitimizing investments in liveable streets" both in the United States and globally. Mexico City, for example, was directly influenced by New York in a number of public space and street redesign initiatives, including pavement-to-plaza projects known locally as "pocket parks" (parques de bolsillo).

In San Francisco, Mexico City, and New York, pavement-to-plaza projects follow a tactical model, typically using inexpensive materials such as paint, planters, blocks of stone, and furniture to quickly create pedestrian-oriented public space, although Times Square was upgraded with more permanent materials, and additional upgrades are planned, depending on capital budgets. More typical sites include where streets meet at oblique angles, creating traffic triangles not needed for traffic movement. As with parklets, nonprofit organizations such as business improvement groups propose projects and, if selected by the city, take on management, maintenance, and insurance costs. New York weighs proposals from lower-income neighborhoods slightly higher in the selection process, although organization capacity and other factors are considered more important (New York City Department of Transportation 2017). There are new plazas in every borough in the city, but most seem to be clustered in lower and Midtown Manhattan, and in the gentrifying neighborhoods of Brooklyn (Figure 46.3). Mexico City's program uses similar

Figure 46.3 A "tactical" pedestrian plaza on Broadway and 24th Street adjoining Madison Square Park created as part of the 2009 Green Light for Midtown project. Note the former curb line delineating the sidewalk, bottom right.

Source: Photo by Paul Hess

management requirements, and pavement-to-plaza projects have been located in central tourist-oriented areas or wealthier neighborhoods. Such efforts fit with competitive cities strategies and political predilection for urban lifestyle preferences associated with the "creative class."

Sadik-Kahn describes these projects as part of the effort to "wrest back" New York's streets from motor vehicles (Sadik-Khan & Solomonow 2016: 5). The Plaza Program has clearly created much-needed public space improvements, but New York has been careful to demonstrate that they do not negatively impact traffic and can even improve it (NYC DOT 2010). The Midtown plan that included Times Square was even titled "Green Light" and was shown to improve most vehicle travel times (NYC DOT 2010). In this sense, these strategies have promoted more and better pedestrian and public space, but they have not challenged the logic of modernist streets.

Conclusion

Despite calls to make pedestrian-oriented public spaces the center of urban design, street design ideas about pedestrians are entangled with ideas about managing and accommodating vehicle traffic. In the early 20th century, a street hierarchy was developed to balance tensions between pedestrians and vehicles and was then broadly implemented in the postwar period. Pedestrianized streets were designed in both North America and Europe, but then abandoned in North America. In Europe the *woonerf* evolved into shared space concepts and broader concepts of traffic calming. These trends have had less impact in North America until the broad adoption of Complete Streets policies. Still, while Complete Streets policies represent a significant change in thinking about street design, they conceptualize separate zones of movement for different modes and largely treat streets as transportation facilities. In contrast, parklet and pavement-to-plazas projects are creating public space, but in spaces previously not used for moving vehicles. Thus, these North American trends continue to replicate basic modernist principles about separate, specialized vehicle and pedestrian spaces.

These strategies however, differently articulate the relationship between pedestrians and public space. North American pedestrian malls and European pedestrian areas treat pedestrianized streets as commercial spaces for shopping and consumption. The *woonerf* focuses on the potential of residential streets to foster community, although the UK home zones also have economic development goals. In Northern Europe, traffic calming has mixed goals for safety, economic development, and social integration. Likewise, shared space is connected to ideas about fostering active social engagement among street users. In the United States and Canada, traffic calming and Complete Streets are conceived more narrowly: targeting traffic safety. In contrast, pavement-to-plaza programs include safety objectives, but are largely understood as public space initiatives. They clearly incorporate economic development goals at the scale of the street, and perhaps at the city scale in fostering competitive and creative city branding.

Just as significant innovations in street design occurred in the last century, this century also faces new technological and political forces that will reshape streets. Here we mention just two that will change streets for pedestrians: the use of motor vehicles for terror, and the mass adoption of automated vehicles. First, the horrific use of vehicles to intentionally kill large numbers of pedestrians, such as recent incidents in Barcelona, Berlin, and Toronto, are likely to encourage more fortified pedestrian zones, especially those in city centers that attract large numbers of people. Whether this will mean expanded pedestrian areas, or stronger physical separations between pedestrian space and vehicle space, is unknown. Second, the advent of autonomous vehicles presents even more unpredictable impacts and will compound the increases in automobility already occurring with the growth of ride-hailing services, especially in places like Mexico City in the

Global South. NACTO has presented the most coherent vision of positive outcomes, with less roadway space devoted to vehicles, more public space, slower traffic, and more opportunities for pedestrians to cross streets (NACTO 2017). Much of the scholarly literature, however, is focused on technical issues, such as the possibility for continuous vehicle flows through intersections, or the development of pedestrian detection systems, without any examination of how vehicles would interact with pedestrians or how streets might be redesigned. There are utopian visions that autonomous vehicles will make cities safer, free of auto-congestion, and more sustainable, but knowledge largely seems to be driven by engineering and economic concerns. This dynamic is reminiscent of the modernists' vision of adapting cities and streets to motor vehicles a century ago. This time we need to be more deliberate in adapting the technology to the cities we want, rather than the other way around.

Note

1 Originating in Sweden, *Vision Zero* is a road safety movement adopted by many North American cities that seeks to eliminate deaths and serious injuries due to traffic. Likewise, *Safe Routes to School* consists of broadly adopted programs that seek to increase the safety and practice of children walking and biking to school.

References

Anzilotti, E. (2017). "No Street Is "Complete" Without Taking Equity into Account." Retrieved February 7, 2018, from www.fastcompany.com/40502981/no-street-is-complete-without-taking-equity-into-account

Appleyard, D. (1980). "Livable Streets: Protected Neighborhoods?" *The Annals of the American Academy of Political and Social Science, 451*(1): 106–117.

Appleyard, D., Gerson, M. S., and Lintell, M. (1981). *Livable Streets.* Berkeley: University of California Press.

Biddulph, M. (2003). "Towards Successful Home Zones in the UK." *Journal of Urban Design, 8*(3): 217–241.

Birdsall, M. (2013). "Parklets: Providing Space for People to Park. Themselves." *Institute of Transportation Engineers: ITE Journal, 83*(5): 36–39.

Brambilla, R., and Longo, G. (1977). *The Rediscovery of the Pedestrian: 12 European Cities.* Washington, DC: Department of Housing and Urban Development.

Brody, J. (2013). "The Neighbourhood Unit Concept and the Shaping of Land Planning in the United States 1912–1968." *Journal of Urban Design, 18*(3): 340–362.

Clarke, A., and Dornfeld, M. J. (1994). *National Bicycling and Walking Study. Case Study No. 19: Traffic Calming, Auto-Restricted Zones and other Traffic Management Techniques: Their Effects on Bicycling and Pedestrians.* Washington, DC: Department of Transportation. Retrieved from www.pedbikeinfo.org/cms/downloads/CS19_TrafficManagementEffects1992.pdf

Ehrenfeucht, R. (2012). "Precursors to Planning: Regulating the Streets of Los Angeles, California, c 1880–1920." *Journal of Planning History, 11*(2): 107–123.

Ewing, R. (2008). "Traffic Calming in the United States: Are we Following Europe's Lead?" *Urban Design International, 13*(2): 90–104.

Ewing, R., Hajrasouliha, A., Neckerman, K. M., and Purciel-Hill, M. (2016). "Streetscape Features Related to Pedestrian Activity." *Journal of Planning Education and Research, 36*(1): 5.

Forsyth, A. (2015). "What Is a Walkable Place? The Walkability Debate in Urban Design." *Urban Design International, 20*(4): 274–292.

Gehl, J. (1987). *Life Between Buildings: Using Public Space.* New York: Van Nostrand Reinhold.

Gillette, H., Jr. (1985). "The Evolution of the Planned Shopping Center in Suburb and City." *Journal of the American Planning Association, 51*: 449–460.

Gregg, K. (2018). "Conceptualizing the Pedestrian Mall in Post-war North America and Understanding its Transatlantic Transfer through the Work and Influence of Victor Gruen." *Planning Perspectives*, 1–27. doi: 10.1080/02665433.2018.1437555

Gregg, K., & Hess, P. (2018). "Complete Streets at the Municipal Level: A Review of American Municipal Complete Street Policy." *International Journal of Sustainable Transportation*, 1–12. doi: 10.1080/15568318.2018.1476995

Guttenberg, A. Z. (1982). "How to Crowd and Still Be Kind: The Dutch Woonerf." *Humboldt Journal of Social Relations, 9*(2): 100–119.

Hamilton-Baillie, B. (2008). "Towards Shared Space." *Urban Design International, 13*(2): 130–138.

Hebbert, M. (2005). "Engineering, Urbanism and the Struggle for Street Design." *Journal of Urban Design, 10*(1): 39–59.

Herrstedt, L. (1992). "Traffic Calming Design: A speed Management Method: Danish Experiences on Environmentally Adapted Through Roads." *Accident Analysis and Prevention, 24*: 3–16.

Hess, P. M. (2009). "Avenues or Arterials: The Struggle to Change Street Building Practices in Toronto, Canada." *Journal of Urban Design, 14*(1): 1–28.

Hoyle, C. (2016, November 29). "Traffic Calming: A Radical Idea in 1995: Alta Planning + Design." Retrieved February 4, 2018, from https://blog.altaplanning.com/traffic-calming-a-radical-idea-in-1995-a4c2a744d332

Institute of Transportation Engineers. (2006). *Context Sensitive Solutions in Designing Major Urban Thoroughfares for Walkable Communities.* Washington, DC: Institute of Transportation Engineers. Retrieved from www.ite.org/css/online/

Jacobs, A., and Appleyard, D. (1987). "Toward an Urban Design Manifesto." *Journal of the American Planning Association, 53*(1): 112–120.

Jacobs, J. (1954). "Northland: A New Yardstick for Shopping Center Planning." *Architectural Forum, 100*(June): 102–118.

Jacobs, J. (1961). *The Death and Life of Great American Cities.* New York, NY: Random House.

Kaparias, I., Bell, M., Dong, W., Sastrawinata, A., Singh, A., Wang, X., and Mount, B. (2013). "Analysis of Pedestrian-vehicle Traffic Conflicts in Street Designs with Elements of Shared Space." *Transportation Research Record, 2393*(1), 1–11.

Karndacharuk, A., Wilson, D. J., and Dunn, R. (2014). "A Review of the Evolution of Shared (Street) Space Concepts in Urban Environments." *Transport Reviews, 34*(2): 190–220.

Kazis, N. (2012, March 19). "How Mexico City Fought and Cajoled to Reclaim Streets for Pedestrians." Retrieved February 13, 2018, from https://nyc.streetsblog.org/2012/03/19/how-mexico-city-fought-and-cajoled-to-reclaim-streets-for-pedestrians/

Knack, R. E. (1982). "Pedestrian Malls: Twenty Years Later." *Planning, 48*: 15–19.

Littke, H. (2016). "Revisiting the San Francisco Parklets Problematizing Publicness, Parks, and Transferability." *Urban Forestry & Urban Greening, 15*: 165–173.

Lockwood, I. M. (1997). "ITE Traffic Calming Definition." *Institute of Transportation Engineers: ITE Journal, 67*(7): 22.

Loukaitou-Sideris, A., and Ehrenfeucht, R. (2009). *Sidewalks: Conflict and Negotiation over Public Space.* Cambridge, MA: MIT Press.

Lydon, M., and Garcia, A. (2015). *Tactical Urbanism: Short-Term Action for Long-Term Change.* Washington, DC: Island Press.

McCann, B. (2013). *Completing Our Streets: The Transition to Safe and Inclusive Transportation Networks.* Washington, DC: Island Press.

Mould, O. (2014). "Tactical Urbanism: The New Vernacular of the Creative City." *Geography Compass, 8*(8): 529–539.

National Association of City Transportation Officials. (2013). *Urban Street Design Guide.* Washington, DC: Imprint: Island Press.

National Association of City Transportation Officials. (2017). *Blueprint for Autonomous Urbanism.* New York, NY: National Association of City Transportation Officials. Retrieved from https://nacto.org/wp-content/uploads/2017/11/BAU_Mod1_raster-sm.pdf

National Complete Street Coalition. (2017). *The Best Complete Street Policies of 2016.* Washington, DC, 27. Retrived from https://smartgrowthamerica.org/app/uploads/2017/06/best-complete-streets-policies-of-2016-1.pdf

National Complete Street Coalition. (n.d.). What Are Complete Streets? Retrieved February 7, 2018, from https://smartgrowthamerica.org/program/national-complete-streets-coalition/what-are-complete-streets/

New York City Department of Transportation. (2010). *Green Light for Midtown Evaluation Report.* New York City Department of Transportation. Retrived from www.nyc.gov/html/dot/downloads/pdf/broadway_report_final2010_web.pdf

New York City Department of Transportation. (2017). *NYC Plaza Program Applications Guidelines 2017.* New York, NY: New York City Department of Transportation. Retrived from www.nyc.gov/html/dot/downloads/pdf/nyc-plaza-program-guidelines-2017.pdf

Northland: A new yardstick for shopping center planning. (1954, June). *Architectural Forum*, *100*: 102–118.

Norton, P. D. (2007). "Street Rivals: Jaywalking and the Invention of the Motor Age Street." *Technology and Culture*, *48*(2): 331–359.

Perry, C. A. (1929). "The Neighborhood Unit, a Scheme of Arrangement for the Family-Life Community. Monograph one, Neighborhood and Community Planning," *Regional Plan of New York and Its Environs*. New York: Committee on Regional Plan of New York and Its Environs, 2–140.

Pharaoh, T. M., and Russell, J. R. E. (1991). "Traffic Calming Policy and Performance: The Netherlands, Denmark and Germany." *Town Planning Review*, *62*(1): 79–105.

Robertson, K. A. (1993). "Pedestrianization Strategies for Downtown Planners: Skywalks versus Pedestrian Malls." *Journal of the American Planning Association*, *59*(3): 361–370.

Royal Dutch Touring Club. (1978). "Woonerf: Residential Precinct." *Ekistics*, *45*(273): 417–423.

Sadik-Khan, J., and Solomonow, S. (2016). *Streetfight: Handbook for an Urban Revolution*. New York: Viking.

San Francisco Planning Department. (n.d.). *Pavement to Parks: Creating and Testing Ideas for New Public Spaces in San Francisco*. Retrieved February 8, 2018, from http://pavementtoparks.org/

Smithies, N. J. (2015). "Appropriate Public-Realm Design for the Blind and Partially Sighted." *Proceedings of the Institution of Civil Engineers*, *168*(6): 43–50.

Southworth, M., and Ben-Joseph, E. (1997). *Streets and the Shaping of Towns and Cities*. New York: McGraw-Hill.

Wagner, P. (2014). "New Life for American Downtowns? The 1958 International Seminar on Urban Renewal and the Travel of Planning Ideas in the North Atlantic World." *Planning Perspectives*, *29*: 189–208.

Zavestoski, S., and Julian, J. (2015). "Towards and Understanding of Complete Streets; Equity, Justice, and Sustainability." In Zavestosk, S., and Agyeman, J. (eds.), *Complete Streets: Processes, Practices and Possibilities*. London: Routledge, Taylor & Francis Group, 306–315.

Further reading

Global Designing Cities Initiative, & National Association of City Transportation Officials. (2016). *Global Street Design Guide*. Island Press. This guidebook presents a diverse array of contemporary multimodal street design treatments from across the globe, including cases from both the Global South and Global North. The guide is useful to urban designers as it articulates the complexity and dynamic nature of contemporary multimodal street design, and provides concrete examples of multiple design solutions.

Hebbert, M. (2005). "Engineering, Urbanism and the Struggle for Street Design." *Journal of Urban Design*, *10*(1): 39–59. This article confronts the tension between the engineering-oriented traffic management approach in street design and the more humanist approach oriented toward creating livable and walkable streets.

Loukaitou-Sideris, A. and Ehrenfeucht, R. (2009). *Sidewalks: Conflict and Negotiation over Public Space*. Cambridge, MA: MIT Press. This book uniquely focuses on the space and culture of sidewalk environments both historically and in contemporary US cities. It highlights competing claims on the space of the sidewalk and the nearly continual conflict between sidewalks as spaces for economic interests versus their role in enabling democratic public space.

47

AERIAL MOBILITY, TRANSPORT INFRASTRUCTURE, AND URBAN DESIGN IN ASIAN CITIES

Max Hirsh

Introduction

This chapter begins with a basic premise. Until the very end of the 20th century, the absolute majority of people living in Asia had never been on a plane, and had never been outside their home country; either because airfares were unaffordable, or because foreign travel was forbidden by their government, or both. Moreover, the average resident of an East or Southeast Asian city didn't see a need to fly: his or her social circle was confined to a geographic area that could comfortably be reached on foot, by rickshaw, or by bus.

Today, flying, and flying abroad, is commonplace. Low-cost airlines advertise $30 flights to a mass audience of middle-class consumers keen on a cheap weekend getaway. Millions of Asian citizens study, work, and retire outside their home country. The annual hajj to Mecca has undergone a leap in scale due to the growth in air passengers from Indonesia, who are now the largest group of pilgrims. And within two short decades, China has become the world's number one source of foreign tourists.

As the number of air travelers expanded exponentially, how have architects and planners redesigned airports – as well as entire cities – to accommodate much larger passenger flows? And how have these flows, in turn, reshaped the spatial organization and typological order of East and Southeast Asian cities?

This chapter examines the urban design implications of that leap in aerial mobility. In so doing, it links broader shifts in migration patterns and the regulation of international mobility to the emergence of innovative transport typologies and urban districts centered around major transit hubs. The chapter begins by outlining the rise of transport megaprojects that conceptualize heavily frequented transport nodes – such as airport hubs and high-speed railway stations – as a tool for realizing large-scale urban expansion and redevelopment plans. These infrastructure-led urban design strategies have yielded costly master-planned urban precincts, designed on a monumental scale, that are poorly integrated into the surrounding built environment, both socially and spatially, and often fail to deliver a substantial return on investment. The chapter then contrasts these projects with transport typologies that are designed for new members of Asia's flying public, whose movement between cities and across borders is limited by their income, citizenship, and technical literacy. While flexible and responsive to rapidly emerging mobility needs, these developments often represent infrastructural "quick fixes" that are lacking in aesthetic quality. By providing a synthetic overview of both bottom-up and top-down engagements with the leap in aerial mobility in Asian cities, the chapter ultimately advances an urban design agenda that incorporates the responsiveness and flexibility of the former with the elevated organizational and aesthetic ambitions of the latter.

Airport islands and airport cities

Arriving in a city like Bangkok, Hong Kong, Seoul, or Shanghai for the first time, visitors from Europe or North America are likely to be struck by three things. First, in contrast to the visitor's home city, the airport infrastructure will be much newer, and built at a much larger scale. Half of the world's 20 busiest airports are located in Asia, and unlike aging hubs like Heathrow and O'Hare, most of them have been built from scratch since the 1990s. Second, visitors will find a much wider range of transportation options for their onward journey, running the gamut from sleek high-speed trains that connect the airport to a downtown check-in terminal, to slightly shabby buses and minivans that transport passengers to dozens of destinations across the surrounding urban region, or even across the country. To the uninitiated, the former will appear orderly and calm, the latter chaotic and confusing. Finally, first-time visitors to an Asian airport hub are likely to notice the contrast between, on the one hand, a much higher level of retail products, services, and architectural finishes, catering to a seasoned crowd of well-heeled frequent travelers; and on the other hand, the palpable presence of thousands of budget tourists and migrant workers, many of whom are traveling in groups and are unfamiliar with basic airport procedures such as security and boarding.

Most of these airport hubs were designed and constructed at the turn of the 21st century. Confronted by exponential growth rates in air traffic, transport planners and urban policymakers across the region endeavored to replace existing inner-city airfields with much larger ones located far from the urban core. In coastal cities, they were built on artificial islands and land reclaimed from the sea (Figure 47.1). In landlocked metropolises, they emerged in rural landscapes on the

Figure 47.1 An aerial photograph of Chek Lap Kok, a human-made island built to house Hong Kong International Airport.

Source: Courtesy of Airport Authority Hong Kong

peri-urban fringe, far from the CBD. These massive relocation projects aimed to enable 24-hour flight operations, to "reduce friction" with the airport's immediate neighbors, and to allow for greater flexibility in future expansion plans.[1] The first city to do so was Osaka, which opened the New Kansai Airport in Japan's Inland Sea in 1994. Similar airport island projects were realized concurrently in Hong Kong, which inaugurated Hong Kong International Airport (HKIA) in 1998; as well as at South Korea's Incheon Airport, which opened three years later (Chang 2010). Other cities that built bigger, more remote airports included Kuala Lumpur (Kuala Lumpur International Airport (KLIA), opened 1998), Shanghai (Pudong, 1999), Guangzhou (New Baiyun, 2004), Bangkok (Suvarnabhumi, 2006), and Kunming (Changshui, 2011).

The case of Hong Kong illustrates this broader paradigm shift in the urban design of aerial mobility: what we might call airport urbanism (Hirsh 2016). Between 1980 and 2010, the city experienced a tenfold increase in passenger air traffic and completely overhauled its airport infrastructure (Figure 47.2). Among transport experts and design critics, the city was lauded for the overwhelming scale and precise execution of its 10-point Airport Core Program, which involved the construction of a brand new airport and a new town located adjacent to it; several suspension bridges; a 34-kilometer road and rail corridor between the airport and downtown Hong Kong; and a series of land reclamation projects in the city's central business district. Tracing the means by which these airport-led urban redevelopment projects were produced – from concept to design to construction – is a complex task. Most design journals offer incomplete accounts that fixate squarely on the "signature architecture" and lead architect of the airport's passenger terminal building. This is typically a telegenic front man – Norman Foster, Helmut Jahn, Renzo Piano – who injects the project with star quality and assigns a human identifier to what are, at

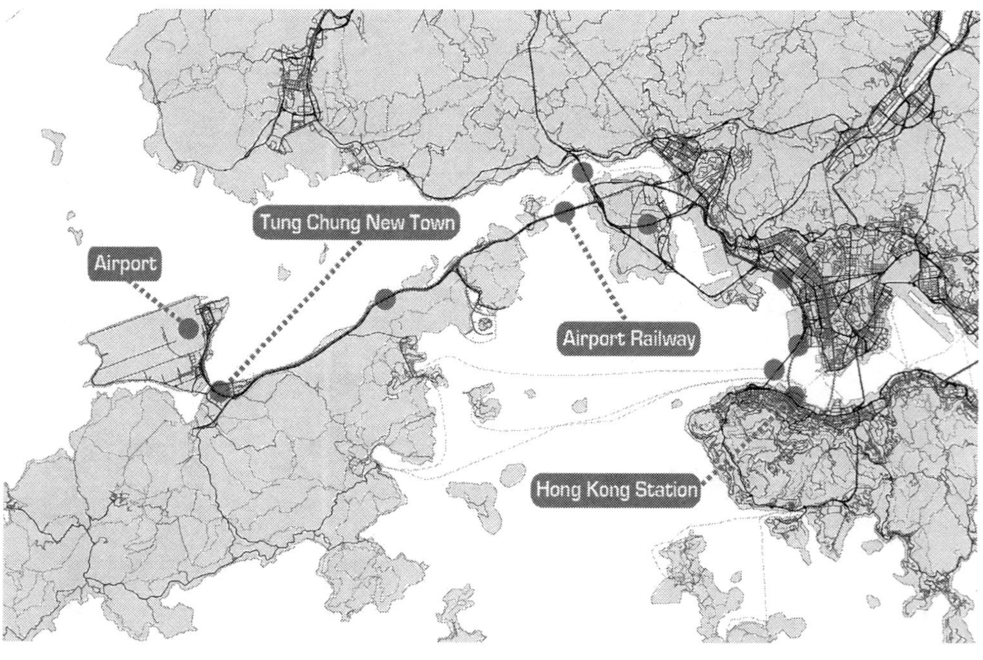

Figure 47.2 A map of Hong Kong's 10-point Airport Core Program. The labels identify the four sites discussed in this chapter.

Source: Map by the author

heart, extremely complicated feats of architecture, planning, and engineering. The fixation on the lead architect has, in turn, led to an unfortunate misunderstanding in the academic literature about the role of "starchitecture" in contemporary Asian cities, one that overlooks the actual processes by which megaprojects are designed and produced.[2] Most of them emerged through the collaboration between local design institutes and planning agencies on the one hand, and foreign technical experts on the other. As such, Asian airport hubs represent an elaboration of vernacular spatial, aesthetic, and programmatic norms that have been hybridized with infrastructural concepts imported from abroad, typically via design professionals from Western Europe, the United States, and Japan. Tasked with harmonizing global infrastructure standards with local building codes and cultural practices, these experts adapted internationally recognized norms to the site-specific design constraints of rapidly developing Asian cities. By planning the overwhelming majority of Asia's airport hubs, they guided the development of the urban infrastructure that critically underpinned the leap in cross-border mobility throughout East and Southeast Asia. Employed by highly specialized planning and technical service firms – notably Landrum & Brown, NACO, and ADPI – these design professionals typically maintain a low public profile, are little known outside of industry circles, and thus remain unaccounted for in the existing academic literature on urbanism, infrastructure, and mobility.

As demonstrated above by the example of Hong Kong, most of these greenfield airport projects doubled as a strategy for large-scale urban development on the ground. Capitalizing on the road and rail infrastructure that was built to connect a new generation of far-flung hubs to the city center, urban planners developed new towns – and sometimes, entire cities – around the airport. Each of these "airport cities" was conceived as a functional or metaphorical complement to the airport itself. For example, Hong Kong's Tung Chung New Town was planned as an "airport support community," designed to house the 60,000 pilots, flight attendants, and ground staff working at HKIA. In Japan, Rinku Town was devised as a shopping and entertainment destination for passengers traveling to the New Kansai airport south of Osaka. Elsewhere, the construction of global airport hubs spurred (or at least justified) infrastructure-led urban development on a much larger scale. South Korea's Songdo International Business District is undoubtedly the most well-known instance. A new city built on reclaimed land, and connected by a suspension bridge to Incheon Airport, Songdo was designed as an Asian gateway for multinational corporations and the expatriate professionals that they employ.[3] In a similar vein, the Malaysian government designated a 50-kilometer stretch of land between the airport (KLIA) and Kuala Lumpur as a "Multimedia Super Corridor," with a new city called Cyberjaya, located at mid-point. The Corridor aimed to attract high-tech industries to Malaysia by combining the financial incentives of a special economic zone with the high-speed road, rail, and air infrastructure that was built to connect the new airport (Bunnell 2004).

All these instances of airport-led urban development were predicated on the firm belief that in a globalizing economy, accessibility to air transport networks would become essential for doing business.[4] Land located near the airport would thus become a desirable place to build corporate headquarters, convention centers, logistics hubs, and research-intensive technology campuses. At the same time, planners believed that these master-planned developments would attract thousands of residents who worked at the airport, or who commuted by air on a regular basis.

The political will to build such projects was strong – they functioned as a handy emblem of modernization via global trade – but they typically rested on a shaky business premise. In fact, most airport cities failed to realize their intended goals as they vastly overestimated the desirability of proximity to the airport. Located remotely on the urban fringe, and lacking the diverse functions and amenities of inner-city locations, these projects struggled to attract both residents and private investment. Hong Kong's airport employees shunned Tung Chung New Town, preferring a longer commute to living on the city's extreme periphery. In fact, many continued to live in the neighborhoods surrounding Hong Kong's former Kai Tak airport. Songdo met a similar fate. As the South Korean government was unable to attract foreign firms, it created incentives for local universities and start-ups to move into the new "international" business district. In effect, these airport-led urban development projects produced a rather naive aestheticization rather than actualization of global connectivity. Yet they have had a significant impact on the peri-urban landscape of East and Southeast Asian cities.

Airports go downtown

By contrast, airport-related development has been a fecund source of urban regeneration in city centers: the result of a broader paradox that confronts urban designers around the world. As more people fly, the less feasible it becomes to site an airport near major population centers, due to the complex operational requirements of much larger airfields. Historic inner-city airports are replaced, or supplemented, by much larger hubs on the exurban fringe, as outlined earlier. This generates considerable traffic between the airport and the inner city, with passengers traversing distances of up to 70 kilometers (44 miles). In order to maintain the essential promise of air travel – that is, quick transport between two cities – Asian airport authorities have made substantial investments in high-speed rail lines that connect airports to dedicated check-in terminals located in central business districts. These terminals outsource basic aviation procedures – such as ticketing and bag check – from the airport itself, and insert them into the urban fabric of downtown. Drawing on the flow of passengers that they create, these terminals form the centerpiece of larger urban renewal projects, a strategy pioneered in the late 1990s by Hong Kong's Airport Express and its two "in-town check-in" terminals.

Combining the flexibility of a remote airfield with the convenience of a downtown air terminal, the Airport Express stations required a formal and programmatic extension of the airport terminal environment into the urban fabric of the inner city. On Hong Kong Island, the in-town check-in terminal anchors the two skyscrapers of the International Finance Centre (IFC), and was planned as the downtown terminus of the Airport Express line. Built on reclaimed land, the eight-hectare plot increased the size of the city's Central district by more than 10%. Following the logic of vertical zoning, the airport train station and check-in hall are located beneath a podium-level shopping mall that supports three office towers, two luxury hotels, and a serviced apartment complex. The station's main hall houses check-in desks, flight departure displays, and luggage trolleys within a cantilevered terminal building, whose pronounced horizontality stands in striking contrast to the extreme verticality of Central Hong Kong's thickly settled forest of skyscrapers (Pearce 2001).

The Hong Kong model has been heavily replicated since its debut in 1998, leading to the construction of similar downtown air-rail terminals in Kuala Lumpur, Seoul, and Taipei. Over time, the in-town terminal concept has expanded both conceptually and typologically, often in unexpected ways. Mainland Chinese cities like Guangzhou and Shenzhen have developed an elaborate network of in-town check-in terminals, sprinkled throughout the urban landscape and managed by local entrepreneurs. Operating out of shopping malls, train stations, urban villages, theaters, and seaports, the companies that manage these routes have developed flexible and loosely organized networks of city-to-airport bus and minivan services that piggyback on existing, non-airport infrastructures in order to shuttle thousands of air travelers to multiple airports located throughout the urban region. In the process, the infrastructure of aerial mobility has been extended far beyond the confines of the two airports and inserted into the everyday urban fabric of Chinese cities.

Typological experiments for the *nouveaux globalisés*

These improvised terminals vary in scale and sophistication: some offer bag drop, ticketing, and even immigration facilities, while others are little more than a makeshift bus stop. Their presence testifies to the radical expansion of the flying public in Asian cities, and to the resulting leap in demand for frequent and affordable transportation to and from the airport. Most of these networks have emerged at the hands of private entrepreneurs rather than through public planning processes.[5] In fact, public transport agencies across the region consistently opted to ignore the democratization of air travel, instead planning high-cost, high-speed transport systems for a narrow segment of government and business elites. The rise of affordable air travel since the turn of the 21st century, which coincided with the emergence of a mass middle class, elicited belated and somewhat lukewarm responses from state transport authorities. In particular, they were unprepared for the rapid emergence of so-called low-cost carriers, or LCCs. In less than a decade, these budget airlines fundamentally reshaped the social and spatial dynamics of East and Southeast Asian cities; and significantly reordered the functional interdependency of those cities by accelerating and expanding cross-border flows of labor, consumption, capital, and knowledge. They did so by developing architectural typologies and low-cost transportation systems that cater to an emerging clientele of what we might call the *nouveaux globalisés*: people who have *just* enough money to travel abroad, but who lack the basic knowledge and technical equipment that is needed to fly, such as a credit card, Internet access, familiarity with check-in procedures, a way to get to the airport.

Budget airlines in Asia owe much of their success to their emphasis on low-tech sales networks – known within the industry as "offline distribution channels" – that are designed for these *nouveaux globalisés*. This is an important consideration among migrant workers in Southeast Asia, where a large proportion of transactions continues to be executed offline and in cash. As the introduction of affordable air travel outpaced the development of online banking systems, low-cost carriers developed creative strategies for extracting payments from their customers. In Thailand, for example, passengers can book a ticket by telephone. The airline replies by SMS with a six-digit reservation code and a request to make a payment in cash at any 7–Eleven within two hours' time. A similar system exists in Malaysia, where customers pay for their plane tickets at petrol stations.

The gap between the demand for budget travel and the inability of customers to purchase tickets online likewise gave rise to a new transport typology – the low-cost travel center – operated by individual airlines in downtown shopping districts, suburban malls, bus terminals, and railway stations (Figure 47.3). At these travel centers, staff members advise passengers on how to book a flight, accept cash payments for existing reservations, and organize affordable bus transfers to the airport. AirAsia, the region's most prominent budget airline, maintains dozens of these travel centers across the region. In Thai cities, for example, they are located inside big-box stores that are the local equivalent of Target or Wal-Mart. As a one-stop shop for groceries, clothing, and electronics catering to consumers on the lower end of the middle class, the warehouses are a logical point of sale for AirAsia's discount plane tickets, sold out of small shopfronts near the check-out lanes.

These travel centers have the look and feel of a busy mobile phone shop. Sales clerks distribute waiting numbers, which are called up on a flatscreen behind the service counter. Customers are then assigned to a kiosk where they select a flight, enter their personal information, and pay for their tickets. Staff members stroll around the store, assisting customers, who are less adept at using touch screens. Visitors to one of these travel centers are likely to encounter a very mixed clientele: migrant workers, young couples, students, retirees, mothers pushing strollers. Most

Figure 47.3 An AirAsia travel center at KL Sentral, Kuala Lumpur, Southeast Asia's largest railway hub.

Source: Photograph by Max Hirsh

come to make advance reservations, but some arrive at the service center packed and ready to go to the airport.

These travel centers are connected by buses and vans to a network of low-cost airports that operate outside the mainstream infrastructure of international air travel, outlined in the first section of this chapter. Budget airlines frequently bypass Asia's megahubs in favor of cheaper airfields: departing from Bangkok's older Don Mueang airport instead of the larger and newer terminal at Suvarnabhumi, for example; or flying out of secondary cities located in the hinterland of primary ones, such as Batam and Johor Bahru in the case of Singapore. And in places where budget airlines do operate out of major international airports, they often do so from low-cost terminals sited on remote parts of the airfield. In so doing, low-cost carriers have produced auxiliary urban transport networks and alternative regional geographies that bypass starchitect-designed airport terminals and displace the flow of travelers into decommissioned airports, retrofitted cargo hangars, and 7–Elevens.

Conclusion

In *The Railway Journey: The Industrialization of Time and Space in the Nineteenth Century*, historian Wolfgang Schivelbusch (1986, 36) argued that:

> a society's space-time perceptions are a function of its social rhythm and its territory. . . . If an essential element of a given sociocultural space-time continuum undergoes change, this will affect the entire structure; our perception of space-time will also lose its accustomed orientation.[6]

While Schivelbusch was referring to the social and cultural shifts ushered in by the introduction of rail transport in 19th-century Europe, his comments resonate with the profound changes in space-time perception engendered by the advent of frequent, affordable air travel in 21st-century Asia. Compared to their European counterparts, the metropoles of East and Southeast Asia are rather far from one another: the distance between Hong Kong and Singapore, for example, is roughly the same as that between Finland and Tunisia. Unlike in Europe, in Asia adversarial foreign relations and impractical topographic conditions have limited the development of high-speed international road and rail connections. Even within the boundaries of a single country, overland trips are often an all-day endeavor. With a paucity of viable alternatives, air travel has thus become the dominant mode of cross-border transportation; and it is not uncommon when boarding a 7:00 p.m. flight from, say, Bangkok to Singapore, or from Manila to Hong Kong, to find that there is another plane flying the exact same route at 7:15, and yet another one at 7:25. Aboard low-cost carriers like AirAsia and Jetstar, journeys of three, four, and five hours feel less like a momentous international voyage and more like a ride aboard a Greyhound bus.

Over the past 30 years, two "essential elements" in East and Southeast Asia's "sociocultural space-time" have undergone significant changes: the dominant mode of transportation, as well as the migration regimes that govern the cross-movement of people. The result is a paradigm shift in the spatial consciousness of a world region where, for much of the 20th century, most people were unable to travel abroad due to intersecting financial, juridical, and infrastructural limitations. Decoupled from its elite connotations, air travel has, in the short span of several decades, become almost banal; and it has fundamentally reshaped the mental map of Asia's urban middle classes. Keeping in mind the kinship ties and historical trade routes that have long bound Asian cities to one another across great distances, it is important to note that contemporary cross-border mobility flows are by no means unprecedented. Indeed, many East and Southeast Asian cities are the product of repeated waves of cultural cross-pollination, evident in a toponymic panorama – Stone Nullah Lane, Jalan Gereja, Muscat Street – that testifies to previous encounters with Europe, the Indian subcontinent, and the Middle East. Yet the volume of contemporary flows – along with their socioeconomic diversity, acceleration by air, and expansion beyond a few select sites explicitly designed for long-distance transportation – are indeed novel phenomena in the region's urban development. By and large, architects, planners, and urban policymakers have been ill-equipped to deal with this shift in the region's space-time continuum, and with the attendant changes in the social and spatial structure of cities. This is evident at the level of public policy, which struggles to grasp the demands of an increasingly large population of tourists, migrants, and transnational commuters; and at the level of urban design, which has not yet developed an aesthetic vocabulary that appropriately reflects the new regime of hypermobility.

Studying cross-border mobility and its physical manifestations in the contemporary Asian city represents a particular challenge for the Western observer, who may not be accustomed to the varying levels of informality, frugality, and entrepreneurialism that penetrate everyday processes of transport and trade in cities such as Bangkok, Hong Kong, and Kuala Lumpur. But it is also a challenge for Asian scholars and urban designers, who have been reared in a context where modernization and cosmopolitanism have invariably been equated with the importation of, and adaptation to, Western models of urbanism. They have likewise been trained to think that informal, flexible, and temporary structural arrangements are embarrassing reminders of a penurious past – and should be replaced by formal, and ideally (for them) megastructural, systems as quickly as possible.

The contrasting examples cited in this chapter – between master-planned, infrastructure-led urban developments on the one hand, and the more spontaneous entrepreneurial mobility practices on the other – illustrate the dialectical relationship between the aestheticization of global connectivity in pricey but often ineffective show projects and its actualization in relatively mundane and pragmatic adaptations of preexisting structures. Across East and Southeast Asia, the everyday experience of cross-border mobility, and the needs of the region's extremely heterogeneous flying public, stands in stark contrast to the fantastical design concepts advanced by airport planners and architects, whose plans rest on the semi-mythological exigencies of an ideal user moving around the city of the future. The result is a pronounced disconnect between utopian designs, which don't address the basic needs of real existing people, and an overtaxed quotidian urban landscape, dominated by improvisational "quick fixes" that are often lacking in aesthetic quality.

How, then, can urban designers move beyond iconic and monumental approaches to transport infrastructure, and instead devise urban design strategies that more accurately reflect the aesthetic experience of cross-border connectivity and hypermobility in the 21st-century Asian city? And how can urban scholars revise their conceptual model of Asian cities – which divides the built environment into exceptional urban enclaves designed for itinerant elites and everyday urban spaces populated by immobilized "subaltern" constituents – in order to account for the democratization of aerial mobility that has taken place over the past three decades?

First and foremost, it is incumbent upon designers to acknowledge the shortcomings of iconic transport megaprojects, and to advance design strategies that are more appropriate for rapidly growing and unpredictably changing urban environments. In so doing, they would address a basic challenge that confronts infrastructure planners not just in East and Southeast Asia, but also in developing regions around the world. Simply stated, the conditions that apply at the start of a given airport's planning process rarely correspond to those that prevail when it is finally opened. As a result, many Asian cities have developed an unfortunate knack for commissioning extremely expensive infrastructure projects that teeter on the edge of obsolescence within months of their inauguration. The pathological displeasure that planners experience when these high-profile installations fail to yield their expected outcomes illuminates the profound limitations of long-term master planning in a rapidly developing urban context. It also reveals the limitations of monumental approaches to transport design, which fixate obsessively upon the construction of visually striking distribution hubs, while ignoring the opportunity to implement more modest interventions and systemic modifications that would greatly improve the overall organization of cross-border mobility flows.[7]

These shortcomings point to the importance of improvisational design strategies that compensate for deficiencies in the master plan: filling in programmatic gaps overlooked by state-sponsored planning practices; and catering more efficiently to end users through flexible and pragmatic urban interventions. These two modes of development can be productively synthesized through the promotion of *incremental design*: an approach that is particularly suited to urban environments that are undergoing long-term changes in their social and spatial organization. Writing on incremental design, urban designer Kevin Lynch argued that:

> In creating a large environment over many years, the perceptual disorder of building – its noise, dust, and inconvenience – can be a serious affair. . . . Processes that may have a desirable conclusion and a well-considered technical order may nevertheless impose frustrating temporary difficulties on the participants and appear to be an incomprehensible chaos. . . . Each brief period in the long process of development must be designed to enhance its quality as a time to be lived in. [T]here must be some sense of completeness at every stage of development so that each generation has some visually stable surroundings, *some feeling of living in today rather than tomorrow*. In any plan,

> the temporary traffic systems, amenities, and patterns of activity must be as carefully planned as the "ultimate" ones.
>
> *(Lynch 1972: 196, 198, italics added)*

Lynch's notion of design by accretion aims to guarantee a degree of built-in flexibility within long-term plans: permitting cities to take advantage of unforeseen opportunities as well as to circumvent unanticipated obstacles. (It also enables urban designers, as well as their municipal clients, to celebrate milestones at more frequent intervals, while solidifying support among the general public for multi-year disruptions to the urban landscape.) Moreover, incremental designs allow the urban landscape to respond more gracefully to unpredictable demographic shifts, as well as to swings in the economic cycle: producing neither an untenable surfeit of grade-A office spaces and luxury condominiums nor a painfully abrupt halt to all construction. Throughout that process, it is incumbent upon designers to deploy building materials, spatial arrangements, and technical systems that are flexible and responsive to change – yet are also aesthetically pleasing and reassuring to the user. Put more bluntly, the challenge is to build infrastructure that, despite its impermanence, looks neither flimsy nor cheap.

The incorporation of incremental design approaches into existing master-planning techniques thus presents a fruitful avenue for addressing the social, political, and economic uncertainties that confront the contemporary Asian city. At the same time, architects and urban planners will be much better equipped to engage with these societal transformations, if scholars begin to participate more actively in public conversations about urban development: drawing attention to broader changes in global mobility patterns, and to their implications for urban form, that are often overlooked by policymakers and designers. This chapter represents a modest attempt to do just that.

Notes

1 Moreover, the development of island airports can be read as a typological reaction to the violent political struggles surrounding the expropriation of land entailed by the construction of Tokyo's Narita airport in the 1970s. For more on the protests at Narita, see Dierikx (2008).
2 On starchitect airport hubs and airport cities in the Asian context, see Rowe (2011). See also Easterling (2005); Graham and Simon (2000), and Ibelings (2002).
3 For a critique of Songdo, see Halpern and Günel (2017).
4 This "aerotropolis" concept was popularized by the economist John Kasarda in the early 1990s and has since been widely discredited.
5 Hong Kong's transportation department, which mandates that each of the city's districts be connected to HKIA via a network of "airbuses," is a notable exception.
6 Here, Schivelbusch is elaborating on Émile Durkheim's argument that "a society's space-time perceptions are a function of its social rhythm and its territory."
7 Reyner Banham outlined the limitations of monumental airport design in a 1962 review of New York's Idlewild Airport. In it, he argued that monumental airport terminals in fact arrested mobility flows, and were "by definition dead, superseded before they were designed" because they could not keep pace with rapid technological changes in the airline industry. A more rational approach to airport design, he contended, would emphasize "the continuity of the process of transportation rather than the monumental halting places along the way." See Banham (1962).

References

Banham, R. (1962). "The Obsolescent Airport." *Architectural Review*. October, 252–253.
Bunnell, T. (2004). *Malaysia, Modernity, and the Multimedia Super Corridor: A Critical Geography of Intelligent Landscapes*. London: Routledge.

Chang, Y. C. (2010). "The Development of Regional Airports in Asia." In Postorino, M.N. (ed.). *Development of Regional Airports: Theoretical Analyses and Case Studies.* Boston: WIT Press.
Dierikx, M. (2008). *Clipping the Clouds: How Air Travel Changed the World.* Westport, CT: Praeger.
Easterling, K. (2005). *Enduring Innocence: Global Architecture and Its Political Masquerades.* Cambridge, MA: MIT Press.
Graham, S. and Simon, M. (2000). *Splintering Urbanism: Networked Infrastructures, Technological Mobilities, and the Urban Condition.* London: Routledge.
Halpern, O. and Günel, G. (2017). "Demoing unto Death: Smart Cities, Environment, and Preemptive Hope." *The Fibreculture Journal* 29. http://fibreculturejournal.org/wp-content/pdfs/FCJ-215Halpern Gunel.pdf
Hirsh, M. (2016). *Airport Urbanism: Infrastructure and Mobility in Asia.* Minneapolis: University of Minnesota Press.
Ibelings, H. (2002). *Supermodernism: Architecture in the Age of Globalization.* Rotterdam: NAi.
Lynch, K. (1972). *What Time Is This Place?* Cambridge, MA: MIT Press.
Pearce, G. (2011). *Arup: Hong Kong Station.* Stuttgart: Axel Menges, 2001.
Rowe, P. (2011). *Emergent Architectural Territories in East Asian Cities.* Basel: Birkhäuser.
Schivelbusch, W. (1986). *The Railway Journey: The Industrialization of Time and Space in the 19th Century.* Berkeley, CA: University of California Press.

Further reading

Hirsh, M. (2016). *Airport Urbanism: Infrastructure and Mobility in Asia.* Minneapolis: University of Minnesota Press. A study of airport infrastructure in five Asian cities that focuses on the interaction between top-down planning methods and user-driven adaptations of cross-border mobility systems.
Lin, W. et al. (eds.). (2017). "Migration Infrastructures and the Production of Migrant Mobilities." *Mobilities* 12:2. A special journal issue that studies the relationship between migration flows and transport infrastructure in East and Southeast Asia.
Nyiri, P. (2010). *Mobility and Cultural Authority in Contemporary China.* Seattle: University of Washington Press. Investigates paradigmatic shifts in migration regimes and cultural attitudes towards mobility in the People's Republic of China.
Solomon, J., Wong, C. and Frampton, A. (2012). *Cities Without Ground: A Hong Kong Guidebook.* New York: ORO. An architectural guide that uses detailed design drawings to illustrate how transport megaprojects have transformed Hong Kong's urban landscape.

PART III.VII

Arts and culture

Introduction

Sharon Zukin was among the first to poignantly observe that "culture is more and more the business of cities" (Zukin 1995: 2). Indeed, in the 21st century, many cities of the Global North have increasingly used the arts and cultural activities and facilities to bolster their image, attract tourism, and even foster economic development. Lured by the economic development potential of attracting the "creative class," cities are designating arts districts, promoting cultural festivals and performances, and commissioning public art. Some of these city-led cultural strategies and neighborhood revitalization efforts valorize formerly disinvested areas, often leading to gentrification and displacement of their residents. Thus, increasingly scholars and activists are asking the question that Renia Ehrenfeucht asks in her chapter: Whose culture? Whose city? Given the market imperatives of the neoliberal city, one could also ask: Is there a role for urban design to redress or obviate displacement, rather than being an agent of such policies of gentrification?

Ehrenfeucht believes that this can happen if urban designers follow four strategies. First, they should try to make or protect space for the ad-hoc, self-organizing activities that constitute the local informal street life of the neighborhood – the street performers, the musicians, the street vendors, or the taco trucks. Such activities are typically the first to be "regulated out" with the insertion of formal design interventions to promote formal cultural activities and facilities. The skills of urban designers are necessary for transforming "lost spaces" (cf. Transick 1986) and those "in between" buildings (cf. Gehl 2011) and creating sidewalk artifacts that can continue to accommodate existing street life. Second, urban designers can also help reduce or soften the rigid edges and boundaries within or between neighborhoods that tend to exclude some groups. Third, designers can help re-envision the ordinary neighborhood infrastructure – the mundane street benches, garbage cans, or bus shelters – as art. Lastly, Ehrenfeucht argues for the co-production of street culture that involves local artists and residents and incorporates the local neighborhood history and culture. Such co-production should allow "for intersecting and conflicting uses, [and] contested representations" in a multicultural pluralistic city.

A different response to the questions "Whose city? Whose art?" comes also from Stefano Bloch, who discusses the contested cultural representations of graffiti art. Such art practices are often displayed at "insurgent artscapes." Bloch defines them as "physical spaces in which unsanctioned, uncommissioned, and often unruly but always expressive art and art practices thrive." They are found at a city's "terrains vogue" – which are exactly the interstitial city spaces that Ehrenfeucht also has in mind. Bloch gives a personal account of the life of one such space in Los Angeles, the Belmont Yard. Until its planned demise, when it became engulfed by the gentrification wave that reached its doors, this former railroad yard served as the artistic terrain for countless graffiti artists. Bloch's chapter is a strong reminder that the "cultures of cities" (cf. Zukin 1995) are not limited to the formally designated arts districts, flagship museums, or outdoor public art sculptures. They also flourish in unsanctioned and unscripted settings. In this case, urban design is not the work of formally trained professionals, but rather the informal and insurgent artists who unwittingly also become place-makers.

What do people think, and how do they react to particular urban design and public art interventions in cities? In the third chapter of this section, Brettany Shannon undertakes "digital fieldwork" to explore exactly this question. More specifically, she mines Instagram's database – the posted images and their captions that accompany the hashtags #urbandesign and #publicart (in English and Spanish). Instagram images are not bounded by the borders of a locale, as they can be reached by any point of the globe that has a wi-fi connection. As such they may be reshaping how the general public views and consumes design and art. As Shannon contends: "Architects and designers have long wished to know more about how the public experiences their projects, and Instagram provides one, albeit limited, view into their perceptions." They now have this opportunity through Instagram, as crowdsourcing, to better understand the affective responses of the public to different design interventions, and how these may shift over time.

References

Gehl, J. (2011). *Life Between Buildings.* Washington, DC: Island Press.
Transick, R. (1986). *Finding Lost Space: Theories of Urban Design.* New York: John Wiley.
Zukin, S. (1995). *The Culture of Cities.* Malden, MA & Oxford, UK: Blackwell Publishers.

48

DESIGNING THE INCLUSIVE CITY

Urban cultures, street arts, and public life

Renia Ehrenfeucht

Introduction

In 2017, Maddox & Co. Realtors presented the event *508 Mural Fest* in the US city of Albuquerque. Produced by arts organizations Warehouse 508 & Worldwide Underground, and sponsored by the city and by corporate and small business owners, the event produced more than 40 murals in a ten-day period along streets in and around downtown (508 Mural Fest n.d.). While the murals reflected Albuquerque's rich diversity, some were tagged before completed, causing outraged passersby to call nearby working-class and mixed-income neighborhoods "a ghetto." The event has been one of many ways that Albuquerque has invested in its creative economy, and the aspirations behind it, as well as the responses, speak to contradictory impulses and effects. The murals in disinvested neighborhoods were intended as a community arts initiative but also implicated gentrification. The "us versus them" responses also raised the critical question: What actions revitalize a city, and for whom?

In the late 20th and early 21st centuries, urban design has played an increasingly important role in reshaping urban experiences as a tool for economic development. Arts and culture have been a fundamental dimension of urbanism throughout history, but during this period, they also played a role in developing urban growth strategies. In a global economy increasingly dependent on creative knowledge workers, or the "creative class," the increased value placed on arts and culture reflects a confluence of urban processes and purposes.

This has led to important shifts in how cities have approached the space of the street, creating openings to use culture for building more inclusive places. Street life, with its varied and intersecting activities, has become more appealing for public life than malls and privatized public spaces. Yet formal street life initiatives often include actions to limit or eliminate existing activities and restrict culture bearers, the people whose daily activities and routines produce local ways of life. Culture bearers can be the myriad street workers whose presence shapes a neighborhood, as well as musicians, craftspeople, or others engaged in street art. At times, art pushes out existing cultural practices, such as when an art market displaces street vendors. Art-oriented development also can fail to recognize conflicting cultural meanings and situations, such as when public-sector mural projects intend to replace street art. This chapter explores these contradictory impulses. I first outline why street art and culture have become widely valued. Then I examine four dimensions of everyday cultures, interrogating their contradictions in order to chart urban design responses that use art and culture for an inclusive urbanism that embraces diversity.

Why street cultures? Why now?

The focus on street art- and culture-driven development entangles everyday life, urban design, regulation, and development. Flagship cultural buildings designed by star architects hope to attract global visitors, but their success also comes from associated retail districts, so much so that at times they reduce the importance of the art (Grodach 2011). Public art – which can range from monuments to sculpture, to well-designed bus stops or bike racks – enlivens cities as well as sparks protest. Neighborhoods with street art and murals in Mexico City and São Paulo attract visitors, but street art can also stimulate gentrification. At times, street art has moved from the street and exhibited in museums and galleries (including the Tate Modern in London and the Museum of Contemporary Art in Los Angeles) (Ehrenfeucht 2014). Street culinary art – from street vending to food trucks or carts – offers convenient food to residents and visitors alike, and its popularity has led some high-end restaurants to serve traditional street food. At the same time, in response to complaints and attempting to retain control, municipal officials still remove street art and vending and arrest street artists and food vendors.

These trends reflect interrelated forces. In the last decades of the 20th century, central city regeneration became visible in the former industrial districts of New York and London. Global economic restructuring re-centered world cities and many second-tier cities; more professionals chose to move to historic neighborhoods that became spaces where a diverse workforce now lives, works, and plays (Sassen 2012).

By the 2000s, gentrification, a process defined by higher-income people displacing lower-income residents in select neighborhoods, became visible in numerous cities and countries in different types of places (although still mainly associated with the central city). At the same time, recreational and business travel increased, generating revenue from visitors. In the 1980s, gentrification was mostly associated with artists who moved to low-rent warehouse districts (Zukin 1982), but by the 2000s higher-income residents were the gentrifiers, and real estate developers renovated and built urban housing to accommodate their market choices. While these new urban residents value working-class cultures and lament their loss, their preferences appropriate and change the very neighborhoods that epitomized such cultures (Zukin 2010).

Urban growth actors have looked favorably upon and attempted to stimulate gentrification. In the early 2000s, Florida's (2002) creative class hypothesis offered a policy pathway to revitalization that used small-scale interventions including promotion of neighborhood-based arts and cultural institutions. Florida argued that creative professionals – who are an increasingly important part of the workforce and have relatively high incomes – are more mobile than other workers and make decisions about where to live based on quality-of-life factors and openness to diversity. Florida's work became highly influential in economic development policies both inside and outside the United States.

During the same era, there has been a renewed interest in localism, cultural diversity, and craft production to counter the homogenizing effects of mass production and consumption (Hess 2009). Localist movements have many strands, as people increasingly seek food and crafts that are made or grown in their region, but also handmade and small batch craft production from other regions. While this aesthetic is highly associated with urban life, it has also influenced some rural areas that have become destinations offering "farm-to-table" experiences. This appreciation extends to independent businesses as well as diverse ethnic and immigrant neighborhoods, foods, and cultural products.

Localism encompasses both the origin and types of goods purchased and the experience of purchasing and living locally. It manifests in local markets, locally owned businesses, and local street vendors. Because global, urban professionals and travelers express preferences for what is unique and authentic to a given locale, the movement's impact receives significant policy attention.

While local governments and other urban growth promoters value arts and cultures for their economic development potential, the benefits extend beyond this goal. Arts and cultures have intrinsic value that enriches daily life and contributes to urban vitality, excitement, and belonging. However, choreographing spaces to cater to consumer desires as alternatives to mass consumption can also lead to displacement and increased inequality (Zukin 2009). Enhancing local distinctiveness simultaneously values, appropriates, and attempts to control the daily lives of people using and working in urban public spaces. Untangling the connections leads to numerous questions. Whose cultures? What art? Where?

Urban cultures in an inclusive society

The following sections explore four dimensions of urban cultures and art. All intersect in the public realm, where urban design interventions can contribute to building culturally rich urban spaces. These are also cautionary tales because, while local businesses, public arts, street vending, local products and food, and neighborhood distinctiveness are valuable, formal interventions often do more to control or manage than to enhance urban life.

Making space for ordinary activities to thrive

Everyday life in public is the basis for urbanism, and good street design can create more spaces for ongoing activities without precluding what might come later. Street life describes activities from standing on a corner, to displaying products outside a shop, or selling jewelry or food from a stand. Watching neighborhood festivals or parades, sitting on a stoop, fixing a car, or waiting for work are also examples of street life. In cities throughout the world, these ordinary expressions of urban life create vibrant neighborhoods of propinquity and social contact, while providing residents with necessary goods and services. They are enjoyable because people like to be around other people. The activities shift in subtle ways and add richness and variation to ordinary routines.

Street cultures develop from both necessity and convenience. As activities become commonplace, they are identified as cultural practices, which then adapt to changing circumstances. While people bring traditions to new places – for example, if someone comes from a place where street vending and informal economy are quite common, they may be more amenable to it – their spatial practices reflect local demand and opportunities (as well as, at times, few alternative options for work).

In the late 19th century, local governments wanted to eliminate street life, and they enacted prohibitions on numerous activities. Throughout the 20th century, the vision for urban streets did not include the messiness of everyday public life. The attempts to control street life intensified in the late 20th century, as cities began to gentrify. Neil Smith (1996) called this "revanchism," when policy and policing aggressively attempted to reclaim public spaces from people who appeared homeless and from others who might bother the middle-class residents moving back to the city. Enforcement of public space regulations has been cyclical and sporadic, however, resulting in widespread street activities, despite the ordinances prohibiting them.

In the 21st century, the approach changed to one of ambivalence. As Gengzhi Huang et al. (2014) have shown for Guangzhou, China, street vendors were targeted by policies in the National Sanitary City campaign that attempted to exclude the 300,000 street vendors from the central city. It was difficult to eliminate vendors entirely because they individually and collectively resisted. At the same time, progressive attempts to address the need for livelihood and promote social resistance led to policies that included spaces for vending.

Street food, goods, and services are convenient in the fast-paced contemporary life. Giving people reasons to frequent an area, street activities also provide "eyes on the street," an idea popularized by Jane Jacobs (1993) that has gained currency in the 21st century. Furthermore, attempts to control street activities are rarely successful. In Guangzhou, however, the "inclusiveness is conditioned" and "the authorized sites are restricted to locations where city landscapes, public space order, formal retail businesses, and residential daily life are not affected," or in other words, they are "located in peripheral or invisible areas," where they are less profitable and convenient (Huang et al. 2014, 185).

The situation in the Guangzhou case illustrates a recurrent theme, which highlights a fundamental tension: When street life is accepted, and even celebrated, it may still be highly controlled and limited to particular (and at times undesirable) spaces. Other examples come from New Orleans, a city celebrated for its public cultures. The city wants artists to sell art in the French Quarter, the historic center and primary tourist destination. Nevertheless, the city's regulations permitting art sales require the art to be "original," preventing artists from selling high-quality prints of original work. Instead, the artists create formulaic pieces to sell to visitors. In another situation, the city decided to offer vending permits to vendors who sell food at weekly neighborhood parades called "second lines." Even though the city took this action in order to recognize the tradition, it adopted regulations that, if followed, would have precluded vending as it has traditionally occurred. However, the regulations are not followed or enforced, indicating ambivalence by all stakeholders (Ehrenfeucht and Croegaert 2017).

How to effectively manage, plan, and design for street activities is a puzzling dilemma. These activities are fundamentally self-organizing, or, in other words, they are activities where participants organize their actions in response to available space, public space users, and stakeholders, including abutting businesses and street-level bureaucrats.[1] Self-organization has been widely documented in the literature about pedestrian behavior as well as street vendors, market vendors, and food truck customers (Whyte 1988; Morales 2010; Ehrenfeucht 2017). In contrast, management practices, including urban design interventions, have an inherently different logic. They are characterized by rigidity and lack of nuance in how relationships and spaces are treated. When spaces are designed for a use such as street vending, too often rigidly programmed kiosks replace traditional activity patterns and limit the activity's interactive, adaptable character.

To make use of the city's interstitial spaces is an alternative to designing separate spaces for street life. Streets have places near utility poles, trash cans, and buildings where people do not walk. Instead of attempting to create zones for public space use, all urban elements can be envisioned for multiple uses. Planters can offer seating, while signs and utility poles can create spaces for street vendors to set up. Figure 48.1 shows a food truck taking advantage of a gap in pedestrian traffic created by a sign.

Reducing boundaries for different worlds to intersect and interact

In the 2000s, Richard Florida's (2002) creative class argument re-focused attention on the importance of amenities. This led to neighborhood-scale interventions to provide spaces where creative workers meet, work, and recreate, including markets, coffeehouses, bars or craft distilleries and breweries, bicycle and pedestrian facilities, and restaurants where wide sidewalks allow for outdoor seating and vending, and trees and awnings provide protection from the weather. Cultural regeneration has a longer history, and this revitalization strategy was used as early as the 1980s and 1990s (Wansborough and Mageean 2010). Cities enacted policies to create special districts

Figure 48.1 Food truck in the Chicago Loop that uses the space near a sign and bike rack.

Source: Photo by Camille Fink

where cultural institutions would flourish. To do this, they also prioritized consumption through restaurants, coffeehouses, high-end shopping, and nightclubs to generate enough foot traffic to also support arts institutions.

However, intentional art districts are only one type of neighborhood culture. Ethnic neighborhoods house diverse residents, and are important sites for residents and others who come to shop for goods unavailable elsewhere. Ethnic diversity is imprinted onto neighborhood landscapes through cultural practices and built environment adaptations as well as intentional urban design interventions (Irazábal 2011). Historically, many neighborhoods such as Chinatowns in the United States and Canada were perceived as exotic, stigmatized as crime ridden, and heavily policed in ways that allowed vice while blaming residents for it (Anderson 1987). In an era that values diversity, these neighborhoods now attract newcomers and face gentrification, and ethnic cultural organizations may also orient themselves towards a broad base of consumers.

People identify with both the social dimensions of the neighborhoods they frequent as well as the qualities of their built environment. Neighborhood character, in particular, is formed from people and activities, as much as from urban form and design. Urban design, nevertheless, can make unique contributions to help define a space, whether it is a neighborhood, street corner, or plaza. Neighborhood-scale urban design interventions can enhance the distinct characteristics generated by people and their activities.

Often, however, as middle-class residents consume heritage through recreation and fixed capital (such as in historic neighborhoods), diverse urbanism disappears (Zukin 2010). In Puebla, Mexico, for example, the municipality's attempts to revitalize the city center focused on the conservation of the built heritage at the expense of the people who lived and used the space. In the 1980s, "the pavements – often broken – were crowded with street traders, many of them indigenous women, often with children, selling fruit and vegetables or preparing food, with steam rising from pots over charcoal stoves" (Jones 2015, 267). Activities spilled onto the streets, and families crowded the often-derelict tenements. After being selected as a UNESCO World Heritage Property, the living city in Puebla's center was forced out as the historic city was restored. The municipality attempted to develop an arts district that included higher-end restaurants, cafes, galleries, and museums. The process displaced many people, and the population dropped from 340,000 to 108,000. When these efforts failed to stimulate a vibrant core, the area was redeveloped once again, replacing historic tenements with modern housing (Jones 2015).

In Puebla, competing views of authenticity were at stake. The city was restored to a designated period of historic significance, and art as business displaced organic daily life. The notion of authenticity can be used as a way to deploy elitist aesthetic sensibility, privileging particular values and cultural forms. Zukin (2009), however, has argued that authenticity may also be used to support rights in the face of widespread neoliberal gentrification. She notes that authenticity has two distinct origins. One is based on long-term life situations and practices for a group attached to their land or place. It is grounded in tradition and local claims and embedded in a "right to the city" that allows people to create and recreate spaces and practices (Lefebvre et al. 1996). The second origin refers to objects that are originally created or innovative, such as original pieces of art, architecture, or cultural trends. These two aspects of authenticity are often in tension with one another as urban spaces change. Developers and gentrifiers seem to value the second aspect of authenticity, which presents a threat to the first.

Integrating new uses into existing landscapes in creative ways – rather than defining new, distinct boundaries – can bridge different neighborhood experiences. Neighborhoods and communities have a core, but the edges are blurry, as residents come and go and life patterns change. Spaces within the neighborhood may also have multiple, overlapping, and contested meanings. For those walking from work or home, the space may feel continuous, but for those riding the bus or driving, there might be more defined beginnings and endings. For some, a clean and policed space feels safe, while others feel unsafe by the presence of police and indicators of heavy surveillance. Vendors might identify boundaries by where they work and replenish goods.

The challenge is to create spaces that continue to support everyday life. Responding to the widespread loss of industrial jobs, the US state of Michigan developed a "cool cities" initiative to invest in amenities to attract young knowledge workers and stimulate economic growth. One investment was in Detroit's Eastern Market, a hub for local agriculture and locally made products. While urban farmers and crafts markets are associated with white aesthetic practices (Alkon and McCullen 2011), the market can also support local agriculture that has diverse participants in Detroit. Cultural distinctiveness can create a sense of belonging for residents or shoppers who find familiar products, foods, and languages, but it is also an explicit attempt to attract diverse residents back to urban neighborhoods. This is the case with the night market at Vancouver's Chinatown. It is a popular attraction for visitors but also serves local residents and sells Chinese products. While it serves various purposes for different groups, some nevertheless question its "authenticity" (Pottie-Sherman 2013).

Neighborhood edges can be intentionally permeable rather than bounded. The hyper-diverse global city of Los Angeles created a process by which residents or local organizations can designate neighborhoods to claim local identities and histories. The street signs designating Koreatown,

Thai Town, Historic Filipinotown and Little Ethiopia reflect the city's rich tapestry. Without trying to resolve boundaries, the city allows diverse groups to visually enact their competing and complementary claims. This also happens organically when shop owners post signs in multiple languages to appeal to different constituents. Facilitating myriad interpretations of a space or neighborhood can honor the core of the neighborhood, or its historical traditions and culture, without creating hard boundaries that exclude some groups or inhibit change.

Designing infrastructure for each element to count

Neighborhood- or destination-oriented interventions have limited reach. They impact small areas and can contribute to gentrification by making local improvements that result in newcomers benefitting from investment in traditionally lower-income areas. At the same time, municipalities regularly invest throughout the city through infrastructure, but little of it is designed to enhance everyday experiences. Public infrastructure, however, can create cultural richness in artistic ways, bringing art into the everyday life. Small improvements can make a difference. Benches and bus stops can move beyond functional design, and play equipment for children can be sculptural.

Well-designed infrastructure transforms experiences, as two transportation systems exemplify. The bus rapid transit (BRT) tubes in Curitiba, Brazil, symbolize the city's revolutionary bus rapid transit system and signify the value placed on pedestrians waiting for this form of motorized mass transit. Curitiba's 357 modernist tube stations were designed as a functional and affordable approach to increase the speed of the transit trip by moving pay stations off the bus, allowing transfers, and protecting people from the weather (Reed 2015). The BRT system is a citywide network, and as Figure 48.2 shows, the stations have given visual identity to an innovative form

Figure 48.2 BRT stop in Curitiba.

Source: Photo by Joni Palmer

Figure 48.3 The University subway station in Naples designed by Karim Rashid.

Source: Photo by Giorgio Galano

of transportation for all people. This trend harkens to an earlier era in urban history, when train stations were important sites that captured designers' and investors' imaginations.

The subway stations in Naples, Italy, offer another example. The stations are intentionally artistic statements that have transformed transit into art. Including both innovative station design and installations by contemporary artists, the experience of the subway system has become an artistic work in progress, as Figure 48.3 shows. With high ridership, residents experience art in their everyday travel (Povoledo 2013).

Distributing investments and benefits among residents, regardless of their income or daily patterns, furthers the goal of building inclusive cities. Re-envisioning infrastructure as art can increase an area's distinctiveness and reflect its character, while enhancing a system or city rather than only select neighborhoods.

Representing and co-producing for people's rights to be realized

Everyday life on the street and intentional design interventions and regulations co-produce street cultures. Two questions arise when deciding what activities should happen, and where they should occur. Who produces the city? How are people, histories, and meanings represented? Urban cultures arise when people live in ways that work for them, and these become life patterns; but street life is regularly contested – whether this involves where people can work, where street art is placed, and whose interests are reflected in new developments.

Street art illustrates the challenges with co-production. Street art refers to art on walls and other public spaces throughout the city. Figure 48.4 shows street art on the infrastructure supporting

Figure 48.4 Street art under an overpass in Toronto.

Source: Photo by author

an overpass in Toronto, Canada. Some street art may be authorized by a governmental agency or property owners, such as commissioned murals. Much of it is painted by artists, where and when they want, without permission. Once created, some unauthorized pieces are appropriated by the property owner, but most are not, and last until they are removed, are painted over, or disintegrate.

As street art became popular and increasingly associated with gentrifying neighborhoods, cities established mural initiatives and art walls to encourage art while controlling where street artists work. Many artists welcome the opportunity to paint without the risks associated with unauthorized action, but others value how they choose a spot and claim the city. Street art extends to other forms of urban culture such as skateboarding, street dance and music, and parkour, an activity where people use buildings, walls, and other dimensions of built form as an obstacle course to climb and traverse. These activities use urban elements in unintended ways (Caldeira 2012; Ehrenfeucht 2014).

Despite its success in museums and galleries, street art is still contested on city streets (Ehrenfeucht 2014). São Paulo, Brazil, is a global street art mecca that has miles of art along freeways, storefronts, and vacant buildings. Although formally embraced by some previous city administrations, in 2017, São Paulo mayor Joao Doria ordered blocks of murals removed under a "pretty city" campaign (DiLorenzo 2017). This contested practice is a struggle not only over aesthetics but also about belonging and right to produce the city (Lefebvre and Nicholson-Smith 1991; Caldeira 2000, 2013). In the situation in Albuquerque mentioned in the introduction, the city and property owners worked with muralists but failed to recognize other stakeholders, who interpreted their actions as a form of exclusion. Murals also enhance properties, but they do not replace street art and the autonomous decision-making involved in the process. In addition, few initiatives recognize how public art may contribute to displacing the lower-income communities of color who started the traditions.[2]

The public dimension of public art makes it a touchstone for societal differences, especially as art commemorates particular histories and perspectives. Many urban sculptures are monuments that commemorate historical events or significant people. As representative of particular times and politics, sculptures exclude as frequently as they include. In the 2010s, many Southern US cities grappled with the question of whether they should remove Confederate sculptures, those associated with the 19th-century Civil War, when southern states seceded in an attempt to uphold slavery. These sculptures adorn parks and other civic plazas, affirming white supremacy. Most were put up by community organizations in the last century; but in the first decade of the 21st century, community organizations are fighting for their removal (even though, as sculptures in public spaces, only local governments have the authority to remove them). This struggle over the Confederate monuments is about not only the history, but also contemporary racism that leads to violence against African American people by white people and the police. Proponents to keep the statues, however, suggest that they represent historic realities, which should not be erased, no matter how wrong-headed. In public, the statues form the iconic boundaries of inclusive and exclusive representation.

Conversely, shallow cultural appropriations, even if intending to celebrate diversity or retain a sense of place, are problematic. To invoke a culturally rich past, many projects invoke diverse people. In US areas with many Native nations and large Indigenous populations, designers use elements that draw from Indigenous crafts or at times depict Indigenous people in native dress. This is occurring in Hawai'i, as new development incorporates Indigenous aesthetic elements and uses Indigenous names for buildings, and native plants in landscape architecture (Koh and Freitas 2018). Design elements and language are appropriated without commissioning Indigenous designers or without working with the community, whose culture is being used. The interest in this past is symbolic. It essentializes particular tradition in Native communities and fails to reflect the myriad ways Indigenous residents live in the city or region. It also fails to address access to the land and how the dominant private property regime is a colonial structure (Koh and Freitas 2018), or to recognize that Indigenous residents disproportionately have low incomes and typically do not benefit from new culturally themed development.

Too often, adopting such cultural themes appropriates people's culture rather than includes the people. It is, therefore, necessary to co-produce imagery as well as urban spaces. Instead of outside designers lightly borrowing images to sell local traditions as unique or exotic, local residents, artists, and artisans can design spaces that incorporate their histories and traditions as formal participants in a process of co-production. This requires more than a participatory process, but rethinking of the roles of designers and reshaping design processes. Formal interventions can enhance new spaces when they reflect and build on traditions, but not if tradition is only used to create an exclusive product.

Conclusion: enhancing inclusive cities

Re-envisioning city streets as primary sites of culture and art can enhance the city, making more spaces that are desirable, usable, and artistic. This is an alternative to displacing living cultures and culture bearers in order to create designated districts for arts and cultural institutions. Too often, urban design has been used to indicate who is out of place, leading to more exclusive neighborhoods. Resulting spaces can lack the complex, vibrant ordinariness that people value, becoming less functional and interesting. In overly choreographed spaces, unauthorized activities are interpreted as a threat rather than an organic part of urban life. Even though people value urban vitality and authenticity, initiatives intending to manage public space replace self-organized activity with programmed areas and narrowly defined events.

Can these contradictions be resolved to ensure that urban design interventions improve urban life? Enhancing existing landscapes and allowing for intersecting and conflicting uses, contested representations, and co-production is an important first step. Interstitial urban spaces can be intentionally designed to allow self-organized public life to thrive and facilitate interaction among people, and between people and the abutting buildings and uses. Design interventions can create more spaces for people rather than replacing some activities with others. Ordinary infrastructure – walls, freeway underpasses, subway stations, signs – can add art into the common spaces. By valuing everyday traditions and street life, urban design can be a force for an inclusive city that incorporates arts and supports diverse cultures.

Notes

1 While not recognized in formal regulations, in practice, people including law enforcement officials respond to public space norms rather than the strict interpretation of the ordinances, and all parties negotiate the spaces and forms that public activities take.

2 Contemporary street art is traced to graffiti that came out of Philadelphia and New York.

References

508 Mural Fest. (n.d.) Available: http://508muralfest.com/

Alkon, A.H., and McCullen, C.G. (2011) "Whiteness and Farmers Markets: Performances, Perpetuations . . . Contestations?" *Antipode* 43 (4): 937–959. doi:10.1111/j.1467-8330.2010.00818.x.

Anderson, K.J. (1987) "The Idea of Chinatown: The Power of Place and Institutional Practice in the Making of a Racial Category." *Annals of the Association of American Geographers* 77 (4): 580–598. doi:10.1111/j.1467-8306.1987.tb00182.x.

Caldeira, T.P.R. (2012) "Imprinting and Moving Around: New Visibilities and Configurations of Public Space in São Paulo." *Public Culture* 24 (2 (67)): 385–419.

Caldeira, T.P.R. (2000) *City of Walls: Crime, Segregation, and Citizenship in São Paulo*. Berkeley: University of California Press.

DiLorenzo, S. (2017) "Sao Paulo Street Art Debate over What Makes Cities Livable." *The Seattle Times* (March 24).

Ehrenfeucht, R. (2014) "Art, Public Spaces, and Private Property along the Streets of New Orleans." *Urban Geography* 35(7): 965–979.

Ehrenfeucht, R. (2017) "Do Food Trucks and Pedestrians Conflict on Urban Streets?" *Journal of Urban Design* 22 (2): 273–290. doi:10.1080/13574809.2017.1281731.

Ehrenfeucht, R. and Croegaert, A. (2017) "Learning from New Orleans: Will Revising or Relaxing Public Space Ordinances Create a Just Environment for Street Commerce?" In *Food Trucks, Cultural Identity, and Social Justice: From Loncheras to Lobsta Love*, edited by J. Agyeman, C. Matthews, and H. Sobel. Cambridge, MA: MIT Press.

Florida, R.L. (2002) *The Rise of the Creative Class: And How It's Transforming Work, Leisure, Community and Everyday Life*. New York, NY: Basic Books.

Grodach, C. (2011) "Cultural Institutions: The Role of Urban Design." In *Companion to Urban Design*, edited by T. Banerjee and A. Loukaitou-Sideris. New York: Routledge.

Hess, D.J. (2009) *Localist Movements in a Global Economy: Sustainability, Justice, and Urban Development in the United States*. Cambridge, MA: MIT Press.

Huang, G., Xue, D. and Li, Z. (2014) "From Revanchism to Ambivalence: The Changing Politics of Street Vending in Guangzhou." *Antipode* 46 (1): 170–189.

Irazábal, C. (2011) "Ethnoscapes." In *Companion to Urban Design*, edited by T. Banerjee and A. Loukaitou-Sideris. New York: Routledge.

Jacobs, J. (1993) *The Death and Life of Great American Cities*. Modern Library ed. New York: Modern Library.

Jones, G. (2015) "Gentrification, Neoliberalism and Loss in Puebla, Mexico." In Lees, L., Shin, H.B., and López-Morales, E. (eds.). *Global Gentrifications: Uneven Development and Displacement,* edited by Bristol: Policy Press.

Koh, A. and Freitas, K. (2018) "Is Honolulu a Hawaiian Place? Decolonizing Cities and the Redefinition of Spatial Legitimacy." In Race and Spatial Imaginary: Planning Otherwise/Introduction. *Planning Theory & Practice* 19 (2): 254–288. doi:10.1080/14649357.2018.1456816.

Lefebvre, H., Kofman, E. and Lebas, E. (1996) *Writings on Cities* (Vol. 63, No. 2). Oxford: Blackwell.

Lefebvre, H. and Nicholson-Smith, D. (1991) *The Production of Space.* Oxford: Blackwell.

Morales, A. 2010. "Planning and the Self-organization of Marketplaces." *Journal of Planning Education and Research,* 30(2), 182–197. DOI 10.1177/0739456X10385561.

Pottie-Sherman, Y. (2013) "Vancouver's Chinatown Night Market: Gentrification and the Perception of Chinatown as a Form of Revitalization." *Built Environment* 39 (2): 172–189.

Povoledo, E. (2013) "Subterranean Naples, Brimming with Art." *The New York Times* (February 15).

Reed, D. (2015) "How Curitiba's BRT Stations Sparked a Transport Revolution." *The Guardian* (May 26).

Sassen, S. (2012) *Cities in a World Economy.* Thousand Oaks, CA: SAGE/Pine Forge Press.

Smith, N. (1996) *The New Urban Frontier: Gentrification and the Revanchist City.* Hoboken, NJ: Taylor & Francis.

Wansborough, M. and Mageean, A. (2010) "The Role of Urban Design in Cultural Regeneration." *Journal of Urban Design* 5 (2): 181–197. doi:10.1080/713683962.

Whyte, W. H. (1988) *Rediscovering the Center.* New York: Doubleday.

Zukin, S. (1982) *Loft Living: Culture and Capital in Urban Change.* Johns Hopkins Studies in Urban Affairs. Baltimore: Johns Hopkins University Press.

Zukin, S. (2009) "Changing Landscapes of Power: Opulence and the Urge for Authenticity." *International Journal of Urban and Regional Research* 33 (2): 543–553.

Zukin, S. (2010) *Naked City: The Death and Life of Authentic Urban Places.* Oxford: Oxford University Press.

Further reading

Kim, A.M. (2015) *Sidewalk City: Remapping Public Space in Ho Chi Minh City.* Chicago: University of Chicago Press. A rich description of sidewalk life in Ho Chi Minh City, which includes activities that take place on the street, how vendors use and adapt space, and their strategies to negotiate regulations and policing. This book includes a countermapping project that makes multiple sidewalk claims visible.

Lees, L., Shin, H.B. and López-Morales, E. (eds.) (2015) *Global Gentrifications: Uneven Development and Displacement.* Bristol, UK: Policy Press. ISBN 9781447313489. An edited volume about contemporary neighborhood-scale change with chapters from cities throughout the world. It shows the multiple and varied ways that arts and cultures are used in development, and the challenges and conflicts that arise among stakeholders.

Loukaitou-Sideris, A. and Ehrenfeucht, R. (2011) *Sidewalks: Conflict and Negotiation Over Public Space.* Cambridge: MIT Press. An examination of wide-ranging sidewalk activities, opportunities, conflicts, and stakeholders in five US cities. This book highlights the significance of sidewalks as spaces where residents meet, see others and engage with their city in spaces that are both public and associated with abutting properties and businesses.

Zukin, S. (2010) *Naked City: The Death and Life of Authentic Urban Places.* Oxford: Oxford University Press. An account of New York that highlights the contradictions in gentrification and redevelopment. The chapters show the varying ways that gentrifiers' preferences change the city despite the fact that newcomers appreciate the diversity that existed prior to their arrival.

49

PAINTING INSURGENT ARTSCAPES

Stefano Bloch

Introduction

I can't quite remember a weekend that I didn't spend in a yard while growing up as a graffiti writer. Yards were illicit and largely hidden spaces located adjacent to decommissioned as well as active rail lines, down freeway embankments, behind industrial complexes, and within washes and tributaries of the LA River. It is where "graffiti artists," or "taggers," depending on who is making the determination, went to practice their styles and hang out with friends for hours on end. These spaces exist all around the city of Los Angeles: the Motor yard, the Whitsett yard, the Venice Pit, Commerce, the Panic Zone, the Duck yard, Sanborn, and of course, Belmont. They were (and some still are) free and open spaces, where sometimes drinkers went to drink and drug users went to use dope, but for the vast majority of us, these yards were the only available art spaces in the city. They were autonomous zones for visual expression.

Nowhere else could someone armed with spray paint, and sometimes brushes or markers, find canvases of such size. Walls were 20 feet high and hundreds of feet long in some yards, and free from a hostile public that conflated our actions as painters with those of violent criminals. We were criminals, too, of course, but made so by the very act of entering those yards and painting on those walls.

I had never consciously connected my occupation of those spaces to a larger political philosophy, but I knew such spaces allowed me to exist and feel alive. That sense of freedom inspired me to search out other ignored, defunct, and derelict spaces where I could simply be without having to pay an entrance fee, beg permission, or feel beholden to some invisible benefactor. Just as I was always reading and scanning the built environment for surfaces on which to write graffiti, I looked even deeper for places that would allow me to slow down, take my time, and produce what some might even consider art.

It wasn't until years later that I started to develop another form of literacy that made it possible to talk about these spaces that dot the urban environment as creative wastelands, alternative sites of social reproduction, or in this chapter, what I call "insurgent artscapes" that are designed from the bottom up.

A spatial taxonomy

Scholarly discussions about bottom-up urban design have tended to gloss over the issue of unsanctioned spaces used for the production of insurgent urban aesthetics. Part of the reason for this is that any discussion about fostering or even acknowledging impromptu and unplanned art spaces is an oxymoron. Nevertheless, I argue that the production of insurgent artscapes should be discussed as a necessary part of theorizing a free and civil society that thrives on variants of informality (Mukhija and Loukaitou-Sideris 2014) and the creative efforts of do-it-yourself place-makers (Iveson 2013; Douglas 2018; Heim LaFrombois 2018). There is a need for a critical discussion about the existence of unsanctioned, uncommissioned, and unfunded art, and how its producers actively occupy and transform public space.

To add to this crucial conversation that has long been going on among artists, activists, and intellectuals in cities around the world, I look back at one of Los Angeles's largest and most successful insurgent artscapes – the Belmont Tunnel. As I show, a brief history of the Belmont Tunnel and its users' demonstrated ability to inhabit such a space highlights Ignasi de Solà-Morales's (1995) concept of "terrain vague" and provides a place-based example of what urbanist Michael Rios (2010) calls "cultural insurgency."

A taxonomy of spaces conceptualized by urban scholars could be applied to Belmont and countless other similar insurgent artscapes: lost space (Trancik 1986), loose space (Franck and Stevens 2007), found space (Rivlin 2006), residual space (Villagomez 2010), interim space (Kamvasinou and Roberts 2013), and spaces of void (Rahmann and Jonas 2013), to name a few. But it is the concept of terrain vague that can best be applied to thinking about how urban artscapes may be found and assertively produced.

According to architectural theorist Ignasi de Solà-Morales (1995), "terrains vague" are those sites left over as underutilized remnants of capitalist development. A decidedly urban feature according to de Solà-Morales, terrains vague facilitate creative potential given their "free, available, unengaged" as well as "indeterminate, imprecise, blurred, uncertain" composition. It is within one of these developmental remainders left over at the periphery of downtown Los Angeles's densely built environment that Belmont existed as a globally recognized graffiti yard and unsanctioned park space. As de Solà-Morales (2013: 26) puts it, such "wastelands," as terrains vague are often called, "exist outside of the city's effective circuits and productive structures, [thereby projecting a] negative image, as much a critique as a possible alternative." Whereas de Solà-Morales's wanderings are largely theoretical, for many urban inhabitants terrain vague is a site of social reproduction and survival for which there is no alternative.

Given their openness and apparent abandonment, terrains vague can be discovered amidst both free-market and state-driven development, lending themselves to quotidian and creative occupation. The insurgent occupation and aesthetic engagement in such voids does not pose an existential challenge to the urban environment or its use values, so much as it provides marginalized groups with the opportunity to freely, even if fleetingly, express themselves.

Unsanctioned and insurgent art practices like graffiti and other forms of street art can visually fill those voids left behind, when property owners and developers keep tracts of land perpetually locked in their portfolios, but with no interest in developing or relinquishing the land to the city for use by a local community. Terrains vague, like voids, residual spaces, and other found and reconceptualized spaces, can be readapted, reimagined, and reused when our collective need for art trumps our dogmatic expectations for visual order. Such was true for Belmont.

A brief history of Belmont

The "Belmont Tunnel," which was also called the "Belmont Yard," or simply "Belmont," became a globally recognized destination for graffiti writers and artists by the early 1990s (Hornish 1993; Docuyanan 2000). The site, located at the confluence of Downtown and the Westlake District of Los Angeles, was a 2½-acre property owned by Pacific Electric for the better part of the 20th century (Figure 49.1). The area was surrounded on three sides by 450-foot-long-by-9-foot-high

Figure 49.1 Belmont site with Downtown Los Angeles buildings in the background.

Source: Photo by the author

concrete retainer walls, a 600-square-foot deco-style substation at the foot of a sloping chaparral-covered hillside, a 13-foot-high tunnel opening, and, at the center of the property, an open space known as the Toluca Yard that sat below street level. A lone scraggly tree sprouted up against one of the walls at the east side of the property, and a small billboard stood at the far north end of the property, at eye level with passing motorists on the Beverly Boulevard viaduct. Despite its size, Belmont was "hiding in plain *site*" (Kamel 2014: 126).

The site, or more accurately, the "industrial ruin" (Edensor 2005), was the terminus and rail yard for Los Angeles's first subway system, which opened in 1924. The "Hollywood Line" that passed through the Belmont Tunnel was designed to bring commuters from the congested streets of Downtown's Bunker Hill district to the burgeoning film studios and growing suburbs in Hollywood and Burbank. By the onset of World War II, the one-mile underground stretch of the extensive transit system was transporting over 60,000 commuters per day. But by 1955, the entire system was taken out of operation, leaving behind an infrastructure whose fate was sealed a little over a decade later, when the south portion of the tunnel was filled in to accommodate the massive redevelopment of Downtown's Bunker Hill district. As such, this terrain vague was both a product and eventual byproduct of urban expansion.

For the next two decades, the tunnel's interior served as a bomb shelter and storage facility for survival rations, including 329,700 pounds of crackers and thousands of drums filled with potable water (Meares 2013). By 1977, and in true LA fashion, the tunnel and its environs became the backdrop for filmmakers, the first film being *MacArthur* (1977), for which the tunnel entrance was dressed up to play the role of a Philippines outpost during the Battle of Bataan. By 1980, the tunnel was sealed with an iron gate (later torn down by graffiti writers) but continued to serve as a filming location for television shows, music videos, and movies looking to depict urban decay, danger, and dereliction with the use of background "gang graffiti."

But Belmont's active second life as an "art yard" began when local graffiti artist Shandu painted the first non-gang graffiti piece there. His piece stylistically replicated the so-called hip hop and "wildstyles" of wall writing that were being imported to Los Angeles from New York City on the glossy pages of the book *Subway Art* (Cooper and Chalfant 1984) as well as in the graffiti documentary *Style Wars* (Silver 1983). As the graffiti scene in LA exploded over the next few years, Belmont, like the city's buses, telephone poles, and freeway walls, would be covered by graffiti. When the movie *Colors* (1988) was filmed, production designers used Belmont's interior walls and existing graffiti as a backdrop for the film's fictitious 21st Street Gang's hangout.

By that point in time there was not a weekend, and hardly a weekday, when Belmont was not filled with dozens of graffiti artists painting its walls and socializing (Figure 49.2). As per existing "yard rules," once someone completed their piece and took a picture of it, it could be covered by another "piecer," or graffiti artist, waiting with spray cans in hand to take their turn. Still, sometimes a piece would last for weeks out of respect for the artist or given the quality or sheer size of the production. The apparent anarchy of the space was anything but anarchic. By learning from what MacGillivray and Curwen (2007: 354) call "tagging as an act of social literacy," I personally received some of my first and most powerful lessons in respect, decorum, and intimidation in Belmont.

Figure 49.2 Graffiti artists at the Belmont Yard, beneath the Beverly Boulevard viaduct.

Source: Photo by the author

By the mid-1990s, Belmont's walls were inches thick with layers of dried aerosol paint that revealed a colorful stratigraphy when peeled away or chipped off. In later years, the layers were added to by street artists, students, and weekend painters, who used paintbrushes on the smoothed walls to the bemusement of long-time graffiti artists.

In addition to lone artists practicing their styles on the Belmont walls, graffiti crews held their meetings in the space and sometimes orchestrated "battles" between two or more crews, who would compete to out-paint the other for the sake of bragging rights. In 1993, I was party to a large-scale production and memorial to the graffiti writer Skate One from the CBS (City Bomb Squad) crew who had died after being struck by a train in the San Fernando Valley while painting in an active train yard. Three graffiti crews came together to paint the production that covered a third of the yard's wall space, taking two days and over 30 artists to complete.

By sundown on both days, cars were pulled into the yard via an access ramp so that their headlights could provide some illumination for night painting. As we painted, local gang members and neighborhood kids sat along the tops of the south-facing walls watching, while women from the neighborhood walked down to sell us homemade food and bottled water.

No matter how many people filled the yard, no matter the outbreak of sporadic fights and gambling, regular drinking, and recreational drug use by some, and despite being trespassers for entering the site and vandals for marking its walls, police rarely disturbed the activities at Belmont. Occasionally a patrol car would stop on one of the side streets above the lip of the yard, and an officer would issue a dispersal order over the car's sound system. The cruiser would then pull away as some of the occupants would begin to exit by walking up the switchbacks that had been carved into the hillside over the years.[1]

As Belmont became an internationally recognized subcultural artscape during the 1990s, it was also becoming the weekend home to *pelota tarasca* players from across the Southern California region – from Tijuana to the south, to Santa Barbara County to the north, and as far inland as the eastern portions of Riverside and San Bernardino Counties (Figure 49.3). The Toluca Yard, which closely approximated the 300-foot-long-by-40-foot-wide playing field for the pre-Columbian ball game, became one of the only playing locations in the United States for the sport-cum-ceremonial competition that was said to provide a spiritual portal to the underworld in both Aztec and Mayan mysticism.

Players, mostly Mexican immigrants from the states of Michoacán and Guerrero, in addition to a few Salvadoran and Guatemalan immigrants, would volley a small rubber ball (usually a shaved tennis ball) across the pitch in an effort to eliminate players, who had missed their turn at striking it with their bare hand. Just like at an ancient *tarasca* match, onlookers at Belmont would line the playing area watching the competition, socializing, playing live music, gambling, and selling food. Combined with the graffiti artists at the walls, Belmont would regularly host over 100 park users on any given Sunday, leaving behind hundreds of empty aerosol cans and plastic bags that had been filled with *papusas* and *curtido de repollo*. As Los Angeles was presenting itself to the world as an established "global city," complete with mounting socio-spatial fragmentation, Belmont provided users with a space to preserve some of the region's connection to its pre-Anglo cultural practices and aesthetics. Part of what makes Belmont an *insurgent* artscape was its users' implicit as well as explicit rejection of what Los Angeles was becoming: a destination for the well-heeled at the expense of its existing working-class communities of color.

Indeed, new-build and residential gentrification was making its way towards Belmont from all directions when, in 2003, the Meta Housing Corporation proposed to build a 276-unit luxury housing complex at the location. Echo Park, the district located just over a mile north of Belmont, experienced a $4,000 rise in median household income as the city's median household

Figure 49.3 *Pelota tarasca* players at Belmont Yard with graffiti writers painting the far wall.

Source: Photo by the author

income dropped by $2,000, just as its white, non-Hispanic population increased to 29%, while the city's overall white, non-Hispanic population dropped to 28% according to US Census data. Just over a mile south of Belmont, in Downtown's South Park district, the Staples Center, which was completed in 1999, was being joined by a bevy of boutique and five-star hotels in addition to the Grammy Museum and Microsoft Theater. They were part of the "LA Live" entertainment complex that broke ground just as the Belmont Yard was being gutted, and existing housing adjacent to the Toluca Yard was being demolished.

For local residents who lived in the neighborhood, the proposed redevelopment was a sure sign that gentrification in Echo Park to the north and in Downtown to the south had finally crept into the long-neglected working-class community of Westlake (Figure 49.4). As Brian Knight (2006: 13) puts it in his brief but brilliant architectural essay on Belmont, the location long existed "in an urban hinterland, and as such it has been immune from the homogenizing effects of mass marketing, advertising, and speculative real estate development." Whereas Belmont had been produced from the bottom up through illicit and insurgent occupation, it had also been preserved from top-down neglect.

Activists, organized in part by local street art enthusiast Stash Maleski, called on the city to force Meta Corp. to file a Mitigated Negative Declaration and full environmental impact report before approving development of the site. In the meantime, Maleski and others formed what they called "Belmont Art Park United," in an effort to get the city to recognize the historical significance of the Belmont Yard. The very use of the word "park" in their title was an attempt to

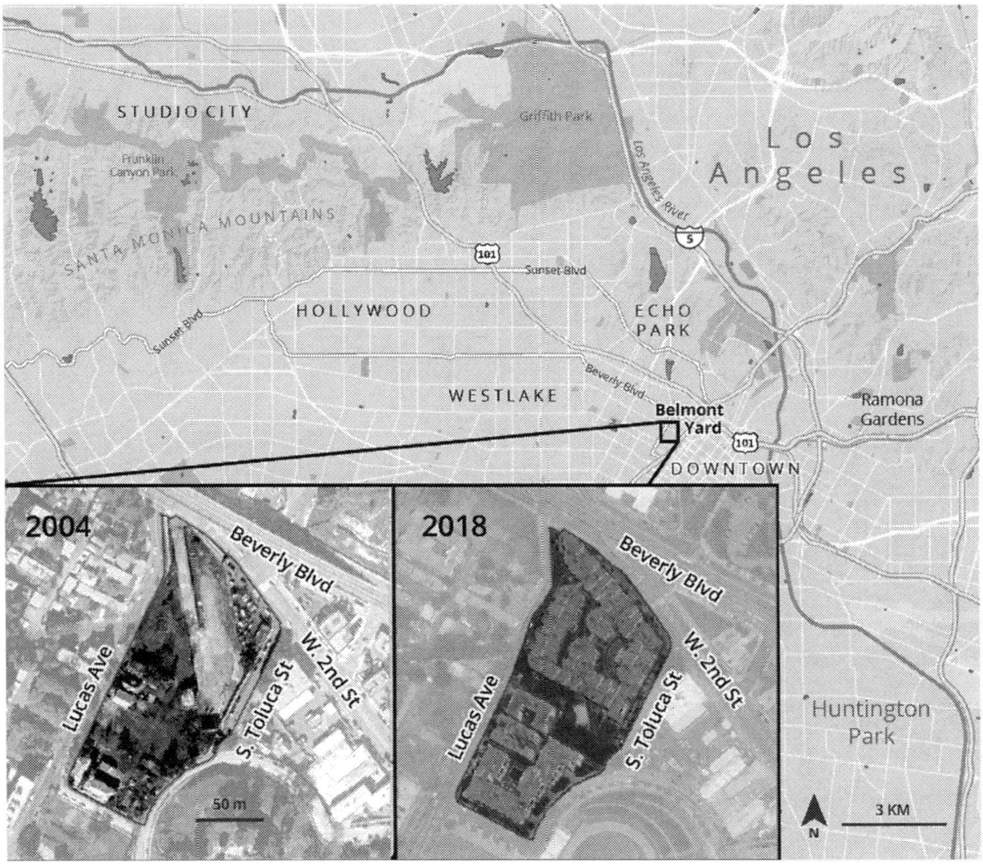

Figure 49.4 Aerial views of Belmont and the Toluca Yard with surrounding residential structures just before redevelopment and today.

Source: Map by Cyrus Maden and the author, 2018

make the illicit artscape and informal "yard" legible, legal, and perhaps worth saving, to those in power. But when the group stood before the Los Angeles City Planning and Land Use Commission asking that historical preservation and landmark status be granted to the site, the commission chaired by Councilmember Ed Reyes dismissed the motion.

In a last-ditch effort to preserve the site, activists made an appeal to the Los Angeles Cultural Heritage Commission, citing the yard's cultural and historical significance as an incubator for graffiti and street art, which had become ubiquitous and globally recognized art forms and practices, thanks in part to Belmont's existence. But Commission President Mary Klaus-Martin said that although graffiti was a recognized art form, the murals around the tunnel were neither old nor unique enough to justify a historic designation on their own (Hernandez 2004a). She then showed the packed hearing room her personal copy of *Spray Can Art*, but added that "an art park would not serve the true interests of those who frequent the site" (Hernandez 2004b), adding "[graffiti writers] like it to be dangerous, so if you designate it an art park and made it legal and validate it, would it become important to them? I don't know. It takes off the edge." This is a valid point that I return to in the next section.

In 2004, Meta Corp. went ahead with the construction of the site, preserving the tunnel entrance and Toluca substation at the center of the large complex at the behest of the Los Angeles Cultural Heritage Commission, which did end up naming part of the site a Historic-Cultural Monument. However, the status only recognizes Belmont's physical infrastructure given its importance to LA's transit history. The permanently sealed tunnel entrance now boasts a mural depicting a Pacific Electric red car, and the fully restored façade of the substation was made free of graffiti for the first time in over three decades.

During Belmont's last days, artists organized an unsanctioned mass painting of the walls as *tarasca* players played one last game. Police watched from the street above but did not move in to secure the space until ground for the newly named "Belmont Station Apartments" had been officially broken. Although this "empty lot," in one of Los Angeles's poorest and predominately Latino neighborhoods, has since been razed and redeveloped, we can still learn from its inception, existence, and contested destruction.

Asserting a space for art

Based on casual observation in addition to the numerous popular books (as well as scholarly articles) published globally on the subject, it is clear that uncommissioned art can be an enduring visual component of the urban environment (e.g. Gastman et al. 2007; McCormick et al. 2010; Schacter 2013). But for publicly produced art to remain in place, it must be granted permission, which puts layers of bureaucracy and patronage in between artists and their creation.

But sanctioned public art, through its funding and granting paradigm, becomes an official urban aesthetic that Sarah Schrank (2009: 4) describes as representing the close relationship between visual arts and capital investment. Consequently, artists looking to paint in public places need only pay lip service to the tenets of community and collaboration to appease public art supporters, who are often looking to promote a sense of place in the interest of selling real estate. Increasingly, sanctioned graffiti and street art also face this existential conundrum, becoming, as Kim Dovey et al. (2012: 39) write, "integral to urban character in places where it helps to construct the legitimation of transgression, yet it cannot do this without also, and at the same time, becoming a form of symbolic capital and place branding." Therefore, artscapes that are not insurgent – i.e. unruly, uncommissioned, and unsanctioned – may more easily be subsumed into the neoliberal landscape.

Since Belmont's closing in 2004, street art has experienced a meteoric rise in global popularity. For example, city officials and scholars in Australia have been at the forefront of the debate about the legitimacy, or even possibility, of free and open art spaces. Criminologist Alison Young (2010) has written about the possibility of there being tacitly tolerated urban art space, where graffiti writers, street artists, and others who wish to create public art will not be criminalized in designated "zones of tolerance." In response, Halsey and Pederick (2010: 97) ask that municipalities consider graffiti and street art "neither publicly sanctioned art nor crime," whereas McAuliffe and Iveson (2011) question if space can even be officially given over to unprofitable art production within the neoliberal city, where the distinction between private sponsorship and public stewardship is blurred beyond recognition.

I have addressed similar issues in my own work (Bloch 2012, 2016a, 2016b), arguing that securing legal spaces for graffiti and street art is tantamount to the confinement of alternative aesthetics. Further, the existence of "legal spots" force graffiti and street artists to be beholden to local government, developers, and property owners, thereby painting only at the pleasure of those in power. Put more bluntly, Dovey et al. (2012: 40) argue that: "nothing will kill graffiti more effectively than promotion and preservation. Graffiti cannot be fully defined or preserved without becoming purified and killed; a quality it shares with urban character and place identity." While supporters of graffiti and street art may have the best intentions in advocating for legal spaces and decriminalization, such advocacy can unwittingly border on enervating romanticization and hastening unsanctioned art's slow erasure.

While sanctioning subcultural space for graffiti and street artists may not be possible given the iconoclastic motivations that both groups possess, perhaps the preservation of art spaces for the general public, whose members do not consider themselves members of a rather parochial community of graffiti and street artists, is in fact feasible. While graffiti writers and street artists continue to assert themselves by reaching out and taking space by force, there remains the need for artscapes produced from the bottom up, which can be found within many of the city's endless supply of terrains vague.

Insurgency

In their discussion of the creative occupation of interim spaces, Kamvasinou and Roberts (2013) advocate for a "light touch," or bottom-up reclamation of underutilized spaces and a gradual shift in thinking about community values when developing neighborhoods. But it may be the more overtly transgressive and insurgent occupation of spaces like Belmont that is necessary for lasting adoption by members of urban art subcultures and larger communities alike.

As Michael Rios (2009: 99) writes, cultural insurgency is the assertive making of place identity, which includes "explicit counter-identities in the urban landscape." He argues that "what makes public space distinctly Latino [and therefore insurgent in a context of institutional white supremacy] are the social and cultural norms that manifest in patterns of use, spatial practice, and material characteristics" (Rios 2010: 101). But new social and cultural norms are hard fought for and won through acts and displays of cultural insurgency. Viewed through a cultural insurgency frame, Belmont provided locals and visitors alike with a space that was decidedly adaptive as used by disparate social groups – graffiti artists, gang members, players of *pelota tarasca*, food vendors, and drifters of various races, classes, gender expressions, sexual orientations, and abilities – but was no less insurgent given its very existence as a space in which to make claims to new kinds of citizenship and belonging.

Insurgency in the form of creative, contentious, and continuous adaptation was key to the production of Belmont, which was maintained as such through near consensus as an artscape. Insurgent occupation and continuous visual adaptation of artscapes are, however, the very type of activities that the stewards of sanctioned "public art" spaces tend to discourage, if not outright forbid.

Conclusion

Insurgent artscapes like the Belmont Yard can, as architect Brian Knight (2006: 14) puts it, be viewed as "laboratories for a new idea of urbanity that can offer an intense, vital experience of the city, [and] as a foil against the standardized and hyper-planned public spaces that increasingly inhabit most of our cities." And Belmont was not alone as a model for an insurgent artscape made by and for the people.

As I write about in the introduction to this chapter, and as Susan Phillips (1999) points out in her rich ethnographic history of graffiti in Los Angeles, writers and artists have maintained several yards around the city over the years. Each of these brightly colored and often cacophonous and semi-hidden spaces is set within a terrain vague, free from the watchful and destructive eyes of authority looking to enforce a moral geographic order.

In addition to the more illicit yards, whose location I will not reveal here, there still exists a lone 100-feet-long-by-6-feet high "free art wall" in Venice Beach, California – the last human-made structure before the land meets the sea. The same activists who could not save Belmont as an "art park" were able to save this wall upon the destruction of a larger "Venice Pit" to which it belonged. However, the institutional stewardship, which includes support from muralist Judith Baca's nonprofit Social and Public Art Resource Center (SPARC), casts a perhaps counterintuitive pall of legitimacy over the protected wall for many graffiti writers and street artists, relegating the space to the status of a legal skate ramp like the one that sits only feet away from it.

Over the years, I have spent time in many of LA's terrains vague-turned-insurgent artscapes, as a member of an illicit subculture and more recently as a still-iconoclastic scholar. I am therefore left ambivalent about advocating for the production of insurgent artscapes, which I think should simply exist without top-down planning and protection. However, free societies can do best to preserve such spaces, indirectly by rebuking premeditated crackdowns, coordinated expulsions, and mass criminalization of the people who do assert themselves spatially.

While insurgent artscapes and sovereign urban spaces cannot be scripted, they can be anticipated and implicitly planned for. Place-makers could be part of the implicit planning and design process by working closely with local government and law enforcement to advocate that state actors resist the impulse to aggressively instill order for incoming residents and ascendant community members.

Acknowledgment

Thank you to Susan Philips for her editing prowess, tireless advocacy for insurgent artscapes, pioneering scholarship on graffiti in Los Angeles, and friendship. And thank you to Gabe 88 for documenting the last days of Belmont.

Note

1 In 2004, while a fight was being waged to save the site, I was stopped, frisked, and questioned by police while in the yard. As I explained to the officer, I was actually doing fieldwork for a class on informal placemaking and park usage taught by Professor Anastasia Loukaitou-Sideris at UCLA's Department of Urban Planning, which was partially true. He told me he had also gone to UCLA, having majored in Chicano/a Studies, and he let me go, ignoring the fresh paint on my fingers.

References

Bloch, S. (2012). "The Illegal Face of Wall Space: Graffiti-murals on the Sunset Boulevard Retaining Walls." *Radical History Review*, 113: 111–126.

Bloch, S. (2016a). "Why Do Graffiti Writers Write on Murals? The Birth, Life, and Slow Death of Freeway Murals in Los Angeles." *International Journal of Urban and Regional Research*, 40 (2): 451–471.

Bloch, S. (2016b). "Challenging the Defense of Graffiti, in Defense of Graffiti." In Ross, J.I. (ed.) *Routledge Handbook of Graffiti and Street Art*. New York: Routledge.

Cooper, M. and Chalfant, H. (1984). *Subway Art*. New York: Macmillan.

de Solà-Morales, I. (1995). "Terrain Vague." In Davidson, C.C. (ed.) *Anyplace*. Boston: MIT Press.

de Solà-Morales, I. (2013). "Terrain Vague." In Barron, P. and Mariani, M. (eds.) *Terrain Vague: Interstices at the Edge of the Pale*. London: Routledge.

Docuyanan, F. (2000). "Governing Graffiti in Contested Urban Spaces." *PoLAR: Political and Legal Anthropology Review*, 23 (1): 103–121.

Douglas, G.C.C. (2018). *The Help-Yourself City: Legitimacy and Inequality in DIY Urbanism.* Oxford: Oxford University Press.

Dovey, K., Wollan, S. and Woodcock, I. (2012). "Placing Graffiti: Creating and Contesting Character in Inner-city Melbourne." *Journal of Urban Design*, 17 (1): 21–41.

Edensor, T. (2005). *Industrial Ruins: Space, Aesthetics and Materiality.* London: Berg Publishers.

Franck, K. and Stevens, Q. (eds.) (2007). *Loose Space: Possibility and Diversity in Urban Life.* London: Routledge.

Gastman, R., Neelon, C. and Smyrski, A. (2007). *Street World: Urban Culture from Five Continents.* London: Thames & Hudson.

Halsey, M. and Pederick, B. (2010). "The Game of Fame: Mural, Graffiti, Erasure." *City*, 14 (1–2): 82–98.

Heim LaFrombois, M.E. (2018). *Reframing the Reclaiming of Urban Space: A Feminist Exploration into Do-It-Yourself Urbanism in Chicago.* New York: Lexington Books.

Hernandez, D. (2004a). "Old Tunnel May Be Tagged as a Landmark." *Los Angeles Times*, September 15.

Hernandez, D. (2004b). "Graffiti Art Park Brushed Aside." *Los Angeles Times*, September 16.

Hornish, M.E. (1993). *Belmont Tunnel: Ethnography of an Empty Lot.* Master's thesis, Department of Anthropology, University of Southern California.

Iveson, K. (2013). "Cities Within the City: Do-It-Yourself Urbanism and the Right to the City." *International Journal of Urban and Regional Research*, 37 (3): 941–956.

Kamel, N. (2014). "Learning from the Margins: Place Making Tactics." In Mukhija, V. and Loukaitou-Sideris, A. (eds.) *The Informal American City: Beyond Taco Trucks and Day Labor.* Boston: MIT Press.

Kamvasinou, K. and Roberts, M. (2013). "Interim Spaces: Vacant Land, Creativity, and Innovation in the Context of Urbanity." In Barron, P. and Mariani, M. (eds.) *Terrain Vague: Interstices at the Edge of the Pale.* London: Routledge.

Knight, B. (2006). "The Belmont Tunnel and Toluca Yard." In Cramer, J.P. and Yankopolos, J.E. (eds.) *Almanac of Architecture & Design.* N.A.: Greenway Communications.

MacGillivray, L. and Curwen, M.S. (2007). "Tagging as a Social Literacy Practice." *Journal of Adolescent & Adult Literacy*, 50 (5): 354–369.

McAuliffe, C. and Iveson, K. (2011). "Art and Crime (and Other Things Besides . . .): Conceptualising Graffiti in the City." *Geography Compass*, 5 (3): 128–143.

McCormick, C., Schiller, M. and Schiller, S., (eds.) (2010). *Trespass: A History of Uncommissioned Urban Art.* New York: Taschen.

Meares, H. (2013). "Stunted Progress: Belmont Tunnel and the Repurposing of a Faded Los Angeles Dream." *KCET.* (www.kcet.org/history-society/stunted-progress-belmont-tunnel-and-the-repurposing-of-a-faded-los-angeles-dream, last accessed May 21, 2018).

Mukhija, V. and Loukaitou-Sideris, A., (eds.) (2014). *The Informal American City: Beyond Taco Trucks and Day Labor.* Boston: MIT Press.

Phillips, S.A. (1999). *Wallbangin': Graffiti and Gangs in LA.* Chicago: University of Chicago Press.

Rahmann, H. and Jonas, M. (2013). "Void Potential: Social Dynamics and Cultural Manifestations of Residual Space." In Barron, P. and Mariani, M. (eds.) *Terrain Vague: Interstices at the Edge of the Pale.* London: Routledge.

Rios, M. (2009). "Public Space Praxis: Cultural Capacity and Political Efficacy in Latina/o Placemaking." *Berkeley Planning Journal*, 22 (1): 92–112.

Rios, M. (2010). "Claiming Latino Space: Cultural Insurgency in the Public Realm." In Hou, J. (ed.) *Insurgent Public Space: Guerilla Urbanism and the Remaking of Contemporary Cities.* New York: Routledge.

Rivlin, L.G. (2006). "Found spaces." In Franck, K. and Stevens, Q. (ed.) *Loose Space: Possibility and Diversity in Urban Life.* London: Routledge.

Schacter, R. (2013). *The World Atlas of Street Art and Graffiti.* New Haven: Yale University Press.

Schrank, S. (2009). *Art and the City: Civic Imagination and Cultural Authority in Los Angeles.* Philadelphia: University of Pennsylvania Press.

Silver, T. dir. (1983). *Style Wars.* Public Art Films.

Trancik, R. (1986). *Finding Lost Space: Theories of Urban Design.* New York: John Wiley & Sons.

Villagomez, E. (2010). "Claiming Residual Spaces in the Heterogeneous City." In Hou, J. (ed.) *Insurgent Public Space: Guerilla Urbanism and the Remaking of Contemporary Cities.* New York: Routledge.

Young, A. (2010). "Negotiated Consent or Zero Tolerance?" *City*, 14 (1–2): 99–114.

Further reading

Barron, P. and Mariani, M. (eds.) (2013). *Terrain Vague: Interstices at the Edge of the Pale.* New York: Routledge. The chapters in this collection speak to the notion and possibilities inherent in urban ruins in a variety of geographical contexts.

Ferrell, J. (2002). *Tearing Down the Streets: Adventures in Urban Anarchy.* New York: Palgrave MacMillan. In this book critical cultural criminologist Jeff Farrell presents a theory-driven and empirically-rich discussion about contestative urban practices and subcultural form of expression in place-making.

Franck, K. and Stevens, Q. (2007). *Loose Space: Possibility and Diversity in Urban Life.* New York: Routledge. Geographers, architects, and urbanists contribute to this thoughtful collection of chapters that discuss the many incarnations and uses for the city's "loose" spaces.

Hou, J. (ed.) (2010). *Insurgent Public Space: Guerrilla Urbanism and the Remaking of Contemporary Cities.* New York: Routledge. This collection of chapters by urbanists addresses the transgressive, insurgent, creative, and sometimes playful reappropriation of public space from a variety of scholarly and practice-based perspectives.

Phillips, S.A. (1999). *Wallbangin': Graffiti and Gangs in LA.* Chicago: University of Chicago Press. This book provides an extensive and authoritative ethnography of the LA graffiti scene of the 1990s, of which the Belmont Yard is a social and geographic centerpiece.

50

URBAN DESIGN AND PUBLIC ART ON INSTAGRAM

Brettany Shannon

Introduction

In December 2017, *CityLab* staff published the article, "Your Entire City Is an Instagram Playground Now." In it, the authors describe the move by boosters to place sculptural installations, "Instagram-ready social tableaus," in heavily trafficked public plazas for city branding and tourism. They note that Amsterdam's tourism increased by 40% since the 2004 installation of the giant letter sculpture spelling "I Amsterdam" and how a similar sign in Toronto has over 120 million social media impressions. *CityLab*'s staff grant that no causal relationship between the signage and tourism can be made, but argue the engagement at least suggests people enjoy using social media to capture public art and urban design interventions. As people congregate around central city plaza sculptures, a parallel practice emerges when they capture and post images of urban spaces and public art installations on social networking sites (SNSs) like Instagram. The *CityLab* article might have a point after all.

From the iconic Hollywood sign in Los Angeles's Hollywood Hills to Robert Indiana's *LOVE* sculpture in Philadelphia's John F. Kennedy Plaza, art objects provoke people to photograph them for their unique aesthetic and sensory attributes. One way to distinguish urban design and public art is to regard the former as the features constituting a city's built form, at the street, block, district, metropolitan, and regional scales (Banerjee and Loukaitou-Sideris 2011), and the latter as art of any medium, genre, size, and duration sited in public space for the purpose of reaching the public (Miles 1997).

But these definitions are not mutually exclusive, and differences blur further when considering how much they have in common. Urban design and public art interventions can be permanent or temporary; may utilize similar objects and materials, and can serve various aims that shift over time, such as placemaking (Lynch 1960), identity- and community-making (Jacobs and Appleyard 1987), and capital investment (Miles 1997; Zukin 2010). Los Angeles's early community builders never dreamed that their temporary "Hollywoodland" billboard would later become the global icon for aspirational Southern California's sunshine, promise, and star power. Meanwhile, other versions of Indiana's *LOVE* sculpture grace corporate headquarter plazas around the country, every bit as handsome as the original but with less social legitimacy, derided as modern "plop art" installations legitimizing late capitalism and obscuring social inequality (Miles 1997). Still, people regard urban design and public art interventions as useful for improving the urban realm's overall appearance and tend to categorize public art as distinct – a subset of urban design in the urban regeneration process (Remesar 2016) or a complementary urban intervention (Radice and Boudreault-Fournier 2017).

Historically, research into how city residents experience urban design and public art has been ethnographic (Lynch 1960; Lynch and Rivkin 1995; Whyte 1980) and case study-based (Giles-Corti, Bull, Knuiman, McCormack, et al. 2013; Hou 2010). Today, SNSs expand opportunities for studying user experiences of urban form and its design, including public art installations. Through the photo-sharing SNS Instagram, for example, we can now conduct large sample-size analyses from its global userbase's data-rich posts to learn how people relate with their cities' built forms, determine if and how they distinguish between urban design and public art, and draw possible lessons for urban design practice.

Instagram and all other SNSs using photography and geo-location technologies underscore the process of globalization that Tomlinson (2007: 94) calls "telemediatization," the "increasing implication of electronic communications and media systems in the constitution of everyday experience." With the emergence of information and communication technologies – particularly smartphones – previously intimate, place-based, and often fleeting interactions with urban design and public art quickly became global, broadly accessible digital artifacts. The meditative wooden bridge in Singapore, the bustling plaza in Guadalajara, the political Banksy mural at the West Bank, and a single night's digital performance in New Delhi now exist anytime, anywhere, for anyone with an Internet connection. In the age of smartphones, anywhere literally means *anywhere*, and of the estimated 2.32 billion smartphone owners in the world (Statista n.d.a), 800 million use Instagram (Aslam 2018).

Instagram's amplification of the cultural significance of telemediatization exemplifies the contemporary urban phenomena of people dedicating their smartphone-enabled SNS engagement towards posting imagery of the urban design and public art they see in their cities. This chapter presents a content and compositional analysis of 800 unique posts gathered from the SNS Instagram bearing the hashtags #urbandesign, #diseñourbano, #publicart, and #artepublico[1] to learn how Instagrammers frame, curate, and share images of urban spaces, and whether and how their posts indicate a perceived difference between urban design and public art. Based on the collected data, I find that Instagram is a rich, if limited, source for studying how people perceive their urban spaces. Analysis of Instagrammers' posted imagery, additional hashtags, and accompanying captions shows that people do see urban design and public art as distinctive; underscores people's desire to engage with the public realm; and suggests Instagram's possible usefulness for urban design research.

Instagram for exploring and making sense of the city

Since 2010, Instagram has become the world's third-largest SNS after Facebook and YouTube (Kallas 2018) – 32% of all Internet users have Instagram, and 80% of Instagrammers live outside the United States (Aslam 2018). Within the United States, Instagrammers constitute 35% of the adult population (Pew Research Center 2018). Sheldon and Bryant (2015) identify Instagram-specific motives as: "surveillance/knowledge about others," "documentation," "coolness," and "creativity," as well as a positive relationship between social activity (e.g., as exploring one's city) and one's increased likelihood to use Instagram.

Urban Instagramming accords with these tendencies. Instagrammers "document" idealized (or "cool" and "creative") imagery of their urban spaces (Boy and Uitermark 2016). MacDowall and de Souza (2018) find that Instagram has a special place in public art's informal street art category because artists can show their hyperlocal projects – artworks otherwise oriented around small geographic scales – to global online community groups. They explain, "Instagram is fundamentally reshaping the practices, aesthetics, and consumption of graffiti and street art" (5), as well as the broader shift of how we experience and interact with urban spaces. Users shoot and share their images *in situ* and in real time, making Instagramming practice a phenomenological one.

Hochman and Manovich (2013) reason that Instagram is popular because its photographs "resonate with more personal, 'authentic' experiences" (para 18). Whereas Google Earth's and Google Street View's "pictorial logics" (para 19) operate in "an objective, elevated, and fixed form," Instagram's logic "present[s] spontaneous and highly personal sentiments that inherently reject the technological pursuit of fine details and accuracy of a 'mechanical' (now digital) eye" (para 19). Google's applications may offer us "objective" data, but their mechanized functionality cannot reflect the nuances of a city's lived street life. Instagram, by contrast, allows people to highlight the urban places and moments *they* see and value, insomuch as they have chosen to share *those* places and moments and not the city's many alternatives.

Instagram and other SNSs are the digital culture's expression of the generations of literature asserting the value of urban exploration, namely, the dynamic, productive act of walking in the city. Baudelaire (1972) introduced us to the *flâneur*, the aloof dandy whose privilege afforded him the opportunity to stroll modern Paris's streets without being subjected to its modern troubles. Benjamin's (1999) *flâneur* was bourgeois and dispassionate like Baudelaire's, but now also the urban spectator, amassing urban observations for transcription into essays about interwar Paris. The Situationist International (1957–1972; Debord 1958) art collective undertook their own psychogeographic *dérives* (Wark 2011), semi-aimless explorations through Paris's urban landscape. Then allied with Lefebvre, the Situationists upheld *dérives* as radical acts of "everyday life" (Lefebvre 1947) for rebelling against contemporary urbanism' s capitalist foundations (Goonewardena 2011). These walks prefigured Lynch's (1960, 1995) and others' (see Appleyard, Lynch, and Myer 1965; Jacobs 1984; Whyte 1980) heuristic mental mapping methodology for recording the sensory experience of the city. De Certeau (1984) argued that urban walks transcend simple observation or "consumption." Rather, walking the city is *writing* the city, an expressly productive act whereby we author our cities just by living in them.

Charles Marville's earliest Paris photographs and the latest Instagram post represent explorations in the aforementioned ways, with a critical exception. Most pre-SNS vernacular photography exists in personal albums for small, familiar audiences. Even Krieger's (2012) Google Street View-based "Re-photographing and Re-viewing Charles Marville's Paris of 1865–1877" project webpage is a "Page Not Found."[2] By contrast, Instagrammers' posts become instantly global, viewable, and shareable by countless friends and strangers.

In addition to supporting the understanding, authoring, and potentially reclaiming our cities, Instagram provides us with empirical units of analysis for studying how people perceive the public realm. For Lynch (1960), the most successful urban design and public art interventions help give even the biggest city a "visible, coherent, and clear" (91) "sensuous form" (119), and "[a] highly developed art of urban design is linked to the creation of a critical and attentive audience" (120). Instagrammers exemplify the "attentive audience": their posts can be broken down into component parts to distinguish "imageability" and the various "qualit[ies]" that give an environment "a high probability of evoking a strong image in any given observer" (1960: 9). "Strong image" works metaphorically and literally here – metaphorically in that urban design and public art resonates enough with an Instagrammer to post it, and literally in that the resultant image is a genuine "strong image" (Figure 50.1).

Instagrammers do not design the places they photograph, but the process of composing, capturing, and sharing images of those places reveals their connection with place, especially urban design and public art. Further, their posts provide us with data points useful for exploring aspects of those connections. With this in mind, the following research project investigates SNS-based "strong image" examples of urban *flânerie* photography, and what lessons they might offer urban design.

Figure 50.1 Instagram pictures imply Lynch's "strong image" thesis and accompanying text generally confirms it. (Left) Like many Instagrammers, this user employs words and emojis to communicate the park's qualities. "☺#amsterdam #holland #weekend #park #landscape #urban #urbandesign #relax #nature #autumn" (Right) and like many others, this Instagrammer accompanies her image with language intimating her approval of public art, allegiance to a neighborhood, and commitment to amplifying the work of a local art and artists. "@cuttyup #jaimemolina for @powwowlongbeach and @thinkspace_art more work on cuttyup.com #cuttyup #powwowlongbeach #streetart #publicart #streetartla #longbeach #longbeachstreetart #powwow #mural #thinkspaceart".

Sources: Left image, Riccardo Celegato, University of Padova (@riccardojstark); right image, Anna Joaquin, California State University, Northridge (@drjoaquinbakes)

Mining Instagram

This chapter discusses a constant-comparative content (Strauss and Corbin 1990) and compositional analysis (Rose 2012) of 800 (200 each) Instagram posts with four different hashtags: #urbandesign, #diseñourbano, #publicart, and #artepublico. Hashtags are words or phrases preceded by # to identify, organize, and search particular topics. The inclusion of #diseñourbano and #artepublico helped escape a purely Anglophone-centric analysis and allowed a look at how people in other parts of the world capture and share examples of urban design and public art. Instagrammers wrote in English, in Spanish, sometimes in both, and occasionally in English and a third language,[3] and posted imagery from all six inhabited continents. In addition to imagery analysis, I analyzed accompanying captions and additional hashtags (i.e., those beyond the sample's qualifying #urbandesign, #publicart, etc., hashtags). The paid Instagram search engine and analytics service Picodash[4] provided the metadata from all hashtagged posts (e.g., dates, captions, comments, likes,[5] geolocation, image URLs, etc.), but not images; therefore, image collection was done by hand.

From September 1, 2016, to September 1, 2017, Instagram users posted 70,155 #urbandesign, 3,044 #diseñourbano, 363,586 #publicart, and #3,838 #artepublico images. To account for the differences between the hashtag sample sizes, I systematically sampled every third #diseñourbano and #artepublico post, and every 69th #urbandesign and every 284th #publicart image to collect 250 examples of each. While collecting, I eliminated interiors, single buildings, and fashion and music imagery. After disqualifying an image, I moved to the next possible image. I kept urban design drawings, renderings, and models because they represent idealized urban design visions. At the end of collection, I pre-coded the first 50 of each for category creation and internal reliability and analyzed the remaining sets of 200. Finally, I collapsed the #urbandesign and #diseñourbano hashtags, and the #publicart and #artepublico hashtags into two categories, #urbandesign and #publicart, comprising 400 observations each.

The project's analysis relied on combining two methods, visual studies' compositional interpretation[6] (Rose 2012) and social science's constant-comparative method[7] (Strauss and Corbin 1990). I conducted a compositional interpretation, or "visual connoisseurship" (Rose 2012: 52), analysis of the Instagram images, identifying the component elements of images and their additional hashtags and captions for later content analysis. By following the constant-comparative method and coding for urban design and public art elements, I uncovered relevant themes and categories. Table 50.1 shows the project's compositional interpretation that includes each

Table 50.1 Image analysis: categories, variables, and definitions coded for image analysis

Category	*Variable*[8]	*Definition*
Object	• Bridges • Buildings • Murals • Nature/landscape • Plazas, public space • Renderings, models • Sculptures • Skylines • Street furniture • Streets and stairs	The primary object within the image
Location (physical placement)	• In district/corridor • In plaza • In street • On building, wall • On ground • On/along water	Where the Object is in the urban fabric
Size	• Large, monumental • Small, intervention • N/A (rendering, model)	Size of Object
Lynchian element (imageability)	• Edge • District • Landmark • Node • Path	The Lynchian element of the city (Lynch 1960) that best describes the Object
Formality	• Impermanent, ephemeral (e.g., performance) • Permanent (formal) • Semi-permanent (e.g., mural) • N/A (rendering, model)	The permanence and durability of Object
Interactivity	• Yes • No	Whether the Object is something with which people can and do physically interact
Expressive content (affect, atmosphere, mood)	• Comical, parodic (e.g., cartoonish, goofy mural) • Grand, imposing (e.g., large plaza) • Insurgent, political (e.g., political performance art) • Menacing, somber (e.g., darkly hued mural with suggested or explicit violence) • Sweet, hopeful (e.g., brightly colored, happy mural) • Serene (e.g., tree-lined plaza, skyline at dawn) • Surreal (e.g., phantasmagoric mural) • Workaday (e.g., bustling commercial street)	The "combined effect of subject matter and visual form" (Taylor 1957: 43–44) Takes the component parts of an image together and reads their combined affective quality

image's primary object, that object's location, size, formality, and interactivity. I also coded for corresponding Lynchian elements (e.g., path, edge, district, node, or landmark; Lynch 1960) as they provide the analytical units for the sensory experience of the city.

Following Rose's *Visual Methodologies: An Introduction to Researching with Visual Materials* (2012), I also coded for each image's expressive content. Referring to an image's overall mood or affect, expressive content is the "combined effect of subject matter and visual form" (Taylor 1957: 43–44). Following Berger's (1972) theory that our "ways of seeing" are socially constructed and Barthes's (1977) "death of the author" thesis that texts have no links to their creators, contemporary visual analysis upholds both the viewer's interpretation as the most meaningful and the corresponding subjective approach as "crucial to compositional interpretation" (Rose 2012: 74). Table 50.1 also shows the expressive content variable names I devised during the open-coding process. For example, I coded the tree-lined plaza in Figure 50.1 as "serene" and the colorful, otherworldly mural to its right as "surreal."

Table 50.2 shows the additional hashtag and caption analysis categories and variables. Here I adapted the expressive content approach to code a post's overall/primary written message, as well as coded both direct references to unique topics. In both cases, I drew the categories from Instagrammers' own language. This analysis proved useful in drawing distinctions between the themes whose definitions and attendant visual representations change depending on context, such as community, social practice, and renewal, and those which concern phenomenological experiences, namely mentions of exploring and travel, personal resonance, life, and light (i.e., dawn or dusk).

After coding, I analyzed the images, additional hashtags, and captions through content analysis. From this, I sought to learn how Instagrammers frame, curate, and share images of urban spaces,

Table 50.2 Hashtag and caption analysis: categories and variables devised and coded for analysis of the additional hashtags and captions

Category	*Variable*[9]
Overall/primary message	• Art/urban design interventions improve urban space • Community, tolerance, peace • Computer technology for design • Historical figures, monuments • Importance of new perspectives for design • Nature, sustainability • Protest, political • Resonance with place or artwork • Social practice • Travel, explore, walk
Direct reference	• Community • Historic preservation • Life • Light • Marginalized groups • Personal resonance • Renewal • Social practice • Street photography • Sustainable design

whether their practice indicates perceived differences between urban design and public art, and how Instagram research may serve urban design.

Instagramming the digital city

Snapping the digital city

Statistically significant[10] relationships between the #urbandesign and #publicart hashtags and particular sets of variables emerge almost immediately, indicating that people do see urban design and public art as distinct. *Objects* have strong relationships with either #urbandesign or #publicart. Within the sample, 88% of bridges, 88% of buildings, 95% of plazas and public spaces, 100% of renderings and models, 100% of skylines, 88% of streets and stairs, and 85% of street furniture were tagged with #urbandesign. Meanwhile, Instagrammers tagged 82% of murals and 93% of sculptures with #publicart. For *location*, images on or at these locations were tagged with #urbandesign: bridges (70%), district or corridor (100%), water (100%). The locations for #publicart images reveal its connection to objects, comprising 81% of building and 77% of street furniture sites. The relationship to *size* is not quite as strong, but still significant: 54% of large and 65% of small objects are #urbandesign and #publicart examples, respectively. *Lynch's elements* also have clear relationships: 89% of paths, 100% of districts, and 97% of nodes bear the #urbandesign hashtag, whereas 67% of landmarks are #publicart examples.[11]

Object *formality* meets expectations in that formal, permanent objects associate with #urbandesign (72%), and #semi-permanent (77%, e.g., murals) or temporary (77%, e.g., performances) correlate with #publicart. For an object to qualify as *interactive*, people must be able to engage with it, such as playgrounds and stairs. Unsurprisingly, 60% of the images deemed interactive are #urbandesign examples. Finally, *expressive content* analysis suggests Instagrammers understand symbolic representations within the urban realm to be #publicart (comical (86%), insurgent/political (85%), somber/menacing (94%), and surreal/phantasmagoric (71%)). At the same time, people record phenomenological objects and places as #urbandesign (grand/imposing (83%), serene/sublime (70%), and workaday (82%)).

The images shown in Figure 50.1 bear out these differences. Earlier, I noted that I coded the pictured plaza and mural in this image as "serene" and "surreal," respectively. With the exception of the mural's size – it does span a building – these images reflect the sample's #urbandesign and #publicart categories. The tree-lined and bench-filled plaza is a large, permanent construction designed for both passage and sitting, and thus typifies Lynch's node. Long Beach's large mural is a landmark people can use for orientation, but they cannot engage with it physically and it remains semi-permanent (subject to new coats of paint).

#Reading The Text

Instagram's primary medium is the image, but users usually add additional hashtags and captions to provide context, commentary, and/or place their photographs within an Internet-wide dialogue. Just 3% of Instagram's users included only the qualifying hashtags, while the remaining 97% wrote additional descriptions, often with a range of ideas. Tellingly, no Instagrammer in the thousand posts collected (200 pre-coding and 800 sample) *ever* used both hashtags together. An image was either #urbandesign/#diseñourbano or it was #publicart/#artepublico, never both.

The primary messages within the written text reveal additional #urbandesign and #publicart differences. Here Instagrammers wrote explicitly about urban design, whether about the

Figure 50.2 An Instagrammer sees hope and beauty in the otherwise routine infrastructure.

Source: Yomara Rodriguez (@yoms_rguez)

usefulness of computer technology (100%) or the goodness of nature and sustainable design (89%). Similarly, #publicart correlates with messages about social issues and public space, such as art's contribution to public space (65%); community, tolerance, and peace (85%); historical figures and monuments (92%); and protest and political action (86%). In Figure 50.1, one Instagrammer praises the plaza's landscape design and uses tree emojis; another celebrates art in the public realm by amplifying both the artist and his local gallery.

Lastly, people express personal resonance in both the #urbandesign and #publicart examples, but the #urbandesign posts' additional hashtags and captions share a strongly significant relationship with action and experience, including exploring and travel, personal resonance, life, and light. The Instagrammer of Figure 50.2 pairs an otherwise workaday image of an elevated train track's construction with notably expressive language: "Todo lo que puedes imaginar es real [Everything you can imagine is real]," with the hashtags #perspectiva [perspective], #enfoque [focus], #luz [light], #sombra [shadow], #luzysombra [lightandshadow], and #azul [blue]. Like the Dutch plaza's Instagrammer, this user sees the urban realm as a place for phenomenological experience, specific objects notwithstanding.

Instagram's place in placemaking

People make sense of urban spaces with the tools they have. This study looked at how we might use Instagram to make sense of place. Precisely because Instagram's 500 million daily users (Aslam 2018) share curated imagery, Instagram photographs reveal the particular versions of our cities we wish to share with others. After collecting, sorting through, and analyzing 800 images, I propose that Instagram is a rich, admittedly limited, source for studying how people perceive and capture their urban spaces, and that Instagrammers' imagery and accompanying text can help inform urban design.

Instagram's limitations: what we cannot see

While this research shows that Instagram's millions of *flâneurs* and *flâneuses* can offer insights into contemporary urban design experience, the SNS has its limitations. First, hashtagging is a haphazard practice. This sample has a robust and broad, but *not* complete, sample of urban design and public art imagery. Not everyone posting pictures of streets, stairs, murals, or street performances during the project timeline used the requisite #urbandesign, #diseñourbano, #publicart, or #artepublico hashtags.

Second, despite the user figures, Instagrammers remain an *un*representative sample. As Hochman and Manovich (2013) note, "Instagram only captures the curated lives of some members of society and not others" (para 26). For one, the age bias is stark. Over 60% of all users are between 18 and 34 years old, whereas just over 2% of users are 65 and older (Statista n.d.b). For another, not all Instagrammers care enough to post about the urban realm. Posting about urban design and public art is less universal than, say, posting pictures of one's family. Self-selection bias persists.

Third, a central weakness of content analysis is that we cannot code what we cannot see – thus in Instagram, critical social commentary. The sample has just two posts with critical primary messages. One shows an empty sidewalk at daybreak with the caption: "#LaVistaUrbanaDelDia #12 de hoy es un ejemplo de una calle por la que nadie desearía caminar [today is an example of a street where no one would want to walk] . . . #unsafestreets #unhealthystreets." The sample has four other examples from this Instagrammer's "#LaVistaUrbanaDelDia" series, but only this image shows an unwelcoming urban environment. The caption accompanying the image of the stone bench in Figure 50.3 refers specifically to hostile urban design: "Diseño urbano sin vagabundos [urban design without homeless]." Kohn (2004) argues that spatial segregation is morally and politically problematic, morally because it reinforces a privileged group's status, and politically because it blinds that group to injustice. Documentation of urban design's weaponization against the most vulnerable exists in social media – the photographer of Figure 50.3 is architect James Furzer and artist Stuart Semple launched the #hostiledesign Instagram campaign in January 2018 (hostiledesign 2018). In the main, though, this research suggests we prefer to show the best and sanitized in cities, not necessarily the nondescript, even shabby neighborhoods (Boy and Uitermark 2016).

Lessons from Instagram

This project does show solid evidence that Instagrammers register urban design and public art as distinct. We cannot say from this research if people view public art as a subset of urban design or as a complementary practice because Instagrammers do not always capture #publicart images' surroundings. Posts showing sculpture, performances, and monumental murals often include their urban contexts in the frame, but the smaller artwork posts lack information about nearby urban design, because capturing the smaller pieces and details requires close, a-contextual photography.

Both urban design and public art afford our cities what Lynch calls a "sensuous form" (1960: 119), but they do so differently. #Urbandesign's relationships with Lynch's (1960) paths, districts, and nodes – spaces characterized by motion, experience, and immersion – illustrate how

Figure 50.3 This image was posted in black and white, reinforcing the message of the bench's hostile design.

Source: James Furzer (@james_furzer)

we register urban design as action-based. By contrast, the project's #publicart posts' correlation with Lynch's landmark category hints at art's symbolic function. That is, in addition to being wayfinding devices, the sample's murals, posters, and sculptures tend also to represent other things and ideas.

Unwittingly, Instagrammers routinely evoke Lynch's work. Sharing their images and ideas about their cities provides us with discrete "strong image" visual and verbal data points for analysis. Some Instagrammers do study urban design (see Figure 50.4), but this sample suggests that a formal urban design education may be unnecessary for identifying quality in built environment interventions and the resulting sensory attributes (e.g., light, color, sound, etc.) that make them special.

People *love* posting images of urban spaces on Instagram. We find consistent evidence that people connect with unique experiences, and specifically with urban spaces and interventions creating those personal moments. Enthusiastic *flâneurs* and *flâneuses* (Apostol, Antoniadis, and Banerjee 2012), who constitute Lynch's (1960) "attentive audience" (120), can help identify the elements that produce a "visible, coherent, and clear" (91) and imageable landscape. From these images, Instagram offers potential sources of information for urban designers.

Architects and designers have long wished to know more about how the public experiences their projects, and Instagram provides one, albeit limited, view into their perceptions. Given that Instagrammers use hashtags and geolocation to identify their imagery, and Picodash offers

Figure 50.4 An architect and urban design student's post. "Porque hoy la clase del Master la damos en el Poblenou 👣👣👣#urbandesign #walkablecity #livablecity #cities4people" [Because today the Master class is given in Poblenou].

Source: Celeste Pereira (@mariacelestepereira)

advanced date-based, geolocated, and hashtag searches, I propose there are at least four useful Instagram-based research approaches available to urban design practice.

1 As in this project, urban design researchers can conduct broad-based, periodic analyses of #urbandesign to glean what people mean when using the term. This project underscores how the urban realm resonates with people and identifies some elements that correlate with urban design and public art. But tastes come and go. Longitudinal comparisons over time of such terms as #urbandesign, #publicart, #publicspace, and so on, might reveal initially exciting but quickly outdated design fads. Instagram surveys could help designers foresee and protect against wasted investments in recreating such projects.
2 Designers could conduct case study analyses of particularly beloved and/or loathed projects to discover what people both see and say about those urban design interventions. How do Instagrammers frame these projects? Are there any repeating visual motifs that designers might be able to replicate (or avoid) in new projects? Here again, researchers could study these sites periodically to see if, when, and how Instagrammers' feelings about those places change.

3 Instagram can also support urban design pedagogy. Students could apply a heuristic methodology to Instagram and other SNSs to learn about the public life in streets, neighborhoods, or entire urban districts. Sample validity and generalizability notwithstanding, this type of social media analysis offers a new, potentially rich, type of digital fieldwork.
4 Finally, urban designers could conduct analyses of urban design practice itself. Within this sample, students and professionals posted renderings, models, and *in situ* images of their work. A critical content analysis of Instagram imagery could discern overlooked social issues (see the aforementioned limitations). Do urban design educators and practitioners use Instagram to shore up urbanism's capitalistic foundations, as the Situationists and Lefebvre feared (Goonewardena 2011)? In a climate where design schools embark on promotional Instagram campaigns (Industry Dive 2018) and "creating Instagrammable moments" is now customary in architectural briefs (Fairs 2018), periodic self-examination could prove consequential.

In Instagram, we see evidence of people perceiving and celebrating the city as a space for profound, personal moments. For urban design, we can potentially use their posts as cues for how better to shape a legible and phenomenologically rich 21st-century city.

Notes

1 I chose Spanish and English because they are the world's second- and third-most popular languages, respectively, and represent a decent proportion of the Global North and Global South.
2 One can find the project through the Internet Archive Wayback Machine (https://archive.org/web/), but the same cannot be said for all webpages.
3 I could not explore cultural differences because English simply dominates the sample. Of the 800 posts, 39% are English-only, 9% are Spanish-only, 43% are both, and the remaining 9% are English and some third language.
4 Picodash offers advanced content, hashtag, and location searches and analyses, and it serves brands, marketing agencies, journalists, and researchers.
5 I did not analyze likes or comments for two reasons: (1) this is research concerned with representation, not participation; and (2) the proliferation of bots, or fake accounts that automatically engage with Instagram imagery to attract followers, renders such analysis largely meaningless.
6 Still image compositional interpretation assesses images themselves and captures such things as content, color, light, spatial organization, and expressive content (Rose 2012). Expressive content coding does assume subjective decision-making on the part of the researcher.
7 The constant-comparative method involves an open- and continuous-coding process where themes, then theories, emerge. This grounded theory approach exists in contrast to the deductive method, which starts with specific, theory-based hypotheses (Strauss and Corbin 1990).
8 In the interest of space, only the most frequently occurring variables are listed.
9 In the interest of space, only the most frequently occurring variables are listed.
10 I used Stata for statistical analysis. Most relationships are significant at the $p < .001$ level, but all are at least at $p < .05$.
11 Lynch's edge does not appear in this sample.

References

Apostol, I., Antoniadis, P. and Banerjee, T. (2012). "*Flânerie* Between Net and Place: Promises and Possibilities for Participation in Planning," *Journal of Planning Education and Research*, 33(1): 20–33.

Appleyard, D., Lynch, K. and Myer, J.R. (1965). *The View from the Road*, Cambridge, MA: MIT Press.

Aslam, S. (2018). *Instagram by the Numbers: Stats, Demographics & Fun Facts*. www.omnicoreagency.com/instagram-statistics/ (accessed January 9, 2018).

Banerjee, T. and Loukaitou-Sideris, A. (eds.). (2011). *A Companion to Urban Design*, Abingdon, Oxon: Routledge.

Barthes, R. (1977). "The Death of the Author." In *Image/Music/Text*, trans. S. Heath, London: Fontana Press.

Baudelaire, C. (1972). "The Painter of Modern Life." In *The Painter of Modern Life and Other Essays (Arts & Letters)*, trans. P.E. Charvet (1863), London: Viking.

Benjamin, W. (1999). *The Arcades Project*, trans. Howard Eiland and Kevin McLaughlin, Cambridge, MA: The Belknap Press of Harvard University Press.

Berger, J. (1972). *Ways of Seeing*, London: Penguin Books.

Boy, J.D. and Uitermark, J. (2016). "How to Study the City on Instagram," *PLoS ONE*, 11(6): 1–16.

CityLab Staff. (2017). "Your Entire City Is an Instagram Playground Now," *CityLab*, December 29. www.citylab.com/design/2017/12/congrats-your-city-is-an-instagram-playground-now/549152/ (accessed May 8, 2018).

Debord, G. (1958). *Theory of the Dérive*. www.bopsecrets.org/SI/2.derive.htm (accessed January 12, 2018).

De Certeau, M. (1984). *The Practice of Everyday Life*, trans. Stephen Randall, Berkeley, CA: University of California Press.

Fairs, M. (2018). "Creating Instagrammable Moments 'Now Part of Architectural Briefs' Says Farshid Moussavi," *Dezeen*, June 5. www.dezeen.com/2018/06/05/instagram-moments-farshid-moussavi-instagram-architecture/ (accessed July 30, 2018).

Giles-Corti, B., Bull, F., Knuiman, M., McCormack, G., Van Niel, K., Timperio, A., Christian, H., Foster, S., Divitini, M., Middleton, N. and Boruff, B. (2013). "The Influence of Urban Design on Neighborhood Walking Following Residential Relocation: Longitudinal Results from the RESIDE Study," *Social Science & Medicine*, 77(2013): 20–30.

Goonewardena, K. (2011). "Critical Urbanism: Space, Design, Revolution," in Banerjee, T. and Loukaitou-Sideris, A. (eds.) *A Companion to Urban Design*, Abingdon, Oxon: Routledge.

Hochman, N. and Manovich, L. (2013). "Zooming into an Instagram City: Reading the Local through Social Media," *First Monday*, 18(7). http://firstmonday.org/article/view/4711/3698

hostiledesign. (2018). *Designs against Humanity*. https://hostiledesign.org/ (accessed July 30, 2018).

Hou, J. (2010). "Taking Place: Rebar's Absurd Tactics in Generous Urbanism," in Hou, J. (Ed.) *Insurgent Public Space: Guerilla Urbanism and the Remaking of Contemporary Cities*, New York, NY: Routledge.

Industry Dive. (2018). "Changing Urban Design with Instagram," *Smart Cities Dive*. www.smartcitiesdive.com/ex/sustainablecitiescollective/user-s-guide-changing-urban-design-instagram/125581/ (accessed July 30, 2018).

Jacobs, A.B. (1984). "Looking at Cities," *Places*, 1(4): 28–37.

Jacobs, A.B. and Appleyard, D. (1987). "Toward an Urban Design Manifesto," *Journal of the American Planning Association*, 53(1): 112–120.

Kallas, P. (2018). *Top 15 Most Popular Social Networking Sites and Apps [January 2018]*. www.dreamgrow.com/top-15-most-popular-social-networking-sites/ (accessed January 9–10, 2018).

Kohn, M. (2004). *Brave New Neighborhoods: The Privatization of Public Space*, New York, NY: Routledge.

Krieger, M. (2012). "Re-Photographing and Re-Viewing Charles Marville's Paris of 1865–1877." https://web.archive.org/web/20150911032855/http://1000eyes.usc.edu/martin/KriegerParisT&C.pdf (accessed July 26, 2018).

Lefebvre, H. (1947). *Critique of Everyday Life*, trans. J. Moore (1992), London: Verso.

Lynch, K. (1960). *The Image of the City*, Cambridge, MA: MIT Press.

Lynch, K. and Rivkin, M. (1995). "A Walk Around the Block," in T. Banerjee and M. Southworth (Eds.) *City Sense and City Design: Writings and Projects of Kevin Lynch*, Cambridge, MA: The MIT Press.

MacDowall, L.J. and de Souza, P. (2018). "I'd Double Tap That! Street Art, Graffiti, and Instagram Research," *Media, Culture & Society*, 40(1): 3–22.

Miles, M. (1997). *Art, Space, and the City: Public Art and Urban Futures*, London: Routledge.

Pew Research Center. (2018). *Social Media Use in 2018*. www.pewinternet.org/2018/03/01/social-media-use-in-2018/ (accessed July 30, 2018).

Radice, M. and Boudreault-Fournier, A. (eds.). (2017). *Urban Encounters: Art and the Public*, Montreal, QC: McGill-Queen's University Press.

Remesar, A. (Ed.). (2016). *The Art of Urban Design in Urban Regeneration*, Barcelona: Edicions de la Universitat de Barcelona.

Rose, G. (2012). *Visual Methodologies: An Introduction to Researching with Visual Methods*, 3rd ed., London: SAGE Publications Ltd.

Sheldon, P. and Bryant, K. (2015). "Instagram: Motives for Its Use and Relationship to Narcissism and Contextual Age," *Computers in Human Behavior*, 58(2016): 89–97.

Statista. (n.d.a). *Number of Smartphone Users Worldwide from 2014 to 2020 (in Billions)*. www.statista.com/statistics/330695/number-of-smartphone-users-worldwide/ (accessed January 9, 2018).

Statista. (n.d.b). *Distribution of Instagram Users Worldwide as of April 2018, by Age and Gender*. www.statista.com/statistics/248769/age-distribution-of-worldwide-instagram-users/ (accessed July 23, 2018).
Strauss, A. and Corbin, J. (1990). *Basics of Qualitative Research: Techniques and Procedures for Developing Grounded Theory*, Newbury Park, CA: SAGE Publications.
Taylor, J.C. (1957). *Learning to Look: A Handbook for the Visual Arts*, Chicago, IL: Chicago University Press.
Tomlinson, J. (2007). "Globalization and Cultural Analysis," in McGrew, A. and Held, D. (eds.) *Globalization Theory: Approaches and* Controversies, Malden, MA: Polity Press.
Wark, M. (2011). *The Beach Beneath the Street: The Everyday Life and Glorious Times of the Situationist International*, London: Verso.
Whyte, W.H. (1980). *The Social Life of Small Urban Spaces*, New York, NY: Project for Public Spaces.
Zukin, S. (2010). *Naked City: The Death and Life of Authentic Urban Places*, New York, NY: Oxford University Press.

Further reading

Hochman, N. and Manovich, L. (2013). "Zooming into an Instagram City: Reading the Local through Social Media," *First Monday*, 18(7). http://firstmonday.org/article/view/4711/3698. An early and still relevant discussion about the empirical possibilities of using Instagram to study the city.
Rose, G. (2012). *Visual Methodologies: An Introduction to Researching with Visual Methods*, 3rd ed., London: SAGE Publications Ltd. Now into its fourth edition, Rose's volume is considered to be an indispensable resource for conducting visual analysis.

EPILOGUE

We began this project with the explicit purpose of addressing most of the lacunae in our previous effort, *Companion to Urban Design*, published by Routledge in 2011. One of the lacunae we were particularly conscious of was the dominance of the Anglo-American axis of scholarship in the previous volume. We felt that the absence of representation and inclusion of voices and examples from the Global South made the definition of the urban design field incomplete. In compiling this edited volume, we accordingly approached many scholars with expertise in comparative urbanism and urban design, as well as many who are situated in countries outside the UK-US axis. We wanted to have a broad coverage of urbanism and urban design as conceived and practiced in other countries, especially those from the Global South, which are going through major urban transformations and explosive urban growth.

As we now take inventory of the contributions of the 50 chapters included in this volume, we are therefore delighted to see the number of countries and cities represented in these writings. The following table shows the list of cities included in this Companion's chapters, classified under either the Global North or the Global South category. We are pleased to see a very balanced representation of cities from both regions. The distribution is even tilting towards the Global South, if we consider the countries represented: 12 from the Global North (albeit dominated by the United States), and 19 from the Global South.

Countries and cities included in the book

Cities from the Global North	*Cities from the Global South*
Baltimore, USA	Abu Dhabi, UAE
Barcelona, SPAIN	Agra, INDIA
Berlin, GERMANY	Ahmedabad, INDIA
Bristol, UK	Amman, JORDAN
Chicago, USA	Bangkok, TAILAND
Cleveland, USA	Beijing, CHINA
Copenhagen, DENMARK	Beirut, LEBANON
Davis, USA	Bogotá, COLOMBIA
Denver, USA	Byblos, LEBANON

(*Continued*)

(Continued)

Cities from the Global North	*Cities from the Global South*
Detroit, USA	Cairo, EGYPT
Flint, USA	Cape Town, SOUTH AFRICA
Herne, GERMANY	Curitiba, BRAZIL
Ise, JAPAN	Damascus, SYRIA
Lakewood, USA	Delhi, INDIA
Los Angeles, USA	Dhaka, BANGLADESH
Melbourne, AUSTRALIA	Dubai, UAE
Milan, ITALY	Guangzhou, CHINA
Naples, ITALY	Hangzhou, CHINA
New Bedford, USA	Hong Kong, CHINA
New Orleans, USA	Jakarta, INDONESIA
New York, USA	Kabul, AFGHANISTAN
Osaka, JAPAN	Kherson, UKRAINE
Rome, ITALY	Kuala Lumpur, MALAYSIA
Rotterdam, NETHERLANDS	Masdar, UAE
San Francisco, USA	Medellín, COLOMBIA
Seattle, USA	Mexico City, MEXICO
Seoul, KOREA	Mumbai, INDIA
Tel Aviv, ISRAEL	Puebla, MEXICO
Tokyo, JAPAN	São Paolo, BRAZIL
Toronto, CANADA	Shanghai, CHINA
Toyama, JAPAN	Shenzhen, CHINA
Vancouver, CANADA	Singapore, SINGAPORE
Washington, D.C., USA	Tijuana, MEXICO
	Yazd, IRAN
	Yichun, CHINA

The previous *Companion* was foundational in content, hence dominated by the works of authors representing the Global North. But this concession need not be interpreted as that the origin of urban design is in the Global North, or that it began at Harvard (see Saunders and Krieger 2009). After all, the canonical and cosmic models of city design go back to ancient cultures of the Global South, as Kevin Lynch (1981) has discussed at length. Here the contributions assembled from the Global South present new dimensions and interpretations of urban design that are embedded in the unique circumstances of the contemporary cities of this region: colonial heritage of exploitation and neoliberal local policies in a background of widespread poverty and major social inequality, degraded environments, informal economy and housing, and also ravages of new urban growth fueled by the unrelenting juggernaut of globalization.

In many ways, these chapters offer new or alternative interpretations of urban design and innovative possibilities for addressing the problems unique to their circumstances, in the context of climate change, global warming, and the imperatives of sustainability of proposed developments. Thus, these articles extend the scope and possibilities of urban design that couldn't have been imagined in the contexts of the Global North under normal circumstances. But as we know, the belief that circumstances such as informality, homelessness, and poverty are present only in the Global South is also a myth. Therefore, we reckon that urban design initiatives and innovations in the Global South may also apply to similar circumstances in the Global North. Thus, many of these essays present new critical perspectives relevant to both hemispheres.

Under the general rubrics of "challenges" and "aspirations," we have assembled a group of articles that address a variety of substantive areas of urban design, which we had not included in the previous book. Thus, the essays about the challenge of designing multicultural cities, or responding to the widespread mimicking of existing built forms, or shaping appropriate design responses to managing growth or decline, or preparing for global warming and natural disasters, and the like, are some of the unique features of this collection. Again, as in our previous *Companion*, we did not expect "how to do" type of responses, as is common in "handbooks." We were pleased to see that the responses from our authors were conceptual, critical, and perhaps anticipatory in nature. However, almost all of the contributions were grounded in specific locations and circumstances, with a whiff of best practices. Some were more normative than others, such as, for example, the essays under "health" and "access and mobility" that include recommendations based on a fairly larger backdrop of relevant literature.

We were also pleasantly surprised to see that many authors who came with scholarly backgrounds in areas other than urban design were quite enthusiastic about the possibilities of urban design and urban designers to intervene and obtain some of the aspirational outcomes in the urbanism of cities, and social and spatial organization of human settlements. We worry, however, that the collective enthusiasm about promises of urban design in solving problems of inequality, justice, exclusion, displacement, etc. – as common leitmotifs in many of the chapters – may be presenting the field as a panacea for all social and environmental problems. While this enthusiasm certainly expands the challenges, promises, and responsibilities for the practitioners of urban design, it may also place untenable and unrealistic expectations on what they can deliver even in the bottom-up and participatory scenarios envisioned and encouraged in many of these writings. After all, as we are aware from the literature on advocacy planning and citizen participation (see Davidoff 1968; Arnstein 1969), such process by itself does not guarantee the kind of changes that require confronting existing power structures and global economic interests. Fundamental changes that are envisioned in this prospectus for urban design are likely to be slow and incremental – a caveat worth noting.

Finally, although our publisher has chosen to call this collection *New Companion to Urban Design*, for reasons of publication protocol, we consider this volume a "sequel" to the earlier publication, as we noted in the Introduction. The word "new" seems to imply that the previous work is "old" and thus dated. Certainly, we would not want the readers of this volume to have that impression. The foundational content of the earlier volume still stands on its own right. Indeed, as we have indicated previously, the content of this volume was shaped by our awareness of several lacunae and professional predilections in the previous volume. In our view, this volume complements the other, and in tandem, they both represent our effort to map the intellectual and professional terrain of urban design as a field, increasingly gaining prominence in urban studies and planning scholarship and beyond.

References

Arnstein, S. (1969). "A Ladder of Citizen Participation." *Journal of the American Institute of Planners*. 35(4): 216–224.

Davidoff, P. (1968). "Advocacy Planning and Pluralism." *Journal of the American Institute of Planner*. 31(4): 331–338.

Lynch, Kevin (1981). *A Theory of Good City Form*. Cambridge, MA: The MIT Press.

Saunders, W. and Krieger, A. (2009). *Urban Design*. Minneapolis, MN: University of Minnesota Press.

INDEX

Note: **Boldface** page references indicate tables. *Italic* references indicate figures and boxed text.

Made in the USA
Monee, IL
14 September 2026

62978581R00391